The Institute of Mathematics
and its Applications
Conference Series

Volumes in the previous series were published by Academic Press to whom all enquiries should
be addressed. The following and all forthcoming titles are published by Oxford University Press
throughout the world.

Continued overleaf

Mixing and Dispersion in Stably Stratified Flows

Based on the proceedings of a conference organized by the Institute of Mathematics and its Applications on Stably Stratified Flows, and held at the University of Dundee in September 1996.

Edited by

PETER A. DAVIES

Department of Civil Engineering
University of Dundee

CLARENDON PRESS • OXFORD • 1999

OXFORD
UNIVERSITY PRESS

Great Clarendon Street, Oxford OX2 6DP

Oxford University Press is a department of the University of Oxford.
It furthers the University's objective of excellence in research, scholarship,
and education by publishing worldwide in.

Oxford New York

Athens Auckland Bangkok Bogotá Buenos Aires Calcutta
Cape Town Chennai Dar es Salaam Delhi Florence Hong Kong Istanbul
Karachi Kuala Lumpur Madrid Melbourne Mexico City Mumbai
Nairobi Paris São Paulo Singapore Taipei Tokyo Toronto Warsaw
with associated companies in Berlin Ibadan

Published in the United States
by Oxford University Press Inc., New York

A catalogue record for this book is available from the British Library

Library of Congress Cataloging in Publication Data
(Data applied for)
ISBN 0 19 850015 7

Typeset using LaTeX

Printed in Great Britain by
Biddles Ltd., Guildford, Surrey

PREFACE

This volume constitutes a record of the Fifth IMA Conference on Stably Stratified Flows, held at the West Park Conference Centre, University of Dundee, during the period 25-27 September 1996. The principal theme of the meeting was Mixing and Dispersion in Stably Stratified Flows, to complement earlier conferences in the IMA Stratified Flows series dealing with Flow and Dispersion over Topography (Guildford, 1992), Waves and Turbulence (Leeds, 1989), Dense Gas Dispersion (Chester, 1986) and Turbulence and Diffusion in Stable Environments (Cambridge, 1983) and to extend a successful Euromech meeting held in 1995 in Lyon, France on the same topic as the Dundee conference. Previous volumes in the IMA series contain papers presented at these earlier meetings, edited by I.P. Castro and N.J. Rockcliffe, S.D. Mobbs and J.C. King, J.S. Puttock and J.C.R. Hunt respectively.

Though the meeting had a well-specified theme, contributions were welcomed from any area of stratified flow research in which buoyancy accelerations were dynamically important. The aim was to bring together researchers from a broad range of backgrounds and disciplines, to consider stratified flow problems having relevance to environmental and industrial processes where the inhibition of vertical mixing of pollutants by ambient density gradients in the atmosphere and oceans is an important constraint. In common with previous meetings, a balanced programme of contributions was achieved, with papers dealing with theoretical, experimental and observational aspects of the subject, particularly from the fields of meteorology, oceanography and geophysical fluid dynamics. The interests of the Organising Committee: Peter Davies (University of Dundee), Paul Linden (University of Cambridge) and Joe Fernando (Arizona State University) were perhaps reflected in the relatively large number of papers dealing with detailed laboratory modelling measurements and associated scalings of stratified flow processes in GFD. As such, these contributions demonstrated well the recent enormous progress made in obtaining precise, automated velocity and density field data from controlled model experiments with turbulent stratified flows.

A number of recurring themes from other meetings in the series continued to be well represented; for example, stratified flow past obstacles, rotating stratified flows, the motion and dispersion of heavy gas on sloping and complex terrain, gravity currents and plume dispersion and pollutant transport flow from power plant discharges. For the first time, the application of plume dynamics to outfall dispersion featured on the programme, together with a number of papers on problems particularly attributable to stratified flow processes influenced by anthropogenic activity. The structure of the meeting was built around a number of invited speakers, each of whom (Don Boyer, Clive Greated, Dieter Etling, Joe Fernando, Ian Castro, Julian Hunt, Peter Baines, Gerhard Jirka and Gert-Jan van Heijst) presented comprehensive reviews dealing with new areas of growth in the subject.

The Organising Committee were particularly pleased to welcome over 80 participants from a wide international constituency, to further support the reputation established by the IMA Stratified Flow conferences for informal, lively and stimulating meetings. The Committee were fortunate to be given excellent administrative support by Pamela Bye and her IMA colleagues and to benefit from generous financial sponsorship of a Conference Dinner Reception by the City of Dundee. The help provided locally by members of the environmental fluid dynamics research group (particularly Dr. Yakun Guo) was invaluable and is gratefully acknowledged here.

Peter A. Davies
University of Dundee

ACKNOWLEDGEMENTS

The Institute of Mathematics and its Applications would like to thank the following who contributed to the organisation of the conference and the production of the proceedings:

Professor Peter Davies (University of Dundee) Editor of the proceedings and Chairman of the Conference Organising Committee, the members of the Conference Organising Committee, the authors of the papers, and the following Institute staff for the organisation of the conference: Dr. Adrian Lepper (Executive Secretary) and Mrs. Pamela Bye (Conference Officer), and for most of the production of the proceedings: Miss Debbie Brown (Publications Officer).

CONTENTS

Contents

Contents

CONTRIBUTORS

K.H.M. ALI; Department of Civil Engineering, University of Liverpool, Brownlow Street, Liverpool, L69 3BX.

P.S. ANDERSON; British Antarctic Survey, High Cross, Madingley Road, Cambridge.

N. AOTA; Sumitomo Heavy Industries Limited, Kitashinagawa 5-9-11, Shinagawa-Ku, Tokyo 141, Japan.

S. ARNOUX-CHIAVASSA; Laboratoire de Sondages Electromagnétiques de l'Environnement Terrestre, Université de Toulon et du Var, B.P. 132, 83957 La Garde Cedex, France.

O. AUBAN; Meteo-France CNMR, 42 avenue Coriolis, 31057 Toulouse, France.

P.G. BAINES; CSIRO Atmospheric Research, Aspendale, Victoria 3195, Australia.

M. BENNETT; Environmental Technology Centre, Department of Chemical Engineering, University of Manchester Institute of Science and Technology, P.O. Box 88, Manchester, M60 1QD.

P. BONNETON; D.G.O., Universite Bordeaux 1, av. des Facultes, 33405 Talence, France.

M.A. BROWN; School of Mechanical and Materials Engineering, University of Surrey, Guildford, Surrey, GU2 5XH.

R. BURROWS; Department of Civil Engineering, University of Liverpool, Brownlow Street, Liverpool, L69 3BX.

I.R. CANTALAPIEDRA; Department de Fisica Aplicada, Universitat Politecnica de Catalunya; B5 Campus Nord, Barcelona 08034, Spain.

J.J. CASARES; Department of Chemical Engineering, University of Santiago de Compostela, E-15706 Santiago de Compostela, Spain.

I.P. CASTRO; Department of Mechanical Engineering, University of Surrey, Guildford, Surrey, GU2 5XH.

C.P. CAULFILED; Centre for Environmental and Geophysical Flows, School of Mathematics, University of Bristol, University Walk, Bristol, BS8 1TW.

L. CHAN; Department of Earth and Atmospheric Science, York University, 4700 Keele Street, North York, Ontario, Canada, M3J 1P3.

H.J.H. CLERCX; Fluid Dynamics Laboratory, Eindhoven University of Technology, Eindhoven 5600 MB, The Netherlands.

I.R. COULBOURN; Department of Civil Engineering, University of Dundee, Dundee, Scotland, DD1 4HN.

M. CRAPPER; Department of Civil Engineering and Environmental Engineering, University of Edinburgh, Mayfield Road, Edinburgh, Scotland, EH9 3JZ.

S.B. DALZIEL; Department of Applied Mathematics and Theoretical Physics, University of Cambridge, Silver Street, Cambridge, CB3 9EW.

P.A. DAVIES; Department of Civil Engineering, University of Dundee, Dundee, Scotland, DD1 4HN.

M. DE CASTRO; Department of Chemical Engineering, University of Santiago de Compostela, E-15706 Santiago de Compostela, Spain.

J.C.W. DENHOLM-PRICE; School of Mathematics and Statistics, University of Sheffield, Hounsfield Road, Sheffield, S3 7RH.

J. DOORSCHOTT; Department of Civil Engineering, University of Dundee, Dundee, Scotland, DD1 4HN.

P.W. DUCK; Department of Mathematics, University of Manchester, Oxford Road, Manchester, M13 9PL.

N. DURAND; Laboratoire de Sondages Electromagnétiques de l'Environnement Terrestre, Université de Toulon et du Var, B.P. 132, 83957 La Garde Cedex, France.

D. ETLING; Institut für Meteorologie und Klimatologie, Universität Hannover, Herrenhäuser Strasse 2, D-30419 Hannover, Germany.

T.K. FANNELØP; Utsikten 6, N-3155 Aasgaardstrand, Norway.

H.J.S. FERNANDO; Department of Mechanical and Aerospace Engineering, Arizona State University, Tempe, AZ 85287-9809, USA.

J.B. FLÓR; Laboratoire des Ecoulements Geophysiques et Industriels, Grenoble, France.

A.M. FOLKARD; Department of Civil Engineering, University of Strathclyde, John Anderson Building, 107 Rottenrow, Glasgow, Scotland, G4 0NG.

M.R. FOSTER; Department of Aerospace Engineering, Applied Mechanics and Aviation, Ohio State University, Columbus, Ohio 43210, USA.

P. FRAUNIÉ; Laboratoire de Sondages Electromagnétiques de l'Environnement Terrestre, Université de Toulon et du Var, B.P. 132, 83957 La Garde Cedex, France.

S.J. GASKIN; Department of Civil Engineering and Applied Mechanics, McGill University, Montreal, Canada.

G. GOUDSMIT; Swiss Federal Institute for Environmental Science and Technology, Ueberlandstrasse 133, CH-8600 Duebendorf, Switzerland.

C.A. GREATED; Room 4201, University of Edinburgh, James Clerk Maxwell Building, The King's Buildings, Mayfield Road, Edinburgh, Scotland, EH9 3JZ.

R.F. GRIFFITHS; Environmental Technology Centre, Department of Chemical Engineering, University of Manchester Institute of Science and Technology, P.O. Box 88, Manchester, M60 1QD.

G.W. HAARLEMMER; Department of Chemical Engineering, University of Manchester Institute of Science and Technology, P.O. Box 88, Manchester, M60 1QD.

G.A. HAMILL; Department of Civil Engineering, Queen's University of Belfast, Belfast, Northern Ireland, BT7 1NN.

D.B. HANN; Room 4201, University of Edinburgh, James Clerk Maxwell Building, The King's Buildings, Mayfield Road, Edinburgh, Scotland, EH9 3JZ.

C. HÄRTEL; Institute of Fluid Dynamics, Swiss Federal Institute of Technology Zürich, ETH Zentrum, CH-8092 Zürich, Switzerland.

R.E. HEWITT; Department of Mathematics, University of Manchester, Oxford Road, Manchester, M13 9PL.

J.M. HOLFORD; Department of Applied Mathematica and Theoretical Physics, University of Cambridge, Silver Street, Cambridge, CB3 9EW.

J.C.R. HUNT; Department of Applied Mathematics and Theoretical Physics, University of Cambridge, Silver Street, Cambridge, CB3 9EW.

R. JIANG; Laboratoire de Recherches Hydrauliques, Ecole Polytechnic Fédérale de Lausanne, CH-1015 Lausanne, Switzerland.

H.T. JOHNSTON; Department of Civil Engineering, Queen's University of Belfast, Belfast, Northern Ireland, BT7 1NN.

H.E. JØRGENSEN; Risø National Laboratory, P.O. Box 49, DK-4000 Roskilde, Denmark.

E.G. KASTRINAKIS; Department of Chemical Engineering, Aristotle University of Thessaloniki, University Box 453, 54006 Thessaloniki, Greece.

J.C. KING; British Antarctic Survey, High Cross, Madingley Road, Cambridge.

E. KIT; Department of Fluid Mechanics and Heat Transfer, Tel-Aviv University, Ramat Aviv 69978, Israel.

C. KOUDELLA; Laboratoire de Physique, Ecole Normale Supérieure de Lyon, 46 Allée d'Italie, 69364 Lyon, Cedex 07, France.

U. LEMMIN; Laboratoire de Recherches Hydrauliques, Ecole Polytechnic Fédérale de Lausanne, CH-1015 Lausanne, Switzerland.

D.M. LEPPINEN; Department of Applied Mathematics and Theoretical Physics, University of Cambridge, Silver Street, Cambridge, CB3 9EW.

R.E. LEWIS; Brixham Environmental Laboratory, Zeneca Limited, Freshwater Quarry, Brixham, Devon, TQ5 8BA.

P.F. LINDEN; Department of Applied Mechanics and Engineering Sciences, University of California, San Diego, 9500 Gilman Drive, La Jolla, CA 92093-0411, USA.

P. LØFSTRØM; National Environmental Research Institute, P.O. Box 358, DK-4000 Roskilde, Denmark.

E. LYCK; National Environmental Research Institute, P.O. Box 358, DK-4000 Roskilde, Denmark.

S.R. MAASSEN; Fluid Dynamics Laboratory, Eindhoven University of Technology, Eindhoven 5600 MB, The Netherlands.

I. MAVROIDIS; 20 Kalogera Street, GR 11 361, Athens, Greece.

M. MICHAUD; Sulzer Innotec AG, Postfach, CH-8401 Winterthur, Switzerland.

T. MIKKELSEN; Risø National Laboratory, P.O. Box 49, DK-4000 Roskilde, Denmark.

J. MÜLLER; SF-Emmen, P.O. Box, CH-6032, Emmen, The Netherlands.

H. NEWTON; Department of Civil Engineering, University of Dundee, Dundee, Scotland, DD1 4HN.

S.G. NYCHAS; Department of Chemical Engineering, Aristotle University of Thessaloniki, University Box 453, 54006 Thessaloniki, Greece.

B.A. O'CONNOR; Department of Civil Engineering, University of Liverpool, Brownlow Street, Liverpool, L69 3BX.

Y. OHYA; Research Institute for Applied Mechanics, Kyushu University, Kasuga 816, Japan.

S. OTT; Risø National Laboratory, P.O. Box 49, DK-4000 Roskilde, Denmark.

S. OUILLON; Laboratoire de Sondages Electromagnétiques de l'Environnement Terrestre, Université de Toulon et du Var, B.P. 132, 83957 La Garde Cedex, France.

S. OZONO; Research Institute for Applied Mechanics, Kyushu University, Kasuga 816, Japan.

M.F. PAISLEY; School of Computing, Staffordshire University, Stafford, ST18 0AD.

P.N. PAPANICOLAOU; 59 Kefallinias Street, 112 51 Athens, Greece.

J.N.E. PAPASPYROS; Department of Chemical Engineering, Aristotle University of Thessaloniki, University Box 453, 54006 Thessaloniki, Greece.

W.R. PELTIER; Department of Physics, University of Toronto, 60 St. George Street, Toronto, Ontario, Canada, M5S 1A7.

V. PÉREZ-MUÑUZURI; Group of Nonlinear Physics, Faculty of Physics, University of Santiago de Compostela, E-15706 Santiago de Compostela, Spain.

M. PERRIER; Meteo-France CNMR, 42 avenue Coriolis, 31057 Toulouse, France.

J.M. REDONDO; Department de Fisica Aplicada, Universitat Politecnica de Catalunya; B5 Campus Nord, Barcelona 08034, Spain.

J.M. REES; Applied Mathematics Section, School of Mathematics and Statistics, University of Sheffield, Hounsfield Road, Sheffield, S3 7RH.

V. REY; Laboratoire de Sondages Electromagnétiques de l'Environnement Terrestre, Université de Toulon et du Var, B.P. 132, 83957 La Garde Cedex, France.

A.M. RIDDLE; Brixham Environmental Laboratory, Zeneca Limited, Freshwater Quarry, Brixham, Devon, TQ5 8BA.

M.A. SANCHEZ; Department de Fisica Aplicada, Universitat Politecnica de Catalunya; B5 Campus Nord, Barcelona 08034, Spain.

V. SCHILLING; Institut für Meteorologie und Klimatologie, Universität Hannover, Herrenhäuser Strasse 2, D-30419 Hannover, Germany.

J.A. SOUTO; Department of Chemical Engineering, University of Santiago de Compostela, E-15706 Santiago de Compostela, Spain.

M.J. SOUTO; Group of Nonlinear Physics, Faculty of Physics, University of Santiago de Compostela, E-15706 Santiago de Compostela, Spain.

K. SPENCE; Department of Civil Engineering, University of Sheffield, Sheffield.

C. STAQUET; Laboratoire des Ecoulements Géophysiques et Industriels, B.P. 53, 38041 Grenoble, Cedex 9, France.

A. STEGNER; LMD, BP 99, Université P. et M. Curie, 4 Place Jussieu, 75252 Paris Cedex 05, France.

E.J. STRANG; Department of Mechanical and Aerospace Engineering, Arizona State University, Tempe AZ 85287-9809, USA.

B.R. SUTHERLAND; Department of Mathematical Sciences, 539 Central Academic Building, University of Alberta, Edmonton, Alberta, Canada, T6G 2G1.

P. TAYLOR; Department of Earth and Atmospheric Science, York University, 4700 Keele Street, North York, Ontario, Canada, M3J 1P3.

G.J.F. VAN HEIJST; Fluid Dynamics Laboratory, Eindhoven University of Technology, Eindhoven 5600 MB, The Netherlands.

S.A. WALKER; Department of Mechanical Engineering, University of Edinburgh, Edinburgh, Scotland, EH9 3JH.

W. WENG; Department of Earth and Atmospheric Science, York University, 4700 Keele Street, North York, Ontario, Canada, M3J 1P3.

I.R. WOOD; Department of Civil Engineering, University of Canterbury, Christchurch, New Zealand.

A. WÜEST; Swiss Federal Institute for Environmental Science and Technology, Ueberlandstrasse 133, CH-8600 Duebendorf, Switzerland.

D. XU; Certicom Corporation, 200 Matheson Boulevard West, Mississauga, Ontario, Canada, L5R 3L7.

G. YU; Department of Civil Engineering, University of Dundee, Dundee, Scotland, DD1 4HN.

V. ZEITLIN; LMD, BP 99, Université P. et M. Curie, 4 Place Jussieu, 75252 Paris Cedex 05, France.

W.B.J. ZIMMERMAN; Department of Chemical and Process Engineering, University of Sheffield, Newcastle Street, Sheffield, S1 3JD.

Downslope Flows into a Stratified Environment - Structure and Detrainment

Peter G. Baines

CSIRO Atmospheric Research, Aspendale, Australia

Abstract

Observations of dense downslope flows into a density-stratified environment are described, and these observations are interpreted quantitatively in terms of dynamical processes. The system is two-dimensional $(x - z)$, the slope is at 6° to the horizontal, and the dense fluid is released at a uniform rate at the top of the slope for a finite period of time. The main downflow has the form of a gravity current with a distinct upper interface, with some similarities to the flow into a homogeneous environment described by Ellison and Turner (1959), but with major differences. The flow may be characterised by a parameter $M = Q_0 N^3 / g_0'^2$, which is empirically related to a bulk Richardson number. Here Q_0 is the volume flux per unit width of tank, N is the buoyancy frequency of the environment and g_0' is the buoyancy of the inflow. The downflow has three discernible regions down the slope:

1. an uppermost region where the inflow adjusts to a state with small downslope variations;

2. a central region with small downslope variations where the depth of the downflowing layer is approximately uniform and significant detrainment is seen to occur;

3. a lowest region where the remaining fluid in the downflow reaches its ambient density and spreads horizontally in a conspicuous pancake or plume.

In the central region both entrainment of environmental fluid into the downflow and detrainment from it occur, across the interface, and these processes are associated with Holmboe instability. The main quantitative results are obtained from measurements of density profiles taken before and after the downflow, which enable the net volume flux and the inflow and outflow to be inferred as functions of downslope position. From these we may obtain the entrainment and detrainment coefficients, with the latter being generally somewhat the larger, but both are much less than the entrainment coefficients of Ellison and Turner. Implications for oceanic and atmospheric flows are discussed.

1

1 Introduction

Dense flows down slopes at large Reynolds numbers occur quite commonly in nature. Prominent examples include nocturnal flows down hillsides due to radiative cooling of air, powder snow avalanches, overflows in the ocean such as the Mediterranean outflow and the Bass Strait overflow, and the flow of cold river water into lakes, to name only a few. The flow of dense fluid whose motion is driven primarily by its own buoyancy is known as a gravity current (Simpson 1997), and these phenomena are highly turbulent. Consequently, the most effective way to study them is usually in the laboratory, where they may be generated relatively easily. The flow of a gravity current down a slope that is produced by the sudden onset of a continuous source of dense fluid at the top is led by a relatively large "head" of dense fluid, which is associated with considerable mixing with the enviromental fluid. This is followed by a turbulent stream that is uniform in the mean, but has conspicuous small-scale mixing taking place at its upper boundary.

Observations of the motion of the head of a two-dimensional downslope gravity current from a line source flowing into a homogeneous environment by Britter and Linden(1980) have shown that the head moves at a uniform speed U_f that is approximately independent of the slope angle ϑ, and depends on the initial mass flux Q_0 (per unit slope width) and buoyancy g_0' in the form $U_f \approx 1.5(g_0' Q_0)^{1/3}$, for slope angles ϑ in the range $5^o < \vartheta \leq 90^o$. Here

$$g_0' = 2g(\rho_0 - \rho_i)/(\rho_0 + \rho_i), \tag{1.1}$$

where g is acceleration due to gravity, ρ_i is the inflowing density, and ρ_0 the density of the environment. The following downslope flow travels more rapidly by a factor of approximately 1.7, and the size of the head increases with time due to the supply of this fluid and to inherent mixing with the environment. This rate of increase itself increases approximately linearly with slope angle, but is not significant for the experiments described in this paper.

Observations of the flow following the head have been described by Ellison and Turner (1959), who derived a dynamical model for the bulk properties of the flow, based on measurements over a broad range of parameters. They showed that the mean fluid velocity was independent of downslope distance, and was dependent on the slope angle, or equivalently, on the local Richardson number, R_i. Following the work of Morton et al. (1956) on vertical buoyant plumes, Ellison and Turner introduced the concept of *turbulent entrainment* for downslope flows. This involved the assumption that the turbulence and associated mixing in the current would cause a net motion of environmental fluid towards and into the current, at a local speed that was proportional to the local mean downslope velocity of the current. This is essentially a dynamical similarity assumption, and the constant of proportionality is known as the *entrainment constant, E*. They showed that E is dependent on the local Richardson number, and incorporated this effect into the mathematical model for the flow, as described in more detail in the next section. This kind of model has been widely used to describe downslope flow in the atmosphere and ocean, with varying success (for example Manins and Sawford 1979, Smith 1975, Price and Baringer 1994).

The experiments of Ellison and Turner were for a homogeneous environment, but the model equations from these experiments may be rescaled to include a stratified environment (Turner 1986) if one assumes that the same physical processes are operating, in the manner that was done for plumes by Morton et al. (1956). On this basis, the model has been used to describe flows into stratified environments.

The first experiments of downslope flows into stratified environments were described by Mitsudera and Baines (1992). Here there is an additional parameter, namely N, the buoyancy frequency of the initial stratification, or equivalently, D, the depth below the source where the initial density of the fluid in the tank equals the inflow density. They are related by $N^2 = g_0'/D$, where g_0' is defined as in (1.1) with ρ_0 taking the value of the initial ambient density at the level of the source. This system has two dimensionless parameters, M and the Reynolds number R_e, defined by

$$M \equiv \frac{Q_0}{(g_0' D^3)^{1/2}} = \frac{Q_0 N^3}{g_0'^2}, \quad R_e \equiv Q_0/\nu, \qquad (1.2)$$

where ν is the kinematic viscosity. $M = 0$ corresponds to a homogeneous environment, and the effect of the stratification increases with M, although $M < 1$ for realistic flows. Mitsudera and Baines carried out experiments with a slope angle $\vartheta = 6^o$, and $R_e \geq 300$, and described some properties of the turbulent downflow and the mixing process that occurred above it. In particular, it was noted that, in addition to the main plume of fluid leaving the slope near the bottom where the inflow reached its ambient density, a second, broader, weaker plume was present higher up, fed by a complex three-dimensional circulation emanating from the mixing region above the downflow (these observations will be reported in more detail elsewhere).

The vertical distance over which the downslope flows of Mitsudera and Baines occurred was relatively small (20 cm from shelf to bottom of tank), and this made some aspects of the interpretation uncertain. The experiments described in the present paper were carried out with a longer slope having a larger vertical distance (25 cm from the shelf to the bottom of the tank). Attention here is focussed on the bulk properties of the main current, the effect of the stratification on entrainment into the current and its converse, "detrainment" from it. As shown below, the mean properties of the current including these properties may be inferred from density soundings before and after the experiment. In this paper, I next describe the experiment and the techniques involved, and then proceed to the analysis and interpretation. These results are also only for a 6° slope, and hence are not quantitative for downslope flows on all slopes. However, they do show that the notion of entrainment needs to be reassessed for stratified fluids, and that detrainment be considered as a process of equal significance in the bulk behaviour of downslope flows. At the time of writing, similar experiments covering the whole range of slope angles from 3° to 90° are currently in progress, and the results from all of these experiments, analysed in a similar manner to that for the experiments described here, will be reported elsewhere.

2   The experiment

The experiments were carried out in the tank illustrated in Figure 1. The glass-sided tank was approximately 80 cm high and rectangular in cross-section, with internal dimensions of 299 cm in length and 38 cm in width, open at the top and with a solid horizontal bottom. A thin vertical partition made of perspex was placed in the tank extending from one end along approximately 80% of its length, with a uniform gap of 23 cm on one side and 15 cm on the other. The main working region of the tank was in the wider region of width 23 cm, and the experiment was made two-dimensional in this portion as much as possible. A horizontal platform 25 cm above the floor of the tank and extending 40 cm from one end (on the right in Figure 1) was inserted. This platform was continued by a plane slope at 6° to the horizontal, extending downward to within about 5 cm above the floor of the tank. A permanent gate or weir consisting of a horizontal cylinder leaving a gap of 1 cm between it and the platform was placed near the end of the horizontal platform, at a distance of 31 cm from the end of the tank. During the experiments the tank was filled to a level above this gap, and dense fluid was supplied to the region behind the weir by a hose from an external container. This fluid was released into a region of wire mesh, which was instrumental in making the supply approximately uniform across the width of the tank. An additional barrier (see Figure 1) was used at the start and then withdrawn, to enable a sudden commencement at a given time to the release of continuously supplied dense fluid. This fluid then flowed through the gap underneath the weir, and down the slope as a gravity current.

At the start of an experiment, the tank was filled with density-stratified fluid that was produced by using the familiar two-tank mixing procedure with salt water. This filled the whole of the tank including the space underneath the slope and shelf region, and gaps into this region were present during filling: a horizontal gap at the bottom of the slope of approximately 5 cm, and a vertical one at the end of the tank. The purpose of this was to obtain a uniform stratification in the tank, by having a uniform vertical cross-sectional area during filling. After filling, the two gaps into the region underneath the shelf and slope were closed and sealed, so that the fluid there was isolated and not involved in the subsequent experiment. After all motion due to filling had subsided, the density profile $\rho_0(z)$ with depth z was measured by a conductivity probe, calibrated at the top and bottom by samples measured in an Anton Paar Densitometer. Apart from a small kink in the profile at the level of the shelf at the top of the slope, this profile was very nearly linear in depth.

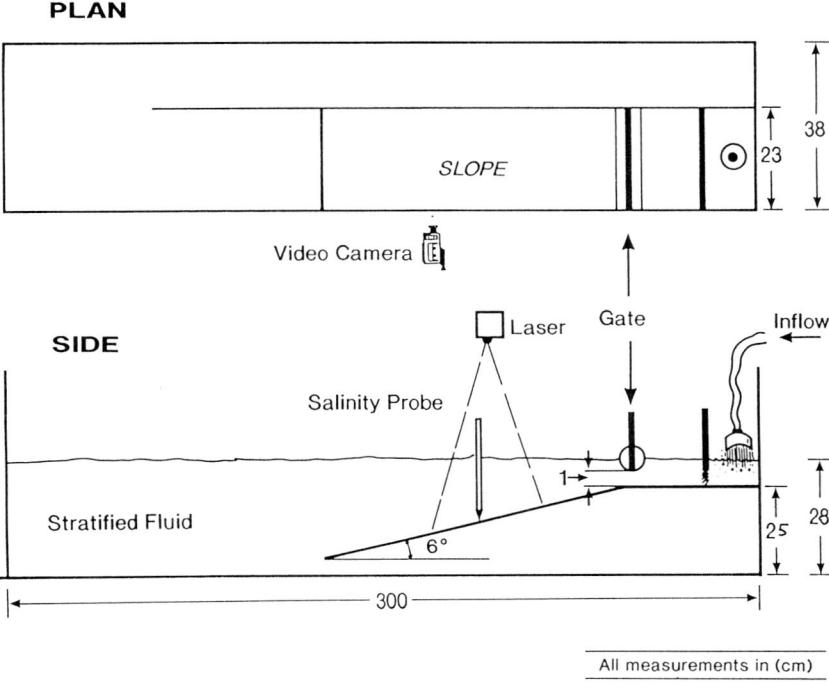

Figure 1. Plan and side views of the tank as used in these experiments. A central vertical section of the flow on the slope was illuminated by a thin sheet of laser light from above, and video-recorded by viewing from the side

The main experiment was then performed by suddenly releasing dense fluid of density ρ_i at the top of the slope in the manner described above. The density of this inflow was chosen to equal that of the undisturbed environmental fluid at a level in the vicinity of the lower part of the tank, more than 5 cm above the bottom, and at a vertical distance D below the level of the shelf. This inflow was continued at a constant rate that was set and monitored by a flow meter in the inflow hose, and lasted for a fixed time that depended on the flow rate and ranged between two and six minutes. Various values of the inflow rate and density and the initial density gradient were used in thirteen separate experimental runs, and the details of these are given in Table 1. The inflowing fluid was dyed with fluorescene, and illuminated in a thin central vertical section by a scanned beam from an Argon ion laser, which gave a clear picture of a two-dimensional cross-section of the motion. This cross-section was recorded on videotape for at least part of each run. The flow had the general character of a gravity current flowing down the slope with a broadly two-dimensional form, with turbulent mixing

P.G. Baines

Table 1. Parameter values for the thirteen experimental runs

M	Q_0	g_0'	N	D	R_e	d_0	R_i
	cm^2 s^{-1}	cm s^{-2}	rad s^{-1}	cm	Q_0/ν	cm	
0.0013	0.463	27.0	1.29	16.3	46.3	0.4±0.1	7.77
0.0026	1.11	27.38	1.19	19.2	111	0.5±0.1	3.51
0.0035	1.82	30.07	1.21	20.5	182	0.5±0.1	1.13
0.0060	2.64	29.3	1.26	18.6	264	0.5±0.1	0.51
0.0067	3.44	31.6	1.27	19.6	344	0.5±0.1	0.33
0.0106	5.56	32.34	1.27	20	556	0.7±0.1	0.36
0.0183	8.39	23.8	1.07	20.75	839	0.9±0.2	0.248
0.0365	6.26	9.31	0.80	14.7	626	0.9±0.1	0.172
0.0453	8.39	18.35	1.22	12.3	839	0.8±0.2	0.127
0.0648	11.21	8.85	0.768	15.0	1121	0.9±0.2	0.051
0.0714	8.39	5.54	0.638	13.6	839	0.9±0.2	0.056
0.0915	8.39	7.67	0.863	10.3	839	0.83±.1	0.061
0.0935	11.21	5.75	0.65	13.5	1121	1.08±.1	0.056

occurring at its upper boundary, and an example is shown in Figure 2. This current had the familiar initial head structure (Britter and Linden 1980) and following current, with the difference that the fluid only penetrated to a level slightly above the point where $\rho_0(z)$ equals ρ_i. During the downflow, inflowing dyed fluid mixed with some of the undyed environmental fluid of the tank. At the end of an experimental run, the extent of dyed fluid consisted of all fluid that had come in contact with the inflowing fluid. When all motion had ceased, the dyed fluid was generally concentrated near a level which was located slightly above a vertical distance D below the shelf, but was also present thoughout the depth range between the shelf and this level. This implied that some inflowing

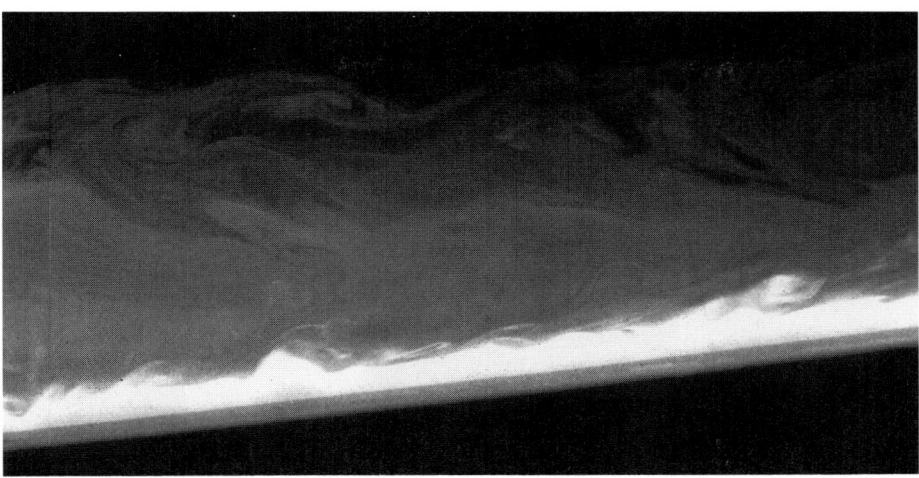

Figure 2. A photograph showing a typical downflow during a run, taken in the central part of the slope (region 2), some time after the passage of the head. The inflowing fluid has been dyed with fluorescene, and the laser illumination shows a vertical section. Note the sharp upper boundary of the main current, the wisps of detrained fluid consistent with the Holmboe instability process, and the presence of detrained fluid in the environment

fluid had become mixed with its environment and found its own level somewhere above its initial density, and the process causing this was continuous in ρ and z. This process is termed *detrainment*.

After the inflow ceased, some motion persisted in the form of low frequency internal waves in the tank, which took up to 30 minutes for viscous decay. When all this motion after a downflow event had died away, the density profile was again measured (several times) by the same techniques and with the same accuracy as the initial density profile, and the comparisons between these two constitute the main quantitative data from the experiment. An example of initial and final density profiles from an experiment is shown in Figure 3. The conductivity probe was again calibrated at the end of the experiment. Generally this showed slight differences from that at the beginning of the experiment, so that the mean of the two could be used for all initial and final profiles, except for a small constant offset which implied a uniform drift in the probe apparatus. This was removed by equating the signal in the before and after profiles in the fluid near the bottom of the tank, well below the active levels of the experiment, where the density was not affected by the

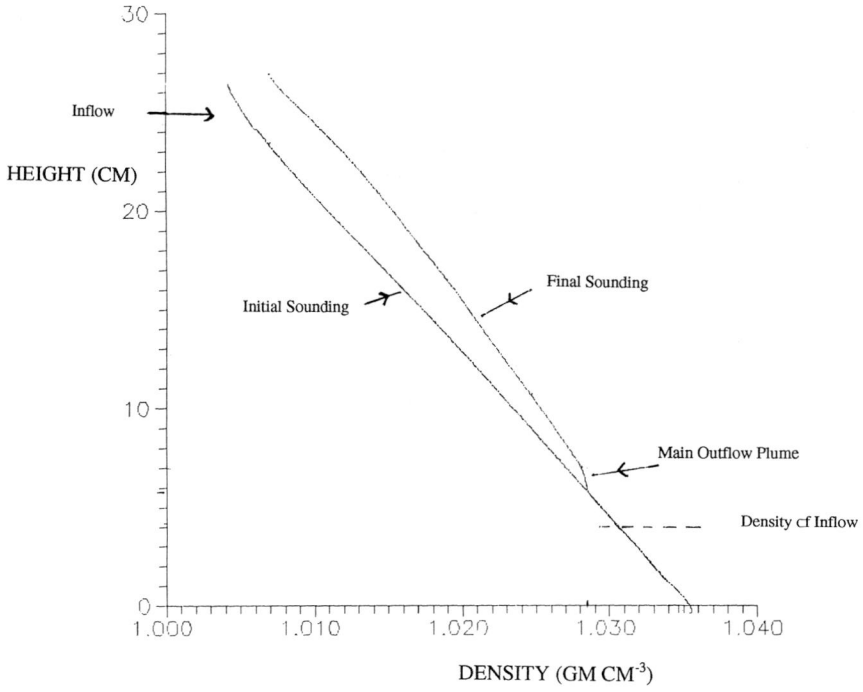

Figure 3. Representative initial and final density soundings for the run with $M = 0.0183$, taken when the fluid in the tank was effectively at rest. The height coordinate denotes height above the floor of the tank

downflow. The time interval of downflow was chosen to be sufficiently long to give measurable differences between the two profiles, and such that these differences were primarily due to the steady downslope flow following the initial head of the gravity current, rather than to processes associated with the head itself (as judged by the relative amount of resulting dyed fluid transferred to the environment). It was for this reason also that the volume of the host fluid in the tank was made as large as possible, to increase the running time, by including the region of the tank behind the partition, as described above.

Observations of the flow during the runs showed that three distinct regions of flow may be identified, as follows. In region (1), occupying the uppermost 30 cm or so of slope length, or 3 cm in depth, the flow emerges from under the weir in an approximately laminar state for the seven runs with $R_e \leq 626$, but is turbulent immediately on exit for the six runs where $R_e \geq 839$. When the flow is initally laminar, it accelerates down the slope under gravity and two-dimensional Kelvin-Helmholtz billows form. The interface then becomes more irregular with three-dimensional wave patterns, leading to turbulent motion. When this has become fully established the flow enters region (2), where the mean layer thickness remains approximately constant (as in Figure 2), and which occupies the main part of the downflow. For the initially turbulent flows, the flow again adjusts to a state of uniform mean layer thickness in region (1), and this includes significant turbulent entrainment that occurs immediately, as discussed below. The mean thickness d of the downflowing current in region (2) was measured by eye, from direct observations and also from videotape records. The interface was unsteady, with continual wave motion and turbulence on long and short length and time scales (see Figure 2 for an example), so that such measurements represented a local mean. However, they showed very little variation in d with downslope distance in region (2), with no systematic trend given the inherent uncertainties in the observations, and mean values are given in Table 1. Net detrainment is visible in this region. At the bottom of region (2) the fluid remaining in the downflow appoaches its ambient density, and spreads from the vicinity of the slope. This denotes region (3). In summary, region (1) denotes the region of adjustment from the weir to the "steady-state" of region (2) - the main region of the downflow, and region (3) denotes the termination of this region, where d becomes large.

In region (2) there are three identifiable regions of fluid motion as one moves outward in a direction normal to the slope. Firstly, there is the downslope flowing region on the surface, with the mean thickness d. Secondly, immediately above this current the flow is in a weakly turbulent state with relatively small velocities, but with a discernible three-dimensional mean flow structure (Mitsudera and Baines 1992). Here the mean density surfaces are observed to have a small upward slope toward the current on the slope (see Section 4). Further away from the slope, the flow is laminar and the motion nearly horizontal. The motion in these regions outside the main downflow is due to buoyancy fluxes at the interface, which cause the fluid to move under gravity to its equilibrium density level in the environment.

The observed value of d shown in Table 1 enables the calculation of a Richardson number based on the volume flux Q_0 and density difference (relative to the environmental value at $z = 0$) of the inflow at the top of the slope. This "bulk" Richardson number is given by

$$R_i = \frac{g_0' d^3 \cos \vartheta}{Q_0^2}, \tag{2.1}$$

and these values are also given in Table 1. A plot of R_i versus M is shown in Figure 4, which shows a clear-cut relationship, R_i decreasing with increasing M.

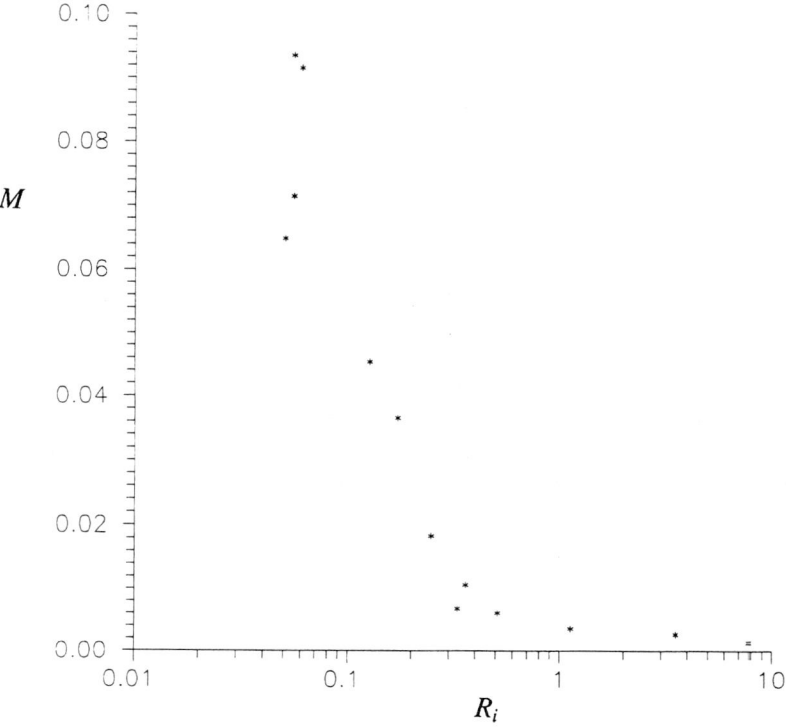

Figure 4. The observed relationship between M and the bulk Richardson number R_i defined by Equation (2.1)

3   Models for downslope flow

In general, the volume flux per unit width down the slope, Q, may be written $Q = Ud$, where U is a mean velocity and d the corresponding thickness of the current. The corresponding momentum flux is then U^2d, and the buoyancy flux $B = GQ$, where $G = g\triangle\rho/\bar{\rho}$, where $\bar{\rho}$ is a mean density and $\triangle\rho(s)$ is the difference in density between the mean density in the downflow and the local environmental density. All of the above quantities are functions of the downslope distance s, with $G = g_0'$ (as given in Table 1) at $s = 0$. In all of the experiments under discussion, the total density variation is only a few percent, so that the Boussinesq approximation is generally applicable. Ellison and Turner developed equations for steady-state flow into a homogeneous fluid of the form

$$\frac{dQ}{ds} = EU, \qquad\qquad (3.1)$$

$$\frac{d(U^2d)}{ds} = S_2 Gd \sin \vartheta - \frac{1}{2}\frac{d}{ds}(S_1 Gd^2 \cos \vartheta) - C_D U^2, \qquad (3.2)$$

$$\frac{dB}{ds} = 0, \qquad (3.3)$$

where s is distance in the downslope direction from the source, S_1 and S_2 are profile factors of order unity, C_D is the drag coefficient on the slope, and E is the entrainment coefficient. In the experiments of Ellison and Turner, the last two terms of (3.2) were small, so that (3.2) could be approximated by (Turner 1973, 1986)

$$\frac{d(U^2d)}{ds} = S_2 Gd \sin \vartheta. \qquad (3.4)$$

For the case of flow into homogeneous fluid, the total fluxes of buoyancy, volume and momentum may be defined by integrals normal to the slope. Environmental fluid that becomes mixed with the downflowing fluid in even the smallest of proportions becomes heavier than its environment, and joins the net flow down the slope. Consequently, it is regarded as entrained, and the value of the entrainment coefficient reflects this. As one moves downslope, the total downslope transport increases, and the affected region (for both velocity and density) extends further out from the boundary (see Figure 4 of Ellison and Turner 1959, or Figure 6.12 of Turner 1973). Close to the boundary there is a rapidly moving region of dense fluid, which can be identified as the partly diluted core of the initially released dense fluid.

The situation with downslope flows into a stratified environment is quite different. Here the principal downflowing stream is observed to have a well-defined upper boundary (see Figure 2), at which there is a sudden change in density. The appearance of the flow is shown schematically in Figure 5, which is representative of the main central region along the downflow. The velocity and density profiles close to the boundary are similar to those just described for a homogeneous environment, but further away from the boundary the growing downslope flow described above is destroyed by the ambient stratification. Waves are present on the boundary, and from the cusped peaks there are wisps of dense fluid that are swept from the lower layer into the ambient fluid above. The process can be readily identified as a manifestation of Holmboe instability (Lawrence et al. 1990, Baines and Mitsudera 1994, Baines 1995), as depicted in Figure 5, occurring on a continuous basis. This process visibly enables dense fluid to be detrained from the main current into the environment above. There, it generates a weak three-dimensional circulation, whereby the fluid settles to its neutral density level in the environment (Mitsudera and Baines 1992).

We may extend the above Equations (3.1–3.4) to describe the motion of the downslope flow into stratified fluid with the process of detrainment included. Here U and d quite naturally become the mean velocity and thickness of the dense layer, below the interface. At this stage, it seems natural to model detrainment

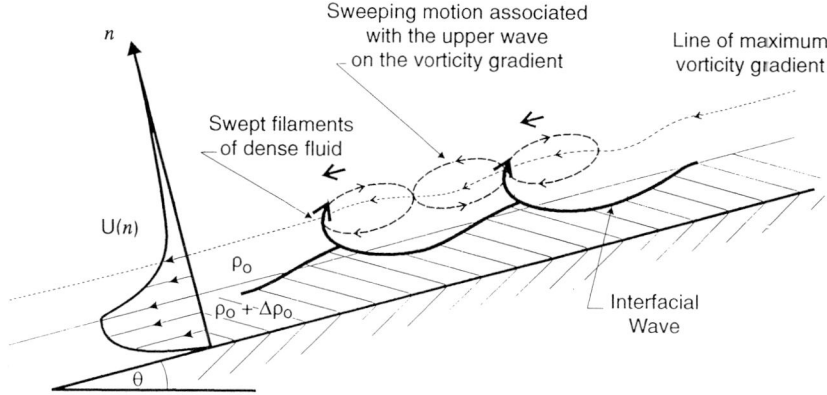

Holmboe instability in downslope flow

Figure 5. A schematic diagram showing the mean velocity profile in region (2), and the process of Holmboe instability, which is due to the mutual interaction between a gravity wave on the interface and a vorticity wave on the vorticity gradient above it (see Baines and Mitsudera 1994, Baines 1995). The main dynamical forces maintain the mean velocity and density profiles, so that the instability process keeps recurring, and wisps of detrained fluid result. n here denotes the coordinate normal to the slope

in the same way as entrainment, with the flux equal to $E_d U$. Entrainment into this current must still be included, and this is again represented by $E_e U$, where E_e and E_d are the entrainment and detrainment coefficients respectively. E_e and E_d are expected to be dependent on the properties of the mean flow, and the main object of the remainder of this paper is to describe this dependence. The relevant physical processes involved are shown in Figure 6, and hence for the stratified environment (3.1), (3.3) and (3.4) become

$$\frac{dQ}{ds} = E_e U - E_d U = (E_e - E_d)U, \tag{3.5}$$

$$\frac{d(U^2 d)}{ds} = G d \sin \vartheta - \frac{1}{2}\frac{d}{ds}(G d^2 \cos \vartheta) - (E_d + C_D + C_{Du})U^2, \tag{3.6}$$

$$\frac{dB}{ds} = -Q N^2 \sin \vartheta - E_d G U. \tag{3.7}$$

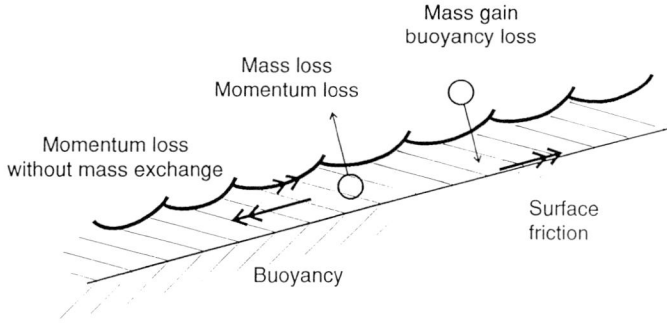

Important Processes

Figure 6. Physical processes associated with the derivation of Equations (3.5–3.8)

Note that the drag term is included here in the momentum equation with two drag coefficients: the first relating to the drag of the rigid bottom surface as before, and the second (C_{Du}), drag due to momentum loss to the overlying fluid that is not associated with mass exchange. The value of C_{Du} has to be inferred from observations. The terms on the right hand side of (3.7) represent the effect of the density gradient in the ambient stratification, and the loss of (negatively) buoyant fluid due to detrainment. From (3.5) and (3.7) we may obtain an equation for G, which is

$$\frac{dG}{ds} = -N^2 \sin \vartheta - G E_e / d. \tag{3.8}$$

These equations are recommended as an improved model for stratified environments over (3.1–3.4), and (3.5) and (3.8) are employed in the next section in the evaluation of E_e and E_d.

4 Analysis and interpretation of observations

The observations described above yield the initial and final states of the stratification, before and after the downflow event. In these steady states the fluid properties are horizontally homogeneous. They would be the same if the tank geometry were altered to be two-dimensional, with the same area at each depth and a uniform width of 23 cm, and we assume this shape for the present analysis. Because of the presence of the slope, the length of this two-dimensional tank is

$$l(z) = 216.1 + (25.0 + z)/\tan 6^\circ, \tag{4.1}$$

where z is vertical coordinate measured from the level of the shelf as before, and has negative values in the region of interest. From the initial and final density profiles, such as those shown in Figure 3, the increase in the height of each density surface as a result of the downflow may be measured. First, the initial and final density profiles are smoothed using the Fourier transform based smoothing routine SMOOFT from Numerical Recipes (Press et al. 1986). This removes small-scale noise in these profiles due to slightly irregular motion of the probe during descent (scarcely visible in the unsmoothed profiles in Figure 3) that can have an undesirable prominence when differences are taken. From the initial and final heights for a given density value, a mean value of $l(z)$ may be obtained from (4.1), and hence the increased volume of the fluid below this density surface calculated. Dividing this by the running time of the inflow and the effective tank width then gives the mean downward flux across this density surface during the experiment, per unit width of the tank. Scaling this with Q_0 gives the function $\tilde{Q}(\rho)$, and an example is shown in Figure 7.

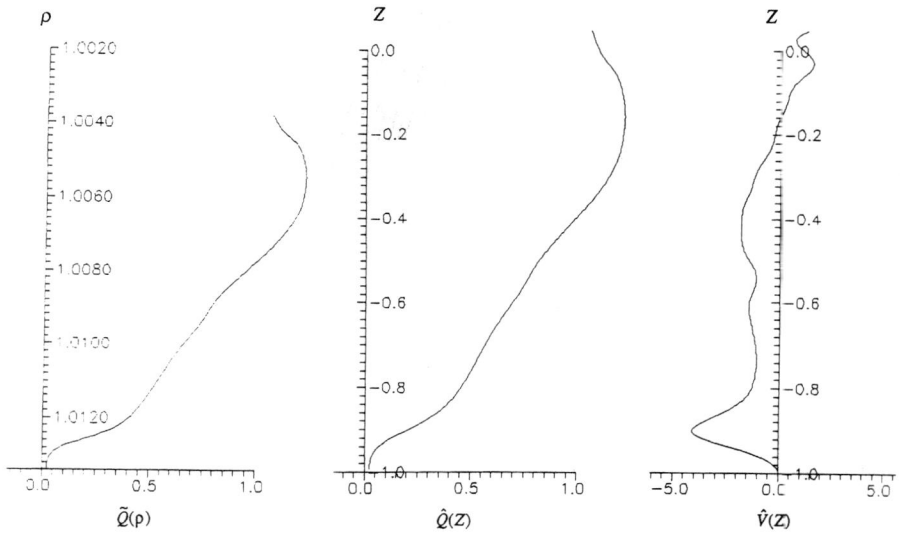

Figure 7. Representative examples of the functions $\tilde{Q}(\rho)$, $\hat{Q}(Z)$ and $\hat{V}(Z)$, for the run with $M = 0.0365$. The first ($\tilde{Q}(\rho)$) shows the downslope volume flux as a function of density, averaged over the whole downflow, and $\hat{Q}(Z)$ and $\hat{V}(Z)$ show the corresponding mean downslope volume flux in the current and outflow velocity respectively, as functions of scaled depth $Z = z/D$. Positive values of $\hat{V}(Z)$ denote flow toward the slope

From $\tilde{Q}(\rho)$ we may calculate the net downward flow at a fixed height z from

$$Q(z) = Q_0 \int_{\rho_i(z)}^{\rho_f(z)} \tilde{Q}(\rho)d\rho, \qquad (4.2)$$

$$= Q_0(\tilde{Q}(\rho_i(z)) + \tilde{Q}(\rho_f(z)))/2, \qquad (4.3)$$

where $\rho_i(z)$ and $\rho_f(z)$ denote the initial and final density values at height z. Expressing this in terms of $Z = z/D$, we have

$$\hat{Q}(Z) = Q(z)/Q_0. \qquad (4.4)$$

$\hat{Q}(Z)$ is therefore an appropriately stretched version of $\hat{Q}(\rho)$. The mean outflow velocity $v(z)$ from the downflow is then given by

$$v(z) = \frac{dQ(z)}{dz}, \qquad (4.5)$$

and in dimensionless form by

$$\hat{V}(Z) = v(z)D/Q_0 = \frac{d\hat{Q}(Z)}{dZ}. \qquad (4.6)$$

Examples of $\hat{Q}(Z)$ and $\hat{V}(Z)$ are also shown in Figure 7.

The initial and final density profiles thus give us the observed dependence of the downslope flux of fluid with depth. There are two complicating factors that should be borne in mind when interpreting and $\hat{Q}$ and $\hat{V}$. Firstly, as the dense fluid flows into the tank, the mean isopycnals rise according to the quantity of fluid that has penetrated below their level. This effect has been neglected in the derivation of (3.5–3.8), where a stationary environment has been assumed. The net rise is small at the lower part of the tank, and increases to a typical value of about 2–3 cm near the top of the slope at the end of a run. Secondly, visual observations during a run indicate that the mean position of the isopycnals near region (2) of the downflow are not exactly horizontal, but slope upward as they approach it. Hence the mean environmental density seen by the main downflow is slightly greater than that given by assuming a stationary environment. The reason for this is clear. In region (2) there is net detrainment (see below) of dense fluid which leaves the current in the form of thin wisps that mix with the environmental fluid, causing a slight increase in the local mean density of the latter. Hence the mean density surfaces slope upward close to the current. This effect has also been neglected in the derivation of Equations (3.5–3.8).

With these provisos, we may infer a great deal about the downslope flow from just the initial and final density profiles. The $\hat{Q}(Z)$ profile in Figure 7 shows that the total downward flux at first increases from unity to a maximum of 1.23 at $Z = -0.16$, and then decreases. This increase is attributed to turbulent entrainment shortly after the inflow, in the manner described by Ellison and Turner. This increased transport takes place outside the main current, and the associated mixed fluid finds its own density levels at Z values above -0.4. This is reflected in the inflow and outflow shown in the $\hat{V}(Z)$ profile. This region of excess transport $\hat{Q}$ helps to define region (1), which extends down to about $Z = -0.4$ where $\hat{Q}(Z)$ returns to its initial value. In region (2) where d is constant there is still some downward transport outside the main current on the sloping mean isopycnals, but this is judged to be small. In Figure 7 this extends from $Z \approx -0.4$ to -0.8. Virtually all experiments show the initial increase in the downflow, with magnitudes of $\hat{Q}$ reaching values ranging from near unity to 1.68. However, no simple relationship between the maxima of $\hat{Q}$ or its location and M, R_e or R_i could be inferred from these experiments. In region (2), $\hat{Q}$ continues to decrease due to detrainment from the main downflow (dyed fluid in Figure 2), and the measurements of detrainment are taken from this region. One noteworthy feature is the "bumpiness" of the outflow curves $\hat{V}$ in region (2). These features appear to be real, and are regarded as due to the three-dimensional flow structure of the mean flow in the region just above (i.e. outside) the downflowing current described by Mitsudera and Baines (1992).

If we make the assumption that E_e and E_d are effectively constant with s in region (2), and use the observation that d is uniform there, we may integrate (3.5) and (3.8) from $s = s_0$ near the top of region (2) to obtain

$$Q(s) = Q(s_0)e^{-(E_d - E_e)(s - s_0)/d_0}, \tag{4.7}$$

$$G(s) = G(s_0)e^{-E_e(s - s_0)/d_0} - \frac{d_0 N^2 \sin \vartheta}{E_e}(1 - e^{-E_e(s - s_0)/d_0}), \tag{4.8}$$

where d_0 is the observed mean value of d (given in Table 1). We may convert (4.7) into $\hat{Q}(Z)$ and compare the result with observations to obtain values for $E_d - E_e$. Specifically, $E_d - E_e$ is given by

$$E_d - E_e = d_0 \sin \vartheta \frac{1}{Q}\frac{dQ}{dz} = \frac{d_0 \sin \vartheta}{D}\frac{1}{\hat{Q}}\frac{d\hat{Q}}{dZ}, \tag{4.9}$$

and all the factors on the right hand side may be measured. E_e may be determined by use of (4.8). The value of G becomes zero when the fluid reaches region (3), and leaves the slope in a conspicuous horizontal plume. The relatively narrow range of densities and depths covered by this plume (at least, compared with D) indicates that the fluid in the downslope current is reasonably well-mixed. We take G to be zero at the depth D_p of the centre of this outflow region, which is identified with the location of the peak in $\hat{V}(Z)$ at the bottom of the profile. Substituting these values into (4.8), and taking $G(s_0) = N^2 D$, $s_0 = 0$, so that (4.8) is assumed to be valid from the top of the slope, through the relatively smaller region (1) as well as region (2), one obtains a transcendental equation that may be solved numerically for E_e in terms of D_p, D and $d_0 \sin \vartheta$.

The values of E_d and E_e obtained by this procedure are shown in Figure 8, as functions of both M and R_i. These constitute the main quantitative results

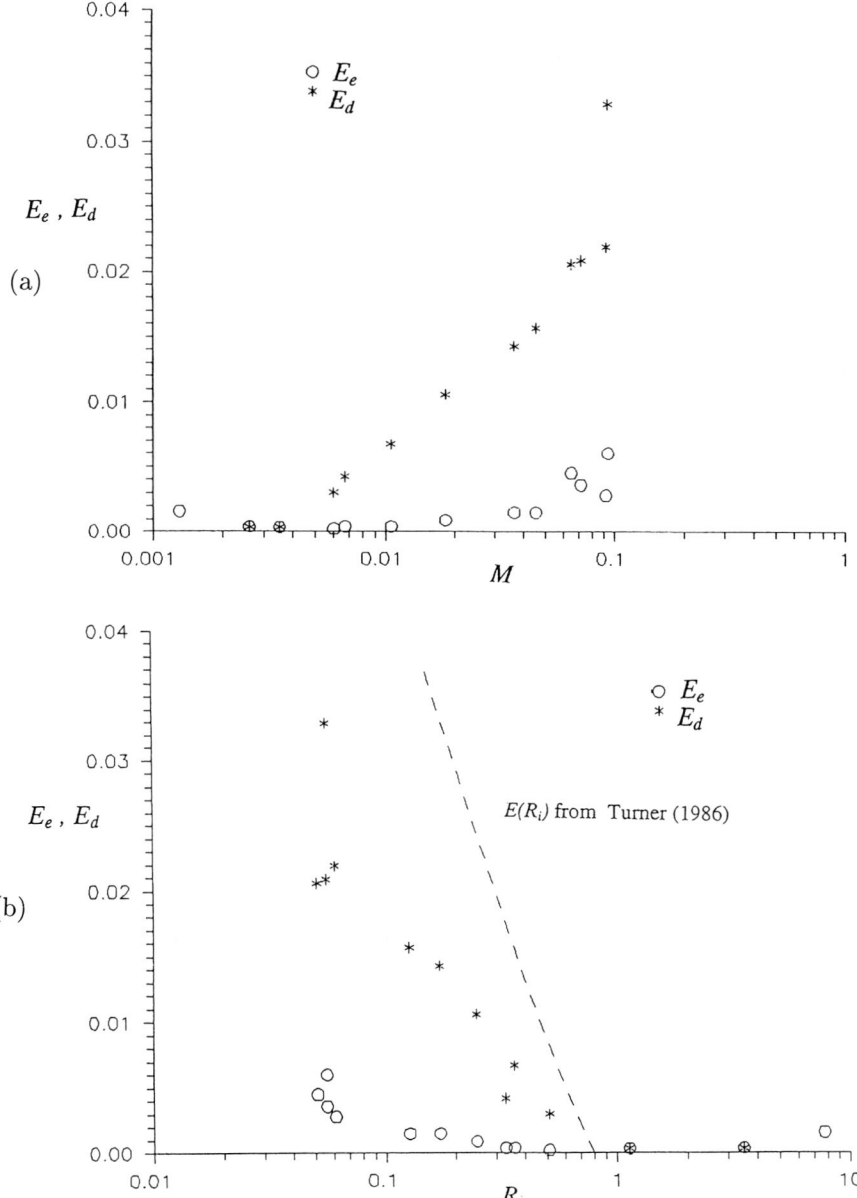

Figure 8. The inferred values of the entrainment E_e and detrainment E_d coefficients, as functions of (a) M, and (b) R_i. The dashed line in the latter also shows the entrainment coefficient E of Ellison and Turner (1959), as given by Equation (23) of Turner (1986). The increased value at small M and large R_i may be attributed to Reynolds number effects (see text)

of this work. Figure 8(a) shows that E_d is much larger than E_e in region (2) for this slope angle, and increases rapidly with M, as does E_e, though from a much smaller value. This indicates that detrainment increases with increasing external stratification, as might be expected. The slight increase in E_e at small $M (= 0.0013)$ may be attributed to Reynolds number effects (here $R_e = 46$), that make this run a special case. Here $\hat{Q}(Z)$ is greater than unity down to the level where region (3) is reached, so that region (2) does not exist in the same sense as for the other runs, and no value of E_d is recorded.

Figure 8(b) shows the corresponding dependence on R_i. This also permits a comparison with the entrainment of Ellison and Turner(1959). In the latter experiments there is no detrainment, and the entrainment coefficient is very much larger than the sum of both the entrainment and detrainment coefficients in the experiments described here. The flows described by Ellison and Turner had structural similarities to the present flows, but with a homogeneous environment all fluid that in the present experiments is regarded as "detrained" still remains part of the flow, and any external fluid that is touched (tainted?) by any part of the dense downflow is regarded as "entrained" by their definition, and joins the general flow downslope. The presence of external stratification destroys this picture as a viable description of the process that occurs. None-the-less, the general dependence of E_e and E_d on R_i closely resembles that of the E of Ellison and Turner, though with smaller magnitudes.

5   Conclusions and Discussion

I have described a series of experiments on the flow of dense fluid down a slope inclined at 6° to the horizontal, into a stratified environment. The flow source is uniform and continuous. The flow can be characterised by the parameter M, which is empirically related to a bulk Richardson number, and the experiments cover the relevant range of M with the Reynolds numbers having mostly large values. The flow is observed to be similar to a gravity current, and three regions of flow type may be identified. These are: an initial adjustment region (region (1)), in which the flow becomes turbulent and adjusts to an approximately uniform thickness. Entrainment of environmental fluid into the downslope current occurs in this region. This leads into region (2), which occupies most of the rest of the flow down the slope. Here the thickness of the downflow is observed to remain approximately uniform, and both entrainment into the current and detrainment from it occur, with the latter being considerably larger in most cases. Hence the volume flux of downflowing fluid decreases with downslope distance in this region, which terminates when the downflowing fluid reaches its ambient density. This fluid then enters region (3), where it accumulates and leaves the slope, spreading horizontally in a conspicuous pancake, or plume. This region covers a small range of densities (relative to the whole range of the downflow), which indicates that the downflow is reasonably turbulent and homogenised. Also, the location of this plume is significantly above the level of the density of the inflowing fluid at the top of the slope, and this difference in levels gives a measure of the entrainment.

The main quantitative observations are made by taking initial and final density profiles in a finite tank, before and after the downslope flow event. From this data, if one assumes that this entrainment and detrainment may each be represented by coefficients in the customary manner of Morton et al. (1956) and Ellison and Turner (1959), one may infer the values of these coefffficients from dynamical considerations. The values of both of these coefficients increase at an approximately exponential rate with M, but their values are always much less than those of the entrainment coefficients of Ellison and Turner, for the same values of the corresponding Richardson numbers. The reason for this is clear - for downslope flows into homogeneous fluid (the case of Ellison and Turner), any environmental fluid that is "touched" by the inflow to even the smallest extent has its density increased, becomes heavier than the environmental density, and thus joins the downflow. Hence it is regarded as "entrained", and the value of the entrainment coefficient reflects this. There is no scope for the concept of detrainment in this picture.

The flow described here is more complicated, and the results less tidy, than those for a homogeneous environment. The arguments of Ellison and Turner for the constancy of the local Richardson number down the slope do not apply here, so that a single initial Richardson number is not adequate to describe the whole flow. We may infer the variation in the local Richardson number, R_{il} (defined as $R_{il} = g'd^3 \cos\vartheta/Q^2$, using the local values of g', d and Q) from Equations (4.7) and (4.8), and these show that, for the bulk R_i less than unity, the local value R_{il} typically varies by a factor of 2 to 3, increasing to a maximum and then decreasing, with downslope distance.

It is possible that there is a better way of representing the detrainment and entrainment processes than the coeffiicient approach that has been used here, which imposes a particular perspective on the observations. One possibility is that much of the entrainment could take place in region (1) at the top of the slope where $\hat{Q}$ increases; however, the magnitudes of this increase are not well correlated with the calculated values of E_e, so that this possibility is not supported by the data. Instead the assumption that the entrainment coefficient is uniform with downslope distance (at least in region (2)) seems to be broadly consistent with most of the data, particularly for the cases with large entrainment where M is large. This paper is a report on progress to date with this series of experiments, and these details should become clearer when more data from other slope angles is analysed.

These results have obvious implications for downslope flows in the environment - the atmosphere, oceans and lakes. Firstly, the dynamical balance of oceanic flows such as the Mediterranean outflow, the Denmark Strait overflow and Antarctic downflows is quite different from the models used so far (see for example Baines and Condie 1998), and the process of detrainment needs to be considered. Similar detrainment is expected from nocturnal katabatic flows down hillsides in the atmosphere. Secondly, detrainment of the downflowing fluid implies that this fluid is distributed over the whole range of depths over which the downflow occurs, excepting a region (1) at the top. This has obvious implications for the distribution of tracers near major downflows in the ocean; evidence exists of detrainment of chemical tracers from downslope flows around Antarctica (Rintoul and Bullister 1997). Thirdly, the presence of both entrainment and detrainment in these flows implies that they provide mechanisms for vertical transport and mixing of the environmental fluid, and constitute a contributory mechanism for "boundary mixing" in the ocean. It is possible that the occurrence of this process at a number of different locations around the world makes a significant contribution to the net vertical mixing in the ocean as a whole.

Acknowledgements

I am most grateful to David Murray, whose constructive participation in all practical aspects and the data analysis for these experiments has made this work possible.

Bibliography

1. Baines, P.G. and Mitsudera, H. (1994). On the mechanism of shear flow instabilities. *J. Fluid Mech.*, **276**, pp. 327–342.

2. Baines, P.G. (1995). *Topographic Effects in Stratified Flows*, Cambridge University Press, p. 482.

3. Baines, P.G. and Condie, S. (1998). Observations and modelling of Antarctic downslope flows. *Ocean, Ice and Atmosphere: Interactions at the Continental Margin*, Editors: S.S. Jacobs and R. Weiss, American Geophysical Union. To appear.

4. Britter, R.E. and Linden, P.F. (1980). The motion of the front of a gravity current travelling down an incline. *J. Fluid Mech.*, **99**, pp. 532–543.

5. Ellison, T.H. and Turner, J.S. (1959). Turbulent entrainment in stratified flows. *J. Fluid Mech.*, **6**, pp. 423–448.

6. Lawrence, G.A., Lasheras, J.C. and Browand, F.K. (1990). Shear instabilities in stratified flow. *Proc. 3rd Int. Symp. on Stratified Flows*, Editors: E.J. List and G.H. Jirka, ASCE, pp. 15–27.

7. Mitsudera, H. and Baines, P.G. (1992). Downslope gravity currents in a continuously stratified environment: a model of the Bass Strait outflow. *Proc. 11th Conf. Australasian Fluid Mechanics*, pp. 1017–1020.

8. Manins, P.C. and Sawford, B.L. (1979). A model of katabatic winds. *J. Atmos. Sci.*, **36**, pp. 619–630.

9. Morton, B.R., Taylor, G.I. and Turner, J.S. (1956). Turbulent gravitational convection from maintained and instantaneous sources. *Proc. Royal Society London*, **A234**, pp. 1–23.

10. Press, W.H., Flannery, B.P., Teukolsky, S.A. and Vetterling, W.T. (1986). *Numerical Recipes*, Cambridge University Press, p. 818.

11. Price, J.F. and Baringer, M.O. (1994). Outflows and deep water production by marginal seas. *Prog. Oceanog.*, **23**, pp. 161–200.

12. Rintoul, S. and Bullister, J.L. (1997). A late winter hydrographic section from Tasmania to Antarctica. *Deep-Sea Res.*. To appear.

13. Simpson, J.E. (1997). *Gravity Currents: in the Environment and the Laboratory*, Cambridge University Press, p. 244.

14. Smith, P.C. (1975). A streamtube model for bottom boundary currents in the ocean. *Deep-Sea Res.*, **22**, pp. 853–873.

15. Turner, J.S. (1973). *Buoyancy Effects in Fluids*, Cambridge University Press, p. 367.

16. Turner, J.S. (1986). Turbulent entrainment: the development of the entrainment assumption, and its application to geophysical flows. *J. Fluid Mech.*, **173**, pp. 431–471.

A Numerical Simulation Study of Two-Dimensional Gravity Currents

Carlos Härtel and Manuel Michaud[1]

Institute of Fluid Dynamics, ETH Zürich, Switzerland

Abstract

A direct numerical simulation study of gravity currents in two-dimensional lock-exchange flow has been conducted. The simulations are based on the Boussinesq equations which are discretized using a mixed spectral/spectral-element scheme in space together with finite differences in time. The influence of viscous forces on the propagation speed of the currents is assessed by a series of simulations with widely different Grashof numbers. From the results it is estimated that a weak Grashof number dependence of the front speed remains up to Grashof numbers of the order of 10^{10}. A direct comparison of simulation results for no-slip and free-slip boundaries reveals that the walls exert significant retarding forces on the propagating currents. The pronounced Kelvin-Helmholtz type instability of the interface between the light and heavy fluid is investigated. It is shown that the onset of this instability is associated with local gradient Richardson numbers below 1/4. An analysis of the structure of the gravity-current head in a moving frame of reference reveals that, in the present case, the foremost point of the gravity current does not coincide with a stagnation point of the flow.

1 Introduction

Gravity currents are flow phenomena which can be observed in many natural and man-made situations (see [1]). These currents are essentially horizontal flows and they are driven by density differences in the fluid which may sometimes amount to merely a few per cent. Common examples of gravity currents encountered in nature are powder-snow avalanches, sea-breeze fronts, thunderstorm outflows or turbidity currents on the ocean floor.

The study of gravity currents is an important issue not only in the geosciences, but also in safety engineering and environmental protection. Such currents may form, for example, after an accidental release of dense industrial gases which may be explosive or poisonous. Another example is the spillage of hydrocarbons on the sea which may spread as a gravity current along the surface. For both risk assessment and the design of efficient counter measures, the understanding of the

---

[1]Present address: Sulzer Innotec AG, Postfach, CH-8401 Winterthur, Switzerland.

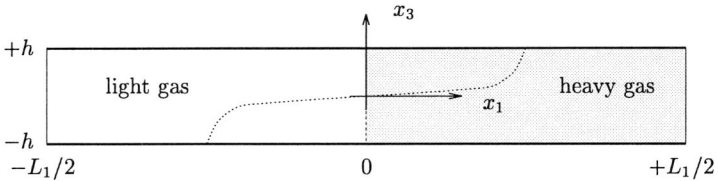

Figure 1. Principle sketch of a lock-exchange flow in a channel of length L_1 and height $2h$. The dotted line gives the interface between the two gases some time after the release (gravity acts in the normal direction x_3)

physical processes involved is essential. Issues which are of particular interest in practice are the speed at which a gravity current spreads and its possible dilution due to mixing with ambient fluid. While extensive theoretical and experimental work has been carried out on gravity currents in the past [2, 3, 4, 5, 6, 7], only few numerical simulations of such flows were presented so far. Moreover, in those simulations which were made, the computational grids employed were too coarse to capture all small-scale structures of the flow [8, 9]. So far, no direct numerical simulations (DNS) of this problem have been attempted, i.e. no simulations where all relevant physical phenomena are thoroughly resolved in space and time.

In the present paper we report about first results of an ongoing research project in which the DNS approach is applied to study fundamental physical properties of gravity currents. The flow considered is the lock-exchange problem, i.e. the mutual intrusion of two fluids of different density in a plane channel [1, 10]. Initially the two gases are separated by a vertical membrane. After the release of the gases a heavy-gas front and a light-gas front develop and propagate along the lower and the upper channel wall, respectively. A sketch of the initial situation and the developing flow is provided in Figure 1. So far we have restricted our attention to strictly two-dimensional flows, but simulations of the full three-dimensional problem are currently being prepared.

2　Governing equations

Since we are primarily interested in flows with small density differences, we adopt the unsteady Boussinesq equations for the mathematical description of the flow [11]. These equations are based on the assumption that the variations in density $\tilde{\rho}$ are small (a tilde denotes a dimensional quantity here). Furthermore, we assume that the variations in density are caused by variations in temperature $\tilde{T}$ only, i.e.

$$\frac{\tilde{\rho} - \tilde{\rho}_r}{\tilde{\rho}_r} = \tilde{\beta}(\tilde{T} - \tilde{T}_r) \ . \tag{2.1}$$

In (2.1) $\tilde{\rho}_r$ and $\tilde{T}_r$ designate the reference values of density and temperature, respectively, and $\tilde{\beta}$ is the coefficient of thermal expansion of the fluid.

In the Boussinesq equations the density variations are only accounted for in the buoyancy term in the momentum equations. If the buoyancy force is assumed to act in the normal direction x_3 only (see Figure 1), the Boussinesq equations take the following dimensionless form

$$\frac{\partial u_k}{\partial x_k} = 0 \qquad (2.2)$$

$$\frac{\partial u_i}{\partial t} + \frac{\partial (u_i u_k)}{\partial x_k} = -\frac{\partial p}{\partial x_i} + \frac{1}{\sqrt{Gr}} \frac{\partial^2 u_i}{\partial x_k \partial x_k} + T\delta_{i3} \qquad (2.3)$$

$$\frac{\partial T}{\partial t} + \frac{\partial (T u_k)}{\partial x_k} = \frac{1}{\sqrt{Gr Pr^2}} \frac{\partial^2 T}{\partial x_k \partial x_k} \quad, \qquad (2.4)$$

where u_i denotes the velocity components, p the pressure and T the temperature. In (2.2-2.4) all terms have been made dimensionless by the channel half-width $\tilde{h}$, the temperature difference $\Delta \tilde{T} = \tilde{T}_{max} - \tilde{T}_{min}$ and the buoyancy velocity $\tilde{u}_b$

$$\tilde{u}_b = \sqrt{\tilde{g}' \, \tilde{h}} \, , \qquad (2.5)$$

where $\tilde{g}'$ denotes the reduced gravity [1] which is computed from the gravitational acceleration $\tilde{g}$ by

$$\tilde{g}' = \tilde{g} \, \tilde{\beta} \, \Delta \tilde{T} \, . \qquad (2.6)$$

The pressure p in (2.3) has been normalized by $\tilde{\rho} \, \tilde{u}_b^2$, and the non-dimensional temperature T is defined as

$$T = \frac{\tilde{T} - \tilde{T}_{min}}{\Delta \tilde{T}} \, . \qquad (2.7)$$

Two dimensionless parameters arise from the normalization discussed above. The first of these is the Prandtl number Pr

$$Pr = \frac{\tilde{\nu}}{\tilde{\kappa}} \qquad (2.8)$$

which is the ratio of kinematic viscosity $\tilde{\nu}$ and molecular diffusivity of temperature $\tilde{\kappa}$. The second parameter is the ratio of buoyancy forces and viscous forces, and it is commonly referred to as the Grashof number Gr [11]

$$Gr = \left(\frac{\tilde{u}_b \, \tilde{h}}{\tilde{\nu}} \right)^2 \, . \qquad (2.9)$$

3   Numerical method

The numerical method employed for the integration of the basic equations is an extension of the scheme used in [12] which was developed for the direct numerical simulation of plane channel flow. The computational box is the domain sketched in Figure 1, where the longitudinal and wall-normal direction of the cartesian coordinate system are denoted by x_1 and x_3, respectively. The third, lateral coordinate direction has been omitted in the figure for simplicity. The numerical scheme is based on a mixed spectral/spectral-element discretization in space, using Fourier-expansions in the wall-parallel directions. In the wall-normal direction a spectral-element collocation technique is employed. The time discretization is done in a semi-implicit manner where a third-order accurate Runge-Kutta scheme is utilized for the nonlinear terms together with a Crank-Nicolson scheme for the diffusive terms and the pressure. More details on the numerical discretization and the solution algorithm are given in [13].

In the lock-exchange flow simulations we impose mirror-symmetry conditions on the flow field at $x_1 = \pm L_1/2$ which corresponds to frictionless end walls at these locations. In the lateral direction x_2 either symmetry or periodicity conditions at $x_2 = \pm L_2/2$ may be utilized. The top and bottom boundaries at $x_3 = \pm 1$ are either rigid no-slip walls or no-stress (i.e. free-slip) boundaries, and they may be either isothermal or adiabatic.

To validate the simulation code we have conducted computations of two-dimensional Rayleigh-Bénard convection, a flow problem for which extensive reference data are available. From the numerical point of view the only differences between the Rayleigh-Bénard flow and the lock-exchange flow are in the boundary conditions for the temperature at the walls: In the Rayleigh-Bénard case the walls are isothermal while we generally employed adiabatic walls in the lock-exchange simulations. Moreover, the initial temperature distribution is different in both cases. In the Rayleigh-Bénard simulations the flow field was initially at rest, and the onset of convection was triggered by minute disturbances (of the order of 10^{-10}) superimposed on the initial linear temperature profile. The flow evolution was then followed in time until a state of steady cellular convection was attained. We have compared the simulation results with both linear-stability theory and reference data for the highly-nonlinear stage of steady convection, and in all cases an excellent agreement was obtained [13].

4 Lock-exchange simulations

4.1 Initial conditions and numerical resolution

In the lock-exchange simulations the flow field was generally initialized with fluid at rest, i.e. $u_i = 0$ everywhere. The initial temperature field $T_0(x_1)$ could in principle be step function with a temperature discontinuity at $x_1 = 0$. However, since we use Fourier expansions in x_1 the temperature profile needs to be continuous and a smooth transient between the temperatures in the left and right channel halves has to be provided. The initial temperature field $T_0(x_1)$ which we specified has the form of an error function

$$T_0(x_1) = \frac{1}{2} - \frac{1}{2} \cdot \mathrm{erf}\left(x_1 \sqrt[4]{GrPr^2}\right) \quad . \tag{4.1}$$

The steepness of the profile (4.1) at the interface depends on $Gr \cdot Pr^2$ which, as seen from (2.4), can be considered as the square of a Péclet number. More details on the derivation of the initial temperature field are given in [13].

The grid size Δx_1 in longitudinal direction which is required to achieve adequate resolution of $T_0(x_1)$ is approximately

$$\Delta x_1 \approx \left(GrPr^2\right)^{-\frac{1}{4}} . \tag{4.2}$$

From our simulations we found that such a grid size is sufficient to obtain a decay of three to four orders of magnitude in the Fourier spectrum of the temperature not only in the initial stage, but also during the further evolution of the flow. A similarly good resolution is found in the velocity field as long as the Prandtl number Pr is not much smaller than one. In the normal direction we employed grid sizes of $\Delta x_3 \approx \Delta x_1$ in the core of the channel, while the mesh was significantly refined near the walls to allow for a thorough resolution of the developing boundary layers.

4.2   Simulation results

An idea of the typical structure of a lock-exchange flow can be gained from Figure 2 where results of a simulation for $Gr=1.25 \cdot 10^6$ are visualized by means of isocontours of the temperature. A numerical mesh with $N_1 \times N_2 = 768 \times 91$ grid points was used in this simulation. The Prandtl number Pr was set to unity which, unless stated otherwise, is the case in all simulations presented here.

From Figure 2 several characteristic features of intrusion fronts become obvious, for example the distinct head of the front with a foremost point, the so-called nose, which is raised above the wall, and the highly unstable interface between the two fluids. The interface is seen to feature pronounced Kelvin-Helmholtz type vortices which are continuously shed from the advancing fronts. Note that the

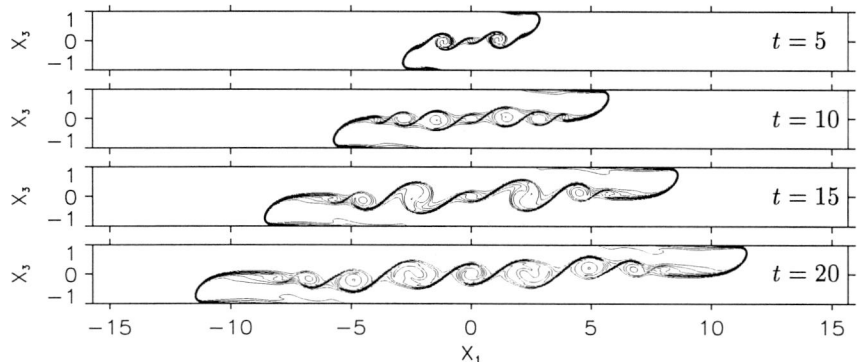

Figure 2. Isocontours of temperature at various times t of the simulation. Results for $Gr=1.25 \cdot 10^6$. No-slip boundary conditions are applied at $x_3=\pm 1$

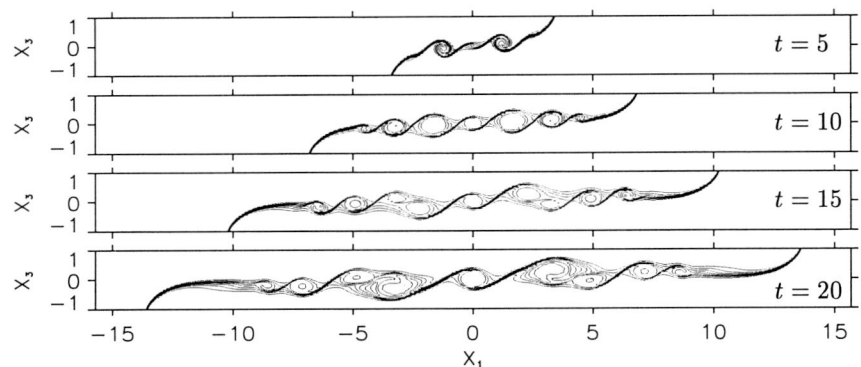

Figure 3. Isocontours of temperature at various time t of the simulation. Results for $Gr=1.25 \cdot 10^6$. Free-slip boundary conditions are applied at $x_3=\pm1$

simulations are strictly two-dimensional and that, hence, the Kelvin-Helmholtz billows may not break up into smaller structures as usually occurs in three-dimensional gravity currents [1, 3]. Therefore, these vortices may thicken until the finite height of the channel inhibits further growth.

Figure 3 shows isocontours of temperature for a flow at the same Grashof number as in Figure 2, but with free-slip boundary conditions applied at $x_3=\pm1$. Similar to the case with no-slip boundaries, intense co-rotating vortices emerge in the interior of the channel, which grow in time and occasionally undergo pairing. From theoretical arguments it is known that for frictionless walls the angle that the front forms with the wall should be 60 degrees right at the wall [2]. Our simulation results are in good agreement with this theoretical finding [14]. Though the general structure of the flow appears to be similar for no-slip and free-slip boundaries, there is a significant difference with respect to the head of the front. While the foremost point of the current is raised above the wall if no-slip conditions are utilized, it is located right on the wall in the case of stress-free walls. Among other things, this implies that no fluid is overrun by the front in the latter case.

Comparing the visualizations shown in Figures 2 and 3 reveals that the fronts spread faster if free-slip boundary conditions are applied at the walls. Such a behavior is generally to be expected, since in this case no boundary layers form beneath the currents, which may exert retarding forces. A more detailed assessment of the influence of these boundary layers is given in the subsequent section.

4.2.1 Speed of the front

The overall speed at which an intrusion front propagates is of particular interest in practice. If one denotes the (time dependent) x_1 position of the nose of the front by x_1^n, the propagation speed u_f is defined as

$$u_f = \frac{dx_1^n}{dt} \ . \tag{4.3}$$

In a lock-exchange flow, the front speed is zero at $t = 0$, but rises rapidly during an initial transient. This is illustrated by Figure 4 where the temporal evolution of the front speed is depicted for the two simulations shown in Figures 2 and 3. It is seen from the curves that the front speed attains a constant value after 5 to 8 dimensionless time units. This constant front speed is about 10–20% lower than the maximum front speed which occurs during the initial transient.

The constant front speed, normalized by the buoyancy velocity (2.5), is usually considered as the Froude number Fr of the gravity current [1]. In order to assess the influence of viscous forces on the propagation speed of the fronts, we have evaluated the respective Froude numbers from a series of simulations with different Grashof numbers. It should be stressed that the examination of such Grashof-number effects is of special interest in DNS. This is due to the fact that in practical applications Grashof numbers are typically very large, whereas direct simulations are constrained to moderate Grashof numbers due to the high resolution requirements.

Figure 5 shows the Grashof-number dependence of the front speed, as obtained from our simulations. Considering the simulations with no-slip boundary conditions it is seen that Fr increases significantly with increasing Grashof number over the whole range examined. Theoretical arguments show that an upper limit for the Froude number is $Fr_{max}=1/\sqrt{2}$ which can be derived under the assumption that potential energy is converted into kinetic energy without losses [1, 2]. From numerous experiments it is known that the high-Grashof-number

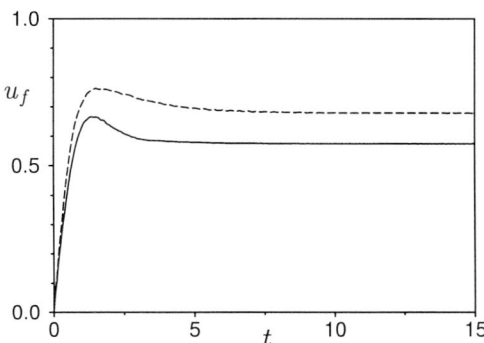

Figure 4. Frontspeed as a function of time after the initial release. Results for no-slip walls (solid line) and no-stress walls, respectively ($Gr=1.25 \cdot 10^6$)

limit of Fr is slightly less than this theoretical value and amounts to about 0.65 to 0.67 [1, 15]. Given this and the results shown in Figure 5, we conjecture that a weak Grashof-number dependence of the front speed may remain up to $Gr \approx 10^{10}$. For comparison, we have included results from recent lock-exchange experiments in Figure 5, and a good agreement between these experiments and our present simulations can be observed. It should be emphasized, that the experimentally examined flow was highly three-dimensional. However, the fact that the Froude number obtained from our strictly two-dimensional simulations agrees well with the experimental result suggests that the front speed is not much influenced by three-dimensional effects.

The curves depicted in Figure 5 again show that at a given Grashof number a front spreads faster, if free-slip conditions are applied at the walls rather than no-slip conditions. The respective differences in the results are more pronounced for lower than for higher Grashof numbers. Particularly noteworthy is the fact that for free-slip boundary conditions the high-Grashof number limit of the Froude number appears to be very close to the theoretical limit of $1/\sqrt{2}$. This indicates that the difference between Fr_{max} and the experimentally established limit for high Grashof numbers is mainly due to wall friction.

At the interface and at no-slip walls there is a continuous transport of fluid away from the head of a gravity current. Consequently, the internal velocity within the advancing front must exceed the front speed in order to provide the mass transport required to balance the losses. Figure 6 shows the longitudinal velocity $u_1(x_1)$ for four different time instants of a simulation for $Gr = 1.25 \cdot 10^6$. The velocities were taken at $x_3 = 0.745$ which is the wall-normal position of the nose of the light-gas front which travels to the right in the upper channel half.

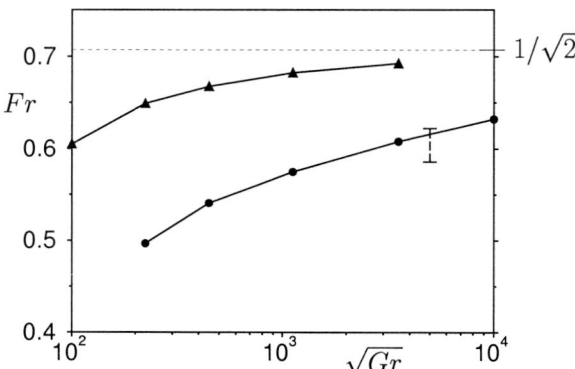

Figure 5. Froude number Fr of the front as a function of Grashof number. Results obtained with free-slip walls (triangles) and no-slip walls (symbols identify the individual simulations). $Fr_{max}=1/\sqrt{2}$ is the theoretical limit for the Froude number [1]. The dashed vertical bar gives the span of results obtained in recent lock-exchange experiments with Ar and CO_2 [16]

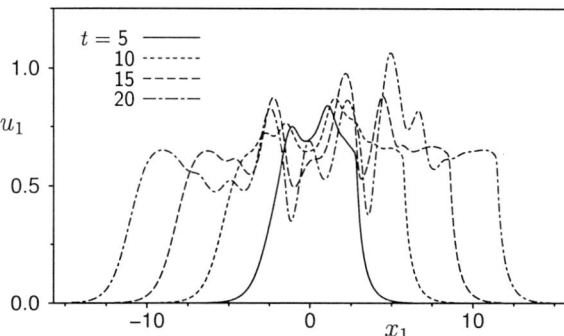

Figure 6. Longitudinal velocity $u_1(x_1)$ at $x_3=0.745$ for four different time instants t. Simulation performed with $L_1=31.4$ and $Gr=1.25 \cdot 10^6$

The actual position of the head of the front is where the steep increase of u_1 occurs. While for this Grashof number the dimensionless front speed is about 0.57 (see Figure 5) u_1 takes much larger values within the gravity current. From $t=10$ on, when the front is fully developed, a zone of almost constant velocity forms immediately behind the front. The velocity in this region is approximately 0.66 which exceeds the front speed by about 16%. This again is in good agreement with experimentally established results [1, 3].

4.2.2 Influence of finite box size

During the course of the simulation, the propagating fronts are approaching the symmetry planes at $x_1=\pm L_1/2$, and it is interesting to know to what extent the velocity of the gravity currents is altered by the presence of these frictionless boundaries. To examine this issue we have conducted three simulations for the same Grashof and Prandtl number ($Gr=6.25\cdot10^5$, $Pr=2$), but with widely different channel lengths. The front speed obtained in the longest channel ($L_1=31.4$) has been taken as the reference solution, and is denoted as $u_{f,\infty}$ in the following. Using this reference solution, the relative "error" ε_f in the front speeds u_f obtained in the other two simulations with channels of length $L_1=10.5$ and $L_1=20.9$, respectively, is given by

$$\varepsilon_f = \frac{|u_{f,\infty} - u_f|}{u_{f,\infty}}. \tag{4.4}$$

In Figure 7 this error ε_f is plotted versus $\Delta x_S = L_1/2 - |x_1^n|$, i.e. versus the distance between front and symmetry plane. It is seen that the two curves essentially collapse which indicates that ε_f is a function of Δx_S only, independent of the actual length L_1 of the channel. Figure 7 also reveals that ε_f decreases exponentially with increasing Δx_S. At $\Delta x_S \approx 4$ the error falls below 10^{-6} which is approximately the noise level set by the numerical accuracy of the finite-difference scheme which we employed to compute u_f. From the curves it can be

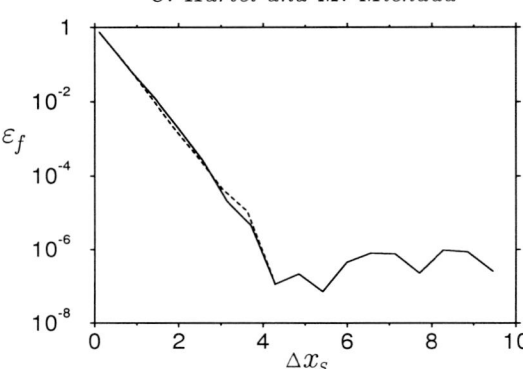

Figure 7. Relative error ε_f in the front speed u_f as a function of the distance Δx_S of the front from the symmetry plane. Results for: $- - -$ $L_1=10.5$, —— $L_1=20.9$. Reference solution obtained in a channel of length $L_1=31.4$. In all simulations $Gr=6.25 \cdot 10^5$ and $Pr=2$

concluded that the propagating fronts are essentially unaffected by the presence of a symmetry plane, provided the distance between front and symmetry plane is more than about one channel height ($\Delta x_S > 2$). In this case ε_f stays well below 1 %. Therefore, in all simulations we performed the box length L_1 was chosen such, that Δx_S was still larger than 2 at the end of each individual computation.

4.2.3 Stability of the interface

From Figures 2 and 3 it was seen that the interface between the light and heavy gas is subject to a pronounced Kelvin-Helmholtz-type instability. For this instability to develop the Grashof number must exceed a certain threshold value, since for very low Grashof numbers no billows can be observed [13]. From our simulations we found that this threshold value is approximately $Gr = 10^5$. In Figure 8 isocontours of the temperature are shown for two simulations having Grashof numbers below and above $Gr=10^5$, respectively. While clear Kelvin-Helmholtz vortices can be seen at $Gr=2.5 \cdot 10^5$, the interface is almost undisturbed at $Gr=9 \cdot 10^4$. We remark that the slight waviness of the interface in the latter case does not indicate the onset of an instability; rather the observed waves are remnants of disturbances which are generated during the initial transient when the fronts start to form. These disturbances, however, are not amplified at this Grashof number and decay during the further evolution of the flow.

At low Grashof numbers, when the interface is stable, the flow in the central part of the channel resembles a parallel stratified shear layer. The stratification

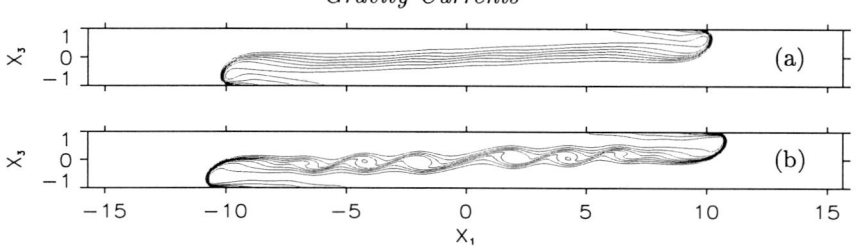

Figure 8. Isocontours of temperature at time $t=20$. Results for (a) $Gr=9 \cdot 10^4$, and (b) $Gr=2.5 \cdot 10^5$

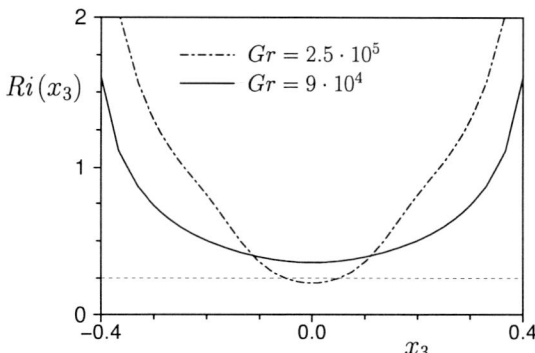

Figure 9. Gradient Richardson number (4.5) at $x_1=0$ as a function of x_3 (only the core flow is shown). Results for two different Grashof numbers at time $t=20$. The dashed line gives the critical Richardson number $Ri=1/4$ of an unbounded stratified shear layer [18]

of shear layers is usually characterized by the gradient Richardson number Ri [17, 18] which, in the present case, is given by

$$Ri = \frac{\partial T/\partial x_3}{(\partial u_1/\partial x_3)^2} \quad . \tag{4.5}$$

In the absence of molecular diffusion the critical Richardson number for an unbounded stratified shear flow is known to be $1/4$, meaning that for $Ri>1/4$ no instabilities can develop [17, 18]. For the flows presented in Figure 8, however, molecular diffusion plays a significant role. Moreover, the thickness of the interface is of the same order of magnitude as the channel height, which means that the influence of the boundaries at $x_3=\pm1$ cannot be neglected. The critical Richardson number should therefore be expected to be somewhat smaller than $1/4$ in the present case.

For the flow at $Gr=9 \cdot 10^4$ shown in Figure 8 the actual values of the gradient Richardson number at $x_1=0$ are given in Figure 9 as a function of the wall-normal coordinate. It is seen that Ri is larger than $1/4$ everywhere in the interior of the channel, consistent with the observation that the interface remains stable. We have evaluated the gradient Richardson number also for the flow at $Gr=2.5 \cdot 10^5$, and the results are included in Figure 9 for comparison. It is seen that in this case Ri falls below the value of $1/4$ in the core flow; note, however, that the flow field at $Gr=2.5 \cdot 10^5$ is rather complex, and that the Richardson number (4.5) is not fully appropriate for the assessment of the stability characteristics of the interface.

4.2.4   Structure of the gravity-current head

Since the propagation speed of the front is constant after the initial transient, the structure of the head of the front can be examined in a frame of reference moving with the front speed u_f. In such a frame of reference the head of the front is at rest and the flow in its neighborhood is essentially stationary. In Figure 10 the structure of the foremost part of the front is illustrated by the streamline pattern in a section of the flow domain around the head. The results were obtained from a simulation at $Gr=1.25 \cdot 10^6$ where the Froude number of the gravity current is about $Fr=0.574$. The streamline pattern in Figure 10 is based on the flow field at time $t=15$ of the simulation, but from an inspection of several other time instants we found that the results are essentially independent of time after the initial transient.

In the moving frame of reference the light fluid approaches the gravity current from the left with speed u_f. It is seen from Figure 10 that the oncoming flow bifurcates and flows either side of the head. In the present case some

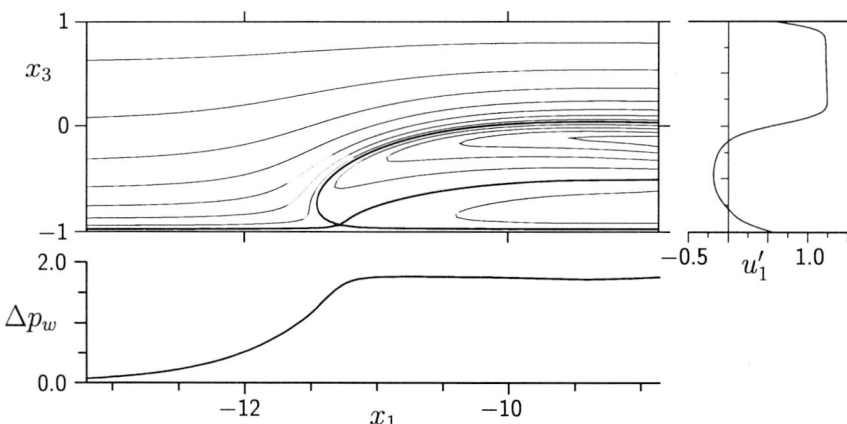

Figure 10. Structure of the gravity-current head in a frame of reference moving with $-u_f$ in x_1. Results for $Gr=1.25 \cdot 10^6$ and $t=15$ ($L_1=31.4$). Above: Streamline pattern in a section of the flow domain (the thick line is the stagnation streamline) and velocity profile at the right margin of this section (note that the walls at $x_3=\pm 1$ are moving to the right with the reference velocity). Below: Increase in wall pressure p_w at $x_3=-1$ normalized by $u_f^2/2$

1.25 % of the oncoming light fluid flows beneath the front, while all the remaining fluid is displaced upwards. The relative speed u_1' in the moving frame of reference is given by

$$u_1' = u_1 + Fr , \tag{4.6}$$

and the profile $u_1'(x_3)$ at the right margin of the section of the flow domain has been included in Figure 10. The graph shows that the fluid displaced upwards is accelerated to about double the reference velocity, and therefore develops a thin boundary layer at the upper channel wall. Since the front is stationary in the moving frame of reference, the total volume flux of heavy fluid is identical to zero at the right margin of the flow domain.

The streamline that separates the bifurcating oncoming fluid is the stagnation streamline, and it meets another streamline from within the gravity current at the stagnation point of the flow. In the present case the stagnation point is located upstream of the nose of the gravity current in the immediate vicinity of the wall. In the neighborhood of the stagnation point a considerable increase in pressure can be observed. The change in wall pressure at $x_3=-1$ is depicted in the bottom graph of Figure 10 where the following normalized pressure difference Δp_w is given

$$\Delta p_w = \frac{p(x_1, x_3=-1) - p(-L_1/2, x_3=-1)}{u_f^2/2} . \tag{4.7}$$

The curve shows that the increase in wall pressure amounts to almost two times the dynamic pressure $u_f^2/2$ of the oncoming light fluid.

We wish to remark that it is usually assumed that the nose of a gravity current is a stagnation point in a moving frame of reference [1, 3]. However, from our simulations we find that the foremost point of the front and the stagnation point do not coincide, as discussed before. Since so far we have studied strictly two-dimensional flows only, we will examine the influence of three-dimensional effects on the location of the stagnation point in the next stage of the project.

5   Concluding remarks

Results from a numerical study of intrusion fronts in two-dimensional lock-exchange flow have been presented. The simulations are based on the Boussinesq equations in which density variations are assumed to be small. For the numerical solution of these equations a mixed spectral/spectral-element scheme is employed.

The numerical simulations reveal the typical characteristics of intrusion fronts which have been observed in numerous experimental studies in the past. Among these is the formation of a pronounced head of the front with a foremost point being raised above the wall. The interface between the light and heavy fluid in the interior of the channel was shown to remain stable at low Grashof numbers, while featuring strong Kelvin-Helmholtz billows at Grashof numbers above 10^5. The occurrence of this Kelvin-Helmholtz type instability was found to be associated with local gradient Richardson numbers below 1/4.

The influence of viscous forces on the propagation speed of the fronts was assessed by a number of simulations with widely different Grashof numbers. It turned out that the propagation speed is affected by viscosity over the whole range examined. We have estimated that a weak Grashof-number dependence of the front speed may remain up to Grashof numbers of the order of 10^{10}. It was also shown that the boundary layers at the walls exert a significant influence on the propagation speed of the front. The Froude numbers obtained from our two-dimensional simulations were compared with recent experimental data and a good agreement was found. Since the flow in the experiments was highly three dimensional, we conjecture that the Froude number in a lock-exchange flow is not much influenced by three-dimensional effects.

The structure of the gravity current head was examined in a frame of reference moving with the front. The streamline pattern revealed that in the present case the foremost point of the front and the stagnation point do not coincide. The latter is located upstream of the nose in the immediate vicinity of the wall. The increase in wall pressure in the neighborhood of the stagnation point was found to be about double the dynamic pressure of the oncoming flow.

In our two-dimensional lock-exchange simulations the Kelvin-Helmholtz billows at the interface may thicken and pair until the finite width of the channel restrains further growth. However, from experiments it is known that propagating gravity currents are subject to pronounced three-dimensional instabilities which break up the large-scale vortices and eventually lead to small-scale turbulence. In the next stage of our research project we will perform three-dimensional simulations to study this breakup process in detail.

Acknowledgments

Computing time on a CRAY J90 has been provided under the collaborative project SuperCluster of ETH Zürich and CRAY/SGI.

Bibliography

1. Simpson, J. E. (1987). *Gravity Currents: in the Environment and the Laboratory*. Ellis Horwood Limited, Chichester.

2. Benjamin, T. B. (1968). Gravity currents and related phenomena. *J. Fluid Mech.*, *31*, 209–248.

3. Simpson, J. E. and Britter, R. E. (1979). The dynamics of the head of a gravity current advancing over a horizontal surface. *J. Fluid Mech.*, *94*, 477–495.

4. Keller, J. J. and Chyou, Y.-P. (1991). On the hydraulic lock-exchange problem. *ZAMP*, *42*, 874–910.

5. Gröbelbauer, H. P., Fanneløp, T. K. and Britter, R. E. (1993). The propagation of intrusion fronts of high density ratios. *J. Fluid Mech.*, *250*, 669–687.

6. Yao, J. and Lundgren, T. S. (1996). Experimental investigation of microbursts. *Experiments in Fluids*, *21*, 17–25.

7. Hacker, J., Linden, P. F. and Dalziel, S. B. (1996). Mixing in lock-release gravity currents. *Dyn. Atmos. Oceans*, *24*, 183–195.

8. Droegemeier, K. K. and Wilhelmson, R. B. (1987). Numerical simulation of thunderstorm outflow dynamics. Part I: Outflow sensitivity experiments and turbulence dynamics. *J. Atmos. Sci.*, *44*, 1180–1210.

9. Klempp, J. Rotunno, R. and Skamarock, W. (1994). On the dynamics of gravity currents in a channel. *J. Fluid Mech.*, *269*, 169–198.

10. Fanneløp, T. K. (1994). *Fluid Mechanics for Industrial Safety and Environmental Protection*. Elsevier Science B.V., Amsterdam.

11. Gebhart, B., Jaluria, Y., Mahajan, R. L. and Sammakia, B. (1979). *Buoyancy-Induced Flows and Transport*. Hemisphere Publishing Corporation, New York.

12. Gilbert, N. and Kleiser, L. (1991) Turbulence model testing with the aid of direct numerical simulation results. *Proc. 8th Symposium on Turbulent Shear Flows*, Munich, September 9-11.

13. Härtel, C., Kleiser, L., Michaud, M. and Stein, C. F. (1997). A direct numerical simulation approach to the study of intrusion fronts. *Journal of Engineering Mathematics* (to appear).

14. Härtel, C., Michaud, M. and Stein, C. F. (1996). Direct numerical simulation of heavy-gas dispersion in a plane channel. *In: Advances in Turbulence VI* (eds. S. Gavrilakis *et al.*). Kluwer Academic Publishers, Dordrecht.

15. Yih, C.-S. (1965). *Dynamics of Nonhomogeneous Fluids*. The McMillan Company, New York.

16. Müller, J. and Fanneløp, T. K. (1996). *Private Communication*.

17. Drazin, P. G. and Reid, W. H. (1981). *Hydrodynamic Stability*. Cambridge University Press, Cambridge.

18. Turner, J. S. (1973). *Buoyancy Effects in Fluids*. Cambridge University Press, Cambridge.

Experimental Study of Heavy-Gas Dispersion on Sloping Surfaces

J. Müller[1] and T.K. Fanneløp[2]

Institute of Fluid Dynamics, Swiss Federal Institute of Technology, Zürich

Abstract

The motion of a dense-gas cloud is known to be strongly influenced by slopes. Various theoretical models have been proposed but the scarcity of experimental data makes their validation difficult. An extensive experimental study of heavy-gas clouds suddenly released on uniform sloping surfaces was therefore undertaken. Both, cylindrical releases on an unobstructed surface and 2-d releases in a channel were investigated. The parameters varied include the gas density, the release volume and aspect ratio, and the slope. A new model combining the slumping motion of the cloud, calculated using a modified box-model, with the slope-dependent motion of its center of gravity, is proposed herein. Good agreement with experimental data is obtained in general as well as in the limit of zero slope and in the early spreading phase.

1 Introduction

For risk assessments of chemical plants, storage facilities and transport systems, knowledge of the spreading and dilution behaviour of hazardous gases is of prime importance. The gases involved often have densities much higher than air, either because of high molecular weight, low temperature or because of suspended particles. The range and the dilution of a heavy-gas cloud is very much affected by obstacles and the geometric features of the terrain including slopes and channels.

Extensive experimental and theoretical studies on the effects of slopes have been performed for underwater turbidity currents and for snow avalanches. Although the flows are similar to the case of present interest, the results are not applicable to gas clouds. For turbidity currents, the density ratios involved are often too small (close to the Boussinesq limit) and the slopes are usually shallow. In the case of avalanche dynamics, the entrainment of snow particles results in a change of cloud density adding yet another unknown. Interesting experiments on a very small scale of heavy gas flow on a sloping surface have been undertaken by Schatzmann et al. [1]. Some theoretical models are available. A model for

---

[1] Present Address: SF-Emmen, P.O. Box, CH-6032, Emmen, The Netherlands.
[2] Present Address: Utsikten 6, N-3155 Aasgaardstrand, Norway.

channel flows based on the theory of thermals is due to Beghin et al. [2]. In
their model, the front velocity is a function of various slope-dependent coefficients. Webber et al. [3] used a shallow-layer formulation to find a similarity
solution, valid for large times. In this model, the cloud, which has the shape
of a wedge, moves down the slope at a constant velocity. The drawback of this
simple model is that it is not valid in the early spreading phase and that it
breaks down for small slopes. Kukkonen and Nikmo [4] proposed another model
which also accounts for wind effects. The cloud is assumed to be circular and the
gravity-driven dispersion can be calculated using any suitable model. A force
equilibrium on the integral cloud is then used to calculate the drift velocity. The
drift due to wind and slope effects does not influence the rate of dispersion.

The lack of adequate experimental data has made the validation of these
theoretical models difficult. The aim of the present work is to fill this gap. To
this end, an extensive experimental study of gas clouds suddenly released on
uniformly sloping surfaces (3-d flow) and in channels (2-d flow) was performed.
The slopes ranged from 0 to 15 degrees. The gases used were Argon, Freon R22
and SF_6, yielding density ratios as high as 5.

2   Experimental setup

The surface on which the experiments are performed consisted of a board 6 meters long and 3.4 meters wide (Figure 1). The slope could be varied between
0 and 15 degrees. Prior to the release, the heavy-gas was contained in a circular cylinder perpendicular to the surface. Three different cylinders were used
(Table 1). A vertical slot was cut into the side of each cylinder. This slot was
partially covered with tape to produce different filling heights. A foam gasket
at the bottom of the cylinder prevented excessive leakage. The gas came from a
pressurized container and was introduced through the top of the cylinder. During the filling process, the open cylinder top was covered by a lid. The gas
concentration was monitored by measuring the oxygen content in the cylinder.
For visualization, an artificial fog was added when the oxygen deficit indicated
a gas concentration of about 90%. The gas supply to the cylinder was continued until the mixing was completed to maintain the concentration. The excess
gas was allowed to drain through the slot until the predetermined height was
reached. The top cover was then removed and the experiment started.

Initially, the cylinder was lifted by hand. This was later changed to a mechanized system, thereby improving the reproducibility of the experiments. The
opening time was short enough so as not to influence the spreading behaviour of
the cloud. Most experiments were conducted with Argon. CO_2, Freon R22 and
SF_6 were used in addition to investigate the influence of density. Table 2 gives
the densities and the density ratio with respect to air of the gases used. Each
experiment was repeated up to 8 times. This number of repetitions was found
necessary when the cylinder was removed manually. Especially for the lighter
dense-gases and for the uphill measurements, where the velocities are small,

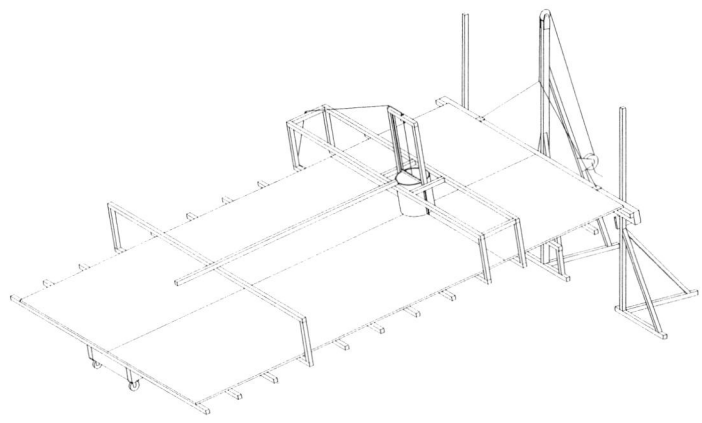

Figure 1. Sloping surface

Table 1. Dimensions in mm of cylindrical gas containers

Cylinder	Internal Diameter	Maximum Height
Small	133	300
Medium	173	440
Large	194	480

variations introduced by the turbulence in the ambient air were noticeable. Outliers were eliminated and the remaining data ensemble averaged. The reproducibility of the experiments was improved considerably by using the mechanized lifting system and only 5 repetitions were found to be necessary for the case when this system was used.

For visual determination of the cloud contour an oil fog was used. The change in gas density due to the presence of the fog particles was estimated to be less than 3%. This result was obtained by weighing a known amount of gas with and without the addition of fog on a highly accurate scale. The visualization experiments were recorded on video. Markings painted on the surface helped to determine the location of the front as a function of time. Some errors in the analysis of the visual record were introduced because of the the camera position and due to the optics of both the camera and the television screen. The visibility

Table 2. Density of gases at 25°C and 0.965 bar

Gas	Density [kg/m^3]	$r = \rho_{gas}/\rho_{air}$
Air	1.12	1.0
Argon	1.54	1.4
CO$_2$	1.71	1.5
Freon R22	3.36	3.0
SF$_6$	5.57	5.0

of the cloud decreased with distance. This could have contributed to the larger scatter at late times.

For the concentration measurements fast aspirated hot-wire probes, developed at the Institute of Fluid Dynamics [5], were used. The data from a maximum of 30 channels was recorded on a PC at a data acquisition rate of 100 Hz. Most of the probes were placed along the centerline, 15 mm above the surface. In addition, vertical rakes of five probes each was used at three stations.

For the 2-d experiments side-walls were added in the direction of maximum slope. The resulting channel had a width of 60 cm and a height of 76 cm. One wall was made out of transparent material to allow visual observations. The length of the channel was about 4 meters. The gate, which could be tilted to remain vertical, was opened with the same pneumatic system used for radial flows. The size and shape of the initial gas volume was determined by the position of the movable rear wall.

3 Theoretical considerations

3.1 Shallow layer model

Webber et al. [3] have made use of a solution of the shallow-layer equations to calculate the spreading behaviour of both a cloud in a channel and a cylindrical release. Appropriate boundary conditions which include the effect of drag are applied. Entrainment is neglected. For late times they found an analytical similarity solution. For channel flows in particular, the predicted cloud consists of a triangular wedge moving down the slope at a constant velocity given by:

$$U = 2^{1/4} k_f \Gamma^{1/4} \left(g'^2 V \right)^{1/4} \tag{3.1}$$

with $\Gamma = \tan \alpha$ the gradient of the surface, V the volume of the cloud and k_f the front Froude-number based on the height of the front. For radial releases they have shown that a solution in the form of a cloud with a constant uniform velocity, a flat top and a fixed shape exists (Figure 2). The velocity is

$$U = \Omega^{-1/6} k_f \Gamma^{1/3} \sqrt{g' V^{1/3}} \tag{3.2}$$

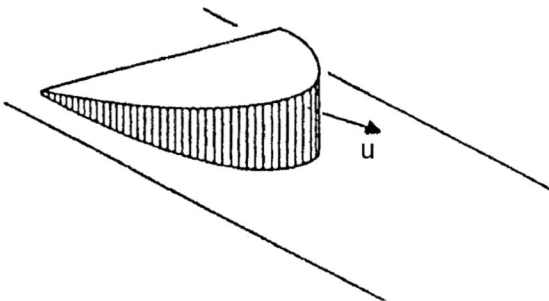

Figure 2. Three-dimensional cloud moving down a slope according to Webber et al.'s shallow-layer model

where $\Omega = 5\pi/16$. Subsequently, similarity solutions for other shapes of the surface have been found. Nielsen [6] has presented solutions for clouds in parabolic and v-shaped valleys of constant slope.

Webber et al.'s model can be expected to give reasonable results only for moderate slopes. For very small angles α the front velocity goes to zero. For high angles the assumptions used in the derivation of the shallow layer equations are violated. Also, since only the late-time solution is obtained from the similarity solution, it is not very useful if one is interested in what happens near the point of release. In laboratory experiments, like those presented here, the spreading distance appears often too short to reach the predicted asymptotic steady state. Also, the validity of the no-entrainment assumption is doubtful, especially at later times in the spreading process.

An improvement of the model that takes into account the dilution of the cloud by the ambient air is presented in a recent paper by Tickle [7]. He has assumed that the wedge shape predicted by the similarity solution without entrainment is conserved also in the case with entrainment. The air entrainment rate is assumed to be proportional to the front velocity and the top area of the cloud. Some of the drawbacks of Webber et al.'s model are retained though. The model breaks down for slopes approaching zero degrees and it is not valid in the early phase of the spreading process.

3.2 Box-model on slopes

The basic assumption used in the present model is that the gravity-driven slumping of the cloud and the motion of its center of gravity can be uncoupled and considered separately (Figure 3). The slumping is assumed to be adequately described by the box-model and a force equilibrium gives the equation of motion for the downslope movement of the cloud. Similar models have been proposed by de Nevers [8] and by Kukkonen and Nikmo [4], who in addition included the effect of wind. The main difference of the present approach from these earlier

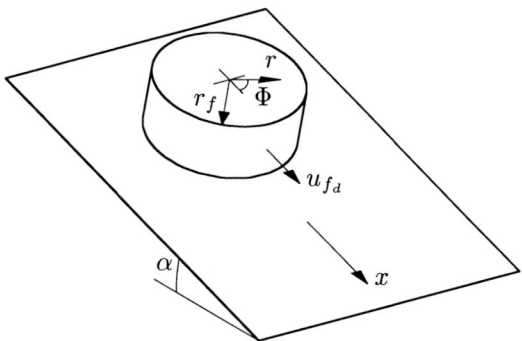

Figure 3. Box-model on sloping surface

models pertains to the way the drag is modeled. Zero ambient wind speed is assumed in accord with the laboratory condition.

In the following considerations, dimensionless variables are denoted with small letters and capital letters are used for dimensional values. An exception to this convention is the dimensional time t (non-dimensional τ) and the gravity g. The reference length is defined as a function of the initial gas volume $L_{ref} = V_0^{1/(2+j)}$ where $j = 1$ is for radial flow and $j = 0$ for channel flows. For the reference time $t_{ref} = \sqrt{g'/L_{ref}}$ is used and for the reference velocity $U_{ref} = \sqrt{g'L_{ref}}$. The reduced gravity constant g' is defined as:

$$g' = \frac{\rho_g - \rho_a}{\rho_a} g. \tag{3.3}$$

Other definitions of the reduced gravity like the $g^* = g\,\Delta\rho/(\rho_g + \rho_a)$ used by Gröbelbauer et al. [9] or $g'' = g\,\Delta\rho/\rho_g$ were found to give less satisfying data correlation for the range of densities considered.

A conventional box-model (for example Fanneløp [10]) is used to calculate the slumping motion of the cloud. The gravity g is replaced by its component perpendicular to the surface ($g\cos\alpha$). The position of the front x_f and its velocity u_f are given by:

$$x_f = \left(\frac{3+j}{2}k\pi^{-j/2}\cos(\alpha)\tau + x_0^{(3+j)/2}\right)^{2/(3+j)} \tag{3.4}$$

$$u_f = k\pi^{-j/2}\left(\frac{3+j}{2}k\pi^{-j/2}\cos(\alpha)\tau + x_0^{(3+j)/2}\right)^{-\frac{1+j}{3+j}} \tag{3.5}$$

where x_0 is the radius of the initial gas volume and $k = U_f/\sqrt{g'H}$ the front Froude-number. Let us first consider the case of radial releases (i.e. $j = 1$).

Top entrainment is assumed. The entrainment velocity is defined as $v_e = \beta u$ with u the magnitude of the local velocity vector:

$$u = \frac{r}{r_f} u_f. \tag{3.6}$$

The average entrainment over the cloud is given by:

$$\bar{v}_e = \frac{1}{\pi r_f^2} \int_0^{2\pi} \int_0^{r_f} \beta |\vec{u}_{cg} + \vec{u}| \, r \, dr \, d\phi \tag{3.7}$$

$$= \frac{\beta u_f}{\pi} \int_0^{2\pi} \int_0^1 \sqrt{\gamma^2 + 2\tilde{r}\gamma \cos\phi + \tilde{r}^2} \, \tilde{r} \, d\tilde{r} \, d\phi \tag{3.8}$$

where $\gamma = u_f / u_{cg}$ and $\tilde{r} = r/r_f$ is the normalized radial coordinate. This integral can not be solved analytically. For values of γ not too large (i.e. up to about $\gamma = 1$), the average entrainment velocity can be approximated by:

$$\bar{v}_e \approx \frac{2}{3}\beta u_f + \frac{\beta}{2} \frac{u_{cg}^2}{u_f}. \tag{3.9}$$

The entrainment is slightly over-estimated when this approximation is used. The second term in the above expression gives the increase of the entrainment due to the downward motion of the cloud. The average concentration in the box-model is:

$$c = \frac{h_g}{h_g + h_e} \tag{3.10}$$

where the height of the entrained air h_e and the height of the gas h_g are given by:

$$h_e = \int_0^\tau \bar{v}_e \, dt \qquad h_g = \frac{1}{\pi r_f^2}. \tag{3.11}$$

A momentum balance on the whole cloud gives an equation for the motion of its center of gravity. The negative buoyancy of the cloud (proportional to $\sin\alpha$) is opposed by a drag force and the inertial force. This yields the following differential equation:

$$m\dot{U}_{cg} = V_0(\rho_g - \rho_a)g \sin\alpha - F_{drag} \tag{3.12}$$

where m is the mass of the fluid which has to be accelerated, i.e. the mass of the dense-gas plus a virtual mass:

$$m = m_g + m_{virt} = \rho_g V_0 + c\rho_a V = L_{ref}^3 \left(\rho_g + c\rho_a(1 + \pi r_f^2 h_e)\right) \tag{3.13}$$

where the virtual mass coefficient c is a constant between 0 and 1. The sensitivity to this parameter is found to be very small. The model can thus be further simplified by neglecting the virtual mass force.

The drag F_{drag} is modeled as a shear force over an area proportional to the footprint of the cloud. Frictional effects at the interface and at the bottom surface are lumped into this term. The drag component in the x-direction is given by:

$$F_{drag} = g' L_{ref}^3 c_T \rho \pi r_f u_f^2 \int_0^1 \int_0^{2\pi} \sqrt{\tilde{r}^2 + 2\tilde{r}\alpha \cos\phi + \alpha^2} \, (\tilde{r}\cos\phi + \alpha)\, \tilde{r} \, d\tilde{r} \, d\phi$$

$$\approx g' L_{ref}^3 c_T \rho \pi r_f^2 u_f^2 \left(\frac{u_{cg}}{u_f} + \frac{3}{8}\frac{u_{cg}^3}{u_f^3} \right) \tag{3.14}$$

where c_T is a drag coefficient and ρ the average density of the cloud:

$$\rho = \frac{h_g \rho_g + h_e \rho_a}{h_g + h_e}. \tag{3.15}$$

Other contributions to the drag proportional to the cross section (form drag) or the perimeter area were found to be negligible. Substituting Equations (3.14) and (3.13) into (3.12) and using dimensionless notations, we get:

$$\left(\rho_g + c\rho_a(1 + \pi r_f^2 h_e) \right) \dot{u}_{cg} = \rho_a \sin\alpha - c_T \rho \pi r_f^2 u_f^2 \left(\frac{u_{cg}}{u_f} + \frac{3}{8}\frac{u_{cg}^3}{u_f^3} \right). \tag{3.16}$$

Together with Equations (3.4), (3.5) and (3.11), Equation (3.16) can be integrated numerically to find the velocity of the center of gravity of the cloud.

The velocity at any point in the cloud is obtained by vector addition of the radial flow velocity u (Equation (3.6)) and the velocity of the center of gravity u_{cg} (Equation (3.16)). In particular for the downward front velocity, we obtain:

$$u_{f_d} = u_f + u_{cg}. \tag{3.17}$$

The same model, appropriately modified can be used for flows in channels. The complete derivation will not be given here. The differential equation to be solved is:

$$\rho_g \dot{u}_{cg} = \rho_a \sin(\alpha) - f_{drag} \tag{3.18}$$

where the friction force and the average entrainment velocity are given for $\gamma = u_f/u_{cg} < 1$ by:

$$\bar{v}_e = \beta u_{cg} \tag{3.19}$$

$$F_{top} = \frac{2}{3}\rho c_T x_f (u_f^2 + 3u_{cg}^2) \tag{3.20}$$

and for $\gamma > 1$:

$$\bar{v}_e = \frac{1}{2}\beta\left(u_f + \frac{u_{cg}^2}{u_f}\right) \tag{3.21}$$

$$F_{top} = \frac{2}{3}\rho c_T x_f \frac{u_{cg}}{u_f}(3u_f^2 + u_{cg}^2). \tag{3.22}$$

The integration of the differential Equation (3.18) is performed numerically also in this case.

An alternate model for releases in channels was proposed by D. Webber [11]. It has the advantage of being valid also for the horizontal case and it is easily modified so as to include the fixed rear wall present in the experiment (see Figure 4). From the assumed geometry, we obtain for the volume of the cloud:

$$v = lh - \frac{1}{2}\tan(\alpha)l^2. \tag{3.23}$$

Making use of the fact that the buoyancy is conserved, we get for the front velocity:

$$u_f = \frac{dl}{d\tau} = k\sqrt{\frac{h}{v}}. \tag{3.24}$$

The cloud dilution is calculated using the front entrainment model:

$$\frac{dv}{d\tau} = \kappa h u_f = \kappa h k\sqrt{\frac{h}{v}}. \tag{3.25}$$

This model is valid as long as the top of the cloud has not touched the ground $(h - \Gamma l > 0)$. Afterwards the shape of the cloud is assumed to be a triangular wedge moving down the slope. For this stage, the model is analogous to the one proposed by Tickle [7] for the 3-d case.

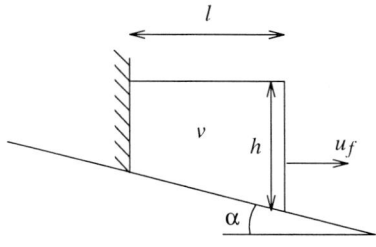

Figure 4. 2-d box-model with rear wall on slope

4   Results and discussion

4.1   Horizontal and lateral spreading

Experiments on a horizontal surface are used to determine the relevant front
Froude-number k and the entrainment coefficient β. Figure 5 shows that for
radial releases the front position is accurately predicted with a value of $k = 1.1$.
For channel flows, a value of $k = 1.3$ is found. This is consistent with the results
of Billeter [5].

The resolution of the concentration measurements is not good enough to
calculate the reliable average needed for a comparison with the box-model. If
we assume the concentration distribution in the cloud to be self-similar then
the average concentration is proportional to the concentration measured at any
one selected point within the cloud. In the horizontal case, the spreading rate
given by the box-model does not depend on the entrainment, the dilution of the
cloud being compensated by its increase in height. For sloping surfaces this is
not true anymore but the sensitivity of the front velocity to the entrainment
parameter is very small. Therefore the entrainment parameter β can be set to
any level desired. Figure 6 shows the values of the entrainment coefficient as a
function of the spreading distance if top entrainment (β) or front entrainment
(κ) is assumed. The front entrainment coefficient κ is seen to be a function
of the spreading distance while the top entrainment coefficient remains almost

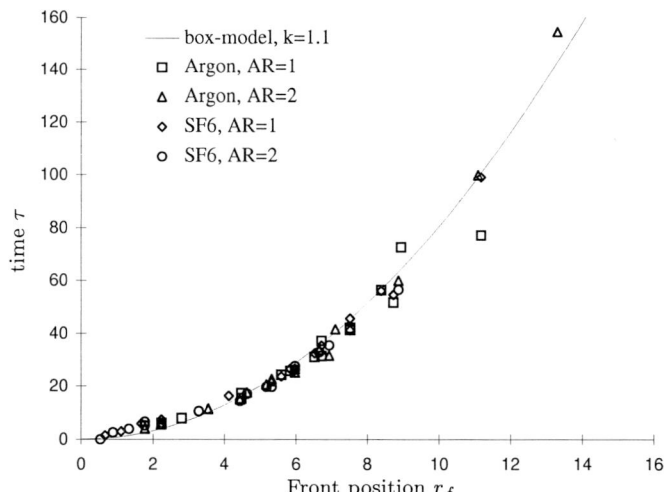

Figure 5. Front position r_f as a function of time τ for radial releases on horizontal
surface

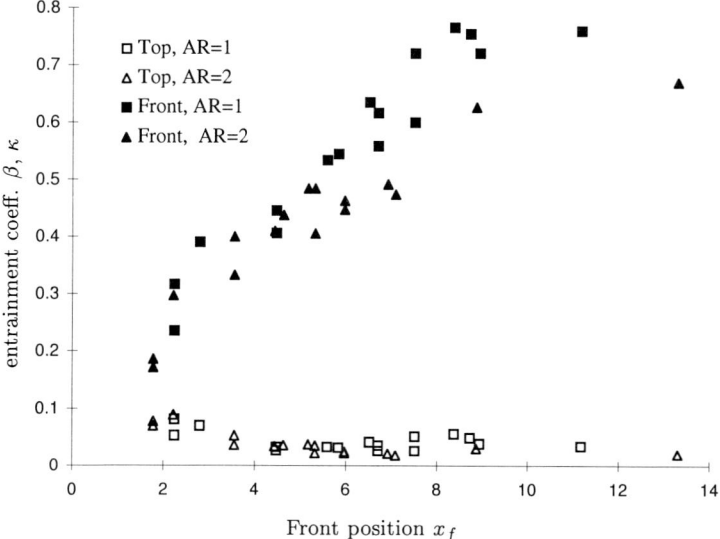

Figure 6. Entrainment coefficient calculated from experimental data for top and front entrainment model

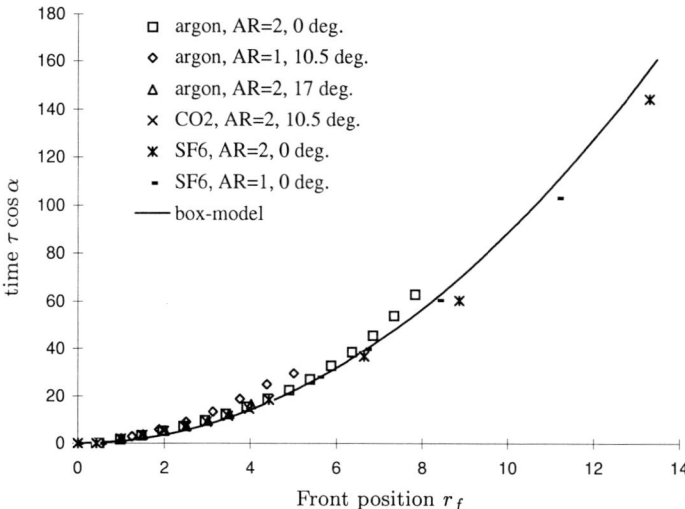

Figure 7. Lateral extent of cloud as a function of the scaled time for various gases, slope angles and aspect ratios

constant. Top entrainment therefore seems to be more suitable, at least for this early phases of the spreading process.

In the proposed model for the motion of the cloud on sloping terrain, the lateral spreading is not affected by the longitudinal motion of the cloud. The only slope dependence is a result of the reduction of the gravity $g \cos(\alpha)$. The maximum lateral extent as a function of time is given by Equation (3.4). Figure 7 shows this independence of slope to be the case for all gases and slopes considered herein.

4.2   Slopes

The sequence of photographs (Figure 8) shows the motion of an Argon cloud on a slope of 10.5 degrees. Initially the shape of the cloud is very nearly circular. A clearly defined frontal region is formed. On the downslope side, the front grows slowly larger as it is fed from behind. The upward moving front loses material end gets smaller. For later times, the resulting asymmetry is in contradiction to the assumptions made in the derivation of our flow model. The wedge predicted by Webber et al. [3] could not be observed on this relatively small surface but it can not be excluded at a later stage of the spreading process.

The entrainment coefficient β and the front Froude-number k required in the model are obtained from the experiments on the horizontal surface. The top friction coefficient c_t is found empirically from the experiments on the sloping surface. Good agreement between the experimental data and the theory is obtained with $c_t = 0.12$ (Figure 9). In general the front position is well predicted by the present model. For angles greater than 6 degrees and dimensionless spreading distances above $x \approx 12$, the predicted velocity is too high. For these high slopes the shape of the cloud differs considerably from the box-model assumption (see Figure 8). Most of the gas is concentrated in the frontal region. The cloud front is thus much higher than predicted and the frontal drag more important.

The agreement between the model and the measurements is less satisfactory for channel flows (Figure 10). The reason for this can be found in the initial condition. The model assumes that both the downslope and upslope limiting walls are removed. In the experiment, the upward motion of the cloud is inhibited by the fixed rear wall. This results in a greater cloud height and leads to a higher downward velocity. This error is expected to be most significant for small slopes where the cloud remains in contact with the rear wall for the longest time.

The cloud model by Webber et al. [3] predicts that the front velocity goes with $(\tan \alpha)^{1/4}$ for channel flows (Equation 3.1) and $(\tan \alpha)^{1/3}$ for radial flows (Equation 3.2). Using the new reference lengths $L_{ref} = (\tan \alpha)^{1/3} V_0^{1/3}$ and $L_{ref} = (\tan \alpha)^{1/4} V_0^{1/2}$ for radial and channel flows respectively, the dependence of the front velocity on the angle can be eliminated. Figure 11 shows the data reduced with this new reference length. For radial flows, especially for small angles, the spreading distance covered is not large enough to reach a constant front velocity. Nevertheless, a value of $k_f = 0.38$ gives reasonable results for later times. For smaller angles the predicted velocities are somewhat too high. For

Figure 8. Release of Argon, slope 10.5^o, $t =0.4$, 1.4, 2.6, 4.1 seconds

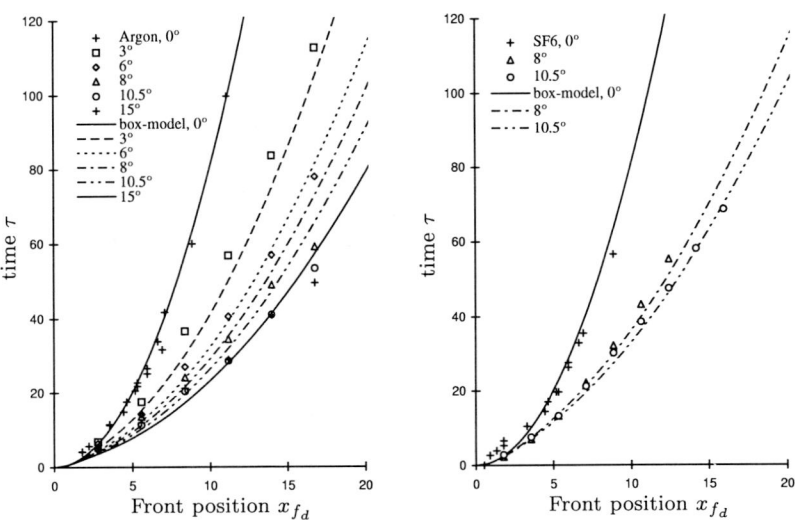

Figure 9. Downward front position x_{f_d} as a function of time for radial releases of Argon (left) and SF$_6$ (right) with $k = 1.1$, $\beta = 0.03$, $c = 0.5$ and $c_T = 0.12$

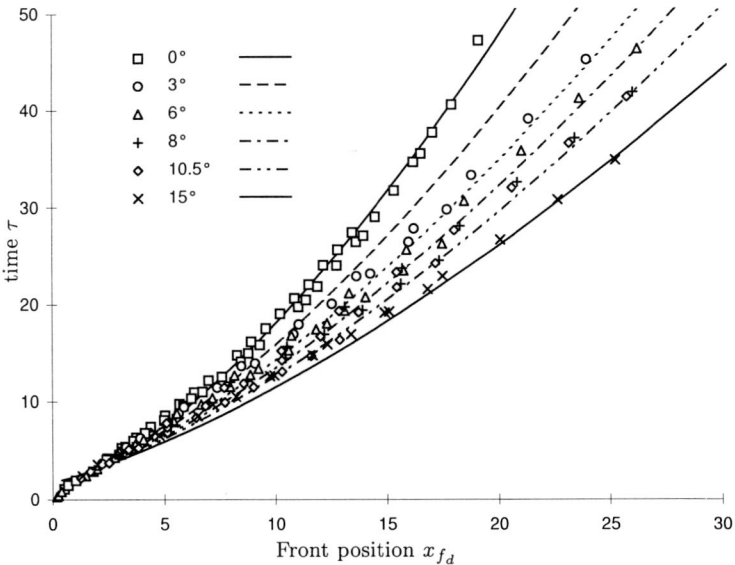

Figure 10. Downward front position for Argon releases in a channel. Symbols represent measurements, lines results from the box-model analyses.

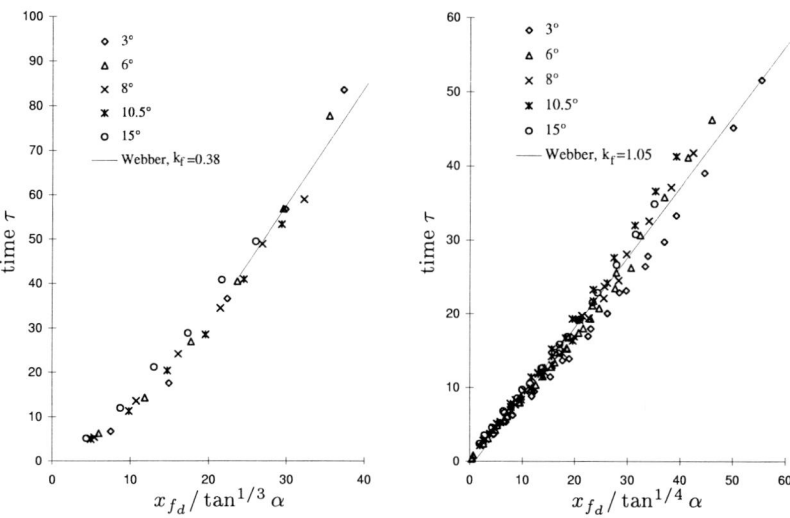

Figure 11. Data regression using Webber et al.'s slope dependence for radial flows (left) and channel flows (right)

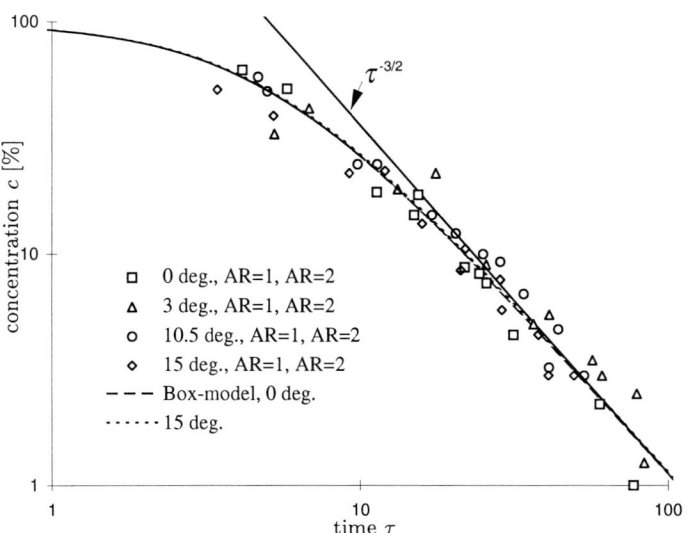

Figure 12. Maximum concentrations, measured 15 mm above the surface, on the centerline as a function of time

channel flows, the non-dimensional front velocity is almost constant ($k_f = 1.05$). In a range of moderate slopes, between 3 and 15 degrees and possibly higher, the model of Webber et al. can thus be used to get a reasonable estimate of the front velocity.

Figure 12 shows the maximum concentrations measured for a 3-d release, on the centerline of the cloud, 15 mm above the surface. The dependence of the concentration on the slope is seen to be rather weak for both the measurements and for the model. An asymptotic time dependence in the form of $\tau^{-3/2}$ is found. This is a consequence of scaling the entrainment with the top surface of the cloud and the front velocity.

4.3   Influence of the initial aspect ratio

For practical reasons, many of the radial experiments had initial aspect ratios of $h/d = 2$. Even though the effect of the aspect ratio is small, it can be seen to be present in the data. Usually, a reference length defined as $L_{ref} = V_0^{1/3}$ is used in the reduction of the experimental data. A reference length proportional to the initial potential energy can be defined:

$$L_{ref} = (V_0 H_0)^{1/4} \tag{4.1}$$

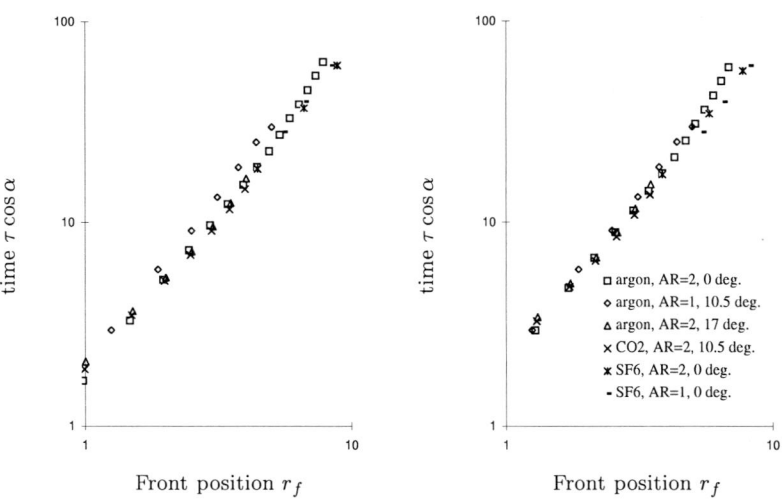

Figure 13. Lateral spreading, non-dimensionalized using reference length based on volume (left) and on initial potential energy (right)

where H_0 is the initial height of the cloud. This reference length accounts implicitly for aspect ratio effects. Figure 13 shows the lateral spreading non-dimensionalized using this new reference length. A slightly better data correlation is obtained.

For releases in channels, the similarly defined reference length is $L_{ref} = (X_0^2 H_0)^{1/3}$. This leads to a front position as function of time which is proportional to $(H/X)^{1/6}$. This dependence on the aspect ratio is confirmed by the experiments of Billeter [5].

5 Conclusion

An extensive experimental study of the behaviour of negatively buoyant gas clouds suddenly released on sloping surfaces has been undertaken. The amount of gas, the aspect ratio of the initial volume, the density of the gas and the slope of the surface were varied. Measurements were performed using concentration probes and by smoke visualizations.

For the horizontal case, a conventional box-model was found to predict accurately the front position as a function of time. The concentration measurements showed that for the early part of the spreading process, the top entrainment assumption gives better results. A small influence on the spreading behaviour of the initial aspect ratio could be found. It confirms the dependence found by Billeter [5] for two-dimensional releases in channels also for radial releases.

At early times in the spreading process on slopes, the cloud can be modeled accurately with a box-model combined with a model for the movement of the cloud center of gravity. The equation of motion of the center of gravity is obtained by considering all the forces acting on the cloud. The primary drag is the shear force on the top surface. In addition to the entrainment parameter β and the front Froude-number k determined from the experiments on the horizontal surface, a new drag coefficient c_T is introduced. c_T is determined empirically from experiments on the sloping surface. The front velocity and the concentrations predicted by the present model are found to be in good agreement with the measured data.

For radial releases, the similarity solution proposed by Webber et al. could not be confirmed. The spreading distances in the present experiment were too small for the cloud to reach the asymptotic constant velocity state. Visualizations showed that the downward moving front gained material from the central regions of the cloud growing progressively larger. At later times, the shape of the cloud looks more like the wedge predicted by Webber et al. than like the circular cylinder assumed in the present model. For both 3-d and channel flows and for slopes greater than 3 degrees, the predicted dependence of front velocity on slope is well reproduced. The similarity model is therefore a good tool for rapid estimates of front velocity for clouds on slopes. The modified box-model is more complicated and requires a numerical integration. It is more accurate, however, in the early spreading phase and doesn't break down in the limit of vanishing slope. Therefore, by combining the box-model for the initial phase with a similarity solution including entrainment for later times, one obtains a useful, simple and reasonably accurate tool to estimate the spreading and dilution of heavy-gas clouds on uniformly sloping terrain.

Bibliography

1. Schatzmann, M., Marotzke, K. and Donat, J. (1990). Research on continuous and instantaneous heavy gas clouds—contribution of sub-project EV4T-00210D to the final report of the joint CEC project, University of Hamburg, Meteorolical Institute Report.

2. Beghin, P., Hopfinger, E.J. and Britter, R.E. (1981). Gravitational convection from instantaneous sources on inclined boundaries. *Journal of Fluid Mechanics*, **107**, 407-422.

3. Webber, D.M., Jones, S.J. and Martin D. (1993). A model of the motion of a heavy-gas cloud released on a uniform slope. *Journal of Hazardous Materials*, **33**, 101-122.

4. Kukkonen J., Nikmo J. (1992). Modelling heavy-gas cloud transport in sloping terrain. *Journal of Hazardous Materials*, **31**, 155-176.

5. Billeter, L., Fanneløp, T.K. (1997). Concentration measurements in dense isothermal gas clouds with different starting conditions. *Atmospheric Environment*, **31(5)**, 755-771.

6. Nielsen, M. (1996). Comment on 'A model of the motion of a heavy-gas cloud released on a uniform slope'. *Journal of Hazardous Materials*, **48**, 251-258.

7. Tickle, G.A. (1996). A model of the motion and dilution of a heavy-gas cloud on a uniform slope in calm conditions. *Journal of Hazardous Materials*, **49**, 29-47.

8. de Nevers, N. (1984). Spread and down-slope flow of negatively buoyant clouds. *Atmospheric Environment*, **18**, 2023-2027.

9. Gröbelbauer, H.P., Fanneløp,T.K. and Britter, R.E. (1993). The propagation of intrusion fronts of high density ratios. *Journal of Fluid Mechanics*, **250**, 669-687.

10. Fanneløp, T.K. (1994). Fluid Mechanics for Industrial Safety and Environmental Protection. Industrial Safety Series, Elsevier, Amsterdam, The Netherlands.

11. Webber, D.M. (1996) personal communications.

The Effect of Uphill Slopes on Mixing in Turbulent Gravity Currents

Matthew A. Brown

Department of Applied Mathematics and Theoretical Physics, University of Cambridge[1]

Abstract

Experiments are performed to study the impact of two-dimensional turbulent gravity currents on uphill slopes, with particular reference to mixing between the current and the ambient caused by the reflection. An optical techique is used to measure cross-stream averaged densities in the current, in an attempt to quantify this mixing. Quantities constructed from the vertical density profiles indicate that for finite-volume currents more mixing occurs during reflection from small slopes than from steep ones, although results for continuous-flux currents are less conclusive. The mechanisms of the reflection process are discussed in order to explain the observations.

1 Introduction

Horizontal flows driven by buoyancy forces in the presence of a surface or interface, known as gravity currents or density currents, are a common occurrence in many environmental contexts. The sea-breeze, turbidity currents and salt-water intrusions in estuaries are typical examples, and many others are given by Simpson [1]. In many cases, such as the spread of a toxic dense gas over land, mixing between the current and ambient is of crucial importance—in this case to determine the time before the concentration is diluted to safe levels. Also in these cases the topography over which the current spreads plays a large role. This paper will examine a particular case where these two effects interact so as to change the nature of the flow.

When a turbulent gravity current propagates over a rigid surface, there are two main processes which contribute to the mixing. Wave breaking behind the head due to shear instability leads to the formation of Kelvin-Helmholtz billows and a mixed layer above the following current. Additionally, the front of the current is broken up into a complex structure of lobes and clefts. Ambient fluid is entrained through the clefts directly into the head of the current, where it is

---

[1]Present address: School of Mechanical and Materials Engineering, University of Surrey, Guildford.

convectively unstable. This is due to the no-slip condition at the rigid interface—experiments where the effect of bottom friction has been removed have shown the suppression of the three-dimensional structure [2, 3]—and the process is often known as "over-running".

Recent experiments designed to quantify the mixing in gravity currents have focussed on the case of a rigid horizontal surface. Hacker et al. [4] looked at lock-release gravity currents during the initial slumping phase using the same technique as in this paper. They calculated entrainment rates into the current, which were shown to depend on the initial aspect ratio. The internal density structure was observed to be consistent with mixing due to detrainment from the head by Kelvin-Helmholtz billows. A similar explanation was provided by Hallworth et al. [5], who studied both two-dimensional and axisymmetric currents, of lock-release and continuous-release types. They used a neutralization technique which could quantify bulk entrainment rates (in the case of a finite release) but not internal structure.

A number of authors have looked at various aspects of gravity current interactions with topography, particularly down-slope flows, but the case of up-slope flows has been relatively neglected. Edwards et al. [6] presented some experiments similar to those in this paper, making only general observations of the flow field, and Lane-Serff et al. [7] looked at the reflection of continuous-flux gravity currents from slopes and overhangs. In neither case was mixing in the current the area of focus.

In this paper experiments are presented where a gravity current is established on a horizontal surface, then allowed to run up a slope. An optical technique is then used to obtain the density structure of the current, and so obtain details of the mixing. Particular attention is paid to the reflection of the current from the slope, the mixing during this phase, and the effect the mixing has on the structure of the current. The quantification of the density structure allows the construction of a measure of the local mixing, and the variation of this in time and space is described.

2   Experiments

Gravity currents are produced in a channel-like tank with dimensions 3.5 m × 0.5 m × 0.2 m, using fresh-water and saline solution to produce the density difference. The experiments can be divided into two types—finite-volume and continuous-flux currents. A finite-volume current involves, as its name suggests, the instantaneous release of a finite volume of dense fluid, whereas a continuous-flux current involves a continuous release of a constant volume and mass flux of dense fluid.

The finite-volume currents are produced using a lock-release mechanism, the simplest and most commonly used method of producing gravity currents in the laboratory (see Figure 1). A short section of the channel (length L) is isolated from the rest by a gate. Fresh water is filled to a depth H in the main tank,

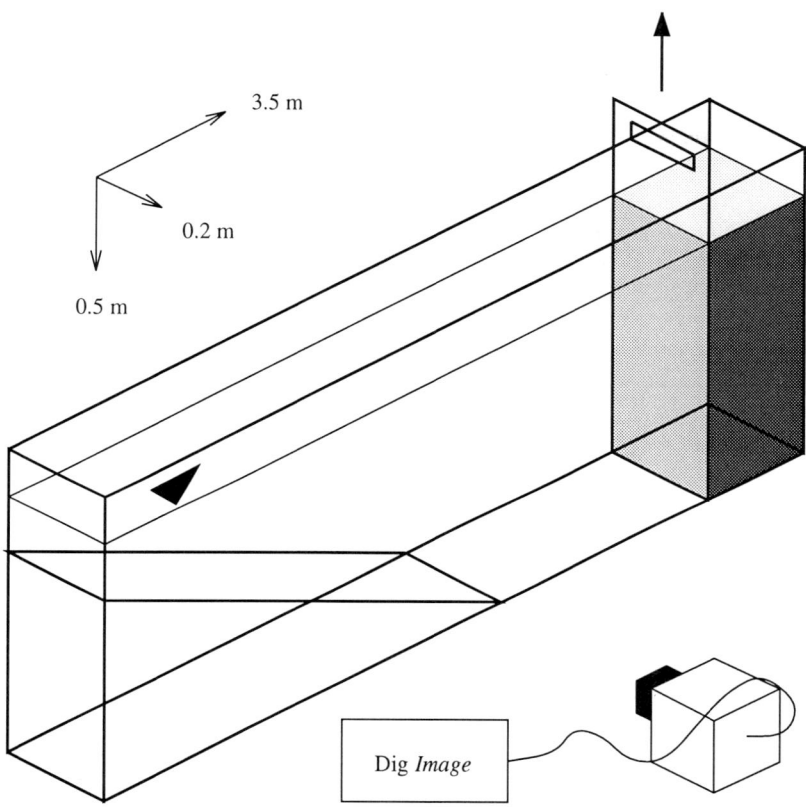

3.5 m

0.2 m

0.5 m

Dig *Image*

Figure 1. Experimental configuration for producing finite-volume gravity currents

and saline solution filled to the same depth in the lock. When the lock-gate is removed, smoothly so as not to disrupt the flow, the dense fluid slumps along the bottom of the channel. Initially, the current front will move at a velocity $\alpha(g'H)^{1/2}$ (g' is the usual reduced gravity), where $\alpha \approx 0.5$, and the velocity will stay constant for a distance of approximately $10L$ [8], if the bottom of the tank is horizontal. In the majority of experiments presented here a ramp of variable angle is set at a distance $5L$ from the back of the lock, but the qualitative behaviour described is not sensitive to this distance.

To produce a continuous-flux current, a similar configuration is used, but only a small gap is opened (height H_g) at the bottom of the gate. An exchange flow is initiated through the gap (Figure 2). Although the pressure head driving the flow does not remain constant throughout the experiment, the exchange flow is governed by a two-layer hydraulic control at the gap and the geometry of the

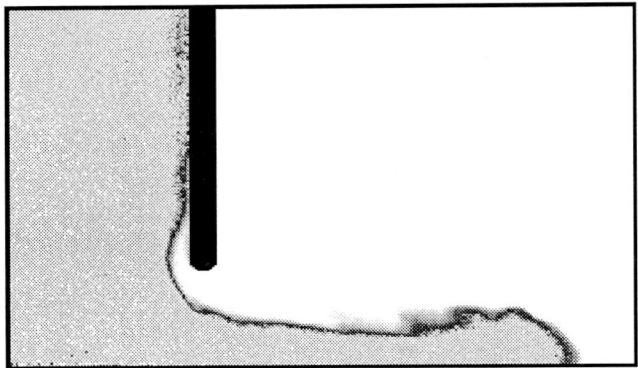

Figure 2. Exchange flow producing constant-flux gravity current. The image is a digitally altered capture from video tape

reservoir into which the dense fluid flows, and so the flux of dense fluid through the gap remains constant for a significant time [9].

Density measurements are taken using the optical technique described by Hacker et al. [4]. The dense fluid is visualized using a known quantity of coloured dye, and the tank is lit from behind. The resulting image is recorded on videotape and then digitally captured for processing by the Dig*Image* image processing system. The attenuation of light passing through the tank is measured and related to the cross-tank-averaged concentration of dye at that point. This in turn is directly related to the density.

3   Results and discussion

3.1   General observations

Previous studies on gravity currents meeting walls or steep slopes have noted a reflection, in the form of a turbulent bore generally about twice the height of the incoming current, and a "splash" of less dense material up the wall to a height of approximately twice this height again [10, 7]. As far as many applications are concerned, however, much shallower slopes are likely to be of interest. In these cases it is difficult to make such a distinction between the splash and the bulk of the current.

The front of the current, then, is decelerated as it moves up the slope. As it does so, fluid is drained from the back of the head, so that the size also decreases. The head of the current stays distinguishable from the following current until it becomes so small that local viscous effects are important, but it does not separate from the rest of the current.

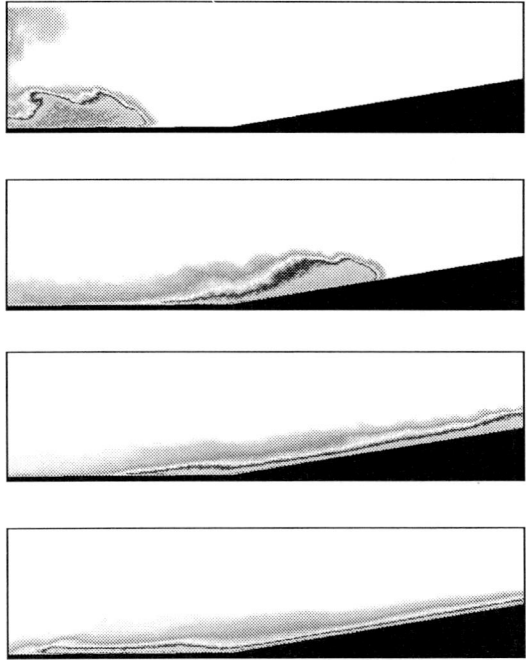

Figure 3. A sequence of images from a finite-volume gravity current interacting with a ramp. The ramp angle is 10°, and the time step between successive images is 8 s. Note that in this, and all other video captured images in this paper, a monochrome rendering of a false-colour scheme has been used which corresponds to densities in the current

The reflection appears to take the form of an internal bore. This is either a turbulent or undular bore, depending on the magnitude of the reflection, which in turn depends on the angle of slope—larger angles causing larger reflections, and a bore of sufficient relative ampitude (the ratio of the two depths) will become turbulent. The bore is followed by a number of waves on the interface.

Examination of the density structure indicates that the reflections can have a more complex structure than this description. This is particularly evident for a finite-volume current. As the current propagates onto the ramp, there is already an inhomogeneous density stratification due to mixing between the current and the ambient. The action of a downslope component of gravity on the stratification leads to the baroclinic production of vorticity, of the opposite sign to the existing vorticity in the current. This induces a shear in the current,

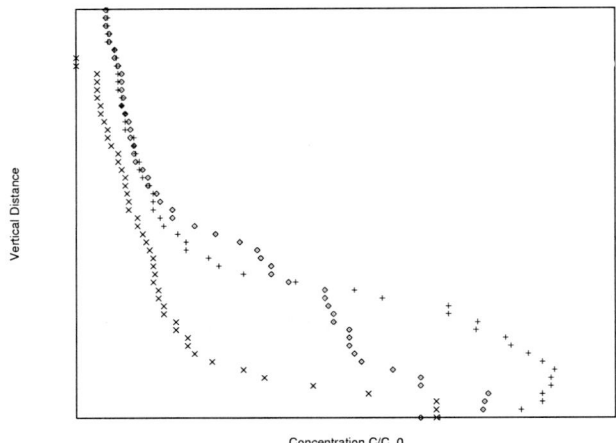

Concentration C/C_0

Figure 4. Vertical density profile of a finite-volume current at a fixed point away from a ramp of angle 15° obtained from optical measurements. The profiles correspond to: × after the passage of the head but before the reflection has arrived; + during the passage of the reflection; ◇ immediately after the passage of the reflection

superimposed on the mean velocity, with downslope velocities highest at the rigid surface beyond the viscous boundary layer. As a result, when the current begins to flow back down the slope it does so first adjacent to the ramp. The nature of the stratification means that the fluid here is densest. This densest fluid accumulates at the foot of the slope until it can overcome the effect of the following current, and so propagates upstream, underneath the following current. Figure 3 shows such a reflection as a series of images.

The resulting flow can be described as another gravity current moving beneath an existing two-layer stratification. Examination of vertical density profiles as the reflection passes (Figure 4) indicates that a description of the stratification as being three layers with constant density in each layer, with a sharp interface between them, is a reasonable approximation, especially for larger reflections. In such a scenario the gravity current in the bottom layer can force an internal solitary wave or bore at the upper interface [11], and it is this which is commonly observed as the reflection. This current structure (without the density structure) was observed, but not fully explained, by Edwards et al. [6].

This effect is not so important in continuous-flux gravity currents. The reversal occurs in a similar way, but since there is a constant flow of densest fluid from upstream the bulge cannot propagate underneath the following stream. There is a complex interaction which cannot be determined from visualizations of the density alone, and the description of the reflection as an internal bore in a two layer fluid is more accurate. Figure 5 shows a typical reflection in this case.

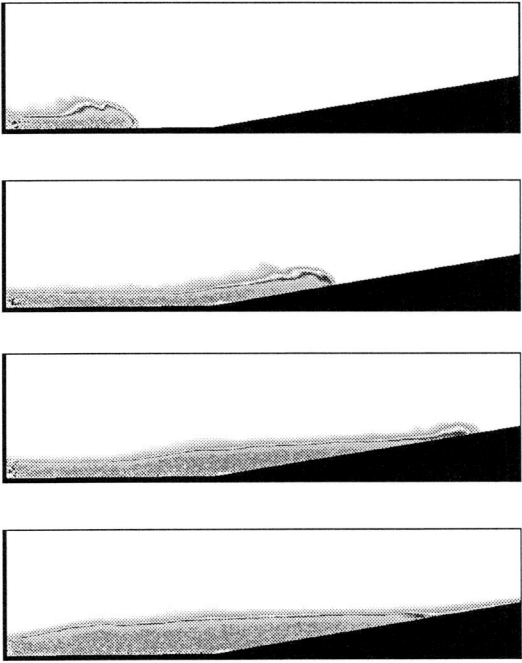

Figure 5. A sequence of images from a continuous-flux current interacting with a ramp. Again, $\theta = 10°$, and the timestep is 8 s

3.2 Mixing diagnostics

A certain amount of qualitative information can be gained from simply observing the structure of the density field. It would be preferable, however, to obtain quantitative information about the amount of mixing that has occurred. The following presents the derivation of two quantities based on the density field which have physical significance, and which can be interpreted as a local measure of the mixing.

We will initially consider the Shallow-Water Equations, in one horizontal dimension, as a model for gravity current flow in a channel. The choice of this model is common in applications such as dense-gas dispersion. The relevant assumptions are that horizontal lengthscales are much greater than vertical lengthscales, and hence that accelerations normal to the slope are small, leading to a hydrostatic pressure. The equations then assume that flow is confined to a thin layer of depth $h(x,t)$, and can be classified by a density $\rho(x,t) = \rho_0 + \Delta\rho(x,t)$ and velocity $u(x,t)$.

Realistic flows do not have velocities and densities which are homogeneous in depth, and neither do they have a well defined height. The formulation can, however, still be used by vertically integrating the Navier-Stokes equations and choosing appropriate characteristic values for $\bar{\rho}$, $\bar{u}$ and h. The exact formulation is not unique; the following choice is after Ellison and Turner [12].

The horizontal momentum equation (with the Boussinesq approximation) becomes:

$$\frac{\partial}{\partial t}(h\bar{u}) = \frac{\partial}{\partial x}(h\bar{u}^2 + \frac{1}{2}S_1 g' h^2 \cos\theta) + S_2 g' h \sin\theta + \tau, \qquad (3.1)$$

where $g' \equiv g\Delta\rho/\rho_0$, and θ is the local gradient of the terrain. τ represents various stress-like terms including turbulent and viscous drag.

The various quantities in this equation are formed from the primitive variables in the following fashion:

$$h\bar{u} = \int_0^\infty u \, dz, \qquad (3.2)$$

$$h\bar{u}^2 = \int_0^\infty u^2 \, dz, \qquad (3.3)$$

$$h\Delta\bar{\rho}\bar{u} = \int_0^\infty \Delta\rho u \, dz, \qquad (3.4)$$

and the shape parameters are defined as

$$S_1 \equiv \frac{2}{h^2\Delta\bar{\rho}} \int_0^\infty \Delta\rho z \, dz, \qquad (3.5)$$

$$S_2 \equiv \frac{1}{h\Delta\bar{\rho}} \int_0^\infty \Delta\rho \, dz. \qquad (3.6)$$

In the case of a uniform layer of depth h, then $S_1 = S_2 = 1$. To pay particular attention to S_1: this parameter represents the modification of the hydrostatic driving force due to the inhomogeneity of the vertical density profile. Since this inhomogeneity is a result of mixing between the current and the ambient, S_1 can be regarded as a measure of the mixing, with particular significance since it represents the way in which the mixing subsequently affects the dynamics of the flow.

Notice that the definition of h and $\Delta\bar{\rho}$, and hence S_1, depends on the velocity structure. This is information which is not available from the experiments. The assumption is therefore made that these quantities can instead be defined using only the density structure—in essence treating $\Delta\rho$ in a similar way to u in (3.2) and (3.3). Therefore:

$$h\Delta\bar{\rho} \equiv \int_0^\infty \Delta\rho \, dz, \qquad (3.7)$$

$$h\Delta\bar{\rho}^2 \equiv \int_0^\infty \Delta\rho^2 \, dz. \qquad (3.8)$$

This is, in fact, more likely to give an appropriate measure of the height in those situations where there is a large shear in the current itself, as observed during the reflection process. If the velocity changes sign with depth then unrealistically small values of h will be obtained using the previous method.

Using (3.7) and (3.8), S_1 can be expressed solely in terms of vertical integrals of the density.

$$S_1 = \frac{2 \left(\int_0^\infty \Delta\rho z \, dz \right) \left(\int_0^\infty \Delta\rho^2 \, dz \right)}{\left(\int_0^\infty \Delta\rho \, dz \right)^3}. \tag{3.9}$$

A value $S_1 = 1$ corresponds to a homogeneous layer of a well-defined depth. Inhomogeneities lead to values $S_1 < 1$. Values $S_1 > 1$ are impossible if $\Delta\rho$ is monotonic decreasing with z, which must be the case for a stable stratification, and so $S_1 > 1$ can only occur for short timescales due to overturning events.

Although S_1 can, then, be used as a measure of the inhomogeneity of the dense fluid layer, it can not be related to the departure of the profile from the initial density ρ_0. A similar quantity can, however, be constructed which does contain this information. A "mixing parameter" Λ is defined in a similar way to S_1, but based on a height and density from an unmixed state.

$$\Lambda = \frac{2}{h_*^2 \Delta\rho_0} \int_0^\infty \Delta\rho z \, dz. \tag{3.10}$$

h_* is defined as the height of the current in the absence of mixing, so as to conserve the vertically integrated mass.

$$h_* \Delta\rho_0 \equiv h \Delta\bar{\rho} \equiv \int_0^\infty \Delta\rho \, dz, \tag{3.11}$$

and so Λ can also be expressed in measurable terms, again using (3.7) and (3.8):

$$\Lambda = \frac{2\Delta\rho_0 \left(\int_0^\infty \Delta\rho z \, dz \right)}{\left(\int_0^\infty \Delta\rho \, dz \right)^2}. \tag{3.12}$$

The quantity Λ can be interpreted as the local ratio of the excess potential energies in the observed current and the hypothetical unmixed one. This is similar to the notion of a "mixedness" used by several authors introduced by De Silva and Fernando [13] to describe a mixed patch in a stratified fluid, the difference being that the mixedness is constrained to lie in the interval $[0, 1]$ by the inclusion of a "fully-mixed" potential energy in the definition. In the case of interest here, the current often has a depth much less than the total fluid depth, and indeed in some applications this total depth can be effectively infinite, so that the fully-mixed potential energy is poorly defined. Therefore Λ is only limited to values $\Lambda \geq 1$, with increasing Λ corresponding to more mixing. Notice that although S_1 and Λ are both based on the same integral $\int \Delta\rho z \, dz$, departures from an unmixed state are signified by variations in different directions—S_1 will decrease while Λ will increase.

3.3   Mixing

The quantities derived above can now be applied to the experiments in order to examine the mixing in the current.

Figure 6(a) shows a typical instance during a reflection of a finite-volume gravity current from a slope of angle $\theta = 21°$. The front of the current is still travelling up the slope, while the reflection, with the structure described in Section 3.1, is moving away from the slope. As is clear, substantial mixing has taken place between the current and the ambient.

In Figure 6(b), the S_1 profile corresponding to the instant in Figure 6(a) is shown. There is some variation, particularly where the depth is small (and so errors in the calculation of S_1 are likely to be at a maximum) but on the whole the value is insensitive to the exact details of the current. Regardless of the internal density structure, and the presence of waves on the interface, a value $S_1 \approx 0.9$ is typical of the current as a whole. It can be concluded that S_1 is not a useful parameter for quantifying the local mixing in the current. However, this does verify that a constant value of S_1 can be used for modelling purposes. In other words, the approximate two-layer nature of the density stratification does not significantly affect the dynamics of the current, although the same can not be said for the associated shear in velocity.

More successful is the use of the mixing parameter Λ to describe the current. Figure 7 contains two representations of the same gravity current being reflected from a slope of angle $47°$. In Figure 7(a) the intensity corresponds to the quantity $I = \int \Delta \rho \, dz$, and shows its variations throughout the current in space and time. In this representation the motion of the front of the current is clearly shown, as is the primary reflection from the slope, which is seen to bounce back and forward between the slope and the back wall of the loch. The bulk of the dense material is concentrated in the head of the current before impact, and in the primary reflection afterwards.

Figure 7(b) shows the same experiment, but this time Λ is displayed. Unfortunately the false intensity scale used to emphasize structure obscures the visualization of specific values, but it is clear that the behaviour of Λ mirrors that of I, but in an inverse manner. Those parts of the current which have low values of I (generally the shallower parts of the current) are precisely those areas with high values of Λ, while high values of I correspond to values of Λ close to unity. This confirms the idea that the head of the gravity current (and the primary reflection) is a core of dense fluid, and fluid which is mixed with the ambient is then left behind in a thin layer.

In the case shown, it is also noticeable that the value of Λ in the core does not significantly increase. Before the first impact, inside the core $\Lambda \approx 1.2$, and this has only increased to around $\Lambda \approx 1.5$ by $t = 100\,\text{s}$—an interval which contains three reflections from the boundaries. What mixing that has occurred has had little effect on the core. For reflections from a smaller slope the variation in Λ is more marked. Reducing the slope to $\theta = 9°$ leads to a value of $\Lambda \approx 2.0$ over the same interval. More mixing takes place between the current and the ambient

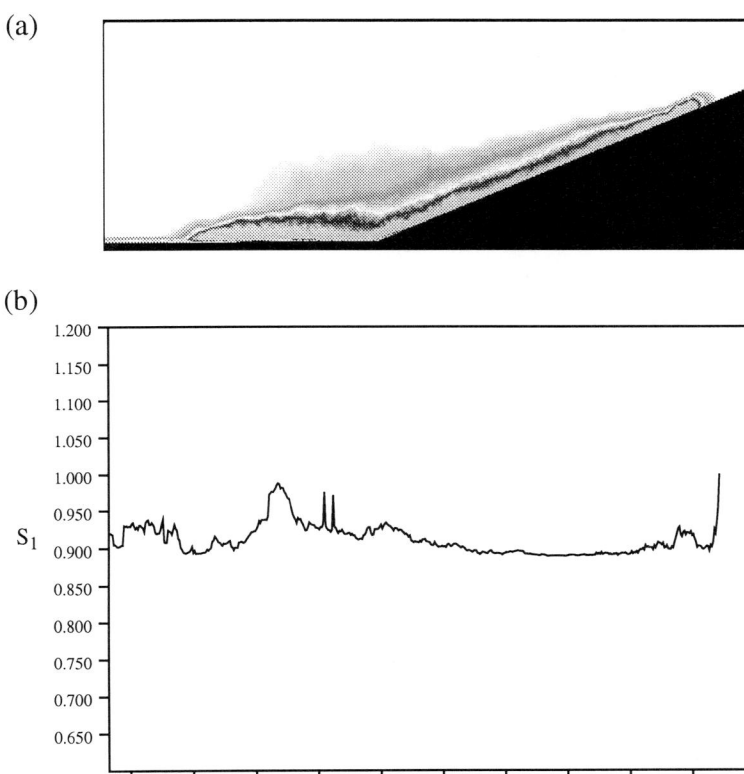

Figure 6. (a) shows a typical reflection of a finite-volume gravity current from a slope with angle $\theta = 21°$, and (b) shows the S_1 profile as calculated from this image. The horizontal scales are the same, with distance x measured from the back of the lock. $L_0 = 10\,\text{cm}$

a)

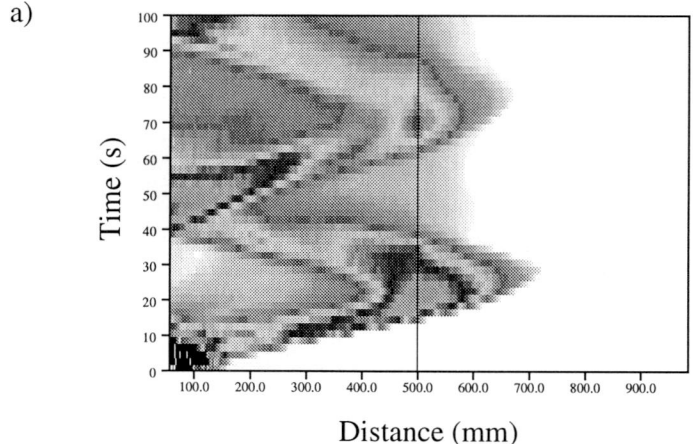

Distance (mm)

b)

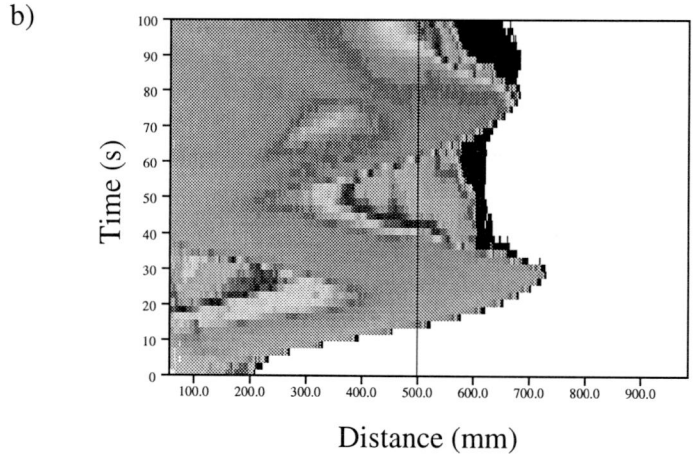

Distance (mm)

Figure 7. Shows the variation of the structure of a typical finite-volume current in space and time. The slope angle here is $47°$. (a) intensities represent the quantity $\int \Delta\rho \, dz$, and (b) intensities represent Λ. Again, a false intensity scale is used

during a reflection from a smaller slope. This is despite the amplitude of the reflection being smaller, and so the motion being less vigourous.

An explanation of this observation lies in the "coherence" of the reflection. At a large slope, or at a wall, the timescale of the interaction of the current head with the boundary is short. There is a large reflection, but relatively little mixing occurs, and the dense core is compact. On smaller slopes, however, the interaction takes place over a longer timescale. The amplitude of the reflection is smaller, but it has a greater wavelength. This, combined with the shear in the current produced during the reflection (described in Section 3.1), produces more significant mixing during the interaction. A simple way of interpreting the process is as follows. If we consider mixing to be the dilution of a dense inner-core of the current, then this mixing will be potentially more effective if the perimeter of this inner-core is large, since mixing can take place all along this contour. A small slope produces a reflection with a larger aspect ratio, or lower "coherence", with a larger perimeter to the inner-core.

This argument can be extended to a continuous-flux gravity current. Here the reflection can always be considered to be coherent, since the constant supply of dense fluid from upstream effectively creates a thick core everywhere upstream from the ramp. It is then expected that the mixing parameter will be close to unity everywhere, and measurements confirm $\Lambda \approx 1.2$ over the period of measurement for a large range of angles.

The mixing parameter Λ is, then, an effective measure of the local mixing. Its use is limited by the relatively small changes in the value, and so in practice variations in the behaviour only become apparent if large differences in slope angle are used. The sensitivity is also not good enough to determine a trend for continuous-flux currents, although this trend is likely to be similar if less pronounced.

4 Conclusions

This paper has presented an experimental study of a two-dimensional gravity current impinging on an uphill slope. Density measurements have been made via an optical technique allowing quantitative examination of the mixing between the current and the lighter ambient fluid.

This technique has allowed the calculation of the shape parameter S_1, a quantity important in shallow-water modelling of such flows, and it is found that S_1 is relatively insensitive to the various reflections, waves and mixing processes occurring in the flow—a value $S_1 \approx 0.9$ is generally appropriate. However, there is a complex reflection structure which perhaps is more significant to modelling efforts. The reflection can sometimes more accurately be described as a two-layer flow, a feature which cannot be captured by traditional models.

A measure of the local mixedness of the flow can be constructed from the potential energy of the current, and this indicates that mixing during the reflection process is greater at small slope angles. This is related to the shorter timescale involved in a reflection at larger slopes resulting in a more coherent mass of the densest fluid, which makes turbulent entrainment a less effective mixing process.

A more detailed description of the flow field is required in order to further evaluate the importance of the effects observed.

Acknowledgements

I would especially like to thank Dr. Paul Linden for his supervison during this work, and various members of the GFD group and DAMTP technical staff for their advice and assistance.

The author has been funded by the EPSRC and the UK Health and Safety Executive, under the CASE award scheme.

Bibliography

[1] Simpson, J. E. (1987). *Gravity currents in the environment and the laboratory.* Ellis Horwood, Chichester.

[2] Simpson, J. E. (1969). A comparison between laboratory and atmospheric density currents. *Q. J. R. Meteorol. Soc.*, **95**, 758–65.

[3] Britter, R. E. and Simpson, J.E. (1978). Experiments on the dynamics of a gravity current head. *J. Fluid Mech.*, **88**, 223–40.

[4] Hacker, J., Linden, P. F. and Dalziel, S. B. (1996). Mixing in lock-release gravity currents. *Dyn. Atmos. and Oceans*, **24**, 183–95.

[5] Hallworth, M. A., Huppert, H. E., Phillips, J. C. and Sparks, R. S. J. (1996). Entrainment into two-dimensional and axisymmetric turbulent gravity currents. *J. Fluid Mech.*, **308**, 289–311.

[6] Edwards, D. A., Leeder, M. R., Best, J. L. and Pantin, H. M. (1994). On experimental reflected density currents and the interpretation of certain turbidites. *Sedimentology*, **41**, 437–61.

[7] Lane-Serff, G. F., Beal, L. M. and Hadfield, T. D. (1995). Reflection of gravity currents from slopes and beaches. Submitted to *J. Fluid Mech.*

[8] Rottman, J. W. and Simpson, J. E. (1983). Gravity currents produced by instantaneous releases of a heavy fluid in a rectangular channel. *J. Fluid Mech.*, **135**, 95–110.

[9] Dalziel, S. B. and Lane-Serff, G. F. (1991). The hydraulics of doorway exchange flows. *Building and Environment*, **26**, 121–35.

[10] Rottman, J. W., Simpson, J. E., Hunt, J. C. R. and Britter, R. E. (1985). Unsteady gravity current flows over obstacles: some observations and analysis related to the phase II trials. *J. Hazard. Mater.*, **11**, 325–40.

[11] Rottman, J. W. and Simpson, J. E. (1989). The formation of internal bores in the atmosphere: a laboratory model. *Q. J. R. Meteorol. Soc.*, **115**, 941–63.

[12] Ellison, T. H. and Turner, J. S. (1959) Turbulent entrainment in stratified flows. *J. Fluid Mech.*, **6**, 423–48.

[13] De Silva, I. P. D. and Fernando, H. J. S. (1992). Some aspects of mixing in a stratified turbulent patch. *J. Fluid Mech.*, **240**, 601–25.

Richardson Number Dependence of Three–Dimensional Motions in Shear Flows

C.P. Caulfield* and W.R. Peltier**

*Department of Mathematics, University of Bristol, and **Department of Physics, University of Toronto, Canada*

Abstract

Spontaneous development of spatially incoherent "turbulent" motion within an initially laminar state of fluid flow is a problem of fundamental importance [1–2] in physics, with applications in both the geophysical sciences [3–4], and engineering [5–6]. An archetypical specific example concerns the onset of "mixing" in a free shear flow within which the background horizontal velocity varies solely in the vertical direction, and over a finite length scale [7]. Such mixing layers are primarily susceptible to the two–dimensional "Kelvin–Helmholtz" instability [8], and are observed to become turbulent only after the occurence of a secondary bifurcation has induced the appearance of three–dimensional motion, which takes the form of streamwise vortex streaks, in both unstratified ([5], [9–16]) and stratified ([4], [17–21]) flows. Using theoretical and numerical analyses, we show that the presence of density variations within the flow qualitatively changes the growth mechanism and physical location of the streamwise vortex streaks, due principally to the presence of (locally) statically unstable regions induced by the "roll–up" of the primary Kelvin–Helmholtz billow.

We confirm that, for fluids of constant density, the mode of secondary instability which precedes the transition to turbulence leads to the development of rib–like [5] streamwise vortical structures, which are initially localized within the "braids" between adjacent billow cores. The "elliptical" instability [22–25] of the Kelvin–Helmholtz billow, though present, appears to play an insignificant role. In density stratified fluids, the behaviour is qualitatively different. Transition in such flows is dominated by a shear aligned convective instability [26–28], [18–19] which develops into a spanwise periodic array of streamwise convective roll vortices which are initially localized within the statically unstable regions around the rim of the billow core.

1 Introduction

The processes by which stratified shear flows become "turbulent" have been much studied, in an attempt to understand how mixing events lead to the exchange of heat and momentum in the atmosphere and oceans [3–4]. In this paper, we have investigated the causes of the various mechanisms of turbulent breakdown, and the effect of stratification and initial perturbation on the subsequent turbulent evolution and mixing characteristics of the flow.

Throughout this paper, we study a temporally evolving shear layer, where the variations in density are sufficiently small so that we can make the Boussinesq approximation. We choose a coordinate system so that x is the spanwise direction, y is the alongstream or streamwise direction, and z is the vertical direction. Two–dimensional shear layers, (where the background laminar velocity $\bar{\mathbf{U}} = (0, V(z), 0)$ only has a streamwise component which varies with height alone) may be thought of as strips of spanwise vorticity, where the vorticity ω is of course the curl of the velocity field:

$$\omega = \nabla \times \mathbf{u} \,. \tag{1.1}$$

Typically, such a shear flow is primarily susceptible to a two–dimensional instability, the "Kelvin–Helmholtz billow" [8]. This instability is essentially an instability of inflectional velocity profiles. The Kelvin–Helmholtz instability "rolls up" this vorticity into elliptical billow cores, connected by thin "braids" [5].

If the fluid is statically stably stratified, with background density profile $\bar{\rho}(z)$, this "rolling up" process is suppressed somewhat. As the billow develops, the potential energy of the flow increases due to heavy fluid being lifted up, and lighter fluid being lowered from its initial, equilibrium position. Therefore, less kinetic energy is available to the perturbation for growth. A measure of the significance of the background stratification is the (local or gradient) Richardson number, $Ri(z)$, defined as

$$Ri(z) \equiv -\frac{g}{\rho_{\text{ref}}} \frac{d\rho/dz}{(dV/dz)^2} \,, \tag{1.2}$$

where ρ_{ref} is some reference density. If the Richardson number is greater than $1/4$ *everywhere* within the flow Miles and Howard showed that the stratification totally stabilised the flow, in the inviscid limit [29–30]. If the characteristic length scale of variation of the velocity profile and the density profile are of the same order, numerical and experimental studies of stratified shear flows have been shown to be susceptible to instabilities of Kelvin–Helmholtz type. For such flows the effect of the stratification on the growth rate of the primary Kelvin–Helmholtz instability appears to be purely stabilising [3–4].

A primary concern is to understand the mixing induced by flow instabilities, and here the rôle of the density distribution is much more complex. From experimental and field observations [3–5], [17], [20–21], [31] mixing within shear flows has been observed to onset only after the appearance of three–dimensional motions within the flow. As we show in this paper, the effect of the background density stratification on the characteristics, growth mechanisms and properties of the three–dimensional motions which lead to flow breakdown and mixing is profound. We demonstrate this by a combination of theoretical calculation and numerical simulation. In Section 2 of this paper, we present detailed theoretical analyses of the stability of two–dimensional Kelvin–Helmholtz billows against three–dimensional perturbations in flows with two different characteristic values of the Richardson number at the midpoint of the shear layer (namely $Ri(0) = 0$ and $Ri(0) = 0.05$). These analyses predict that in general the primary, two–dimensional instability is subject to a range of distinct secondary, three–dimensional instabilities. The growth mechanisms and finite amplitude characters of the various secondary instabilities are strongly dependent on the initial flow parameters. The results of these theoretical analyses are then compared with the results of direct numerical simulations of fully three–dimensional mixing layers.

Provided the initial Richardson number is high enough, the theoretical analyses predict that the three–dimensional spectrum is dominated by a shear aligned convective instability, which is localized around the edge of the billow core where the process of "rolling up" of the (initially one–dimensional) vorticity in the shear layer has induced (locally) statically unstable density distributions. On the other hand, for unstratified shear flows, the theoretical calculations predict that, at sufficiently high Reynolds number, the dominant mode of secondary instability is located primarily in the braids between the billows.

In Section 3, we present results of fully three–dimensional numerical simulations, which agree very well with our theoretical predictions. In the unstratified simulation, rib–like vortex structures in the braids are observed to dominate the breakdown of the primary Kelvin–Helmholtz core in the numerical simulations. Our results suggest that the core–centred instability of the elliptical (finite amplitude) vortex arising from Kelvin–Helmholtz instability, though present, is far less important in the transition to turbulence, except at low Reynolds number. In the stratified case, the simulations verify the dominance of this convective mode of three–dimensionalization, which at finite amplitude is manifested as convective roll–like streaks of streamwise vorticity which alternate in sign in the spanwise direction, and the spanwise wavelength of these streaks is well predicted by the theory. We draw some brief conclusions in Section 4.

2 Theoretical calculations

We consider the behaviour of a Boussinesq incompressible fluid. The initial background profiles of velocity and density are chosen such that

$$\bar{V}(z,0) = V_0 \tanh\frac{z}{d}, \tag{2.1}$$

$$\bar{\rho}(z,0) = \rho_a + \rho_0 \tanh\frac{Rz}{d}, \tag{2.2}$$

where R is the ratio of the characteristic scale of velocity variation to the characteristic scale of density variation. We use the experimentally realistic value of $R = 1.1$ throughout [17]. For these initial profiles, the initial Richardson number is minimal at the midpoint of the shear layer (i.e. at $z = 0$) with value

$$Ri(0) \equiv \frac{gR\rho_0 d}{\rho_a V_0^2}. \tag{2.3}$$

Using V_0, ρ_0 and d as characteristic velocity variation, density variation and length scales, we non–dimensionalize the Navier–Stokes equations to obtain

$$\frac{Du_i^*}{Dt^*} = -\frac{\partial p'^*}{\partial x_i^*} - \frac{Ri(0)\rho'^*}{R}\delta_{i3} + \frac{1}{Re}\frac{\partial^2 u_i^*}{\partial x_j^{*2}} \ , \tag{2.4}$$

$$\frac{\partial u_j^*}{\partial x_j^*} = 0 \ , \tag{2.5}$$

$$\frac{D\rho'^*}{Dt^*} = \frac{1}{RePr}\frac{\partial^2 \rho'^*}{\partial x_j^{*2}} \ , \tag{2.6}$$

in which asterisks denote nondimensional quantities, and summation is implied by repeated subscripts. Henceforth, we will only consider nondimensional quantities, and hence we will drop the asterisks. In the governing equations, the dependence of the flow on the three nondimensional parameters, i.e. the Reynolds number Re, the Prandtl number Pr and the minimum initial Richardson number $Ri(0)$ is now made explicit. For simplicity, we have chosen the Prandtl number $Pr \equiv \nu/\kappa = 1$ in all simulations. We chose the Reynolds number to be 750 (a typical experimental value), where $Re = V_0 d/\nu$.

We solved these equations numerically, using a modified, anelastic finite–difference model [32]. This model is based upon the use of second–order accurate finite differences to solve the equations on a staggered grid, while the equations are stepped forward in time using an explicit leapfrog scheme. The streamwise extent L_y of the computational domain was taken in all cases to be equal to one wavelength of the most unstable mode of linear theory, i.e. $L_y \simeq 14.7$. We consider the flow within a channel which is sufficiently wide so that the upper and lower boundaries have no effect on the flow dynamics. In nondimensional units, this means that the vertical domain is $-10 < z < 10$.

In the first instance, we restrict the simulation to two–dimensional (i.e. restricted to the y–z plane) flow. In this case, a Kelvin–Helmholtz billow develops as the shear layer, "rolls up".

To investigate the development of such a flow in time, we decompose the velocity field into mean and perturbation parts. In two dimensions, the total velocity field, $\mathbf{U}_T$, has two elements

$$\mathbf{U}_T(y, z, t) = (v_T(y, z, t), w_{KH}(y, z, t)) \ . \tag{2.7}$$

Clearly, there is no mean component of the vertical velocity, and so any component of vertical velocity must be associated with the development of the Kelvin–Helmholtz billow, and is thus labelled with the subscript "KH". On the other hand, the total streamwise velocity may be written as

$$v_T = \bar{V}(z, t) + v_{KH}(y, z, t) \ , \tag{2.8}$$

where

$$\bar{V}(z, t) = <v_T>_y \ , \tag{2.9}$$

$$<v_{KH}>_y = 0 \ , \tag{2.10}$$

and $< . >_y$ denotes an average in the streamwise direction.

Similarly, the density may be decomposed into

$$\rho_T(y, z, t) = \bar{\rho}(z, t) + \rho_{KH}(y, z, t) , \tag{2.11}$$

where

$$\bar{\rho}(z, t) = < \rho_T >_y , \tag{2.12}$$

and ρ_{KH} is the density perturbation associated with the development of the Kelvin–Helmholtz billow.

Then, in the Boussinesq approximation, we may identify the total kinetic energy of the flow over the computational domain as

$$KE_T(t) = \frac{1}{2} \int_{-10}^{10} \int_0^{L_y} (v_T{}^2 + w_{KH}{}^2) \, dy dz . \tag{2.13}$$

The perturbation kinetic energy over the computational domain of the two–dimensional perturbation associated with the Kelvin–Helmholtz billow may be deinfed as

$$KE_{KH} = \frac{1}{2} \int_{-10}^{10} \int_0^{L_y} (v_{KH}{}^2 + w_{KH}{}^2) \, dy dz . \tag{2.14}$$

We define the billow to be "fully developed" when KE_{KH} is maximal.

We then consider the stability of both the fully developed unstratified Kelvin–Helmholtz billow, and the stratified Kelvin–Helmholtz billow against three–dimensional perturbations, using a previously elaborated theoretical methodology [27–28]. Fundamentally, this method assumes that the fully developed Kelvin–Helmholtz billow is "steady" (which it is to a good approximation), and that it is reasonable to consider a three–dimensional perturbation from this two–dimensional, (but definitely not laminar) "steady" flow, and to investigate whether this perturbation grows with time. We thus decompose the (total) three–dimensional velocity fields $\mathbf{u}_T$ as follows:

$$
\begin{align}
\mathbf{u}_T(x, y, z, t) &= [u_T(x, y, z, t), v_T(x, y, z, t), w_T(x, y, z, t)] , \tag{2.15} \\
\bar{\mathbf{V}}(z, t) &= << \mathbf{u}_T >_x >_y = [0, \bar{V}(z, t), 0] , \tag{2.16} \\
\mathbf{u}_{KH}(y, z, t) &= < \mathbf{u}_T - \bar{\mathbf{U}} >_x = [0, v_{KH}, w_{KH}] , \tag{2.17} \\
\mathbf{u}'(x, y, z, t) &= \mathbf{u}_T - \mathbf{u}_{KH} - \bar{\mathbf{V}} = [u', v', w'] , \tag{2.18}
\end{align}
$$

in which $< . >_x$ denotes spanwise averaging, over the total extent of the domain of interest. (It is possible to decompose the total density field ρ_T in a similar way, using ρ_0 to non–dimensionalize.) From experimental evidence, we are interested in three–dimensional perturbations which are localized in the vicinity of a two–dimensional billow. From Floquet theory, we may suppose that the three–dimensional perturbations $\mathbf{u}'$ and ρ' take the form

$$
\begin{align}
\mathbf{u}'(x, y, z, t) &= \hat{\mathbf{u}}(y, z, t) \exp[i\gamma x] , \tag{2.19} \\
\rho'(x, y, z, t) &= \hat{\rho}(y, z, t) \exp[i\gamma x] , \tag{2.20}
\end{align}
$$

where the components of $\hat{\mathbf{u}}$ and $\hat{\rho}$ are also spatially periodic with lengthscale L_y. Since we assume that the Kelvin–Helmholtz billow is steady, or at least slowly varying in time compared to the three–dimensional perturbation, we assume that

$$\hat{\mathbf{u}}(y,z,t) \;=\; \mathbf{u}^\dagger(y,z)e^{st} , \tag{2.21}$$

where $s \equiv \sigma + i\omega$, and instability corresponds to $\sigma > 0$. It is possible to use the Galerkin method to calculate the structure (for example $\mathbf{u}^\dagger(y,z)$) spanwise wavenumber (γ) and growth rate (σ) of three–dimensional instabilities by standard matrix techniques.

The growth mechanisms of the various, distinct secondary instabilities which the billows support can be best understood by considering the spatial distribution of perturbation kinetic energy for the three–dimensional perturbation ($KE'(x,y,z)$) from the spanwise–averaged flow. An appropriate measure of the perturbation kinetic energy of these three–dimensional motions is

$$KE'(x,y,z,t) \equiv \frac{1}{2}\left(u'^2 + v'^2 + w'^2 \right) . \tag{2.22}$$

To understand the significance of various processes to the (initially small) three–dimensional instabilities, it is very instructive to derive an evolution equation for KE', integrated over the whole domain. After some manipulation, we obtain

$$\int_{-10}^{10}\int_0^{L_y} \frac{1}{2KE'}\frac{d}{dt}KE'\,dydz \;=\; \left[-\frac{1}{Re}\int_{-10}^{10}\int_0^{L_y}\left(\frac{\partial u_i'}{\partial x_j}\right)^2 dydz \right.$$

$$-\int_{-10}^{10}\int_0^{L_y} v'w'\frac{\partial \bar{V}}{\partial z}\,dydz \tag{2.23}$$

$$-\int_{-10}^{10}\int_0^{L_y} v'w'\left(\frac{\partial v_{\mathrm{KH}}}{\partial z}+\frac{\partial w_{\mathrm{KH}}}{\partial y}\right)dydz$$

$$-\int_{-10}^{10}\int_0^{L_y}\left(v'^2\frac{\partial v_{\mathrm{KH}}}{\partial y}+w'^2\frac{\partial w_{\mathrm{KH}}}{\partial z}\right)dydz$$

$$\left. -\frac{Ri(0)}{R}\int_{-10}^{10}\int_0^{L_y}\rho'w'\,dydz \right]/(2KE') ,$$

$$\sigma' \;\equiv\; \mathcal{D}' + \mathcal{SH}_{\mathrm{B}}' + \mathcal{SH}_{\mathrm{KH}}' + \mathcal{ST}_{\mathrm{KH}}' + \mathcal{H}' . \tag{2.24}$$

where σ' is a numerical measure of the "growth rate" of the (general) perturbation. Here, $\mathcal{D}'$ is (negative definite) dissipation, $\mathcal{SH}_{\mathrm{B}}'$ quantifies extraction of energy through shear stress from the background field, $\mathcal{SH}_{\mathrm{KH}}'$ quantifies extraction of energy through shear stress from the two–dimensional perturbation field, $\mathcal{ST}_{\mathrm{KH}}'$ is extraction through straining motions, and $\mathcal{H}'$ is conversion of potential energy (through buoyancy flux) into KE'.

From our stability calculations, we verified that both unstratified and stratified Kelvin–Helmholtz billows are subject to three–dimensional, spanwise periodic instabilities. For an instability with structure defined by (2.19) and (2.20), it is clear that

$$\sigma = \sigma' = \mathcal{D}' + \mathcal{SH}'_{\mathrm{B}} + \mathcal{SH}'_{\mathrm{KH}} + \mathcal{ST}'_{\mathrm{KH}} + \mathcal{H}' , \tag{2.25}$$

exactly, and so it is possible to determine which physical processes are dominant in the growth of the three–dimensional instabilities. Furthermore, since our calculations yield the eigenstructure of the instabilities, it is also possible to determine the locations within the flow where we expect the instabilities to be most energetic. The results of our calculations are summarized in Table 1.

For both flows, the instability was found to be relatively broadband, in the sense that the growth rate σ varied relatively slowly over a relatively wide range of values of γ. For the unstratified flow, the dominant contribution to growth is through shear extraction of energy from the mean (streamwise independent) component of the flow, and the component of the flow associated with the Kelvin–Helmholtz billow has an appreciably weaker effect. The most unstable wavenumber $\gamma = 2.5$ implies that the spanwise wavelength of the most unstable mode is about five or six times smaller than the wavelength of the primary Kelvin–Helmholtz billow (which has wavenumber $\alpha \simeq .43$). For the stratified flow, we see that the wavelength of the most unstable mode is smaller still (about seven or eight times smaller than the wavelength of the Kelvin–Helmholtz billow). Although shear extraction from the background flow is still very important as a growth mechanism, the buoyancy flux term $\mathcal{H}'$ is of comparable magnitude, and is clearly very significant for the (predicted) growth of the most unstable spanwise periodic three–dimensional perturbation.

The difference between stratified and unstratified flows is even more marked when we consider the spatial distribution of KE'. Our theoretical analyses predict that for the unstratified flow (see Figure 1(a)) KE' should be concentrated in the braid region. Indeed, similar calculations at various Reynolds numbers suggest that the core–centred instability [22–25] of the elliptical (finite amplitude) vortex arising from the primary Kelvin–Helmholtz instability, though present, is far less important in the transition to turbulence, except at low Reynolds number. The equivalent analysis for the flow with $Ri(0) = 0.05$ (Figure 1(b)) shows a markedly different behaviour. The theory predicts that the spatial distribution of KE' should be concentrated in the regions around the core, the very loca-

Table 1. Properties of most unstable three–dimensional modes

$Ri(0)$	γ	σ	$\mathcal{SH}'_{\mathrm{B}}$	$\mathcal{SH}'_{\mathrm{KH}}$	$\mathcal{ST}'_{\mathrm{KH}}$	$\mathcal{H}'$	$\mathcal{D}'$
0.00	2.5	0.147	0.127	0.017	0.017	0	-0.014
0.05	3.4	0.179	0.115	-0.024	0.019	0.091	-0.022

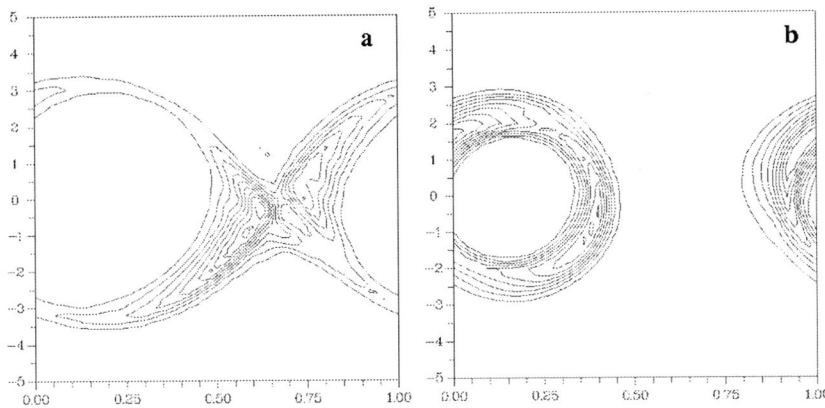

Figure 1. Theoretical isolines of KE' for the most unstable mode for flows with $Re = 750$ and, (a) $Ri(0) = 0$, and (b) $Ri(0) = 0.05$, when the primary two–dimensional billow has achieved maximal amplitude

tions the density distribution has become statically unstable. Therefore, there is an available source of potential energy, which through the term $\mathcal{H}'$ can be converted into perturbation kinetic energy. For this particular value of $Ri(0)$, both the shear conversion terms (which now are most significant away from the braid region and towards the edges of the billow core) and the release of potential energy through heat flux terms should be significant in the increase in the KE'. This instability may be thought of as being of the so–called "shear–aligned convective" type [19,26–28].

3   Numerical simulations

We investigated the validity of our theory by conducting fully three–dimensional simulations of shear layer evolution, with $Ri(0) = 0$, and $Ri(0) = 0.05$. Both calculations were initialized by superimposing on the background flow a small amplitude perturbation with structure determined by the most unstable mode of linear theory and a small amplitude isotropic noise field for the spanwise velocity. Spanwise periodicity was imposed, and the width of the computational domain was $0 < x < 10$, i.e. of the order of four or five spanwise wavelengths of the most unstable modes predicted by our theoretical calculations.

In Figures 2(a)–(c) we show, at three different times in the evolution of the unstratified shear flow, an iso–surface of spanwise vorticity (in light grey) denoting the primary Kelvin–Helmholtz billow core and two iso–surfaces of (oppositely signed) streamwise vorticity (in darker shades of grey). In nondimensional time units, the times are $t = 35, 101.2$, and 126 respectively. In Figure 2(a), the

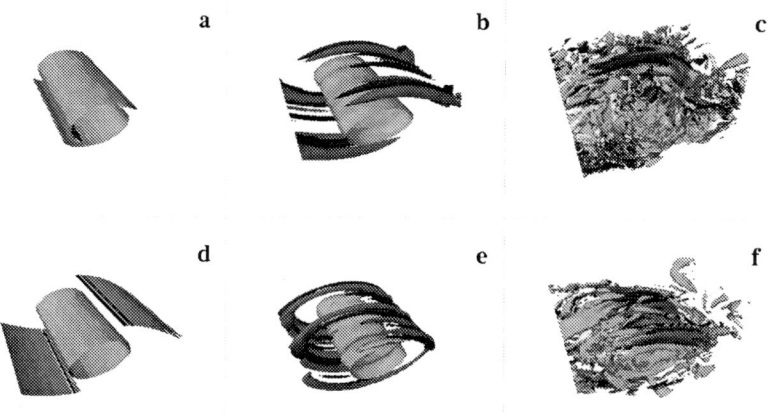

Figure 2. Positive (darkest grey) and negative (intermediate grey) streamwise vorticity iso–surfaces with a spanwise vorticity iso–surface (light grey) which delineates the billow core from three–dimensional simulations of an unstratified shear flow (2(a)–(c)) and stratified shear flow with $Ri(0) = 0.05$ (2(d)–(f)) at three characteristic times: when KE_{KH} is maximal; when the integrated value of KE' is 1% of KE_{KH}; and when the flow is "turbulent"

Kelvin–Helmholtz billow has reached its maximum amplitude, and the vorticity of the initial shear layer has been "rolled up" into the billow core in a highly two–dimensional manner. This state is essentially identical to the two–dimensional state whose stability we considered in the previous section. During the growth of this instability, the three–dimensional perturbation (seeded by the isotropic noise field) is actually strongly suppressed. However, following saturation of the primary instability, three–dimensional perturbations grow rapidly, and manifest themselves as spanwise periodic pairs of streamwise vortex tubes, which are strongly localized in the "braids" between the billow cores (Figure 2(b)). Immediately following the appearance of these highly coherent structures, the Kelvin–Helmholtz billow cores collapse (Figure 2(c)) and the flow becomes fully turbulent.

The behaviour of a stratified shear flow is markedly different from this, even for quite small Richardson numbers. Figures 2(d)–(f) show the same iso–surfaces of vorticity at equivalent times to the unstratified simulation for a flow with a (Boussinesq) density distribution, such that $Ri(0) = 0.05$. The equivalent times for Figures 2(d)–(f) are $t = 42, 127.2$, and 156 respectively. The primary instability still develops into a Kelvin–Helmholtz billow. In this case, however, there is significant baroclinic generation of vorticity in the braid regions during the two–dimensional evolution of the Kelvin–Helmholtz billow (Figure 1(d)). Furthermore, the process of vortex "roll-up" causes the development of statically unstable regions within the billow core. The secondary bifurcation that ensues (Figure 2(e)) is markedly different from that which developed in the unstratified flow, and is concentrated around the periphery of the billow core, precisely overlapping these statically unstable regions. The appearance of these secondary structures is also rapidly followed by the collapse of the primary billow core and the onset of truly turbulent flow (Figure 2(f)), though the stratification is clearly flattening the vertical extent of the disordered motion. In both cases, it is clear that the development of the streamwise vortex tubes is localized in the vicinity of a single Kelvin–Helmholtz billow, though the actual part of the flow where the amplitude of the perturbation is largest is exceedingly dependent on $Ri(0)$.

Clearly, there is qualitative agreement in the properties of the nonlinear flow with our theoretical predictions. In Figure 3, we compare the (linear) prediction of our theory for the KE' field (i.e. Figure 1), with the numerically calculated spanwise averaged nonlinear KE' field realised at the times shown in Figures 2(b) and 2(e).

There is excellent agreement. For the unstratified flow, (Figure 3(a)) the nonlinear process of vortex stretching has smeared the distribution over the entire braid region, but still the braid localization of the perturbation is very apparent. Similar good agreement is found in the stratified case (through comparison of Figures 1(b) and 3(b), evaluated at the time shown in Figure 2(e)), though even more so in this case, the nonlinear process of vortex stretching has smeared

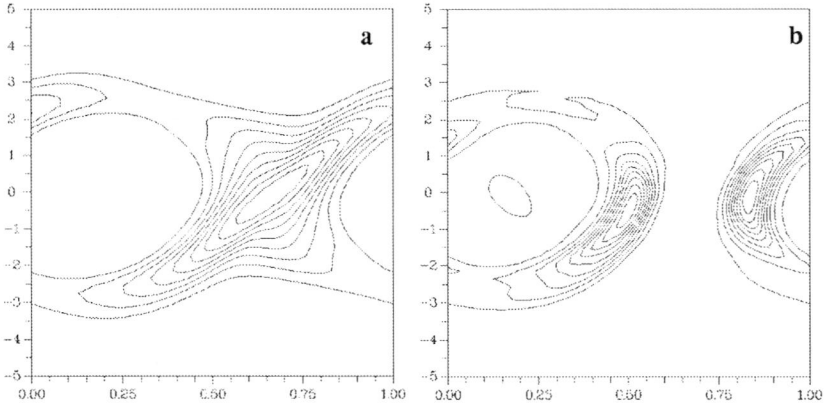

Figure 3. Numerically calculated isolines of KE' for flows with $Re = 750$ at the times shown in, (a) Figure 1(b) for $Ri(0) = 0$, i.e. $t = 35$, and (b) Figure 1(e) for $Ri(0) = 0.05$, i.e. $t = 42$

the distribution such that it extends further towards the braids. Not only is the localization of the dominant three–dimensional instabilities in each case accurately predicted by the theory, but also the spanwise wavenumber is as predicted. Clearly, from Figure 2, the spanwise wavelength of the three–dimensional perturbation is appreciably smaller than the wavelength of the primary Kelvin–Helmholtz billow. Spanwise powerspectra of the streamwise vorticity field in each case show strong peaks at the value of the predicted value of the most unstable wavenumber.

The streamwise vortex tubes which develop are clearly crucial in the process of transition to turbulence. To understand how the tubes have developed, it is appropriate to look at the temporal evolution of the perturbation kinetic energy. In Figure 4, we plot the numerically determined growth rate σ' (as a thick solid line) and the various terms on the left hand side of (2.23) and (2.24) against time for both our simulations. It is particularly interesting that the three–dimensional motion starts to grow **exactly** at the moment when the Kelvin–Helmholtz billow saturates.

For the unstratified simulation, the theory predicts that the dominant mechanism for increase in KE' should be through the action of Reynolds stresses against the shearing deformation applied in the vicinity of the braids by the two–dimensional billow–related flow. This is exactly what is observed to occur in the direct three–dimensional numerical simulation, as can be seen in Figure 4(a). This growth mechanism appears to be totally localized within the braid region.

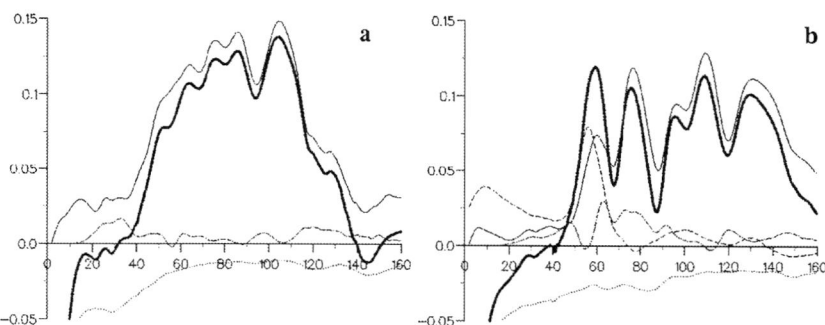

Figure 4. Growth rate σ' (thick solid line) of three–dimensional perturbations and the shearing deformations $\mathcal{SH}'_B + \mathcal{SH}'_{KH}$ (thin solid line), buoyancy flux contributions $\mathcal{H}'$ (dashed line), straining deformations $\mathcal{ST}'_{KH}$ (dot-dashed line) and dissipation $\mathcal{D}'$ (dotted line) plotted against nondimensional time for, (a) $Ri(0) = 0$, and (b) $Ri(0) = 0.05$. Note that in both cases the three–dimensional perturbations are initially suppressed (i.e. $\sigma < 0$) until exactly the time when the Kelvin–Helmholtz billow has maximal amplitude (i.e. 35 and 42 respectively)

In the stratified flow, as the flow evolves, and the three–dimensional structures attain finite amplitude, shear conversion effects through vortex stretching become increasingly important, as can be seen in Figure 4(b). However the initial mechanism of growth of the three–dimensional perturbation, which to a large degree determines its location and structure at finite amplitude, is associated with a release of potential energy through the buoyancy flux term associated with a (single) convective overturning, localized in the statically unstable regions which surround the billow core. The flow here is convectively unstable, but the background shear acts to suppress "rolls" aligned in the spanwise direction [26], thus leading naturally to alternating-sign streamwise vortices, which are not shear–suppressed. This conversion of potential energy to KE' appears to be the most important trigger for the onset of the three–dimensional instability as it precedes the increase in shear conversion, as can be clearly seen in Figure 4(b). This would appear to confirm that the streamwise vortex streaks which precede the onset of mixing in stratified parallel shear flow are largely governed at small amplitude by shear–aligned convection. In additional analyses at larger $Ri(0)$, the importance of convective processes unsurprisingly increases. In all of these stratified cases, the wavenumber of the observed instability also agrees very well with the theoretical analysis which predicts localization within the region immediately surrounding the billow core.

4   Conclusions

We find that the onset of turbulence within both stratified and unstratified shear flows is triggered by the development of streamwise vortex tubes. The particular growth mechanisms and the location of maximal amplitude of these streamwise vortices which spontaneously appear immediately upon saturation of the primary Kelvin–Helmholtz billow are, however, strongly dependent on the initial Richardson number. For the unstratified flow, the dominant mode of secondary instability is found to be located primarily in the "braid" region of strain between adjacent billow cores, and the "elliptical" instability appears to be largely insignificant at the Reynolds numbers we consider. In the stratified case, the three–dimensional motions appear to be triggered by a single convective overturning, which converts potential energy into kinetic energy for the three–dimensional perturbation. This perturbation is then amplified by the background flow, but the fundamental structure (streamwise vortex tubes), location (around the periphery of the vortex core, where there are statically unstable density distributions) and length scale are determined by this initial overturning event.

In both flows, these streamwise–aligned vortex tubes lead rapidly to disordered turbulent flow, but the processes by which this turbulent breakdown is reached are fundamentally affected by stratification.

Bibliography

1. Hinze, J. O. (1975). *Turbulence* (2nd edn). McGraw-Hill.

2. Lesieur, M. (1990). *Turbulence in Fluids* (2nd revised edn) Kluwer.

3. Thorpe, S. A. (1987). Transitional phenomena and the development of turbulence in stratified fluids: a review. *J. Geophys. Res.* **92** C 5231–5248.

4. Fernando, H. J. S. (1991). Turbulent mixing in stratified flows. *Ann. Rev. Fluid Mech.* **23** 455–493.

5. Hussain, A. K. M. F. (1983). In *Turbulence and Chaotic Phenomena in Fluids* (ed. T. Tatsumi) North Holland 453–460.

6. Liu, J. T. C. (1989). Coherent structures in transitional and turbulent free shear flows. *Ann. Rev. Fluid Mech.* **21** 285–315.

7. Brown, G. L. & Roshko, A. (1974). On density effects and large structure in turbulent mixing layers. *J. Fluid Mech.* **64** 775–816.

8. Drazin, P. G. & Reid, W. H. (1981). *Hydrodynamic Stability* C. U. P.

9. Corcos, G. M. & Lin, J. S. (1984). The mixing layer: deterministic models of a turbulent flow. Part 2. The origin of the three-dimensional motion. *J. Fluid Mech.* **139** 67–95.

10. Metcalfe, R. W., Orszag, S. A., Brachet, M. E., Menon, S. & Riley, J. J. (1987). Secondary instability of a temporally growing mixing layer. *J. Fluid Mech.* **184** 207–243.

11. Lasheras, J. C. & Choi, H. (1988). Three–dimensional instability of a plane free shear layer: an experimental study of the formation and evolution of streamwise vortices. *J. Fluid Mech.* **189** 53–86.

12. Ashurst, W. T. & Meiburg, E. (1988). Three–dimensional shear layers via vortex dynamics. *J. Fluid Mech.* **189** 87–116.

13. Nygaard, K. J. & Glezer, A. (1991). Evolution of streamwise vortices and the generation of smallscale motion in a plane mixing layer. *J. Fluid Mech.* **231** 257–301.

14. Rogers, M. M. & Moser, R. D. (1992). The three-dimensional evolution of a plane mixing layer: the Kelvin-Helmholtz rollup. *J. Fluid Mech.* **243** 183–226.

15. Comte P., Lesieur, M. & Lamballais, E. (1992). Large and small–scale stirring of vorticity and a passive scalar in a 3D temporal mixing layer. *Phys. Fluids* **A4** 2761–2778.

16. Lesieur, M., Comte P. & Métais, O. (1995). Numerical simulation of coherent vortices in turbulence. *Appl. Mech. Rev.* **48** 121–148.

17. Thorpe, S. A. (1985). Laboratory observations of secondary structures in Kelvin-Helmholtz billows and consequences for ocean mixing. *Geophys. Astrophys. Fluid Dyn.* **34** 175–199.

18. Palmer, T. L., Fritts, D. C., Andreassen, Ø. & Lie, I. (1994). Three–dimensional evolution of Kelvin–Helmholtz billows in stratified compressible flow. *Geophys. Res. Let.* **21** 2287–2290.

19. Caulfield, C. P. & Peltier, W. R. (1994). Three dimensionalization of the stratified mixing layer. *Phys. Fluids* **6** 3803–3805.

20. Schowalter, D. G., Van Atta, C. W. & Lasheras, J. C. (1994). A study of streamwise vortex structure in a stratified shear layer. *J. Fluid Mech.* **281** 247–292.

21. Caulfield, C. P., Yoshida, S. & Peltier, W. R. (1996). Secondary instability and three dimensionalization in a laboratory accelerating shear layer with varying density differences. *Dyn. Atmos. Oceans* **23** 125–138.

22. Pierrehumbert, R. T. & Widnall, S. E. (1982). The two- and three-dimensional instabilities of a spatially periodic shear layer. *J. Fluid Mech.* **114** 59–82.

23. Pierrehumbert, R. T. (1986). Universal short-wave instability of two-dimensional eddies in an inviscid fluid. *Phys. Rev. Lett.* **57** 2157–2159.

24. Waleffe, F. (1990). On the three–dimensional instability of strained vortices. *Phys. Fluids* **A2** 76–80.

25. Nygaard, K. J. & Glezer, A. (1990). Core instability of the spanwise vortices in a plane mixing layer. *Phys. Fluids* **A2** 461–464.

26. Kelly, R. E. (1977). The onset and development of Rayleigh-Bénard convection in shear flows: a review. In *Physicochemical Hydrodynamics* Volume 1, (ed. V. G. Levich). Advance.

27. Klaassen, G. P. & Peltier, W. R. (1985). The evolution of turbulence in finite amplitude Kelvin-Helmholtz billows. *J. Fluid Mech.* **155** 1–35.

28. Klaassen, G. P. & Peltier, W. R. (1991). The influence of stratification on secondary instability in free shear layers. *J. Fluid Mech.* **227** 71–106.

29. Miles, J. W. (1961). On the stability of heterogeneous shear flows. *J. Fluid Mech.* **10** 496–508.

30. Howard, L. N. (1961). Note on a paper of John W. Miles. *J. Fluid Mech.* **10** 509–512.

31. Clark, T. L. (1977). A small-scale dynamic model using a terrain-following coordinate transformation. *J. Comp. Phys.* **24** 186–215.

32. De Silva, I. P. D., Fernando, H. J. S., Eaton, F & Hebert D. (1996). Evolution of Kelvin–Helmholtz billows in nature and the laboratory. *Earth and Planetary Science Letters* **143** 217–231.

Confined and Unbounded Mixing in Stratified Flows

J.M. Redondo, M.A. Sanchez and I.R. Cantalapiedra

Department Fisica Aplicada, Universitat Politecnica de Catalunya, Barcelona, Spain

Abstract

Several laboratory experiments have been used to study the vertical structure and the mixing efficiency across sharp and linear density interfaces. The two laboratory experiments described here exhibit no mean velocity shear; in the first one the turbulence is generated by a horizontally oscillating grid or rod system, in the other experiment a line of air bubbles injected in the base of a stratified tank produces localized mixing. We will consider here mixing in zero mean flows with a constant input of external kinetic energy, as well as the decay of the turbulence.

The relaminarization process is discussed showing the effect of molecular difusivity and stirring characteristics on the layering, horizontal flows are forced by intrusion type flows. Most of the experiments show that there is a maximum of mixing efficiency for confined mixing, when a large fraction of base area is stirred, but no maximum is apparent when the stirred area is only a small fraction of the tank.

1 Introduction

A large part of the mixing processes that take place in real geophysical situations have strong temporal and spatial variation of energy inputs. Typical examples are sea and mountain breezes, tidal stirring and coastal mixing by waves. This is true as well as many industrial applications such as reactor stirring where the paddles or stirrers are localized, or in thermal storage systems.

Phillips (1972) and Posmentier (1997) argued that a mechanism for producing sharp interfaces would take place if for high density gradients the vertical mass fluxes diminished, then large gradients would support smaller fluxes and the density interfaces would become sharper. The behaviour of the Flux Richardson number, R_f or mixing efficiency versus the gradient Richardson number, Ri was used by Linden (1979,1980) to describe a variety of experiments wich showed a maximum in $Rf(Ri)$.

Previous experiments of mixing of a two layer stably stratified fluid in a confined space were performed by Yague and Redondo (1990), and Redondo and Cantalapiedra (1993) with air bubble generated turbulence. Grid stirred mixing of stratified flows has been used in various configurations, see Fernando (1991) for a review. Here we discuss the effect of horizontal heterogeneity on mixing structure and efficiency in linear stratifications and sharp density interfaces, following the results of Browand and Hopfinger (1984) on lateral grid stirring. This work also addresses the problem of mixing across a density interface and the conflicting views on the shape of the mixing efficiency versus Richardson number curve; some authors, Linden (1979,1980), Redondo (1987), see Fernando (1991), have found a decrease in mixing efficiency at high Richardson numbers while others McEwan (1983), Britter (1984), found a monotonic relationship between mixing efficiency and Richardson number.

There is some difference in the amount of internal wave generation between the vertically and the horizontally oscillated grids; these also produce lateral intrusions as in Browand et al. (1987), and Redondo and Cantalapiedra (1993). Horizontally oscillating rods have been used in a similar fashion as in the experiments of Ruddick et al. (1989), but concentrating the stirring in a section of the tank.

The length of the tank has been varied in the experiments in order to investigate different behaviours of $\eta(Ri)$ as gravitational transport of mixed material starts to play an important role in the horizontal and vertical transports. The transport of mass in the region far away from the source of turbulent kinetic energy is mostly due to intrusion and gravity currents. These are controlled by the geometry of the reservoir and the secondary flows generated by the intrusion penetrating at the interface level. This geometrical constraint on the localized mixing is similar to that ocurring in a closed tunnel as described by Linden et. al. (1991). There, the hot gases produced by a fire inside a tunnel and the mixing produced by a Water Spray Barrier, were modelled in the laboratory adjusting the buoyancy fluxes of brine to the real situation.

The structure of the paper is as follows, next a description of the experiment and the characteristics of mixing generated by bubbles and vertical rods are presented, in Section 3 the definitions of Richardson number and mixing efficiency are discussed, in Section 4 the experimental results are presented and in Section 5 the results are discussed and some general conclusions presented.

2   Description of the experiments

A comparison between experiments where mixing is produced uniformly and those where mixing is localized and there are sideways intrusions is performed with two experimental apparatus. Visualization methods were used to investigate the flows and measure the small scale structure. Shadowgraphs were analysed in a similar way as in Pearson and Linden (1983). A microconductivity probe described in Redondo (1987) with a DANTEC micropositioner were used to measure the evolution of the density profiles, and from those the mixing efficiency.

2.1 Oscillating grid and rod experiments

A box made of 1 cm perspex plate, 0.255×0.255 m in base and 25 cm in height was used, after filling up to 18 cm of the box with either a stepwise sharp density interface or a linear stratification, with constant N. The experiments covered the range $N = 0.01 rads^{-1}$ to $N = 1 rads^{-1}$.

The turbulence in the midst of the tank was produced by oscillating laterally an array of vertical bars 0.003 m in diameter separated 0.03 m. The region stirred was varied as a fraction of the base of the tank.

The stratification, whether linear or produced by a sharp interface defines several overturning or buoyancy scales. When the integral turbulence scale ℓ and the buoyancy scale are of a similar size the stratification stops further vertical displacements of fluid and internal waves are usually generated. Ozmidov's lengthscale is $l_o = (\epsilon/N^3)^{1/2}$, where ϵ is the mean rate of viscous dissipation and N is the buoyancy frequency $N = (-(g/\rho)\partial\rho/\partial z)^{1/2}$. Any motion larger than this scale would, in principle, not be able to overturn and would be restricted to wave-like motions. See in Figure 1(a) a description of the experiment.

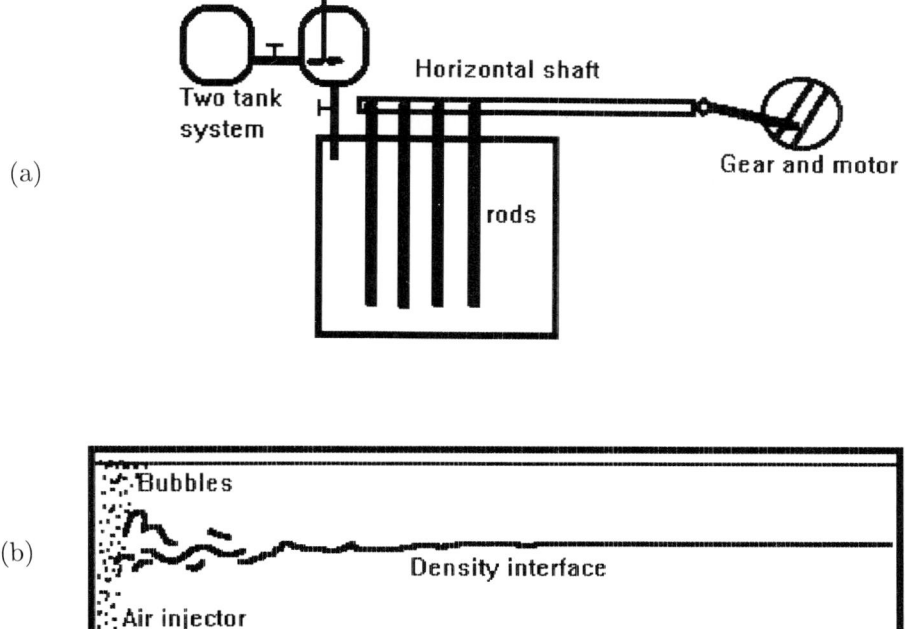

Figure 1. Description of the experimental apparatus, (a) Oscillating Rod system, and (b) Bubble induced mixing

2.2 Mixing by a line of air bubbles

A box made of 1 cm perspex plate, 120 × 15 cm in base and 35 cm in height was used. The fluid near the side of the tank was stirred by bubbling air from a feeder built from a circular perspex tube, 1 cm in diameter spanning the base of the tank (13 cm) with holes placed every cm, see Figure 1(b), Yague and Redondo (1990), Yague (1992), and Redondo and Cantalapiedra (1993). Provided the air flow rate was not too large, the bubbles rose almost vertically producing vigorous mixing near the corner of the tank. The experiments of Yague and Redondo (1991) were performed in a tank of dimensions 27.5 cm × 27.5 cm in base and 40 cm in height, and we will consider these experiments as taking place in a confined space when compared with the present experiments.

A reason for using bubbles in order to generate the turbulence, instead of an oscillating grid, as in the experiments of Browand et. al (1987) is to avoid the lateral dependence of the grid-stirred turbulence paramenters, $\ell = 0.1\,x$ and $u' = c\,\omega\,x^{-1}$, being ℓ and u' the integral lengthscale and the r.m.s turbulent velocity, ω the frecuency of oscillation of the grid and x the distance from the grid. Another reason is that bubbles are simple to produce and there are no temporal oscillations of the turbulence intensity associated with the grid oscillation. Considering the effective turbulent velocities associated to the bubble flow stirring as $u' = 0.1V$, by analogy with a plume entrainment model, Turner (1973), we define a local Richardson number associated with the density interface as described below, the integral lengthscale of the turbulence, ℓ, is modelled by the separation between bubble generating holes, which gives an approximate measure of the small eddies generated by the air bubble wakes.

3 Analysis of the experiments

Conductivity probe recordings and visual observations were used to measure the total mixing time, t_m, which is defined as the time taken to mix the whole of the fluid column throughout the length of the tank; this occurs when the density is equal in all points. A visual indication of the duration of the mixing could also be obtained by placing dye in one of the layers; another indication that total mixing had occurred was the absence of small scale structure in a shadowgraph of the tank.

Density profiles were taken at horizontal positions every 10 cm until the end of tank. The 11 traverses were completed in less than a minute, and this process was repeated every 5 minutes until the end of the experiment (i.e. when the fluid was homogeneous). Thus we can measure both the spatial and temporal evolution of density at different positions inside the box. The vertically traversing probes were used to obtain the evolution of density profiles during the mixing process.

From videos of a shadowgraph of the experiments the evolution of the vertical scales and some features of the structure of the stratified turbulence were found, see in Figure 2 an example of the vertical rod induced turbulence. In

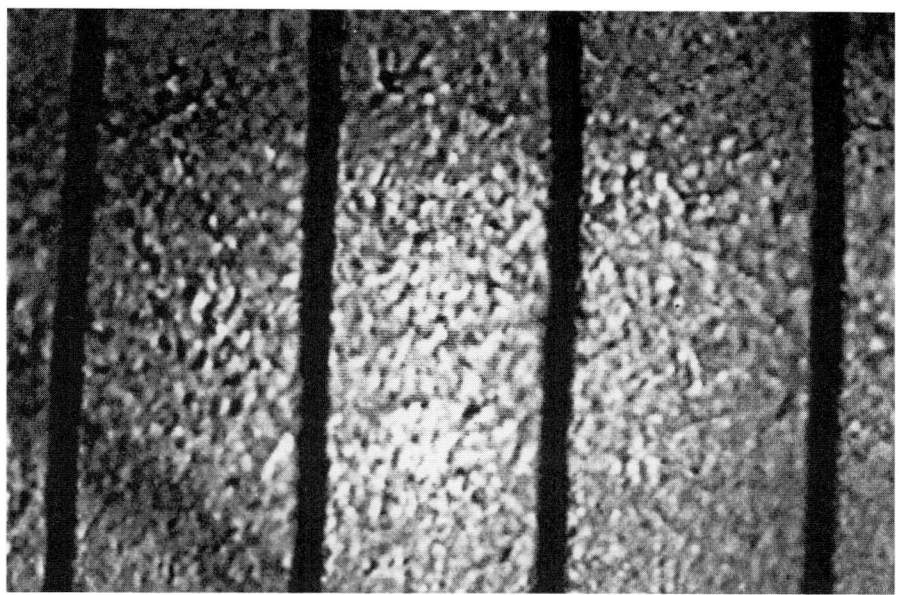

Figure 2. Shadowgraph of the mixing by oscillating rods, digitized image

Figures 3(a) and 3(b) digitized vertical sections in the spacing between the bars were fourier transformed (Figure 3(c)) in order to investigate the pattern selection of vertical microstructure.

3.1 Calculation of the mixing efficiency

The mixing efficiency is defined as the relation between the increase in potential energy of the fluid, ΔPE and the kinetic energy that produces the mixing, ΔKE.

$$\eta = \frac{\Delta PE}{\Delta KE}. \tag{3.1}$$

We compare the density profiles before and after mixing. Initially there is a sharp density interface before the start of the experiment and after a time t_m the whole column of fluid is well mixed; this is checked both through shadowgraph visualization and by means of conductivity probes, thus the final density profile will be constant with height and is given (using mass conservation) by $\Delta \rho_f = \Delta \rho (1 - h_2/L)$.

The increase in potential energy is then

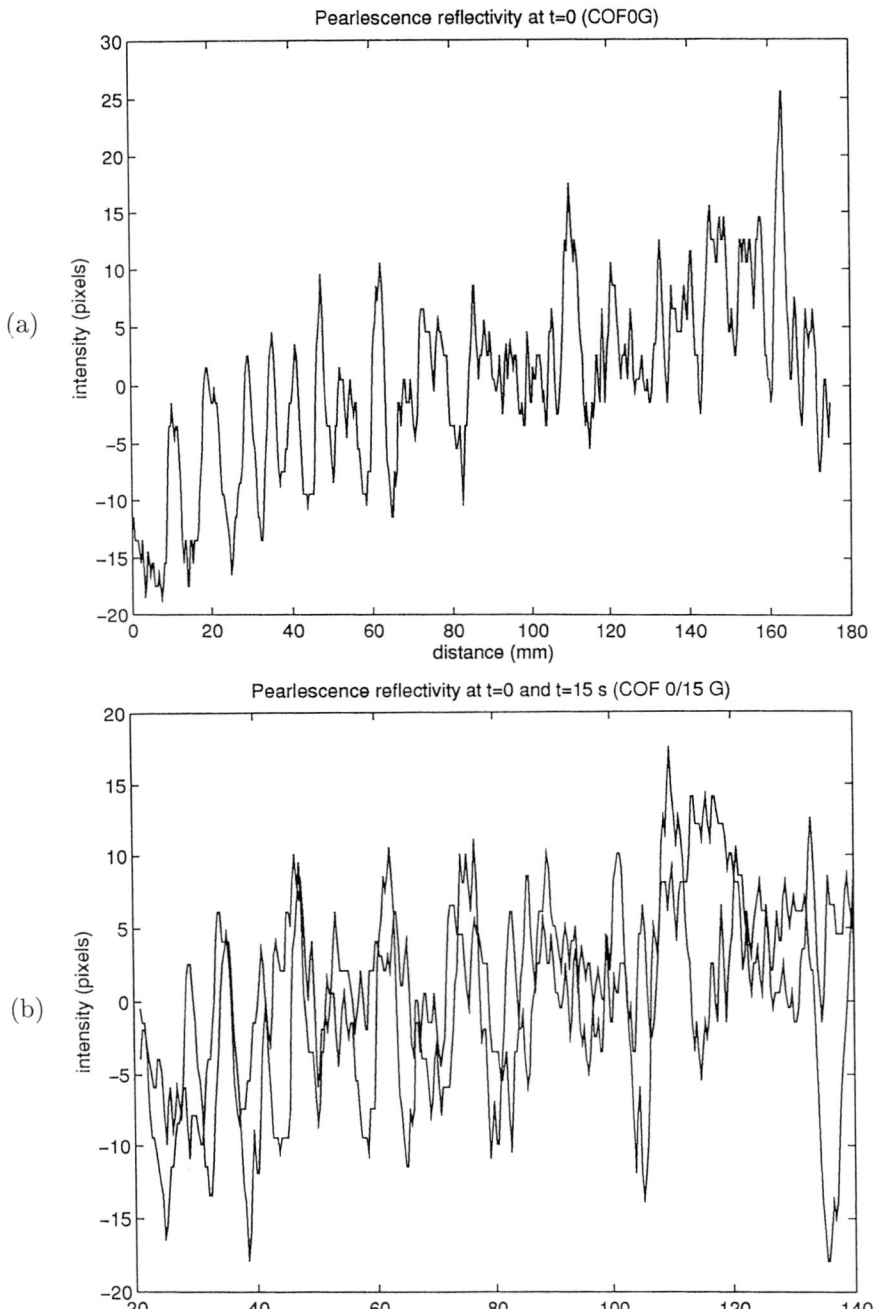

Figure 3. (a) Example of digitized vertical line from a shadowgraph, and (b) Intensity signals at two different times

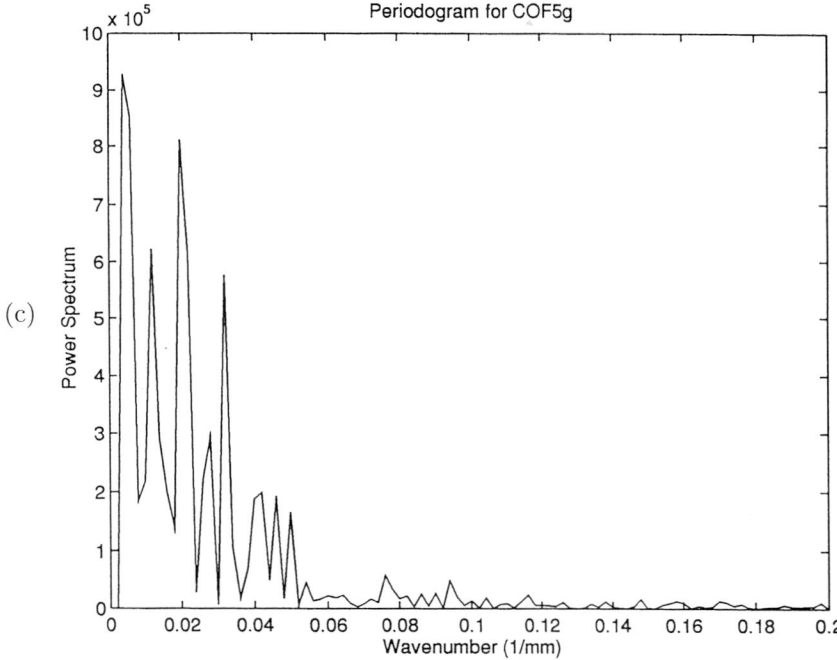

Figure 3. (c) Periodogram obtained from FFT

$$\Delta PE = gS \int_0^L [\rho_{final}(z) - \rho_{initial}(z)] \, z \, dz, \tag{3.2}$$

$$\Delta PE = \frac{1}{2} gS \Delta \rho h_1 h_2. \tag{3.3}$$

In the case of equals heights, as in our case, $h = h_1 = h_2$ where h is the heights of both the brine layer and initial fresh layer (15 cm). S is the surface of the tank's base and $L = 2h$ its total height.

The kinetic energy for the whole mixing process is estimated as the energy dissipated by the drag on an air bubble times the number of bubbles.

The drag of a single bubble is

$$F_d = \frac{1}{2} \rho C_d V^2 \pi \left(\frac{d}{2}\right)^2. \tag{3.4}$$

Thus the work done by the number of bubbles formed in time t, which is $n^o = V \, N \, t \, /2d$, can be given as

$$\Delta KE = \frac{1}{2} \rho C_d \pi \left(\frac{d}{2}\right)^2 V^2 \frac{V}{2d} N t_m L, \tag{3.5}$$

where C_d is the drag coefficient of a single bubble of diameter d rising at a terminal speed V, N is the number of holes producing bubbles and $\pi = 3.1416$.

If we now substitute the values of ΔPE and ΔKE in η, described above, for the total mixing process then we obtain an expression inversely proportional to the total mixing time

$$\eta = C \, Ri \, t_m^{-1}, \tag{3.6}$$

with C defined as: $C = 2LS/VdNC_d\pi100\ell$ where all the terms have been described above.

4 Experimental results

4.1 Mixing by horizontal oscillations of vertical rods

For oscillating grid experiments Turner (1968, 1973) proposed that the entrainment velocity V_e, defined as $V_e = dD/dt$, where D is the depth of the turbulent layer, is given by a simple law of the form $E \propto Ri^{-n}$ where E, the entrainment rate is defined as $E = V_e/V$, V being some global or local reference velocity. The relevant Richardson number in terms of local parameters is:

$$Ri = \frac{g \frac{\partial \rho}{\partial z} \ell^2}{\rho \, u'^2} = \frac{g \, \Delta \rho \, \ell}{\rho \, u'^2}, \tag{4.1}$$

for linearly or stepwise stratified flows, where $\Delta\rho$ is the buoyancy jump across a density interface, u' is the r.m.s turbulent velocity and ℓ is an integral length scale of the turbulence.

Turner (1968) found that the value of n was 3/2 when the stratification was due to salt. When the density-stratification resulted from a temperature gradient, the value of n was found to be close to 1. Using $R_f = ERi$ together with the experimental values of E gives $R_f \propto Ri^{1-n}$, and we can see that for salt experiments, there is a decrease in mixing efficiency with increasing Ri, while for heat, the mixing efficiency should be constant.

The thickness of the interfacial layer h both in heat and in salt-stratified experiments was determined by Crapper and Linden (1974), who used both a travelling conductivity probe and shadowgraph techniques. They found in their experimental range that for salt (high Pe) h/ℓ seemed to be independent of Ri and given by $h/\ell \approx 1.5$.

In experiments involving heat as a stratifying agent, Crapper and Linden (1974) found that molecular diffusion is important in the determination of the interfacial structure at $Pe \leq 200$. In this case they found that a diffusive core is formed at the centre of the interfacial layer, across which most transport occurs by molecular diffusion.

The final vertical structure after a period of decay of the internal waves was obtained for stratifications formed by different solutes: salt, sugar and temperature, Redondo (1996). As an example the evolution in time, during 100s of a horizontal line of the shadowgraph is shown in Figure 4(a) and the evolution of a vertical line in Figure 4(b). The internal waves are clearly seen.

The frequency of oscillations ω was kept greater than N in order to avoid the radiation of internal waves. The amplitude of the oscillation was 0.015 m, producing localized mixing as in the the experiments of Browand et al. (1987) with grid stirred turbulence causing the mixing.

For the mixing process we obtain an expression for the buoyancy flux as a function of the entrainment exponent, n, and the angular frequency of the oscillating rods, ω,

$$B = \frac{g}{\rho}\overline{w'\rho'} \propto \omega^{(1/2-2n)}, \qquad (4.2)$$

where n varies with the type of solute used, 1 for heat, 3/2 for salt and 5/3 for sugar, reflecting their respective diffusivities κ_ρ.

The evolution of the microlayer formation as the turbulent decays is different for the different solutes, showing vertical scales between the rod diameters and the Ozmidov scale, which are diffused on a timescale of $l_o^2\kappa_\rho^{-1}$.

The mean vertical spacing of the striations in the shadowgraph are presented in Figure 5 versus the non dimensional parameter $\nu\kappa_\rho/N^2D^4$, using as a relevant scale D, the rod diameter used for stirring, Pearson and Linden (1983) used a similar parameter with D the diagonal of the tank.

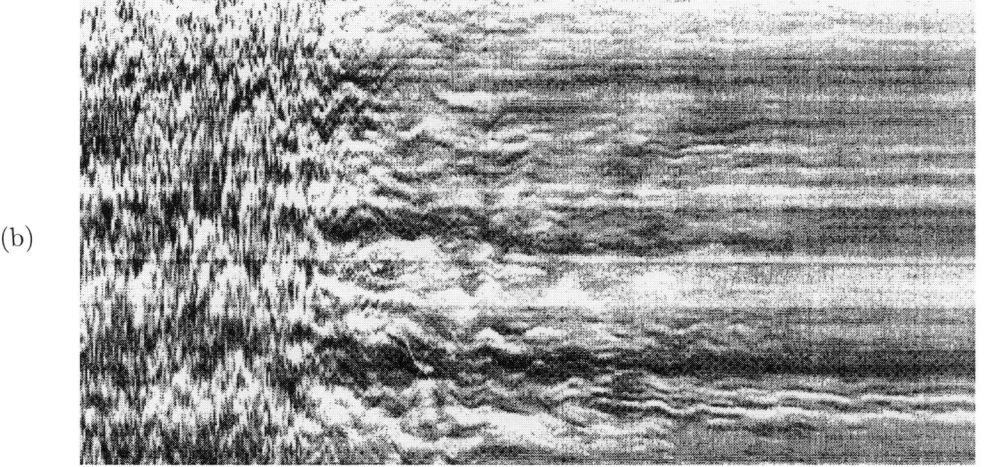

Figure 4. Decay of turbulence, (a) Time (down) evolution of a horizontal line, and (b) Time (right) evolution of a vertical line during 100s

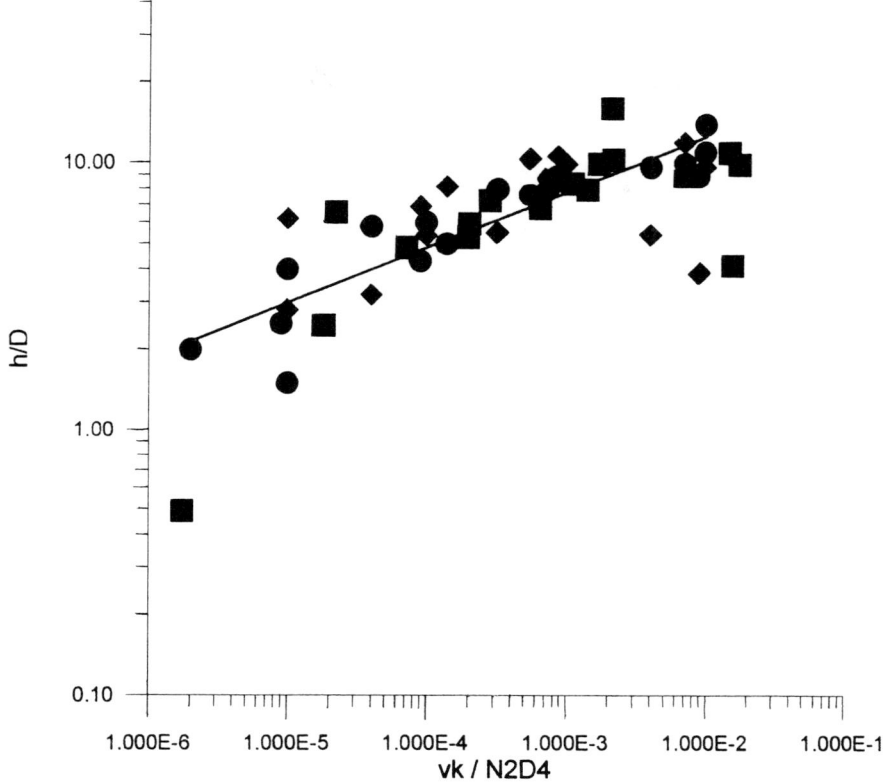

Figure 5. Plot of the nondimensional average striation thickness vs. the non dimensional parameter $\frac{\nu\kappa}{N^2 D^4}$. Different symbols indicate different frequencies ω

4.2 Mixing by a line of air bubbles

Mixing time increases as Ri increases, although when a critical value of the Richardson number is reached ($Ri \geq 2$) we can observe by means of shadowgraph the structure of the intrusions and the generation of internal waves showing some qualitative differences, (i.e. more waves, different patterns, fractal dimensions, etc...) which seem to reduce vertical mixing efficiency for large initial Richardson numbers and confined spaces.

Local Richardson numbers vary both spatially and temporaly throughout the mixing process because the Richardson number throughout the experiment decreases with time until Ri approaches zero, as the density interface is destroyed by mixing processes. Due to the laterally non-uniform turbulent r.m.s. velocities

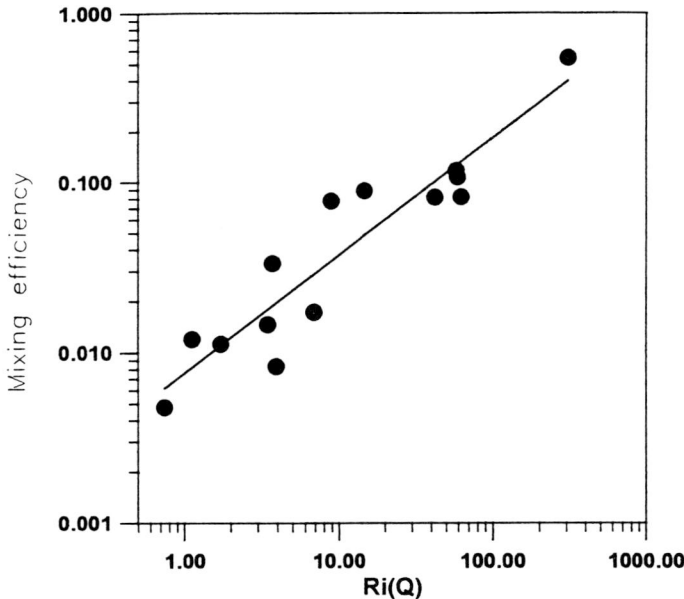

Figure 6. Mixing efficiency vs. Richardson number for unbounded mixing, where the length of the tank is more than 20 times the stirred area

produced by the air flow there is an increase in the local Richardson number as we move away from the source of turbulent kinetic energy.

Some differences in the behaviour of η versus Ri, and difficulties in collapsing the experimental results, have been observed when we have used different air flows. This led us to use a Richardson number depending on the air flow, in order to compare experiments with different air flow rates. The data can be expressed as $t\,V/L \propto Ri^n$, giving a mixing efficiency $\eta \propto Ri^{1-n}$. In Figure 6 the mixing efficiency versus the Richardson number is shown for the range of experiments performed. We should compare this results with the $\eta(Ri)$ curves found by Yague and Redondo(1991), Figure 7, which gave values of n between 0.7 - 1.2 with a strong dependence on Ri. In both cases a similar behaviour is observed, showing different local power law behaviour of the mixing efficiency with the Richardson number. Only in the cases where $n \geq 1$, which occur only for confined mixing there is an important change in behaviour of the system as indicated by Phillips (1972) and Posmentier (1977) noting then the differences between confined and non confined mixing processes.

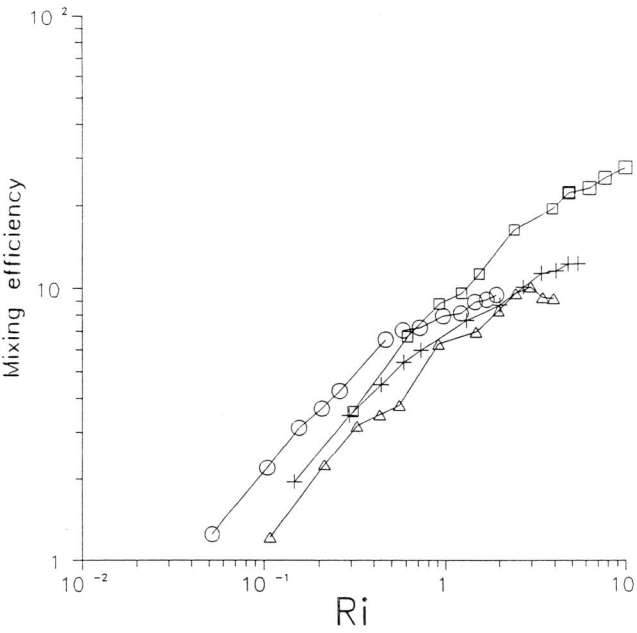

Figure 7. Mixing efficiency vs. Richardson number for confined mixing from Yague and Redondo (1990)

5 Discussion and concluding remarks

Different methods of stirring, such as a vertically oscillating grid (Turner 1968, 1973), bubble injection (Redondo and Cantalapiedra 1993) show different characteristics which need to be studied in detail and compared. The effect of the stratification on the zero-mean-turbulence is to restrict the typical vertical distance travelled by the fluid particles. There is no evidence of a critical local Richardson number, where mixing efficiency ceases to increases monotonically if the tank is large enough.

We find empirical relationships between mixing efficiency and Richardson number of the type $\eta = k \, Ri^{1-n}$, and $\eta = k'(Ri)e^{-k''x}$, where η is the mixing efficiency, n is an exponent which depends on the Richardson number Ri, x is the distance from the energy source and k, k', k'' are constants. The value of n is greater than 1 when mixing efficiency decreases with Ri, this situation is only observed in confined flows. The same observations have been done for mixing in the linearly stratified experiments, where the mixing was produced by horizontal oscillations of an array of vertical rods. Layering in weak turbulence forms if $B \geq 0.2u'/\ell$ with $B = 2\pi \, \omega^{1/2-2n}$.

In the experiments presented with air bubble mixing, efficiencies have been measured for a range of Ri and tank lengths. Changes in the air flux, Q, producing the bubbles alter the shape of the mixing efficiency versus Richardson number curves. The most important factor seems to be the ratio between the turbulent zone and the size of the tank, showing both consistency with Linden (1979, 1980) and McEwan (1983), Britter (1984). Only for very strong density interfaces and mixing taking place in a confined space a decrease in mixing efficiency with Ri can be appreciated. The Reynolds number based on a single bubble is the same, as the air flux changes, only reduce the rate of kinetic energy input at the interface, regardless where it is sharp or diffuse. The mixing efficiency for bubble generated turbulence can be expressed as

$$\eta \propto Ri^{1-n(Ri,Q)} \qquad (5.1)$$

with $n(Ri, Q)$ a variable exponent found from the time a two layer stratified system takes to mix. We find that $1/3 < n < 1/2$ in the free end experiments but $7/8 < n < 10/7$ for the confined experiments. A maximum in η may only be seen for the confined mixing process while the interface remains sharp enough to dissipate enough energy in internal waves, for small Ri the internal wave field is very weak and the results show a monotonic increase of mixing efficiency indicating that a saturation process has not taken place.

Acknowledgements

We would like to thank Dr. C. Yague from the National Institute of Meteorology in Madrid for helpful discussions and G. Van der Graaf for his technical help in the laboratory and in the preparation of the paper. This work has been supported by the European Union and DGES grants MAS3-CT96-0049, MAS-CT95-0016 and EU-950016.

Bibliography

1. Britter R.E. (1984) "Diffusion and decay in stably-stratified turbulent flows" in "Turbulence and Diffusion in Stable Environments" IMA conference proceedings, Edited by J.C.R. Hunt. p 15-27, Clarendon, Oxford, England.

2. Browand F.K. and Hopfinger, E. (1984) "The inhibition of vertical turbulence scale by stable stratification" in "Turbulence and Diffusion in Stable Environments" IMA conference proceedings, Edited by J.C.R. Hunt. p 15-27, Clarendon, Oxford, England.

3. Browand F.K., Guyomar D. and S.C. Yoon (1987) "The Behavior of a Turbulent Front in a Stratified Fluid: Experiments With an Oscillating Grid" *J. Geophysical Res.* **92**, 5329-5341.

4. Crapper R. and Linden P.F. (1974) The structure of density interfaces, *J. Fluid Mech.* **85**, 65-90.

5. Derbyshire, S. and Redondo J.M. (1990) "Fractals and waves, some geometrical approaches to stably stratified turbulence". *Anales de Fisica, Real socieda española de Fisica y Quimica.* 1990 Serie A. **86**, 67-76.

6. Fernando H.J.S. (1991) Turbulent mixing in stratified fluids, *Ann. Rev. Fluid Mech.* **23**, 455-493.

7. Hannoun I.A. and List J.E. (1988) Turbulent mixing at a shear free density interface, *J. Fluid Mech.* **189**, 209-227.

8. Koop,C.G. & Browand F.K. (1979) "Instability and turbulence in a stratified fluid with shear" *J.Fluid Mech.* **135**, 135-147.

9. Linden,P.F. (1979) Mixing in stratified fluids, *Geophysical and Astrophysical Fluid Dynamics* **13** 3-23.

10. Linden, P.F. (1980) Mixing across a density interface produced by grid turbulence, *J. Fluid Mech.* **100**, 691-703.

11. Linden, P.F.,Jagger S.F., Redondo J.M., Britter R.E. and Moodie K (1991) The effect of a water spray barrier on a tunnel fire, *ASME.* , 1-6.

12. McEwan A.D.(1983) "Internal mixing in stratified fluids" *J. Fluid Mech.* **128**, 59-80.

13. Pearson H.J. and Linden P.F. (1983) The final stage of decay of turbulence in stably stratified fluid , *J. Fluid Mech.* **134**, 195-203.

14. Phillips O.M. (1972) Turbulence in a strongly stratified fluid - is it unstable?, *Deep Sea Research* **19**, 79-81.

15. Posmentier E.S. (1977) The generation of salinity finestructure by vertical diffusion *Journal of Physical Oceanography* **7**, 292-300.

16. Redondo J.M. (1987) Difusion turbulenta en fluidos estratificados, *PhD. Thesis* , Univ. Barcelona.

17. Redondo J.M. (1996) Vertical microstructure and mixing in stratified flows. S. Gavrilakis et al. (eds.), *Advances in Turbulence VI* , 605-608.

18. Redondo J.M. and Cantalapiedra I.R. (1993) Mixing in horizontally heterogeneous flows, *Applied Scientific Research* **51**, 217-222.

19. Ruddick B.R., McDougall T.J. and Turner J.S. (1989) The formation of layers in a uniformly stirred density gradient, *Deep-Sea Research* **36**, 597-609.

20. Turner J.S. (1968) The influence of molecular diffusivity on turbulent entrainment across a density interface, *J. Fluid Mech.* **33**, 639-656.

21. Turner S.T. (1973), "Buoyancy effects in fluids" Cambridge University Press.

22. Yague C. (1992) "Estudio de la mezcla turbulenta a traves de experimentos de laboratorio y datos micrometeorologicos" Ph.D. Thesis. Univ. Complutense Madrid.

23. Yague C. & Redondo J.M. (1990) "Mezcla convectiva a traves de una interfase de densidad" Revista de Geofísica. **46**, 147-160.

Breaking Progressive Internal Gravity Waves: Two-Dimensional and Preliminary Three-Dimensional Numerical Experiments

C. Koudella and C. Staquet[1]

Laboratoire de Physique, Ecole Normale Supérieure de Lyon, France

Abstract

The dynamics of progressive internal gravity waves of small amplitude are studied by means of two-dimensional (512^2) and preliminary three-dimensional (64^3) direct numerical simulations of the Boussinesq equations. Our results show that an internal wave always breaks (for a viscosity low enough), this event being initiated by resonant interactions. A turbulent regime sets in after breaking. This regime is characterized by kinetic and potential energy spectra that behave as $\alpha N^2 k^{-3}$, the level α for potential energy being identical as that found in the oceanic thermocline (for example Gregg, 1987). The overall dynamics are analogous to that found for a two-dimensional standing internal gravity wave by Bouruet-Aubertot *et al.* (1995, 1996). This suggests that small scale geophysical flows, which mainly involve progressive internal waves, may be studied by laboratory experiments, where standing waves are mainly encountered.

In three dimensions, breaking of the progressive wave is initiated by two- and three-dimensional interactions among gravity waves, whereof resonant nature needs to be investigated. Vertical vorticity is produced within this weakly nonlinear regime of the flow, due to three-dimensional interactions among gravity waves, but no potential vorticity is produced. By contrast, potential vorticity is generated by the breaking event. In the present case where the resolution is low, the motion associated with potential vorticity is a horizontal mean shear with large scale vertical variability. It is noteworthy that, from an energetic point of view, this motion dominates the flow after breaking.

1   Introduction

In the oceanic thermocline and in the lower stratosphere, *in situ* measurements of vertical gradient of temperature and horizontal velocity yield spectra with a universal dependence in the vertical wavenumber, of the form k_z^{-3} (for example Gregg 1987, Sidi & Dalaudier 1989). The vertical scales for which this dependency holds range between a few meters to a few tens of meters in the ocean,

---

[1]Present address: LEGI, BP 53, 38041 Grenoble, Cedex 9, France.

and between a few meters to a few hundred meters in the stratosphere. Several theoretical models have been proposed to account for this behavior. The main idea (actually subjected to recent controversy, see for example Sidi 1995, Gregg et al. 1996) is that these spectra would result from *breaking* internal gravity waves. Remarkably, these velocity and temperature spectra have been reproduced by two-dimensional numerical experiments in a vertical plane of a large scale standing internal gravity wave (Bouruet-Aubertot et al. 1996): whatever its initial amplitude, the wave always breaks (for a low enough viscosity), yielding a "wave turbulence" regime characterized by these spectra. One purpose of the present paper is to examine the validity of these conclusions for a large scale progressive wave, which is most often encountered in geophysical media.

Wave breaking is also known to play an essential role in atmospheric motions, in its producing a mean flow. The latter motion is associated with potential vorticity. Generally speaking, dissipative processes (such as breaking) can change irreversibly the potential vorticity distribution of the flow (for example McIntyre & Norton 1990). In the ocean, it is a challenge for observationalists to distinguish motions associated with potential vorticity from internal waves (Müller et al. 1986, Kunze & Sanford 1993). Both motions have a very different dynamics and thus contribute very differently to dissipative events of the flow. Potential vorticity is zero in a vertical plane and a second part of this paper is to present preliminary numerical simulations of a breaking progressive internal gravity wave in three dimensions.

Numerous two-dimensional numerical studies of progressive and standing internal gravity waves have been performed since almost thirty years. A crucial aspect of breaking studies lies in the resolution of the numerical calculations, because small dissipative vertical scales are produced during breaking. As shown by Bouruet-Aubertot et al. (1996), a high resolution together with a numerical method of high precision is a prerequisite to generate a turbulent regime with clear statistics. Only a few studies have been performed in three dimensions. In Ramsden & Holloway (1992), low resolution (32^3) simulations of forced progressive random waves are presented and the properties of the statistical regime are investigated. A qualitative study of the dynamics of a single progressive wave is presented in Lombard & Riley (1996), for different amplitudes of the wave. Only in Riley *et al.* (1991) is the generation of potential vorticity addressed. The purpose of the latter paper is to examine the efficiency of resonant interactions in energy transfers, as predicted by Lelong & Riley (1991), between two waves and one vortex mode (i.e. a motion with potential vorticity) or among vortex modes only.

The equations of motion and the numerical model are presented in the next section. Two-dimensional and three-dimensional direct numerical simulations of a progressive wave are described in Sections 3 and 4 respectively. Conclusions are drawn in the final section.

2 Equations of motion and numerical model

2.1 Equations of motion

We consider a stably stratified Boussinesq fluid with constant density gradient. Making a decomposition of the density ρ and pressure p into a hydrostatic time independent and a fluctuating part:

$$\rho(x, y, z, t) = \rho_0 + \bar{\rho}(z) + \tilde{\rho}(x, y, z, t) \tag{2.1}$$
$$p(x, y, z, t) = p_0(z) + \tilde{p}(x, y, z, t), \tag{2.2}$$

and invoking the hydrostatic approximation, the Navier-Stokes equations in the Boussinesq approximation read:

$$\frac{\partial \vec{u}}{\partial t} + \vec{u} \cdot \nabla \vec{u} = -\nabla p' - \rho' \vec{i}_z + \nu \nabla^2 \vec{u} \tag{2.3}$$

$$\nabla \cdot \vec{u} = 0 \tag{2.4}$$

$$\frac{\partial \rho'}{\partial t} + \vec{u} \cdot \nabla \rho' = N^2 w + \frac{\nu}{Pr} \nabla^2 \rho'. \tag{2.5}$$

$\vec{u} = (u, v, w)$ is the velocity field, $p' = \tilde{p}/\rho_0$ are normalized pressure fluctuations and $\rho' = \tilde{\rho} g/\rho_0$ are reduced density fluctuations. Further $\vec{i}_z$ is a vertical unit vector pointing upwards; ν is the kinematic viscosity, $Pr = \nu/\kappa$ is the Prandtl number, where κ is the thermal diffusivity and

$$N^2 = -(g/\rho_0) \, d\bar{\rho}/dz, \tag{2.6}$$

is the square of the Brunt-Väisälä frequency.

2.2 Numerical method

The equations of motion (2.3) to (2.5) have been integrated directly in space and time using a pseudo-spectral method (Orszag 1971, Canuto *et al.* 1988). The method approximates the Eulerian fields to be solved as truncated Fourier series. This representation of the fields transforms the equations of motion into a set of coupled first order in time integro-differential equations which are advanced in time by a third order Adams-Bashforth scheme. The equations are solved in a square domain (in two dimensions) or in a cube (in three dimensions) of side length 2π. The fields all verify 2π-periodic boundary conditions in space.

2.3 Initial condition and physical parameters

The initial condition is a single internal wave, propagating in the vertical (x, z) plane:

$$\begin{aligned}
u(x, z, t = 0) &= -a \, cos(x + z) \\
v(x, z, t = 0) &= 0 \\
w(x, z, t = 0) &= a \, cos(x + z) \\
\rho'(x, z, t = 0) &= \sqrt{2} \, a \, N^2 \, sin(x + z).
\end{aligned} \tag{2.7}$$

This initial condition is an unstable solution of the Boussinesq equations and a (either two- or three-dimensional) white noise with uniform distribution is used to destabilize it, of amplitude equal to 10^{-5} a. Parameter a varies from 0.256 to 0.75. For convenience the Brunt-Väisälä frequency N is set equal to one and time is scaled by the Brunt-Väisälä period $T_{BV} = 2\pi/N$. The Prandtl number has the constant value of 0.72 in two dimensions and of 1 in three dimensions.

The resolution of the two-dimensional simulations is fairly high, equal to 512^2. Thus energy dissipation can be modelled by a simple Laplacian operator and the flow dynamics still be driven by nonlinear effects. For any particular physical situation the viscosity is adjusted to its minimum by checking that the energy and enstrophy balances are verified within a precision of 1%. The value of the viscosity varies from $1.5\ 10^{-4}$ (for $a = 0.256$) to $3\ 10^{-4}$ (for $a = 0.75$).

By contrast, the resolution of the three-dimensional calculations is low, equal to 64^3 only. As we shall show, each simulation requires to integrate the equations of motion over a long time, of more than $100\ T_{BV}$, to cover all stages of the wave dynamics. In order not have the flow dynamics soon dominated by viscous effects, we use a time varying (but uniform) viscosity. At each time step, this viscosity is determined by the Kolmogoroff scale η, at which dissipative effects arise, to be a constant fraction of the grid size ds. This yields:

$$\nu(t) = (\alpha\ ds)^2\ [2\ Z(t)]^{1/2} \quad \text{with} \quad \alpha < 1. \tag{2.8}$$

Z is the volume averaged enstrophy, equal to $\langle \Omega^2 \rangle/2$ where $\Omega = \nabla \wedge \vec{u}$ is the vorticity. In the present study, $\alpha = 1/2$.

3 Breaking of a two-dimensional progressive internal gravity wave

3.1 Weakly nonlinear dynamics

We have investigated the transient dynamics of a two-dimensional (in a vertical plane) progressive internal gravity wave. This wave is referred to as the primary wave hereafter. Whatever its initial amplitude, an initial stage is observed, during which the dynamics of the internal wave field are controlled by the growth of secondary waves feeding on the energy of the primary wave. In the case of low or intermediate initial amplitudes, this stage may be appropriately termed weakly nonlinear, since then, as we have checked, the growth of these secondary waves is predicted by resonant interaction theory (Phillips 1966): two secondary waves interact with the primary wave, forming an interacting triad, if the wavevectors $\vec{k}_1, \vec{k}_2$ and $\vec{k}_3$ and associated angular frequencies ω_1, ω_2 and ω_3, the index 3 identifying the primary wave, satisfy the following resonance conditions: $\vec{k}_1 + \vec{k}_2 = \vec{k}_3$ and $\omega_1 + \omega_2 = \omega_3$. The theory predicts that secondary waves grow exponentially in time, where the growth rate is a function of the triad configuration and of the initial amplitude of the primary wave.

Due to the finite size of the computational domain and of the grid mesh, only a finite number of discrete wavevectors are present within the flow fields.

Nevertheless, we have observed that the secondary waves that are selected for growth, verify the resonance conditions within a few percent. Figure 1 depicts the energy history of the two secondary waves belonging to the resonant triad $\vec{k}_1 = (3,7)$, $\vec{k}_2 = (-2,-6)$ and $\vec{k}_3 = (1,1)$ as a function of time. This triad meets the resonance conditions within a precision of 0.4%. The numerical results for the exponential growth rate of the secondary waves underestimate the theoretical predictions by 10 to 15%. This effect may be due to the fact that the numerical simulation uses a viscous fluid. The quantitative properties of this are under study, but since the excited secondary waves are not of very small scale, dissipation is probably weak. A second argument is that the theoretical calculation of the growth rate proceeds from an isolated triad of internal waves. This is not so in a numerical simulation, where an internal wave simultaneously interacts with other resonant (and possibly non-resonant) internal waves.

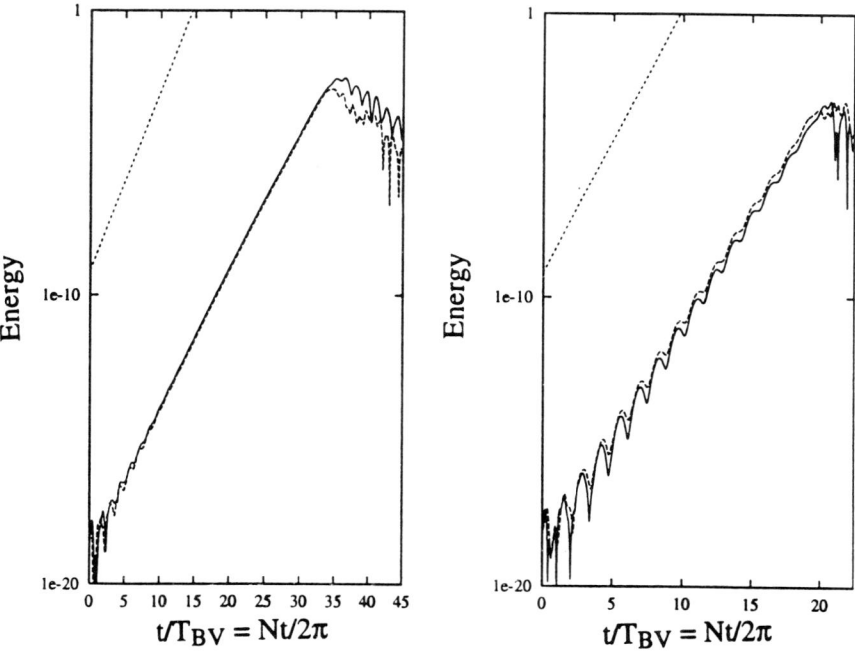

Figure 1. Energy versus time of triadic secondary modes $\vec{k}_1 = (3,7)$ and $\vec{k}_2 = (-2,-6)$ for initial amplitude wave amplitudes 0.256 (left) and 0.382 (right). The numerically computed growth rate for this triad are a) 0.074 and b) 0.137, while the theoretical predictions (upper left corner) are a) 0.01 and b) 0.149

The resonant triads we observe in the numerical simulations are close to a parametric instability. Indeed, the frequencies of the secondary waves belonging to the resonant triad mentioned above are $\omega_1 = 0.39\,N$ and $\omega_2 = 0.31\,N$, which should be compared to $\omega_3/2 = 0.35\,N$. Also, the excited secondary waves are of smaller scales than the primary wave. This resonant behavior is consistent with the findings of Drazin (1977) and Mied (1976), who performed a linear stability analysis for progressive internal waves and showed that these waves are parametrically unstable.

As the initial amplitude of the primary wave is increased, the growth of secondary modes is more disorganized. Beside the existence of resonant triads, other non-resonant modes also grow substantially.

3.2   Evolution in physical space

We present the evolution of the total density and vorticity fields in the course of time for the internal wave defined by (2.7) at $t = 0$. Initial configurations are represented in Figures 2(a) and 2(b), respectively. Vorticity isolines, which are lines of constant phase, make an angle $\pi/4$ to the vertical. After $17\,T_{BV}$, a significant fraction of the energy remaining in the primary mode has been redistributed among secondary modes and the effect of the instability becomes visible on the total fields. At that stage, the deformation of the initially smooth isopycnals can clearly be seen (Figure 2(c)). The vorticity field is characterised by vorticity bands superimposed on the primary wave vorticity (Figure 2(d)). These bands are the manifestation of growing secondary wave packets. Further, the vorticity bands make an approximate angle of $21°$ to the horizontal, which is consistent with these wave packets to result from a parametric instability. At that stage however, interactions between the waves have become strong. At about $21\,T_{BV}$ (Figure 2(e)), the deformation of the isopycnals produces locally unstable density gradients, leading to overturning. The vorticity bands are stretched and rotated, under the action of the shear associated with the background primary wave. As can be seen on Figure 2(f), they form a parallel set of shear layers that become unstable through Kelvin-Helmholtz instability. The primary wave thus breaks. Note that the overturning of isopycnals and the Kelvin-Helmholtz instability of the vorticity bands coincide exactly. This indicates that the internal wave breaks by a shear instability rather than by a buoyancy induced instability. The breaking of the wave field generates a freely decaying turbulent flow whose spectral properties will be presented in the next section.

A plot of the total energy against time is displayed in Figure 3, for different amplitudes of the primary wave. The breaking event corresponds to a sudden decrease of the energy in all cases. The breaking process rapidly transfers energy to small vertical scales, where it is dissipated. The breaking time, and thus the life time of the wave, depends on the amplitude of the primary wave: a large amplitude yields stronger interactions with secondary waves, which promotes breaking. The breaking time also depends upon the amplitude of the white noise superposed at initial time upon the primary wave. Indeed, the wave is

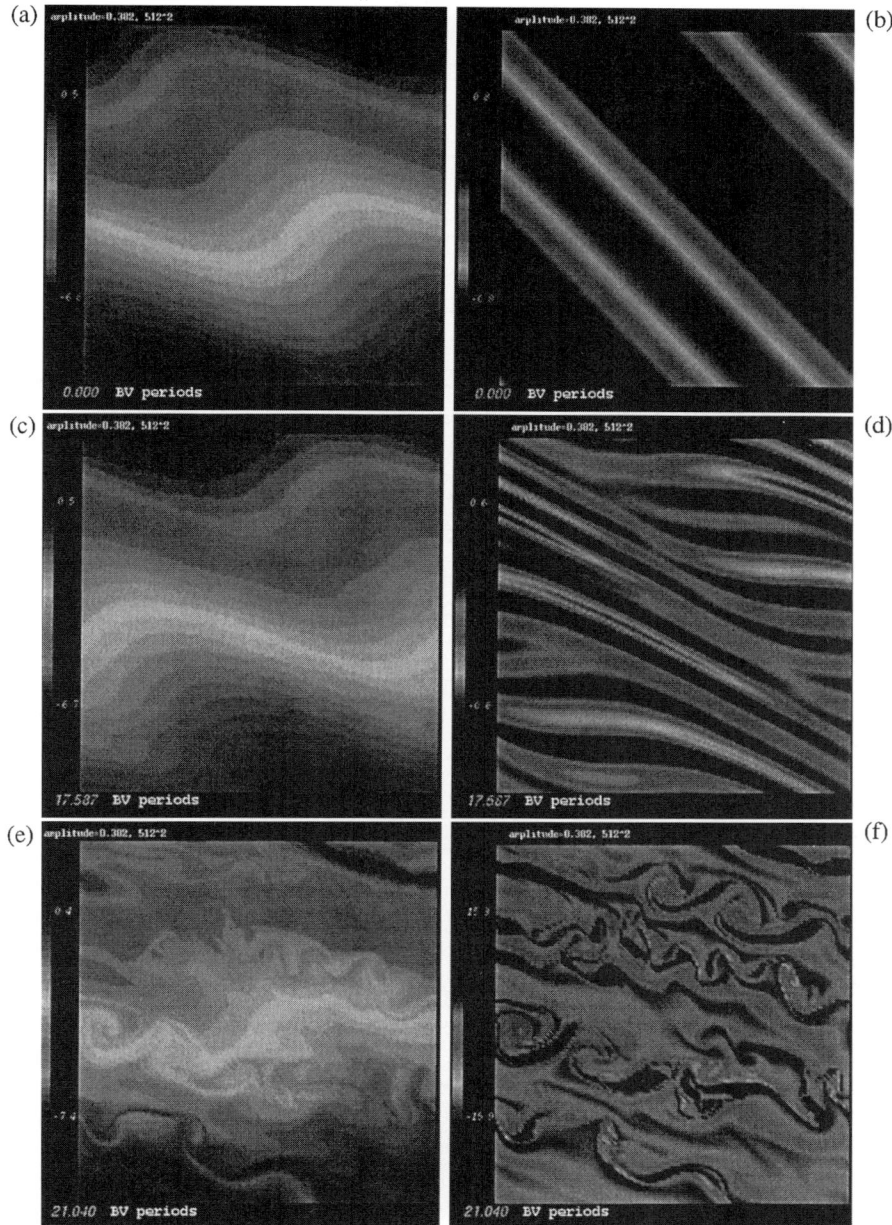

Figure 2. The total density field (left column) and vorticity field (right column) at various times for the primary wave $\vec{k}_3 = (1,1)$ with an initial amplitude 0.382. (a) and (b) $t = 0$; (c) and (d) $t = 17$; (e) and (f) $t = 21$, in units of Brunt-Vaisala periods

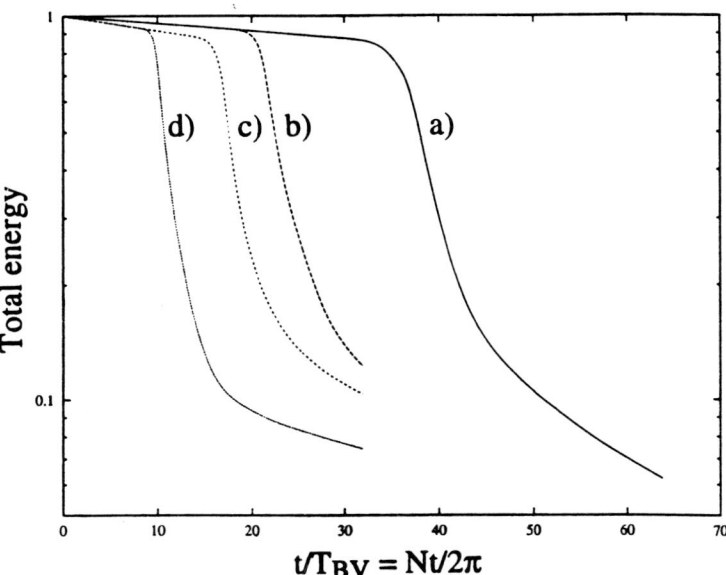

Figure 3. Total normalised energy as a function of time (scaled in Brunt-Väisälä periods) for the primary wave $\vec{k}_3 = (1,1)$ for the initial amplitudes: a) 0.256, b) 0.382, c) 0.5, and d) 0.75

stable (i.e. the breaking time is infinite) if the amplitude of the white noise is zero.

3.3   Energy spectra

The deterministic approach we have followed to investigate the mechanisms leading to breaking cannot be pursued anylonger when the wave has broken. A statistical approach has to be used. Thus, instantaneous kinetic and potential energy spectra as a function of the vertical wavenumber k_z are plotted in Figure 4, every 5 T_{BV}, for one amplitude of the primary wave. Both spectra exhibit a k_z^{-3} law, as in geophysical flows at small scales. Moreover, the potential energy spectra display a level close to the value of 0.2, as in the oceanic thermocline. The remarkable point is that these results are obtained whatever the amplitude of the primary wave, from $a = 0.256$ up to $a = 0.75$.

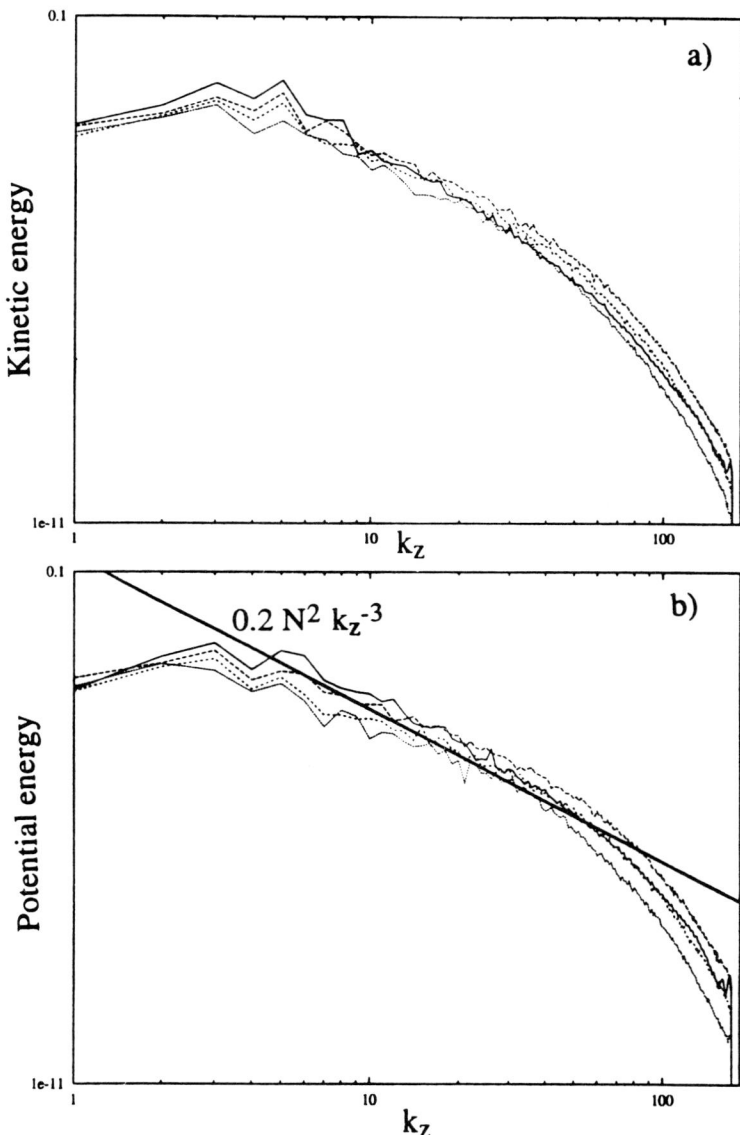

Figure 4. (a) Kinetic; and (b) potential, energy density spectra as a function of vertical wave number for an initial amplitude of 0.256 shortly after breaking time at $t = 38$ Brunt-Väisälä periods

4   Breaking of a three-dimensional progressive internal gravity wave: preliminary results

4.1   Decay of the total energy

To the previous two-dimensional internal wave has now been added a three-dimensional white noise. We recall that a 64^3 resolution is employed. Three calculations will be presented in this section, which solely differ by the viscosity: ν is either constant, equal to 0 or $5\ 10^{-4}$, or varies in time according to definition (2.8). In the three runs, the amplitude of the primary wave is equal to 0.256 and that of the white noise to $0.256\ 10^{-5}$.

The time varying viscosity is plotted in Figure 5. It decreases monotonically by more than a factor 4, except around $t \simeq 55$ where it strongly increases. As shown by Figure 6 where the total energy is plotted, breaking occurs at this time:

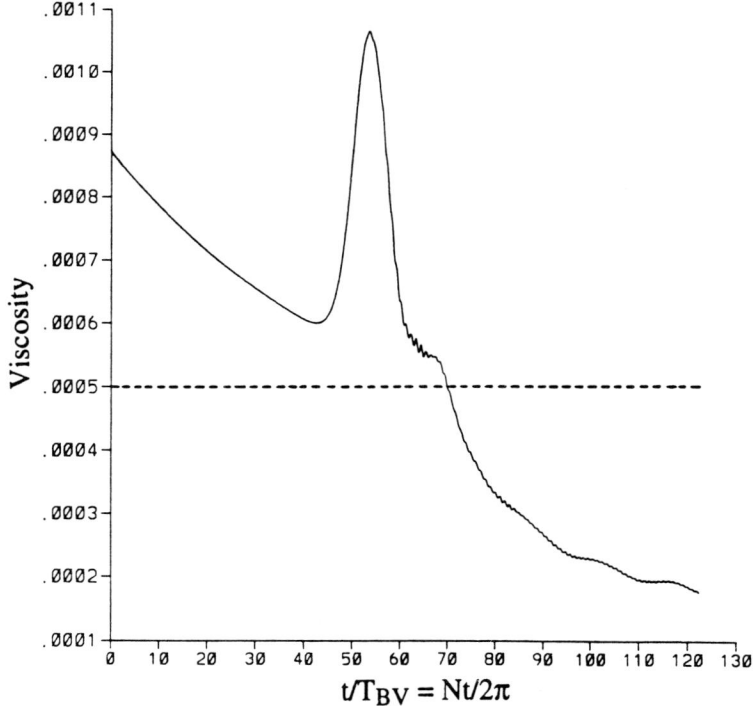

Figure 5.   Three-dimensional numerical simulations (primary wave amplitude $a = 0.256$). *Full line*: Time varying viscosity defined by (2.8). *Dashed line*: Constant viscosity, equal to $5\ 10^{-4}$

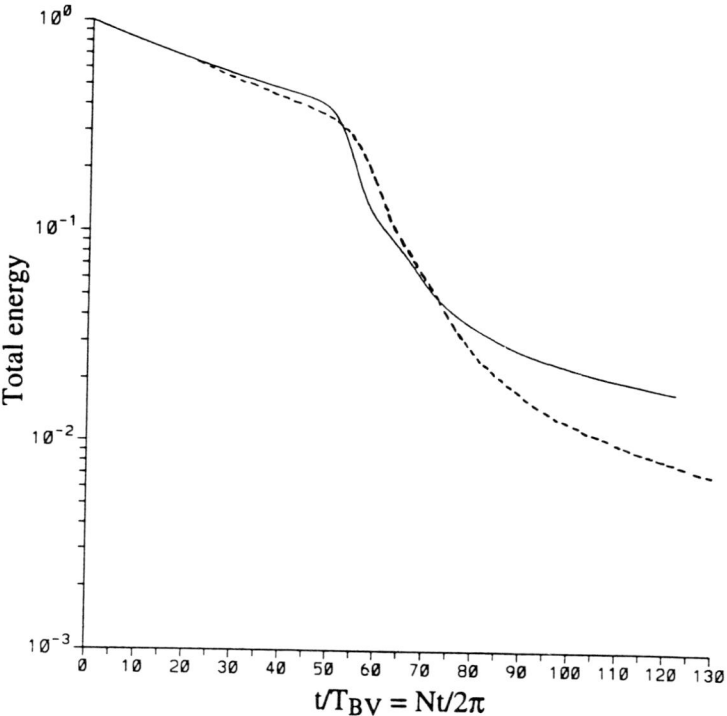

Figure 6. Three-dimensional numerical simulations ($a = 0.256$). *Full line*: Total energy of the flow with a time varying viscosity. *Dashed line*: Total energy of the flow with constant viscosity $5 \; 10^{-4}$

the total energy displays a sudden decay. Hence, the same behavior as in the two-dimensional wave is observed. However, breaking occurs later in time than in two dimensions, possibly because of the larger value of the viscosity in three dimensions than in two dimensions (the resolution per direction being lower in the former case). The total energy for the simulation with constant non zero viscosity is also plotted in Figure 6: only during the final regime, when the primary wave has lost 95% of its energy, do the two curves definitely separate. The key event of the wave evolution, namely breaking, is only weakly sensitive to the time varying viscosity in this figure. The interest in this viscosity is thus to prevent the later stage of the flow from being dominated by viscous effects.

4.1.1 Background

We recalled in Introduction that, in three dimensions, dissipative internal waves may generate another type of motion entirely, which is non propagating and characterized by its containing the potential vorticity of the flow. In the Boussinesq approximation, the potential vorticity Q is defined by:

$$Q = \frac{\vec{\Omega} \cdot \nabla \rho}{\rho_0} \tag{4.1}$$

(ρ being the total density). When molecular effects are neglected, each fluid particle conserves its potential vorticity (Ertel 1942). The potential vorticity is therefore not exchanged between fluid particles in an inviscid fluid, as noted by Lelong & Riley (1991). Thus, the motion characterized by potential vorticity does not propagate. This motion is forbidden in two dimensions because the potential vorticity is zero by definition in a vertical plane.

One way to extract the non propagating motion from the velocity field is to use the decomposition proposed by Staquet & Riley (1989): this motion is supposed to be incompressible, to lie on the isopycnals and to contain all of the potential vorticity of the flow. We shall refer to it as the potential-vortex part of the flow. As a first step however and for simplicity, we rather use the so-called Craya-Herring decomposition to which the former decomposition reduces when isopycnals are horizontal (Craya 1958, Herring 1974). In the latter decomposition, the non propagating part of the flow is horizontal and contains all of the vertical vorticity of the flow. We shall refer to it as the horizontal-vortex part of the flow.

The Craya-Herring decomposition is best described in Fourier space. The idea is that the incompressible velocity field has two components only in a plane perpendicular to $\vec{k}$; one of this component is taken in the vertical plane while the other one lies along a horizontal direction. This decomposition is a mathematical one but it takes a physical meaning for a stably stratified flow, in the linear limit (Riley *et al.* 1981): the former component then coincides with the velocity field of fluid particles in a gravity wave field. By contrast, the latter component is associated with a zero frequency and thus, with a non propagating motion. This decomposition reads, in Fourier space:

$$\hat{\vec{u}}(\vec{k}, t) = \phi_1(\vec{k}, t) \; \vec{e}^1(\vec{k}) + \phi_2(\vec{k}, t) \; \vec{e}^2(\vec{k}), \tag{4.2}$$

where $\hat{}$ refers to the Fourier space. $\vec{e}_1$ and $\vec{e}_2$ are two unit vectors in the plane perpendicular to $\vec{k}$, which are respectively horizontal and lying in a vertical plane:

$$\begin{aligned}
\vec{e}^1(\vec{k}) &= (\vec{k} \times \vec{i}_z)/|\vec{k} \times \vec{i}_z| \\
\vec{e}^2(\vec{k}) &= (\vec{k} \times \vec{e}^1)/|\vec{k} \times \vec{e}^1|.
\end{aligned} \tag{4.3}$$

An important point is that decomposition (4.2) is undefined when $\vec{k}$ is vertical. Motions associated with a vertical $\vec{k}$ are of zero frequency; also, because a vertical $\vec{k}$ writes $(0, 0, k_z)$, these motions are mean motions along horizontal planes, with a vertical variability. Such motions, of horizontal-vortex type, should be mostly contributed by the mean flow generated by dissipative waves that we mentioned in Introduction.

4.1.2 Energy in the wave and horizontal-vortex components of the velocity field

The kinetic energy has been decomposed into the energy in the horizontal-vortex part for non vertical wavevectors, denoted $\Phi_1(t)$, into the internal gravity wave kinetic energy, denoted $\Phi_2(t)$ and into the energy of the motions associated with a vertical wavevector, denoted $\Phi_0(t)$. Note that, because a white noise is added to the primary wave at $t = 0$, the horizontal-vortex component (as well as the potential-vortex component) are non zero initially. These three kinetic energy components are plotted in Figure 7, together with the potential energy E_p.

Figure 7 shows that, from $5\,T_{BV}$, $\Phi_0(t)$ exhibits a nearly exponential growth, and saturates shortly after breaking. An instability mechanism very likely underlies this behavior. The energy in the horizontal-vortex component $\Phi_1(t)$ also

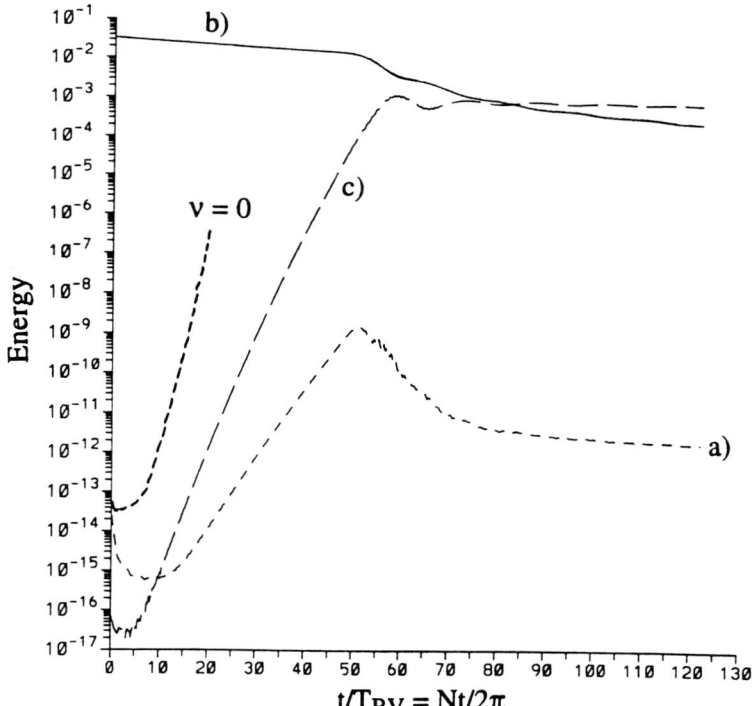

Figure 7. Three-dimensional numerical simulations ($a = 0.256$). (a) Kinetic energy of the horizontal-vortex component of the flow for non vertical wavevectors, $\Phi_1(t)$; $\Phi_1(t)$ is also plotted for a simulation with no viscosity ($\nu = 0$). (b) Kinetic and potential energy of the internal wave field, $\Phi_2(t)$ and $E_p(t)$ (both curves superpose). (c) Kinetic energy of the horizontal-vortex component of the flow for vertical wavevectors, $\Phi_0(t)$

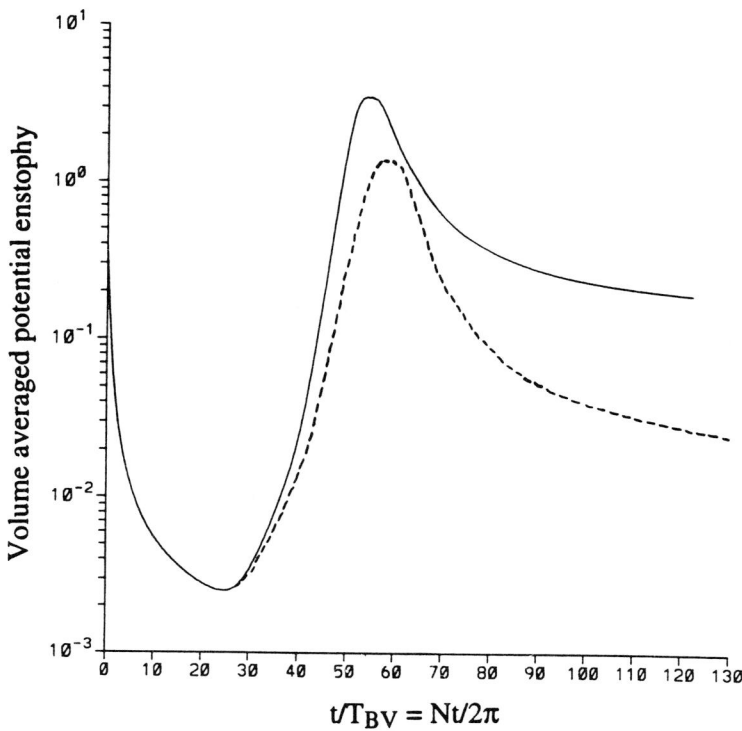

Figure 8. Three-dimensional numerical simulations ($a = 0.256$). *Full line*: Volume averaged potential vorticity squared (defined by (4.1)) for the flow with a time varying viscosity. *Dashed line*: Same quantity for the flow with constant viscosity $5 \ 10^{-4}$

starts growing nearly exponentially but saturates at the breaking time at a very small level. This energy is contributed by smaller scale motions than Φ_0. Its energetic insignificance thus indicates that no small scale energetic motions of the horizontal-vortex type occurs in the flow. The large value of the viscosity very likely accounts for this behavior. Indeed, when investigating the Fourier spectrum of the motions with vertical $\vec{k}$, we have found that they are contributed by large scales only. The influence of the viscosity upon this behavior is confirmed by the evolution of Φ_0 and Φ_1 for the $\nu = 0$ simulation, which is also plotted in Figure 7: both energies grow at the same rate in this case. This growth rate is twice larger than that of Φ_0 in the non zero viscosity calculation.

To interpret the growth of $\Phi_0(t)$ and $\Phi_1(t)$, it is necessary to investigate the behavior of the potential vorticity. The volume averaged potential enstrophy $\langle Q^2 \rangle$ is thus plotted versus time in Figure 8. $\langle Q^2 \rangle$ is non zero at initial time because of the white noise, as already said. The evolution of $\langle Q^2 \rangle$ displays three stages: up to $\simeq 25 \ T_{BV}$, it decreases by two orders of magnitudes. The reason is very likely that the white noise, which solely contributes to the potential vorticity at early times, is damped by viscous effects at small scales. This behavior is crucial to the interpretation of Figure 7: no potential-vortex mode is thus produced during the stage of the flow preceding breaking. Hence, the use of the Craya-Herring decomposition incorrectly predicts the growth of a non propagating component in the flow. The growth of $\Phi_0(t)$ and $\Phi_1(t)$ in Figure 7 soley attests of the production of vertical vorticity in the flow and should result from three-dimensional interactions among gravity waves. The very likely exponential growth rate suggests that three-dimensional resonant interactions occur among waves. It is expected however that two-dimensional resonant interactions should dominate the transfers of energy toward small scales, because the primary wave amplitude is very small (Klostermeyer 1991). These questions are currently investigated.

Figure 8 shows that, from $\simeq 25\ T_{BV}$, $\langle Q^2 \rangle$ starts growing, attesting that nonlinear interactions have transferred energy toward dissipative scales. The breaking event that occurs at about $t \simeq 55\ T_{BV}$ is responsible for the dramatic increase of $\langle Q^2 \rangle$. This quantity next decreases again but the time varying viscosity prevents it from decaying below its initial level.

5 Conclusions

We have presented two-dimensional and preliminary three-dimensional numerical experiments of a progressive standing wave.

This study has been preceded by a two-dimensional numerical study of standing internal gravity waves by Bouruet-Aubertot *et al.* (1995, 1996) and by an associated laboratory experiment by Benielli & Sommeria (1996). The dynamical evolution of progressive and standing internal waves display striking analogies. As Bouruet-Aubertot *et al.* (1995), we observe that the primary wave breaks whatever its initial amplitude. In the standing wave case, it was found that the breaking time is proportional to the inverse of the initial amplitude squared. The initial stage of the evolution is also qualitatively similar in both cases, since the dynamics of the standing wave were observed to be governed by resonant interactions of the parametric instability type. Finally we have seen that the energy spectra of the progressive wave case display a characteristic behavior encountered in geophysical flows. The same behavior was observed by Bouruet-Aubertot *et al.* (1996). There is however an important difference. In the standing wave study, the turbulence generated by the breaking process persists for a longer time than in the progressive wave study and a quasi-stationary regime is reached after breaking. This difference in behavior is essentially due to the fact that growth of instabilities and subsequent breaking in the standing wave case is confined to the center of the computational domain. Consequently energy is dissipated in a smaller region of the computational domain, than in the progressive wave case.

C. Koudella and C. Staquet

This difference apart, the standing wave and progressive wave cases display analogous dynamical behavior, while the numerical and laboratory experiments of standing waves have shown excellent agreement. This feature suggests the usefulness of laboratory waves, which tend to be standing, to simulate and study internal waves in geophysical flows, which tend to be progressive.

The three-dimensional numerical experiments have the novelty, compared to two-dimensional simulations, to contain another type of motion entirely, with respect to the gravity waves. This motion does not propagate and contains all of the potential vorticity of the flow (which is zero in a vertical plane). Our calculations have shown that, in the stage of the flow that precedes breaking, no such potential-vortex motion is produced in the flow. By contrast, vertical vorticity is produced, very likely through three-dimensional interactions among internal waves. Thus, the use of the Craya-Herring decomposition would lead to the erroneous conclusion that a non propagating part is produced in the flow during this early regime. The primary wave next breaks, which produces potential enstrophy (square of the potential vorticity). The residual non propagating motion appears to consist of a horizontal mean motion with large scale variability, because of the low resolution, and thus low viscosity, of our simulations. An important result is that this horizontal mean motion dominates the later stage of the flow, from an energetic point of view. Higher resolution calculations need to be performed in order to investigate the incidence of the potential-vortex motion upon the turbulent regime generated by breaking.

Acknowledgements

This work has been supported by a DRET contract (no. 951143/A000) and by a DRET-CNRS grant. Calculations have been performed on the Cray C98 of IDRIS (CNRS computer center), thanks to computing time allocated by the Scientific Council of IDRIS.

Bibliography

1. BENIELLI, D. & SOMMERIA, J. (1996). Excitation of internal waves and stratified turbulence by parametric instability. *Dyn. Atmos. Oceans*, **23**, 335-343.

2. BOURUET-AUBERTOT, P., SOMMERIA, J. & STAQUET, C. (1995). Breaking of standing internal gravity waves through two-dimensional instabilities. *J. Fluid Mech.*, **285**, 265-301.

3. BOURUET-AUBERTOT, P., SOMMERIA, J. & STAQUET C. (1996). Stratified turbulence produced by internal wave breaking: two-dimensional numerical experiments. *Dyn. Atmos. Oceans*, **23**, 357-369.

4. CANUTO, C., HUSSAINI, M.Y., QUARTERONI, A. & ZANG, T.A. (1988). Spectral methods in fluid dynamics. *Springer Series in Computational Physics*, Springer-Verlag. Berlin.

5. CRAYA, A. (1958). Contribution à l'analyse de la turbulence associée à vitesses moyennes. *P.S.T. Ministère de l'Air (Fr)*, 345 pp.

6. DRAZIN, P. G. (1977). On the instability of an internal gravity wave. *Proc. R. Soc. Lond. A.* **356**, 411-432.

7. ERTEL, H. (1942). Ein neuer hydrodynamischer Wirbelsatz. *Meteorol. Z.*, **59**, 271-281.

8. GREGG, M.C. (1987). Diapycnal mixing in the thermocline : a review. *Journal of Geophysical Research*, **92**-C5, 5249-5286.

9. GREGG, M.C., WINKEL, D.P., SANFORD, T.B. & PETERS, H. (1996). Turbulence produced by internal waves in the oceanic thermocline at mid and low latitudes. *Dyn. Atmos. Oceans*, **24**, 1-14.

10. HERRING, J.R. (1974). Approach of axisymmetric turbulence to isotropy. *Phys. Fluids*, **17**, 859-872.

11. KLOSTERMEYER, J. (1991). Two- and three-dimensional parametric instabilities in finite-amplitude internal gravity waves. *Geophys. Astrophys. Fluid Dynamics*, **61**, 1–25.

12. KUNZE, E. & SANFORD, T.B. (1993). Submesoscale dynamics near a seamount. Part I: Measurements of Ertel vorticity. *J. Phys. Ocean.*, **23**, 2567-2588.

13. LELONG, M.-P. & RILEY, J.J. (1991). Internal wave/vortical mode interactions in a stably stratified flow. *J. Fluid Mech*, **232**, 1-19.

14. LOMBARD, P. N. & RILEY, J.J. (1996). On the breakdown into turbulence of propagating internal waves. **23**, 345-355.

15. MCINTYRE, M.E. & NORTON, W.A. (1990). Dissipative wave-mean interactions and the transport of vorticity or potential vorticity. *J. Fluid Mech.*, **212**, 403-435.

16. MIED, R. P. (1976). The occurence of parametric instabilities in finite-amplitude internal gravity waves. *J. Fluid Mech.* **78**, part 4, 763-784.

17. MÜLLER, P., HOLLOWAY, G., HENYEY, F. & POMPHREY, N. (1986). Nonlinear interactions among internal gravity waves. *Reviews of Geophysics.*, **24**, 3, 493-536.

18. ORSZAG, S.A. (1971). Numerical simulation of incompressible flows within simple boundaries. I. Galerkin (spectral) representations. *Studies in Applied Maths*, vol. **L**, no. 4, 293-327.

19. PHILLIPS, O. M. (1966). The dynamics of the upper ocean. Cambridge University Press.

20. RAMSDEN, D.& HOLLOWAY, G. (1992). Energy transfers across an internal wave-vortical mode spectrum. *J. Geophys. Res.*, **97**, C3, 3659-3668.

21. RILEY, J.J., LELONG, M.-P. G. & SLINN, D.N. (1991). Organized structures in strongly stratified flows. *Turbulence and coherent structures*, Métais & Lesieur eds., Kluwer, 413-428.

22. RILEY J.J., METCALFE R.W. & WEISSMAN M.A. (1981). Direct numerical simulations of homogeneous turbulence in density stratified fluids. In *Nonlinear properties of internal waves*, AIP conf. proc., B.J. West ed., 79-112.

23. SIDI, C. & DALAUDIER, C. (1989). Temperature and heat flux spectra in the turbulent buoyancy range. *Pure and Applied Geophysics*, **130**, 547-569.

24. SIDI, C. (1995). Some observed properties of small scales ($o[1...10^2 m]$) fluctuations in the stratosphere. *Proc. of the Euromech colloquium "Internal waves, turbulence and mixing in stratified flows"*, Lyon, Sept 6-9.

25. STAQUET, C. & RILEY, J.J. (1989). On the velocity field associated with potential vorticity. *Dyn. Atmos. Oceans*,**14**, 93-123.

Vertical Velocity Structure and Eddy Scale Distribution in the Water Column of the Stratified Lake of Geneva

U. Lemmin and R. Jiang

Laboratoire de Recherches Hydrauliques, Ecole Polytechnique Fédérale de Lausanne, Switzerland

Abstract

An acoustic Doppler velocity profiler is used to study vertical velocity dynamics in the stratified lake. Profiles were taken by lowering the instrument combined with a CTD at constant speed from a boat. The vertical velocity fluctuation amplitude is directly linked to the mean stratification in the water column. It is largest in the epilimnion, strongly reduced in the thermocline and homogeneous below the thermocline in the constant N layer. In the thermocline a strong correlation between vertical velocity, stratification, light transmissivity and acoustic backscattering is found. Turbulence scales which have been calculated from the vertical velocity data and the corresponding N-profile are about 0.5 to 1 m. An active layer is observed around 40 m where these scales correspond well with the Thorpe displacements.

1 Introduction

In a stratified lake, external wind forcing will simultaneously produce a large scale advection field, a field of internal waves and turbulence. Superimposed on the wind forcing, heat exchange between the atmosphere and the lake will cause convective cooling influencing stratification. These conditions are characterized by a different distribution of the heat flux $c_p \langle \rho' w' \rangle$ or $c_p \langle w' T' \rangle$. ($c_p$ = specific heat; ρ' = density fluctuation, T' = temperature fluctuation and w' = vertical velocity fluctuation). It is therefore of interest to measure the correlation $\langle \rho' w' \rangle$ or $\langle w' T' \rangle$ directly in order to be able to distinguish between the different processes. Understanding how the heat flux is driven and how it is associated with different stages of turbulence production and dissipation appears to be essential for obtaining reliable predictions of mixing in lakes.

The practical application of this diagnostic tool in lakes has been hindered by the difficulty to measure the vertical velocity with sufficient spatial and temporal resolution. Instead, heat flux has been inferred from turbulent dissipation rates of temperature variance and kinetic energy. In this context the measurement of length scales has been proved to be useful. Thorpe [1] devised an objective method to estimate the length scales of turbulent overturning in flow with overall stable stratification. The Thorpe scale L_T is defined as the root mean square Thorpe displacement d_T which is found by reordering a measured temperature profile into a stable one. Thorpe further proposed that this overturning scale should be related to the Ozmidov scale, L_o, which links the buoyancy force and the inertial force $L_o = \left(\dfrac{\varepsilon}{N^3}\right)^{1/2}$; ε is the kinetic energy dissipation rate and N is the buoyancy frequency $N^2 = \dfrac{g d\rho}{\rho dz}$ (z is downward vertical and ρ the mean density). Dillon [2] has given evidence that the Thorpe scale L_T is indeed comparable to the Ozmidov scale L_o when the mean horizontal density gradient is much smaller than the vertical gradient. This is most probably true in layers far from the lake surface. The definition of a buoyancy length scale which is equivalent to the Ozmidov scale L_o is given as [3],

$$L_b = \frac{\langle w'^2 \rangle^{\frac{1}{2}}}{N}. \tag{1.1}$$

In the ocean, however, great discrepancies have been found recently between the rather scant direct measurements of heat flux and the indirectly determined ones. Moum [4] showed that the heat flux inferred from the turbulent kinetic energy dissipation rate ε was an order of magnitude larger than the directly measured heat flux.

In preparation for heat flux measurements we have tested the feasibility of an acoustic velocity profiler (ADVP) to obtain vertical velocities. In particular we investigated the reliability of that technique in estimating the fall velocity of the instrument, the variance of the vertical velocity fluctuation and the detection of instabilities in the stratification by buoyancy length scale determination. We report on an analysis of a set of profiles obtained during the summer of 1995 and discuss the relationship between the stratification and observed vertical turbulence characteristics.

2   Instrumentation and experimental setup

Temperature stratification and vertical velocity have to be measured to make the analysis outlined above. In the present study these data are obtained by a CTD combined with an acoustic velocity profiler. The CTD samples at 4.5 Hz and has a temperature resolution of 0.001°C. The acoustic transducer of the velocity profiler is clamped to the CTD (Figure 1) and connected to the hardware and the data acquisition on deck. The weight and the buoyancy of the package are distributed to opposite ends and the vertical orientation of the acoustic transducer is assured. The system is adjusted to provide for a stable free-fall velocity close to 10 cms^{-1}. Effects of the boat motion on the descending

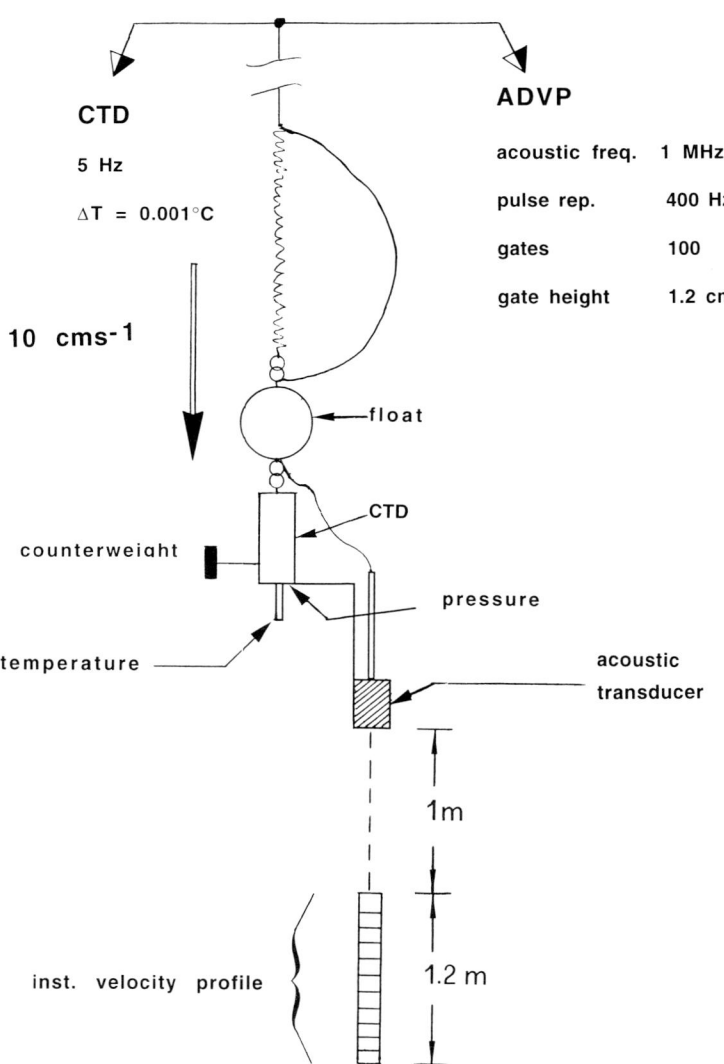

Figure 1. Instrument package consisting of a CTD and the acoustic transducer of the ADVP lowered on a cable and bungee cord from a moored boat. The range of the instantaneous velocity profile is indicated. Drawing not to scale

instrument have been minimized by mounting a set of 1 m long bungee cords between the lowering cable and the instrument. The number of parallel bungee cords was optimized to minimize boat interference. The cable to the package is paid out at a controlled, constant speed of 10 cms^{-1} by a winch from a boat which is held in place by three anchors.

A high resolution ADVP (Acoustic Doppler Velocity Profiler) is used to measure vertical velocity. Details of the technique are given in [5]. An important feature of the ADVP is that it simultaneously takes instantaneous velocity profiles at a certain number of consecutively spaced "gates" or "bins" over a predetermined depth range. The ADVP has previously been used in a bottom boundary layer study in Lake Geneva [6] where it provided time series of a section of the boundary layer while being fixed on the lake bottom.

The present ADVP works at an acoustic frequency of 1 MHz using a transducer with an opening angle of 1.3^o (7.5 cm diameter). The instrument was set to sample simultaneously at $M_g = 100$ consecutive gates at a pulse repetition frequency f_{prf} of 300 Hz. Each gate had a vertical length of $l_g = 1.2$ cm which resulted in an instantaneous velocity profile length $L_{\max}$ of 1.2 m. The near field of the transducer has been omitted in the present study. Therefore the first gate is placed 1 m ahead of the transducer and the remotely sensed profile covers the distance from 1 m to 2.2 m ahead of the transducer face (Figure 1). This arrangement will also avoid any physical interference of the instrument with the velocity measurements. The vertical velocity in each gate was determined by the pulse-pair algorithm [5] averaged over $N_{pp} = 32$ consecutive Doppler frequency samples sampled at f_{prf}. Thus vertical velocity profiles are obtained at a rate of $N_{pp}/f_{prf} = 300/32 = 9.375$ complete profiles at 100 gates per second with a nominal spatial resolution of 1.2 cm (at a fall speed of 10 cms^{-1}). Tests of the system have shown [7] that it is stable to < 3 mms^{-1}.

Compared to Laser Doppler anemometers, the ADVP has the advantage that it takes instantaneous velocity profiles. Thus a given water volume, represented by a gate, is repeatedly sampled by consecutive gates while the instrument is lowered through the water column. This feature, unique to the ADVP technique, allows the establishment of (short) velocity time series at any given depth from the instantaneous profiles. In that way some information on the time stability of the instantaneous velocity samples is provided.

The vertical velocity of the water can be obtained from

$$w(t, z, m_g) = w_m(t, z, m_g) - \overline{W}(t, z) \qquad (2.1)$$

with

$$\overline{W}(t, z) = \frac{1}{L_{\max}} \int_z^{z+L_{\max}} w_m(t, z, l_g) dl_g = \frac{1}{M_g} \sum_{m_g=1}^{M_g} w_m(t, z, m_g) \qquad (2.2)$$

where $w_m(z, t, m_g)$ is the measured vertical velocity (the speed relative to the falling instrument) at ADVP gate number m_g

- l_g is the gate length, t time, z instantaneous reference depth;

- M_g is the total number of gates;

- $L_{\max}$ is the maximum length of the instantaneous velocity profile corresponding to M_g;

- $\overline{W}(t, z)$ is the fall velocity at time t and depth z.

The advantage of calculating the mean speed of descent $\overline{W}(t, z)$ by Equation (2.2) is that the mean is taken over all gates ($M_g = 100$ gates in the present study) which provides for a good statistical stability of this estimate. A drawback of this method is that in the case where the whole layer of thickness $L_{\max}$ moves at a physically real and constant vertical velocity this movement is integrated into the fall velocity $\overline{W}(t, z)$ and therefore lost for the analysis of the motion field. Such constant vertical velocities over layers of several meters thickness can be produced by internal waves [8]. Thus the larger $L_{\max}$ the more precise the calculated $\overline{W}(t, z)$. However, because of the range-velocity ambiguity of the ADVP [5], $L_{\max}$ is limited by the pulse repetition frequency f_{prf}. The velocity $w(t, z, m_g)$ obtained by the method described here limits the study of turbulence to scales smaller than $L_{\max}$. A longer range $L_{\max}$ could be obtained by lowering the acoustic frequency.

In order to verify the fall velocity measured by the ADVP, the simultaneously registered time series of CTD depth can be used to calculate the true fall velocity of the package. We will compare the two results.

3   Experiments and results

During the summer of 1995 the instrument package was tested several times in the Lake of Geneva. Typical examples which will be presented here were taken on 2nd August 1995. The weather was calm and sunny. The water surface was flat thus eliminating wave induced boat motion. The experimental site was near Buchillon at a depth of about 60 m about 1 km off-shore on the sloping side of the lake basin. Profiles (taken between 09:55 and 15:15) started at about 2 m depth and were continued to below 30 m depth.

An ensemble of the temperature profiles (Figure 2) shows that the mean stratification remained nearly stable throughout the day. The water was most strongly stratified in the depth range of the diurnal thermocline from 5 to 7 m. A slight stratification remains in the mixed layer above. Below the thermocline the temperature changes at a nearly constant rate resulting in a layer of almost constant buoyancy frequency N (Figure 2(b)).

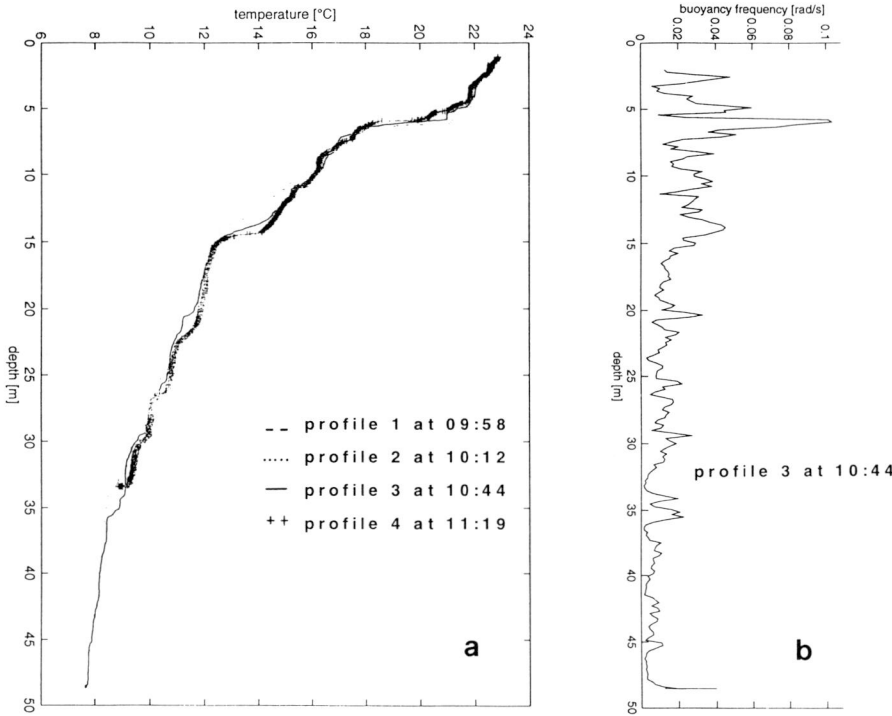

Figure 2. (a) Four CTD temperature profiles on 2nd August 1995 near Buchillon. Times indicate when each profile was started, and (b) Profile of the Brunt-Vaisalla frequency for profile 3

3.1 Mean fall velocity

The mean fall velocity of the package was calculated in three different ways. A comparison of the results of all three methods is given in Table 1. First, the mean of each gate over the whole profile was calculated. These mean velocities are centered around 10 cms^{-1}, the fall speed of the instruments and fluctuate in the range of ± 5 cms^{-1} being smaller most of the time. The general appearance of the recorded signal is similar at all gates. Variations of mean parameters between gates of one profile were always less than ± 5%. Next, a profile of $\overline{W}(t,z)$ was established and finally the true fall velocity profile was calculated from the CTD pressure data. The mean of $\overline{W}(t,z)$ and the true fall velocity are very close. The same holds true for the two standard deviations. The standard deviations of the individual gates give higher values because they include the

Table 1. Comparison of mean velocities and standard deviations for profile 3

	Gate 1; Mean Whole Profile	Gate 100; Whole Profile	Mean of $\overline{W}(t,z)$	Mean of True Fall Velocity
Mean Velocity (cms^{-1})	10.39	10.79	10.62	10.68
Standard Deviation (cms^{-1})	2.17	2.11	1.85	1.26

true vertical velocity fluctuations as well. Similar results were obtained for all other profiles indicating that the ADVP functions correctly.

An excerpt of the record of $\overline{W}(t,z)$ and the true fall velocity is given in Figure 3. It can be observed that the fall velocity varies smoothly with depth indicating that free fall conditions are approached. Superimposed, a regular oscillation is visible in both curves. Bursts of oscillations are observed throughout the profiles and spectral analysis reveals a broad peak in the range between 4.5 and 6 s. This signal cannot be the result of surface waves induced boat motion which should be at 2 to 3 s; instead, this oscillation is attributed to vortex shedding of the instrument package. Due to the robust estimation of $\overline{W}(t,z)$ this signal can be eliminated from the vertical velocity profile.

3.2 Vertical velocity dynamics

The mean value of the vertical velocity obtained by Equation (2.1) was calculated over the full profile and for all gates within each profile. In both cases it is always < 5 mms^{-1}. The standard deviation for the full profiles varies from 1.9 to 2.6 cms^{-1}.

3.2.1 The thermocline layer

Inspection of the data reveals some systematic structure in all profiles. The CTD/ADVP profile number 3 is taken here as an example to present the typical structure of the vertical velocity. For the range around the diurnal thermocline from 2 to 10 m, Figure 4 shows the profiles of temperature and light transmissivity obtained by the CTD and of the corresponding vertical velocity and backscattered acoustic signal intensity obtained from the ADVP. Vertical velocity is strongest above the thermocline and is organized in layers of O(50cm) vertical extent. The most obvious feature is the strong reduction in vertical velocity in the thermocline layer. This has been observed in all profiles taken.

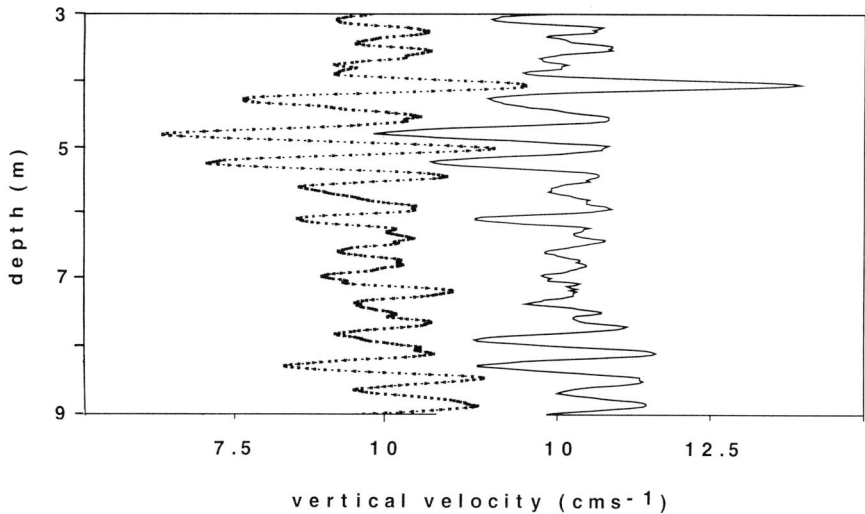

Figure 3. Example of the mean velocity calculated over the 100 gates of the instantaneous profile (left; measured points) compared with the instrument fall velocity calculated from the pressure data of the CTD (right; solid line). Data from profile 3

Strongest fluctuations in light transmissivity are found in the upper 5 m followed by a layer from 5.8 m to 6.8 m depth in which the smallest transmissivity values coincide with the strongest stratification. Below that depth, variability increases again. In the hypolimnion (not shown here) transmissivity eventually reaches higher values than in the epilimnion. Acoustic backscattering intensity is highest in the layer of lowest vertical velocity which occurs in the thermocline. It varies on vertical scales similar to those of the velocity. While the maximum of the backscattering intensity corresponds to a peak of low light transmissivity, the two signals do not coincide but show again significant variability on similar vertical scales.

3.2.2 Time series

The structure of the observed instantaneous vertical velocities can be further investigated by constructing time series of the vertical velocity at constant depths. The combination of fall speed, $\overline{W}(t, z)$, pulse repetition frequency, f_{prf}, and instantaneous profile length, $L_{\max}$, allows for a time series length of about 12 s for a given depth which is taken here as a layer of about 1.2 cm thickness. Examples of these time series from three different sections of the water column are given in Figure 5. It is obvious that the overall structure of the velocity fluctuations does not change with time during the period of the time series.

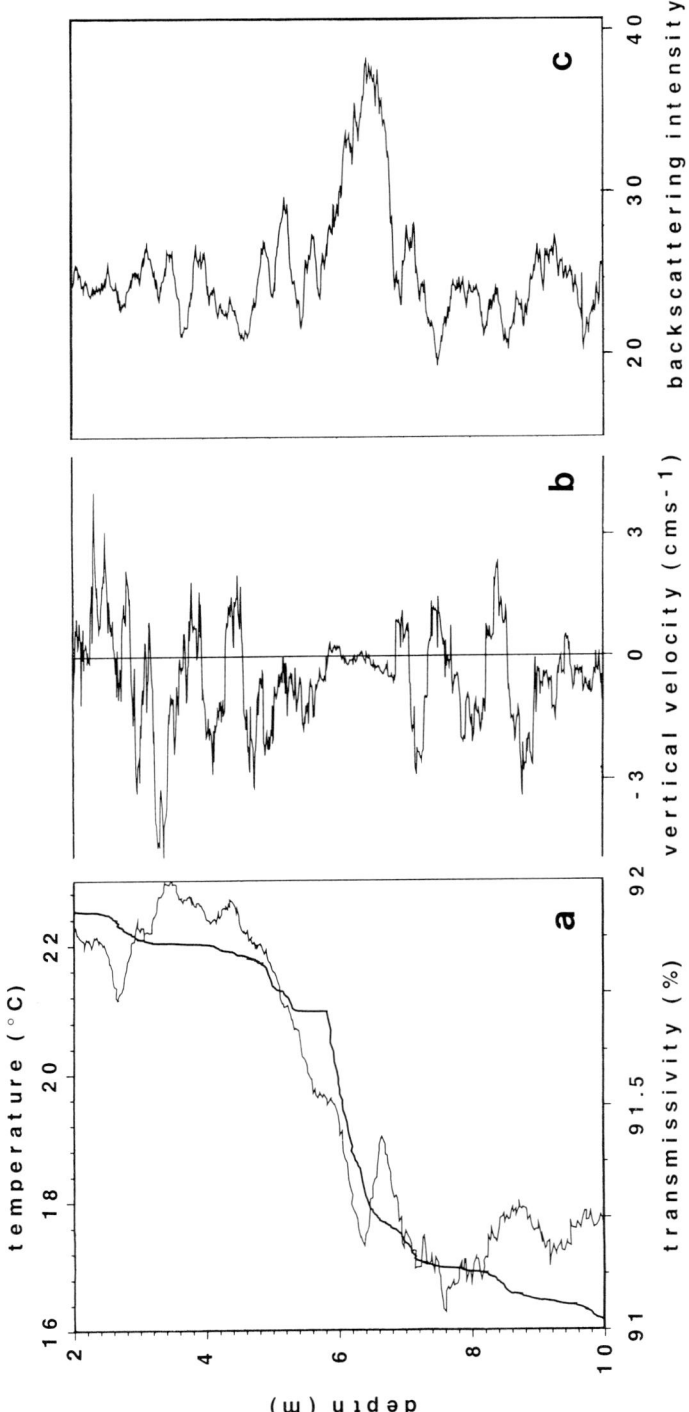

Figure 4. For the layer from 2 m to 10 m depth: (a) profile of temperature and light transmissivity (CTD), (b) profile of vertical velocity (ADVP), and (c) profile of acoustic backscattering intensity (ADVP). Data from profile 3

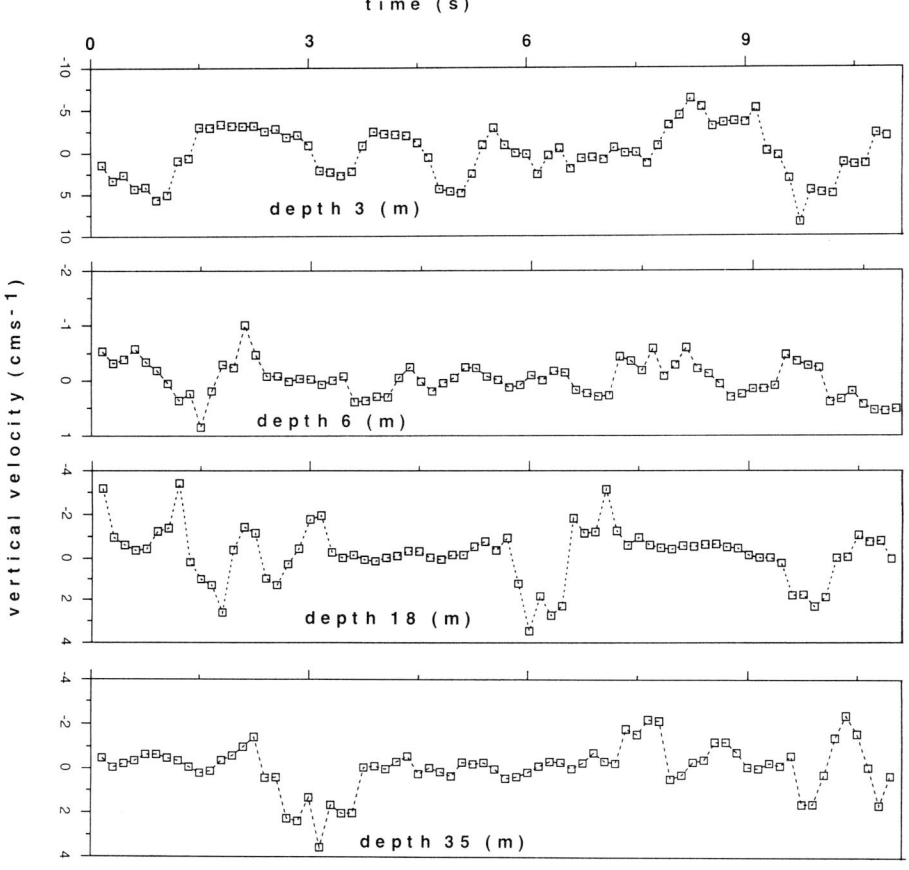

Figure 5. Time series of vertical velocity at selected depths. The depth is indicated on the diagram. Data from profile 3. Note the change in velocity scales

Above the thermocline, amplitudes are large and change regularly. The amplitude of fluctuation decreases systematically in the thermocline and increases again below, but the amplitude is smaller than above the thermocline. In the hypolimnion with weak stratification the amplitude of the fluctuations remains similar throughout the layer. In Figure 6, a time series obtained from consecutive gates of the instantaneous profiles is compared to a time series recorded by gate one falling with the time delay corresponding to the fall velocity through the same water column which is located in the thermocline. Thus time separation between the two series increases progressively. It can be seen that

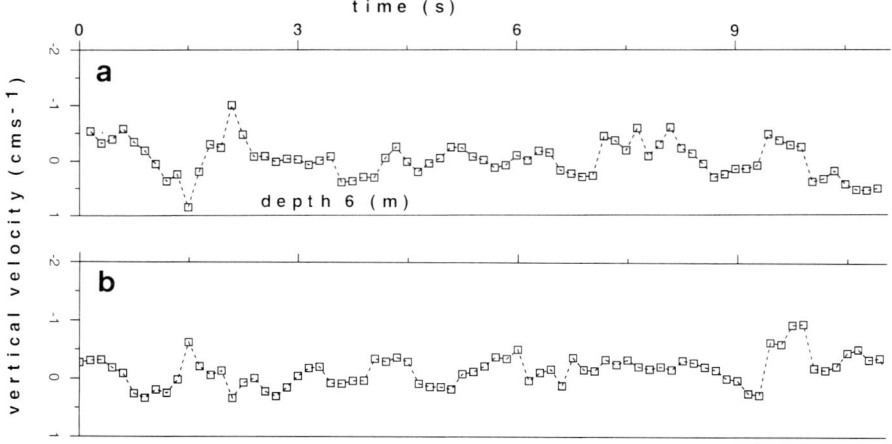

Figure 6. Comparison of: (a) Time series at fixed depth in the thermocline with, and (b) Instantaneous profile for the same depth range. Data from profile 3

the overall structure of the two series is similar but certain changes in the time delay of their appearance do occur.

3.2.3 Variance profiles

Profiles of the variance of the vertical velocity as function of depth which is a measure for the strength of the vertical turbulence are presented in Figure 7 for the four profiles taken that day. The variance has been averaged over layers of about 50 cm thickness. The variance is strong in the upper mixed layer and decreases systematically to its minimum in the diurnal thermocline (5 m to 7 m). Below that depth it increases again gradually to reach a nearly constant value in the layer of constant N (Figure 2(b)). Overall, the variance decreases progressively towards the end of the observation period (profile 4).

3.2.4 Statistical analysis

The statistical significance of the depth variability of the variance has been further investigated. The first four moments of the vertical velocity have been calculated for depth segments of profile 3 corresponding to the upper layer of constant variance (epilimnion), the minimum of the variance (thermocline), the layer of increasing variance (below the thermocline) and two consecutive layers in the hypolimnion (constant N-layer). Results are summarized in Table 2. It can be seen that except for the thermocline layer, the vertical velocity approaches

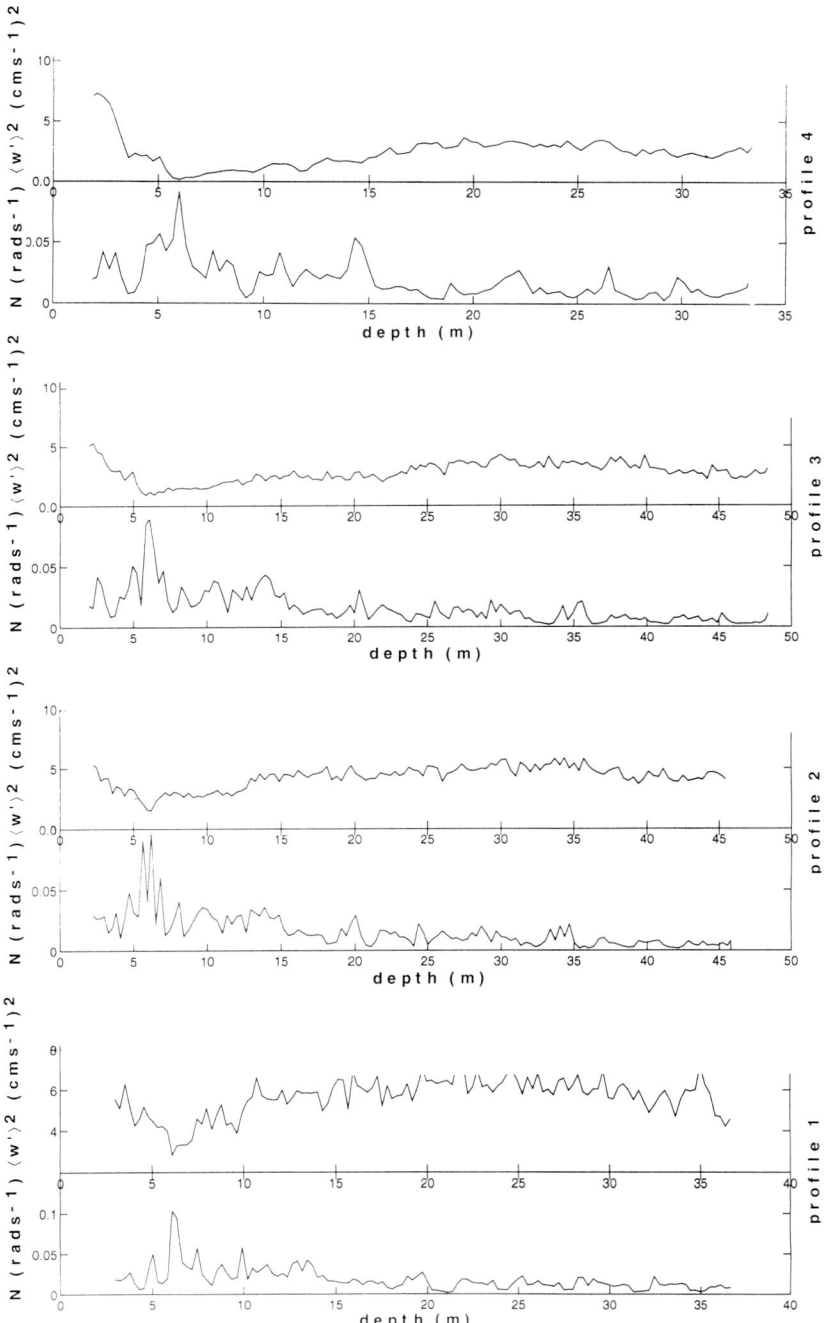

Figure 7. Ensemble mean variance of vertical velocity as function of depth for four profiles taken 2nd August 1995. For profiles times see Figure 2

Table 2. Depth variability of the standard parameters for profile three

Depth (m)	2 - 5.8	5.8 - 6.8	6.8 - 10.5	10.5 - 25	25 - 48
Mean (cms-1)	-0.33	-0.04	-0.31	-0.12	-0.08
Standard Deviation (cms-1)	3.86	1.61	2.65	3.79	3.48
Skewness	0.72	3.28	0.92	0.72	0.56
Kurtosis	2.76	34.55	8.59	3.23	3.28

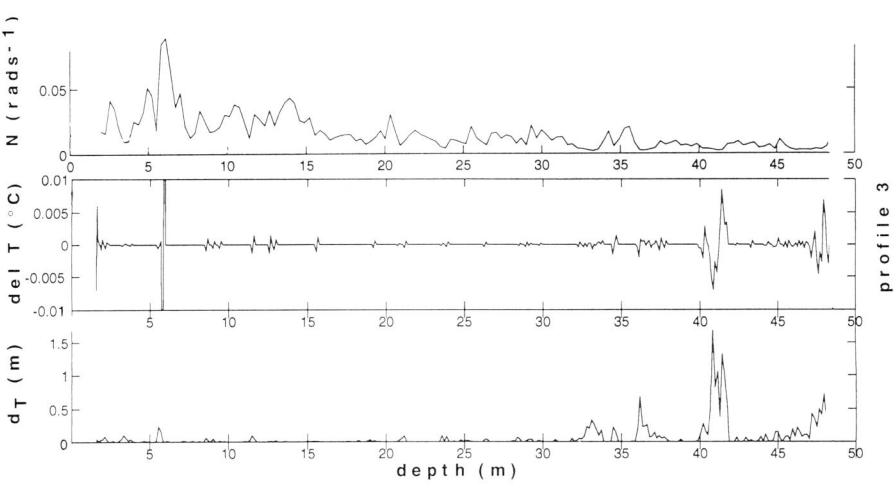

Figure 8. Temperature instabilities in profile 3

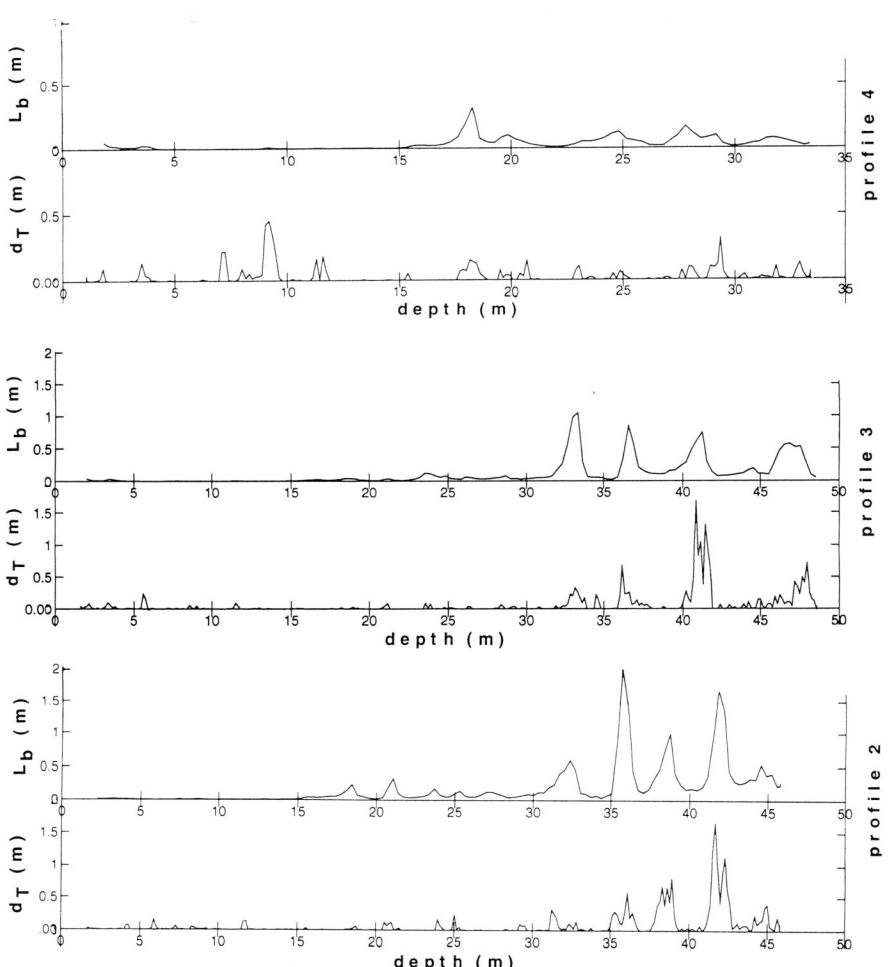

Figure 9. Buoyancy scales L_b (Equation 1.1) and Thorpe displacements d_T for four profiles taken on 2nd August 1995

a Gaussian distribution expressed by small skewness values. However, an over-representation of small velocity fluctuations expressed by the high kurtosis values is found everywhere. It is most pronounced in the layer below the thermocline. This can be made evident by smoothing the data with a running mean over twenty profiles. In that case, skewness and kurtosis fall both below 0.3 for all four layers in question. In the thermocline the velocity structure is left leaning and very pointed approaching characteristics of a log-normal distribution.

3.3 Length scale dynamics

Instabilities in the temperature profiles have been identified and an example is shown in Figure 8. While instabilities are found throughout the hypolimnion in all profiles, the largest ones are observed in the layer between 35 m and 45 m depth. From these instabilities, Thorpe displacement scales d_T have been calculated by restructuring the profiles [1] for all four profiles of 2nd August 1995 (Figure 9). It can be seen that in all profiles the largest displacements occur in the same layer situated in the constant N-section of the profile. Some smaller displacements are randomly distributed. No significant instability is seen in the thermocline or above.

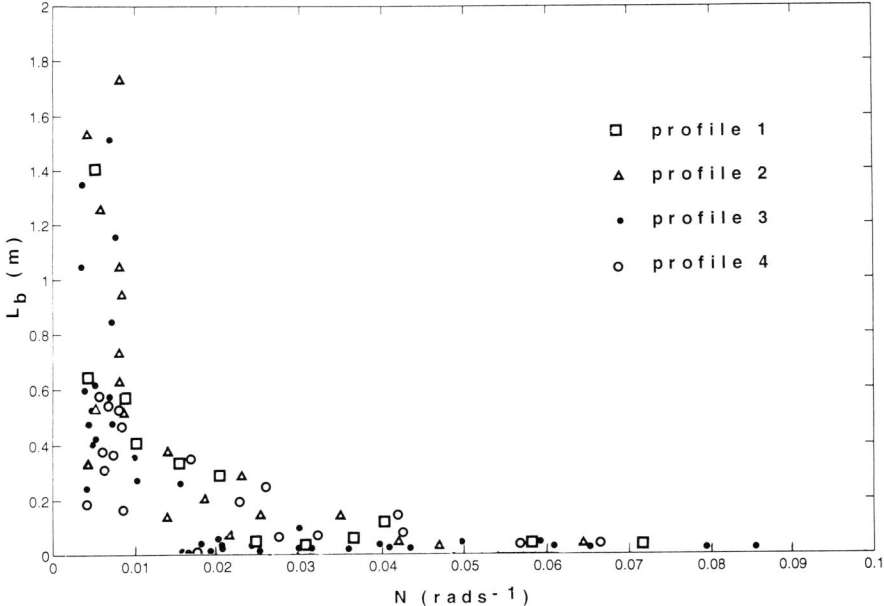

Figure 10. Buoyancy scales L_b as function of buoyancy frequency N

Instabilities in the water column may also be detected by the buoyancy scale L_b which can be calculated from the vertical velocities and the stratification using Equation (1.1). This calculation was carried out for all four profiles and results are also shown in Figure 9. In all profiles, the layers in which large Thorpe displacements are found coincide with those of large buoyancy scales. Least coincidence is seen in profile 4 when variance had overall decreased (Figure 7). Smaller buoyancy scales are seen in most cases where the smaller Thorpe displacements have been observed. The buoyancy scale estimates are typically larger than the Thorpe displacements by 10% to 50%.

The observed buoyancy length scales have been plotted in Figure 10 as function of the mean buoyancy frequency N. The trend of large buoyancy scales at small buoyancy frequencies N is clearly visible.

4   Discussion

In this study we have applied the ADVP technology to lake hydrodynamics in open water. Previously we have obtained excellent results with the ADVP in laboratory open channel flow [5,7] where we have investigated the performance and the resolution of the system and have proven under controlled laboratory conditions that this instrument is a bona fide research tool. We have used it in the lake to study bottom boundary layer dynamics [6]. In the present study we observe that the fall speed of instrument package calculated from ADVP data is consistent with the one calculated from the pressure of the CTD. Mean value, standard deviation and velocity pattern details closely coincide. This is an indication that the system functions correctly also under free-fall lake conditions.

A further indication is the low mean vertical water velocity. The decrease in mean velocity with depth observed in all layers outside the thermocline indicates that there is no net vertical transport in the whole water column. Nevertheless, a vertical velocity of 0.3 cms^{-1} measured above the thermocline implies a vertical transport of more than 200 m a day. This is higher than the typical fall velocity of particles estimated by budget methods from particle trap data. A fall velocity of about 0.3 cms^{-1} is typical for very fine sand particles (radius 0.033 mm) and is just above that of silt particles which may flocculate [9]. The reduction in mean velocity from the epilimnion to the thermocline may explain the accumulation of particles in the thermocline which was evidenced by the reduction in light transmissivity in that layer. The increase in light transmissivity below the thermocline can be correlated with the increase in the mean vertical velocity observed in that layer. No corresponding observations appear to have been made before.

The increase in particulate matter in the thermocline is also correlated with the increase in acoustic backscattering intensity in that layer. As yet, little is known about the nature of the backscattering targets in lakes. It can be assumed, though, that the backscattering mechanism in the lake may be somewhat different from that in open-channel flow where under conditions of high turbulent dissipation, small air bubble-microstructures have been found to contribute most strongly [10]. In the lake, under much lower levels of turbulent dissipation, particles can be expected to contribute to the backscattering. The difference between the variation in light transmissivity and acoustic backscattering indicates that the particles or particle clusters of the size corresponding to the 1 MHz acoustic frequency (i.e. of radius 0.375 mm) are not a dominant member of the particle community in the lake. There is an order of magnitude size difference between the particles which may have the observed mean fall velocity in the epilimnion and those to which the ADVP is sensitive. Thus we do not believe that our measurements are dominated by particle fall velocities. Low turbulent dissipation and low particle density combined may explain why the backscattering intensity recorded in the lake is by two orders lower than that found in open channel flow. Following the calculations by Shen and Lemmin [10] this sensitivity may be improved by reducing the acoustic frequency for lake measurements to 200 kHz. The variation of light transmissivity and acoustic backscattering on vertical scales of O(10cm) indicates a vertical layering which appears to be correlated with the vertical velocity distribution varying on the same scale size.

The magnitude of vertical velocity fluctuations measured by the ADVP are in agreement with those reported from time series measurements under oceanic conditions with much less spatial and temporal resolution [11–13]. A similar layering of alternate positive and negative velocities in the profile of vertical velocity can be seen in the data of [14]. The magnitude of their velocities though never exceeds 1 cms^{-1}. The observed structure of our data is physically consistent with the stratification dynamics. The turbulence is strongest in the upper mixed layer due to the proximity of the wind forcing and the effects of convective cooling. It is nearly suppressed by the strong stratification in the diurnal thermocline In our previous measurements we have rarely observed buoyancy instabilities in this layer around the thermocline [15] except near the sides of the lake. It can therefore be expected that the net vertical mixing in the thermocline will be smaller than that in the other parts of the water column. In this thermocline layer we have previously observed strong horizontal shear and short period internal waves [8].

The ADVP is the only instrument which allows profiling and time series measurements at the same time. From the examples of these time series (Figure 5) it is seen that the vertical velocities change very little in their characteristics over times scales of more than 10 s. They can therefore be considered as being typical for their respective depth. At the same time the structures of vertical velocity fluctuations are similar for adjacent layers as long as stratification does not change. The data show that the vertical velocity varies smoothly between consecutive points indicating that the vertical velocity is organized at scales larger than the presently chosen resolution of the ADVP system of 1.2 cm. It should be noted that the resolution of the present ADVP can be increased without system modification to 0.6 cm. The instrument is therefore well suited for the study of vertical velocity dynamics.

We have found that the small scale variations are over-represented in the data when no smoothing is applied. At present it is not possible to determine whether this is physically real or introduced by the measurement technique. The instrument itself is stable to smaller variations but low backscattering intensity conditions encountered in the lake may have introduced some noise. This can also be seen from a spectral analysis carried out over the four layers of profile 3 described above. When the analysis is carried out on the velocity data without smoothing, a white noise spectrum results. When the data are first treated with a running mean over 6 samples all spectra show a slope which is close to -5/3 in the two constant N hypolimnion layers. The reduction in variance amplitude with time (Figure 7) is reflected in the spectra as well. In the hypolimnion, the energy level of the spectra is shifted down by more than two orders of magnitude between profile 3 and profile 4. Since the stratification is also small in those layers, isotropic conditions which we had previously already seen in the bottom boundary layer [6] may be approached in the open waters of the hypolimnion.

The temperature instabilities observed all show a symmetric pattern over scales comparable to the Thorpe displacements d_T indicating that the instability is the result of local shear [2]. With both methods of calculation we find the largest scales in the same layer. The systematic difference between the Thorpe displacement obtained from the instabilities of the temperature profiles alone and buoyancy scales obtained from the combined velocity and temperature information is in part due to the layout of the present instrument package. The two sensors are not measuring in the same volume because their vertical paths are separated by about 10 cm and because they are separated in the vertical by 1 m or more. Furthermore their response times are slightly different. In future measurements both velocity and temperature will be measured in the same volume.

In all cases, the ratio L_o/d_T is <2. If the buoyancy scale is taken as representative for the Ozmidov scale L_o then a turbulent Froude number can be calculated as $Fr_t \propto (L_o/d_T)^{2/3}$. Ivey and Imberger [16] have shown that mixing efficiency is greatest (25%) when $Fr_t = 1$. Thus in the layer from 35 m to 45 m where we have found the strongest activity persisting throughout the observation period, mixing efficiency is also close to optimal. The sensitivity of the buoyancy scale to the stratification is demonstrated in Figure 10. In the present example, the buoyancy scale size increases rapidly when the buoyancy frequency falls below 0.005 [rad s^{-1}]. The general trend of the variation of the buoyancy scale coincides with the observations made by [15] who have shown that the Thorpe scale calculated from temperature profile instabilities increases from a minimum in the thermocline to large values in the hypolimnion.

Throughout the measurements we observe the largest L_o and d_T values in the same depth layer from 35 m to 45 m. It is therefore likely that this is an active layer and not a layer of decaying turbulence [17]. Some indication for the reason of this activity may be found in the last temperature profile (profile 4) where some small steps are visible in that same depth range. Caldwell et al. [18] have interpreted these steps as indications of intrusion penetrating from the side of the lake. Our own measurements [6] indicate that they are reflected internal waves. Both mechanisms could play a roll in the present measurements which were made only 1 km from the sloping side of the lake. Either mechanism will produce shear and mixing in that layer.

From the temperature measurements alone it appears that the ambient condition remained stable during this time. At the same time the variance of the turbulence decreases with time. This indicates that either an important horizontal advective motion occurs which displaces the water mass or an overall reduction of turbulence takes place because of a decaying current field. No clear answer can be given without knowledge of the horizontal current vector. Nevertheless, a permanently present layer between 35 m and 45 m in which instabilities occur is found to be superimposed on a changing vertical turbulence field. This indicates that the concept of homogeneity of a water mass which is frequently invoked when dissipation is calculated from a series of temperature profiles over a certain time lapse may be more difficult to attain in lakes than in oceans.

5 Summary and Conclusions

From the results of the combined CTD / ADVP profiling study carried in the stratified Lake of Geneva the following conclusions can be drawn:

We have demonstrated that the ADVP works well under free-fall conditions. The simultaneous multigate profiling capacity and the possibility to establish short time series at any depth give new insight which cannot be obtained with other instruments. Both indicate that the vertical velocity is predominately organized on scales of O(10cm).

Vertical turbulence is systematically suppressed in the strongly stratified thermocline and the turbulence most likely becomes non-isotropic. In the thermocline, correlation of low vertical velocity variance with increase in light transmissivity loss and acoustic backscattering points towards a direct effect of vertical velocity variance on particle transport. The turbulence intensity is larger and more homogeneous below the thermocline in the constant N layer. A decrease in time of the variance in the whole water column indicates that a large-scale water-mass process is superimposed which cannot be detected from temperature profiling alone.

Largest instabilities are found in each profile in the same depth layer between 35 m and 45 m in the hypolimnion despite the change of overall variance intensity of the vertical velocity. Lateral boundary effects may explain this active layer.

The observations in the lake presented here are first results obtained with a new velocity measuring technology. They have already given some insight into the vertical velocity dynamics which have not been observed before. In the next stage, the buoyancy flux will be measured combining two high resolution temperature sensors with the ADVP. We expect that the direct measurements will provide better insight into heat flux and vertical mixing dynamics in lakes. The range of measurements and the resolution of the ADVP span the range of transition from finescale internal waves to outer scale turbulence. This is the range where internal waves, potential vorticity carrying finestructure, growing instabilities and turbulence all coexist [11]. Measurements in this range using the new ADVP, either by profiling or by time series at fixed depth, are suitable for studying the still largely missing links between internal waves and turbulence in lakes.

Acknowledgements

C. Perrinjaquet was responsible for the field work and the data preprocessing. We appreciate his care and efficiency. Many critical comments and helpful suggestions by S.A. Thorpe on an earlier version of the manuscript were greatly appreciated. This work was funded by the Swiss National Science Foundation grant Nr. 21-34'116.92. We are grateful for the support.

Bibliography

1. Thorpe, S.A. (1977). Turbulence and mixing in a Scottish loch. *Phil. Trans. Roy. Soc. Lond. A*, **286**, pp. 125–181.

2. Dillon, T.M. (1982). Vertical overturns: a comparison of Thorpe and Ozmidov length scales. *J. Geophys. Res.*, **87**, pp. 9601–9613.

3. Stillinger, D.C. et al. (1983). Experiments on the transition of homogeneous turbulence to internal waves in a stratified fluid. *J. Fluid Mech.*, **131**, pp. 91–122.

4. Moum, J.N. (1990). The quest for $K\rho$ - preliminary results from direct measurements of turbulent fluxes in the ocean. *J. Phys. Oceanogr.*, **20**, pp. 1980–1984.

5. Lhermitte, R. and Lemmin, U. (1994). Open-channel flow and turbulence measurement by high-resolution Doppler sonar. *J. Atm. and Oceanic Tech.*, **11**, pp. 1295–1308.

6. Lemmin, U., Jiang, R. and Thorpe, S.A. (1995). Finescale dynamics in the thermocline of Lake Geneva. *IUTAM Symposium on Physical Limnology*, pp. 409–422.

7. Lemmin, U. and Rolland, Th. (1997). A monostatic velocity profiler for laboratory and field studies of turbulent flow. *J. Hydr. Eng.*, ASCE, **123**, pp. 1089–1098.

8. Thorpe, S.A., Keen, J.M., Jiang, R. and Lemmin, U. (1996). High frequency internal waves in Lake Geneva. *Phil. Trans. Roy. Soc. Lond. A*, **354**, pp. 237–257.

9. Julien, P.Y. (1995). Erosion and sedimentation. Cambridge University Press, p. 280.

10. Shen, C. and Lemmin, U. (1996). Ultrasonic scattering in highly turbulent clear water flow. *Ultrasonics*, **35**, pp. 57–64.

11. Sun, H., Kunze, E. and Williams, A.J. III (1996). Vertical heat flux measurements from a neutrally buoyant float. *J. Phys. Oceanogr.*, **26**, pp. 984–1001.

12. McPhee, M.G. and Stanton, T.P. (1996). Turbulence in the statically unstable oceanic boundary layer under Arctic leads. *J. Geophys. Res.*, **101**, pp. 6409–6428.

13. Gargett, A.E. and Moum, J.N. (1995). Mixing efficiencies in the turbulent tidal fronts: Results from direct and indirect measurements of density flux. *J. Phys. Oceanogr.*, **25**, pp. 2583–2608.

14. Lemckert, C. and Imberger, J. (1995). Turbulent benthic boundary layers in fresh water lakes. *IUTAM Symposium on Physical Limnology*, pp. 115–124.

15. Zhang, J., Lemmin, U. and Hopfinger, E. (1994). The variability of vertical finescale structures in Lake Geneva, Switzerland. *Fourth International Symposium on Stratified Flows*, **3**.

16. Ivey, G.N. and Imberger, J. (1991). On the nature of turbulence in a stratified fluid. Part I: The energetics of mixing. *J. Phys. Oceanogr.*, **21**, pp. 650–658.

17. Gibson, C.H. (1987). Fossil turbulence and intermittency in sampling oceanic mixing processes. *J. Geophys. Res.*, **92**, pp. 5383–5404.

18. Caldwell, D.R., Brubaker, J.M. and Neal, V.T. (1978). Thermal microstructure on a lake slope. *Limnol Oceangr.*, **15**, pp. 372–374.

Interior and Basin-Wide Diapycnal Mixing in Stratified Water: A Comparison of Dissipation and Diffusivity

G.-H. Goudsmit and A. Wüest

Swiss Federal Institute for Environmental Science and Technology (EAWAG) and Swiss Federal Institute of Technology (ETH), Dübendorf, Switzerland

Abstract

Turbulence measurements and basin-scale tracer balances can be employed to yield direct and indirect estimates of diapycnal diffusivity in stratified water bodies. As values inferred from these two fundamentally different approaches differ by an order of magnitude in the ocean thermocline, a series of experiments was carried out in order to quantify and to understand the mechanism of diapycnal mixing in a stratified freshwater basin.

Temperature microstructure measurements in the hypolimnion showed that about 90% of the turbulent kinetic energy was dissipated within the bottom boundary layer, whereas only 10% was lost in the interior of the stratified water body (i.e. far from the boundary). These findings were corroborated by three fluorescent dye tracer release experiments. Injection of the tracer into the center of the hypolimnion revealed that diapycnal diffusivity in the interior was low ($(0.2 \pm 0.04) \cdot 10^{-6} m^2 s^{-1}$), but increased by an order of magnitude (to $(4 \pm 1) \cdot 10^{-6} m^2 s^{-1}$) upon reaching the sediment boundary. The latter value agreed well with the basin-wide diapycnal diffusivities determined from heat flux measurements. As a consequence, diapycnal mixing in the stratified hypolimnion can be considered as a composite of weak turbulence in the interior and strong turbulence within the bottom boundary, leading to a net buoyancy flux compatible with the basin-wide tracer diffusion.

The results of the two completely independent methods coincide perfectly within the margin of error demonstrating that diapycnal fluxes - at least in small to medium-sized basins - are generated predominantly by mixing within the boundary layer above the sediment. This results consistently support equivalent findings from ocean basins.

1 Introduction

Vertical exchange in stratified natural waters is of great interest both because of its ecological implications and because of its relevance to fundamental fluid dynamics. In the past, different methods have been used to determine the diapycnal diffusivity of stratified water bodies. In particular, two completely different methods have been applied in the ocean thermocline. Indirect methods provide estimates based on fits to large-scale property distributions by assuming a balance between upward advection and turbulent downward diffusion (Munk, 1966; Toggweiler et al., 1989). Direct estimates of diapycnal diffusivity can be obtained from turbulence measurements resolving the small-scale fluctuations of temperature and shear (currents) at which viscous and diffusive dissipation complete the action of turbulent mixing (Gregg, 1987). Independent of which microstructure technique is employed (temperature or shear) and of the type of profiler used, the values determined by the direct method have been found to be approximately 10 times smaller than Munk's (1966) value (Gregg and Sanford, 1988; Toole et al., 1994; Davis, 1994).

It is possible that this discrepancy results from incorrect assumptions adopted in the small-scale turbulence models (Osborn and Cox, 1972; Osborn, 1980). Another scenario is that diapycnal diffusivity as measured by micro-structure profilers in the *interior* water of the thermocline is much smaller than the *basin-wide* diffusivity determined by large-scale tracer studies. Although many possibilities of resolving this discrepancy have been considered (Sherman and Davis, 1995), the situation has still not been clarified fully.

In an attempt to explain these ostensibly paradoxical results, a set of measurements was conducted in a medium-sized freshwater basin. These measurements included temperature microstructure profiles, artificial tracer spreading and heat budgets, as well as observations of currents and seiching. The goal of this series of experiments was

1. to quantify diapycnal diffusivities as exactly as possible,

2. to explain the above-mentioned paradox, and

3. to understand the mechanism by which mixing is generated in the stratified water below the surface layer.

Lake Alpnach was chosen for these studies since diapycnal diffusivities in this density-stratified lake are similar to oceanic thermocline values, and since tracer balances are relatively well defined in enclosed basins.

Interestingly enough, the measurements revealed the same discrepancy, in the sense that basin-wide mixing was an order of magnitude larger than mixing in the interior bulk water body. We therefore began to evaluate systematically the difference between *boundary mixing*, which occurs directly above the sediment, and *interior mixing*, which occurs in the inner part of the hypolimnion far from the boundary. The purpose of this paper is to provide a coherent summary of the results of these different experiments with respect to this question, summarising internal seiching (Münnich et al., 1992; Münnich, 1996), bottom boundary dynamics (Gloor et al., 1994), microstructure measurements (Gloor, 1995; Wüest et al., 1996), artificial tracer studies (Goudsmit et al., 1997) and the relationship between boundary turbulence and basin-wide diffusivity (Wüest and Gloor, 1997).

The limnological background of the lake basin under study is presented in Section 2. The results of the measurements of dissipation of turbulent kinetic energy ε and diapycnal diffusivity K follow in Sections 3 and 4, respectively. After comparing mixing in the interior with mixing in the bottom boundary layer, conclusions related to the basin-wide diapycnal diffusivity and the discrepancy between the results obtained using direct and indirect methods are drawn.

2 Characteristics of the fresh-water basin

The experimental work was carried out in Lake Alpnach (Figure 1), the smallest (surface area 4.8 km^2) and almost separated basin of Lake Lucerne, situated in

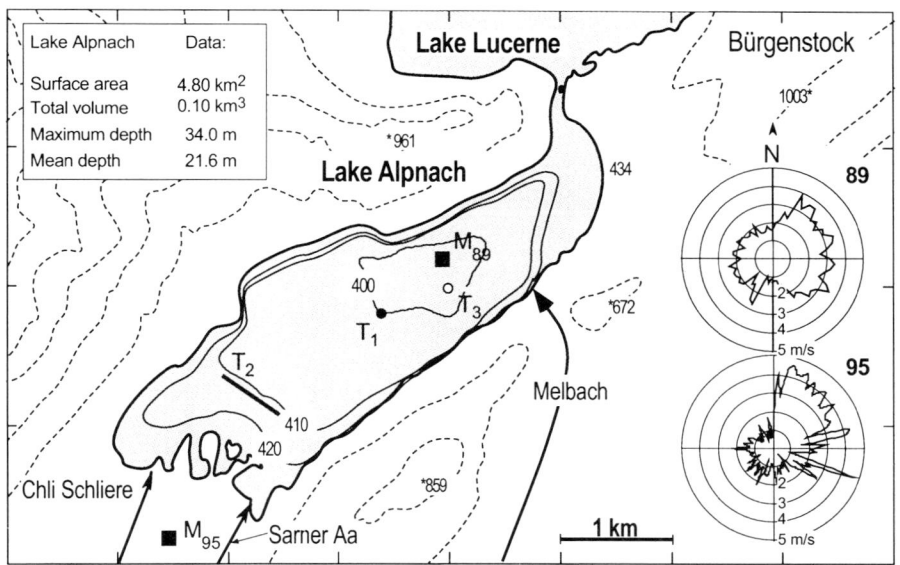

Figure 1. Map of Lake Alpnach (a side basin of Lake Lucerne, Switzerland). The locations of the tracer releases on 12th October 1992 (experiment I), 13th June 1994 (experiment II), and 19th June 1995 (experiment III), labelled T1, T2, and T3, are marked by •, \, and o, respectively. Depth contours, lake surface elevation, and heights of nearby mountains are given in m a.s.l. The two insets give the wind speed distribution measured in 1989 (upper) and 1995 (lower) at the meteorological buoy M_{89} at the deepest point of the lake and M_{95} at the weather station near the airport at Alpnach (both marked by ■). The locations of the temperature microstructure measurements were homogeneously scattered over the lake

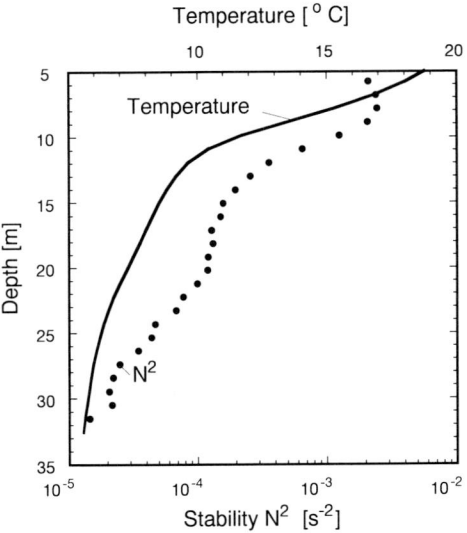

Figure 2. Profiles of temperature, representing the diel average of 28th June 1989, and stability N^2, averaged from 82 CTD profiles recorded during June/July 1989

Central Switzerland. The lake is approximately elliptical and relatively shallow (maximum depth 34 m). In the absence of thunderstorm-induced turbidity currents, the river input (Sarner Aa) merges into the epilimnion during the periods of the experiments in summer. Consequently, wind is the dominant driving force for generation of turbulence in the hypolimnion of Lake Alpnach.

The stratification of the hypolimnion of Lake Alpnach is usually strong during summer and almost unaffected by dissolved solids (salinity about 0.3‰). Figure 2 shows a typical temperature profile and the stability N^2 profile averaged from 82 CTD casts taken during the first study in June/July 1989.

During summer time, atmospheric thermals produced by solar heating along the mountain slopes stimulate a predominant diel wind blowing remarkably regularly (Figure 3) and uniformly parallel to the major axis (inset of Figure 1). As shown by Münnich et al. (1992) the whole hypolimnetic water body is excited by seiches of the first and second vertical modes. The seiche-induced horizontal currents with amplitude typically varying from 3 to 6 cm s^{-1} generate well-mixed bottom layers, which thicken and decay in accordance to the strength of the bottom currents (Gloor et al., 1994; Gloor, 1995). The variation of their thickness is illustrated in Figure 4, where three examples of temperature profiles through the well-mixed bottom layer are shown. The mean thickness of these boundary layers varies depending on depth and season. Unlike the situation in many eu-

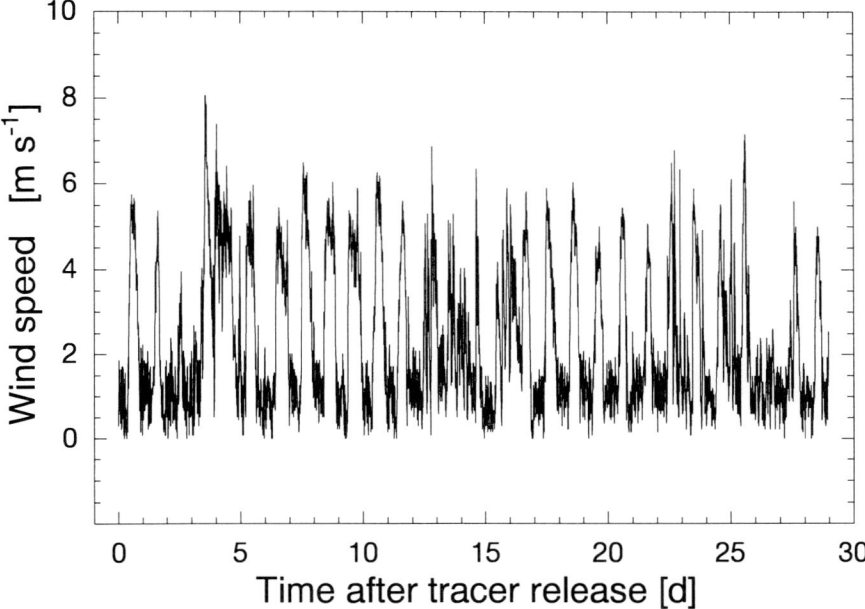

Figure 3. Wind speed from weather station M_{95} at the airport Alpnach (Figure 1) for 19th June - 17th July 1995. The wind shows a remarkably regular diurnal pattern due to the heating/cooling cycle of the adjacent mountains. Adapted from Goudsmit et al. (1997)

trophic lakes in which the occurrence of mixed boundary layers is suppressed by chemically induced stratification, this is not the case for the mesotrophic Lake Alpnach (Wüest and Gloor, 1997). Moreover, due to the regular wind excitation and, therefore, the permanent development of well-mixed boundary layers Lake Alpnach is an ideal basin for those investigations.

3 Dissipation measurements

The first objectives of the study were

1. to verify the agreement of diffusivities determined by tracers with diffusivities estimated by microstructure measurements,

2. to evaluate the discrepancy between them, and

3. to make inferences on the mixing mechanism in an enclosed water body (Wüest et al., 1996).

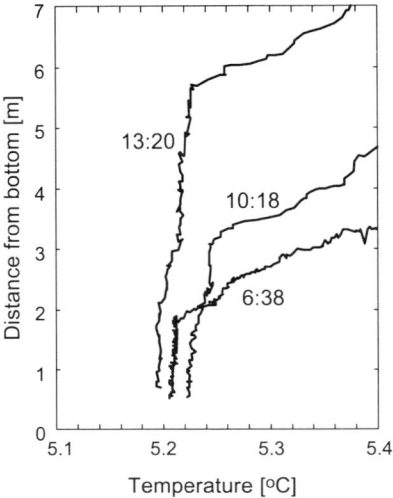

Figure 4. Three examples of temperature profiles through the boundary layer above the sediment at the deepest location of the lake. The profiles are labelled with the times of May 12th, 1992, when the profiles were measured. The varying thickness is due to internal seiching. Adapted from Gloor et al. (1994)

The experiment was carried out in the hypolimnion of Lake Alpnach over the period of one month in June/July 1989. We determined the turbulent kinetic energy dissipation $\varepsilon[W \ kg^{-1}]$ within the stratified hypolimnion by collecting temperature microstructure profiles and estimated tracer-based diapycnal diffusivities $K[m^2 s^{-1}]$ from the vertical spreading of deliberately injected sulfur hexafluoride (SF_6) and from the heat budget (Powell and Jassby, 1974).

In order to determine dissipation, 130 temperature microstructure profiles were measured using a self-contained profiler (Carter and Imberger, 1986), freely rising at about 0.1 m s^{-1} and collecting data at 100 Hz from a pair of fast-response thermistors (FP-07) starting 0.4 m above the sediment (Figure 5). After data filtering (Fozdar et al., 1985), the temperature microstructure profiles were divided into turbulent and non-turbulent segments (Imberger and Ivey, 1991). Dissipation ε was determined by fitting the calculated spectra of the temperature fluctuations to Batchelor's model spectrum (Gibson and Schwartz, 1963; Dillon and Caldwell, 1980). The error introduced due to ambiguous spectra, is far below the statistical uncertainty of the dissipation estimates.

In order to obtain adequate averages, we divided the water column into two regions: *interior* and **boundary layer** (Figure 5). Dissipation was ensemble averaged for the 130 profiles, in "depth bins" beginning at the upper end of the thermocline (6 m depth) and extending downwards to a maximum depth of

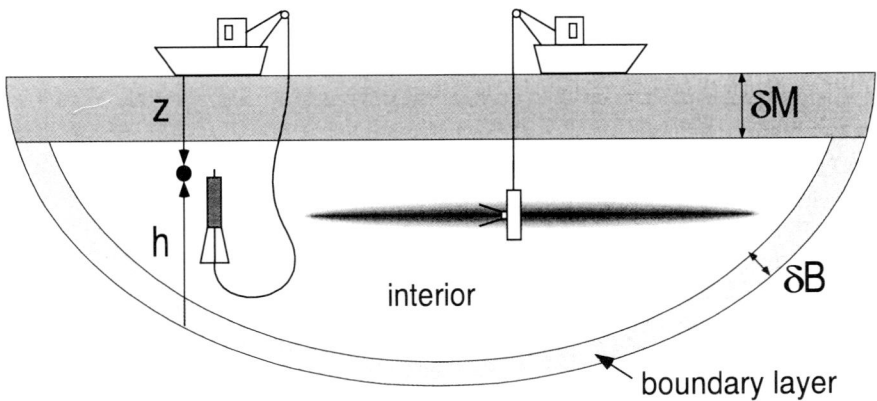

Figure 5. Scheme, illustrating the concept of the experimental set-up of the studies in Lake Alpnach 1989–1995: Temperature microstructure profiles were collected with a self-contained free rising profiler (left). Fluorescence measurements were carried out using an in-situ fluorometer mounted on a CTD probe. The picture illustrates the concept of separation between mixing in the interior and within the bottom boundary layer (with thickness $\delta B = 5$ to 10 m). In the upper layer (thickness $\delta M \approx 6$ m) mixing processes are dominated by direct surface forcing of wind and convection

23 m (distance from sediment > 10 m) in order to obtain dissipation in the interior (Figure 6(a), left scale). Dissipation in the bottom boundary were ensemble averaged for the 130 profiles for "height bins" beginning 0.5 m above the lake bottom and extending upwards (Figure 6(a), right scale). Beside the bin-averaged arithmetic mean the standard deviation of $ln(\varepsilon)$ were calculated in order to estimate the intermittency and the maximum likelihood of the ensemble averages (Baker and Gibson, 1987).

3.1 Interior mixing

Dissipation ε_I in the interior (index "I") of the hypolimnion decreases with depth, rapidly reaching the limit of detection (few times 10^{-11} W kg^{-1}) below about 20 m depth (Figure 6(a)). The decay of $\varepsilon_I(z)$ with depth z follows, within the range of intermittency, the exponential form

$$\varepsilon_I(z) = \varepsilon_{6m} \cdot \exp^{-(z/z_0)} \quad [W\ kg^{-1}] \tag{3.1}$$

A fit of Equation (3.1) to the data yields $\varepsilon_{6m} = 3.6 \cdot 10^{-9} W\ kg^{-1}$ (upper end of thermocline) and $z_0 = 4.0$ m. Above 6 m depth turbulence is subject to

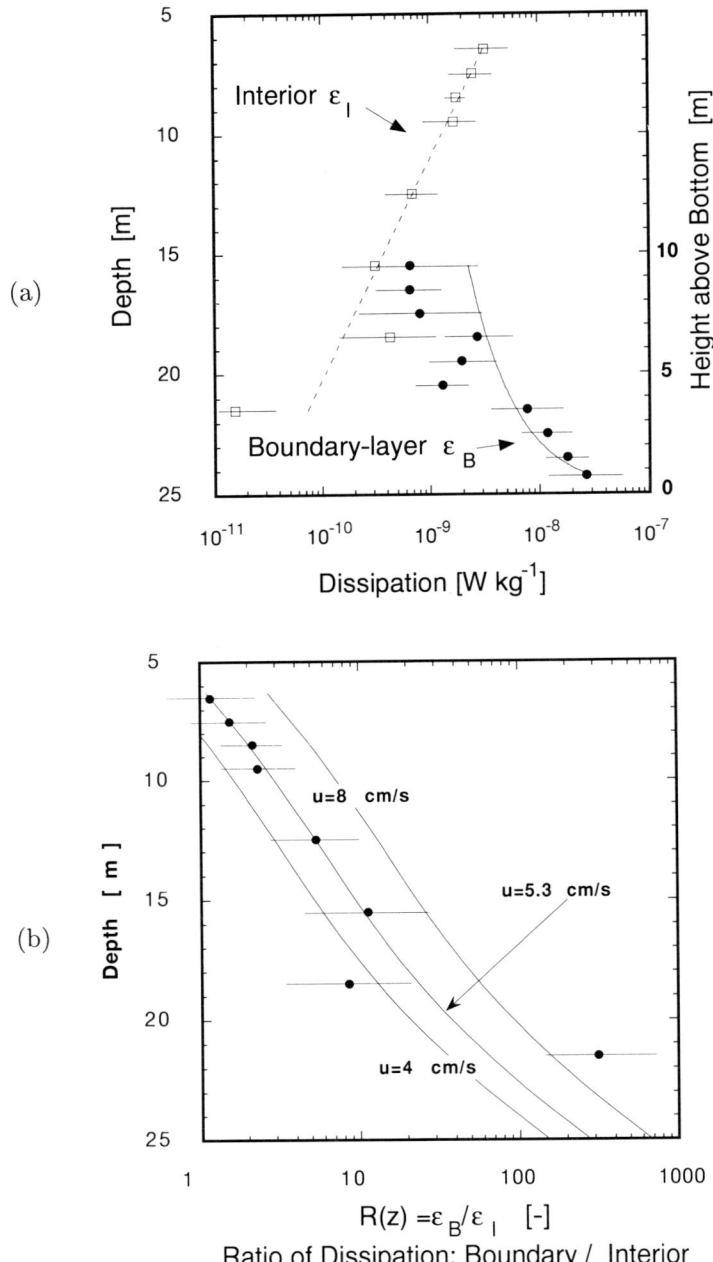

Figure 6. (a) Arithmetic means of dissipation $\bar{\varepsilon}$, averaged in vertical "depth bins" (left scale) and in "bins above the sediment" (right scale). The error bars represent the standard deviations of $ln(\varepsilon)$ due to Baker and Gibson (1987), and (b) Ratio of dissipation within the boundary layer to dissipation in the interior as given by Equation (3.4)

direct surface forcing from heat flux and wind stress and is not considered in this paper. The total area-specific dissipation in the interior of the hypolimnion is given by the vertical integral

$$P_I = \frac{1}{A(6m)} \int_{max.depth-10m}^{6m} \rho \cdot \varepsilon_I(z') \cdot A(z')dz' = 1.5 \cdot 10^{-5} W m^{-2}. \quad (3.2)$$

The rapid vertical decrease of $\varepsilon_I(z)$ is most probably due to lower shear as a function of increasing depth as it is expected for the two dominant seiches corresponding to the first and second vertical modes of the first horizontal seiches (Münnich et al., 1992).

3.2 Boundary mixing

In contrast to the interior, dissipation ε_B within the boundary layer (index "B" increases drastically towards the sediment (Figure 6(a)). As the intermittency is large, we can not decide with enough confidence the functional dependence of $\varepsilon_B(h)$; where h is the height above the local sediment bottom (Figure 5). Within the margin of intermittency, $\varepsilon_B(h)$ follows h^{-1}, in accordance to *law of the wall*, for the lowest few meters of h with a break-off somewhere between 5 and 10 m above sediment. Fitting the *law of the wall* scaling relation $\varepsilon_B(h) = \varepsilon_{1m}/h = u_*^3 \cdot (\kappa h)^{-1}$ to the data in Figure 6(a) yields $\varepsilon_{1m} = 2.1 \cdot 10^{-8} W \ kg^{-1}$ and a friction velocity of $u_* = 0.20$ cm s^{-1}, which corresponds to a typical velocity of $(C_{1m})^{-1/2} \cdot u_* = 5.1$ cm s^{-1} at 1 m above the bottom ($C_{1m} = 1.6 \cdot 10^{-3}$ = drag coefficient (Elliott, 1984); $\kappa = 0.41$ = von Kàrmàn's constant). These values are representative of Lake Alpnach for that time of year (Gloor, 1995). The break-off somewhere between 5 and 10 m above sediment may be related to the thickness of the well-mixed bottom layer which is typically a few m in the deepest part of the hypolimnion (Gloor et al., 1994).

Corresponding to *law of the wall* the integrated area-specific dissipation at the bottom boundary yield

$$P_B = \int_{\delta_\nu}^{\delta_B} \rho \cdot \varepsilon_B(h')dh' = \rho \frac{u_*^3}{\kappa} ln(\delta_B/\delta_\nu) = \rho \cdot \varepsilon_{1m} ln(\delta_B/\delta_\nu) \approx 1.5 \cdot 10^{-4} W \ m^{-2}$$

$$(3.3)$$

where δ_ν and δ_B are the thickness of the viscous and logarithmic layers, respectively. The choice of the temporally varying δ_B is not crucial to the integral (Equation 3.3); taking 5 m instead of 10 m for δ_B would reduce the integrated dissipation within the bottom boundary by only 10%.

3.3 Ratio of boundary to interior mixing

From this bulk consideration we find that the ratio of dissipation at the bottom boundary to dissipation in the interior is $P_B/P_I = 1.5 \cdot 10^{-4} W \ m^{-2}/1.5 \cdot$

$10^{-5}W\ m^{-2} = 10$. It means that in the stratified hypolimnion basin-averaged dissipation at the bottom boundary is 10 times as high as within the interior water body. However, as evident from Figure 6(a), the relative ratio R of the dissipation at the boundary $\left(= \frac{dA}{dV}(z)\frac{P_B}{\rho}\right)$ to dissipation in the interior $\varepsilon_I(z)$

$$R(z) = \left(\frac{dA}{dV}\frac{P_B}{\rho}\right) \bigg/ \varepsilon_I(z) = \left(\frac{dA}{dV}\frac{u_*^3}{k}ln(\delta_B/\delta_v)\right) \bigg/ \varepsilon_I(z) \qquad (3.4)$$

is a function of depth z, as shown in Figure 6(b). R is about 1 in the upper thermocline, and much larger than 1 in the deepest reaches. The relative importance of boundary mixing is steadily increasing with depth due to two factors: $\varepsilon_I(z)$ decays with decreasing interior shear (mode-dependent) whereas dA/dV, the sediment surface per volume of water, increases with depth. To evaluate the relative importance of the two terms in Equation (3.4), $R(z)$ has been calculated for different bottom current velocities and plotted as lines in Figure 6(b). For these calculations $\varepsilon_I(z)$ was parameterized as a function of the gradient Richardson number Ri, i.e. $\varepsilon_I(z) = \varepsilon_I(Ri(z))$, where $Ri = N^2(\partial u/\partial z)^{-2}$, has been taken from Münnich et al. (1992) and P_B was determined using *law of the wall* (Equation 3.3). The comparison in Figure 6(b) shows that the observed ratios R are consistent with horizontal currents u of amplitudes in the range of 5 to 7 cm s^{-1}.

Based on this significant large ratio R we can also expect the diapycnal diffusivity K to increase drastically from the interior bulk water to the bottom boundary layers. The interior diffusivity

$$K_I(z) = \gamma_{\mathrm{mix}} \cdot \varepsilon_I(z) \cdot N^{-2}\quad [m^2 s^{-1}] \qquad (3.5)$$

(Osborn, 1980) can be determined from the measured quantities $\varepsilon_I(z)$ and N^2. As $N^2(z)$ decays exponentially at a slightly lower rate ($z_0 \approx 5m$ than $\varepsilon_I(z)$), $K_I(z)$ is slowly decreasing with depth, averaging at about $K_I = 0.20\cdot10^{-6}m^2s^{-1}$ (Table 1). The basin-wide tracer-based estimates (index "BW") from SF_6 and temperature at 17 m depth (Wüest et al., 1996) yielded, indeed, $K_{BW} = 3.4 \pm 0.4 \cdot 10^{-6}m^2s^{-1}$ (Table 1), which is found to be 17 times larger than K_I. Using Equation (3.5) within the bottom boundary layer (index "B") would lead to an apparent diffusivity $K_B = 20 \pm 10 \cdot 10^{-6}m^2s^{-1}$ (Table 1), which is 100 times lager than K_I.

As we were puzzled, whether this increase is real or the artefact of the microstructure method and their assumptions (for example for γ_{mix}) we repeated the experiment by releasing an artificial tracer in the interior, as reported in the next section.

Table 1. Internal, boundary, and basin-wide diapycnal diffusivities obtained from tracer experiment, temperature profiles, and temperature microstructures. In the two right columns the ratio between basin-wide and internal, and boundary and internal diffusivity, respectively, is given

	Date of Experiment	Depth [m]	Method			Ratio	
			Tracer (Uranin/SF$_6$)	Temperature	Microstructure	Basin-wide / Internal	Boundary / Basin-wide
Internal K$_I$	1992	12	0.17±0.04	-	-	25±7	-
[10^{-6} m^{2}s^{-1}]	1992	23	0.06±0.02	-	-	11±5	-
	1995	25	0.23±0.04	-	-	16±3	-
	1989	17	-	-	0.20±0.08	15±7	-
Boundary K$_B$	1994	22	80±40	-	-	-	10±5
[10^{-6} m^{2}s^{-1}]	1989	17	-	-	20±10	-	7±4
Basin-wide K$_{BW}$	1992	12	4±1	4.6±0.5	-	-	-
[10^{-6} m^{2}s^{-1}]	1992	23	0.4±0.1	0.9±0.4	-	-	-
	1994	22	7±3	8.5±0.5	-	-	-
	1995	25	3.7±0.7	3.5±1.0	-	-	-
	1989	17	3.0±0.4	3.7±0.4	2.5±1.0	-	-

4 Tracer observations

In order to study the contributions from interior and from bottom boundaries to the basin-wide diapycnal transport three tracer release experiments were carried out in 1992, 1994, and 1995 using the fluorescent dye uranin (sodium fluorescein). In the following, we will address them in chronological order as experiment I, II, and III, respectively. Two types of experiments were conducted.

In the first one (experiment I and III), a tracer was injected as a point source in the centre of the hypolimnion close to a predefined isothermal surface (Figures 1 and 5). In the second one (experiment II) the tracer was released close to a pre-defined isothermal surface approximately 2 meters above the sediment-water interface along a track of approximately 700 m length (Figures 1 and 5). For all experiments the horizontal and vertical spreading of the tracer cloud was detected by an *in-situ* fluorometer (Dr. Haardt, Klein-Barkau, Germany) mounted on a conventional CTD probe (Meerestechnik-Elektronik, Trappenkamp, Germany).

For the time following the injection all vertical profiles of an individual day (up to 500) were integrated and interpolated horizontally. As an example of the spreading, the horizontal distribution of the tracer cloud of experiment III after 4 (lower panel) and 28 days (upper panel) of its point-like release is depicted in the insets of Figure 7. By analysing experiment III it was found that the tracer cloud reached part of the bottom boundary at the respective depth after approximately 7 days.

(a)

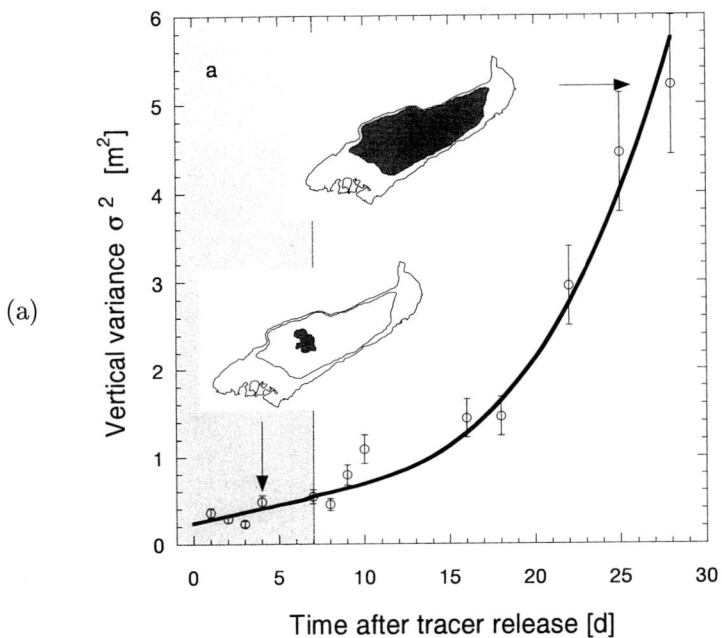

Figure 7.

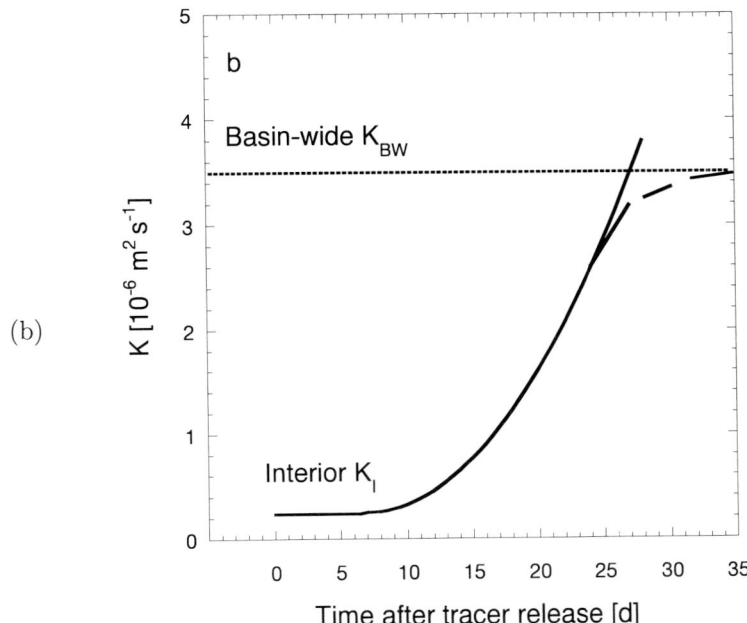

(b)

Figure 7. (a) Vertical variances σ_z^2 of the tracer cloud injected on 19th June 1995 (experiment III) with interpolated line (linear function for day 0 - 7, and third order polynomial function for day 7 - 28, respectively). The grey area represents the period in which the tracer had not yet reached the boundaries of the lake (day 0 - 7). The two insets show the horizontal distribution of the tracer cloud after 4 (lower) and 28 days (upper), respectively, of its point release on 19th June 1995 into the central hypolimnion of Lake Alpnach (Figure 5), and (b) Diapycnal diffusivity K (solid line) calculated from the solid line of Figure 7(a) using Equation (4.2). The horizontal dashed line represents the basin-wide diapycnal diffusivity K_{BW} determined by the heat budget method. During the process of horizontal tracer homogenization K increases from K_I to K_{BW} (dotted line)

In order to obtain genuine, mean vertical tracer profiles free from distortions induced by internal seiching, the tracer concentration profiles of each day were averaged in bins of density instead of bins of depth. Finally, using the mean density profile, the averaged concentrations were retransferred to profiles of depth.

Assuming that the initial distribution of the tracer is a Gaussian profile and that the vertical spread of the tracer obeys Fick's Second Law, the vertical distribution of the tracer is given by:

$$C(z,t) = C(\bar{z},t) \exp^{(-(z-\bar{z})^2/2\sigma_z^2(t))} \quad [g \; m^{-3}]. \tag{4.1}$$

Independent of the initial distribution the vertical variance σ_z^2 is related to the vertical diffusion coefficient K by:

$$K = \frac{1}{2}\frac{\partial \sigma_z^2}{\partial t} \quad [m^2 s^{-1}]. \tag{4.2}$$

In order to determine the vertical variance, all profiles were first normalized (i.e. $\int c_n(z)dz = 1$, were c_n represents the normalized uranin concentration) and then fitted to Gaussian curves.

To reduce the noise induced by temporal differentiation, σ_z^2 was interpolated by a linear or a third order polynomial fit, corresponding to the characteristics of the temporal variation of σ_z^2. For experiment III the vertical variances σ_z^2 are plotted as a function of time in Figure 7(a). Furthermore, vertical eddy diffusion coefficients K were obtained (Figure 7(b)) by applying Equation (4.2). The temporal development of diapycnal diffusivity demonstrates the increasing influence of mixing at the bottom boundary. As long as the tracer is confined to the open water region (0 to 7 days), the diapycnal diffusivity is rather small $(0.23 \cdot 10^{-6} m^2 s^{-1})$ representing the average value of diffusivity in the interior K_I. However, after the tracer patch has reached the bottom of the lake, diapycnal diffusivity increased by more than an order of magnitude (to $3.7 \cdot 10^{-6} m^2 s^{-1}$), i.e. by a factor of 16. The latter value represents the basin-wide diffusivity K_{BW} after the tracer has been homogeneously distributed across the lake.

However, the value of K_{BW} at the end of the experiment is to some extent uncertain, as the polynomial fit would imply the diffusivity to increase, which is for obvious reasons not the case. Therefore, K_{BW} was also calculated by the heat flux method (Powell and Jassby, 1974) over the same time and depth intervals as for the tracer experiment. As can be seen from Figure 7(b), the agreement between the results of both methods is good, although the transition to the basin-wide value can not be determined from the tracer observations (dotted line).

One may argue that the temporal increase of K is due to stronger wind. This, however, can definitely be denied as to be seen from the wind record in Figure 3. There is even a slight decline after 7 days and, consequently, we can reject the possibility of intensified mixing due to stronger wind forcing.

In the second type of tracer experiment, the tracer solution was released near the sediment-water interface. In this case the temporal development of the vertical tracer spreading was completely opposite (Figure 8). During the first seven hours after the tracer release, when the tracer cloud was located close to the bottom boundary, diffusivities K_B (index "B") were large ($K_B = (80 \pm 40) \cdot 10^{-6} m^2 s^{-1}$). After this initial period of intense boundary mixing, the tracer cloud was spread and drifted partially away from the sediment-water interface. Then (7 hours - 2 days) the influence of boundary mixing was reduced, as demonstrated in Figure 8 by the lower K of $(7 \pm 3) \cdot 10^{-6} m^2 s^{-1}$.

All diapycnal diffusivities, measured by using different methods, are compiled in Table 1. The results are classified in interior K_I, boundary K_B, and basin-wide K_{BW} diffusivities, according to the locations of the measurements. From this comparison and from the observations on dissipation in Section 3 we can conclude that basin-wide diffusivity K_{BW} can consistently be interpreted as a super-position of low mixing in the interior and intense mixing at the bottom boundaries.

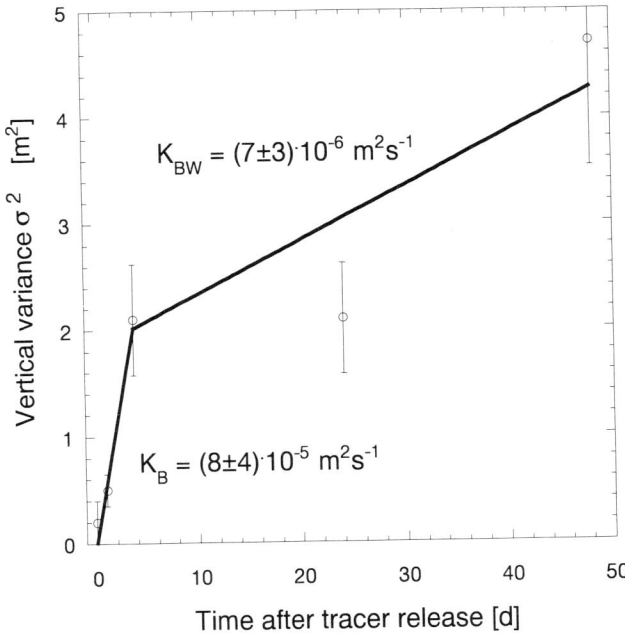

Figure 8. Vertical variances σ_z^2 and diapycnal diffusivities obtained from the tracer release experiment conducted in the bottom boundary layer in June 1994 (experiment II)

5 Discussion and conclusions

The microstructure measurements presented in Section 3 reveal that there are
two distinctly different sources of turbulence: one in the interior (away from the
boundary) and one within the bottom boundary layer (thickness approximately
5 - 10 m). Whereas in the surface layer of the lake, dissipation is dominated by
interior (open water) processes, the main source of turbulent energy in the stratified hypolimnion is located within the bottom boundary. The ratio of boundary
to interior dissipation increases with depth, as both Ri and dA/dV increase. Averaging over the entire hypolimnion during the period of measurements yielded
90% dissipation within the bottom boundary layer and only 10% in the interior.
Using the dissipation method (Equation 3.5) diapycnal diffusivities yield orders
of magnitude differences between interior $K_I = (0.2 \pm 0.08) \cdot 10^{-6} m^2 s^{-1}$, basin-
wide $K_{BW} = (2.5 \pm 1.0) \cdot 10^{-6} m^2 s^{-1}$ and boundary $K_B = (20 \pm 10) \cdot 10^{-6} m^2 s^{-1}$
values, respectively.

However, since dissipation is not an absolute scale for diapycnal mixing, three
tracer release experiments were carried out from which diffusivities were obtained. The tracer revealed a consistently low diapycnal diffusivity of about
$K_I = 0.2 \cdot 10^{-6} m^2 s^{-1}$ until the tracer reached the sediment boundary at the
respective lake depth. There, the tracer indicated a 15 fold increase in diffusivity. In addition, diapycnal diffusivities were found to be large for tracer released
within the bottom boundary ($K_B = (80 \pm 40) \cdot 10^{-6} m^2 s^{-1}$) and decreased by an
order of magnitude (Table 1) after the tracer cloud had partially drifted away
from the lake bottom. Basin-wide diapycnal diffusivities were also measured
applying horizontal homogenous tracer budgets (SF_6 and temperature). All approaches yielded the same result within the margin of uncertainty of the methods
(Table 1).

In Table 1 the results of the various investigations are compiled. The diapycnal diffusivities are classified according to the applied technique (tracer, temperature, and microstructure) and according to the location of the measurements (interior, boundary, and basin-wide). Of course the absolute values of the different
classes of diapycnal diffusivity (interior, boundary, and basin-wide) differ for the
various experiments. This is no surprise, as the measurements were conducted
at different depths (12 to 25 m), for different meteorological conditions (for
example different wind forcing), and for slightly differing density stratification
among the experiments. However, basin-wide diapycnal diffusivity, determined
over week-long periods, was relatively constant at $K_{BW} = (4 \pm 2) \cdot 10^{-6} m^2 s^{-1}$.

Interior diffusivity, $K_I = (0.06 \text{ to } 0.24) \cdot 10^{-6} m^2 s^{-1}$ was found to vary within
a reasonable range. Applying the interior gradient Richardson numbers Ri, based
on the internal seiches (Münnich et al., 1992), to parameterizations of diffusivity as a function of Ri, i.e. $K = K(Ri)$, yields values comparable to the observed K_I (for example Peters et al., 1988). Also the level of dissipation within
the bottom boundary layer is in a reasonable range. Average bottom current
speed, 1 m above sediment, was found to be consistent with *law of the wall* for

$u_{1m} = 5.1 cms^{-1}$ a value to be found representative in Lake Alpnach for that time of year (Gloor, 1995).

The results of these studies revealed that

1. microstructure-based methods are well suited for estimations of diapycnal diffusivities, and

2. that the diapycnal diffusivities calculated from tracer experiments represent basin-wide quantities only after homo-genization.

   The most striking results are the facts that

3. the ratio between basin-wide and interior diapycnal diffusivity was nearly constant (approximately 15) for all experiments, and

4. that the ratio of basin-wide to interior dissipation was about 11.

The high ratio of boundary to basin-wide diffusivity of about a decade consistently supports the model, that diapycnal mixing is dominantly generated at the bottom boundary. These results lead to the general conclusions that

1. for this stratified basin interior mixing is one order of magnitude weaker than basin-wide, and

2. that bottom boundary mixing is the dominant process for diapycnal transport in the hypolimnion of medium-sized lakes.

As the direct estimates, based on microstructure measurements, obviously provide correct values for diapycnal diffusivity, we have to conclude that diapycnal mixing in the interior of the ocean thermocline is indeed much smaller than the basin-wide average. This hypothesis is strongly supported by the tracer experiments from the open ocean (Ledwell et al., 1993). In their work diapycnal diffusion coefficients, calculated from the vertical spread of tracer clouds, were also small compared to typical basin-wide diffusivities. Recently, Ledwell and Bratkovich (1995) have demonstrated that the vertical spread of a tracer increases rapidly (by a factor of approximately 7) when the tracer meets the bottom boundaries of the basin (Ledwell and Hickey, 1995). These results are in excellent agreement with our findings and emphasises the general importance of boundary mixing in stratified natural waters.

Acknowledgements

We are indebted to many members of the Environmental Physics Department at EAWAG for making the collection of field data possible: in particular to F. Peeters, M. Gloor, M. Hofer and J. Schlatter for the tracer measurements; M. Münnich, A. Lück, A. Zwyssig and M. Ulrich for field work and M. Schurter for his excellent technical expertise on Lake Alpnach during the entire sampling program. We thank J. Imberger for providing the microstructure probe, D.C. van Senden for planning and organising the microstructure measurements, G. Piepke for data analysis and computer support and D.M. Imboden for discussions of the concept and the results. This study was supported by Swiss National Science Foundation grants 20-27751.89; 20-32700.91 and 20-36364.92.

Bibliography

1. Baker, M.A. and Gibson, C.M. (1987). Sampling turbulence in the stratified ocean: Statistical consequences of strong intermittency. *J. Phys. Oceanogr.*, **17**, pp. 1817–1836.

2. Carter, G.D. and Imberger, J. (1986). Vertically rising microstructure profiler. *J. Atoms. Oceanic Technol.*, **3**, pp. 462–471.

3. Davis, R.E. (1994). Diapycnal mixing in the ocean: The Osborn-Cox model. *J. Phys. Oceanogr.*, **24**, pp. 2560–2576.

4. Dillon, T.M. and Caldwell, D.R. (1980). The Batchelor spectrum and dissipation in the upper ocean. *J. Geophys. Res.*, **85**, pp. 1910–1916.

5. Fozdar, F.M., Parker, G.J. and Imberger, J. (1985). Matching temperature and conductivity sensor response characteristics. *J. Phys. Oceanogr.*, **15**, pp. 1557–1569.

6. Gibson, C.H. and Schwartz, W.H. (1963). The universal equilibrium spectra of turbulent velocity and scalar fields. *J. Fluid Mech.*, **16**, pp. 365–384.

7. Gloor, M. (1995). Methode der temperaturmikrostruktur und deren anwendung auf die bodengrenzschicht in geschichteten wasserkörpern. *Diss. ETH Nr. 11'336*, p. 159.

8. Gloor, M., Wüest, A. and Münnich, M. (1994). Benthic boundary mixing and resuspension induced by internal seiches. *Hydrobiologia*, **284**, pp. 59–68.

9. Goudsmit, G.-H., Peeters, F., Gloor, E. and Wüest, A. (1997). Boundary versus internal mixing in stratified natural waters. *J. Geophys. Res.*, **102**, pp. 27903–27914.

10. Gregg, M.C. (1987). Diapycnal mixing in the thermocline: A review. *J. Geophys. Res.*, **92**, pp. 5249–5286.

11. Gregg, M.C. and Sanford, T.B. (1988). The dependence of turbulent dissipation on stratification in a diffusively stable thermocline. *J. Geophys. Res.*, **93**, pp. 12381–12392.

12. Imberger, J. and Ivey, G.N. (1991). On the nature of turbulence in a stratified fluid - Part II: Application to lakes. *J. Phys. Oceanogr.*, **21**, pp. 659–679.

13. Ledwell, J.R. and Bratkovich, A. (1995). A tracer study of mixing in the Santa Cruz Basin. *J. Geophys. Res.*, **100**, pp. 20681–20704.

14. Ledwell, J.R. and Hickey, B.M. (1995). Evidence for enhanced mixing in the Santa Monica Basin. *J. Geophys. Res.*, **100**, pp. 20665–20679.

15. Ledwell, J.R., Watson, A.J. and Law, C.S. (1993). Evidence for slow mixing across the pycnocline from an open-ocean tracer-release experiment. *Nature*, **364**, pp. 701–703.

16. Munk, W.H. (1966). Abyssal recipes. *Deep-Sea Res.*, **13**, pp. 707–730.

17. Münnich, M. (1996). The influence of bottom topography on internal seiches in stratified media. *Dyn. Atmos. Oceans.*, **23**, pp. 257–266.

18. Münnich, M., Wüest, A. and Imboden, D.M. (1992). Observations of the second vertical mode of the internal seiche in an alpine lake. *Limnol. Oceanogr.*, **37**, pp. 1705–1719.

19. Osborn, T.R. (1980). Estimates of the local rate of vertical diffusion from dissipation measurements. *J. Phys. Oceanogr.*, **10**, pp. 83–89.

20. Osborn, T.R. and Cox, C.S. (1972). Oceanic fine structure. *Geophys. Fluid Dyn.*, **3**, pp. 321–345.

21. Peters, H., Gregg, M.C. and Toole, J.M. (1988). On the parameterization of equatorial turbulence. *J. Geophys. Res.*, **93**, pp. 1199–1218.

22. Powell, T. and Jassby, A. (1974). The estimation of vertical eddy diffusivities below the thermocline in lakes. *Water Resour. Res.*, **10**, pp. 191–198.

23. Sherman, J.T. and Davis, R.E. (1995). Observations of temperature microstructure in NATRE. *J. Phys. Oceanogr.*, **25**, pp. 1913–1929.

24. Toggweiler, J.R., Dixon, K. and Bryan, K. (1989). Simulations of radiocarbon in a coarse-resolution world ocean model: 1. Steady state prebomb distributions - 2. Distribution of bomb-produced carbon 14. *J. Geophys. Res.*, **15**, pp. 8207–8264.

25. Toole, J.M., Polzin, K.L. and Schmitt, R.W. (1994). Estimates of diapycnal mixing in the abyssal ocean. *Science*, **264**, pp. 1120–1123.

26. Wüest, A. and Gloor, M. (1997). Bottom Boundary Mixing: The Role of Near-Sediment Density Stratification. Editor: J. Imberger, "Physical Limnology", Coastal and Estuarine Studies, AGU-Series. In press.

27. Wüest, A., Van Senden, D.C., Imberger, J., Piepke, G. and Gloor, M. (1996). Comparison of diapycnal diffusivities measured by tracer and microstructure techniques. *Dyn. Atmosph. Oceans.*, **24**, pp. 27–39.

The Development of Layers in a Stratified Fluid

Joanne M. Holford and P.F. Linden[1]

Department of Applied Mathematics and Theoretical Physics, University of Cambridge

Abstract

The character of perturbations in a stratified fluid varies with the flow stability, as indicated by a Richardson number Ri. At low Ri, turbulent motions lead to diffusion-like mixing, whereas at high Ri, a combination of horizontal, vortical motions and internal waves lead to intermittent mixing dominated by wave-breaking events. Data from some laboratory experiments show that at high Ri, the mixing efficiency decreases with increasing Ri, allowing the possibility that small perturbations away from a smooth density profile may be enhanced to give interfaces. This implies that strongly stratified fluids have a naturally layered character, and may explain the prevalence of finestructure observed in density profiles of the oceans.

In this paper, results are presented from experiments on mixing at high Ri and low Reynolds number, $Re < 180$. A rake of vertical bars is towed back and forth through a linearly stratified fluid. The bar wakes are dominated by vertically aligned Karman vortices, which interact with the stratification and undergo distortions with a well-defined vertical scale. Well-mixed layers develop at half this vertical scale, separated by interfaces which sharpen with time, according to the tendency of high Ri flow.

1 Introduction

The diapycnal transport of substances through a perturbed stratified fluid is an important component of the dynamics of the oceans and atmosphere. The input of heat, momentum and pollutants is largely confined to the earth and ocean surface, from where fluid motions mix these substances into the body of the fluids. Both large scale, mean overturning motions and smaller scale, turbulent transports contribute to mixing, which always increases the potential energy (PE) of the fluid. The amount of mixing in a confined system can be measured by the mixing efficiency, defined here as the rate of change of PE as a fraction of the rate of input of kinetic energy (KE) of the perturbing flow. For a horizontally isotropic turbulent shear flow with uniform shear and density

---

[1]Present address: Department of Applied Mechanics and Engineering Sciences, University of California, San Diego, 9500 Gilman Drive, La Jolla, CA 92093-0411, USA.

gradient, the mixing efficiency is equivalent to the flux Richardson number, which is the ratio of the buoyancy flux to the work done by the Reynolds stress. As small-scale motions cannot be resolved in global models, it is necessary to be able to parametrise the mixing efficiency in terms of mean flow characteristics.

The fluid flow causing the mixing can be affected by the density distribution. The strength of the stratification, relative to the strength of the stirring, is quantified by the Richardson number Ri. In the limit of very small Ri, the effect of the stratification is weak, and density is transported as a passive tracer. In this regime the mixing is often modelled by a constant turbulent eddy diffusivity. For large Ri the flow is very stable, and a significant fraction of the input energy typically generates internal waves, which do not cause irreversible mixing unless they break. It is therefore possible that the mixing efficiency rises to a maximum at some critical Richardson number Ri_c, and then falls again as Ri increases further. This behaviour was observed in experiments in which a biplanar grid was dropped through the interface between two homogeneous layers [1]. For $Ri > Ri_c$, the Phillips/Posmentier mechanism [2, 3] may allow the development of a series of well-mixed layers separated by sharp density interfaces. Layers developed in experiments in which a linearly stratified fluid was stirred by oscillating arrays of vertical bars [4]. The scale of layers formed through repeated stirring with one vertical bar has also been investigated, and some measurements of the mixing efficiency made [5]. It has been suggested [6] that the change in mixing efficiency as Ri increases is related to the change in the character of stratified flow, as the scale of the energy-containing eddies becomes larger than the Ozmidov scale, which is the scale above which buoyancy forces affect the flow.

Theoretical models of mixing usually assume a homogeneous turbulent field, or a turbulent field that is unaffected by the stratification. In the laboratory, mixing is often produced by the action of oscillating or translating biplanar grids or rakes of bars. The flows generated by these methods are not homogeneous, and involve considerable mean motion and large scale structure. The experiments presented here investigate the evolution of a linear density stratification perturbed by repeated stirring with a rake of vertical bars. The Reynolds number Re, based on bar diameter and towing velocity, is sufficiently low that the wakes of the bars are not turbulent, and vertically aligned vortices of a Karman vortex street are present. Previous experiments [5] concentrated on higher Re regimes in which the perturbing flow is turbulent. In the present experiments, mixing is generated predominantly by the interaction of large-scale vortical structures and stratification.

In Section 2 the experimental apparatus and different forms of visualisation and measurement are described. The characteristics of cylinder wakes in an unstratified fluid are reviewed in Section 3. Results from the present experiments, on the structure of the wake behind a single bar in a stratified fluid, are presented in Section 4. When the fluid is repeatedly stirred at high Ri with a rake of bars of high solidity, the density profile is altered and sharp interfaces form, as described in Section 5. Conclusions are given in Section 6.

2   Experimental apparatus

A tank, of width $W = 25.6$ cm and maximum depth $H = 50$ cm, was filled with a linearly stratified salt solution of density $\rho(z)$, where z is the vertical coordinate (positive upwards). Internal walls allowed the length L of the tank to be varied up to a maximum of $L = 70$ cm. The buoyancy frequency $N = \sqrt{-(g/\rho_0)d\rho/dz}$, where ρ_0 is the mean fluid density and g the acceleration due to gravity, was in the range $0.3\,\mathrm{s}^{-1} < N < 2.2\,\mathrm{s}^{-1}$. A rake of eight vertical bars, with square cross-section and of side length $d = 0.65$ cm, was mounted across the tank. The centrelines of the bars were $M = 3.25$ cm apart, giving a grid solidity $S = d/M = 0.2$. The rake of bars was repeatedly towed back and forth along the tank at a speed U, where $1.0\,\mathrm{cms}^{-1} < U < 3.0\,\mathrm{cms}^{-1}$. A brief acceleration and deceleration occurred at the ends of each tow, and the bars reached to within 3 cm of the ends of the tank.

Experiments to generate layers were made up of a number of stirring sequences, where each sequence comprised a fixed number of tows of the bar, followed by a brief period in which the transient motions were allowed to decay and the changes in $\rho(z)$ observed. In addition, a number of experiments with single tows of one bar mounted near the centre of the tank were carried out, in order to study the wake structure in stratified fluids. Two larger bars, of side length $d = 1.3$ cm and $d = 2.6$ cm, were also used in these experiments.

The Richardson number Ri and Reynolds number Re are defined using the lengthscale of the bars and the towing speed, that is $Ri = N^2d^2/U^2$ and $Re = Ud/\nu$. The kinematic viscosity ν increases slightly with salt concentration, and here the value appropriate for the mean fluid density is used. The results presented here are from experiments in the range $30 < Re < 140$ and $0.3 < Ri < 1.0$.

Several forms of visualisation and quantitative measurement were taken during these experiments, and are described below. The initial linear density profile was measured using a conductivity probe.

2.1   Shadowgraph visualisation

A shadowgraph is a non-intrusive visualisation which uses the refractive index variation of salt solution with salinity to visualise the curvature of the density profile. Parallel light rays are shone horizontally through the tank. Regions where the density profile curvature $d^2\rho/dz^2 > 0$ appear dark, whereas regions where $d^2\rho/dz^2 < 0$ appear bright. The results of this technique are integrals along light rays through the tank at each level. In these experiments the light source used was a slide projector placed 4.5 m from the tank. A sensitive video camera and image enhancement techniques were used to detect small changes in curvature.

2.2   Powder visualisation of vortex cores

The flows in the parameter ranges studied here are dominated by the vertical Karman vortices shed behind the bars. These structures were visualised by marking the vortex cores with a fine grey powder, given off from a solder wire attached to one of the bars during electrolysis with an alternating current. Apart from the slight disturbance caused by the extra wire on the bar, this is also a non-intrusive technique. When the wire was attached to one side of the bar, only vortices of one sign were marked by powder, whereas when the wire was placed on the back face of the bar, vortices of both signs contained powder.

2.3   Velocity in a horizontal plane

The method of stirring excites little vertical motion, and the presence of stratification damps the vertical velocity component, so that fluid motions are largely horizontal. The horizontal components of velocity in one plane were measured by seeding the flow with small, white particles of approximately uniform density and lighting the tank at the level of neutral buoyancy with a thin (0.5 cm) horizontal light sheet. The motion of individual particles was then followed using a video camera mounted above the tank. The particles were tracked between video frames approximately $d/4U$ s apart, and the velocity field calculated by finite differences. Particles can be located to within 0.2 of a pixel, equivalent to 0.01 cm, giving particle velocities accurate to within 3%. The particle tracking algorithm was developed by Dalziel [7]. The Lagrangian data from particle motions can be interpolated onto a regular grid to provide Eulerian data at each timestep. The grid resolution used here was $0.5d$, giving approximately 8 gridpoints across a vortex, so that the horizontal structure of the vortices could be resolved, as well as the overall dynamics.

3   The wake of a circular cylinder in an unstratified fluid

Many studies have been carried out on the wakes of bluff bodies in an unstratified fluid. Attention has concentrated on the wakes of circular cylinders, which exhibit rich behaviour, as the flow pattern changes from laminar to turbulent with increasing Re. There has been relatively little work on the wake of a square cylinder.

   The following description of the flow past a circular cylinder is taken from Tritton [8], and Re is based on the cylinder diameter D. For $Re \lesssim 40$, the flow is steady, but above this value, the wake is unstable. For $40 \lesssim Re \lesssim 100$ attached eddies occur immediately behind the cylinder and the instability develops further downstream, whereas for $Re \gtrsim 100$ eddies are shed at finite amplitude from the cylinder. The wake structure becomes unstable at $Re \approx 200$ to three-dimensional instabilities, but the vortices are not turbulent on formation until $Re \approx 400$. The character of wakes then remains unchanged until $Re \approx 3 \times 10^5$, when the boundary layers on the cylinder undergo a transition to turbulence.

The wake of a finite length cylinder will exhibit some three-dimensional structure even below $Re \approx 200$. Vortices are shed at an oblique angle to the cylinder in order to match the conditions at the ends of the cylinder, giving a spanwise chevron pattern [9]. At $Re \approx 180$, the wake becomes unstable to the formation of loops of vorticity with a vertical wavelength of approximately $3D$, and at $Re \gtrsim 230$ streamwise vortices linking adjacent Karman vortices are formed, with a vertical wavelength of about D [10]. As Re increases still further, vortex dislocations between spanwise regions of different shedding frequency grow rapidly in the spanwise direction [11].

If the flow is constrained to be two-dimensional, as in a soap film [12], a subharmonic instability of the vortex street leading to merging of like-signed vortices is seen for $Re \gtrsim 150$, while for $Re \gtrsim 400$, some vortices move away from the street as vortex dipoles. In general, the vortex street is unstable to finite amplitude three-dimensional disturbances, and precise initial conditions are necessary to see the flow transitions described above.

The sharp corners of a square cylinder ensure that the flow always separates from the forward edges, although for small Re the flow reattaches and separates again from the trailing edges [13]. Two-dimensional numerical simulations [14] give a critical Re for wake instability of $50 < Re_c < 60$, and above this Re the shedding frequency increases in the same overall manner as for a circular cylinder [13]. Although we know of no studies of the three-dimensional flow transitions around a square cylinder, behaviour similar to that of flow around a circular cylinder is expected.

Vortex streets generated in an unstratified fluid behind the square cylinder in our tank were short-lived, undergoing rapid vortex breakdown. Weak residual circulations, vibrations of the bar and towing mechanism and the effects of the initial acceleration of the cylinder probably all contribute to the vortex breakdown.

4   The wake of a cylinder in a stratified fluid

Stratification damps vertical motions and so may be expected to constrain vortex lines to remain vertical. This is indeed the case in these experiments, with vortex structures in the stratified wake of a cylinder remaining coherent for many multiples of the vortex turn-around time. As in an unstratified fluid, the wake structure depends on Re. For $Re \lesssim 40$, Karman vortices are not formed, and a thin vertical vortex sheet is shed from behind the bar. In the range $40 \lesssim Re \lesssim 180$, Karman vortices are formed, although they slowly develop perturbations away from the vertical. At $Re \gtrsim 180$, vertical motions are generated in the wakes. The wakes of larger square cross-section bars in this tank (not discussed here in detail), show the presence of streamwise vortices linking the Karman vortices, with a vertical separation comparable to the bar diameter. A further transition occurs at $Re \approx 700$, when the wakes become more fully turbulent. The transition Re are observed to increase with increasing Ri, and the values given here are appropriate for $0.25 \lesssim Ri \lesssim 1.0$. In stratified circular Taylor-Couette flow, the transition Re also increases as Ri increases [15]. The layering experiments described here are all in the regime in which the cylinder wakes are non-turbulent.

A side view of the wake behind one bar moving from right to left at $Re = 75$ and $Ri = 0.46$ is shown in Figure 1. The cores of the vortices of one sign only are well marked by the solder powder technique, and appear evenly spaced and of a regular size. The horizontal motions associated with weak vertical vortices do not alter the density profile, and therefore give no shadowgraph image. Vertical distortions of the vortex cores are visible in the along-stream direction, in phase in the along-stream direction. The vortices are most sharply bent where they meet the bottom of the tank, but there is no evidence of connections between the vortices or between a vortex and the bar.

The horizontal velocity and vertical vorticity fields for a similar experiment are shown in Figure 2. There is some variation near the start of the vortex street, on the right. Elsewhere, the wavelength of the Karman vortex street is 4.1 cm, and the vortices drift in the direction of the bar at $0.12U$, giving a Strouhal number (non-dimensional shedding frequency based on the bar side length and towing speed) of $St = 0.14$. In general, the vortex pattern varies with Re, and changes with time. For $Re \gtrsim 60$, the vortices are initially elliptical and lie almost in a straight line behind the bar. Subsequently, the vortices develop a more triangular shape, seen in Figure 2, as the staggered double row of the stable vortex street configuration develops. At lower Re, the vortex street is staggered

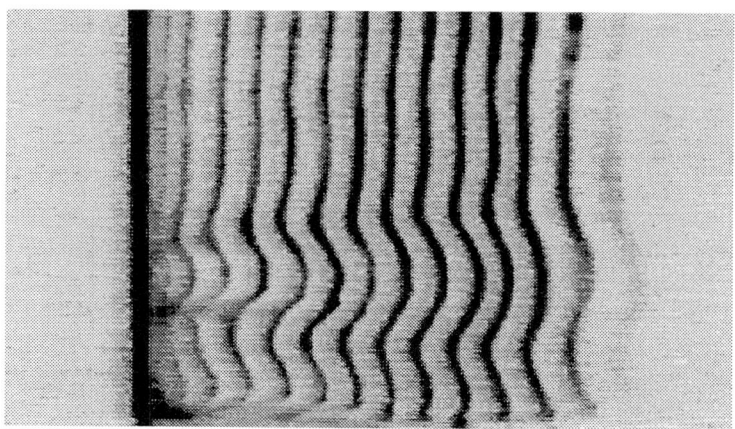

Figure 1. Vertical Karman vortices of one sign, marked by solder powder, shed behind a single bar moving to the left at $Re = 75$, $Ri = 0.46$. Half of the tank depth is shown, and the bottom of the tank is aligned with the bottom of the image

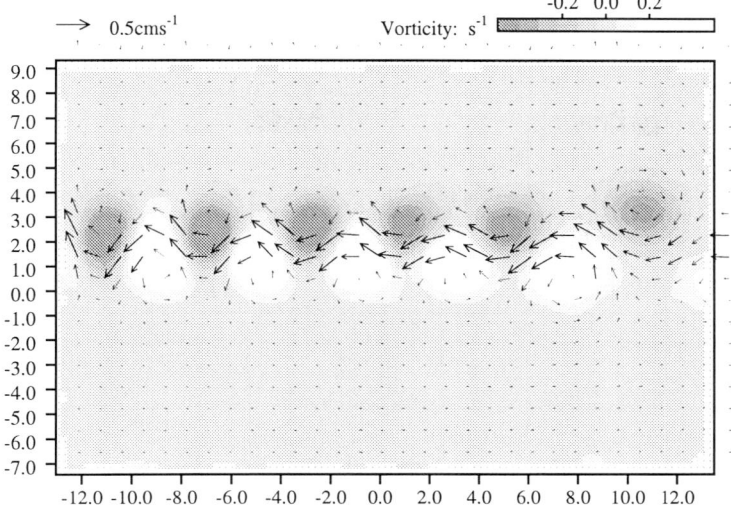

Figure 2. Vorticity (greyscale) and velocity (arrows) in a horizontal plane about a fifth of the way down the tank, in the wake of one bar, which has moved from right to left at $Re = 84$, $Ri = 0.60$. The axes give the scale in cm

at formation. At higher Re and lower Ri, the regular pattern of the street is lost as like-signed vortices merge. From the narrow range of parameter space investigated with the current geometry, the exact boundary in Re, Ri space is unclear, but the data so far are consistent with a critical $Ri = 0.3$ for vortex merging.

The typical ellipticity of vortices was a ratio of major to minor axes of 1.2. In order to follow the main features of the vortex evolutions through time, each vortex was approximated by an axisymmetric vortex. The radial profile of one vortex from Figure 2, $t = 15\,\mathrm{s}$ behind the bar, is shown in Figure 3(a). If the flow is two-dimensional, the evolution equation for the vertical vorticity ω is

$$\frac{D\omega}{Dt} = \nu\nabla^2\omega. \tag{4.1}$$

A point (line) vortex $\omega(r,0) = \Gamma_0\delta(r)$ evolves to the Lamb-Oseen vortex, with vorticity profile

$$\omega(r,t) = \frac{\Gamma_0}{4\pi\nu t}e^{-r^2/4\nu t}, \tag{4.2}$$

where Γ_0 is the total circulation. Therefore, the evolution of an initial Gaussian vorticity profile is

$$\omega(r,t) = \omega_0(t)e^{-r^2/a_0^2(t)}, \tag{4.3}$$

where the square of the radius increases as $a_0^2(t) = a_0^2(0) + 4\nu t$ and the peak vorticity decreases as $\omega_0(t) = \omega_0(0)a_0^2(0)/a_0^2(t)$.

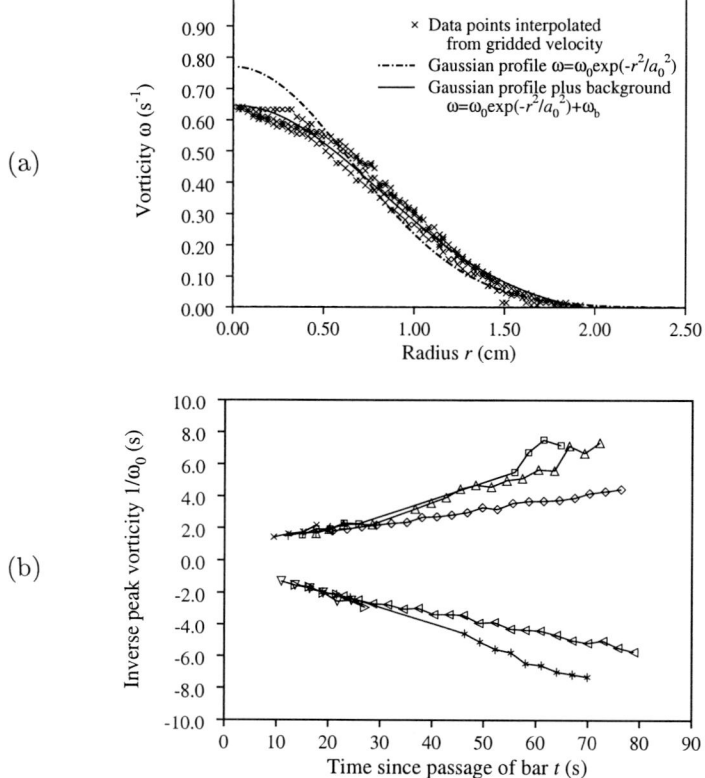

Figure 3. Characteristics of the vortices depicted in Figure 2. (a) Radial vorticity profile through the axisymmetric approximation to one vortex, and (b) Evolution of the inverse peak vorticity of all vortices with time since the passage of the bar. Each vortex is represented by a differemt symbol

A Gaussian vorticity profile is fitted to the data in Figure 3(a). The fit over-estimates the peak vorticity, as this functional form cannot account for the reduction to zero vorticity at the edge of the vortex. This reduction is due to the proximity of vortices of the opposite sign. The observed profile is more successfully fitted with the sum of a Gaussian profile and a constant, where the constant takes a value of opposite sign to the peak vorticity, as shown in Figure 3(a). The circulation of any isolated vortex with vorticity of only one sign remains constant, by Kelvin's circulation theorem. However, if vorticity of both signs is initially present in a flow, then the circulation of any vortex can decrease in magnitude due to the annihilation of positive and negative vorticity. The circulation of the vortices in these experiments decays in this way.

A vortex set in a background vorticity field will diffuse away, leaving the background vorticity field. A better model for the vortices in these experiments is the superposition of two Gaussian vorticity profiles of opposite sign, one being weaker, but having a larger radius, so that

$$\omega(r,t) = \omega_0(t)e^{-r^2/a_0^2(t)} + \omega_1(t)e^{-r^2/a_1^2(t)}. \qquad (4.4)$$

If the total circulation is κ, the peak vorticity $\omega(0,t)$ then evolves according to

$$\frac{1}{\omega(0,t)} = \frac{1}{\omega(0,0)}\left\{1 + \frac{4\nu t}{a_0^2(0)}\right\}\left\{1 + \frac{4\nu t}{a_1^2(0)}\right\}\left\{1 + \frac{4\nu t \kappa}{\pi\omega(0,0)a_0^2(0)a_1^2(0)}\right\}^{-1}. \qquad (4.5)$$

The inverse peak vorticity in an isolated Gaussian vortex increases linearly with time. However, the inverse peak vorticity of the composite vortex in (4.4) may increase faster than linearly with time, for example if $\kappa/\omega(0,0) \le 0$. The time evolution of the inverse peak vorticity of the vortices in Figure 2 is shown in Figure 3(b), and displays such an increase. The data in Figure 3(b) do not collapse because the two vortices at the start of the vortex street decay more slowly than the other vortices. In practice, the level of interaction with neighbouring vortices, which is parametrised by ω_1 and a_1^2, will vary with time. A value of ω_1 of the same sign as ω_0 would represent a vortex interacting with neighbours of the same sign.

This model of vortex evolution does not suggest that the vortices observed are shielded (a shielded vortex has a core of vorticity of one sign surrounded by a ring of opposite-signed vorticity), but suggests that the interaction of the vortices with the surroundings can be modelled in this way. In this model of vortex evolution the advection of vorticity $\mathbf{u}.\nabla\omega$ is identically zero, and viscosity controls the evolution. This is in contrast to previous studies of viscous vortex decay, such as the cancellation of strained vortices [16].

One possible mechanism for the generation of three-dimensional perturbations in the vortices is an instability involving a resonance of inertial waves in the vortex core and internal waves [17]. A purely axisymmetric vortex is stable to this instability mechanism, however the vortices in this flow do have a small ellipticity. Alternatively, the perturbations may be related to the vortex dislocations that develop in unstratified wakes.

5   Stirring with a rake of bars

When stirring the fluid with the rake of vertical bars of solidity 0.2, the wake motions fill the tank, so that the energy density of the perturbations is higher and vortices from neighbouring wakes may interact. Additionally, each pass with the rake may deform vortex structures previously created. This vortex stretching enhances the perturbation vorticity and velocity fields.

Figure 4 shows a shadowgraph picture of stirring with the rake of bars, in which vortices from one side of one of the central bars are marked by solder powder. The stratification is slightly higher than for Figure 1, and the fluid has been stirred for about 90 minutes, or 60 stirs. The solder lines have once

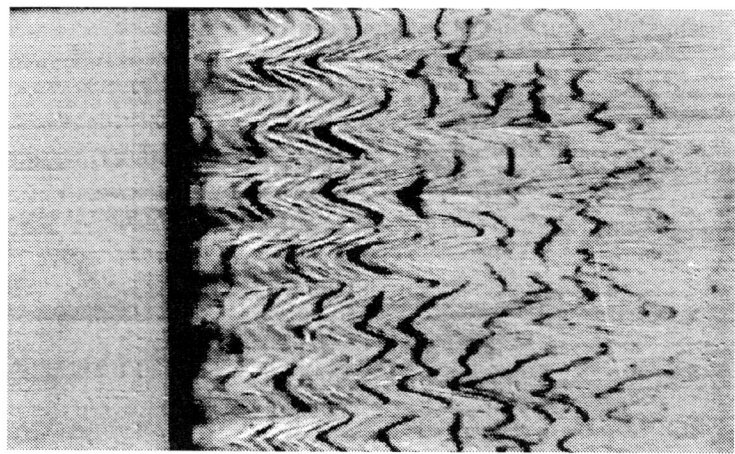

Figure 4. The wakes behind eight bars moving to the left, at $Re = 78$, $Ri = 0.65$. The fine herringbone pattern is a shadowgraph visualisation, and the heavier dark lines are solder powder traces from one side of one of the central bars. The top two-thirds of the tank is shown

again developed vertical perturbations, but there is now a well-defined vertical wavelength. The shadowgraph visualisation shows similar patterns to the lines of solder powder. When the vortices are tilted, overturning motions around the core can mix fluid vertically, as shown in Figure 5(a). If the pattern of vortex perturbation is uniform across the tank, this can result in a significant alteration of the density field, which is visible on the shadowgraph. Experiments in which more than one line of vortices is marked with solder powder show that the pattern of vortex distortions is repeated across most of the width of the tank, with different behaviour at the walls.

The cross-stream coherence in these experiments can be explained by studying the horizontal velocity field. Figure 6 shows the distribution of vortices behind the rake for an experiment with similar parameters to Figure 4. Near the bars, eight vortex streets may be seen, while further downstream, like-signed vortices have merged across the tank leaving only four or five bands of each sign of vorticity. At lower Ri, vortices pair preferentially with the next like-signed vortex in the same vortex street, rather than across the stream, and the regular pattern of vortices breaks down.

As the stirring continues, a pattern of interfaces and well-mixed layers develops in the density profile, throughout the whole depth of the fluid, as shown in the compilation of shadowgraph images in Figure 7. The time evolution of the density profile, and its relation to the shadowgraph image, is sketched in Figure 5(b). Interfaces form initially at the upstream limit of the vortex distortions, as the tilted vortices mix fluid at either side. Sharp kinks in the density

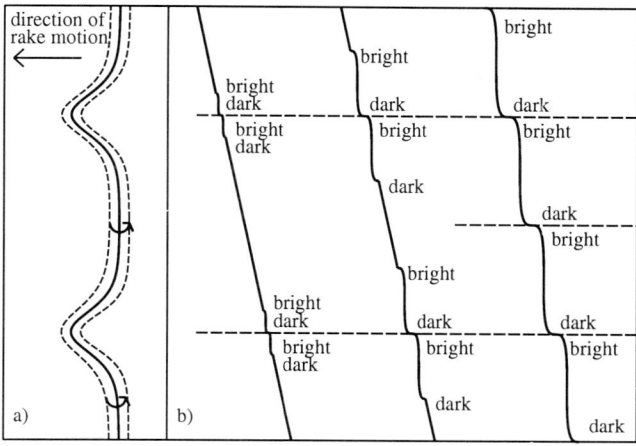

Figure 5. A sketch of the mechanism by which tilted vortex lines alter the density profile. (a) A distorted vortex, and (b) three stages in the evolution of the density profile with time, indicating the shading on an equivalent shadowgraph image

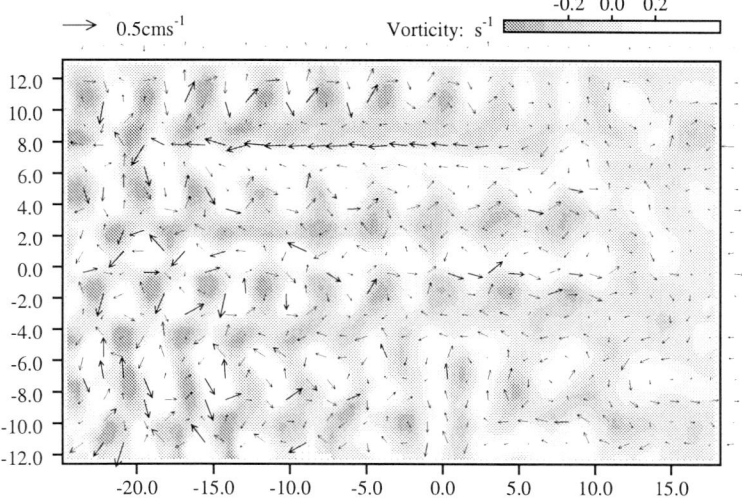

Figure 6. Vorticity (greyscale) and velocity (arrows) behind eight bars, which have moved from right to left at $Re = 71$, $Ri = 0.84$. The axes give the scale in cm

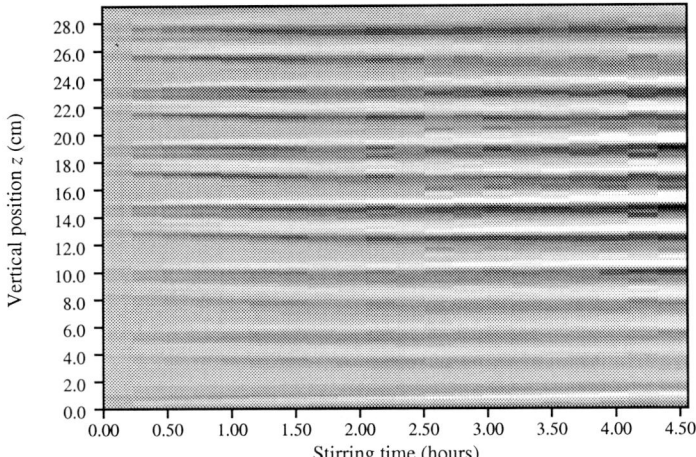

Figure 7. The development of layers with time at $Re = 93$, $Ri = 0.75$ by repeated stirring with the rake of bars. At the end of each sequence, the intensities from a shadowgraph image are averaged along the tank, giving the variation of the curvature of the density profile with z

profile occur at the limits of the mixing, while a linear stratification is retained between the mixing regions. The linearly stratified regions decrease in thickness as fluid is mixed into the less stratified regions above and below. Finally, the mixed regions meet forming other interfaces, midway between the initial interfaces. Once this stage is reached, all interfaces develop to an equal strength. The intensification of density perturbations is typical of high Ri flow. The vertical lengthscale of the layering is now half the wavelength of the original vortex distortions. Throughout the experiments, the layers remain weakly stratified, since the patterns of vortex distortions are always visible in shadowgraph images.

At this stage, the pattern of vortex distortions may suddenly shift vertically by half a wavelength. Often this change in vertical phase is driven by the advancement of the well-mixed boundary layers at the top and bottom boundaries. In a diffusive or turbulently mixed fluid, the boundary layer thickness increases smoothly like $t^{1/2}$. In these experiments, the boundary layers grow by discrete amounts, as the density jump across the interface nearest the boundary falls to zero and the outermost layer is absorbed into the boundary layer. The experiments are too short to determine whether, on average, the boundary layer thickness is still proportional to $t^{1/2}$.

The non-dimensional lengthscale l/d of layers formed by this mechanism is given in Figure 8, as a function of the Froude number $F = 1/\sqrt{Ri}$. The data is well approximated by the linear relation $l = 3.1U/N$. This scaling for l has been justified by a general energetics argument [4], since moving a fluid parcel up a distance l in a stratified fluid requires the conversion of KE $\propto U^2$ into PE $\propto l^2N^2$,

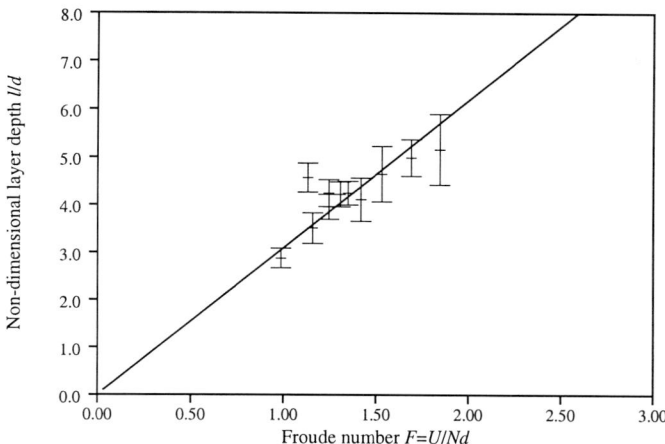

Figure 8. Variation of the non-dimensional scale l/d of layers formed by the mechanism of vortex distortion with Froude number, and a linear least squares fit to the data. Error bars are based on the requirement that there must be an integer number of layers

implying that $l \propto U/N$. A non-dimensional layer scale that varied linearly with Froude number was seen in stratified circular Taylor-Couette flow [15], and a layer scale $l = 2.6U/N + 1.0\,\text{cm}$ was measured in more turbulent towing bar experiments [5]. The form of layering discussed in this paper, associated with vortex distortions, only occurs for $Ri > 0.3$ and, presumably, sufficiently low Re for Karman vortices to dominate the wakes.

6 Conclusions

Results have been presented from experiments investigating the development of a linear stratification repeatedly stirred with a rake of vertical bars. For the stratification used here and $40 \lesssim Re \lesssim 180$, the bar wakes are dominated by vertically aligned Karman vortices. These vortices have an approximately Gaussian vorticity profile, and are typically slightly elliptical or triangular. The circulation of each vortex decays due to interactions with neighbouring vortices of the opposite sign, and the decay may be modelled by the evolution of the central portion of a shielded vortex.

These vortices are distorted in the along-stream direction and, at high grid solidities, the distortions develop a clear vertical wavelength. For high Ri, like-signed vortices merge between neighbouring vortex streets, giving a cross-stream coherence to the vortex distortions. Where the vortex lines are tilted away from the vertical, circulations around the vortex cores mix the fluid. Interfaces develop throughout the whole depth, with a vertical lengthscale of half the wavelength of the vortex distortions. With this grid geometry, the layer scale is found to satisfy $l = 3.1U/N$, for $Ri > 0.3$. Once formed, the interfaces strengthen, as the increased stratification further inhibits mixing across the interface. A tendency to layering is typical of flows at high Ri, but here the layer scale appears to be set by a dynamic interaction between the vertical vortices and the stratification.

The flow in a high Ri fluid may be described by the combination of an internal wave field and a vortical field of horizontal motions [18]. The breaking of internal waves is known to be an important mixing process at high Ri. The mixing mechanism identified in these experiments, that is the tilting of vertical vortices, may prove to have equal generality.

Acknowledgements

The authors wish to thank David Page-Croft, Brian Dean and David Lipman for the construction and maintenance of the laboratory apparatus. This work is supported by the Natural Environment Research Council, grant reference number GR3/8891.

Bibliography

1. Linden, P. F. (1980). Mixing across a density interface produced by grid turbulence. *J. Fluid Mech.*, **100**, 691–703.

2. Phillips, O. M. (1972). Turbulence in a strongly stratified fluid - is it unstable? *Deep-Sea Res.*, **19**, 79–81.

3. Posmentier, E. S. (1977). The generation of salinity finestructure by vertical diffusion. *J. Phys. Oceanogr.*, **7**, 298–300.

4. Ruddick, B. R. , McDougall, T. J. , and Turner, J. S. (1989). The formation of layers in a uniformly stirred density gradient. *Deep-Sea Res.*, **36**, 597–609.

5. Park, Y.-G. , Whitehead, J. A. , and Gnanadesikan, A. (1995). Turbulent mixing in stratified fluids: Layer formation and energetics. *J. Fluid Mech.*, **279**, 279–311.

6. Ivey, G. N. and Imberger, J. (1991). On the nature of turbulence in a stratified fluid. Part I: The energetics of mixing. *J. Phys. Oceanogr.*, **21**, 650–658.

7. Dalziel, S. B. (1992). Decay of rotating turbulence: Some particle tracking experiments. *App. Sci. Res.*, **49**, 217–244.

8. Tritton, D. J. (1988). *Physical Fluid Dynamics*. Oxford Science Publications, 519 pp.

9. Williamson, C. H. K. (1989). Oblique and parallel modes of vortex shedding in the wake of a circular cylinder at low Reynolds numbers. *J. Fluid Mech.*, **206**, 579–627.

10. Williamson, C. H. K. (1988). The existence of two stages in the transition to three-dimensionality of a cylinder wake. *Phys. Fluids*, **31**, 3165–3168.

11. Williamson, C. H. K. (1992). The natural and forced formation of spot-like 'vortex dislocations' in the transition of a wake. *J. Fluid Mech.*, **243**, 393–441.

12. Couder, Y. and Basdevant, C. (1986). Experimental and numerical study of vortex couples in two-dimensional flows. *J. Fluid Mech.*, **173**, 225–251.

13. Okajima, A. (1982). Strouhal numbers for rectangular cylinders. *J. Fluid Mech.*, **123**, 379–398.

14. Kelkar, K. M. and Patankar, S. V. (1992). Numerical prediction of vortex shedding behind a square cylinder. *Int. Journ. Numer. Methods Fluids*, **14**, 327–341.

15. Boubnov, B. M. , Gledzer, E. B. , and Hopfinger, E. J. (1995). Stratified circular Couette flow: Instability and flow regimes. *J. Fluid Mech.*, **292**, 333–358.

16. Buntine, J. D. and Pullin, D. I. (1989). Merger and cancellation of strained vortices. *J. Fluid Mech.*, **205**, 263–295.

17. Miyazaki, T. and Fukumoto, Y. (1992). Three-dimensional instability of strained vortices in a stably stratified fluid. *Phys. Fluids A*, **4**, 2515–2522.

18. Lelong, M.-P. and Riley, J. J. (1991). Internal wave-vortical mode interactions in strongly-stratified flows. *J. Fluid Mech.*, **232**, 1–19.

.

The Erosion of a Salinity Step by a Localized Heat Source

D.M. Leppinen and S.B. Dalziel

Department of Applied Mathematics and Theoretical Physics, University of Cambridge

Abstract

In this paper we compare the erosion of a salinity step by a localized heat source with the erosion of the same salinity step by a distributed heat source. Upper and lower bounds are obtained for the time required for the merger of the two layers and it is shown that a localized heat source is more efficient at eroding a salinity step than a distributed heat source providing the same total heat input. The underlying physical mechanisms controlling the erosion of a salinity step by a localized heat source are identified through the use of laboratory experiments and numerical simulations and these mechanisms are compared with those involved in the erosion of a salinity step by a distributed heat source.

1 Introduction

If two components contribute to a fluid's density then it is possible for convection to result even if the fluid is stably stratified. This convection, known as double diffusive convection, only occurs if the two components have differing diffusivities and one of the components contributes to the density field in a statically unstable sense. Double diffusive convection was examined first in the oceanographic context with heat and salt acting as the two stratifying components, and is now recognized to be an important feature of convection in the earth's mantle and is also relevant to the dynamics of salt gradient solar ponds and many solidification processes. Reviews of the geophysical and industrial applications of double diffusive convection are available in [1–3].

In this paper we examine the double diffusive convection which results when a two–layer salinity stratification is heated from below. Previous studies have considered the case of a distributed heat source [4–7]. In this paper we consider the case of a localized heat source. We begin in Section 2 by detailing the configuration of our two–layer system. In Section 3 we present a time scale analysis to predict the time required to erode the two–layer salinity stratification. This time scale analysis is supported using data obtained from laboratory experiments and numerical simulations. In Section 4, we provide a qualitative description of the processes involved in the erosion of a salinity step by a localized heat source.

This description is the outcome of a series of numerical simulations described in [8] and a series of laboratory experiments described in [9, 10].

2   Configuration

The basic configuration of the two–layer system we consider in this paper is depicted in Figure 1.

The system is assumed to be two–dimensional with all gradients in the direction normal to the page being zero. The total height of the system is h, and the total width is w. The system is composed of a lower layer of salty water which is overlayed with a layer of fresh water. Initially, at time $t = 0$, the lower layer has thickness $h_1 = \lambda h$, salinity $S_1 = S_o + \Delta S$, with $\Delta S > 0$, and temperature $T_1 = T_o$. The upper layer has thickness $h_2 = (1 - \lambda)h$, salinity $S_2 = S_o$ and temperature $T_2 = T_o$. In our system, gravity (g) acts downwards as indicated. We make the Boussinesq approximation and use α and β to represent the thermal expansion and salinity contraction coefficients respectively. The reference density for our system is ρ_o, the initial density of the upper layer. The fluid is initially motionless and has kinematic viscosity ν and a specific heat capacity c_p. The diffusivity of temperature is κ_T and the diffusivity of salt is κ_S. All

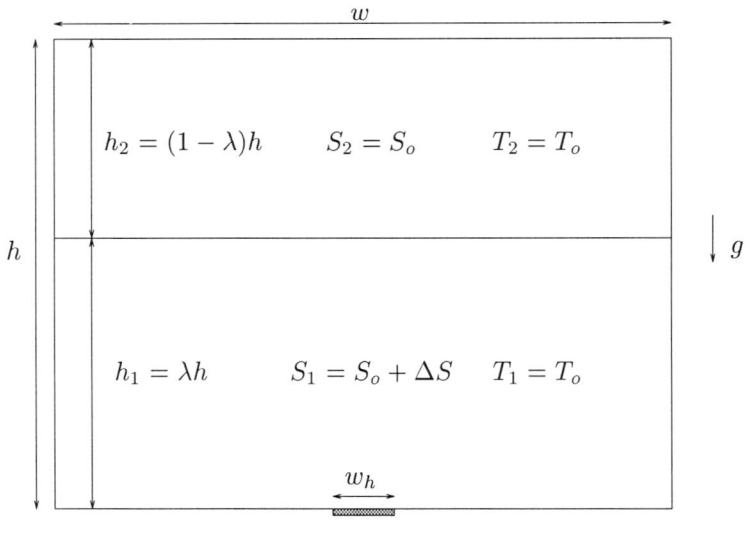

Figure 1. Layout of two–layer system

boundaries are assumed to be perfectly insulating to salt and heat except for the heat source at the bottom of the lower layer which is only insulating to salt. From time $t = 0$, a constant heat flux is supplied at the bottom of the layer through a heat source of width w_h. The total heat input is q per unit length into the page. If $w_h \ll w$ we say that the heat source is localized. If $w_h = w$ we say that the heat source is distributed. Our goal in this paper is to compare the the erosion of a salinity step by a localized heat source with the erosion of the same salinity step by a distributed heat source.

3   Time scale analysis

An absolute upper bound on the time required for the two layers in our system to merge into a single layer is given by the limiting case of no heating. The layers will equilibriate in density due to molecular diffusion over a time scale $t \sim t_{\kappa_S}$ where

$$t_{\kappa_S} = h^2 / \kappa_S. \tag{3.1}$$

Moreover, we expect the merging process to occur over this time scale for small finite values of heating if the resulting increases in temperature contribute negligibly to the density field.

3.1   Time bounds for the case of large heating rates

We now consider the case of large heating rates and we assume that throughout the merger process: (1) the interface separating the two layers remains horizontal and of negligible thickness, (2) the two layers remain well mixed at all times, (3) there is no entrainment of fluid across the interface, so that the interface remains at the same height, and (4) the values of κ_T, κ_S, α, β and c_p remain constant. Subject to these constraints, the two layers will merge when they have equilibriated in density.

Assumptions 1 and 2 are in reasonable agreement with experimental results [7, 9] and correspond to using layer averaged quantities. The validity of assumption 3 is dependent on the experimental parameters. For the experiments discussed in Section 3.2, this assumption is approximately true except during the latest stages of the layer merger process. A discussion on the rate of entrainment across double diffusive interfaces is given in [11]. In the case of a salt–heat double diffusive system, assumption 4 cannot be justified: some properties can vary by a factor of 3 during an experiment. Nevertheless, we proceed by using the average value of the physical properties.

Suppose that the interface separating the lower salty layer from the upper fresh layer is perfectly insulating to both heat and salt. Upon the start of heating the lower layer temperature rises continually while the temperature in the upper layer and the salinity in both layers remain constant. When the temperature in the lower layer has increased by an amount $\Delta T = \beta \Delta S/\alpha$, the two layers will be of equal density and merge. This occurs after time

$$t_q = \lambda h w \rho_o c_p \beta \Delta S / \alpha q. \tag{3.2}$$

We interpret t_q as the bulk heating time scale because it is related to the convective forcing of the heat source in the lower layer.

By considering the ratio, R, of the bulk heating and diffusive time scales we may determine the situations under which the merger process is controlled by convective heating:

$$R = t_q / t_{\kappa_S} = \lambda A \tau / q^*, \tag{3.3}$$

where $A = w/h$ is the aspect ratio of the system, $\tau = \kappa_S / \kappa_T$ is the ratio of the diffusivity of salinity to heat, and $q^* = \alpha q / \rho_o c_p \kappa_T \beta \Delta S$ is the dimensionless heating rate. When $R \ll 1$ we conclude that the heating of the lower layer is more important to the merger process than the direct diffusion of salt across the interface. For the heat–salt system we are studying $\tau \approx 1/70$ and it is clear that convective heating will dominate the merger process for most practical situations.

In this $R \ll 1$ regime, t_q provides a lower limit for the time required for layer merger. We justify this statement by noting that in real systems, double diffusive convection across a diffusive interface results in a net upwards transfer of buoyancy [12]. Hence if the interface between the two layers is non-insulating, t_q will underestimate the time required before the densities in the two layers equilibrate. It is also worth noting that t_q is independent of the depth of the upper layer and that the assumption of an insulating interface can be relaxed provided the time required for the growth of an unstable thermal boundary layer at interface is larger than t_q.

We now seek a revised upper bound for the time required for merger in the limit when $R \ll 1$. We obtain this upper bound by explicitly accounting for the fact that double diffusive convection across a diffusive interface results in a net upwards flux of buoyancy. This upwards flux is given by $\alpha F_T(1 - R_F)$ where $R_F = \beta F_S / \alpha F_T$ and αF_T and βF_S are the double diffusive fluxes of buoyancy across the interface due to heat and salt respectively. Since it has been shown by Stern [13] that the minimum possible value of R_F is $\tau^{1/2}$, we obtain our upper bound for the time required for layer merger by considering the limiting case of $F_S = \alpha F_T \tau^{1/2} / \beta$.

We proceed by writing the equations of conservation of energy and salt for our two–layer system (see [7] for example),

$$\lambda h \frac{dT_1}{dt} = Q - F_T \tag{3.4}$$

$$(1 - \lambda) h \frac{dT_2}{dt} = F_T \tag{3.5}$$

$$\lambda h \frac{dS_1}{dt} = -F_S \tag{3.6}$$

$$(1 - \lambda) h \frac{dS_2}{dt} = F_S, \tag{3.7}$$

where $Q = q/\rho_o c_p w$. In practice F_T varies with time but instead of explicitly accounting for this variation we write $F_T = \epsilon(t)Q$. The only restrictions we place on ϵ is that it remains positive and varies in such a way that $T_1(t) \geq T_2(t)$, which are both physical constraints of the system.

The temperature difference between the upper and lower layers may be calculated from Equations (3.4) and (3.5) as

$$T_1 - T_2 = \frac{Q}{h} \int_0^t \left(\frac{1 - \epsilon(t')}{\lambda} - \frac{\epsilon(t')}{1 - \lambda} \right) dt' \tag{3.8}$$

$$= \frac{Q}{h} \left(\frac{(1 - \bar{\epsilon})t}{\lambda} - \frac{\bar{\epsilon}t}{1 - \lambda} \right), \tag{3.9}$$

where $\bar{\epsilon} = 1/t \int_0^t \epsilon(t')dt'$ is the average value of ϵ from the initiation of heating until time t. Since we require $T_1(t) \geq T_2(t)$, it follows that $\bar{\epsilon} \leq 1 - \lambda$. Similarly we obtain

$$S_1 - S_2 = \Delta S - \frac{\alpha \bar{\epsilon} Q \tau^{1/2}}{h\beta} \left(\frac{1}{\lambda} + \frac{1}{1 - \lambda} \right) t, \tag{3.10}$$

from Equations (3.6) and (3.7) upon substituting $F_S = \alpha F_T \tau^{1/2}/\beta$. Recalling that the two layers merge when they equilibriate in density, we obtain our estimate for the upper bound on the time until layer merger by equating Equation (3.9) times α with Equation (3.10) times β to obtain

$$t = \frac{\lambda h \beta \Delta S}{\alpha Q \left(\tau^{1/2} \bar{\epsilon}/(1 - \lambda) + (1 - \lambda - \bar{\epsilon})/(1 - \lambda) \right)}. \tag{3.11}$$

Finally, we note that the maximum value of (3.11), subject to the constraints $0 \leq \bar{\epsilon} \leq 1 - \lambda$, is obtained when $\bar{\epsilon} = 1 - \lambda$. Thus an upper bound on the time required for layer merger in the limit $R \ll 1$ is

$$t_{up} = \frac{\lambda h \beta \Delta S}{\alpha Q \tau^{1/2}}. \tag{3.12}$$

Equation (3.12) can be rearranged to give $t_{up} = t_q/\tau^{1/2}$. Thus, in the limit of $R \ll 1$, the time required for layer merger to occur is bounded by:

$$1 \leq \frac{t}{t_q} \leq \frac{1}{\tau^{1/2}}. \tag{3.13}$$

This relationship shows the importance of the bulk heating time scale t_q in two–layer double diffusive systems. It must be stressed, however, that Equation (3.13) is obtained under the assumption of no entrainment across the interface separating the upper and lower layers. Since entrainment provides a mechanism for the downwards transport of buoyancy, $t = t_q$ should only be thought of as an approximate lower bound. In cases of strong entrainment, $t = t_q$ may well overestimate the time required for layer merger.

3.2 Experimental and numerical results

In the previous section we obtained bounds for the time required for layer merger in the convectively dominated limit. In this section we present results of laboratory experiments and numerical simulations which were performed to confirm this time scale analysis and also to investigate whether the time required for layer merger depends on the exact nature of the convective heating of the lower layer. In particular, we wish to examine the hypothesis that a localized heat source is more effective than a distributed heat source at precipitating layer merger.

3.2.1 Experimental setup

The laboratory experiments were performed in a 40 cm by 40 cm by 50 cm (deep) tank constructed from 1 cm thick perspex. Experiments were only performed with a localized heat source which was constructed by enclosing a length of high resistance nichrome wire in a 3 mm diameter stainless steel tube. The tube was electrically insulated from the wire to prevent electrolysis. The heat source was 35 cm long and was positioned in the centre of the tank, raised by 0.5 cm from the bottom. The two–layer salinity stratification was obtained by filling the tank to the required depth with a salt solution of the desired concentration, and then gently overlaying a layer of fresh water through a floating sponge. A syphon was then used to selectively withdraw fluid in order to sharpen the interface. In all cases the thickness of the interface was less than 1 cm as measured by vertically traversing a conductivity probe. To minimize heat losses, the tank was insulated with 10 cm of polystyrene insulation on all sides and on the top. The bottom of the tank was insulated with 5 cm of polystyrene. The tank was left for an hour after filling to allow residual motions to die down before initiating heating.

In order to monitor the merger process, three thermistors were used to measure temperatures. The first was located at the initial mid-depth of the lower layer, the second was located at the initial mid-depth of the upper layer, and the third was located 1.5 cm from the top of the upper layer. All three thermistors were located 10 cm to one side of the heat source, and were monitored every 30 seconds by a BBC microcomputer. By examining the temperature traces it was possible to determine when the interface rose above the initial mid-height of the upper layer. Shortly after this, the interface reached the thermistor located 1.5 cm from the top of the tank. At this this time, it was concluded that the layers had merged.

3.2.2 Numerical simulations

A control volume based numerical scheme employing a stream function–vorticity formulation has been used to examine layer merger for both the case of a distributed and a localized heat source. Details of the numerical implementation are given in [8] and [9], and are not repeated here. The criterion used to determine the time required for layer merger reflects the fact that layer merger is associated with the conversion of an initial step salinity distribution into a homogeneous salt distribution. Physically, this redistribution of salt represents an increase in the potential energy of our system due to salt. Hence, we introduce the quantity

$$E = \frac{1}{\Delta S h^2} \int_0^h (\bar{S} - S_o) z \, dz, \tag{3.14}$$

where $\bar{S}$ is the horizontally averaged salinity. The quantity E is proportional to the potential energy due to salt in our system. Initially, $E = E_o = \lambda^2/2$ and after complete mixing, $E = E_\infty = 1/4$. We consider the layers to be completely merged when

$$E = E_o + c(E_\infty - E_o) = (1 - c)E_o + cE_\infty. \tag{3.15}$$

The value of $c = 0.9$ in Equation (3.15) was chosen so that the merging requirement was not limited by fine scale mixing yet was stringent enough that no layered structure persisted. Provided c is sufficiently close to unity, the results presented below are not sensitive to the precise value of c.

3.2.3   Comparison with time scale analysis

The time required for layer merger as a function of heating rate for the numerical and laboratory experiments is plotted in Figure 2. Time has been non-dimensionalized with the bulk heating time scale, $t^* = t/t_q$, and the heating rate is nondimensionalized as $q^* = \alpha q / \rho_o c_p \kappa_T \beta \Delta S$. The numerical results for a localized heat source are plotted with a dashed line and the numerical results for the distributed source are plotted with a solid line. For the numerical results the following parameters were used: $Ra_S = g\beta\Delta S h^3/\nu\kappa_T = 7\times 10^6$, $Pr = \nu/\kappa_T = 7$, $A = 2$, $\tau = 1/70$ and $\lambda = 1/2$ where Ra_S is the salinity Rayleigh number and Pr is the Prandtl number. In the case of a localized heat source $w_h/w = 1/32$ and for the distributed heat source $w_h/w = 1$. All numerical simulations were performed using a uniform 256×256 grid. An axis of reflectional symmetry was imposed about the centre of the heat source.

The experimental results are plotted using symbols and the details of these experiments are summarized in Table 1 where the column labelled T_f refers to the mean temperature at the time of layer merger. The experimental runs are grouped according to depth of the lower layer, and total depth of the tank with different symbols used to identify each group. In all cases, κ_T, κ_S, α, β and c_p were evaluated at the mean salinity and temperature during the experiments, using the correlations presented in [14, 15].

From Figure 2 we conclude that there is good qualitative agreement with the trends observed in the data and the time bounds given in Equation (3.13). We note that t^* decreases monotonically as q^* increases and approaches unity at large heating rates in line with the suggestion that at large heating rates the interface may be considered insulating. For small heating rates the data is consistent with the assertion that t^* is bounded from above by $1/\tau^{1/2} \approx 8.4$, and in reality the upper bound may be even lower. We also conclude that t_q is the correct scaling for the time associated with layer merger. In particular, the

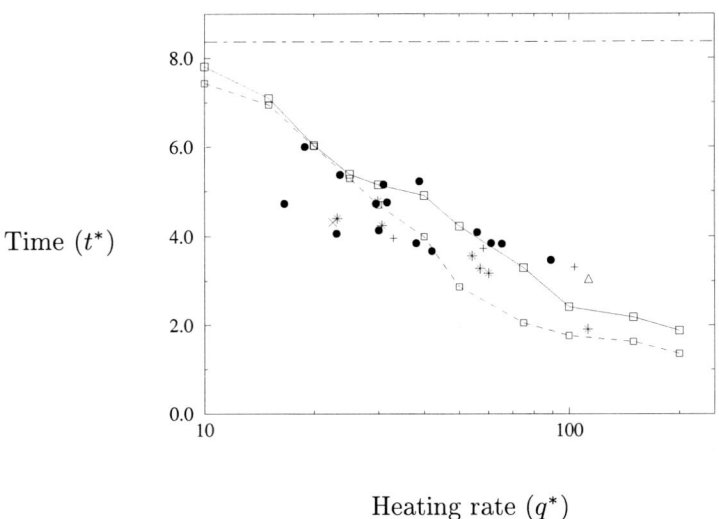

Heating rate (q^*)

Figure 2. Time required for layer merger versus heating rate: numerical simulation with a localized heat source (dashed line), numerical simulation with a distributed heat source (solid line), experimental data (symbols – see Table 1). The dot-dashed line is $t^*/t_q = 1/\tau^{1/2} \approx 8.4$

time required for layer merger is only dependent on the thickness of the lower layer and is independent of the thickness of the upper layer. While there is a great deal of scatter in the experimental data, the scatter is not correlated with differences in layer thicknesses.

For the parameter values considered in this study, the numerical simulations indicate that a localized heat source is up to 30% more efficient than a distributed heat source when $q^* \gtrsim 50$, while the erosion times differ by less than 10% when $q^* \lesssim 25$. As noted in [9], the enhanced efficiency of a localized heat source at high heating rates is related to the ability of a plume to penetrate deeply into the upper layer and effectively mix the system and is not a double diffusive phenomena. An experimental comparison of the efficiency of a localized versus a distributed heat source in terms of the erosion of a salinity step is not possible due to the lack of experimental data for the case of a distributed heat source. Such data was unattainable using the current experimental set up and data from previous studies [4–7] was unsuitable since it has not been possible to unambiguously determine t^* and q^*. These previous studies were primarily concerned with measuring the fluxes of heat and salt across diffusive interfaces. Fernando [7] does provide a history of layer merger experiments, however he does not provide the necessary data to determine t^* and q^*.

Table 1. Summary of two–layer experiments

q w/m	h_1 cm	h_2 cm	ΔS g/ℓ	T_1 °C	T_2 °C	T_f °C	t min	
204.0	10.0	10.0	6.20	21.1	20.8	43.5	833	•
210.0	10.0	10.0	3.62	20.9	21.0	34.1	405	•
215.0	10.0	10.0	2.25	20.2	20.3	29.2	272	•
100.8	10.0	10.0	3.91	20.0	20.0	38.0	1350	•
100.8	10.0	10.0	2.92	23.8	22.9	37.1	1075	•
99.0	10.0	10.0	1.77	20.8	20.5	29.9	625	•
100.8	10.0	10.0	0.96	22.6	22.1	26.5	297	•
52.5	10.0	10.0	1.88	25.1	23.9	34.0	1503	•
52.5	10.0	10.0	1.07	24.2	23.1	27.9	650	•
52.5	10.0	10.0	3.92	22.3	21.7	40.0	3256	•
125.0	12.2	11.8	2.58	19.1	19.1	32.4	878	•
125.0	11.4	11.7	2.59	20.0	20.1	33.6	891	•
92.8	11.1	11.9	0.53	18.4	18.4	20.9	203	•
103.3	12.3	11.7	1.96	19.9	19.8	29.2	829	•
103.3	10.5	12.5	1.07	20.5	20.4	25.0	362	•
133.0	12.0	11.0	1.90	17.8	17.8	25.8	571	•
199.8	10.0	20.0	5.94	22.5	22.1	39.0	900	*
199.8	9.0	21.0	1.96	22.8	22.1	27.1	224	*
199.8	10.0	20.0	1.97	20.9	20.9	26.7	272	*
199.8	10.0	20.0	3.89	19.8	19.8	32.7	653	*
199.8	10.0	20.0	0.96	21.7	21.2	23.4	80	*
106.3	10.0	20.0	0.99	19.6	19.7	22.5	310	*
106.3	10.0	19.5	1.97	19.8	19.8	27.7	755	*
199.8	20.0	10.0	1.02	19.7	19.3	25.5	302	+
199.8	20.0	10.0	1.99	19.4	19.7	31.3	605	+
106.3	20.0	10.0	1.99	21.5	21.3	32.6	1133	+
125.5	6.0	15.0	3.00	17.2	17.2	25.8	540	△
192.5	9.6	14.6	1.19	22.5	22.5	26.2	120	×

Broadly speaking, the experimental data lies somewhat above the dashed curve obtained from the numerical simulations for the case of a localized source for values of $q^* \gtrsim 40$, and somewhat below for smaller values of q^*. An exact agreement between the numerical and experimental is not expected however, since the erosion time is a function of Rayleigh number, as well as a function of heating rate. By necessity, the Rayleigh numbers of the numerical simulations were three orders of magnitude smaller than the Rayleigh number during the laboratory experiments (i.e. a prohibitively fine grid would have been required to perform simulations using typical experimental values of the Rayleigh numbers).

There are two undesirable features of the experimental data presented in Figure 2 which need to be addressed. First, the scatter in the data is unacceptably large. The measurement errors associated with determining q, h_1, h_2 and ΔS are small and cannot explain the scatter. We believe the scatter is due to: (1) differences in the initial layer temperatures, (2) differences in the amount of heat loss through the sidewalls in each experiment, (3) differences in the thickness of the initial salt interface, and (4) the temperature dependence of the various physical properties. The second undesirable feature of the experimental data is the relatively small spread in q^*. In practice q^* was bounded from above by the maximum heating rate of the heat source and the ability to establish a small salinity jump across the two–layer system. The lower bound on q^* was dictated by the desire to keep heat losses, as a fraction of heat input rate, relatively small throughout the experiments.

4 A qualitative description of the erosion process

In the previous section we concentrated on the time required for the two layers in our system to merge. In this section we look in greater detail at the dynamics of the erosion process and we compare the case of a localized heat source with the case of a distributed heat source.

4.1 Localized heat source

Shortly after the initiation of heating a narrow plume rises from the heat source. The plume is initially laminar, and for the heating rates and layer depths considered in this paper, the plume remains laminar until it reaches the interface separating the salty lower layer from the fresh upper layer. If the plume is sufficiently buoyant when it reaches the interface, it will penetrate through the interface and rise to the top of the domain. We shall not consider this situation here. Instead we concentrate on the case when the plume impinges upon the interface and then spreads laterally towards the sidewalls.

As the plume impinges upon the interface, it entrains a small amount of fluid from the upper layer into the lower. With time, the lower layer heats up and this ultimately leads to the double diffusive transport of heat and a small amount of salt from the lower layer to the upper layer. This double diffusive transport results in the net upwards transport of buoyancy, however the continual heating of the lower layer reduces the density jump across the interface. As the density jump across the interface becomes smaller, the plume is able to entrain more and more fluid as it impinges, and ultimately the plume is able to penetrate through the interface and rise to the top of the upper layer. Soon afterwards, the two layers become well mixed and the merger process is complete.

A distinguishing feature of the erosion process for the case of a localized heat source is the formation of a stable thermal stratification in the lower layer. This stable thermal stratification develops as a result of the filling box mechanism [16] which has been extended by Baines [17] and Kumagai [18] to the case of a

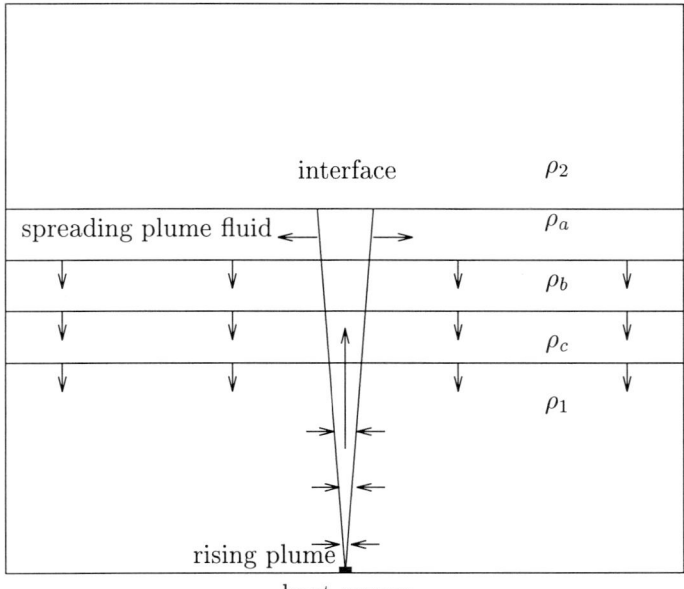

interface ρ_2

spreading plume fluid ρ_a

ρ_b

ρ_c

ρ_1

rising plume

heat source

Figure 3. The filling box mechanism establishing a stable density stratification in our two–layer system: $\rho_2 < \rho_a < \rho_b < \rho_c < \rho_1$

two–layer density stratification. The mechanism is best described with reference to Figure 3. A narrow plume rises from the buoyancy source, impinges upon the interface, and then spreads horizontally towards the side of the domain. The plume provides a conduit to carry buoyancy to the height of the interface. By continuity, there is a slow uniform downward flow of the ambient fluid away from the plume. The key feature of the filling box mechanism is that the plume is rising through an environment which is continually increasing in buoyancy. Therefore as time passes, the plume loses less of its buoyancy due to interaction with the environment and consequently deposits increasingly buoyant fluid at the level of the interface, thus establishing a stable stratification.

Baines [17] and Kumagai [18] considered single diffusive convection: their buoyancy source and density step were both due to the same stratifying component. The problem we consider here is double diffusive, and as we discuss below, the stable thermal stratification which develops in the lower layer has a profound influence on the nature of the double diffusive convection which occurs at the interface.

4.2   Distributed heat source

When a salinity step is heated from below by a distributed heat source, the erosion process is similar to that described above. The lower layer gradually heats up and eventually there is double diffusive transport of heat and salt across the interface separating the layers. The interface rises slowly at first due to entrainment, and then when the density jump across the interface becomes small, the interface begins to rise rapidly. The key distinction however, occurs in the lower layer. For sufficiently high heating rates, the convection in the lower layer is comparable to high Rayleigh number Benard convection: the lower layer is continuously stirred by randomly rising and falling thermals. This stirring action ensures that the lower layer temperature and salinity fields remain well mixed. This is in contrast with the filling box circulation and stable thermal stratification which develops when the heating is due to a localized source. It is this difference in the thermal stratification within the lower layer which leads a different character to the double diffusive convection at the interface.

4.3   Transport mechanisms across a diffusive interface

A number of different mechanisms have been proposed for the double diffusive transport of heat and salt across an interface separating a lower layer of hot, salty water from an upper layer of relatively cool and fresh water. In order to highlight the characteristics of the double diffusive convection which results when a salinity step is heated from below by a localized heat, we first discuss two of these mechanisms.

Linden and Shirtcliffe [19] have proposed a "thermals" mechanism in which the transport of heat and salt is controlled by the growth of thermal boundary layers on either side of a diffusive core separating the upper and lower layers. This model is symmetrical in that the mechanisms for the double diffusive convection is the same on either side of the interface. They suggest that when the boundary layers become unstable, buoyant thermals are released, which carry heat and a small amount of salt away from the interface. The basis of this model is that the transport of heat and salt across the interface occurs in a quasi-steady manner through the repetitive breakdown of unstable thermal boundary layers. This model is only appropriate if the upper and lower layers are much thicker than the diffusive core, and if there is no external forcing which might inhibit the formation of thermal boundary layers at the edge of the diffusive core. Moreover, the Linden–Shirtcliffe model breaks down when the stability ratio $R_\rho = \beta \Delta S / \alpha \Delta T$ exceeds $1/\tau^{1/2}$, where ΔS and ΔT are the salinity and temperature jumps across the interface.

Fernando [7] has proposed a model for the transport of heat and salt across a diffusive interface for the case when a salinity step is heated strongly from below by a distributed heat source. In his model, turbulent eddies in the lower layer which are driven by the strong heating from below scour fluid away from the interface, limiting the diffusive thickening of the interface and preventing the growth of thermals in the manner described by Linden and Shirtcliffe [19]. The double diffusive convection described in this model is asymmetric in that the convective eddies in the lower layer are more energetic than the convective eddies in the upper layer and hence are more effective at scouring the diffusive interface. Fernando argues that if the turnover time scale of the convective eddies in the lower layer is much smaller than the time scale required for the growth of unstable thermal boundary layers, then his "scouring" mechanism will control the growth of the thickness of the interface and will be the limiting factor controlling the transport of heat and salt across the interface. Fernando's model breaks down when the turnover time scale of convective eddies in the lower layer becomes comparable to the time required for the growth of unstable thermal boundary layers. This occurs in the case of a high heating rate.

4.4   Double diffusive convection due to localized heating

The processes which result when a salinity step is heated from below by a localized heat source are sketched in Figure 4. In the upper layer, the double diffusive convection takes the form of thermals which rise from the interface. As discussed by Linden and Shirtcliffe [19], the relatively hot lower layer induces the

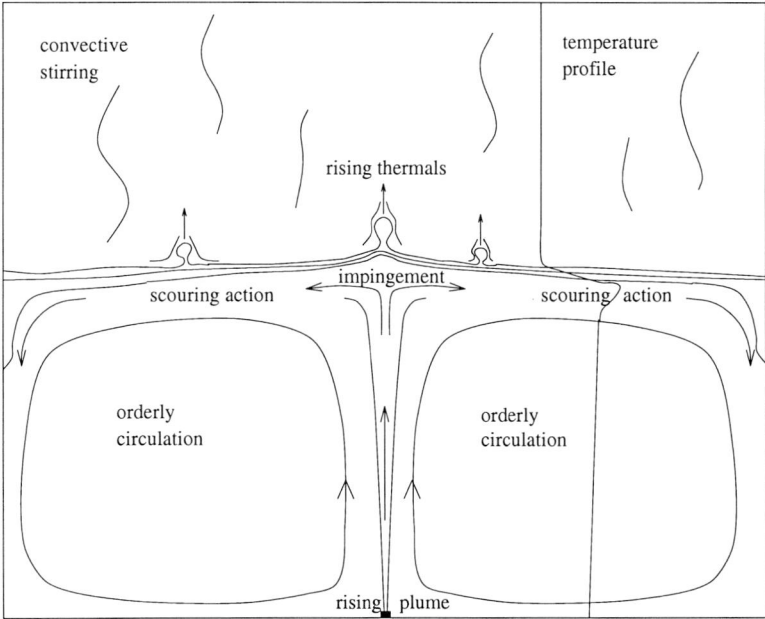

Figure 4. The processes involved in the erosion of a salinity step by a localized heat source

formation of unstable thermal boundary layers along the upper edge of the diffusive interface. When these boundary layers collapse, they release thermals which carry heat and salt upwards. These thermals are released intermittently across the whole interface, however, they are most prevalent near the point of plume impingement since this is where there is the maximum temperature gradient between the upper and lower layer. The rising thermals result in a disorganized velocity field and the upper layer remains well mixed. For the heating rates considered in this paper, the "scouring" mechanism discussed by Fernando [7] does not inhibit the formation of thermals in the upper layer.

The double diffusive convection in the lower layer is quite distinct from that of the upper layer. As the plume rises from the heat source, impinges upon the interface and then spreads laterally, it establishes an orderly circulation in the lower layer. This circulation is similar to the filling box circulation discussed above and establishes a stable thermal stratification in the upper portion of the lower layer. A typical temperature profile is shown in Figure 4 corresponding to a vertical traverse midway between the heat source and the sidewalls. We see that the temperature is constant in the upper layer and increases rapidly across the interface separating the upper and lower layers. The temperature reaches a maximum in the upper portion of the lower layer, and then decreases slowly throughout the rest of the layer. The region of strong static stability in the lower layer is confined to a thin region in the upper portion of the lower layer. It is this thin region of stability and the relatively strong flow associated with it, which prevents the formation of thermals along the lower edge of the diffusive interface. Instead, the heat and salt which diffuses through the interface from the upper layer to the lower layer, are quickly and efficiently scoured away by the laterally spreading plume fluid towards the sidewalls. The circulation in the lower layer then drags this relatively cool fresh fluid downwards along the sidewalls. The scouring action of the laterally spreading plume fluid is similar to Fernando's "scouring" mechanism in the sense that fluid is carried away from the interface before unstable thermal boundary layers can develop by molecular diffusion. Given time, the statically stable temperature profile in the lower layer would diffuse in a statically unstable sense. The scouring action of the spreading plume fluid prevents this from happening.

5   Conclusions

We have shown that in the limit of a convection dominated system, the time required for the complete erosion of a two–layer salinity stratification is bounded from below by the bulk heating time scale $t_q = \lambda h w \rho_o c_p \beta \Delta S / \alpha q$. An upper bounds on the merger process is $t_q / \tau^{1/2}$ where τ is the ratio of the diffusivity of salt to the diffusivity of heat. These time bounds show that the time required for layer merger depends only on the thickness of the lower layer, independent of the thickness of the upper layer. Numerical simulations show that a localized heat source is up to 30% more efficient than a distributed heat source at eroding a two–layer salinity stratification. In the case of a localized heat source, a stable thermal stratification is established in the lower layer during the erosion process due to the filling box mechanism while in the case of a distributed heat source, such a stratification is not established. The stable thermal stratification and the well organized velocity field in the lower layer inhibit the formation and downwards propagation of thermals from the double diffusive interface. Instead, heat and salt which diffuse from the upper to the lower layer, are efficiently scoured away from the interface by the orderly circulation in the lower layer which is established as a result of heating from a localized source.

Bibliography

1. Huppert, H.E. and Turner, J.S. (1981). Double diffusive convection. *J. Fluid Mech.*, *106*, 299–329.

2. Turner, J.S. (1985). Multicomponent convection. *Ann. Rev. Fluid Mech.*, *17*, 11–44.

3. Schmitt, R.W. (1994). Double diffusive oceanography. *Ann. Rev. Fluid Mech.*, *26*, 255–285.

4. Turner, J.S. (1965). The coupled turbulent transport of salt and heat across a sharp density step. *Int. J. Heat Mass Trans.*, *8*, 759–767.

5. Crapper, P.F. (1975). Measurements across a diffusive interface. *Deep-Sea Res.*, *22*, 537–545.

6. Marmorino, G.O. and Caldwell, D.R. (1976). Heat and salt transport through a diffusive thermohaline interface. *Deep-Sea Res.*, *23*, 59–67.

7. Fernando, H.J.S. (1989). Buoyancy transfer across a diffusive interface. *J. Fluid Mech.*, *209*, 1–34.

8. Leppinen, D.M. (1996). The erosion of a density step by a localized source of buoyancy: single and double diffusive effects. *Proceedings of the 4th Annual Conference of the CFD Society of Canada*. 707–714.

9. Leppinen, D.M. (1997). Aspects of convection. *PhD. Thesis*, University of Cambridge.

10. Leppinen, D.M. and Dalziel S.B. An experimental study of the erosion of a salinity step by a line heat source. *In preparation*.

11. Fernando, H.J.S. (1990). Comments on "Interfacial migration in thermohaline staircases". *J. Phys. Oceanogr.*, *20*, 1994–1996.

12. Turner, J.S. (1979). Buoyancy effects in fluids. Cambridge University Press.

13. Stern, M.E. (1982). Inequalities and variational principles in double diffusive turbulence. *J. Fluid Mech.*, *114*, 105–129.

14. Ruddick, B.R. and Shirtcliffe, T.G.L. (1979). Data for double diffusers: physical properties of aqueous salt–sugar solutions. *Deep-Sea Res.*, *26*, 775-787.

15. Gill, A.E. (1982). Atmosphere–ocean dynamics. Academic Press, London.

16. Baines, W.D. and Turner, J.S. (1969). Turbulent buoyant convection from a source in a confined region. *J. Fluid Mech.*, *37*, 51–80.

17. Baines, W.D. (1975). Entrainment by a plume or jet at a density interface. *J. Fluid Mech.*, *68*, 309–320.

18. Kumagai, M. (1984). Turbulent buoyant convection from a source in a confined two layer region. *J. Fluid Mech.*, *147*, 105–131.

19. Linden, P.F. and Shirtcliffe, T.G. (1978). The diffusive interface in double diffusive convection. *J. Fluid Mech.*, *87*, 417–432.

Experimental Studies of Thorpe Scale Decay Rate

Andrew M. Folkard[1] and Harindra J.S. Fernando

Environmental Fluid Dynamics Program, Department of Mechanical and Aerospace Engineering, Arizona State University, Tempe, USA

Abstract

The accurate parameterization of turbulent diffusion coefficients is a critical pre-requisite for the development of useful models of ocean circulation and budgets. Determination of these coefficients rely on field measurements of microstructure in both vector and scalar fields. Understanding these measurements depends upon knowing how the state observed relates to the history of the turbulent event which caused it. A central, unresolved problem in the effort to understand these matters is that of how the decay rate of scalar anomalies caused by turbulent stirring relates to external measurable parameters. Such a relationship would allow field oceanographers to gain a fuller insight into the nature of their microstructure data. The present work addresses this problem. A set of laboratory experiments was carried out using the standard vertically oscillating, horizontal grid positioned within a horizontally constrained container, which was filled with a linearly-stratified fluid. Measurements of the decay rate of density anomalies after the cessation of grid oscillations (quantified in terms of both the r.m.s. and maximum Thorpe scales, L_T and L_M respectively (Thorpe, 1977), and the available potential energy function A (Dillon, 1984)) were carried out in a set of experimental runs during which were varied the grid oscillation frequency (quantified in terms of the grid action, K), the initial buoyancy frequency N_0 and the time of grid oscillation t_{osc} (which played a role in determining the ultimate buoyancy frequency of the mixed patch N). Although there was no apparent dependence of decay rate on the grid oscillation frequency, significant and consistent dependencies on the other two parameters were observed. The observed dependence on the final buoyancy frequency N tenuously corroborated the $N^{2/3}$ law deduced by Pearson and Linden (1983), although the best fit curves were found to be $N^{0.52}$ (for L_T) and $N^{0.71}$ (for A). The dependence on the initial buoyancy frequency N_0 was found to follow an N_0^{-2} law for both parameters. Implications of these findings are discussed.

---

[1]Present address: Department of Civil Engineering, University of Strathclyde, John Andreson Building, 107 Rottenrow, Glasgow, Scotland, United Kingdom, G4 0NG.

1   Introduction

A knowledge of oceanic mixing processes is vital in developing atmospheric and oceanic general circulation models (GCM's) and nested regional models, which have immediate practical applications, for example, in the prediction of optical properties and currents of coastal waters and large scale oceanic circulations.

Turbulent mixing is known to produce a first-order effect on ocean dynamics, as exemplified by the studies of Wijesekera and Dillon (1991). In current weather forecasting models (for example the European Centre for Medium Range Weather Forecasting, ECMWF), the mixing at the sheared inversion layer is parameterized by assuming a uniform entrainment rate across the interface, irrespective of local conditions. These models can be improved, if proper entrainment parameterizations can be incorporated into the model.

Thus the necessity of studying the behaviour of turbulence in a stratified regime is clear, and a large body of research has been carried out in this area over the past several years (see, for example, Fernando, 1991, for a review). The present work makes a detailed analysis of the behaviour of density anomalies produced by turbulence during their decay, that is, once the source of turbulent energy has been removed. The turbulence is quantified in terms both of the Thorpe scale (Thorpe, 1977) and the available potential energy function (Dillon, 1984). The former is essentially equivalent to the overturning turbulent length scale $L_B = \rho'/(\partial\rho/\partial z)$ where ρ' is the root mean square density fluctuation and $\partial\rho/\partial z$ is the ambient density gradient (for example Hopfinger, 1987). The turbulence is assumed for the purposes of the theoretical part of this work to be in the "final stages of turbulent decay" (Batchelor and Townsend, 1948), that is, the turbulent velocity field is uniformly zero, and only the density anomalies remain. The processes then occurring are

1. buoyant return of anomalies to their level of neutral buoyancy;

2. diffusion of the density anomaly into the ambient fluid.

The presence of density anomalies in the absence of the turbulent velocity field which caused them has been referred to as "fossil turbulence" (for example Woods et al., 1969; Gibson, 1986), although the precise nature and relevance of such a term is a matter of debate.

2   Experimental procedure

The experimental apparatus is depicted schematically in Figure 1. Experiments were carried out in a rectilinear tank of dimensions 0.45 m (height) × 0.25 m (width) × 0.25 m (depth). A horizontal grid was positioned 0.16 m above the bottom of the tank and consisted of limbs 1 cm^2 in cross-section separated by 0.05 m. The grid extended to cover the entire horizontal cross-section of the tank, with a gap of ~3 mm separating it from the tank-wall. The grid was

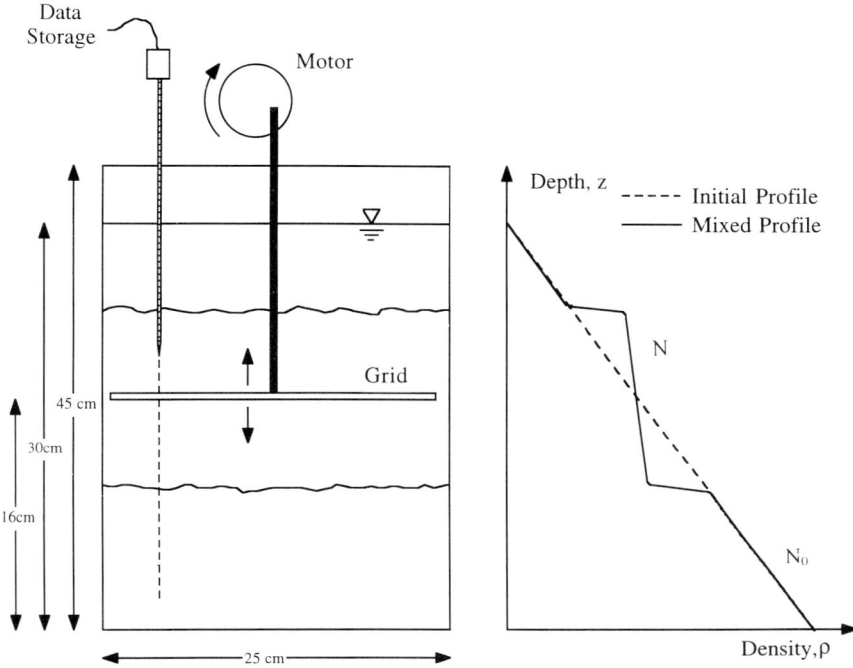

Figure 1. Schematic diagram of the experimental apparatus. A plot of the generic density profile before and after turbulent mixing and decay is shown to the right

supported from above by a rod which was attached off-centre to a wheel driven by a motor, the speed of which was adjustable in the range 2-20 rad/s.

The tank was filled with a linearly-stratified salt solution using the classical two-bucket method of Oster (1965). The inlet for the solution was situated horizontally at the bottom of one of the vertical sides of the tank, close to a vertex. A baffle was placed diagonally across this vertex, a few centimetres from the inlet and with a narrow ($\sim$2 mm) gap beneath it. The flow from the inlet was arrested by this baffle and forced into a shallow sheet of flow which then entered the main body of the tank at its very base. A valve was attached to the inlet to ensure that the inflow was neither too fast (this would have caused undesired mixing of the inflow with the solution already in the tank) nor unnecessarily slow. The tank was filled to a depth of 0.30 m.

In order to measure vertical density profiles during the experimental runs, a rapid-response conductivity probe (Head, 1983) was arranged to shoot through the fluid column at operator-defined times. This was realised by mounting the probe on a carriage attached to a screw-rod. The rod was rotated by a stepping motor, which caused the carriage to move down the screw-thread on the rod, thus propelling the probe vertically through the fluid column. The probe was positioned so that it passed through one of the gaps in the grid approximately halfway between the centre and the edge of the tank. The density probe was connected, via processing hardware to one channel of a data acquisition card within a computer. Simultaneously with taking readings from the probe, the card also took readings via another channel from a potentiometer attached to the screw rod. As the rod turned, it varied the resistance across the potentiometer. A constant current was applied across the potentiometer, thus the voltage read by the data acquisition card was proportional to the depth of the probe. Hence time series of data pairs representing depth and density were acquired. This arrangement has been used successfully many times in the past (for example De Silva and Fernando, 1992; Folkard et al., 1997).

Since only salt solution was used in these experiments, the effects of varying the viscosity and diffusivity of the fluid were not tested. The three parameters which were varied during the experiments, in order to determine their effects on the Thorpe scale decay rate, were K - the grid action (Long, 1977), N_0 - the initial buoyancy frequency, and t - the time for which the grid was oscillated before decay was allowed to start. These three between them determine a fourth parameter, namely N - the buoyancy frequency of the Thorpe-ordered turbulent patch at the onset of decay. Since $N = N(N_0, K, t)$, if follows that $t = t(N_0, K, N)$ and hence N_0, K and N can be considered as the three independent variables which determine the evolution of the fluid. This inversion is performed since it is easier to think of the physical effects of N on the turbulence than it is to think of those of t.

A standard set of parameters was chosen, namely $K = 6$ cm^{2}s^{-1}; $N_0 = 1.0$ rad s^{-1}; and $t_{osc} = 100$ seconds, corresponding to $N = 0.26$ rad s^{-1}. Runs were carried out by changing one parameter at a time from this set of values. The sets of values used is shown in Table 1. A total of six runs was

Table 1. Summary of the parameter values used in each of the sets of runs of the experimental study

Run Number	Turbulent Reynolds Number Re	Initial Buoyancy Frequency N_0 (rad/s)	Grid Oscillation Time t (seconds)
1	500	1.0	100
2	300	1.0	100
3	700	1.0	100
4	500	0.66	100
5	500	1.5	100
6	500	1.0	15
7	500	1.0	600

carried out for each parameter set, one each with a delay between turning off
the grid and taking the first density profile of 0, 5, 10, 15, 20 and 25 seconds. A
delay of 30 seconds was used between density profiles, this having been found in
preliminary runs to give the best trade-off between minimum delay time and lack
of background disturbances from the wake of the previous profiling. A total of
eight profiles was carried out in each run, giving a data set for each parameter set
of 48 profiles, at spacings of 5 seconds from 0 to 235 seconds after the cessation
of grid oscillations.

The profiles were analysed, and the r.m.s. Thorpe displacement L_T, the maximum Thorpe displacement L_M, and the available potential energy function A
calculated. These were smoothed using a five-point rolling mean and plotted to
give a decay curve of which Figure 2 is a typical example. Note that the value
of each parameter does not decay to zero, but stays at some value that is due to
noise in the data. In order to remove this, it was noted that in none of the 336
profiles obtained did the decay continue beyond 100 buoyancy periods (indeed,
in most of them the decay had ceased by 50 buoyancy periods). Thus for each
profile, the mean of the values obtained between 100 buoyancy periods and the
end of the run was calculated and taken to be the contribution to the parameter
of noise. This was then subtracted from all the values, and the new parameter
thus obtained, L_T^* say, was replotted from $t' = 0$ to the data value immediately
before the first value of L_T^* below zero. This was then plotted on semi-log
axes, data sets in which a particular parameter had been varied being grouped

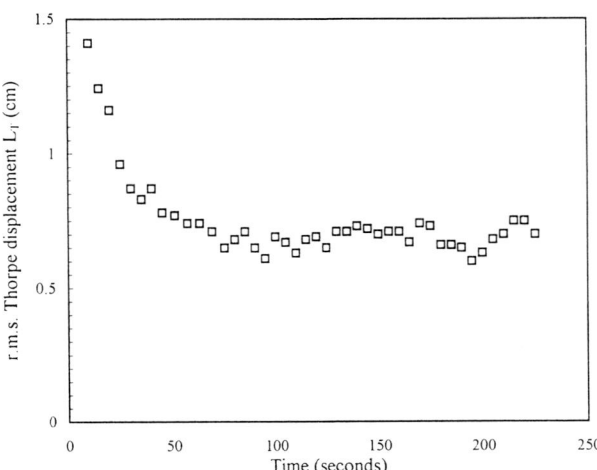

Figure 2. Typical plot of r.m.s. Thorpe displacement L_T against time. The data
have been smoothed from their raw state by applying a rolling 5-point mean

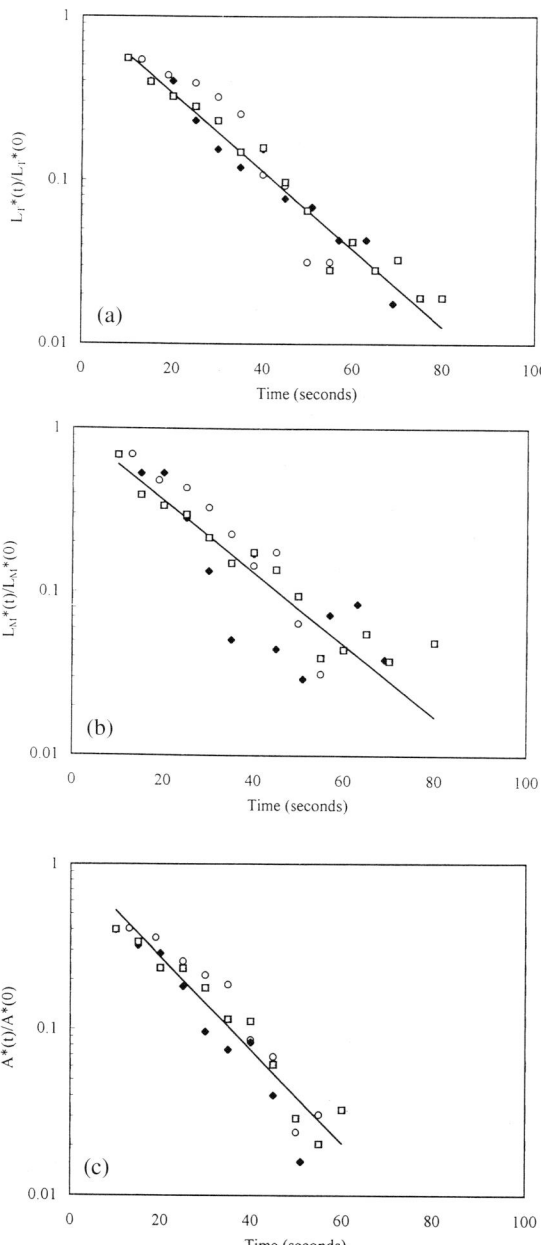

Figure 3. (a) Plot showing decay of r.m.s. Thorpe displacement L_T for $N_0 = 1.0$ rad/s, $t_{osc} = 100$ seconds and $K = 3$ cm^{2}s^{-1} (circles); $K = 5$ cm^{2}s^{-1} (diamonds); and $K = 7$ cm^{2}s^{-1} (squares). The best fit line to these combined data is shown; (b) as for Figure 3(a) for maximum Thorpe displacement L_M; and (c) as for Figure 3(a) for available potential energy function A

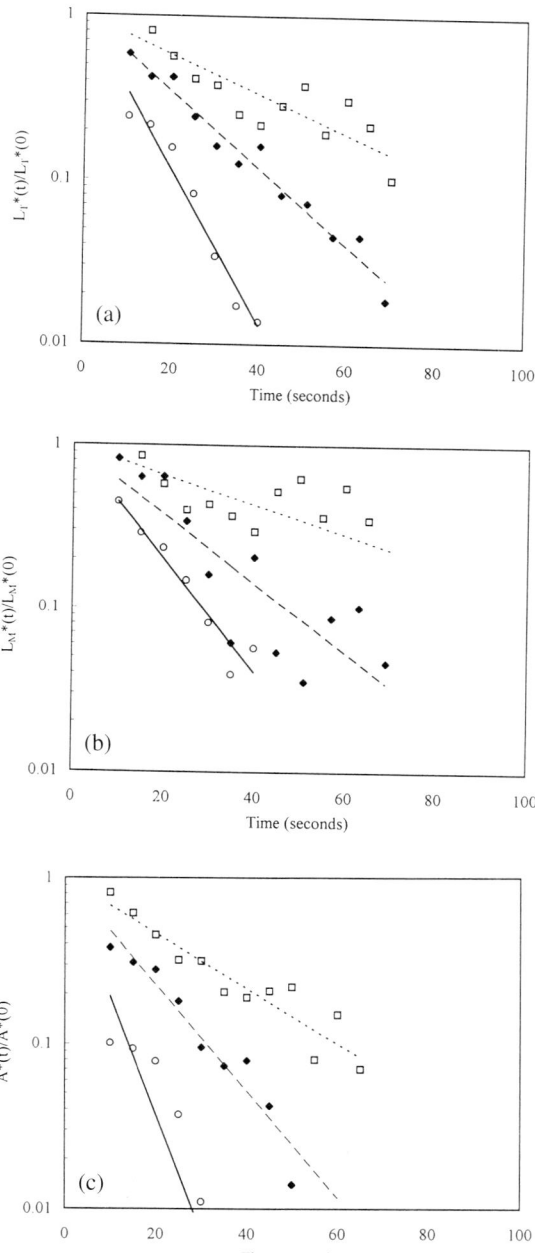

Figure 4. (a) Plot showing decay of r.m.s. Thorpe displacement L_T for $K = 5$ cm^{2}s^{-1}, $t_{osc} = 100$ seconds and $N_0 = 0.66$ rad/s (circles); $N_0 = 1.0$ rad/s (diamonds); and $N_0 = 1.5$ rad/s (squares). Best fit lines are shown; (b) as for Figure 4(a) for maximum Thorpe displacement L_M; and (c) as for Figure 4(a) for available potential energy function A

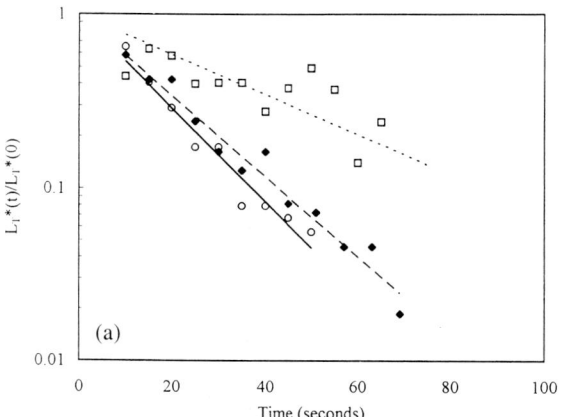

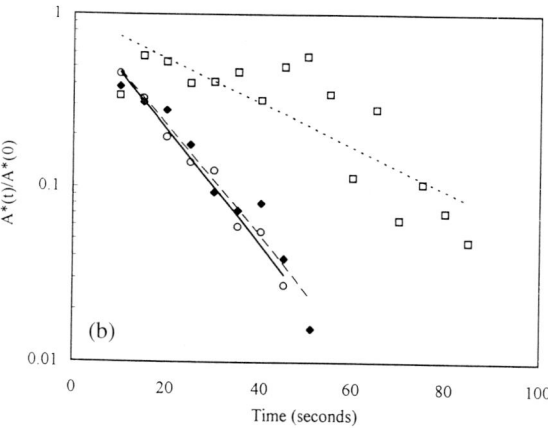

Figure 5. (a) Plot showing decay of r.m.s. Thorpe displacement L_T for $K = 5$ cm^{2}s^{-1}, $N_0 = 1.0$ rad/s and $t_{osc} = 15$ seconds ($N = 0.45$ rad/s - circles); $t_{osc} = 100$ seconds ($N = 0.26$ rad/s - diamonds); and $t_{osc} = 600$ seconds ($N = 0.17$ rad/s - squares). Best fit lines are shown; and (b) as for Figure 5(a) for available potential energy function A

together on one plot, in order to determine the dependence of the exponent on each of the parameters considered. Plots for L_T*, L_M*, and $A*$ are shown in Figures 3–5 respectively. Hereinafter, the asterisks are dropped from this notation for clarity.

3   Results

The most striking aspect of the plots in Figure 3 is the lack of dependence of any of the three measured parameters L_T, L_M and A on the grid action K. The logarithmic ordinate axis results in a spreading of the data at its lower end, particularly in the L_M plot, but otherwise the adherence to a straight line is clear, the coefficient of correlation r^2 being above 0.9 in all three cases. The plots indicate that the decay rate of A is 18% greater than that of L_T whereas that of L_M is 7% less. The former of these differences is found to be significant at the 95% level, but the latter is not.

Figure 4 shows very clearly the strong influence N_0 has on the decay rate of L_T, L_M and A. As in Figure 3, the data are least correlated with a straight line relationship in the L_M plot, although the data are also somewhat scattered in places in the other plots. The data, however, show consistently that the decay rate of all three parameters decreases with increasing N_0. Once again, these plots indicate that the decay of A proceeds most rapidly, and that of L_M most slowly.

Figure 5 shows a consistent, though less convincing, dependence on N than the dependence on N_0 shown in Figure 4. The greater scatter in the $t = 600$ seconds case is evidently due to the more completely-mixed patch which results, since smaller errors in the density data will correspond to larger Thorpe displacements in a patch of lower buoyancy frequency. The data for L_M were found to be dominated by scatter, and hence are not presented.

Linear regression is applied to the data in Figure 5(a), and the gradients of the best-fit straight lines obtained are plotted against N in Figure 6(a). The gradients here correspond to $1/t_E$. A remarkably good correlation is observed between these data and the $N \log N$ curve, which is shown in the plot. There is no apparent physical explanation, however, for the global maximum which this function implies in the data at $N \sim 0.35$ rad/s, hence the correlation is assumed to be coincidental. Assuming a monotonic (power law) relationship between N and the decay time scale yields a best fit line with a power index of 0.52. Within the range of N considered, this is shown in Figure 6(a) to be in good agreement with the 2/3 power law deduced analytically by Pearson and Linden (1983) for the decay of a density anomaly in the final stages of turbulent decay. These results can therefore be claimed to corroborate the findings of Pearson and Linden, and the 2/3 power law is assumed in the following.

A similar plot is shown in Figure 6(b), these data having been taken from the plot of A in Figure 5(b). Although the agreement between the best fit power law curve and the 2/3 power law curve here is not so good as in Figure 6(a),

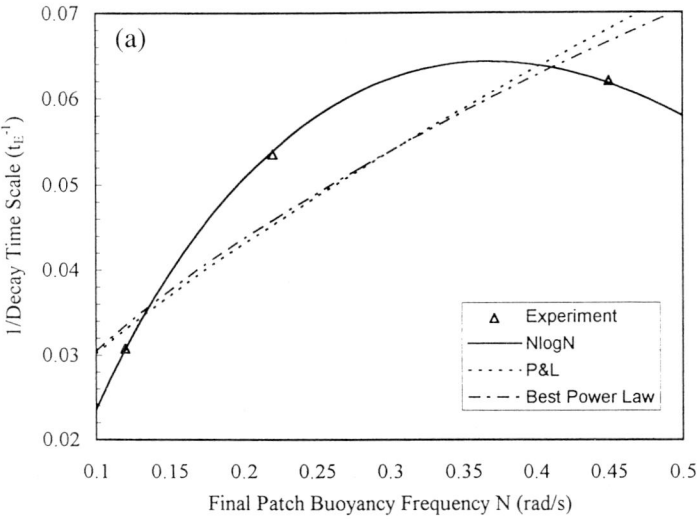

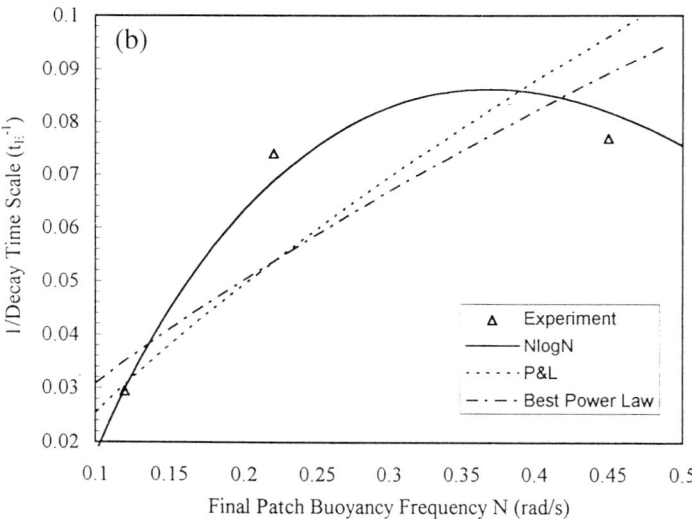

Figure 6. (a) Decay time scale t_E of r.m.s. Thorpe displacement L_T against final patch buoyancy frequency. Curves representing best fit function ($N \log N$), the best fit power law ($N^{0.52}$) and the 2/3 power law deduced by Pearson and Linden (1983) are shown; and (b) as for Figure 6(a) for available potential energy function A. Here the best fit power law curve is $N^{0.71}$

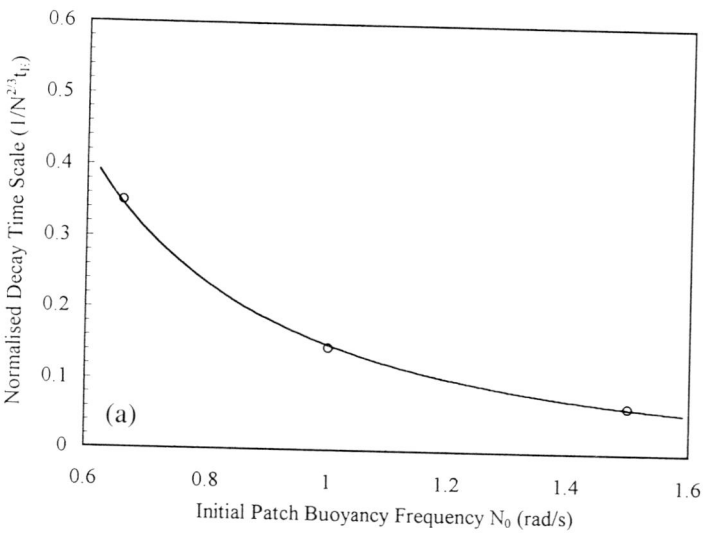

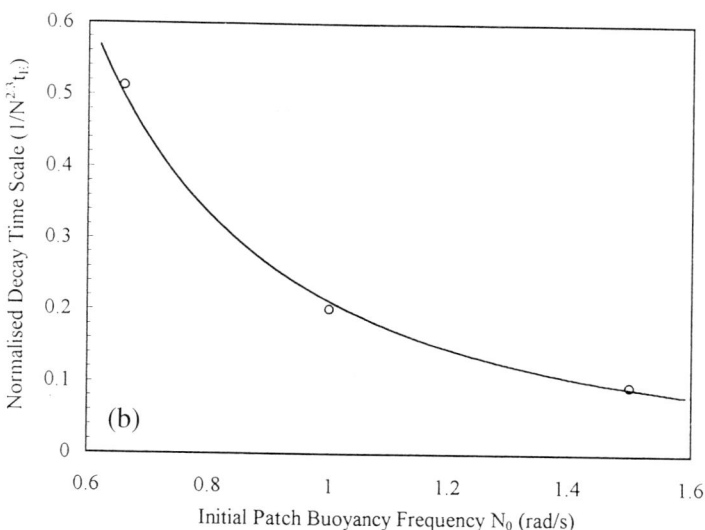

Figure 7. (a) Plot of decay time scale for r.m.s. Thorpe displacement L_T normalised by $N^{2/3}$ against the initial patch buoyancy frequency N_0. The curve shown is N_0^{-2}; and (b) as for Figure 7(b), for available potential energy function A. Again, the curve shown is N_0^{-2}

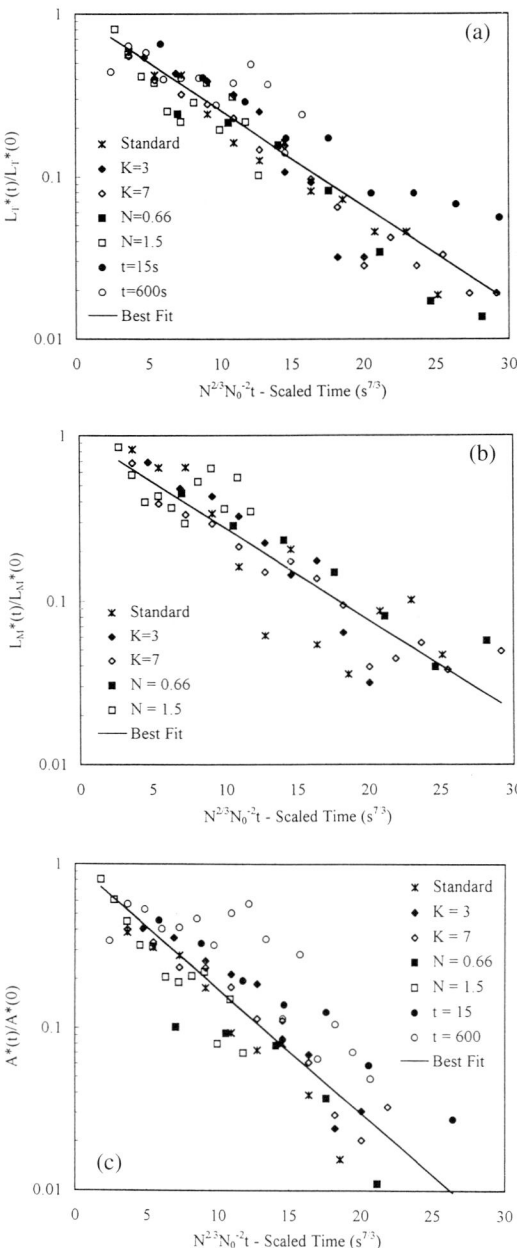

Figure 8. (a) Normalised r.m.s. Thorpe displacement L_T plotted against time scaled according to dependencies described in text. The best fit line, which has a gradient of -0.059, is shown ($r^2 = 0.85$); (b) as for Figure 8(a) for maximum Thorpe displacement L_M. The best fit line here has a gradient of -0.056 and $r^2 = 0.82$; and (c) as for Figure 8(a) for available potential energy function A. The best fit line here has a gradient of -0.076 and $r^2 = 0.80$

it is considered to be sufficiently close, particularly considering the scatter of the data around the best fit curve, to allow the 2/3 law to be assumed for this relationship also.

In Figure 4(a), the results from experiments in which both N and N_0 have been varied are shown. Having deduced that the $N^{2/3}$ law can be assumed above, this dependence is now taken out of the data in Figure 4(a), so that their dependence on N_0 can be deduced. This relationship is shown in Figure 7(a), which demonstrates a very good agreement between the experimental data and the N_0^{-2} curve. This dependence is also shown to hold for A in Figure 7(b).

These analyses together imply that the decay of L_T and A can be described as

$$\frac{L_T(t')}{L_T(0)} = \exp\left(-\alpha_1 \frac{N^{2/3}t'}{N_0^2}\right); \quad \frac{A(t')}{A(0)} = \exp\left(-\alpha_2 \frac{N^{2/3}t'}{N_0^2}\right) \quad (3.1)$$

where α_1 and α_2 are functions of the viscosity and difussivity of the fluid and a characteristic length scale relevant to the system being considered, and have dimensions $T^{-7/3}$. Plots showing the collapse of data caused by this scaling are shown in Figure 8. In Figure 8(b), L_M is assumed to have the same dependence on N and N_0 as L_T, an assumption which is borne out by the collapse of data shown. Comparisons of these plots demonstrate clearly the more rapid rate of decay of A than of either L_T or L_M. Put another way, this implies $\alpha_2 > \alpha_1$.

4 Discussion

The data obtained here for both L_T and A, as described in Section 3 appear to tenuously corroborate the $N^{2/3}$ law - deduced in the analytical study of Pearson and Linden - of the decay of density anomalies caused by turbulent mixing once the turbulent velocity field has decayed away. The best fit power law curves for these two parameters, however, are notably different (having indices of 0.52 and 0.71 respectively). Further work is required to test the nature of these results.

Pearson and Linden's analyses begin from the state after the turbulent kinetic energy has decayed and therefore do not take into account the situation in which the turbulence was created and how the environment of this stage of the process impacts upon the decay stage. Furthermore, they assume an infinite patch of uniform N, and do not take into account the effect on the decay rate of the presence of sharp density jumps at the edge of the patch of mixed fluid observed here. Nor do they consider the presence of an ambient fluid above and below the mixed patch of a different buoyancy frequency N_0. Of the two parameters which determine this state, the grid action (i.e. a quantification of the nature of the source of turbulent kinetic energy) was found to have no effect on the decay rate. The grid action is closely related to the turbulent Reynolds number of the fluid, which can be formulated as K/ν where ν is the viscosity of the fluid. This would be expected to determine the length scales of the turbulent motions within the patch, and it is therefore somewhat surprising that it appears to have no effect on the final decay rate. The initial buoyancy frequency would be expected to affect the size (in terms of density) of the buoyancy anomalies, the rate of mixing processes during both the stages of turbulent production and decay, and the size of the buoyancy jumps which bound the turbulent patch both above and below. These factors are shown to have a large effect on the final decay rate, reflected in the approximately inverse-square relationship between the decay time scale and the initial buoyancy frequency. This relationship is shown to be adhered to very closely by the experimental data for both L_T and A.

5   Conclusions

The data obtained in the experiments described above have been analysed and the following deduction made regarding the decay rate of L_T, L_M and A:

In general, the decay rate of the r.m.s. Thorpe scale L_T during the final stages of turbulent decay is found to be significantly slower than that of the available potential energy function A, but not significantly different from that of the maximum Thorpe displacement L_M.

The initial buoyancy frequency of the fluid in which the turbulence is produced is found to have a strong effect on the decay rate of density anomalies once turbulent production is increased; the relationship $t_E \propto N_0^{-2}$ for both Thorpe scale and available potential energy function is suggested by these results.

The rate of energy production of the mechanism producing the turbulence has no effect on the final rate of decay of any of the dependent variables studied.

The dependency of the time scale for the final stages of decay of density anomalies in a stratified fluid on the background buoyancy frequency N is found to be consistent with that found analytically and experimentally by Pearson and Linden (1983), although the correlation between their prediction ($t_E \propto N^{2/3}$) and the results obtained here is relatively-poor. Work is required to further investigate this correlation.

Acknowledgements

Work in environmental turbulence at ASU is supported by the Office of Naval Research (small-scale mixing and Arctic sciences), the National Science Foundation and the Environmental Protection Agency (Office of Exploratory Research). The authors express their thanks to Ryan Provencio for his considerable assistance in this work.

Bibliography

1. Batchelor, G.K. and Townsend, A.A. (1948). Decay of isotropic turbulence in the final period. *Proc. R. Soc. Lond.*, **A194**, pp. 527–543.

2. De Silva, I.P.D. and Fernando, H.J.S. (1992). Some aspects of mixing in a stratified turbulent patch. *J. Fluid Mechanics*, **240**, pp. 601–625.

3. Dillon, T.M. (1984). The energetics of overturning structures: implications for the theory of fossil turbulence. *J. Phys. Oceangr.*, **14(3)**, pp. 541–549.

4. Fernando, H.J.S. (1991). Turbulent mixing in stratified flows. *Ann. Rev. of Fluid Mechanics*, **23**, pp. 455–493.

5. Folkard, A.M., Davies, P.A. and Fernando, H.J.S. (1997). Measurements in a turbulent patch in a rotating linearly-stratied fluid. *Dyn. Atmos. Oceans*, **26**, pp. 27–51.

6. Gibson, C.H. (1980). Fossil temperature, salinity, and vorticity turbulence in the ocean. *Marine Turbulence*, Editor: J.C.J. Nihoul, Elsevier, Amsterdam, pp. 221–257.

7. Head, M.J. (1983). The use of miniature four-electrode conductivity probes for high resolution measurement of turbulent density or temperature variations in salt-stratified water flows. *Ph.D. Thesis*, University of California, San Diego.

8. Hopfinger, E.J. (1987). Turbulence in stratified fluids: A review. *J. Geophysical Research*, **92(C5)**, pp. 5287–5303.

9. Long, R.R. (1978). A theory of mixing in stably stratified fluids. *J. Fluid Mechanics*, **84**, pp. 113–124.

10. Oster, G. (1965). Density gradients. *Scientific American*, **70**.

11. Pearson, H.J. and Linden, P.F. (1983). Decay of turbulence in a stably stratified fluid. *J. Fluid Mechanics*, **134**, pp. 195–203.

12. Thorpe, S.A. (1977). Turbulence and mixing in a Scottish loch. *Phil. Trans. R. Soc. London Series A*, **286**, pp. 125–181.

13. Wijesekera, H. and Dillon, T.M. (1991). Internal waves and mixing in the upper equatorial Pacific Ocean. *J. Geophysical Research (Oceans)*, **96**, NC4, pp. 7115–7125.

14. Woods, J.D. (Editor), (1969). Report of working group: Fossil turbulence. *Radio Science*, **4**, pp. 65–67.

Aspects of Turbulent Mixing at Density Interfaces

E.J. Strang[*], H.J.S. Fernando[*] and E. Kit[**]

[*]*Environmental Fluid Dynamics Program, Department of Mechanical and Aerospace Engineering, Arizona State University, USA, and [**]Department of Fluid Mechanics and Heat Transfer, Tel-Aviv University, Israel*

Abstract

A summary of an experimental program designed to study the nature of zero-mean-shear turbulence near a density interface is presented. The experimental results are discussed vis-a-vis previous theoretical and direct numerical simulation work on the distortion of homogeneous turbulence by a solid plate with different boundary conditions at the plate. The nature of turbulence near the interface, entrainment at the interface and the fate of entrained parcels as they move in the turbulent environment are discussed.

1   Introduction

Fluid masses in our environment such as oceans, lakes and the atmosphere are stably stratified at least part of the time, and usually they are in turbulent motion. Because of the highly diffusive nature of turbulence, fluid parcels in stratified turbulent fluids are advected randomly to environments of different densities where they distort and sometimes exchange heat, salt and other species with the background. Random advection does not change densities of the fluid parcels, and when their kinetic energy is expended they return to their original equilibrium density levels (the so called stirring motions). On the other hand, if the fluid parcels change their density via molecular transfer, with a density flux of

$$F_i = -\kappa \frac{\partial \rho}{\partial x_i}, \tag{1.1}$$

where κ is molecular diffusivity and ρ is the density, then the parcels change their densities irreversibly and will not return to their original positions upon the decay of turbulent energy. The homogenization between adjacent fluid parcels is a signature of mixing, and the rate of mixing can be defined by the Fickian diffusion formula (1.1). In reality, for effective mixing to occur between two fluid parcels separated by a distance δ, the fluid velocities should be small enough to transfer the densities between the two parcels before they are separated by advection. Diffusion velocities are typically of the order κ/δ, and only among the

213

fluid parcels that have separation velocities of this order or less can the molecular diffusion take place. If the fluid parcels belong to the inertial subrange,
then their velocities are of the order $(\epsilon \delta)^{1/3}$, and hence molecular-scale mixing
can take place at scales smaller than the Obukhov-Corrsin scale $\delta \sim (\kappa^3/\epsilon)^{1/4}$,
where ϵ is the rate of turbulent kinetic energy dissipation. Consistency requirements imply that δ should be greater than the Kolmogorov scale $(\nu^3/\epsilon)^{1/4}$ or
$Sc = \nu/\kappa < 1$, where Sc is the Schmidt number and ν is the kinematic viscosity.
On the other hand, if $Sc > 1$, then the molecular-diffusive effects are manifested
at scales smaller than the Kolmogorov scale. As the adjacent fluid parcels separated by a distance δ move apart with a time scale of $(\nu/\epsilon)^{1/2}$, molecular-scale
homogenization can occur only if this time scale is larger that the diffusion time
scale δ^2/κ or at lengthscales smaller than the Batchelor scale $(\nu\kappa^2/\epsilon)^{1/4}$.

Although the mixing rate can be defined by (1.1), it is very difficult to measure since the scales over which the diffusion takes place are very small. For
example in oceans, $\epsilon \sim 10^{-8} \ m^2/s^3$ and, hence, the scales where heat diffusion
occurs are of the order $10^{-3} \ m$, which are unresolvable by typical microstructure
instruments. The difficulty of measuring mixing rates is common to both laboratory and natural flows, and obtaining sufficient resolution to estimate scalar
dissipation currently requires highly specialized measurement techniques. As
such the mixing rates are estimated by implicit methods, for example, by measuring or estimating the turbulent density flux $\overline{w'\rho'}$ and assuming that under
equilibrium conditions the mixing at small scales is balanced by the buoyancy
flux cascading down to smaller scales. The density flux is typically divided by
the mean density gradient to define an equivalent "eddy diffusivity". A popular
paradigm for the role of ocean mixing (specified by the eddy diffusivity K_ρ) in
determining the large-scale circulation is that there is a balance between vertical
(z-direction) advection and turbulent diffusion,

$$w\frac{\partial \rho}{\partial z} = K_\rho \frac{d^2\rho}{dz^2}, \tag{1.2}$$

or

$$\frac{K_\rho}{w} = \frac{(\partial \rho/\partial z)}{(\partial^2 \rho/\partial z^2)} \approx 1300 \ \text{m}, \tag{1.3}$$

[1,2]. Assuming a vertical velocity of $w \approx 0.5 - 1$ cm/day, $K_\rho \approx 10^{-4} \ m^2/s$ can
be inferred. Oceanic measurements, on the other hand, are confined to local or
regional scales and yield widely different values from the above bulk estimates.
Measurements are typically in the range $(0.15 - 1) \times 10^{-5} \ m^2/s$, and attuning of
such disparities is at the heart of oceanic microstructure research. Laboratory
experiments such as those described here are expected to shed light on various
facets of the problem of mixing parameterizations.

This paper is concerned with the transport of buoyant fluid parcels across
density interfaces and their mixing in the new environment. Fluid parcels are
advected across the interface by a stirring mechanism such as interfacial wave
breaking or scouring of the interface by large-scale eddies. Upon arrival, these

fluid parcels find themselves either positively or negatively buoyant and, to remain in the new environment, molecular-scale mixing should be accomplished before buoyancy forces rebound them back to their original levels. To study stirring and mixing at density interfaces adjacent to turbulent layers, we will consider a simple flow configuration wherein a density interface (with a density jump $\Delta\rho$) is subjected to shear free turbulence induced by oscillating grids, as in the experiments of Turner [3] and others. In some experiments, the turbulence is forced in the upper layer whereas in others both the top and bottom layers are maintained turbulent with identical r.m.s. velocity and integral length scales. The flow was observed using laser-induced fluorescence (LIF) with proper refractive index matching between upper and lower layers. Velocities were measured using laser-Doppler velocimetry and hot-film anemometry; in the latter case, the hot-film was rotated, as in [4], to generate a mean flow at the tip. In Section 2, we will describe the qualitative properties of such interfaces whereas in Section 3 the fate of fluid parcels that cross the interface will be discussed, paying particular attention to the evolution of the spectra of scalar fluctuations. Section 4 discusses the role of molecular diffusivity on mixing of fluid parcels entrained into turbulent environments and the critical role effective diffusivities play in ocean models. Section 5 is devoted to the presentation of turbulence measurements near density interfaces and comparison of results with previous direct numerical simulations and analytical results.

2 Entrainment at density interfaces

Ever present turbulent stirring in environmental flows causes exchange of fluid elements carrying species such as humidity, salinity or temperature across natural density stratified layers. If the species contribute to the density of fluid parcels, then the density changes associated with crossing the interface determine the fate of these fluid elements and hence the resulting concentration distribution. If the density interfaces are strong enough, eddies on one side of the interface cannot simply cross to the other side and vice versa, because of the opposing influence of their buoyancy. Yet, if the fluid parcels can change their density to values closer to the ambient density rapid enough (via turbulent mixing), then the exchange of fluid parcels is irreversible with a net density flux through the interface.

  If the turbulence is confined to one side of the interface, then the transfer of parcels occurs from the non-turbulent side to the turbulent side. The simplest mechanism of this transfer is the engulfment and advection of non-turbulent fluid by energy-containing eddies with characteristic velocity and length scales u_H and L_H, respectively, which work against the interfacial buoyancy jump Δb to carry interfacial fluid into the turbulent layer. This is possible when the inertial forces u_H^2/L_H of the eddies are strong enough to lift the fluid parcels against the buoyancy forces Δb or when the Richardson number

(a)

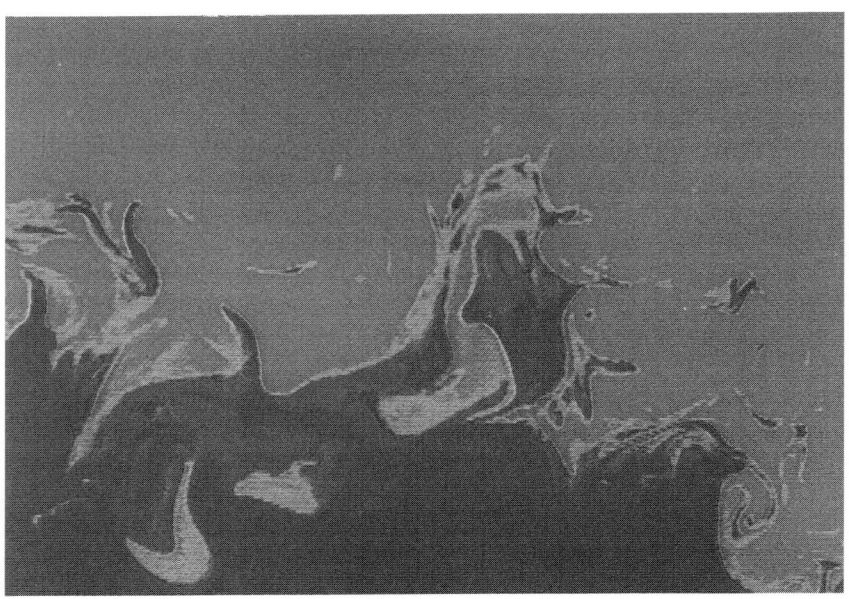

(b)

Figure 1. (a) A LIF picture of a shear-free entraining interface, which shows the severing off of a heavy fluid lump by a turbulent eddy approaching from the left; $Ri \approx 20$, and (b) An interface at $Ri \approx 12$, but with turbulence on either side of the interface

$Ri = \Delta b L_H / u_H^2$ is sufficiently low. Figure 1(a) shows a case with $Ri \approx 20$, where eddies penetrate oblique to the interface and sever off a lump of fluid from the interface; in this photo, the upper layer is turbulent with forcing by an oscillating grid, the bottom layer is non-turbulent and laser-induced fluorescence has been used for flow visualization. As the separated heavy fluid lump enters the upper layer it is distorted, crumbled and homogenized with the surrounding fluid. Near the interface, the distortions are biased toward horizontal stretching and wrinkling due to strong anisotropy of turbulence; the fluid parcels are still negatively buoyant thus creating a stably stratified intermediate zone above the interface (in an averaged sense). However, as they are lifted up, the strong directional nature of the distortions disappears, smaller scales are generated (with homogenization at molecular scales) and the fluid parcels lose their buoyancy via dilution; under these conditions the scalar fluctuations become passive. Further away from the interface, the passive scalar fluctuations are gradually weakened and dominated by the advection of already diluted fluid parcels by the eddies.

Figure 1(b) shows an interface at $Ri \approx 12$, where turbulence is present on both sides of the interface. The behavior of the entrained fluid parcels appears to be qualitatively the same as in the earlier case of a single turbulent layer, although the entrained fluid parcels themselves are already in a turbulent state. LIF observations indicate that the buoyant fluid parcels complete their dilution at a distance about $\xi \sim (0.5 - 1.0) L_H$ from the interface.

3   Spectral measurements of scalar fluctuations

As discussed above, the scalar gradient close to the interface is expected to be affected by buoyancy forces, as the entrained fluid blobs have not yet lost their buoyancy completely. The buoyancy gradient in this region $(db/dz)_a$ is expected to depend on Δb, L_H, h and u_H, where h is the interfacial thickness; since the interfacial buoyancy gradient (i.e., buoyancy frequency) $N^2 = \Delta b / h \sim \Delta b / L_H$, it is possible to surmise that, in the limit $h/L_H = $ constant (which is typical for this type of flows), $(db/dz)_a \sim N^2 f_1 (N^2 L_H^2 / u_H^2) \sim N^2 f_2(Ri)$, where $f_1, f_2, \ldots$ are functions. If this region is disturbed by eddies of wave number k, having an Eulerian frequency of $\omega (= u_H k)$, then the resulting buoyancy fluctuations should be of the order $b \sim N^2/k$, and hence the Eulerian frequency spectrum $E_b(\omega)$ should show the behavior

$$E_b(\omega) \sim b^2/\omega \sim N^4 u_H^2 \omega^{-3}. \tag{3.1}$$

Figure 2 shows two buoyancy fluctuation spectra taken at average distances of $\xi/L_H = 0.08$ and 0.1 from an entraining interface of an oscillating-grid, salt-stratified experiment. The Richardson number Ri for this experiment is about 100. Microscale conductivity probes were used for the measurements; a buck and gain amplifier and a high quality Kron-Hite low-pass filter were used to minimize the noise and aliasing problems during data acquisition. Note that the spectra

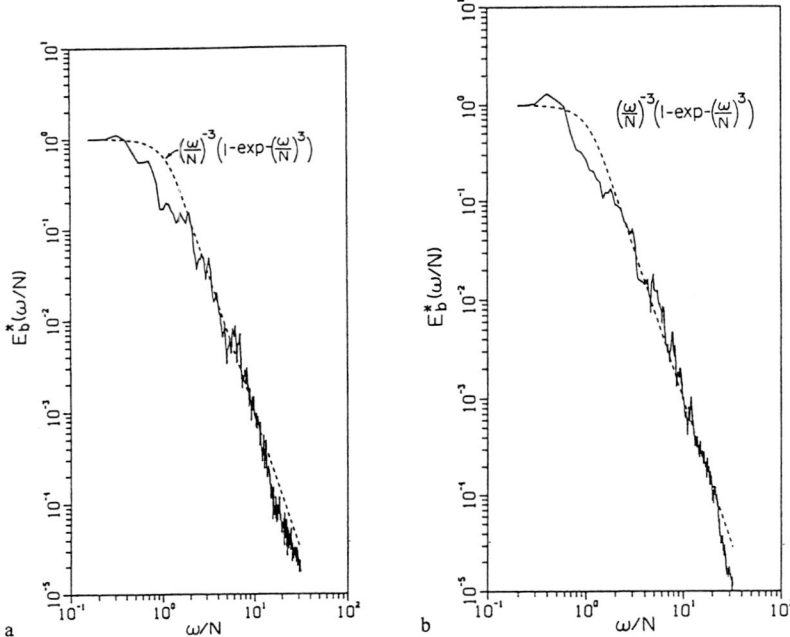

Figure 2. Salinity fluctuation spectrum $E_b^*(\omega/N)$ measured by the conductivity probe. Distances between the density interface and the probes are: (a) $h_{pi} = 0.4$ cm, and (b) $h_{pi} = 0.5$ cm

have been normalized so that $E_b^*(\omega/N \to 0) = 1$. The spectral shape obeys the form,

$$E_b^*(\omega/N) = \left(\frac{\omega}{N}\right)^{-3}\left[1 - \exp\left(-\frac{\omega^3}{N^3}\right)\right], \tag{3.2}$$

which is consistent with (3.1) at large ω/N values. The agreement of the measurements with (3.1) implies that the fluid parcels entrained at the interface are agitated by eddies that are advected by the integral scale motions, which eventually carry fluid parcels to the turbulent interior.

Figures 3(a) and 3(b) shows two spectra taken at $\xi/L_H \approx 0.36$ and 0.40, respectively. The data clearly show that, as the fluid parcels are advected away from the interface, the typical passive scalar spectral form $\omega^{-5/3}$ develops in the inertial subranges (note the locations of the Kolmogorov ω_k and Batchelor $\omega_{\nu c}$ scales). Then, further away from the interface, around $\xi/L_H \approx 0.56$, as these

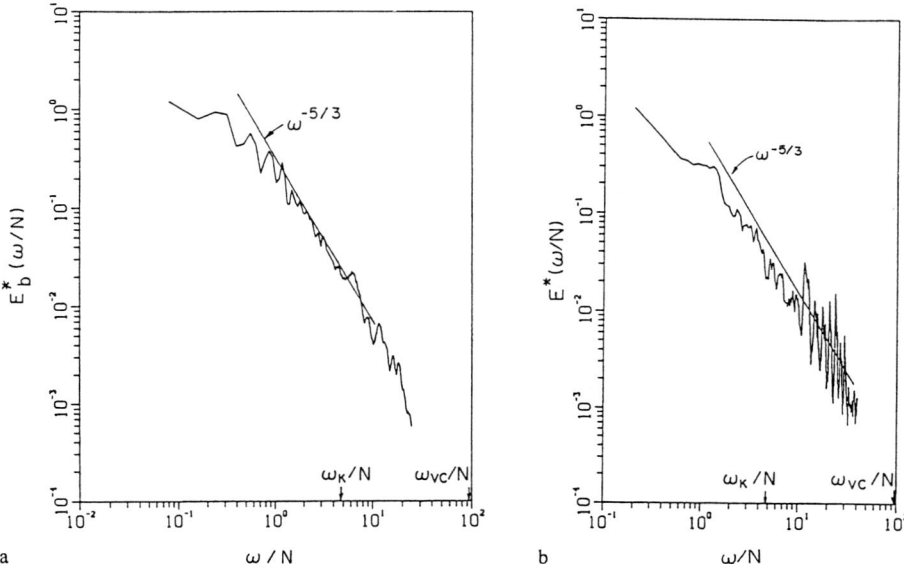

Figure 3. Same as Figure 2, but with, (a) $h_{pi} = 1.8$ cm, and (b) $h_{pi} = 2.0$ cm

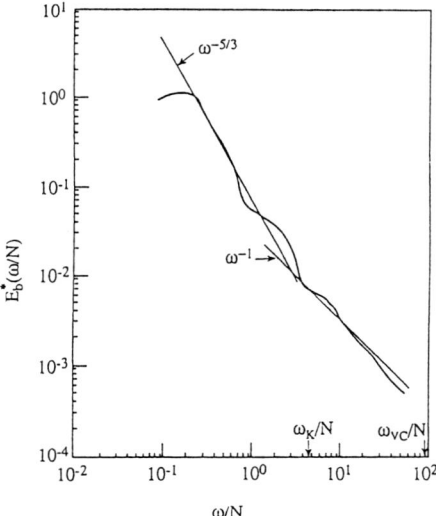

Figure 4. Normalized salinity fluctuation spectrum $E_b^*(\omega/N)$ at $h_{pi} = 2.8$ cm

scalar fluctuations are broken down to smaller scales, the viscous-convective part of the spectra appears to be developed, producing a ω^{-1} spectral subrange between ω_k and $\omega_{\nu c}$ (Figure 4). Thus, it appears that the entrained fluid parcels are degraded to molecular scales at distances around $z/L_H \approx 0.5$, and beyond that the entrainment process is completed via mixing.

4   Are the molecular diffusivities important?

Turner [3] pointed out, based on his grid experiments, that the entrainment rates in heat-stratified experiments are greater than that of salt-stratified experiments at a certain range of Ri. Given that all other experimental conditions remained the same, the only factor that can manifest the observed disparity was the differences in molecular diffusivities of heat and salt. Further experiments in this regard illustrated that the Peclet number Pe may play a determining role in mixing across the interface [5], in that when $Pe = u_H L_H/\kappa$ is less than about 200 the interface becomes purely molecular diffusive and local stirring events such as interfacial wave breaking are untenable. Noh and Fernando [6] also showed that the rapid decay of interfacial motions under low Pe conditions leads to retarded growth of interfacial waves, thus leading to low mixing rates.

Despite numerous studies of the type described above, the increased rate of mixing seen in heat stratified experiments remains to be explained using convincing arguments. Estimates show that, at high Pe, the entrained fluid mixes down to molecular scales in several eddy overturn times, independent of Pe [7], but the question is whether the entrained parcels can stay in the turbulent layer until the molecular diffusion completes mixing, without falling back on the interface. Consider the simple example of a fluid blob of radius R_o entrained into a turbulent fluid, which is stretched into a cylindrical shape of radius r by mean straining motions having a characteristic velocity u and lengthscale L. If the length of the blob at a given time t is l, then $dl/dt \sim u$, $r^2 l \sim R_o^3$, and at large times the radius changes as $r \alpha \sqrt{R_o^3/ut}$. When the diffusion velocity scale κ/r becomes of the same order as the radial velocity of the fluid element due to stretching, molecular diffusion is expected to take over the evolution of fluid mass. This time scale can be evaluated as $t_H \sim (R_o^3/\kappa u)^{1/2}$ or $t_H u/L \sim (R_o/L)^{3/2}(Pe)^{1/2}$. For the fluid parcel to complete the entrainment process, without being subjected to restratification, it should remain there at least for a time of order t_H. Since the rebounding time scale of the fluid parcels is of the order $u/\Delta b$, it can be inferred that the stirred fluid parcels by the entrainment can complete its mixing when,

$$Ri < \left(\frac{R_0}{L_H}\right)^{-3/2} Pe^{-1/2}, \tag{4.1}$$

where $u \sim u_H$ and $L \sim L_H$ have been assumed. This shows that at large Pe, the fall back of entrained fluid parcels are more likely even at relatively low Richardson numbers. The critical Richardson numbers above which incomplete entrainment occurs for given R_0, u_H and L_H values are about ten times larger for heat-stratified fluids than for salt-stratified ones.

Overall, an understanding of the disparities between heat and salt mixing rates remains a central issue in stratified turbulent mixing studies, from laboratory to geophysical scales. Gargett and Holloway [8] used the GFDL ocean model [9] to investigate the role of "effective diffusivities" of heat and salt in the evolution of oceanic circulation, meridional heat transport and vertical scales. Typically, in ocean models, the effective diffusivities of heat and salt are assumed to be the same, but in their study the diffusivity ratio was changed over a selected range. They found that the ratio of diffusivities indeed plays a very sensitive role in determining the steady state magnitude and direction as

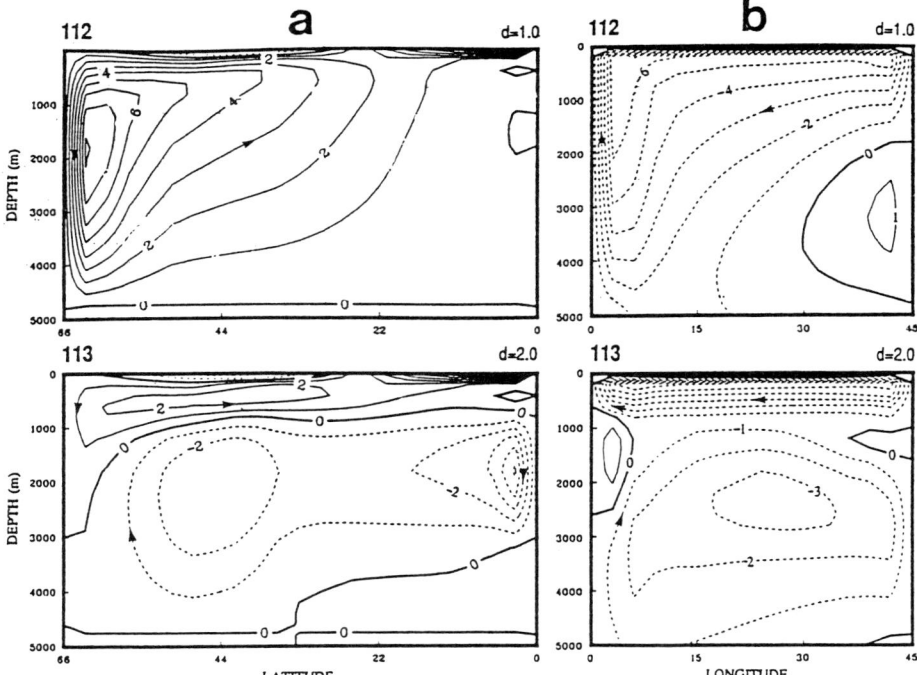

Figure 5. (a) Meridional, and (b) zonal overturning stream functions for model runs reported by Gargett and Holloway [8]. GFDL ocean model was used for the simulation of a oceanic region extending 45° in longitude and from 0° to 66° N in latitude. The resolution was 3° in both meridional and zonal directions with 15 levels in the vertical and a flat bottom. Two cases with the ratio of effective diffusivities between salt and heat $d = 1$ and 2 are shown

well as T/S characteristics of the thermohaline circulation (Figure 5). Since the differences of "effective" diffusivities ought to be related to the disparities of molecular diffusivities, the foregoing discussion points to the possibility that molecular effects can be an important ingredient for the parameterization of ocean mixing. Detailed studies on the dependence of effective diffusivities on molecular parameters, however, are yet to be reported.

5   Turbulence structure near density interfaces

The motion field within a density interface is driven by the nearby turbulence field, and therefore the understanding of the nature of mixing at an interface requires a knowledge of turbulence surrounding the interface. In the following discussion, we will focus on the distortion of homogeneous turbulence by a density interface and compare the past theoretical and numerical predictions with the experimental results obtained with shear-free oscillating-grid mixing experiments (when the grid system is properly designed, oscillating grids can yield approximately homogeneous turbulence). On the theoretical side, Hunt and Graham [10] and Hunt [11] have made a Rapid Distortion Theory (RDT) based analysis to predict how the homogeneous turbulence is affected by the presence of a solid surface. The method consists of the introduction of a solid plate to a homogeneous turbulent flow, which is specified by its three-dimensional energy spectral tensor. The condition of zero velocity normal to the surface instantaneously distorts the motion field, and through the pressure fluctuations this effect is immediately felt by the other components of the velocity. The resulting flow field is so adjusted as to satisfy the continuity near the surface, and hence the resulting distortion velocity field can be considered as due to a kinematic effect. If the undistorted homogeneous velocity field is given by $\underset{\sim}{u}^H(\underset{\sim}{x}, t)$ and the flow field after the distortion is represented by $\underset{\sim}{u}(\underset{\sim}{x}, t)$, then it is possible to write

$$\underset{\sim}{u}(\underset{\sim}{x}, t) = \underset{\sim}{u}^H(\underset{\sim}{x}, t) + \underset{\sim}{u}^D(\underset{\sim}{x}, t), \tag{5.1}$$

where $\underset{\sim}{u}^D$ is the induced velocity field for the flow to adjust to the new boundary condition. Because of the suddenness of the flow distortion at the boundary, it is possible to argue that the distortion field is irrotational, and $\underset{\sim}{u}^D(\underset{\sim}{x}, t) = -\nabla \phi$, where the velocity potential is given by $\nabla^2 \phi = 0$. Although this assumption is good for the initial stages of flow evolution, clearly there will be adjustments to the vorticity field near the interface at large times, via the viscous diffusion of vorticity from the boundary, baroclinic effects and the distortion of eddies by the surrounding velocity field wherein the vortex filaments are stretched and compressed. Thus, as a first approximation, the flow distortion by the boundary can be treated as irrotational for short time periods after the introduction of the interface. A schematic of such a (hypothetical) distorted flow field is given in Figure 6(a), where the development of irrotational motions surrounding the eddies is shown. Figure 6(b) shows the modification of vortex structures via stretching by larger eddies.

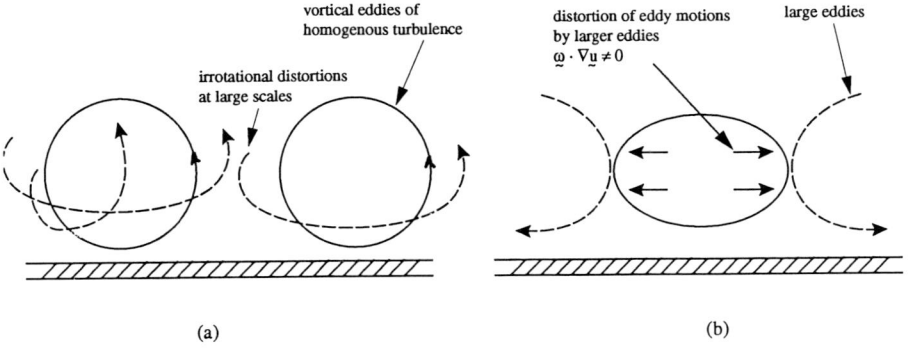

Figure 6. (a) A schematic illustrating the RDT-based assumption that the vorticity of the eddies remains unchanged during the flow adjustment. Calculations show that additional (large-scale) irrotational motions are generated to satisfy the normal boundary condition, and (b) Stretching of eddies by larger eddies, thus amplifying the horizontal velocities. This process has a lesser effect on the w-component because of wall locking

A comparison of horizontal and vertical r.m.s. velocity measurements taken at various distances from the density interface (ξ) in an oscillating grid experiment with the predictions of the linear theory of Hunt and Graham [10] is given in Figure 7. Also included are the measurements of Hannoun et al. [12]. Note that the agreement between the RDT prediction of the vertical component and the data is satisfactory, but the u-component data showed a marked disagreement with the theoretical prediction when $\xi/L_H < 0.5$. The observed amplification of $\overline{u^2}/u_H^2$ at smaller $\frac{\xi}{L_H}$ can be attributed to the adjustment of vortex structures near the boundary as discussed above.

Hunt [11] developed a model to estimate the change of turbulence by nonlinear vorticity distortion processes near the boundary, specifically by systematic vortex stretching mechanism illustrated in Figure 6(b). This stretching causes an amplification of the u-component, but the blocking effect of the wall precludes substantial changes to the w-component. According to the estimates of Hunt [11], the non-linear vortex dynamics increase the r.m.s. horizontal velocity according to

$$u^2 = u_H^2 \left[1 + \frac{1}{3} \left(\frac{L_H}{\xi} \right)^{1/2} \right]. \tag{5.2}$$

In Figure 7, the prediction 5.2 is also plotted. When $\xi/L_H > 0.5$, the expression 5.2 overpredicts the horizontal velocity component, indicating that additional vortex distortions are not significant at such distances. Since the energy

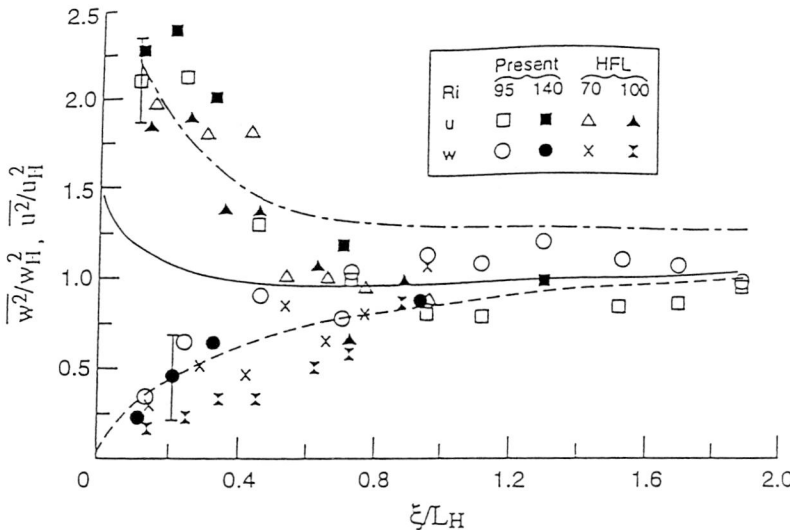

Figure 7. The normalized variance of the vertical and horizontal velocity components versus the normalized distance from the density interface. The r.m.s. velocity and integral lengthscales corresponding to undistorted turbulence have been used for normalization. The dashed and solid lines correspond to the theoretical relations obtained by Hunt and Graham [10] for vertical and horizontal components, respectively. The chain line corresponds to the non-linear correction to the Hunt and Graham [10] solution by Hunt [11]. HFL is the abbreviation for the data from Hannoun, Fernando and List [12] taken at various Ri

flux from the turbulent flow to the interface is governed by the w-component via pressure-velocity correlation and since w is well predicted by RDT, one may further conclude that the RDT-based analysis may adequately predict the energy flux absorbed into the interface from the contiguous turbulent flow (measurements of Hannoun et al. [12] show that the non-linear transport of energy into the interface via energy flux divergence is close to zero).

The RDT-based analysis provides information only on statistically averaged quantities, and does not provide details of turbulence structures near the interface. To this end, the results of direct numerical simulations (DNS) are useful, though they are usually carried out at relatively low Reynolds numbers. DNS calculations on the distortion of homogeneous decaying turbulence by solid surfaces with differing boundary conditions (permeable, free and solid surfaces) have been reported by Perot and Moin [13]. Their results show that the imprints of eddies impinging on the interface (called splats) can cause systematic

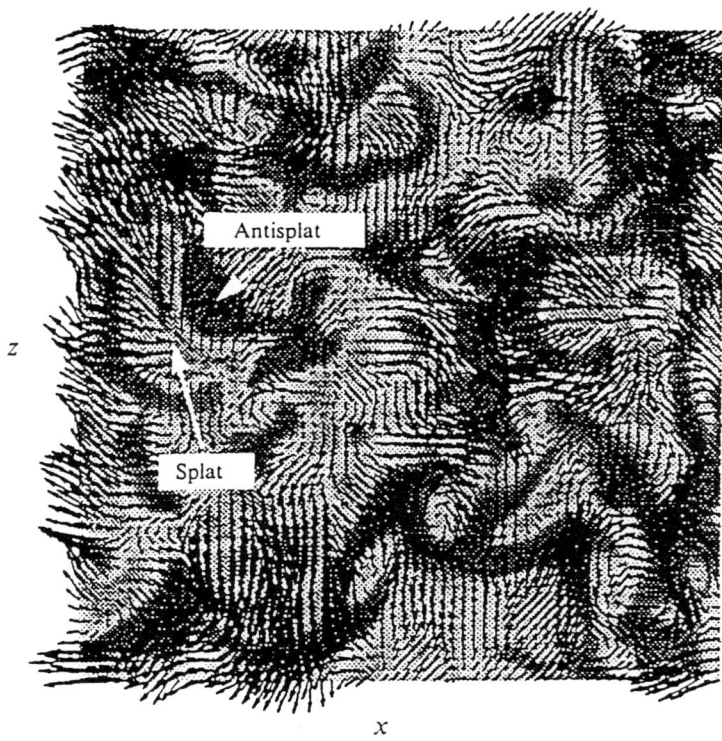

Figure 8. Velocity vectors in a plane parallel to an ideal free surface located contiguous to a decaying homogeneous turbulent fluid layer ($Re = 134$). Shading shows regions of outward flow and lighter regions indicate regions of impinging flow (from Perot and Moin [13])

high-pressure stagnation regions on the interface. Flow diverging at the "splats" is directed horizontally over the interface until they meet a similar flow of another splat. The resultant converging flows meet and then diverge away from the interface, typically in narrower regions called "antisplats." Figure 8 shows a velocity vector plot taken from Perot and Moin [13], showing the presence of splats and antisplats on a free surface.

DNS calculations of pressure-strain correlation near a solid boundary show that it is a dominant term, supporting the possibility that splats, upon impingement on the interfaces, transfer energy from the vertical to the horizontal components. This notion is consistent with the data shown in Figure 7 which indicates a systematic reduction of the w-component and an increase of the

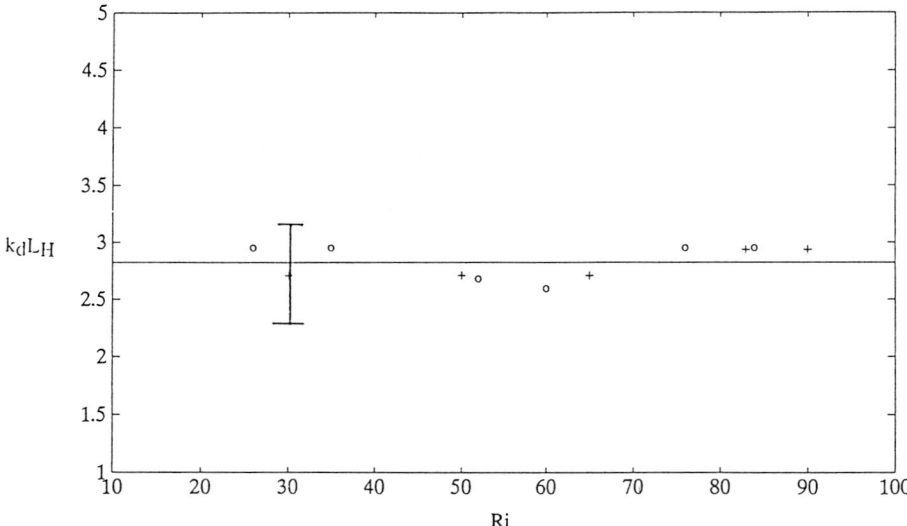

Figure 9. Variation of the normalized (with the integral scale L_H) dominant disturbance wave number at the interface with the Richardson number. o - turbulence is present on one side of the interface; + - both layers above and below the interface are turbulent

u-component. The results of the numerical simulations of Perot and Moin [13], however, did not show an amplification of the u-component, except for a high Reynolds number simulation which showed an increase of $\overline{u^2}/u_H^2$ near the interface. This was attributed to the low decay rate of $\overline{u^2}$ near the interface compared to u_H^2, which may misleadingly show that the increase of $\overline{u^2}/u_H^2$ as an amplification of the $\overline{u^2}$. Perot and Moin [13] asserted that the horizontal motions of splats are strongly influenced by the viscosity, in that a significant fraction of the incoming energy of the splats is lost before the fluid leaves as antisplats, and hence the intercomponent energy transfer is dominated by the viscous frictional effects. The calculation of the rate of dissipation near the surface, however, showed a reduction of dissipation rather than an increase–an observation which is at odds with their assertion of increased viscous influence on horizontal motions near the interface!

The fluctuating pressures so created on an interface are expected to excite internal waves (if the splats do not have enough vertical inertia to break up the interface), with the scale of the splat (or the horizontal scale of excited waves) being the scale of energy containing eddies near the interface. Figure 9 shows the measurement of the dominant horizontal wavelength k_d at the interface. The results indeed show that the scale of the wave disturbance is of the order of the integral scale of turbulence near the interface.

6   Discussion

The results of laboratory studies described herein show that the buoyant fluid parcels crossing from a stratified layer to a turbulent layer through a density interface can lose buoyancy rather rapidly when the Peclet number is relatively low. At large Ri and high Pe, such fluid parcels may return back to their original fluid layers (detrainment), thus reducing the efficiency of mixing. Typically, the exchange of fluid parcels across the density interface occurs by local instability mechanisms. First, partially mixed fluid blobs are generated in the vicinity of the interface, and these are then advected over the interface by larger integral-scale eddies. During this advection, the fluid parcels are broken down into smaller scales, and at a distance about $\xi/L_H = 0.5$ from the interface the molecular-diffusive effects can complete the homogenization of fluid parcels with the ambient fluid. In this region, the concentration fluctuations behave similarly to that of a passive scalar in unstratified turbulence (see Kit and Fernando [14]).

It was found that the vertical velocity fluctuations near the interface are well predicted by the Rapid Distortion Theory-based analysis of Hunt and Graham [10], but the horizontal velocity is underestimated by this theory. A modification that accounts for the stretching of small-scale eddies by the large scales [11] predicted satisfactorily the r.m.s. horizontal velocities near the interface. Despite indications from the direct numerical simulations to the contrary, it appears that there is intercomponent energy transfer from the vertical to the horizontal components near the interface, and this energy transfer occurs at large scales covering the integral scale and the low wave number end of the inertial subrange (also see [15]).

Acknowledgements

Stratified and rotating turbulent flow research at Arizona State University is supported by the Office of Naval Research, Army Research Office and the National Science Foundation.

Bibliography

1. Munk, W.H. (1966). "Abyssal recipes". *Deep-Sea Res.*, **13**, pp. 707–730.

2. Kunze, E. (1996). "Abyssal mixing: where it is not". *J. Phys. Oceanogr.*, **26**, pp. 2286–2296.

3. Turner, J.S. (1968). "The influence of molecular diffusivity on turbulent entrainment across a density interface". *J. Fluid Mech.*, **33**, pp. 639–656.

4. Thompson, S.M. and Turner, J.S. (1975). "Mixing across an interface due to turbulence generated by an oscillating grid". *J. Fluid Mech.*, **67(2)**, pp. 349–368.

5. Crapper, P.F. and Linden, P.F. (1974). "The structure of the turbulent density interfaces". *J. Fluid Mech.*, **65**, pp. 45–83.

6. Noh, Y. and Fernando, H.J.S. (1993). "The role of molecular diffusion in the deepening of the mixed layer". *Dyn. Atmos. Oceans*, **17**, pp. 187–215.

7. Fernando, H.J.S. and Hunt, J.C.R. (1996). "Some aspects of turbulence and mixing in stably stratified layers". *Dyn. Atmos. Oceans*, **23**, pp. 55–61.

8. Gargett, A.E. and Holloway, G. (1992). "Sensitivity of the GFDL ocean model to different diffusivities for heat and salt". *J. Phys. Oceanogr.*, **22**, pp. 1158–1177.

9. Bryan, K. and Cox, M.D. (1967). "A numerical investigation of the oceanic general circulation". *Tellus*, **19**, pp. 54–80.

10. Hunt, J.C.R. and Graham, J.M.R. (1978). "Free stream turbulence near plane boundaries". *J. Fluid Mech.*, **84**, pp. 209–235.

11. Hunt, J.C.R. (1984). "Turbulence structure in thermal convection and shear-free boundary layers". *J. Fluid Mech.*, **138**, pp. 161–184.

12. Hannoun, I.A., Fernando, H.J.S. and List, E.J. (1988). "Turbulence structure near a sharp density interface". *J. Fluid Mech.*, **189**, pp. 189–209.

13. Perot, B. and Moin, P. (1995). "Shear-free turbulent boundary layers". Part 1: Physical insights into near-wall turbulence. *J. Fluid Mech.*, **295**, pp. 199–227.

14. Kit, E. and Fernando, H.J.S. (1997). "Frequency spectra of scalar fluctuations at entraining stratified interfaces". *Fluid Dyn. Res.*, **19**, pp. 65–75.

15. Kit, E., Strang, E. and Fernando, H.J.S. (1997). "Measurement of turbulence near shear-free density interfaces". *J. Fluid Mech.*, **334**, pp. 293–314.

Development and Erosion of a Mixing Layer in a Stratified Flow

S.A. Walker, G.A. Hamill and H.T. Johnston

Department of Civil Engineering, Queen's University of Belfast, Northern Ireland

Abstract

The temporal behaviour of a salt water impoundment due to a buoyant shear flow is presented. Mixing has been found to result in the formation of a third distinct layer, the characteristics of which change with time. Various fresh water velocities and depths combined with various salt water densities, were used to obtain a range of bulk Richardson numbers. Two regimes of interfacial layer behaviour have been identified. In the first regime the density of the layer decreases to a constant value, while in the second regime the density and thickness of the layer fluctuates as the layer is removed by sequential stripping. Transport rates across both boundaries of the interfacial layer are assessed from the change in density profile. Average transport rates for each test were obtained and found to best related to the bulk Richardson number.

1   Background

The most fundamental characteristic of an estuary is the interaction between fresh and salt water. When the river flow is high, the salt water and fresh water mix. When the river flow is low, a stratification forms as the less dense fresh water from the river overrides the salt water. Due to the frictional forces developed along the interface between the layers, the lower saline mass will develop into a wedge. The salt wedge remains quasi-stationary, advancing or retreating due to variations in river flow rates and tidal cycles. This condition maybe be repeated where water of different densities meet in conditions that do not induce significant mixing. Density stratification in fluids can also occur as a result of differences in temperature and concentration of suspended sediment. Thermally stratified flows may be generated by solar heating or by the discharge of power plant or industrial plant cooling water. A special case of highly stratified flow is an entrapped wedge, which occurs when a pond of saline water becomes retained behind a barrier of some form. The saline water is isolated from the ocean by the barrage and is dependent upon the transfer of oxygen from the fresh water. When vertical mixing is predominately of a diffusive nature, there is little transfer of oxygen between the fresh and salt water layers, and a stagnant layer with low dissolved oxygen content results.

229

The behaviour of arrested and entrapped wedges differ in some important ways, the main differences being the lack of circulation within the wedge. New barrages may be constructed as part of estuary rehabilitation, examples of which are the Tees (Cardiff Bay) and Lagan (Belfast) barrages, or to prevent flooding, as in the Thames, the Dutch North Sea Defence and the Venice Barrages [1].

The interface of a saline wedge in an estuary can have a major influence on the ecology of the area and an understanding of its behaviour is therefore of considerable importance to river managers/engineers, environmental agencies and industries. Discharge to, and water supply extraction from, the estuary are of particular interest to industrial users and river managers. Thermal plumes discharging to the estuary may create stratification problems if there are insufficient currents to disperse the plume.

The behaviour of stratified flows has been studied experimentally in laboratory flume tests by many previous researchers [2–6]. Dimensionless parameters which influence the mixing processes, entrainment rates, and the shape of the wedge have been found from these experiments. The Richardson number is a measure of the ratio of gravity forces to inertial forces in a shear flow having a vertical density gradient. It is widely used as a measure of the stability of the flow with regards to vertical mixing and has two forms; the gradient Richardson number (Ri_g) defined as:

$$Ri_g = -g/\rho \frac{\partial \rho/\partial z}{\{\partial v/\partial z\}^2} \tag{1.1}$$

where $\partial \rho/\partial z$ is the vertical density gradient, $\partial v/\partial z$ is the vertical gradient of horizontal velocity, g is the acceleration due to gravity and ρ is the fresh water density.

The bulk Richardson number (Ri_b) is defined as:

$$Ri_b = g \frac{\Delta \rho H}{\rho V^2} \tag{1.2}$$

where $\Delta \rho$ is the density difference, H is the depth of fresh water and V is the average fresh water velocity.

Other dimensionless parameters such as the Reynolds number, used to define the degree of turbulence [7,8], and the inverse of the Keulegan number, used to investigate interfacial shear stress [9,10], are also relevant to entrainment.

One of the earliest studies of entrainment rates was conducted by Ellison and Turner [11] who conducted two series of laboratory experiments in a channel 5 m long and 15 cm wide. The entrainment rate ($E = u_e/V$) found was observed to fall rapidly with increasing Richardson number, where u_e is the entrainment velocity and V is the average turbulent layer velocity. The range of Ri covered was low (< 1) and the experimental set-up was considerably different to that expected in an estuary, which limited the usefulness of the results. A number of model tests were performed by Suga [12] on an arrested saltwater wedge. The tests were conducted in two channels, the larger being 100 m long and 0.8 m wide and the other was 30 m long and 0.3 m wide. A mixing equation was obtained for the entrainment rate over the range of $1 < Ri < 300$. This equation is defined as:

$$E = 6 \times 10^{-3} Ri^{-5/3} \quad 1 < Ri < 300. \tag{1.3}$$

Field studies were conducted by Buch [13], in several stratified fjords. He concluded that the vertical mixing was due essentially to the production of turbulent energy originating from the interfacial shear stresses. The equation which was obtained by Buch for the entrainment rate over the range $50 < Ri < 250$ is defined as:

$$E = 5.2 \times 10^{-4} Ri^{-1} \quad 50 < Ri < 250. \tag{1.4}$$

Various experimental set-ups have been used by many researchers from which empirical equations of entrainment rates have been obtained. Entrainment rates for various differing experimental set-ups were reviewed by Christodolou [3]. Data from previous work [11,13–20] was re-examined. Christodolou developed governing laws for four types of stratified flows by looking at the turbulent interfacial mixing in flows due to surface stress, buoyant overflows, density currents and counter-flows. Christodolou concluded that for low Ri numbers (< 1) mixing occurs by vortex entrainment and for larger Ri values, mixing occurs by cusp entrainment. By plotting the results from the re-analysed data, Christodolou was able to relate the mixing phenomenon to the equations:

$$E = 0.007 Ri^{-1/2} \quad 0.001 < Ri < 1 \tag{1.5}$$

$$E = 0.007 Ri^{-3/2} \quad 1 < Ri < 100. \tag{1.6}$$

For some range around $Ri = 1$, the two mechanisms seem to co-exist, an intermediate expression described by:

$$E = 0.002 Ri^{-1} \quad 0.1 < Ri < 10 \tag{1.7}$$

was proposed. At $Ri < 0.01$, the mixing rate was considered to be approximately constant at 0.07.

The entrainment rates for an arrested saline wedge were examined by Grubert [5]. Experiments were conducted in two open channels, one 0.31 m wide and 15 m long and the other 0.61 m wide and 30 m long. With fresh water passing over a pool 5 m long, mixing rates were measured and several equations obtained. The work of Lofquist [14] and Keulegan [21] was extended by using a larger channel which increased the degree of accuracy of the rates measured and, by investigating the influence of Re and channel friction on the mixing rates for comparison with rates found in an estuary. The rate of mixing was obtained by monitoring the level of the interface in the channel at regular time intervals. Salt water was added at a constant rate with a constant discharge of fresh water. By computing the difference in volume of salt water supplied to, and lost from the trough, estimations were made of the interfacial mixing. Considerable differences were found in the rates of mixing observed and those found in an estuary. These were explained by the differing values of Re found in each circumstance. Grubert suggested that "at the Reynolds numbers of the experiments conducted, the interfacial layer was mainly in a quasi-laminar state, similar to that of lakes, whereas in the prototype estuaries the interfacial layer was turbulent". Grubert concluded that an accurate prediction of the interfacial mixing was not dependent on Ri alone but depended on accurate calculation of the interfacial friction.

A large number of entrainment experiments have been performed, most of which concentrate on the entrainment laws. The type of turbulence produced, the experimental set-up and the Ri range have been shown to influence the equations produced. Little work has been carried out on the temporal erosion of the saline layer and the parameters which influence it, with the most recent and relevant work being carried out by Grubert [5]. In this work it has been shown that Re has an influence on the entrainment rate but the main dimensionless parameter used to define entrainment is the bulk Richardson number, which combines the velocity, depth and density difference.

One of the main aims of this study was to develop a more complete understanding of the behaviour of the entrapped wedge with particular attention to the development and characteristics of the interfacial or mixed layer.

2   Apparatus and procedure

Experiments were conducted in a laboratory flume which was 20 m long, 0.75 m wide and 0.75 m deep, a schematic diagram of which is shown in Figure 1. The salt water was retained between two baffles which were 17.5 m apart. The salt water depth was kept constant at 0.25 m whilst the fresh water depths were varied by means of an adjustable weir. The fresh water was pumped over the top of the stationary salt water impoundment and various velocities and depths, combined with differing salt water densities were chosen to investigate the effect of a range of Richardson numbers.

Fresh water was mixed with dried salt and dye in the storage tanks and then pumped into the salt water impoundment to a depth of 0.25 m. Fresh water was then filled slowly on top, over the course of 12 to 15 hours. Flow was initiated slowly in the upper layer by starting the fresh water pump. Once the flow had reached the required rate, the water in the storage tanks was changed slowly by opening the waste valve and filling from the supply. This was necessary due to the contamination of the fresh water which occurred as a result of mixing. The degree of contamination was relatively small, but large enough to change the salinity of the fresh water over a substantial time period. The scanning process was commenced after flow was initiated. Velocity and density profiles were obtained at the centre of the channel. Velocity measurements were obtained using LDA with a fibre optic attachment to limit the effect of differing refrac-

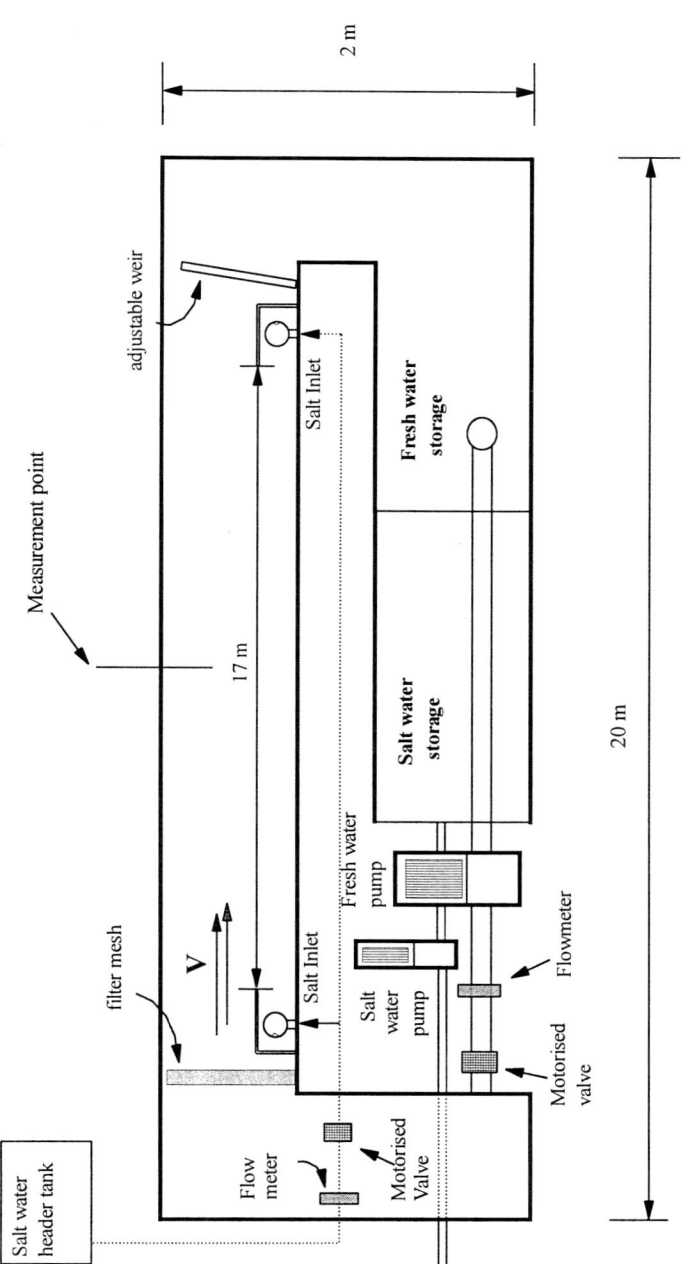

Figure 1. Experimental facility

tive indices. Density measurements were obtained using a conductivity probe with small measuring volume. The probe consisted of two wires 5 mm thick which were completely covered except for 2 rings, 2 mm thick, near the bottom. Both probes were mounted on a linear actuator, the position of which was computer automated. The probes were placed in the salt water, usually about 20 to 30 mm below the interface in order to obtain a reading of maximum salinity. An average was taken over a period of 30 seconds and the probes were then moved up 5 mm and another point measurement obtained.

Due to the temporal nature of the experimental investigation it was necessary to obtain as many profiles per hour as possible. The profile obtained was therefore not a full vertical traverse but was limited to 120 mm around the interface. With time the interface moved downwards, as salt water was removed by entrainment from the impoundment, and it was necessary to incorporate movement of the base profile position into the scanning program.

3   Temporal characteristics of interfacial layer

Initially the interface between the two layers is clearly defined and a typical density profile is plotted in Figure 2(a). The profile is plotted in a dimensionless form by plotting S/S max against Z_o/H, where S is the local density difference, S max is the initial density difference, Z_o is the position from the original interface and H is the depth of fresh water. On the vertical scale 0 is the original interface and 1 is the water surface and on the horizontal scale 0 is the fresh water and 1 indicates fully saline water. Also plotted on the figure is the actual position in the channel. The depth of salt water was kept constant at 250 mm for all tests. The profile obtained after 15 minutes in most tests was similar to that expected, with the same basic shape as the initial condition but with a more sudden change to fully fresh than to fully saline. The density profile was continually monitored with time and profiles were obtained after approximately 1 hour 30 minutes. Due to the time taken to obtain a vertical traverse of the probes it was only possible to monitor tests with a sufficiently slow rate of salt water removal. This ensured that the density profile did not alter significantly during the traverse.

With time a third distinct layer was visible, as shown in Figure 2(b). The density profile had changed and two boundaries can be seen. When examined closely it was observed that the interfacial layer formed at the downstream end of the flume and elongated with time. The density of the layer at the measuring point was found to change with time, as shown in Figure 2(c). When the thickness and density of the layer, at the centre of the channel, were plotted with time, two mechanisms of erosion were identified. These were classified as oscillating and non-oscillating. Significant differences in the temporal characteristics of the velocity and density profiles of the two mechanisms were found. In the first mechanism, the density of the layer decreased rapidly and then reached a constant value, as shown in Figure 3. In the second mechanism the density and

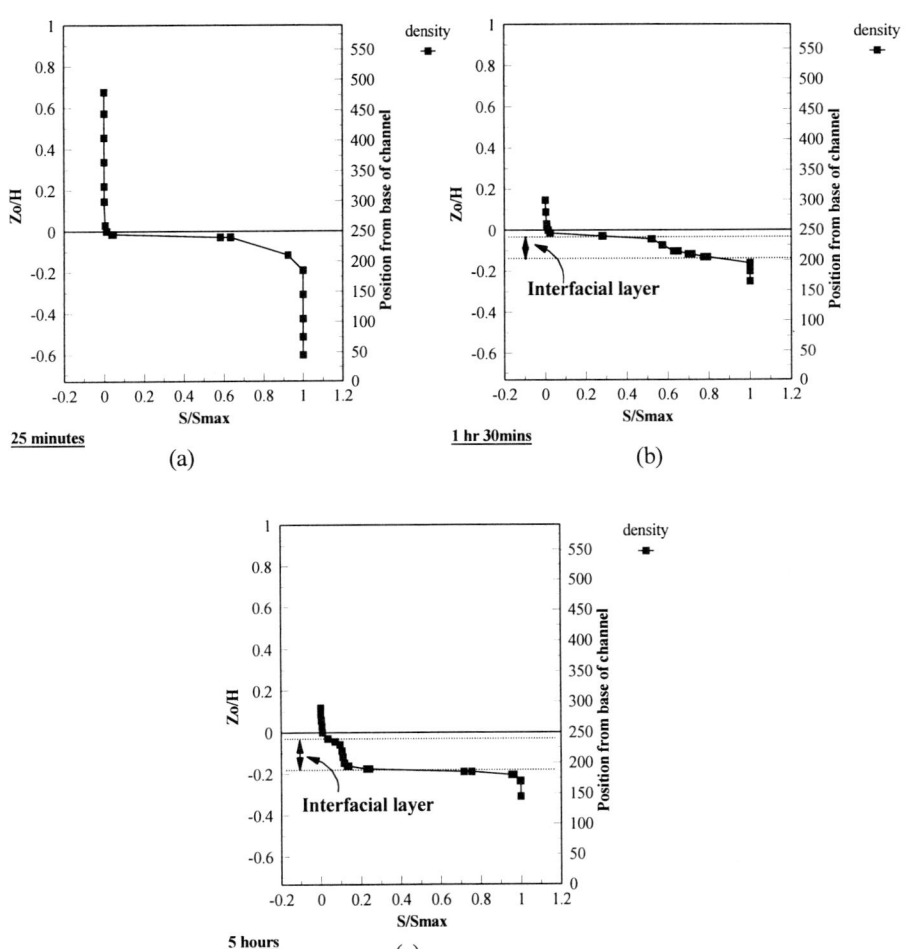

Figure 2. Density profiles, (a) After 25 minutes, (b) After 1 hour 30 minutes, and (c) After 5 hours

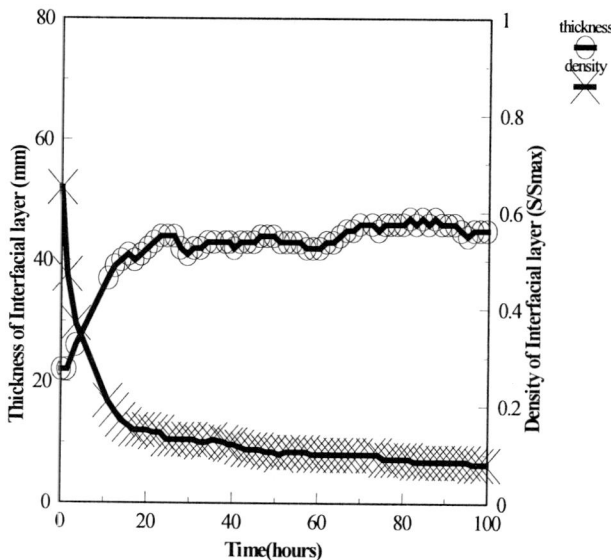

Figure 3. Relative interfacial layer density and thickness (non oscillating)

thickness of the layer showed a cyclic pattern as the layer developed and eroded. Figure 4 shows the temporal density and thickness of a typical test which showed this behaviour.

3.1 Non-oscillating

During the tests which reached a steady state condition, there were times when a small circulation velocity within the layer was observed and the impoundment eroded slowly. In other cases the velocity within the layer was large and additional circulation cells could be seen in the lower fully saline layer. This resulted in rapid erosion of the impoundment whilst the density within the layer remained constant. In all cases, the layer formed downstream next to the weir and extended the full length of the channel. The time taken to reach the length of the channel depended upon the velocity of the circulation within the layer. The values of velocity and density gradient obtained from the profiles, remained constant when the layer became established, as shown in Figure 5. As a result, the value of the gradient Richardson number (Ri_g) remained constant for the duration of the test, as shown in Figure 6. The layer was in an equilibrium condition and the value of Ri_g was above the critical number ($Ri_g = 0.25$) for instability. Relative turbulence intensity measurements showed an approximately uniform

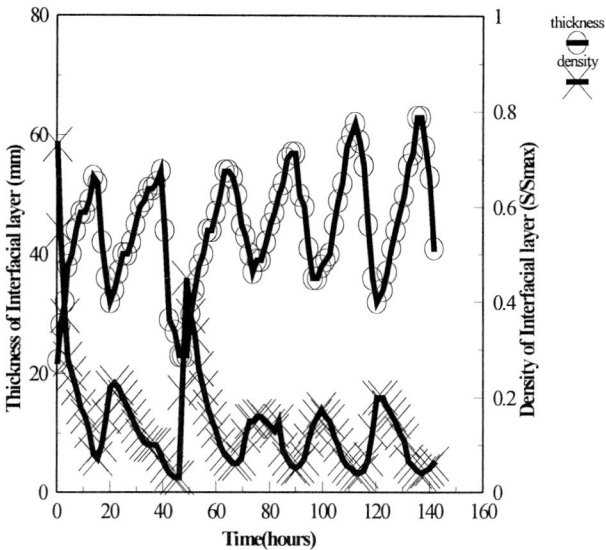

Figure 4. Relative interfacial layer density and thickness (oscillating)

value at both the upper and lower boundaries, the value at the upper boundary
being slightly larger.

3.2 Oscillating

For those tests which displayed a cyclic behaviour, the velocity and density profiles showed considerable velocities within the layer. Using the density profiles,
the layer position was plotted with time. The boundaries were defined by the
location of the points of maximum density gradient. Periods of rapid upper
boundary movement were found corresponding to times when part of the mixed
layer was washed out.

Associated with this, when the layer was thinnest, there appeared to be a
corresponding increase in lower boundary movement as the free stream velocity
was now much closer to this boundary. The layer also showed considerable
differences in upper boundary slope. This slope was not uniform along the
length of the channel but changed as the density within the layer varied along the
channel. From this it would appear that as the density of the layer decreases, the
interfacial slope increases in order to resist the interfacial shear stress. Velocity
and density gradients for a typical test are shown in Figure 7. The interfacial
velocity and density gradients showed a definite cyclic behaviour, with maximum
velocity gradients occurring when the layer was most dense. This suggested an
effect of the density difference on the velocity gradient and the position of the

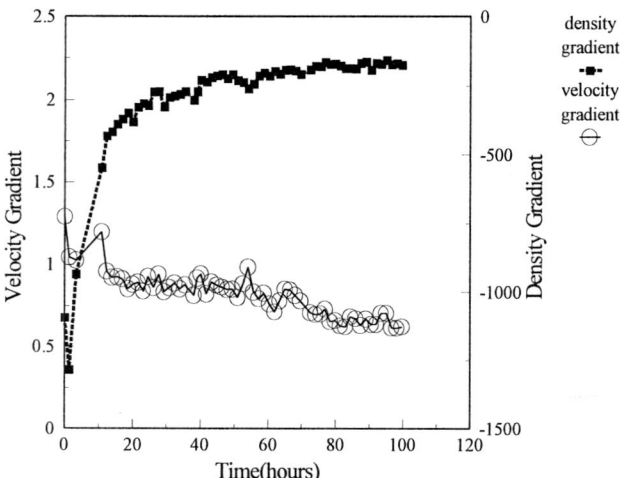

Figure 5. Interfacial velocity and density gradients (non oscillating)

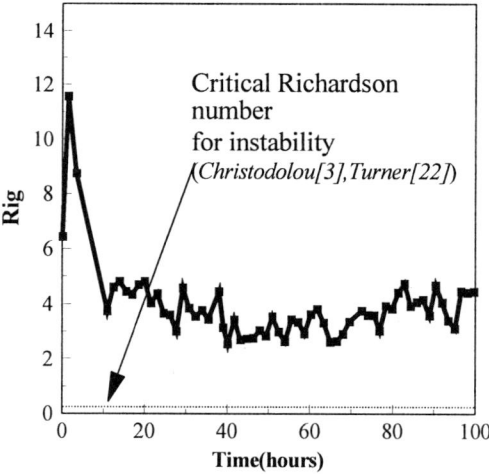

Figure 6. Variation of gradient Richardson number (non oscillating)

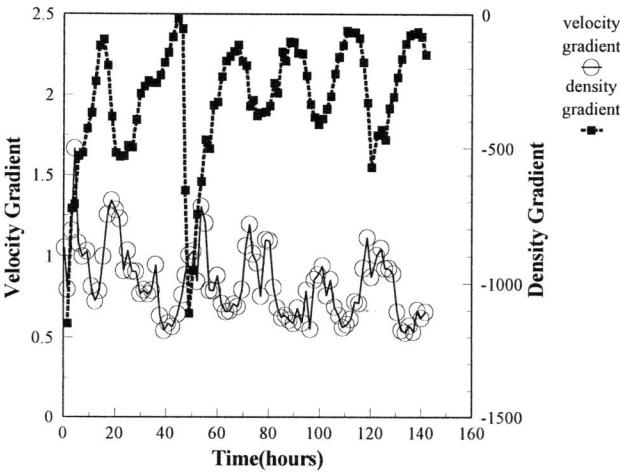

Figure 7. Interfacial velocity and density gradients (oscillating)

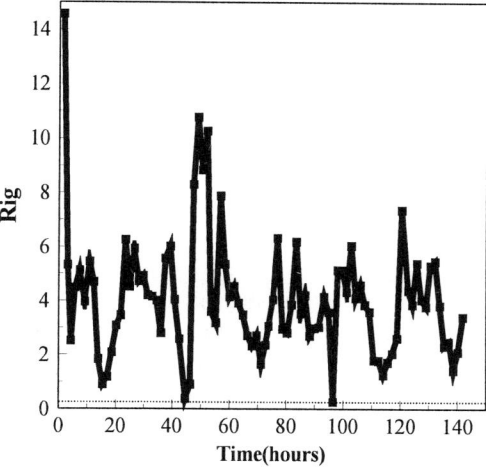

Figure 8. Variation of gradient Richardson number (oscillating)

maximum velocity in the profile. From these gradients, the gradient Richardson number was plotted with time. Figure 8 shows the variation of Ri_g with time for a typical oscillating test. The value of Ri_g dropped to the order of 0.25 in most cases when the upper part of the layer was removed. This suggested that the flow was unstable at this point in the test and rapid mixing occurred as a result of large amplitude breaking internal waves. The turbulence intensity at both interfaces showed a corresponding cyclic pattern. An initially high value of turbulence at the upper boundary decreased as the layer formed and the value at the lower boundary increased as the circulation cell became established in the layer. Greater turbulence was found to occur at the upper interface when the velocity gradient was high, which corresponded to times when the density within the layer was also high.

4   Transport rates

The density profiles were used to determine the rate of erosion of the salt water impoundment. By comparing the position of the boundaries, and the density and thickness of the interfacial layer, it was possible to estimate the rate of salt water removal. Transport across the upper boundary was a measure of the salt water removed during each test, as any salt water crossing this boundary entered the fresh water and was removed from the system. Movement across the lower boundary was different in that the movement was into the mixed layer.

Profiles were obtained, on average, every 1 hour 30 minutes and it was assumed that the transport rate was constant during this period. Each profile was approximated to a square profile based on the position of the boundaries and density of the layer. Any change in density of the interfacial layer, or drop in position of the upper boundary, was deemed to be due to transport across the upper boundary. Any change in position of the lower boundary was calculated as transport across this boundary. The rate of transport across both boundaries of the interfacial layer for a typical test which showed the non-oscillating interfacial density behaviour is shown in Figure 9. Also plotted on the figure is the density within the layer. This enabled key points within the test to be compared with the rate of transport. The figure shows that the rate of transport across the upper boundary was initially considerably higher than that at the lower boundary. Transport at the lower boundary showed a slight increase as the layer formed, and after 10 hours a constant value of 0.5 kg/hr had been reached. Transport at the upper boundary continued to decrease slowly during the test until at around 20 hours, when it remained constant with a value of 0.7 kg/hr. The fact that the upper value was higher than that at the lower interface showed that a greater amount of salt was being removed from the interfacial layer. This was confirmed by the relative density values within the layer which showed a continual decrease during the test.

The transport rates across both boundaries for a typical test which showed oscillating behaviour are plotted in Figure 10. A rapid movement of salt was

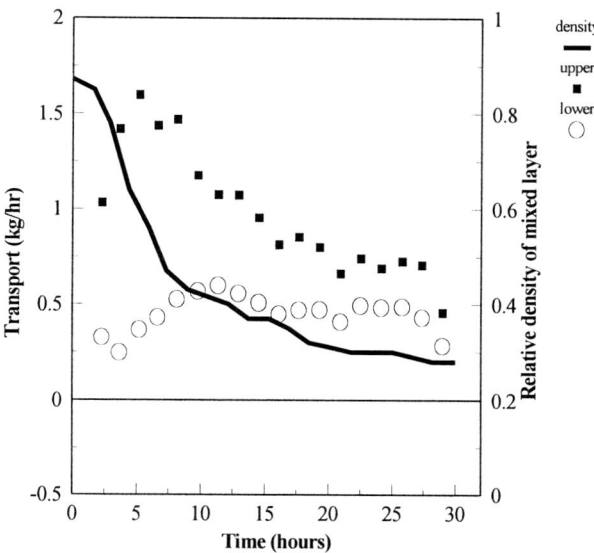

Figure 9. Temporal transport rates across both boundaries (non oscillating)

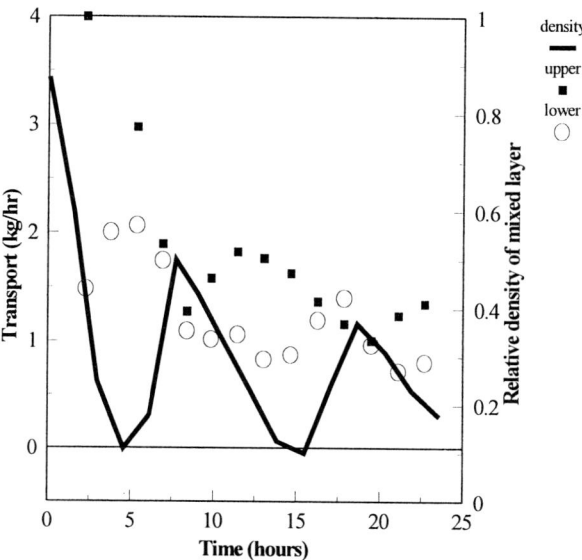

Figure 10. Temporal transport rates across both boundaries (oscillating)

observed initially as the layer developed and the resulting rate of transport at the upper boundary was high (not shown on Figure 10 at 2 hours 30 minutes and 4 hours due to the limited scale). Transport across the lower boundary increased and then decreased again as that across the upper decreased until 10 hours when both were equal. A cyclic behaviour is evident from the figure with greater rates being observed at the upper boundary when the density within the layer was decreasing. At certain times during the cycle, the transport rate at the lower boundary was higher than that at the upper boundary. This increased rate of movement at the lower boundary corresponded to times when the density within the layer was increasing. To allow rates to be compared easily, the values for each test were averaged to give an overall average rate for each test. For some tests up to 100 values were averaged while in others only 5 were used depending upon the duration of the test. The values obtained had dimensions of kg/hr and, in order to compare values with other researchers, the rates were then non-dimensionalised in a manner similar to that of Christodolou [3]. The values were divided by the surface area of salt water in contact with the fresh water to give an entrainment velocity which was then non-dimensionalised with respect to the average fresh water velocity. The entrainment rate was plotted against the bulk Richardson number (Ri_b). This form of transport rate allowed direct comparison of the data with existing transport formulae. Figure 11 shows the entrainment rate plotted against Ri_b for both boundaries. The figure shows a basic trend of increasing entrainment rates with decreasing Ri_b. The figure also shows that the oscillating tests (labelled as "O" on the figure) follow a clearly defined trend. The non-oscillating tests (labelled as "non-O" on the figure) show a less well defined pattern but the same overall trend of decreasing transport with increasing Ri_b.

In all cases the rate of transport across the upper boundary (or the rate of entrainment) was greater than that across the lower boundary. A best fit line was obtained for each boundary and each mechanism. The equations obtained for the oscillating tests, upper boundary (Eu_o) and lower boundary (El_o) are defined as:

$$Eu_o = 9.7 \times 10^{-4} Ri_b^{-0.944} \tag{4.1}$$

$$El_o = 3.5 \times 10^{-4} Ri_b^{-0.798}. \tag{4.2}$$

The form of Equation (4.1) is comparable to the rate of entrainment found in previous work [3,12,13] for the given range of Richardson numbers. Figure 12 presents a comparison of the transport rates obtained in the current study with the rates of other researchers. When compared with Christodolou's, the equation presented produces a smaller transport prediction for the lower range of Ri_b and a larger prediction at the higher values of Ri_b. The equation presented by Christodolou was valid up to a maximum Ri_b of 100, which was the

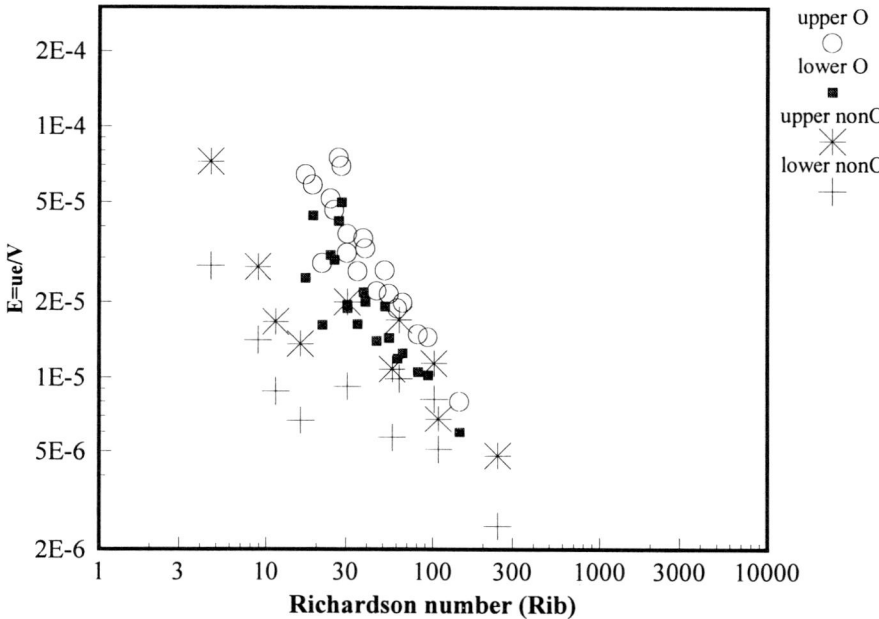

Figure 11. Dimensionless transport rates across each boundary, for each mechanism

limit of the experimental investigation. The equation presented by Suga results in a similar approximation although the range covered is larger with Ri_b values up to 300. The equation presented by Buch, as a result of field studies, is the best approximation for the results obtained in this study. The equations for entrainment at both boundaries for the non-oscillating tests are defined as:

$$Eu_{no} = 9 \times 10^{-5} Ri_b^{-0.519} \qquad (4.3)$$

$$El_{no} = 3 \times 10^{-5} Ri_b^{-0.415}. \qquad (4.4)$$

Using the initial flow conditions it is now possible to predict the rate of erosion of the salt water impoundment. It is hoped that the equations developed can be used by practising engineers to predict transport rates under similar circumstances. However, the presence of scale effects, in the equations presented, is a possibility which needs further investigation.

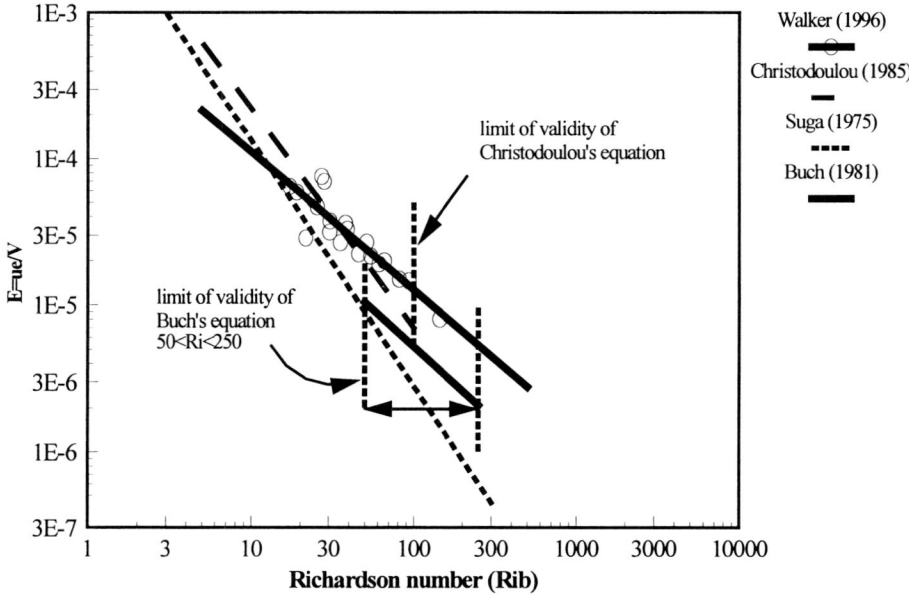

Figure 12. Comparison of upper boundary rates

5 Physical behaviour

At the start of each test interface between the fresh and salt water layers was clearly defined, and once flow was initiated interfacial shear waves appeared forming slope. This change in surface slope caused a circulation cell to develop in a manner similar to that observed in stratified lakes [22,23]. The breaking of interfacial shear waves resulted in the transport of salt water across the boundary. This mixed water collected as a wedge shape at the downstream end of the channel. The circulation cell became established in this wedge, and the upper boundary slope changed due to the lower density within the layer.

With time the layer elongated within the channel. This initial stage of layer formation took on average about 5 minutes, with major variations arising in the time taken for the layer to elongate along the channel. After some time the layer was visible at the centre of the channel. In all cases the layer had a thickness of around 30 mm when first observed at the centre of the channel. An increase in layer thickness was observed at the downstream end at this stage, with depths of up to 100 mm. The layer continued to elongate and eventually reached the entrance of the channel. The slope from the centre of the channel to the downstream end was much greater than that from the upstream end to the centre as a density differential existed along the channel. With time the layer continued to elongate and increase in thickness at the centre of the channel. The density of the layer also decreased with time, as salt water was lost from the system.

When the layer reached the start of the channel, one of two things could have happened. A flow chart of events is given in Figure 13. The chart shows that the initial stage of layer formation is similar for both mechanisms, with a difference being observed when the layer has reached the full length of the channel. Whilst the layer formation was similar, the flow patterns within the layer were not. In the first case the layer remained quasi-static and the density of the layer changed with time. The position of both boundaries, however, were continually moving downwards suggesting equal volumes of transport into and out of the layer. The tests which showed this behaviour could either erode the impoundment very quickly (in less than 5 hours) or relatively slowly, depending upon the velocity observed within the layer. For those tests which eroded the impoundment very quickly, the circulation cell tended to penetrate the fully saline water. Additional circulation cells were observed in the saltwater below the layer suggesting movement within the impoundment as a whole. For those tests which took a considerable time to erode the impoundment, the circulation was confined to the interfacial layer. Small movements were observed in the underlying salt water by dropping potassium permanganate through the flow.

When the second mechanism was observed in the channel, much more movement of the layer was observed. The initial formation of the layer was similar as shown on the chart in Figure 13, in that the layer elongated over the entire length of the channel (the time taken to reach the upstream end depended upon the velocity within the layer). During the test, the layer density decreased to a point where it could no longer be sustained and the slope of the interface needed for stability could not be achieved for the flow conditions. The layer was washed out slowly from the toe of the slope, so that the initial condition was again observed at the start of the channel with two layers separated by interfacial waves. As the layer washed out, the profile measured at the centre of the channel, showed a decrease in layer thickness. The density of the layer then increased due to greater transport of salt water across the lower boundary into the layer. As time progressed the layer formation process repeated. The layer observed at the centre of the channel became thicker and the density started to reduce again as the transport across the upper boundary became greater than that across the lower boundary. The process of layer formation and erosion then repeated but at a lower level in the channel. The fresh water had a greater depth and therefore a lower average fresh water velocity, as the flow was constant. It was found that a drop in average velocity was not the key to the constant cycle times observed but that the velocity gradients were important.

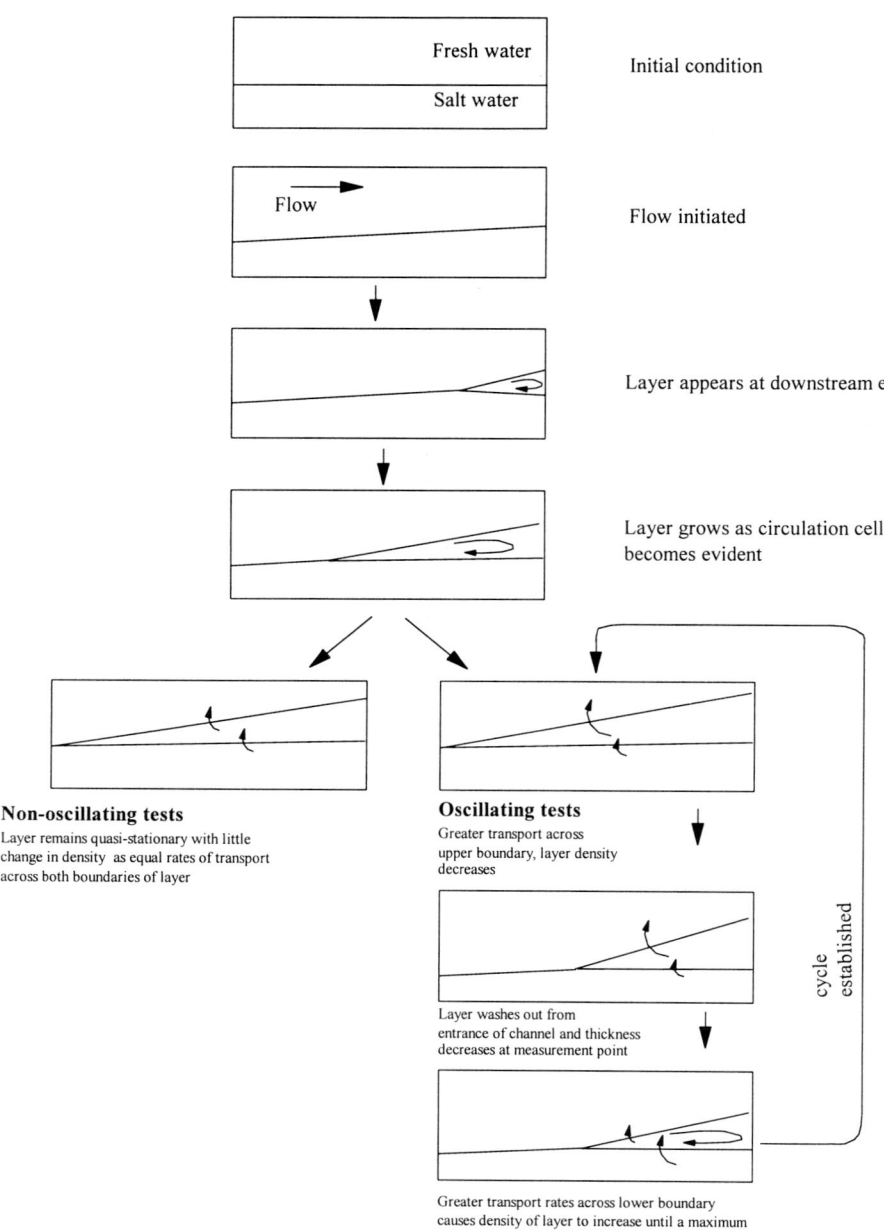

Figure 13. Flowchart of events

6 Conclusions

In all cases interfacial mixing resulted in the formation of an interfacial or mixed layer which was around 30 to 40 mm thick. Only those tests conducted with low values of Re i.e. < 2000 showed no mixing.

Two mechanisms of erosion have been observed. These were classified as oscillating and non-oscillating, depending on the behaviour of the layer in terms of its relative density and thickness. In some tests the density of the layer decreased rapidly, and then remained constant and these were classified as non-oscillating. In a number of cases the density within the layer decreased initially and then increased again after the upper part of the layer was removed. These tests were classified as oscillating as a cyclic behaviour was observed in the layer density.

The gradient Richardson number remained approximately constant during those tests which did not oscillate, indicating a stable flow. For those tests which showed the oscillating behaviour, the value of Ri_g was found to show a cyclic behaviour. When the layer was washed out, the value of Ri_g dropped towards 0.25, indicating unstable flow. Mixing occurred as a result of large amplitude breaking interfacial waves.

After the start of a test greater rates of transport were observed at the upper boundary as the layer formed. This resulted in a decrease in density until the layer washed out and a peak rate was observed at the upper boundary. Greater rates of transport were observed at the lower boundary when the layer was thinnest. This caused the density of the layer to increase with time.

Average transport rates were calculated during each test from the change in density profiles. Rates were averaged over the duration of the test and calculated for both boundaries. Dimensionless transport rates were found to be best related to the bulk Richardson number (Ri_b) with higher rates of transport for low values of Ri_b.

Bibliography

1. Lewin, J. (1996). Mechanical aspects of water control structures. *Proceedings of the Institution of Civil Engineers, Water Maritime and Energy*, **118**, pp. 29–38.

2. Keulegan, G.H. (1966). The mechanism of an arrested saline wedge. *Hydrodynamics of Estuaries and Fjords, Proceedings of the 9th International Liege Colloquium on Ocean Hydrodynamics*, Elsiever Scientific Publishing Company, pp. 546–574.

3. Christodolou (1985). Interfacial mixing in stratified flows. *Journal of Hydraulic Research*, **24**, pp. 77–92

4. Dermissis, V. (1990). Velocity distribution in arrested saline wedges. *Journal of Waterway, Port, Coastal and Ocean Engineering*, **116**, pp. 21–42.

5. Grubert, J.P. (1989). Interfacial mixing in stratified channel flows. *Journal of Hydraulic Engineering*, **115**, pp. 887–905.

6. Foo, M.H., Shuy, E.B. and Chen, C.N. (1995). Entrainment across a density interface inside a flume compartment. *Journal of Hydraulic Research*, **33**, pp. 181–196.

7. Grubert, J.P. (1980). Experiments on arrested saline wedge. *Journal of Hydraulic Engineering*, **106**, pp. 945–960.

8. Curran, G. (1993). The effect of interfacial layer development on fresh water flows over an entrapped saline intrusion. Ph.D. Thesis submitted to The Queen's University of Belfast, Northern Ireland.

9. Dermissis, V. and Parthenaides, E. (1985). Dominant shear stresses in arrested saline wedges. *Journal of Waterway, Port, Coastal and Ocean Engineering*, **111**, pp. 733–752.

10. Arita, M. and Jirka, G.H. (1987). Two-layer model of saline wedge. I: Entrainment and interfacial friction. *Journal of Hydraulic Engineering*, **113**, pp. 1229–1248.

11. Ellison, T.H. and Turner, J.S. (1959). Turbulent entrainment in stratified flows. *Journal of Fluid Mechanics*, **6**, pp. 423–448.

12. Suga, K. (1975). Salt water intrusion with entrainment. *Proceedings of the 16th International Association of Hydraulic Research Congress*, pp. 172–179.

13. Buch, E. (1981). On entrainment and vertical mixing in stably stratified fjords. *Estuarine Coastal and Shelf Science*, **12**, pp. 461–469.

14. Lofquist, K. (1960). Flow and stress near an interface between stratified liquids. *National Bureau of Standards, The Physics of Fluids*, **3**, pp. 158–175.

15. Kato, H. and Phillips, O.M. (1969). On the penetration of a turbulent layer into stratified fluid. *Journal of Fluid Mechanics*, **37**, pp. 643–655.

16. Moore, M.J. and Long, R.R. (1971). An experimental investigation of turbulent stratified shearing flow. *Journal of Fluid Mechanics*, **49**, pp. 635–655.

17. Wu, J. (1973). Wind induced turbulent entrainment across a stable density interface. *Journal of Fluid Mechanics*, **63**, pp. 275–287.

18. Chu, V.H. and Vanvari, M.R. (1976). Experimental study of turbulent stratified shearing flow. *Journal of Hydraulic Division*, ASCE, **102**, pp. 691–706.

19. Kantha, L.H., Phillips, O.M. and Azad, R.S. (1977). On turbulent entrainment at a stable density interface. *Journal of Fluid Mechanics*, **79**, pp. 753–768.

20. Kit, E., Berent, E. and Vajda, M. (1980). Vertical mixing induced by wind and a rotating screen. *Journal of Hydraulic Research*, **18**, pp. 35–58.

21. Keulegan, G.H. (1949). Interfacial instability and mixing in stratified flows. *National Bureau of Standards*, **32**, pp. 487–500.

22. Turner, J.S. (1973). Buoyancy effects in fluids. Cambridge University Press.

23. Stevens, C. and Imberger, J. (1996). The initial response of a stratified lake to a surface shear stress. *Journal of Fluid Mechanics*, **312**, pp. 39–66.

An Experimental/Numerical Study of Internal Wave Transmission Across an Evanescent Level

B.R. Sutherland[1] and P.F. Linden

Department of Applied Mathematics and Theoretical Physics, University of Cambridge

Abstract

Numerical and experimental studies show that a wavepacket of internal gravity waves may be partially reflected from and transmitted into a region where the wave intrinsic frequency is comparable to the background buoyancy frequency. The departure of the propagation behaviour from that predicted by ray theory is examined in detail. In particular, weakly nonlinear effects are shown to play a significant role when horizontally periodic waves are of moderately large amplitude, though such effects are inhibited for waves that are horizontally as well as vertically compact. Some qualitative results of these analyses are presented here.

1 Introduction

Ray theory is often employed to predict the path followed by a packet of internal gravity waves (IGW) in a background with variable flow and stratification. The theory itself employs the WKBJ approximation and consequently is valid only if the vertical wavelength of the waves is much smaller than the length scale of the background variations (for example, see Lighthill [1]). Nonetheless, if a wavepacket propagates into weakly stratified fluid or if the background flow speed changes with height so that the intrinsic frequency of the waves approaches the value of the background buoyancy frequency, then the vertical wavelength increases to infinity and the assumptions of ray theory are violated. Such circumstances occur frequently in nature. For example, by way of numerical simulations, Laprise [2] has demonstrated that considerations of non-WKBJ effects may be crucial to appropriately model the propagation and reflection of mountain waves, and Sutherland [3] has proposed that a potential momentum source for the deep zonal countercurrents in the equatorial oceans may be internal waves generated near the surface that are partially transmitted below the thermocline by a mechanism enhanced by non-WKBJ effects.

In an idealised study of IGW in stationary but non-uniformly stratified fluid in which the profile of the buoyancy frequency, $N(z)$, decreases from N_0 to a

---

[1]Present address: Department of Mathematical Sciences, CAB 539, University of Alberta, Edmonton, Alberta, Canada, T6G 2G1.

value N_+ comparable with the initial wavepacket frequency, ω_0, it was shown that a horizontally periodic wavepacket partially reflected from and transmitted into the weakly stratified region [4]. Reflection coefficients predicted from linear theory allowing for transient effects agreed well with those determined for a simulated small amplitude wavepacket. For waves of large amplitude, however, IGW reflection was shown to be enhanced for non-evanescent waves (i.e. $\omega_0 < N_+$) and it was shown that enhanced IGW transmission of evanescent waves could likewise occur. This nonlinear effect was shown to occur due to the change in phase speed of the waves at their reflecting levels: the phase speed changes as the mean flow is rapidly accelerated then decelerated by the incident waves that first increase then decrease in amplitude as the wavepacket reflects. In their study of non-WKBJ effects upon transient IGW incident upon a critical level, Fritts and Dunkerton [5] referred to this weakly nonlinear wave-mean flow interaction as the "self-acceleration" of the waves. In their study, dissipative processes near the critical level result in permanent changes to the mean flow, in part, due to transient effects. Here, however, dissipative processes are negligible and transient accelerations therefore induce no permanent changes to the mean flow [6, 7, 8], but serve only to modify the characteristics of the waves themselves.

The numerical study of IGW propagation in non-uniform stratification is extended here to include the effects of background shear. In Section 2 we describe the numerical model used and report some of the more salient results. Experiments have been performed to examine IGW generated by a vertically oscillating cylinder in salt stratified water. The path followed by these horizontally and vertically compact wavepackets is compared in Section 3 with the predictions of ray theory. It is shown that when the buoyancy frequency varies rapidly with height compared with the vertical wavelength of the waves the waves propagate more vertically than predicted. If the buoyancy frequency far from the cylinder reduces to values comparable with the IGW frequency, partial reflection and transmission occurs. The implications of this work are discussed in Section 4.

2   Numerical simulations

Numerical simulations of the fully nonlinear evolution of an IGW wavepacket in a horizontally periodic channel with free-slip upper and lower boundaries are performed by solving the primitive equations for Boussinesq flow restricted to two dimensions, using a code based upon that developed by Smyth and Peltier [9].

In studies with variable background flow, the stratification is assumed constant with $N^2 = 1$ everywhere and the background flow $U(z) = T(z)$ is assumed to have a hyperbolic tangent form such that the flow decreases with height from zero to $-\Delta U$ over a distance characterised by the length scale H:

$$T(z) = -\frac{\Delta U}{2}\left[1 + \tanh\left(\frac{z}{H}\right)\right]. \tag{2.1}$$

In studies with variable stratification the background flow is assumed to be stationary and the stratification is characterised by

$$N^2(z) = 1/[1 - |\vec{k}|T(z)]^2. \tag{2.2}$$

N^2 is defined in this way so that the profile of the ratio, $\Omega(z)/N(z)$, of the wave intrinsic frequency ($\Omega = \omega_0 - k_x U$) to the buoyancy frequency is the same if ΔU is the same for simulations either with shear or with non-uniform stratification. In particular, whether for large z the value of this ratio, $\Omega_+/N_+ = \omega_0 + k_x \Delta U$, is greater than or less than one determines whether or not, respectively, the incident waves from below are evanescent in the upper region.

A vertically compact wavepacket centred about $z_0 = -30$ is superimposed initially on the basic state prescribed by a streamfunction of the form

$$\psi(x, z) = A \exp\left(-|z - z_0|/D)\right) \cos\left(\vec{k} \cdot \vec{x}\right) \tag{2.3}$$

in which the vertical extent of the wavepacket $D = 5$ is chosen to be sufficiently large that the wavepacket supports many vertical waves, but it is sufficiently small that the IGW amplitude is negligible near $z = 0$. The wavenumber vector $\vec{k} = (1, -0.7071)$ corresponds approximately to that of a wavepacket with the largest positive vertical group velocity where $N^2 = 1$. From the dispersion relationship of IGW, the frequency and, with $k_x = 1$, the horizontal phase speed of the wavepacket are approximately $\sqrt{2/3}$.

The effect of nonlinearity is illustrated in Figure 1 which shows profiles of the perturbation kinetic energy (PKE) and the perturbation density field at the end of two simulations of IGW in uniformly stratified shear flow: in diagrams (a) and (b) for small amplitude waves ($A = 0.02$) and in diagrams (c) and (d) for large amplitude waves ($A = 0.3$). In both simulations the background flow well above $z = 0$ is set so that the incident wavepacket is evanescent in the region, specifically, where $\Omega_+/N_+ \simeq 1.02$.

In corresponding simulations of IGW in stationary, but variable N^2 fluid the same qualitative behaviour is observed if $\Omega_+/N_+ \simeq 1.02$, though in this case the transmitted waves propagate with smaller vertical group velocity and a larger proportion of the wavepacket is reflected. A measure of the proportion that is reflected is determined at the end of the simulation by comparing the perturbation kinetic energy of the wavepacket associated with a downward momentum flux to the perturbation kinetic energy of the the entire wave field. Figure 2 shows the reflection coefficients as a function of Ω_+/N_+ for small (dashed lines) and large (solid lines) amplitude waves in simulations with (a) uniform stratification and shear, and (b) non-uniform stratification and stationary flow.

As mentioned earlier, the transmission and reflection of waves is nonlinearly enhanced due to the effect of the wave-induced mean flow upon the phase speed of the waves. If the waves are horizontally compact, however, the effect is significantly retarded. Simulations have been performed for vertically and horizontally

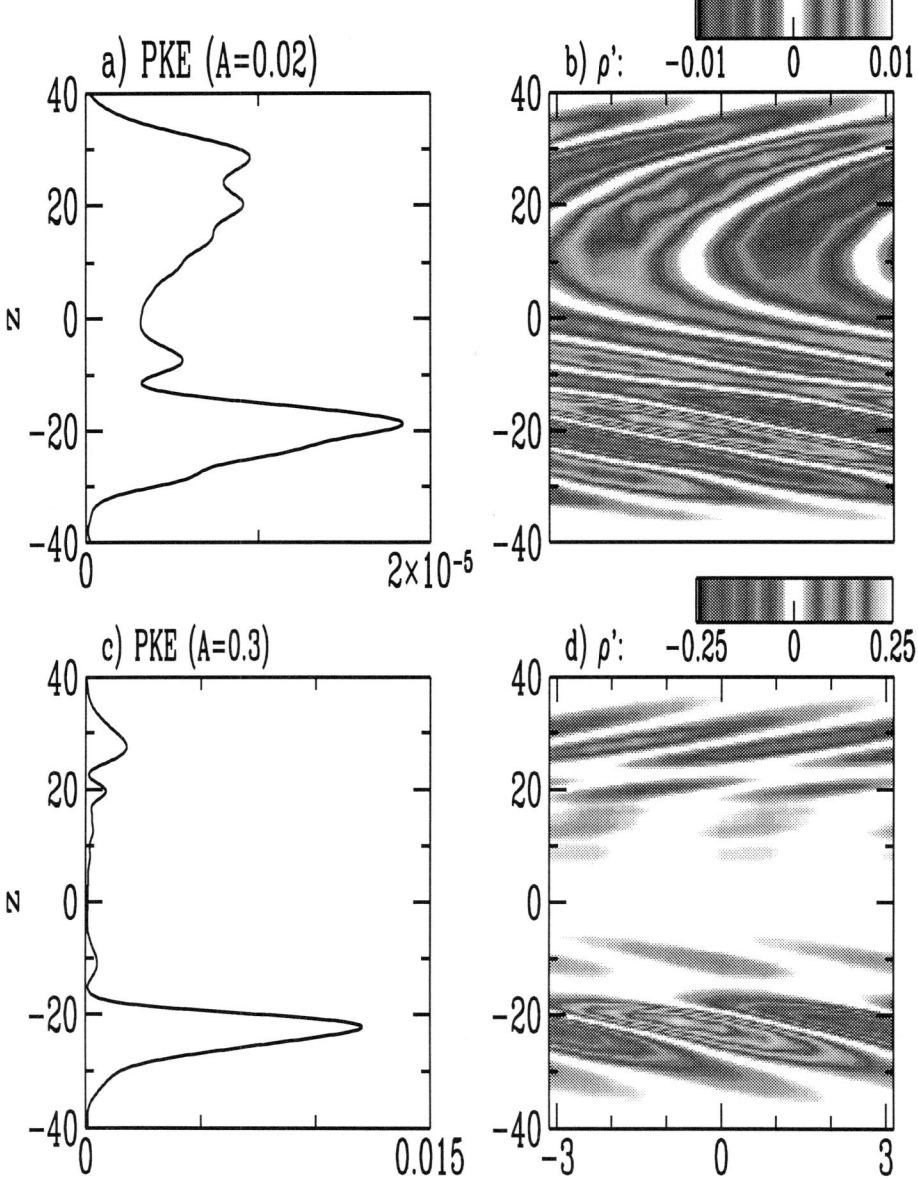

Figure 1. (a) and (c) perturbation kinetic energy profiles, and (b) and (d) perturbation density fields are shown at time $t = 150$ in simulations with $\Omega_+/N_+ \simeq 1.02$. The waves in (a) and (b) are of small amplitude ($A = 0.02$), and (c) and (d) they are of large amplitude ($A = 0.3$). Note that contours of the perturbation density are shown on a scale 25 times smaller for small amplitude waves than for large amplitude waves

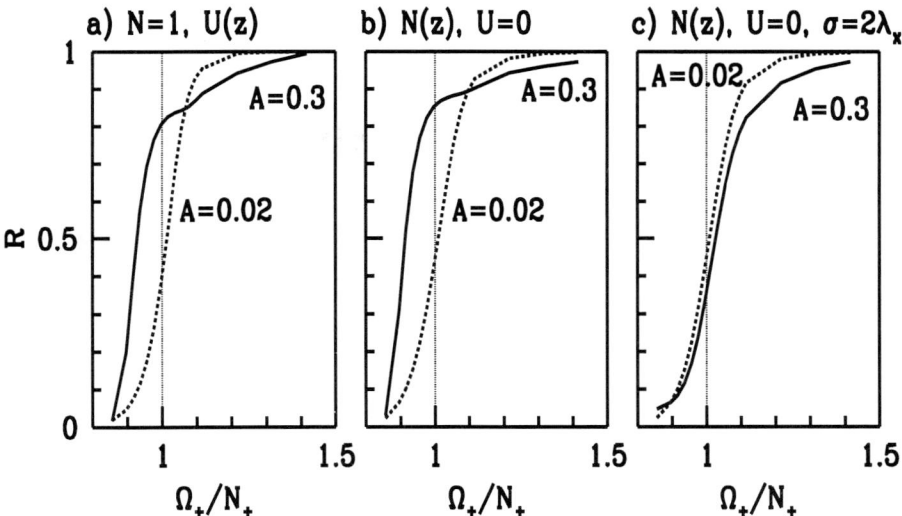

Figure 2. Reflection coefficients determined at time $t = 150$ for large (solid lines) and small (dashed lines) amplitude waves in simulations of horizontally periodic, vertically compact IGW with (a) uniform stratification and shear, and (b) non-uniform stratification and stationary flow. (c) the coefficients are calculated similarly for a simulation of horizontally and vertically compact IGW in non-uniformly stratified stationary flow. The vertical dotted line in each diagram indicates the critical value of the ratio of the wave intrinsic frequency to the background buoyancy frequency for $z \gg 0$, Ω_+/N_+. When this ratio is greater than 1, the incident wavepacket is evanescent

compact wavepackets whose envelopes in the horizontal are Gaussian with standard deviation σ:

$$\psi(x, z) = A \exp\left(-|z - z_0|/D\right) \exp\left(-x^2/2\sigma^2\right) \cos\left(\vec{k} \cdot \vec{x}\right) \qquad (2.4)$$

Figure 2(c) shows the reflection coefficient for small and large amplitude waves in stationary but variable N^2 fluid for such a wavepacket with $\sigma = 2\lambda_x$, twice the horizontal wavelength of the initial waves. The graph shows that the proportion of reflected waves is similar for both large and small amplitude waves when the incident waves are not evanescent in the region well above $z = 0$ (i.e. $\Omega_+/N_+ < 1$). Nonetheless, wave transmission is nonlinearly enhanced when the incident waves are evanescent (i.e. $\Omega_+/N_+ > 1$). Future work will quantify these results further.

3   Laboratory experiments

Non-WKBJ effects are examined in laboratory experiments for IGW generated by a vertically oscillating cylinder in salt stratified water. Waves so generated have frequently been studied [10, 11, 12, 13, 14]. In particular, an oscillating cylinder in uniformly stratified fluid is well known to produce a wave pattern in the shape of a Saint Andrew's Cross [1, 15]. Our experiments are performed in a racetrack shaped tank with test section 240 cm long, 40 cm deep, and 20 cm wide [16]. Uniform stratification is set up using a "double-bucket" system, and variable stratification is established by changing the salinity of one bucket during the filling process. Typically the tank is filled to a depth of 35 cm. The density profile is measured between 9 and 31 cm from the bottom of the tank using a mechanically traversed conductivity probe which, in one pass samples the local conductivity every 0.2 millimetre at a rate of 100 samples per second. A PVC cylinder of radius 1.6 cm generates IGW by oscillating vertically with its axis horizontal and spanning the width of the test section, the axis in its equilibrium position being situated 11.4 cm above the bottom of the tank. The cylinder oscillates with a peak to peak amplitude of up to 0.7 cm for a range of periods between 1 and 50 seconds. We examine the upward propagating waves emanating to the right of the cylinder. An angled barrier spanning the width of the test section with one end resting on the bottom of the tank is inserted with its other end near the cylinder to block bottom reflections from the right and downward propagating waves. For some experiments (not reported here) a shear flow is established near the surface by an Odell-Kovasznay drive [17]: two sets of intermeshed horizontal disks situated between 30 and 35 cm height rotate in opposite directions, thus driving fluid by viscosity with a minimum of vertical mixing.

Visualisation of the waves is accomplished using a "synthetic schlieren" system. The visualisation is relatively simple and inexpensive to set up: a back-illuminated grid of horizontal black lines is positioned well behind the test section of the tank and a CCD camera is positioned on the opposite side of the tank focussed on the grid lines through the salt stratified water, as shown schematically in Figure 3. Because the index of refraction of salt water varies with salinity the stretching and compressing of isopycnals due to the passage of IGW deflects light rays passing through the tank between the grid and the camera. Stretched isopycnals deflect light upward; compressed isopycnals deflect light downward. By comparing the initial position of the grid lines with their position when waves are present, waves with amplitudes as small as tenths of millimeters can be visualised. The comparison is done by instantaneously digitising and processing the images from a CCD camera using "DigImage", a image processing system developed by Dr. Stuart Dalziel [18]. A powerful feature of this technique is that it provides a method by which to measure non-intrusively and quantitatively the amplitude of the two-dimensional wave field everywhere in space and time to the resolution of the camera and video.

With this set-up, we have studied the propagation of IGW into weakly stratified fluid. As expected, IGW have larger vertical tilt as they propagate into more weakly stratified fluid and reflect from a mixed region near the surface. Figure 4 shows the results of three experiments of IGW propagating into weakly stratified, stationary fluid. In each experiment, the cylinder oscillates at a frequency 0.47 rad/s and with peak to peak amplitude 0.7 cm. The background N^2 profiles

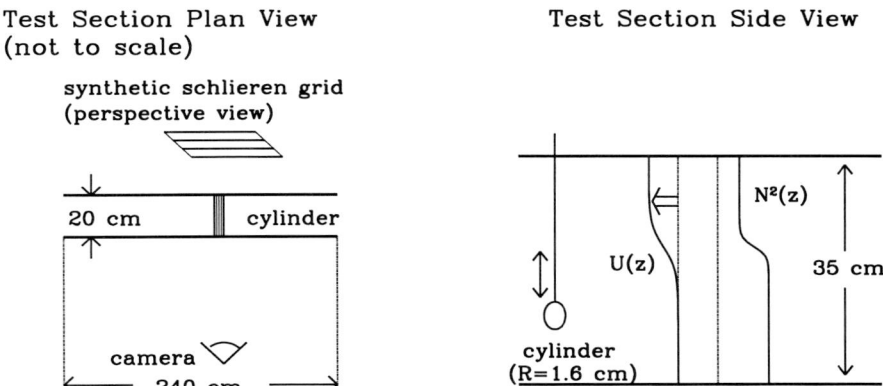

Figure 3. Schematic of the experimental set-up

are shown for each experiment with $z = 0$ corresponding to the position of the cylinder axis at equilibrium. To the right of these are shown the fields of the time rate of change of N^2 after the cylinder has oscillated continuously for 100 s. A comparison of the path of the waves with that predicted by ray theory (the dashed line superimposed on each field) shows that the waves tend to propagate more vertically than predicted, particularly when N^2 varies rapidly with height compared with the vertical wavelength of the waves. In particular, Figure 4(b) demonstrates the limitations of the WKBJ approximation.

In experiments for which the buoyancy frequency reduces with height to a value comparable to the wave frequency, IGW are observed to undergo partial transmission and reflection. For example, Figure 4(c) shows weak partial IGW reflection well below the evanescent level. This reflection occurs continuously and does not result from transient effects in the usual sense.

The numerical model described in Section 2 has been adapted to simulate the experimental set-up by adding a local external forcing term to the vertical velocity equation. The forcing is uniform over a circular patch centred at the origin with the same radius as the cylinder in the experiment, and the forcing magnitude varies sinusoidally in time with the same frequency. The simulated waves follow the same path as those generated experimentally, partially reflecting from and transmitting into weakly stratified fluid. After calibrating for the amplitude of the simulated waves, the average vertical flux of horizontal momentum is determined for waves near the cylinder (τ_0), at $z = 3.6$ cm averaged between $x = 0$ and 16 cm, for transmitted waves (τ_+) at $z = 31.6$ cm, and for reflected waves (τ_-) at $z = 3.6$ cm averaged between $x = 16$ and 35 cm.

Figure 5 shows that an approximately constant positive momentum flux is associated with waves near the cylinder shortly after it begins to oscillate. The

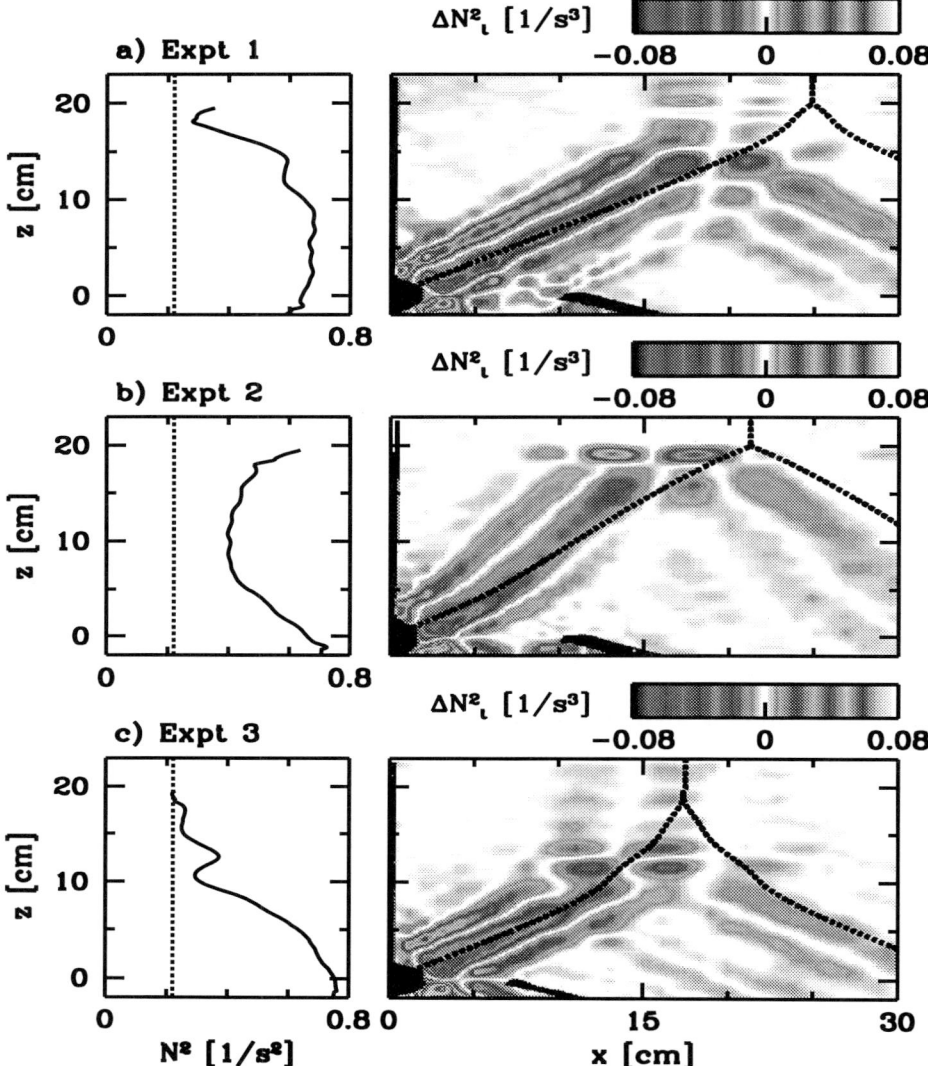

Figure 4. Three laboratory experiments of IGW propagation in non-uniformly stratified fluid. In each case the cylinder oscillates at the origin with frequency 0.47 rad/s
and amplitude 0.7 cm. (a) the background stratification varies slowly with height near
the source, (b) it reduces rapidly with height, and (c) the buoyancy frequency far from
the source is comparable to the wave frequency. Each diagram shows the background
N^2 profile with the vertical dashed line representing the squared frequency of the IGW.
To the right of each profile are fields of the time rate of change of N^2 shown 100 s after
the cylinder begins oscillating. The bold dashed lines superimposed on each field is the
path followed by an IGW wavepacket as predicted by ray theory

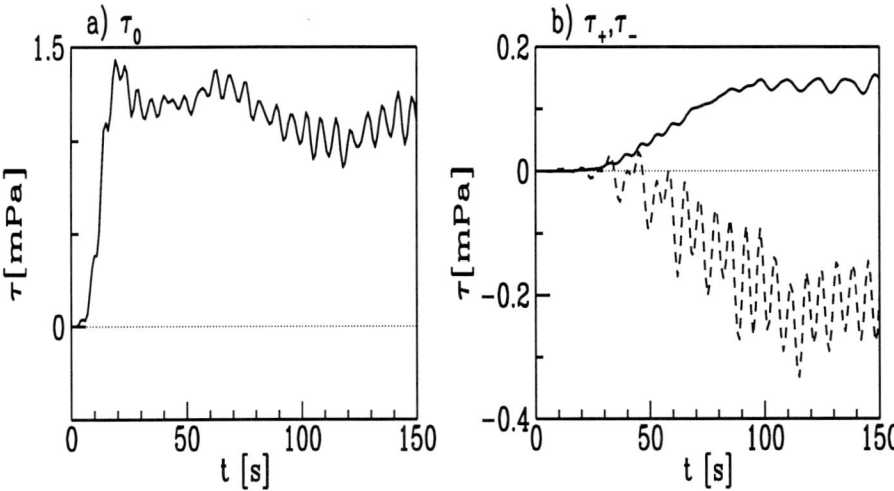

Figure 5. From a numerical simulation of the laboratory experiment shown in Figure 4(c), vertical fluxes of horizontal momentum ($\tau = \rho_0 \langle u'w' \rangle_x$) are determined over time for (a) incident, and (b) transmitted (solid line) and reflected (dashed line) IGW

fluctuations about the mean flux occur on the same frequency as the oscillation frequency. As the waves propagate into the weakly stratified fluid a positive momentum flux is associated with the transmitted waves and a negative momentum flux is associated with the reflected waves. The magnitude of the mean reflected wave momentum flux is approximately double in magnitude that of the transmitted waves and exhibits larger fluctuations about the mean.

4   Discussion and conclusions

We have shown numerically and experimentally that the behaviour of IGW deviates significantly from that predicted by ray theory when the assumptions of the WKBJ approximation are violated due to either rapid background variations or weakly nonlinear effects. Of course, there is no reason to expect ray theory to apply to IGW under such circumstances. Nonetheless, general circulation models of the atmosphere presently adopt its predictions in order to parametrise subgrid-scale gravity wave drag, even under circumstances in which the theory fails (for example, see McFarlane [19], Palmer et al. [20]). This work introduces a programme of study that will attempt to lead to improved parametrisation schemes. In order to model simply the effect of partial reflection and transmission, reflection coefficients have been calculated for wavepackets of different geometries (vertically compact and either horizontally periodic or horizontally compact) and amplitudes. In either shear flow with uniform stratification or stationary flow with non-uniform stratification, the coefficients differ by only a small amount if the profiles of $\Omega(z)/N(z)$ are the same in both cases. If $\Omega_+/N_+ \simeq 1$, however, the reflection coefficient for large amplitude IGW is almost double that for small amplitude IGW if they are horizontally periodic. This nonlinearly enhanced reflection is inhibited for horizontally compact wavepackets of width comparable to the horizontal wavelength.

The numerical work leading to these conclusions requires that the wavepacket to be vertically compact and that the transient self-acceleration of the waves plays a crucial role. However, in laboratory experiments of IGW generated by an oscillating cylinder and incident upon a weakly stratified region, partial reflection and transmission occurs with steady periodic IGW. This may result from transience when the horizontal wavelength is comparable with the wavepacket width. The same study shows that IGW tend to tilt more vertically than predicted by ray theory when the stratification varies rapidly with height, compared with the vertical wavelength. In future work the numerical model will be adapted to include non-Boussinesq effects appropriate to the middle atmosphere.

Acknowledgement

The authors would like to thank Stuart Dalziel and Graham Hughes who developed and helped adapt the synthetic schlieren system used extensively here. This work has been supported by the Natural Environment Research Council (NERC) grant GR3/09399.

Bibliography

1. Lighthill, M. J. (1978). *Waves in Fluids*. Cambridge University Press, Cambridge, England, 504 pp.

2. Laprise, J. P. R. (1993). An assessment of the WKBJ approximation to the vertical structure of linear mountain waves: implications for gravity wave drag parameterization. *J. Atmos. Sci.*, **50**, 1469–1487.

3. Sutherland, B. R. (1996). The dynamic excitation of internal gravity waves in the equatorial oceans. *J. Phys. Oceanogr.*, **26**, 3214–3235.

4. Sutherland, B. R. (1996). Internal gravity wave radiation into weakly stratified fluid. *Phys. Fluids*, **8**, 430–441.

5. Fritts, D. C. and Dunkerton, T. J. (1984). A quasi-linear study of gravity-wave saturation and self-acceleration. *J. Atmos. Sci.*, **41**, 3272–3289.

6. Eliassen, A. and Palm, E. (1960). On the transfer of energy in stationary mountain waves. *Geofys. Publ.*, **22**, 1–23.

7. Andrews, D. G. and McIntyre, M. E. (1976). Planetary waves in horizontal and vertical shear: The generalized Eliassen-Palm relation and the mean flow acceleration. *J. Atmos. Sci.*, **33**, 2031–2048.

8. Andrews, D. G. and McIntyre, M. E. (1978). An exact theory of nonlinear waves on a Lagrangian-mean flow. *J. Fluid Mech.*, **89**, 609–646.

9. Smyth, W. D. and Peltier, W. R. (1989). The transition between Kelvin-Helmholtz and Holmboe instability: An investigation of the overreflection hypothesis. *J. Atmos. Sci.*, **46**, 3698–3720.

10. Görtler, H. (1943). Über eine schwingungserscheinung in flüssigkeiten mit stabiler dichteschichtung. *Z. angew. Math. Mech.*, **23**, 65–71.

11. Mowbray, D. E. and Rarity, B. S. H. (1967). A theoretical and experimental investigation of the phase configuration of internal waves of small amplitude in a density stratified liquid. *J. Fluid Mech.*, **28**, 1–16.

12. Stevenson, T. N. and Thomas, N. H. (1969). Two-dimensional internal waves generated by a travelling oscillating cylinder. *J. Fluid Mech.*, **36**, 505–511.

13. Koop, C. G. (1981). A preliminary investigation of the interaction of internal gravity waves with a steady shearing motion. *J. Fluid Mech.*, **113**, 347–386.

14. Nicolaou, D. , Liu, R. , and Stevenson, T. N. (1993). The evolution of thermocline waves from an oscillatory disturbance. *J. Fluid Mech.*, **254**, 401–416.

15. Voisin, B. (1991). Internal wave generation in uniformly stratified fluids. Part 1. Green's function and point sources. *J. Fluid Mech.*, **231**, 439–480.

16. Redondo, J. M. R. (1989). Internal and external mixing in a stratified-shear flow. In Fernholz, H. H. and Fiedler, H. E. , editors, *Advances in Turbulence 2* 198. Springer-Verlag.

17. Odell, G. M. and Kovasznay, L. S. G. (1971). A new type of water channel with density stratification. *J. Fluid Mech.*, **50**, 535–543.

18. Dalziel, S. B. (1993). Rayleigh-Taylor instability: experiments with image analysis. *Dyn. Atmos. Oceans*, **20**, 127–153.

19. McFarlane, N. A. (1987). The effect of orographically excited gravity wave drag on the general circulation of the lower stratosphere and troposphere. *J. Atmos. Sci.*, **44**, 1775–1800.

20. Palmer, T. N. , Shutts, G. J. , and Swinbank, R. (1986). Alleviation of a systematic westerly bias in general circulation and numerical weather prediction models through an orographic gravity drag parametrization. *Quart. J. Roy. Meteor. Soc.*, **112**, 1001–1039.

The Temporal Evolution of the Boundary Layer in a Rotating Stratified Fluid

R.E. Hewitt*, P.W. Duck* and M.R. Foster**

*Department of Mathematics, University of Manchester, and **Department of Aerospace Engineering, Applied Mechanics and Aviation, Ohio State University, USA

Abstract

We consider the boundary layer that forms on the wall of a rotating container of stratified fluid that is altered impulsively from a state of rigid body rotation. We generalise the boundary–layer equations, as given by Duck, Foster and Hewitt [1], for an axisymmetric container of cross sectional shape $z = R^\alpha$ in a cylindrical polar coordinate system. The governing equations (valid near the container wall and away from the axis of rotation) are obtained for the three distinct cases of $\alpha < 1$, $\alpha = 1$ and $\alpha > 1$. We introduce a similarity–type solution and solve the resulting unsteady problem numerically. Computational results are compared with asymptotic solutions for a number of cases. In general the system may evolve to a steady state, a growing boundary layer, or a finite–time singularity depending upon α and the initial configuration.

1 Introduction

The transient response of a contained fluid undergoing a spin–up from a state of rigid body rotation has been discussed for both homogeneous and stratified fluids (for example, Greenspan and Howard [2], Pedlosky [3], Walin [4], and Spence, Foster and Davies [5]). When a stratified fluid is considered, the time scale for the readjustment to the new conditions has been shown to be much longer than that appropriate in the homogeneous case. However, conclusions drawn from investigations into the stratified spin–up problem have been based on containers that do not have sloping walls.

If the spin–up problem is considered for a stratified fluid in a container with angled walls, then a non–normal component of buoyancy is introduced into the unsteady boundary–layer problem. The coupling of this buoyancy effect with the velocity boundary–layer can alter the unsteady problem in this region considerably.

Duck, Foster and Hewitt [1] (subsequently referred to as DFH) considered the boundary–layer problem for a conical container of viscous, stratified fluid, showing that the evolution for a fixed Schmidt number could be entirely classified according to two parameters. The parameters involved were shown to be the initial rotation rate (denoted by $\hat{W}_e$, which was taken to be positive, with a final rotation rate nondimensionalised to unity) and a "modified Burger number" that was defined in terms of the Burger number, the angle of the sidewalls to the horizontal, and the boundary–layer edge conditions. The parameter space was shown to have well defined boundaries that separated three distinct regions in which the governing equations evolved to either a steady state, a growing boundary layer or a finite–time singularity. The DFH analysis allowed for non–linear changes in the rotation rate of the container and showed that a linear analysis (for example, MacCready and Rhines [6] and Thorpe [7]) is only valid in a restricted region of parameter space, namely when the change in rotation rate is much smaller than the square of the Ekman number for the flow.

In this paper we extend the analysis presented by DFH to a more general form of axisymmetric container. Figures 1(a)–(c) show the three distinct cases that we consider; they are containers that have a cross–sectional shape defined by $z = R^\alpha$ in a cylindrical polar coordinate system (R, z, λ). The case considered by DFH corresponds to Figure 1(b) (their analysis uses α to denote the quantity shown as $\hat{\alpha}$ in the figure). We distinguish the cases according to the sign of $(1-\alpha)$ since it is this quantity that determines the appropriate solution expansion away from the "apex" of the container.

The governing boundary–layer equations are as given by DFH,

$$\frac{1}{r^2}\frac{\partial(r^2 v_r)}{\partial r} + \frac{1}{r}\frac{\partial v_\theta}{\partial \theta} = 0, \tag{1.1}$$

$$\frac{\partial v_r}{\partial t} + (\mathbf{q}\cdot\nabla)v_r - \frac{v_\phi^2}{r} + \frac{dp}{dr} = \frac{1}{r^2}\frac{\partial^2 v_r}{\partial\theta^2} - B\sin\hat{\alpha}, \tag{1.2}$$

$$\frac{\partial v_\phi}{\partial t} + (\mathbf{q}\cdot\nabla)v_\phi + \frac{v_\phi v_r}{r} = \frac{1}{r^2}\frac{\partial^2 v_\phi}{\partial\theta^2}, \tag{1.3}$$

$$\frac{\partial B}{\partial t} + (\mathbf{q}\cdot\nabla)B - Sv_r\sin\hat{\alpha} = \frac{1}{\sigma}\frac{1}{r^2}\frac{\partial^2 B}{\partial\theta^2}, \tag{1.4}$$

$$\mathbf{q}\cdot\nabla \equiv v_r\frac{\partial}{\partial r} + \frac{v_\theta}{r}\frac{\partial}{\partial\theta}, \tag{1.5}$$

with boundary conditions $\partial B/\partial\theta = v_r = v_\theta = 0$, $v_\phi = r\omega(t)$ on $\theta = 0$, and with prescribed edge conditions as $\theta \to -\infty$. The reader is referred to DFH for details of the non–dimensionalisation. We have used a spherical polar coordinate system (r, θ^*, ϕ) centred on the "apex", in which r is a radial coordinate relative to the axis of rotation, ϕ is an azimuthal coordinate and θ is a scaled boundary–layer coordinate (where $\theta(\nu/\Omega h^2)^{\frac{1}{2}} = \theta^* - (\pi/2 - \tan^{-1} R^{\alpha-1})$). Here h and Ω are a characteristic length scale and rotation rate, and ν is the kinematic viscosity of the fluid. In (1.2)–(1.5), $\hat{\alpha}$ denotes the local angle between the container and the horizontal, B is the buoyancy, S is a Burger number and $\omega(t)$ is the angular frequency of the container (which we take to be an impulsive change from $\hat{W}_e$ to unity at $t = 0$).

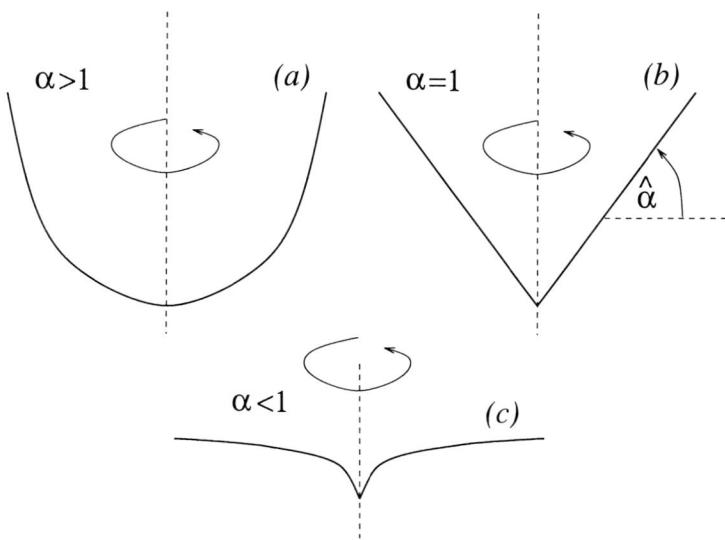

Figure 1. The container geometries

In the following sections we discuss the $\alpha > 1$ case (in Section 2) and the $\alpha < 1$ case (in Section 3), which contains two sub-cases that are discussed in Sections 3.1 and 3.2. Finally, in Section 4, we compare the evolution type found in each case with that found for the conical container (corresponding to $\alpha = 1$) considered by DFH. The notation used in each section is defined independently.

2　The $\alpha > 1$ container

A sensible balancing of terms in the boundary–layer Equations (1.2–1.5), for a container of the type shown in Figure 1(a) leads to the following scalings as $r \to \infty$

$$v_r = r^{\frac{2}{\alpha}-1}\hat{U}(\Theta,t) + \dots \,, \quad v_\phi = r^{\frac{1}{\alpha}}\hat{W}(\Theta,t) + \dots \,, \tag{2.1}$$

$$v_\theta = r^{\frac{2}{\alpha}-2}\left(\hat{V}(\Theta,t) - \Theta\hat{U}(\Theta,t)\right) + \dots \,, \tag{2.2}$$

$$B = r^{\frac{2}{\alpha}-1}\hat{B}(\Theta,t) + \dots \,, \quad P = r^{\frac{2}{\alpha}}\hat{P} + \dots \,, \tag{2.3}$$

where $\Theta = r\theta$. The above expansions follow directly from the scaling $R \sim r^{1/\alpha}$ with $\hat{\alpha} \to \pi/2$. Substitution of these expansions yields the following governing equations within the boundary layer

$$\frac{\partial \hat{U}}{\partial t} - \hat{W}^2 + \frac{2}{\alpha}\hat{P} = \frac{\partial^2 \hat{U}}{\partial \Theta^2} - \hat{B}, \tag{2.4}$$

$$\frac{\partial \hat{W}}{\partial t} = \frac{\partial^2 \hat{W}}{\partial \Theta^2}, \tag{2.5}$$

$$\frac{\partial \hat{B}}{\partial t} - S\hat{U} = \frac{1}{\sigma}\frac{\partial^2 \hat{B}}{\partial \Theta^2}, \tag{2.6}$$

$$\frac{2}{\alpha}\hat{U} + \frac{\partial \hat{V}}{\partial \Theta} = 0. \tag{2.7}$$

We can simplify this system by introducing

$$\hat{B}(\Theta) = B^*(\bar{\Theta}) - (\hat{W}_e^2 - \hat{B}_e) \ , \ \ \hat{U}(\Theta) = \alpha U^*(\bar{\Theta}), \tag{2.8}$$

$$\hat{V}(\Theta) = \sqrt{\alpha}V^*(\bar{\Theta}) \ , \ \ \hat{W}(\Theta) = W^*(\bar{\Theta}), \tag{2.9}$$

$$S = \alpha^2 S^*, \tag{2.10}$$

where

$$\bar{\Theta} = \frac{\Theta}{\sqrt{\alpha}}, \ \ \text{and} \ \ \bar{t} = \frac{t}{\alpha}. \tag{2.11}$$

The governing equations are thus reduced to the form shown in (2.4)–(2.7) but with $\hat{P} = 0$, $\alpha = 1$, $(\hat{U}, \hat{V}, \hat{W}, \hat{B}, S)$ replaced by $(U^*, V^*, W^*, B^*, S^*)$ and a bar–notation for the time and boundary–layer coordinate. The boundary conditions for this system are

$$U^* = V^* = B_{\bar{\Theta}}^* = 0, W^* = 1 \ \ \text{on} \ \bar{\Theta} = 0, \tag{2.12}$$

and

$$U^* \to 0, W^* \to \hat{W}_e, B^* \to \hat{W}_e^2 \ \ \text{as} \ \bar{\Theta} \to -\infty. \tag{2.13}$$

Obviously there are no steady state solutions to this problem for a general $\hat{W}_e$; this can be seen from the form of (2.5). Numerical investigations for $S^* > 0$, $\hat{W}_e > 0$ suggest that the system always evolves to a growing boundary layer. The evolution is characterized by an overall thickening of the boundary layer which is easily observed when the velocity or buoyancy profiles are examined. The analysis presented by DFH concerning the double structure of the growing boundary layer can also be applied here.

In an outer layer, $\bar{\Theta} = O(\sqrt{t})$, we can introduce the following expansions

$$\eta = \frac{\bar{\Theta}}{\sqrt{t}}, \tag{2.14}$$

$$W^* = \bar{W}_0(\eta) + \dots \ , \ \ B^* = \bar{B}_0(\eta) + \dots, \tag{2.15}$$

$$U^* = \frac{\bar{U}_0(\eta)}{t} + \dots \ , \ \ V^* = \frac{\bar{V}_0(\eta)}{\sqrt{t}} + \dots. \tag{2.16}$$

The boundary conditions reduce to

$$\bar{W}_0 = \hat{W}_e \, , \bar{B}_0 = \hat{W}_e^2 \, , \bar{U}_0 \to 0 \quad \text{as } \eta \to -\infty, \tag{2.17}$$

and the solution must be matched with an inner layer (considered shortly) at $\eta = 0$. The solution in this outer layer is

$$\bar{W}_0 = (\hat{W}_e - 1)\mathrm{erfc}\left(\frac{\eta}{2}\right) + (2 - \hat{W}_e), \tag{2.18}$$

$$\bar{U}_0 = \frac{\bar{W}_0'}{-\sigma S^*} \{\eta \bar{W}_0(\sigma - 1) + 2\bar{W}_0'\}, \tag{2.19}$$

where $\bar{B}_0 = \bar{W}_0^2$.

In the inner layer, which is $O(1)$ and immediately next to the container wall, the relevent expansions are

$$W^* = 1 + \frac{W_0(\bar{\Theta})}{\sqrt{t}} + \dots, \; B^* = 1 + \frac{B_0(\bar{\Theta})}{\sqrt{t}} + \dots, \tag{2.20}$$

$$U^* = \frac{U_0(\bar{\Theta})}{\sqrt{t}} + \dots, \; V^* = \frac{V_0(\bar{\Theta})}{\sqrt{t}} + \dots. \tag{2.21}$$

Matching conditions must be applied as $\bar{\Theta} \to -\infty$ and

$$U_0 = V_0 = B_0' = W_0 = 0 \quad \text{on } \bar{\Theta} = 0. \tag{2.22}$$

The solution in this inner layer is easily shown to be

$$W_0 = \bar{W}_0'(\eta = 0)\,\bar{\Theta}, \tag{2.23}$$
$$U_0 = D\sin(\hat{\lambda}\bar{\Theta})\exp(\hat{\lambda}\bar{\Theta}), \tag{2.24}$$
$$B_0 = 2\hat{\lambda}^2 D\cos(\hat{\lambda}\bar{\Theta})\exp(\hat{\lambda}\bar{\Theta}) + 2\bar{W}_0'(\eta = 0)\bar{\Theta}, \tag{2.25}$$

where $\hat{\lambda} = \left(\frac{\sigma S^*}{4}\right)^{\frac{1}{4}}$, and the boundary condition $B_0'(\bar{\Theta} = 0) = 0$ requires that

$$D = -\frac{\bar{W}_0'(\eta = 0)}{\hat{\lambda}^3}. \tag{2.26}$$

If we consider the value of B^* at the container wall we see that, from (2.18),

$$\bar{W}_0'(\eta = 0) = -\frac{1}{\sqrt{\pi}}(W_e^* - 1). \tag{2.27}$$

Therefore a comparison may be made with the numerical results using the asymptotic results

$$B^*(\bar{\Theta} = 0, t) = 1 + \frac{1}{\sqrt{t}}\left(\frac{2}{\sqrt{\pi}\hat{\lambda}}\right)(W_e^* - 1) + \dots, \tag{2.28}$$

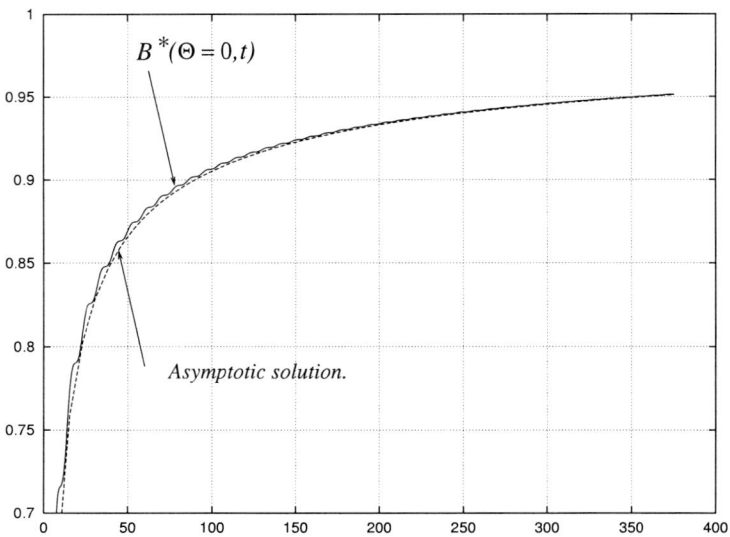

Figure 2. Comparison of $B^*(\Theta = 0, t)$ with the expansion Equation (2.28); $\hat{W}_e = .5$, $S^* = .5$

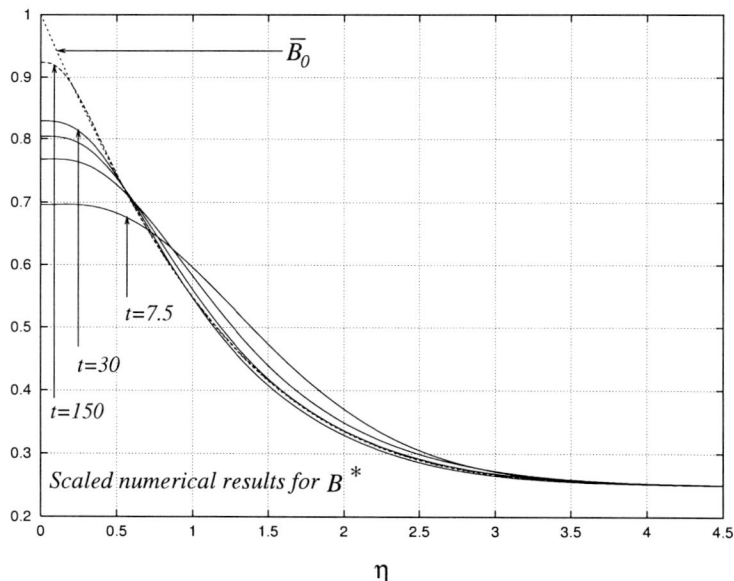

Figure 3. Comparison of the scaled numerical results with the leading order asymptotic profile for $\hat{W}_e = .5$, $S^* = .5$

and

$$W_{\bar{\Theta}}^*(\bar{\Theta} = 0, t) = \frac{1}{\sqrt{t}} \left(-\frac{1}{\sqrt{\pi}} (W_e^* - 1) \right) + \dots . \tag{2.29}$$

Figure 2 compares the value of B^* at the container wall with that predicted by the first two terms in the expansion (2.28) for $W_e^* = 0.5$, $S^* = 0.5$. The oscillatory behaviour shown in the figure is not an artifact of the numerical method. Figure 3 compares the scaled B^*–profiles generated by the numerical scheme to the leading order outer solution obtained from the above asymptotic approach; the profiles are shown at times $t = 7.5, 15, 22.5, 30$ and $t = 150$.

3   The $\alpha < 1$ container

The relevant expansions for a container of the type shown in Figure 1(c) are

$$v_r = r\hat{U}(\Theta, t) + \dots \ , \quad v_\phi = r\hat{W}(\Theta, t) + \dots \ , \tag{3.1}$$

$$v_\theta = \hat{V}(\Theta, t) - \Theta\hat{U}(\Theta, t) + \dots \ , \tag{3.2}$$

$$P = r^2\hat{P} + \dots \ , \quad \Theta = r\theta \ , \tag{3.3}$$

which follow from the scaling $R \sim r$ with $\hat{\alpha} \to 0$. However, in this case there are two possible scalings for the buoyancy term B, namely

$$B = r^{2-\alpha}\hat{B}(\Theta, t) + \dots \ , \quad \text{or} \ \ B = r^\alpha \hat{B}(\Theta, t) + \dots . \tag{3.4}$$

In the first of the above scalings, B is of sufficient magnitude for the density transport equation to remain coupled with the momentum equations. However, the Burger number, S, does not appear in the density transport equation in this case. The second scaling shown in (3.4) corresponds to balancing the $\hat{B}$ terms in the density transport equation with the inhomogeneous Burger–number term of $S\hat{U}$. In this case the buoyancy is not of sufficient magnitude for the coupling term to be present in the $\hat{U}$–momentum equation and the density transport equation remains decoupled. We examine both cases individually in the following sections.

3.1   The $B \sim r^{2-\alpha}$ case

When this choice of scaling is made for B the governing equations reduce to

$$\frac{\partial\hat{U}}{\partial t} + \hat{U}^2 + \hat{V}\frac{\partial\hat{U}}{\partial\Theta} - \hat{W}^2 + 2\hat{P} = \frac{\partial^2\hat{U}}{\partial\Theta^2} - \hat{B} \ , \tag{3.5}$$

$$\frac{\partial\hat{W}}{\partial t} + 2\hat{U}\hat{W} + \hat{V}\frac{\partial\hat{W}}{\partial\Theta} = \frac{\partial^2\hat{W}}{\partial\Theta^2} \ , \tag{3.6}$$

$$\frac{\partial\hat{B}}{\partial t} + \hat{V}\frac{\partial\hat{B}}{\partial\Theta} + (2 - \alpha)\hat{U}\hat{B} = \frac{1}{\sigma}\frac{\partial^2\hat{B}}{\partial\Theta^2} \ , \tag{3.7}$$

$$2\hat{U} + \frac{\partial\hat{V}}{\partial\Theta} = 0 \ . \tag{3.8}$$

We note that a substitution of $B^* = \hat{B} + (\hat{W}_e^2 - \hat{B}_e)$ simplifies the radial momentum equation and the density transport equation to

$$\frac{\partial \hat{U}}{\partial t} + \hat{U}^2 + \hat{V}\frac{\partial \hat{U}}{\partial \Theta} - \hat{W}^2 = \frac{\partial^2 \hat{U}}{\partial \Theta^2} - B^*, \tag{3.9}$$

$$\frac{\partial B^*}{\partial t} + \hat{V}\frac{\partial B^*}{\partial \Theta} + (2 - \alpha)\hat{U}B^* - S^*\hat{U} = \frac{1}{\sigma}\frac{\partial^2 B^*}{\partial \Theta^2}, \tag{3.10}$$

where the term S^* plays the role of the Burger number but is now defined solely in terms of the edge conditions as

$$S^* = (\hat{W}_e^2 - \hat{B}_e)(2 - \alpha). \tag{3.11}$$

The boundary conditions for this system are

$$\hat{U} = \hat{V} = \frac{\partial B^*}{\partial \Theta} = 0, \ \hat{W} = 1 \quad \text{on } \Theta = 0, \tag{3.12}$$

and

$$\hat{U} \to 0, \ \hat{W} \to \hat{W}_e, \ B^* \to \hat{W}_e^2 \quad \text{as } \Theta \to -\infty. \tag{3.13}$$

We observe that the only difference between this case and that of the conical container (as discussed by DFH) are the definitions of B^* and S^*, together with the $2 - \alpha$ coefficient of the $\hat{U}B^*$ term in (3.10).

3.1.1 The parameter space

Numerical investigation of this system reveals that the same three types of evolution that are found for the conical case of DFH are also obtained with this geometry of container. The asymptotic analyses of the growing boundary layer and the finite–time breakdown for this case are detailed below. We can categorise the behaviour of the solution according to two parameters (as can be done for the $\alpha = 1$ case). For this particular geometry we observe that the relevant parameters are $\hat{W}_e$ and $\tilde{S}$ (defined below). A schematic of this parameter space is shown in Figure 4, and displays three distinct regions corresponding to the three evolutionary types.

Region (i)

The region (i) shown in Figure 4 represents those parameter values for which an evolution to a steady state is achieved. There is a critical value of S^*, denoted by S^*_{crit}, above which a steady solution cannot be located. We can consider the steady problem for a perturbation about this critical value by defining

$$S^* = S^*_{crit} + \delta, \tag{3.14}$$

which leads to the following expansions in a $\Theta = O(1)$ layer adjacent to the container wall

$$\hat{U} = \delta U_1(\Theta) + \dots, \ \hat{V} = \delta V_1(\Theta) + \dots, \tag{3.15}$$
$$\hat{W} = 1 + \delta W_1(\Theta) + \dots, \ \hat{B} = 1 + \delta B_1(\Theta) + \dots. \tag{3.16}$$

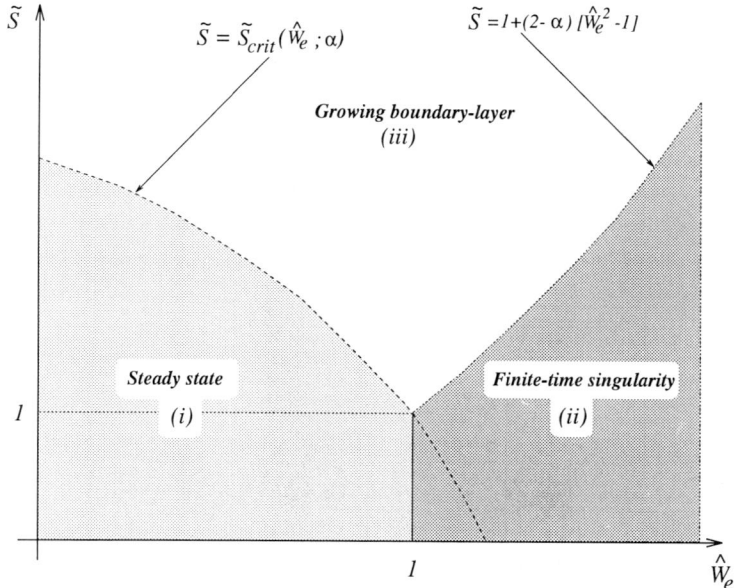

Figure 4. Different flow regimes for $\sigma = 1$

The leading order equations are therefore

$$-2W_1 = U_1'' - B_1 \ , \ 2U_1 = W_1'' \ , \tag{3.17}$$

$$\frac{1}{\sigma}B_1'' = \left((2 - \alpha) - S_{crit}^*\right)U_1 \ , \ 2U_1 + V_1' = 0 \ , \tag{3.18}$$

with boundary conditions

$$B_1' = U_1 = V_1 = W_1 = 0 \ , \quad \text{on} \ \ \Theta = 0 \ , \tag{3.19}$$

and matching conditions as $\Theta \to -\infty$.

We note that there is a natural substitution that simplifies the parameter space diagram, namely

$$\tilde{S}_{crit} = S_{crit}^* - (1 - \alpha) \ ; \tag{3.20}$$

which also reduces the inner–layer problem to that considered by DFH but with S_{crit}^* replaced by $\tilde{S}_{crit}$ in their analysis. Therefore, as $\Theta \to -\infty$, we must have

$$U_1 \to 0 \ , \quad V_1 \to \gamma \ , \quad W_1 \to \gamma\Theta[\bar{\lambda}^4 - 1] \ , \tag{3.21}$$

where $\bar{\lambda}$ is defined by

$$\bar{\lambda}^4 = \frac{1}{4}\left[4 - \sigma(1 - \tilde{S}_{crit})\right] \ . \tag{3.22}$$

Thus an outer layer is required, defined by $\tilde{\Theta} = \delta\Theta = O(1)$, as discussed by DFH.

An analysis of this outer layer yields a problem that, when combined with the conditions (3.21), can be solved by a simple numerical algorithm to determine $\tilde{S}_{crit}$. The results of such an analysis are shown in Figure 5. The sign of δ, as was noted by DFH, must be negative in the above expansion. A positive δ leads to a solution domain of $0 \leq \tilde{\Theta} < \infty$ for the outer problem (because matching with the numerical results requires that $\gamma < 0$) and this leads to exponentially growing solutions. Therefore steady state solutions can only be located for $\tilde{S} < \tilde{S}_{crit}$ where

$$\tilde{S} = S^* - (1 - \alpha). \tag{3.23}$$

We note here that when $\hat{W}_e > 1$ unsteady calculations evolved to a finite–time breakdown even in the region for which a steady solution existed.

Region (ii)

The region (ii) in Figure 4 consists of those values of $\tilde{S}$ and $\hat{W}_e$ for which the solution fails at a finite time. The scalings involved in the breakdown process are as given by DFH with only a minor difference in the coefficients involved in the governing equations. The character of the breakdown process is therefore the same in this region of the parameter space as that presented for the conical container.

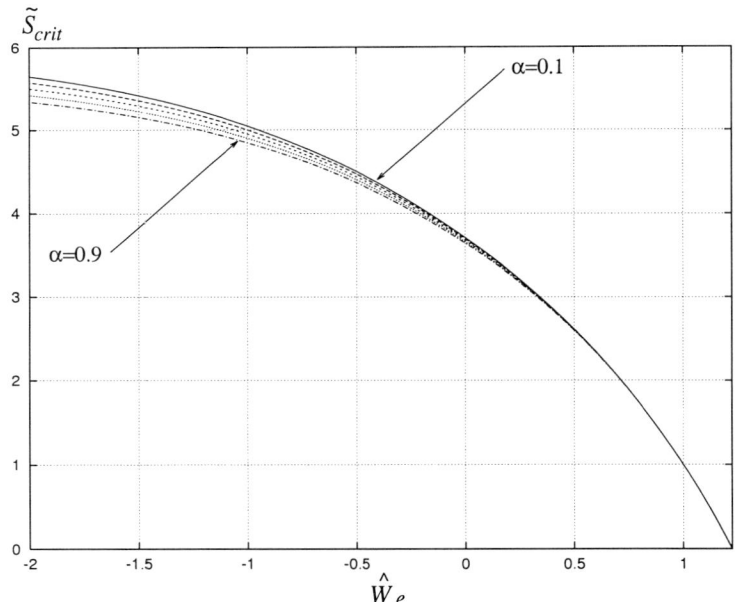

Figure 5. The critical line $\tilde{S}_{crit}$ for varying $\alpha < 1$

A balancing of terms leads to an inner layer with a lengthscale of $\Theta = O((t_0 - t)^{\frac{1}{2}})$ as $t \to t_0$ (the time at breakdown). An analysis of this inner layer yields the following scalings

$$\hat{U} = (t_0 - t)^{-1} \hat{U}_1(\tilde{\eta}) + \dots, \quad \hat{V} = (t_0 - t)^{-\frac{1}{2}} \hat{V}_1(\tilde{\eta}) + \dots, \tag{3.24}$$
$$\hat{W} = (t_0 - t)^{-1} \hat{W}_1(\tilde{\eta}) + \dots, \quad B^* = (t_0 - t)^{-2} \hat{B}_1(\tilde{\eta}) + \dots, \tag{3.25}$$

where $\tilde{\eta} = \Theta/(t_0 - t)^{\frac{1}{2}}$. Substitution of these expansions into the governing equations yields a system of equations that form a nonlinear eigenvalue problem. These equations can be solved numerically by a finite–difference method with Newton iteration used to obtain the correct behaviour at $\tilde{\eta} = 0$. We found that the simplest way to iterate to a non–trivial solution was to use an appropriately scaled solution to the full unsteady numerical problem as a starting point in the iterative procedure. The scaled velocity components and buoyancy in the $\Theta = O((t_0 - t)^{\frac{1}{2}})$ region decay algebraically like

$$\hat{U}_1, \hat{W}_1 = O(\tilde{\eta}^{-2}), \quad \hat{V}_1 = O(1), \quad \hat{B}_1 = O(\tilde{\eta}^{-4}), \tag{3.26}$$

as $\tilde{\eta} \to -\infty$.

There are also two other layers, a passive inner (where $\Theta \sim (t_0 - t)^{-1}$) in which the no–slip conditions are satisified, and an outer layer (where $\Theta = O(1)$) in which

$$\hat{U}, \hat{W}, B^* = O(1), \quad \text{and} \quad \hat{V} = O((t_0 - t)^{-\frac{1}{2}}). \tag{3.27}$$

A comparison of the numerical solution to the system of equations valid in the $\tilde{\eta} = O(1)$ region and scaled numerical solutions to the full problem, (3.9), (3.6), (3.10), (3.8), is not presented here. As noted above, there is very little difference between the inner–layer equations for this geometry of container and those for the conical container. Therefore, the reader is referred to the figures of DFH for examples of the agreement between the asymptotic solution and the unsteady solution as $t \to t_0$ for $\alpha = 1$; similar results are found in this $\alpha < 1$ case.

Numerical results indicate that the dividing boundary between the finite–time breakdown behaviour and the growing boundary layer type of evolution is the curve $\tilde{S} = (2 - \alpha)[\hat{W}_e^2 - 1] + 1$. The appearance of this particular curve through parameter space as a boundary between evolution types is not arbitrary. We observe that for these parameter values there exist uniform B^* solutions to the problem, which have a velocity field that is equivalent to that found for the corresponding infinite rotating disk problem.

Region (iii)

An analysis of the growing boundary layer follows in the manner of Section 2. In the outer, $\eta = \Theta/\sqrt{t} = O(1)$, region we have the same scalings (2.14)–(2.16) with the leading order equations

$$\bar{W}_0'' = -\frac{1}{2}\eta\bar{W}_0' - \bar{V}_0'\bar{W}_0 + \bar{V}_0\bar{W}_0',\tag{3.28}$$

$$\frac{1}{\sigma}\bar{B}_0'' = -\frac{1}{2}\eta\bar{B}_0' - \frac{(2-\alpha)}{2}\bar{V}_0' + \bar{V}_0\bar{B}_0' + \left(\tilde{S} + (1-\alpha)\right)\bar{V}_0',\tag{3.29}$$

$$\bar{W}_0^2 = \bar{B}_0,\quad 2\bar{U}_0 + \bar{V}_0' = 0.\tag{3.30}$$

This system has matching conditions at $\eta = 0$ and

$$\bar{W}_0 = \hat{W}_e,\ \bar{B}_0 = \hat{W}_e^2,\ \bar{U}_0 \to 0\quad\text{as }\eta \to -\infty.\tag{3.31}$$

The inner solution is equivalent to that presented by DFH when S^* is replaced by $\tilde{S}$ in their analysis. The same fourth–order shooting method that was applied in DFH can be applied here, showing excellent agreement with the numerics, although in the interests of brevity we do not present any figures of comparison here.

The growing boundary layer scenario is found for those parameter values in the region $\tilde{S} > \max\{\tilde{S}_{crit}(\hat{W}_e;\alpha), (2-\alpha)[\hat{W}_e^2 - 1] + 1\}$ when $\hat{W}_e > 0$. This region is denoted by (iii) in Figure 4.

3.2   The $B \sim r^\alpha$ case

When this choice of scaling is made for B we note that the momentum equations become decoupled from the density transport equation, which now includes a term that is dependent on the Burger number. The governing system of equations is therefore formed from (3.6) and (3.8) together with

$$\frac{\partial\hat{U}}{\partial t} + \hat{U}^2 + \hat{V}\frac{\partial\hat{U}}{\partial\Theta} - \hat{W}^2 = \frac{\partial^2\hat{U}}{\partial\Theta^2} - \hat{W}_e^2,\tag{3.32}$$

$$\frac{\partial\hat{B}}{\partial t} + \hat{V}\frac{\partial\hat{B}}{\partial\Theta} + \alpha\hat{U}\hat{B} - S\hat{U} = \frac{1}{\sigma}\frac{\partial^2\hat{B}}{\partial\Theta^2}.\tag{3.33}$$

The boundary conditions on the velocity components are given in (3.12) and (3.13), but the buoyancy must now satisfy $\hat{B}_\Theta(\Theta = 0, t) = 0$ and $\hat{B}(\Theta, t) \to \hat{B}_e$ as $\Theta \to -\infty$. In this case the edge conditions for the azimuthal rotation and buoyancy can be prescribed independently since the large Θ form of the radial momentum equation determines the pressure constant to be $\hat{P} = \hat{W}_e^2/2$.

Obviously the initial value problem, from the point of view of the decoupled velocity components, is equivalent to that for the infinite rotating disk problem (for example, see Von Kármán [8], Bödewadt [9], Bodonyi and Stewartson [10], Bodonyi [11]). However, a numerical investigation of the whole system suggests that two types of behaviour can be found for the decoupled buoyancy. Eventually, for the $\hat{B}$ term, we find that a steady state solution is achieved, or an exponential growth is obtained depending on the the sign of $1 - \hat{W}_e$ (for $\hat{W}_e > 0$). When the system undergoes a spin up, a steady state is achieved, but in the spin–down case an exponential growth of the buoyancy is found and presumably, in this case, the density transport equation will eventually re–couple with the momentum equations.

Since the numerical results suggest that $\hat{W}_e = 1$ is a dividing boundary between evolution types, we begin by investigating the steady solution to the density transport Equation (3.33) when $\hat{W}_e = 1 + \delta$. In this case the velocity components

$$\hat{U} = \delta u + \dots, \quad \hat{V} = \delta v + \dots, \quad \hat{W} = 1 + \delta w + \dots, \tag{3.34}$$

are given by the linearised steady solution

$$u = \exp(\Theta)\sin(\Theta), \quad v = -\{\exp(\Theta)[\sin(\Theta) - \cos(\Theta)] + 1\}, \tag{3.35}$$
$$w = 1 - \exp(\Theta)\cos(\Theta). \tag{3.36}$$

Now, considering the form of the Equation (3.33) we see that it may be simplified by the substitutions

$$S^* = S + \alpha \hat{B}_e, \quad \text{and } B^* = \frac{\hat{B} - \hat{B}_e}{S^*}, \tag{3.37}$$

to give

$$\frac{\partial B^*}{\partial t} + \alpha \hat{U} B^* + \hat{V}\frac{\partial B^*}{\partial \Theta} - \hat{U} = \frac{1}{\sigma}\frac{\partial^2 B^*}{\partial \Theta^2}. \tag{3.38}$$

The boundary conditions for this transformed buoyancy are $B^*_\Theta(\Theta = 0, t) = 0$ and $B^*(\Theta, t) \to 0$ as $\Theta \to -\infty$.

We can examine the steady state solution to (3.38) when $\hat{W}_e = 1 + \delta$ by using the solutions (3.35)–(3.36). The solution in this linear limit has a double–layer structure consisting of a $\Theta = O(1)$ inner layer and a $\Theta = O(\delta^{-1})$ outer layer.

In the inner layer we see that the appropriate expansion for $\hat{B}$ is

$$\hat{B} = b_0 + \delta b_1(\Theta) + \dots, \tag{3.39}$$

where the leading order term b_0 is a constant. Substitution of the above linearised solution (3.34) to the rotating disk problem leads to

$$b_1(\Theta) = \frac{\sigma(\alpha b_0 - 1)}{2}\{\Theta - \exp(\Theta)\cos(\Theta)\} + C, \tag{3.40}$$

where C is a constant.

Similarly, in an outer layer defined by $\bar{\Theta} = \delta\Theta = O(1)$, we see that

$$\hat{B} = \bar{b}_0(\bar{\Theta}) + \dots, \tag{3.41}$$

where
$$\bar{b}_0(\bar{\Theta}) = b_0 \exp(-\sigma\bar{\Theta}) . \tag{3.42}$$

We also note that matching the inner and outer layers determines the leading order constant, $b_0 = 1/(2+\alpha)$, in the inner layer expansion.

Thus the steady solution to the density transport equation involves a lengthening boundary–layer scale as $\hat{W}_e \to 1$, and cannot be continued to $\hat{W}_e > 1$ (i.e., $\delta > 0$) because we find that $\bar{\Theta} \in [0, -\infty)$ and (3.42) leads to an exponentially growing solution. Thus a steady solution is eventually attained for $\hat{W}_e < 1$, however, for $\hat{W}_e \geq 1$ no such steady solution is possible. Numerical results for the spin–down case show that the solution to the density transport equation grows with time. When t is sufficiently large for $B^* \gg \hat{U}$, we can find a solution to (3.38) in the form
$$B^*(\Theta, t) = \tilde{B}(\Theta) \exp(\mu t) , \tag{3.43}$$

where μ and $\tilde{B}(\Theta)$ are determined by the eigenvalue problem
$$\mu\tilde{B} + \alpha\hat{U}_s\tilde{B} + \hat{V}_s\tilde{B}' = \frac{1}{\sigma}\tilde{B}'' , \tag{3.44}$$

with $\tilde{B}'(0) = 0$ and $\tilde{B} \to 0$ as $\Theta \to -\infty$. Here we have used $\hat{U}_s$ and $\hat{V}_s$ to denote the appropriate steady solutions (from the Von Kármán family) to the rotating disk problem.

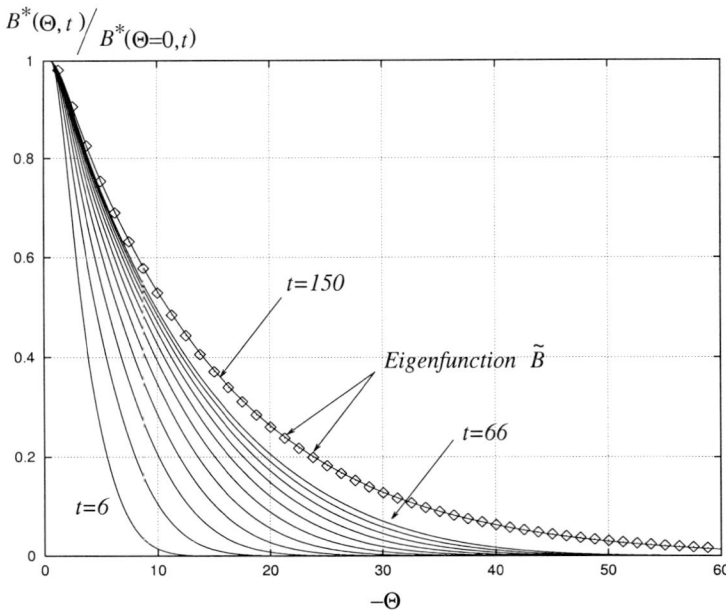

Figure 6. A comparison of the eigenfunction with the numerical results for $\hat{W}_e = 1.5$, $\alpha = 0.5$

In Figure 6 we compare the growing buoyancy profile obtained from a numerical solution to the governing system of Equations (3.32), (3.6), (3.8) and (3.38) to the eigenfunction associated with the largest eigenvalue of (3.44). The results shown in Figure 6 are for $\hat{W}_e = 1.5$ and $\alpha = 0.5$ ($\sigma = 1.0$), which leads to only real eigenvalues; the profiles are shown for $t = 6, 12, ..., 60, 66$, and $t = 150$. For this case there is only one positive eigenvalue, namely $\mu \approx 0.0372$.

4 Conclusions

We have extended the analysis of the boundary–layer problem discussed by DFH to include a more general class of axisymmetric container. The analysis is valid near the container walls and away from the axis of rotation. We place no bounds on the size of the change in rotation rate for the container, but do require that the final rotation is in the same sense as the initial rotation (i.e., $\hat{W}_e > 0$). The change in rotation rate is assumed to be impulsive in our analysis, however numerical investigations into other methods of deceleration have revealed no qualitative differences in the overall character of the evolution types obtained. Indeed, the boundaries (in the appropriate parameter space) that separate the different evolution types seem to hold even for gradual decelerations, as might be expected since these boundaries are either all known explicitly or are obtained from an analysis of the steady problem.

For the container type shown in Figure 1(a) we have shown how the boundary–layer equations can be reduced to a form similar to those obtained in the $\alpha = 1$ case but without the convection terms. There are no non–trivial steady solutions and at all points in parameter space (S^*, $\hat{W}_e > 0$, for σ fixed) we find a growing boundary layer that can be described in terms of a double–layer structure for large times.

For the container type shown in Figure 1(c) we have shown that there are two possible scalings that can be applied. If we scale the buoyancy in such a way that the density transport equation remains coupled to the momentum equations then we find an overall behaviour that can be described in the same manner as the conical container (after an appropriate redefinition of the parameters involved). In this case we find all three evolution types; a steady state, a growing boundary layer and a finite–time singularity. When scaling the buoyancy to balance the Burger number term in the density transport equation we find that an increase in the rotation rate of the container leads to both steady velocity profiles and a steady buoyancy profile across the boundary layer. However, when the rotation rate of the container is decreased we find that an exponential growth of the buoyancy term can occur and, presumably, after sufficient growth has occurred we cannot neglect the coupling effect between the momentum equations and the density transport equation. Obviously we can not comment on which scaling for the buoyancy will be appropriate in any practical situation.

We must also note that when developing the eigenproblem description for the ultimate behaviour of the buoyancy in Section 3.2 (3.44), we have assumed that the decoupled momentum equations will eventually evolve to a steady state (referred to as $\hat{U}_s$ and $\hat{V}_s$ in (3.44)). However, as discussed in DFH these momentum equations (which essentially form the similarity equations for the flow above a rotating disk) do not evolve to a steady state when $\hat{W}_e > 1$. The eventual state is in fact the appropriate Von Kármán type solution with a superimposed "wave–packet" travelling away from the boundary, however the use of the steady state may be justified after a sufficiently long time as the unsteady "wave–packet" part moves away from the container wall.

Bibliography

1. Duck, P.W., Foster, M.R. and Hewitt, R.E. (1997) On the boundary layer arising in the spin–up of a stratified fluid in a container with sloping walls. *to appear in J. Fluid Mech.*

2. Greenspan, H.P. & Howard, L.N. (1963) On a time–dependent motion of a rotating fluid. *J. Fluid Mech.* **17**, 385.

3. Pedlosky, J. (1967) The spin up of a stratified fluid. *J. Fluid Mech.* **28**, 463.

4. Walin, G. (1969) Some aspects of time–dependent motion of a stratified fluid. *J. Fluid Mech.* **36**, 289.

5. Spence, G.S.M., Foster, M.R., & Davies, P.A. (1990) The transient response of a contained rotating stratified fluid to impulsively started surface forcing. *J. Fluid Mech.* **243**, 33.

6. MacCready, P. & Rhines, P.B. (1991) Buoyant inhibition of Ekman transport on a slope and its effect on stratified spin up. *J. Fluid Mech.* **223**, 631.

7. Thorpe, S.A. (1987) Current and temperature variability on the continental slope. *Phil. Trans. Roy. Soc. Lond.* **A323**, 471.

8. Kármán, T. Von (1921) Über laminare und turbulente Reibung. *ZAMP* **1**, 244.

9. Bödewadt, U.T. (1940) Die Drehströmung über festem Grund. *ZAMM* **20**, 241.

10. Bodonyi, R.J. & Stewartson, K. (1977) The unsteady boundary layer on a rotating disk in a counter–rotating fluid. *J. Fluid Mech.* **79**, 669.

11. Bodonyi, R.J. (1978) On the unsteady similarity equations for the flow above a rotating disc in a rotating fluid. *Quart. Appl. Math. and Mech.* **31**, 461.

The Use of Particle Image Velocimetry for the Measurement of the Effect of Post Breaking Turbulence on Stratified Fluids

D.B. Hann*, C.A. Greated* and P.A. Davies**

**Department of Physics and Astronomy, University of Edinburgh, and*
***Department of Civil Engineering, University of Dundee*

Abstract

Measurements are presented of the post breaking turbulent velocities in a stratified fluid, recorded using particle image velocimetry. They show that there is a significant recirculation of the fluid set up by the breaking process which penetrates deeper than the turbulence. This will mix the interface by more than the turbulence component alone.

1   Introduction

There are many indications in current research that the effect of surface breaking waves on a stratified fluid is more important than previously thought [1, 2]. The effect of breaking processes has been shown to extend to depths of the order of the wavelength, rather than the amplitude of the waves as previously expected [3, 4]. This means that the environmental effect of breaking waves in the deep ocean needs to be re-addressed, since this will increase the transfer of heat, gas and energy between the upper and lower layers in the ocean [5]. In order to investigate this more thoroughly, this research has been concerned with modelling the effect of surface breaking waves on an interface between two fluids of different density using a 2D wave tank.

2   The 2d wave tank

The experiments were performed in a 2D wave tank at the University of Edinburgh (see Figure 1). This tank is 8 metres long, 0.4 m wide, 1 m deep and filled to a level of 0.70 m. One end of the tank has a flap type wave paddle and the other is filled with a sponge beach to absorb the waves. The wave paddle can generate frequency dispersed waves [6], which means that a wave train may be produced where each wave frequency is given a phase so that all the wave components are focussed at a point. At this focus point, the wave will become unstable and will break, thereby generating turbulence. There are two types of breaking waves which can be generated, spilling and plunging breakers. We are

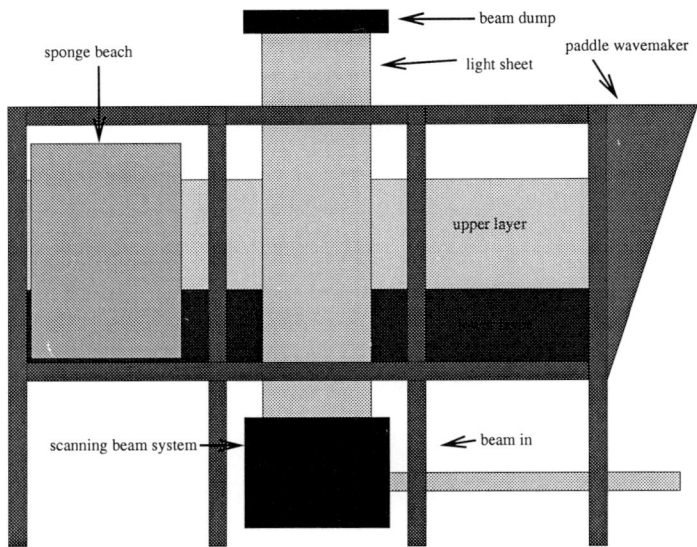

Figure 1. The schematics of the 2D wave tank

more interested in the spilling breakers as these are more common in the deep ocean.

A pump assembly is connected to the tank to allow a second layer of more dense fluid to be introduced in the tank. The lower layer in this case was a salt water solution with a refractive index of about 1.2% sugar, the upper layer being tap water. The relative heights were 0.30 m and 0.4 m for the lower and upper layers respectively. Breaking waves were generated with a critical frequency of 1.375 Hz, a frequency spread of 0.75 Hz and an amplitude of 0.05 m.

3   The Particle Image Velocimetry technique

The particle image velocimetry (PIV) technique has many variations. The basic setup involves a light sheet illuminating seeding particles suspended within the fluid. Several exposures of these particles are recorded photographically and the displacement of their positions between exposures is measured to give the velocity maps of the fluid flow. In this case the light sheet was produced by a rotating multi-faceted mirror [7, 8] which will scan the laser beam across an area of approximately 1m^2 in 5ns. The seeding used was conifer pollen particles, which are highly reflective and only slightly buoyant in water so they will only rise to the surface in a matter of hours. The images of the particles were recorded using an interlaced video camera connected to a high quality video recorder. This would record the image at a resolution of 512 x 750 pixels at a rate of 25 frames a second. The turbulent velocities were quite low, so the particles did not move more than one pixel in 1/25th of a second. This meant that it was not necessary to use a more expensive non-interlaced camera for these experiments. The image frames were grabbed 13 at a time from the playing video at approximately 3 second intervals. Adjacent frames were then analyzed using the cross correlation technique to find the particle displacements between successive frames to produce 12 vector-maps. These were edited to remove spurious vectors and then interpolated to fill all points. This provided 6 sets of 12 vector-maps which could be analyzed for turbulent and mean characteristics.

The results were analyzed by averaging the 12 vector-maps to obtain the mean velocity over the 12/25th of a second, and then finding the variance from that for the turbulent velocity. Thus a mean velocity map and a turbulent velocity map for each set of measurements could be constructed.

4 Results

The mean velocity maps are shown in Figures 2–7, with the wave travelling from right to left and breaking slightly to the right of the picture. The picture is of an area slightly below the surface. The pictures start 4 seconds after breaking which means that the main wave packet has already passed and the only velocities present are those generated by the breaking wave. The figures show two things: a mean flow and a developing vorticity structure. The mean flow is evident in the first picture as it comes in from the right 4 seconds after breaking. It is still present 17 seconds after breaking. The flow must be formed as a response to the surface flow generated by a breaking wave. Since momentum must be conserved within the tank, there must be a return flow. Figure 8 shows the averaged horizontal velocity as a function of depth which illustrates the steady nature of the flow over the time period of the experiment.

The other main factor which can be ascertained from these vector-maps is the development of the turbulence. The vorticity shows the regions of turbulence shifting and deepening in the time frame shown. If we make an estimation of the depth of the turbulence from these vector maps, then we can plot the depth of turbulence as a function of time. This will be expected to be of the form $depth \propto t^{0.25}$. Figure 9 shows that this relationship holds true for these results.

By averaging the energy spectrum over all depths, the averaged energy spectrum can be calculated (Figure 10). This shows that at 4 seconds after breaking the dissipation region of the energy spectrum decays as k^{-3} which is what is expected for 2-dimensional turbulence. After 17 seconds this has changed to a decay rate of $k^{-5/3}$ which is the expected value for 3-dimensional homogeneous turbulence.

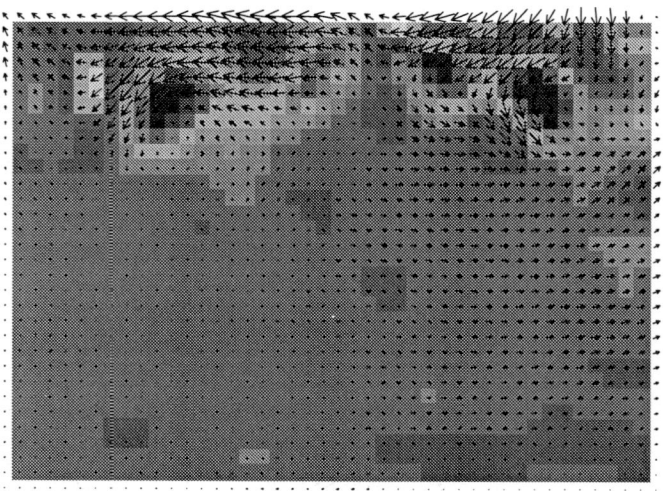

Figure 2. The mean velocity averaged over 0.48 seconds at 4 seconds after breaking

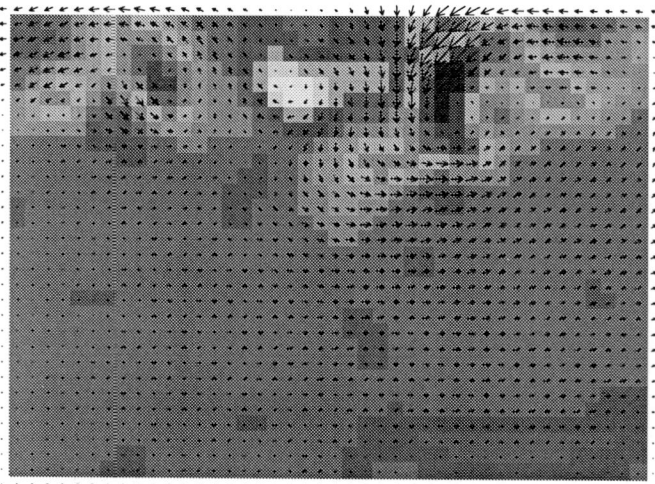

Figure 3. The mean velocity averaged over 0.48 seconds at 8 seconds after breaking

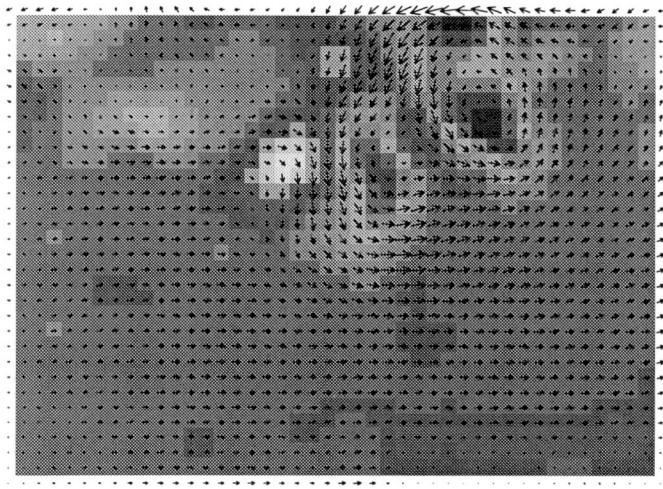

Figure 4. The mean velocity averaged over 0.48 seconds at 10 seconds after breaking

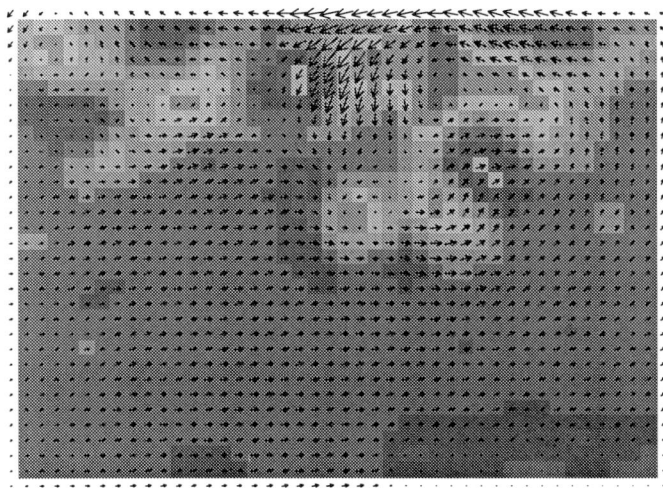

Figure 5. The mean velocity averaged over 0.48 seconds at 13 seconds after breaking

D.B. Hann et al.

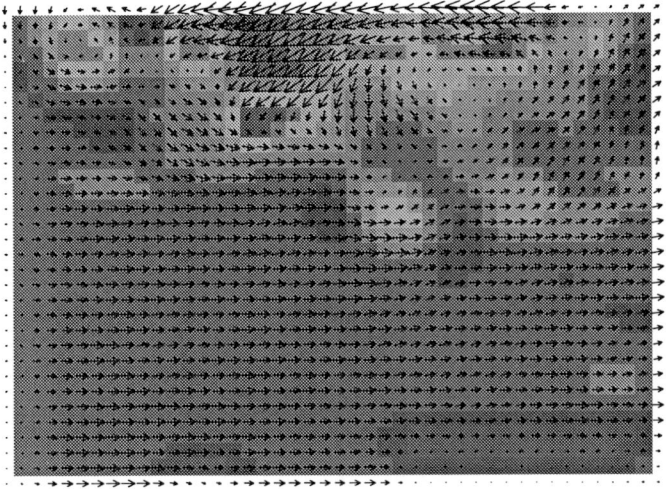

Figure 6. The mean velocity averaged over 0.48 seconds at 15 seconds after breaking

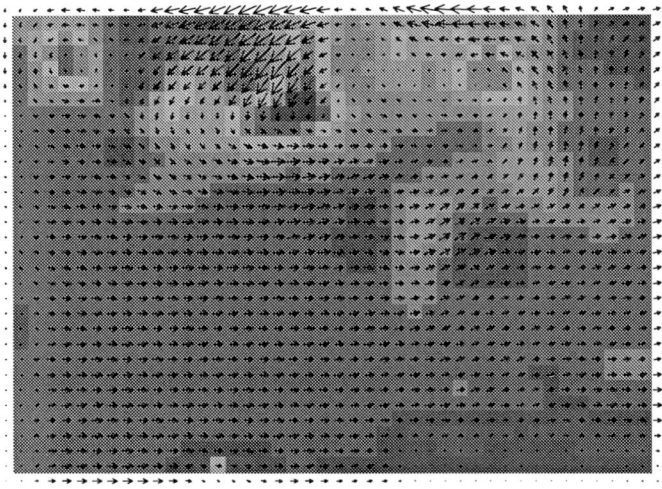

Figure 7. The mean velocity averaged over 0.48 seconds at 17 seconds after breaking

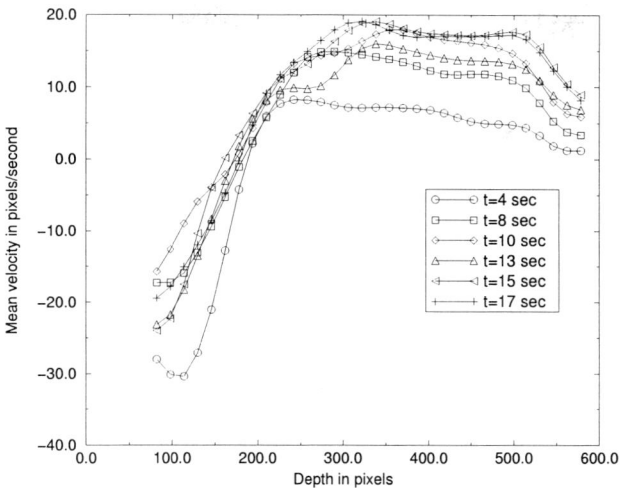

Figure 8. The horizontal velocity with depth showing the return flow developed by the breaking wave

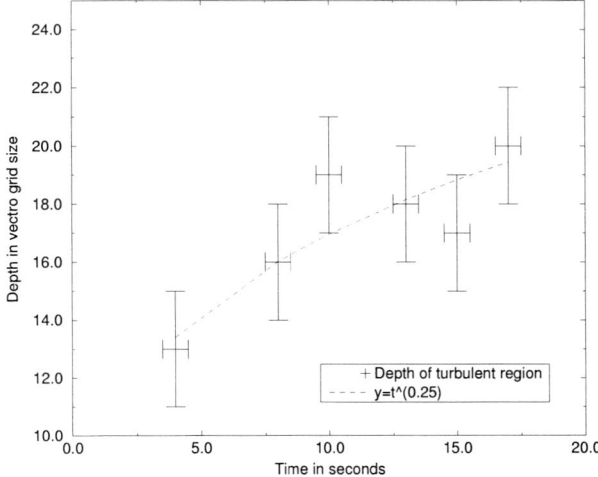

Figure 9. A graph showing that the depth of the turbulence in the averaged vector maps increases as $t^{0.25}$ as expected

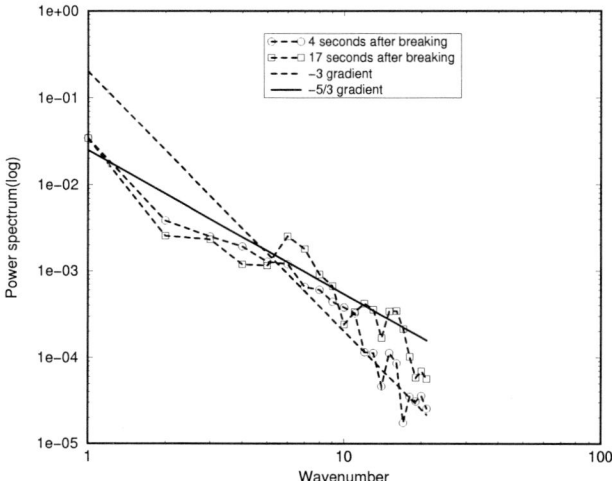

Figure 10. A graph showing that the dissipation region gradient changes from 2-D (-3) to 3-D (-5/3)

5   Conclusion

These results show that the breaking wave generates a mean recirculation in the 2-D wave tank which is fairly steady over the duration of the measurements. The turbulence generated is 2D initially, but becomes more 3D at longer times after breaking.

The measurements shown were carried out only in the upper layer but further measurements will be made in the area around the interface.

Bibliography

1. Y. C. Agrawal, E. A. Terray, M.A. Donelan, P.A. Hwang, A. J. Williams III, W. M. Drennan, K. K. Kahma, and S.A. Kitaigorodskii. Enhanced dissipation of kinetic energy beneath surface waves. *Nature*, 359:219–220, 1992.

2. W.M Drennan, K. K. Kahma, E. A. Terray, M. A. Donelan, and S. A. Kitaigorodskii. Observations of the enhancement of kinetic energy dissipation beneath breaking wind waves. In M. L. Banner and R. H. J. Grimshaw, editors, *Breaking Waves*, pages 95–101. IUTAM Symposium Sydney, 1991.

3. S. A. Kitaigorodskii and J. L. Lumley. Wave-turbulence interactions in the upper ocean. part 1: The energy balance of the interacting fields of surface wind waves and wind–induced three–dimensional turbulence. *Journal of Physical Oceanography*, 13:1977–87, 1983.

4. S. A. Kitaigorodskii, M. A. Donelan, J. L. Lumley, and E. A. Terray. Wave-turbulence interactions in the upper ocean. part ii: Statistical characteristics of wave and turbulent components in the random velocity fields in the marine surface. *Journal of Physical oceanography.*, 13:1988–1999, 1983.

5. A. Anis and J.N Moum. Surface wave-turbulence interactions: Scaling $\epsilon(z)$ near the sea surface. *Journal of Physical Oceanography*, 25:2025–2045, 1995.

6. D. J. Skyner. *The mechanics of extreme water waves.* PhD thesis, The University of Edinburgh, 1992.

7. C. Gray, C. A. Greated, W Easson, and N. E. Fancey. The application of particle image velocimetry to the study of waver waves. *Optics and lasers in Engineering*, 9:265–276, 1988.

8. C. Gray, C. A. Greated, D. R. McClusky, and W. J. Easson. An analysis of the scanning beam PIV illumination system. *Journal of Physics E. Measuring Science and Technology*, 2:717–723, 1991.

Stably Stratified Flow in a Rectangular Container: Cell Pattern Formation and Anomalous Diffusion

G.J.F. van Heijst*, H.J.H. Clercx*, S.R. Maassen* and J.B. Flór**

*Fluid Dynamics Laboratory, Eindhoven University of Technology, The Netherlands, and **Laboratoire des Ecoulements Geophysiques et Industriels, Grenoble, France

Abstract

Laboratory experiments on decaying, stably-stratified flows in slender rectangular containers have demonstrated that the selforganisation process - a manifestation of the inverse energy cascade acting in quasi-2D flows - results in the formation of a linear array of cells with alternating circulations. Advection of viscously-produced vorticity from the lateral domain boundaries gives rise to small perturbations in the cellular flow pattern. Owing to the time-dependence thus introduced, passive scalars may migrate from one cell to a neighbouring cell. Compared to molecular diffusion, this mechanism results in an efficient tracer transport along the array, a process commonly referred to as *anomalous diffusion*. The laboratory observations of the cell pattern formation and the tracer transport are supported by numerical simulations based on a spectral method.

1 Introduction

In contrast to three-dimensional (3D) turbulence, two-dimensional (2D) turbulence is characterised by an inverse energy cascade, i.e. a spectral transfer of kinetic energy to larger scales of motion. This process may lead to selforganisation of the 2D flow, as can be observed from the formation of coherent vortex structures. This process has been demonstrated in numerical simulations [1, 2], and also by laboratory experiments in soap films [3], in rotating fluids [4, 5] and in density-stratified fluids [6, 7, 8]. Strictly speaking, the flow in these experiments was not exactly 2D, owing to boundary effects (tank bottom) and other disturbing features. In the stratified flow experiments, for example, the vortices take on the appearance of thin pancake-shaped structures, in which the motion is planar. Because of vertical gradients in the planar flow, the motion is not purely 2D. Notwithstanding these "mild" 3D effects, the flow evolution shows the emergence of coherent vortex structures as in purely 2D flows, and is hence referred to as quasi-2D.

In the cited experiments, the flow was not affected by any nearby lateral boundaries of the domain. The cited simulations were carried out in a doubly-periodic domain, so that physical boundaries were absent. When the initial flow is forced locally, like for example by a pulsed jet or a moving obstacle as in the experiments, one usually observes the formation of one single vortex or a few coherent vortex structures that may interact or translate away (as in the case of dipole vortices). When the flow is initially forced over the entire domain (as in the cited numerical simulations), a cloud of structures emerges. Through complicated interaction processes, which may lead to vortex stripping and merging, one finally observes a dilute cloud of vortices, which drift around slowly, without any strong interactions.

The question that we address in the present study concerns the effect of lateral no-slip boundaries on the flow evolution. This effect will be studied in one specific configuration, viz. a rectangular container. As will be discussed further on in more detail, a randomly generated initial (quasi-) 2D flow on a rectangular domain shows a transition into a regular linear array of flow cells with alternating circulations. This phenomenon was also observed during the spin-up of a non-stratified fluid in a rectangular tank [9], in which the background rotation provides the two-dimensionalising mechanism. Although many interesting similarities exist, in the present study we will mainly consider the evolution of stratified, non-rotating flow in a rectangular tank. In order to investigate whether the ultimate, organised flow pattern is dependent on details of the initial state, two different types of forcing were applied in the experiment: jet forcing and mechanical stirring by a rake.

Along with the laboratory experiments, numerical 2D flow simulations were performed by using a spectral method. For a detailed discussion of this high-resolution numerical method the reader is referred to Clercx [10]. Apart from investigating the evolution of the decaying flow and the structure of the final cellular pattern, the laboratory experiments and the simulations were also aimed at studying the transport of tracers in the organised, cellular flow. Many of the laboratory experiments discussed in the present paper have been discussed in detail by Flór [11].

The remainder of the paper is organised as follows. The laboratory arrangement is described in Section 2. Laboratory observations and results of numerical simulations of the flow evolution and the ultimate cell pattern are presented in Section 3. Next, aspects of tracer transport along the array of cells are discussed in Section 4. Finally, some general conclusions are formulated in Section 5.

2 Laboratory arrangement

The laboratory experiments were carried out in two different rectangular tanks with horizontal dimensions 40×200 cm^2 and 90×115 cm^2, and in a square tank of 80×80 cm^2. In all cases the tank was filled with a density-stratified fluid with a total depth of 20 cm. In a number of experiments in the rectangular tank the ratio of length (L) to width (D) of the working section was reduced by mounting a false wall in the tank; the aspect ratio $\delta = L/D$ was varied in the range 1 to 8.

The density stratification was established by using salt as a stratifying agent. Linear stratifications were made by applying the well-known two-bucket technique, while two-layer stratifications were obtained by carefully adding salt water underneath a layer of fresh water. Owing to mixing during the filling process and to diffusion at later stages, a relatively thick interface (typically 5 cm) was obtained. The initial flow was generated by two different types of forcing (i) by two horizontal jets along opposite (long) walls, working during some time δt at the same level but in opposite directions, and (ii) by moving a rake of slender vertical bars horizontally through the fluid. After the flow was initiated the forcing devices were removed by lifting them carefully out of the tank. The Reynolds number of the jet-generated flow (based on the flow rate and the nozzle diameter) varied from 600 to 1200, while the Reynolds number of the stir-forcing (based on the bar size and the translation velocity of the rake) measured typically 1500.

The experiments reported on in this paper were all performed in the two-layer stratification with a pycnocline (with a depth of typically 5 cm or more) between both uniform-density layers. The forcing jets were positioned at the centre of the pycnocline. Although with both types of forcing initially motion was induced both in the pycnocline and in the upper and lower layers, the motion outside the pycnocline was observed to decay relatively quickly, presumably as a consequence of the 3D nature of the flow in these unstratified layers. In contrast, the flow in the pycnocline soon became planar (quasi-2D) and was observed to persist for a long while (typically more than one hour).

Small polystyrene tracer particles were used to obtain information about the horizontal motion in the pycnocline. The density structure of the stratified fluid column was chosen in such a way that the tracer particles floated in a thin layer exactly at the mid-level of the interfacial layer. Their motion was recorded from above both by a photo camera (streak photography) and by a video camera. The digital image analysis software package DigImage [12] was used for the tracking of the particles. In this way quantitative information was obtained about the flow evolution, in terms of the horizontal velocity field (u, v) and the spatial distribution of the vertical vorticity component $\omega = \frac{\partial v}{\partial x} - \frac{\partial u}{\partial y}$ and the stream function ψ, which is defined by $(u, v) = \left(\frac{\partial \psi}{\partial y}, -\frac{\partial \psi}{\partial x} \right)$. In addition to particle tracking, in a few experiments dye was added to the fluid, which allowed for a study of the transport of passive scalars.

3 Cell pattern formation

A typical example of the flow evolution, observed after forcing by moving the rake back and forth through the tank, is shown by the streak recordings in Figure 1. Initially, just after the forcing was stopped, the flow had still an irregular character, with mainly smaller vortices interacting in a complicated way (Figure 1(a)). As time proceeds the size of the vortices increases (see Figures 1(b) and 1(c)), until eventually a regular cell pattern is formed

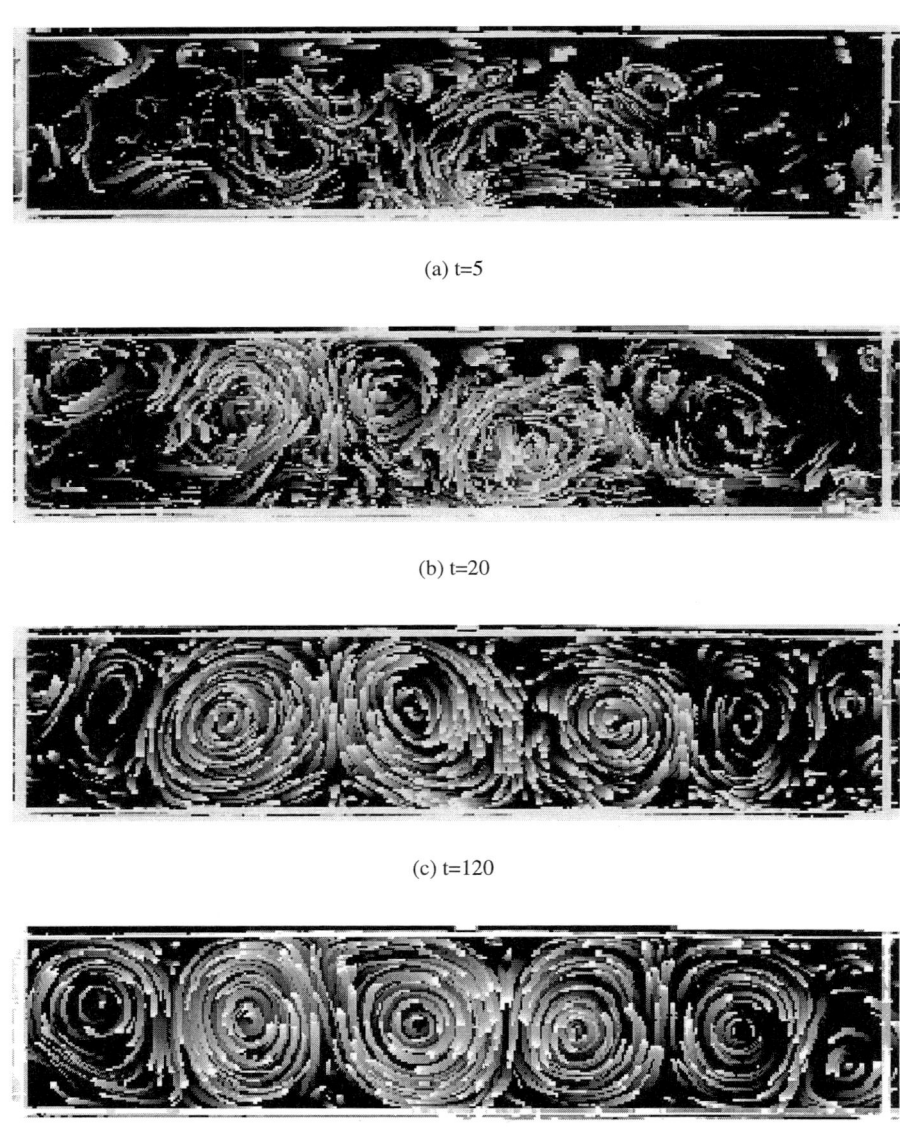

(a) t=5

(b) t=20

(c) t=120

(d) t=360

Figure 1. Sequence of plan-view video streak recordings showing the evolution of the flow, initiated by grid stirring, in a rectangular tank, with aspect ratio $\delta = L/D = 5$, with $L = 200$ cm. The recordings were taken at, (a) $t = 5$ s, (b) $t = 20$ s, (c) $t = 120$ s, and (d) $t = 360$ s. The length of the streaks is (a) 5 s or (b)–(d) 10 s

(Figure 1(d)). In this experiment the tank had an aspect ratio $\delta = 5$, and the number of large cells is roughly in agreement with this value; the small cell in the right-hand end of the tank is very weak. From the lengths of the streak tails, one can infer from Figure 1(d) that the cells in the centre of the tank are stronger than those at the ends. This is most likely caused by the initial stirring (done by hand) not having been uniform over the whole domain: the stirring generated the strongest motion in the central part of the container (see Figure 1(a)).

A similar evolution towards a cellular pattern takes place when the flow is initially forced by horizontal jets, oppositely directed along the longer side walls. The cellular pattern in the quasi-final stage of the flow evolution was also observed in the spin-up experiments by van Heijst et al. [9]. In such experiments the fluid is initially in solid-body rotation at angular velocity Ω, and after the rotation speed of the tank was increased to $\Omega + \Delta\Omega$ (at time $t = 0$), the fluid was observed to adjust to the new rotation speed through a complicated spin-up process. In a rectangular container the (relative) flow takes on the appearance of a linear array of counter-rotating cells, which slowly decay by the action of Ekman layers at the tank bottom. This spin-up process has dynamic similarities with the stratified jet-forced or stirring-induced flow, although the initial forcing is different: relative to the rotating frame, the fluid acquires at $t = 0$ a uniform vorticity $-2\Delta\Omega$ all over the domain, except in thin viscous boundary layers at the tank walls. Note that the total relative vorticity is zero, since the circulation calculated along the (no-slip) walls is by definition zero. During the spin-up process the vorticity is redistributed over the flow domain.

Figure 2 presents results of a numerical simulation of the 2D flow in a rectangular domain ($\delta = 5$) with no-slip walls, initiated according to the instantaneous spin-up. One clearly observes that initially the flow consists of one cell (with clockwise circulation) that fills the rectangular domain completely. Boundary-layer separation sets in quickly, resulting in cells with opposite circulation in the corners. In Figure 2(b) only the two strongest corner cells are still observed, which are then gradually moving inwards (Figures 2(c)–2(g)). Finally, the flow has reached an organised state of 5 cells: two strong anti-clockwise cells and three weaker clockwise cells. The vorticity contour plots in the right column of Figure 2 show how filaments of large-amplitude vorticity are advected from the wall into the interior throughout the adjustment process. The no-slip walls obviously play an essential role: viscosity leads to large vorticity at the walls, not only in the initial state ($t = 0^+$), but during the whole evolution. The vorticity plot of the quasi-final state (Figure 2(h)) shows that each flow cell is accompanied by bands of oppositely-signed vorticity at the boundaries.

Figure 3 shows contour lines of the stream function ψ and the vorticity ω derived from tracking measurements of the flow in the quasi-final state of a grid-stirring experiment in a stratified fluid. In this case the aspect ratio of the container was $\delta = 4$, and four cells are observed. As in the numerical simulation, each cell is accompanied by oppositely-signed vorticity bands at the tank walls. Apparently, in this particular experiment the negative cells are somewhat

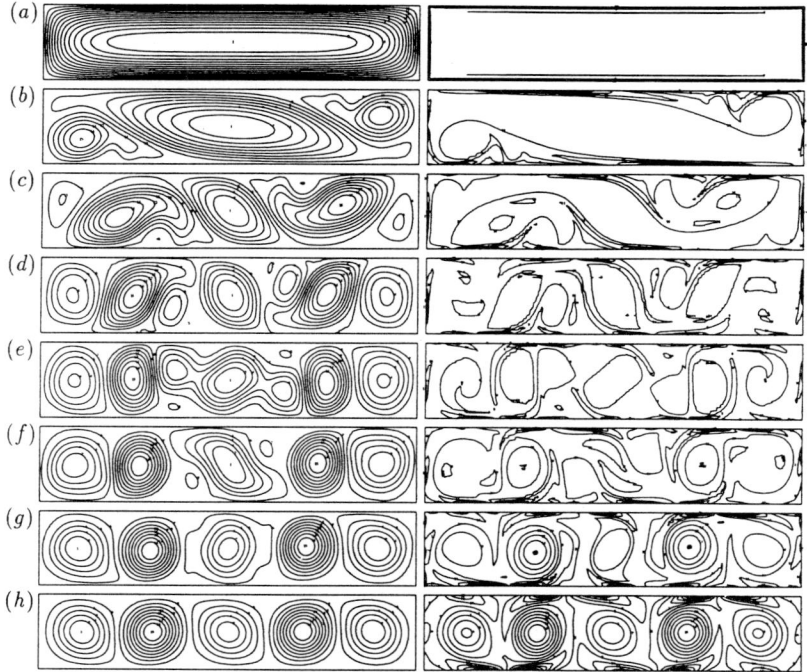

Figure 2. Sequence of stream function and vorticity contour plots calculated numerically by a spectral method for 2D spin-up in a rectangular tank with $\delta = 5$ for non-dimensional times, (a) $t = 0$, (b) 20, (c) 30, (d) 35, (e) 40, (f) 55, (g) 97.5, and (h) 247.5. Time was non-dimensionalised by $\frac{1}{2}\Delta\Omega$. The Reynolds number is $Re = 2D^2\Delta\Omega/\nu = 5000$, where D is the width of the tank, ν the kinematic viscosity of the fluid and $\Delta\Omega$ the change of rotation speed of the container (by courtesy of Frank Tacken)

stronger than the positive ones. This is probably caused by asymmetries in the stirring, which was done by hand. In similar experiments, however, cellular patterns were observed with the cells having approximately equal strengths.

A crucial question is: how can we characterise this organised quasi-final stage of the flow evolution? In the various cases (initial forcing by jets, stirring, impulsive rotation) one observes the eventual formation of an array of cells, but is their dynamical structure similar, i.e. independent of the type of initial forcing?

Inviscid 2D flow is governed by the vorticity equation, which is here written as

$$\frac{\partial \omega}{\partial t} + J(\omega, \psi) = 0 , \tag{3.1}$$

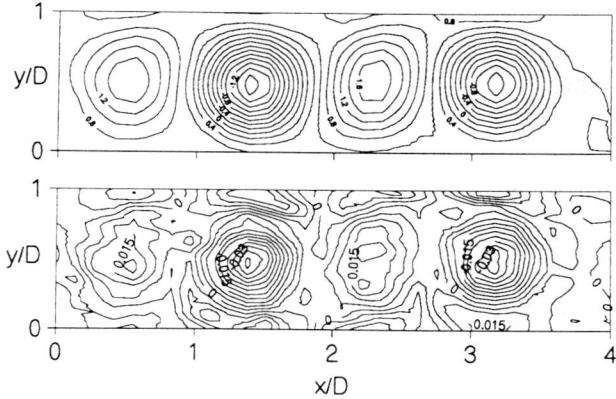

Figure 3. Contour plots of the stream function (upper plot) and the vorticity (lower plot) of a four-cell pattern in a tank with $\delta = L/D = 4$ (with $L = 160$ cm) at $t = 45$ min after forcing by translation of a rake through the fluid

where $J(\omega, \psi) = \frac{\partial \omega}{\partial x}\frac{\partial \psi}{\partial y} - \frac{\partial \omega}{\partial y}\frac{\partial \psi}{\partial x}$ is the Jacobian operator. By virtue of their definitions, the vorticity ω and the stream function ψ satisfy

$$\nabla^2 \psi = -\omega \ . \tag{3.2}$$

For steady flows, (3.1) reduces to

$$J(\omega, \psi) = 0 \ . \tag{3.3}$$

Apparently, *any* functional relationship $\omega = F(\psi)$ satisfies, so that a multitude of solutions of (3.2) is possible. The key to solving this paradoxical problem most likely lies in the action of viscosity, which is entirely neglected in the vorticity Equation (3.1). Many theoretical studies have addressed this fundamental aspect of steady 2D flows, but we will not discuss them here. Instead we follow an alternative approach: the laboratory experiment may also provide information about the relationship between ω and ψ. The particle tracking technique effectively produces a set of velocity vectors associated with the motion of the tracer particles in the flow. As a next step in the data processing, an interpolation is carried out to a regular rectangular grid, which yields velocity vectors in all the grid points. By numerical differentiation and by solving the Poisson Equation (3.2), the vorticity ω and the stream function ψ, respectively, are calculated in each nodal point. Instead of drawing contours of ω and ψ, like in Figure 3, one can also plot each grid point according to its (ω, ψ)-values in a so-called scatter plot. Figure 4(a) shows such an ω, ψ-plot for the central part of the flow domain of Figure 3, namely only those grid points that cover the two

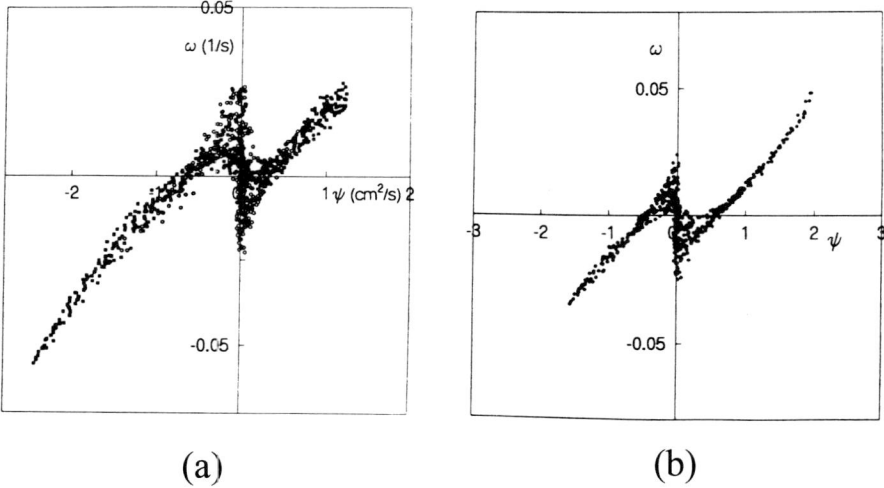

(a)   (b)

Figure 4. Scatter plots of (ω, ψ) for, (a) the two central cells of Figure 3, and (b) two cells arising in an experiment in which the flow was initiated by jet forcing

middle cells. The two branches each represent one cell: the "positive" branch represents the weak cell of positive vorticity with negative vorticity at the wall, while the "negative" branch corresponds with the strong cell of clockwise scirculation. Apparently, the ω, ψ-relationship in each cell is approximately linear, although some curvature is noticed. A similar plot for two cells arising in an experiment with jet-forcing is shown in Figure 4(b). In this case the cells are more or less equally strong; again one observes a weak departure from a linear ω, ψ-relationship. Similar linear dependence of ω and ψ has been observed in the cell pattern that eventually arises during the spin-up of fluid in a rectangular tank [13]. We may thus quite generally conclude that the cells in the decaying organised state are characterised by an approximately linear ω, ψ-relationship, indicating that the flow in this decaying state is not primarily determined by details of the initial (global) forcing.

A specific case of self-organisation of flow in a rectangular configuration was also investigated: the case $\delta = 1$, i.e. a *square* domain. The geometry can nicely accommodate a single cell, and this is exactly what has been observed for initial forcing by two oppositely directed jets. Contour plots of the stream function and the vorticity of the flow during a later stage of the decay are presented in Figure 5. Apparently, the flow has a smooth, continuous vorticity distribution in the centre part of the domain, and oppositely-signed vorticity is again present at the sidewalls of the container. These graphs describe the flow at $t = 145$ min, so

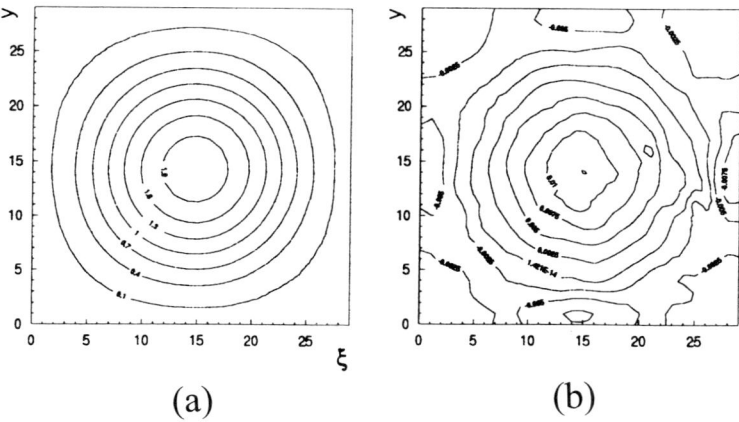

(a)   (b)

Figure 5. (a) Characteristic contour plots of the stream function, and (b) the vorticity of the jet-forced flow in a square tank at $t = 145$ min after the forcing was stopped

a long time after the forcing was stopped (at $t = 0$). A big central cell was formed much earlier in the evolution (after 10 min), but it had the shape of an ellipse, which rotated slowly about its centre. Besides, smaller vortices with oppositely-signed circulation were present in the corners. These corner cells were seen to alternatingly grow and decrease in size, an effect that is presumably directly associated with the rotation of the central, elliptical cell. An ω, ψ-scatter plot of the flow at this stage is shown in Figure 6(a). The scatter in the principal branch is not due to experimental inaccuracies, but hints at non-steadiness of the flow at this stage: in the central part of the domain the flow is thus *not* described by (3.3), so that a well-defined ω, ψ-relationship is not to be expected. Anyhow, the non-steadiness causes only moderate scatter, and the characteristics of the vorticity distribution are clearly visible: an approximately uniform positive vorticity in the centre (implying approximate solid-body rotation) and a band of negative vorticity at the edge of the cell (near the walls and in the corners). In the course of the decay process the shape of the ω, ψ-relationship shows a gradual change into a linear one, see Figure 6(b). The same relaxation towards a linear relationship was observed by Boubnov et al. [14] in an experiment in which a stratified flow in a square tank was continuously forced by a set of horizontal sources and sinks, regularly spaced along the circumference of the container. The sources and sinks were all directed radially, all in the same horizontal plane. Some time after the forcing was turned on a large central cell was observed, with an ω, ψ-relationship very similar to that shown in Figure 6(a). When the forcing

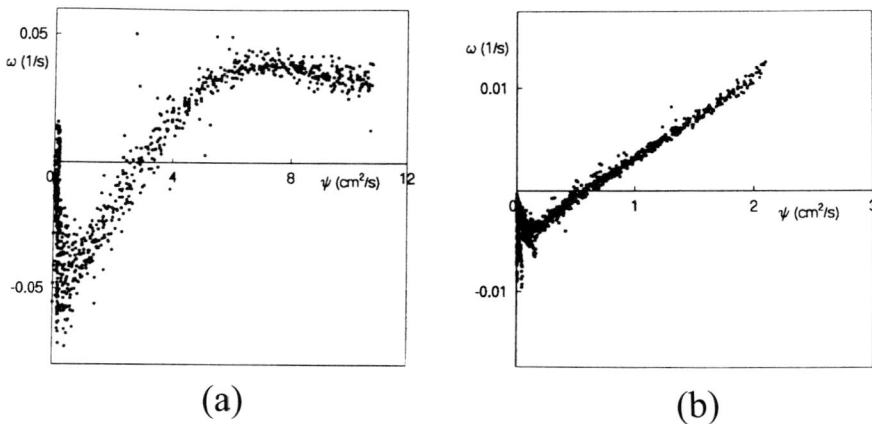

(a)   (b)

Figure 6. Two ω, ψ-scatter plots of the cellular flow in a square tank, generated by jet forcing. The plots represent the flow at (a) $t = 35$ min, and (b) 145 min

was stopped, however, this relationship relaxed towards a linear one, like the one as shown in Figure 6(b).

In their numerical study of the motion of a 2D monopolar vortex in a square domain, Van Geffen et al. [15] observed a similar tendency during the later stages of the decay: a monopolar vortex with a continuous vorticity distribution, released at some off-centre position, was seen to drift along a complicated path but finally ended up in the centre of the domain. The presence of viscosity caused wall-generated vorticity to play a crucial role in the flow evolution, and effected a redistribution of the vorticity by diffusion. Once the vortex centre has arrived at the tank centre, the vorticity has become spread smoothly over the whole domain. The scatter plot of the flow in this viscously decaying state again shows a linear relationship between ω and ψ, except near the walls and in the corners.

Van de Konijnenberg [16] has demonstrated that the relaxation towards a linear relationship can be explained by purely viscous diffusion, i.e. when advective effects have become negligibly small, as is the case during the last stage of the decaying flows described above.

4   Tracer transport

As was seen in the numerical simulation results presented in Figure 2, throughout the flow evolution filaments of viscously generated vorticity are advected away from the wall into the interior. Even in the quasi-final state one observes that - in particular near the domain boundaries - the streamlines do not coincide with the iso-vorticity contours, which implies time-dependence of the flow. Close inspection of the cells - both in the numerical simulations and in the laboratory experiments - has revealed a wiggling motion of their centres with an approximate frequency that is roughly inversely proportional to the turnover time of the cells. Streak photographs showing this phenomenon are presented in Figure 7.

The periodic shifting of the cell centres gives rise to staggered configurations, in which wavy flow patterns between the cells are visible (see Figure 7(b)). This wiggling motion becomes gradually weaker, and eventually a more or less steady cell pattern results.

It was seen in Figure 4 that the flow in the quasi-steady cellular state is characterised by an approximately linear ω, ψ-relationship. By assuming such a relationship $\omega = k^2\psi$, with k a constant, and ignoring viscous effects, it is easy to find a solution of (3.2), which now takes on the appearance of the Helmholtz equation,

$$\nabla^2\psi = -k^2\psi, \tag{4.1}$$

on a rectangular domain ($0 \le x \le L$, $0 \le y \le D$), with $\psi = 0$ at the boundaries (free-slip boundary condition). The gravest mode solution that fits in a rectangle with an integer aspect ratio δ is

$$\psi(x,y) = \psi_0 \sin\frac{\pi x}{D}\sin\frac{\pi y}{D}. \tag{4.2}$$

a)

b)

Figure 7. Streak photographs of the flow in a rectangular tank with $\delta = L/D = 8$, where $L = 115$ cm, initiated by jet forcing. The pictures were taken approximately 2 min after the forcing was stopped, at a time interval of approximately 2.5 min

In order to study the advection induced by the wiggling cell centres, one could take this basic flow solution (4.2) and add a time-periodic perturbation, similar to the approach taken by Gollub and Solomon [17] in a study of chaotic mixing in Rayleigh-Bénard convection rolls. Although this approach is somewhat idealistic (no-slip walls are not included, and the perturbation is prescribed artificially), it provides useful insight in the way passive tracers are advected by the cells. For example, particles initially placed on the separatrix $\psi = 0$ between two neighbouring cells are seen to be advected along the cellular array: some of them are wrapped around the core of one cell, the others get entrained by the other cell. This process repeats, and particles may thus move to the next neighbouring cell, all the way along the array. This phenomenon of advective transport associated with (weak) time-dependent perturbations of an otherwise steady flow is referred to as *chaotic advection*. Compared to molecular diffusion, this mechanism usually results in an efficient transport of tracers, a process known as *anomalous diffusion*. The name is somewhat misleading, because in this model the transport is purely advective, since molecular diffusion is not taken into account. Intuitively, one may expect that the dispersion of tracers in the cellular flow configuration will be enhanced when molecular diffusion of the tracer substance is incorporated in the model. The evolution of a passive scalar is then governed by the dimensionless advection-diffusion equation:

$$\frac{\partial C}{\partial t} + (\boldsymbol{v} \cdot \nabla)C = \frac{1}{Pe}\nabla^2 C \ , \tag{4.3}$$

with $C(x, y, t)$ the tracer concentration and $Pe = UL/\kappa$ the Péclet number based on the velocity scale U, with κ the molecular diffusivity. Impermeable sidewalls imply that the normal gradient of C vanishes at the domain boundaries. The velocity field $\boldsymbol{v}$ is calculated from the Navier-Stokes equation, and serves here merely as an "input" into (4.3).

An example of a numerical tracer transport calculation for the case of a three-cell pattern is shown in Figure 8. Once the flow has reached a weakly non-steady three-cell state, the diffusing substance was released uniformly along one of the end walls, with a Gaussian distribution in the perpendicular direction, at time $t = 0$. The advection by the first cell can be clearly seen in the wrapping of the dark tracer around its core. At later times, the tracer is observed to enter the second, and eventually even the third cell. Comparison of the concentration distribution in the subsequent graphs (the grey scale is indicative of the value of C) reveals how effective the advective mechanism works.

Figure 9 shows a dye visualisation of the tracer transport in a three-cell array. Blobs of orange and green dye were released in the left and right cells, respectively, see Figure 9(a) (on these black and white pictures the orange dye is slightly brighter). After some time the dye has been spread over each of the cells, and thin dye filaments are observed to enter the centre cell (Figure 9(b)). As time proceeds (see Figure 9(c)) these filaments are seen to be stretched and also folded. This feature confirms the non-steady nature of the flow and is indicative

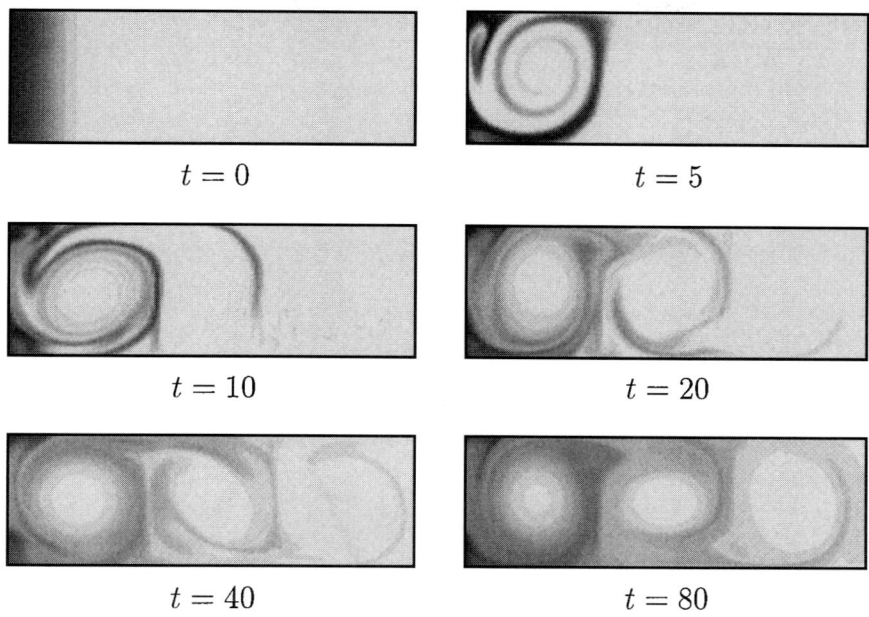

$t = 0$ $t = 5$

$t = 10$ $t = 20$

$t = 40$ $t = 80$

Figure 8. Numerical calculation of tracer transport according to (4.3) for a three-cell flow pattern during impulsive spin-up in a rectangular container. The flow field is computed by a spectral method. The tracer is released at the left end wall, with a Gaussian distribution. Grey scales indicate the value of the tracer concentration; in the last graph the grey levels are rescaled for clarity. $\delta = 3$, $Re = 1250$, $Pe = 5000$ (by courtesy of Jeroen van Oijen)

of chaotic advection: particles initially located close to each other may become separated over a long distance as the flow evolves.

The transport of passive particles in the flow (so ignoring diffusion of the tracer, i.e. for $Pe \to \infty$) can be calculated once the flow field is known. An example of such a simulation is given in Figure 10 for the case of a square domain. The flow is initiated by instantaneous spin-up, and after some time the big central cell behaves as the one in the jet-forced experiment shown in Figures 5 and 6: it is not yet perfectly circular, and smaller cells exist in the corners of the domain. When the central cell is formed, which we assign at $t = 0$, a total of 1000 closely-packed tracer particles are released near the upper right corner cell of the weakly-unsteady flow. The subsequent graphs show how the particles are effectively dispersed: apparently they are not able to enter the main cell, but they reveal an efficient mixing in the outer band and in the corner cells.

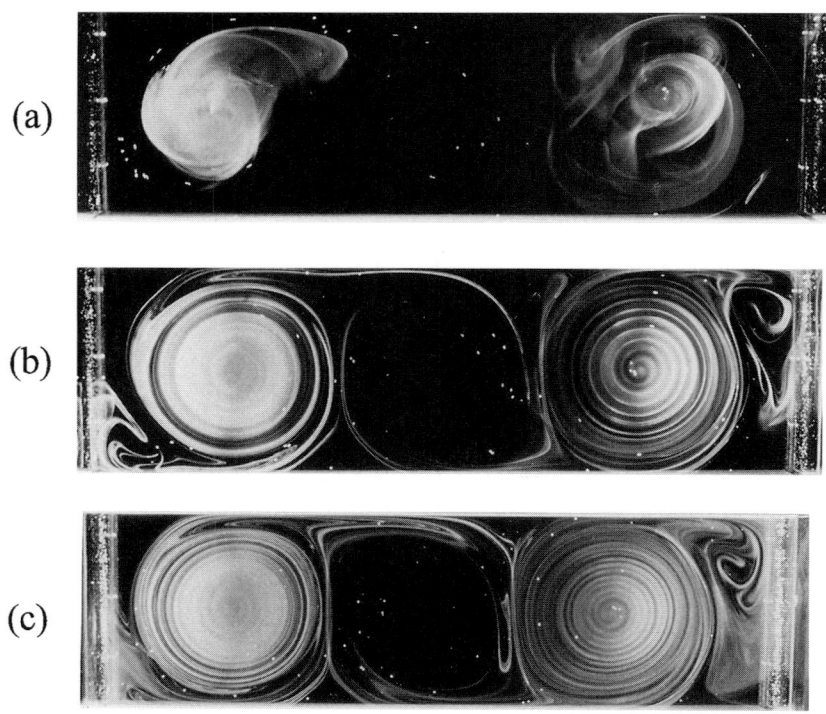

Figure 9. Plan-view photographs of the dye-visualised cell pattern arising in a jet-forced flow in a rectangular tank with aspect ratio $\delta = L/D = 3.5$, with $L = 115$ cm. The "brighter" dye (injected in the left cell) corresponds to orange, the "darker" to green dye. The photographs were taken at, (a) $t = 2$ min, (b) 17.5 min, and (c) 45 min, after the forcing stopped

This behaviour is very similar to what was observed in a laboratory experiment in which the flow was initially forced by grid stirring, upon which a big central cell developed as in the experiment shown in Figure 5. A dye visualisation of the tracer transport by this flow is given in Figure 11. At some stage a blob of dye was released at the mid of the "upper" wall at $t = 30$ min, while another blob was released near the centre at $t = 40$ min, just before the picture of Figure 11(a) was taken. An additional blob of dye was released in the central cell, but at a larger distance from its centre, just before the picture of Figure 11(b) was taken. One clearly observes the complicated deformation of the outer dye blob into a long filament that undergoes considerable stretching and folding (the latter in particular in the corners), while the inner blobs are just

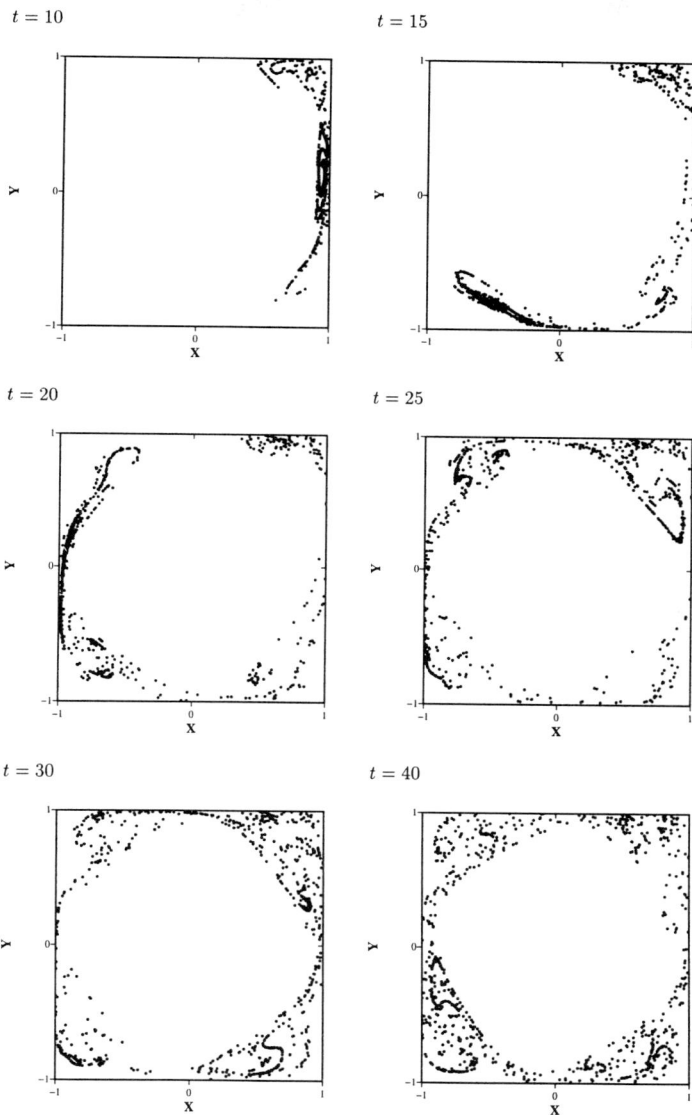

Figure 10. Numerical calculation of the dispersion of passive particles in 2D flow in a square tank forced by instantaneous spin-up. The particles (total of 1000) were released at a stage in which the flow had acquired the shape of one big central cell with clockwise circulation and four smaller cells of opposite circulation in the corners. The particles were released with a Gaussian distribution in a circular area with centre at $(x, y) = (0.7, 0.7)$ and variance of 0.1. The Reynolds number of the flow was 2500 (by courtesy of Jeroen van Oijen)

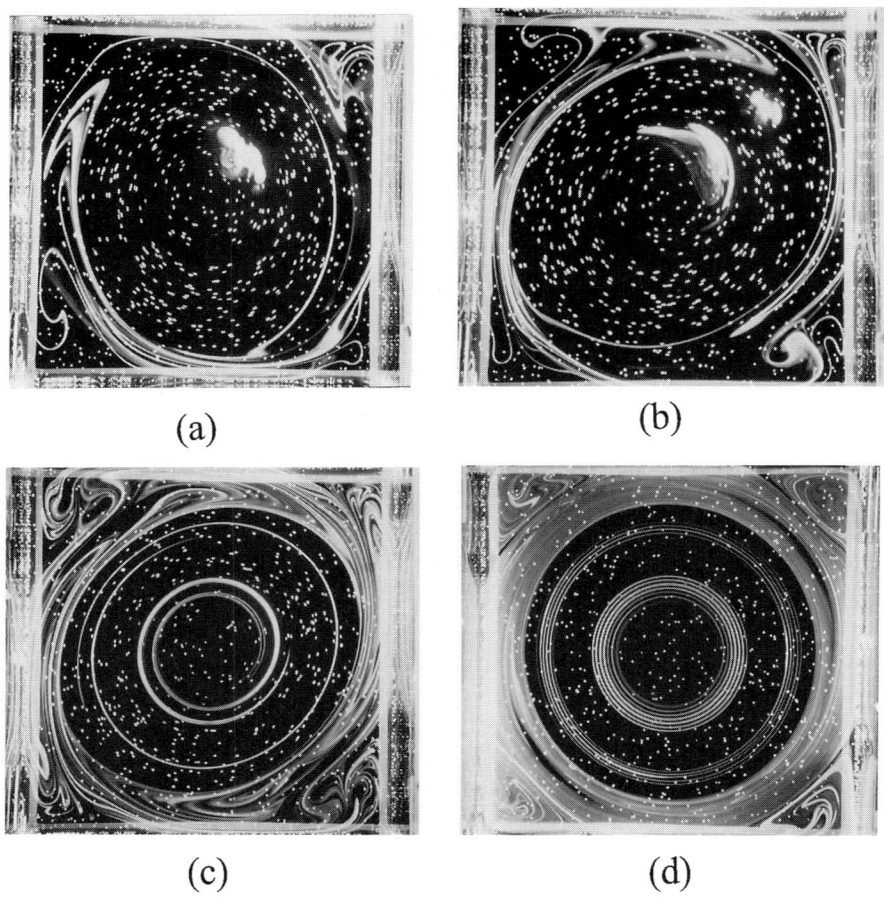

(a)   (b)

(c)   (d)

Figure 11. Dye visualisations of the rake-forced flow in a square tank with $L = 80$ cm. The dye in the outer region was released at $t = 30$ min at the upper wall; the dye in the middle at $t = 40$ min and at $t = 42.5$ min. The pictures were taken at (a) $t = 41$ min, (b) 43 min, (c) 70 min, and (d) 154 min

stretched into long spiral-shaped filaments, owing to the differential azimuthal velocity of the central cell. After a long time ($t = 154$ min, see Figure 11(d)) the dye appears to have become mixed very well in the outer band of the central cell. Some additional dye was injected in the upper left corner cell, some time before the photograph of Figure 11(d) was taken. Close inspection of this photograph reveals that some of this dye has leaked along the wall into the upper right cell, and also - although to a lesser extent - into the bottom cells. This indicates that flow in the outer band of the central cell has become more or less steady at this stage.

5   Conclusion

Laboratory experiments on stably stratified flows in a rectangular container have demonstrated that the selforganisation mechanism acting in quasi-2D flows results in the formation of a linear array of counter-rotating cells that fill the domain completely. This quasi-final state is characterised by an approximately linear ω, ψ-relationship, similar to what has been observed in experiments on spin-up of a homogeneous fluid in a rectangular tank. A tentative conclusion that can be formulated on the basis of these observations is that the quasi-final state of the decaying quasi-2D flow in this geometry is not dependent on the details of the initial (global) forcing.

Numerical simulations have revealed the crucial role of viscously-produced vorticity at the lateral domain boundaries: filaments of vorticity are observed to be advected away from the walls into the interior throughout the flow evolution. This advection of filamentary vorticity structures continues even after the cellular pattern has established, which leads to small perturbations of the cells. The time-dependence thus introduced provides the mechanism for chaotic advection of scalars. Both the numerical simulations and the laboratory experiments have demonstrated this transport mechanism: tracer particles were observed to migrate from one cell to a neighbouring one and so on.

In the case of a square container only a single large cell arises in the organised state, with small cells in the corners of the domain. The time-dependence associated with the advection of filaments of wall-generated vorticity results in enhanced mixing in only the outer band of the central cell.

It is clear that dye visualisations (Figures 9 and 11) and numerical computations of particle and scalar dispersion (Figures 8 and 10) provide important information about the transport characteristics of the cellular flow, in particular about the location of areas of weak and strong mixing. In addition to this, there is a need to quantify the efficiency of mixing. Work on this is still in progress.

Acknowledgements

One of the authors (S.R.M.) gratefully acknowledges support from the Netherlands Geosciences Foundation (NWO-GOA), while J.B.F. was financially supported by the Netherlands Foundation for Fundamental Research on Matter (FOM) under project SW 88.650.

Bibliography

1. McWilliams, J.C. - The emergence of isolated coherent vortices in turbulent flows. *J. Fluid Mech.* **146**, 21-43 (1984).

2. Legras, B., P. Santangelo & R. Benzi - High-resolution numerical experiments for forced two-dimensional turbulence. *Europhys. Lett.* **5**, 37-42 (1988).

3. Couder, Y., & C. Basdevant - Experimental and numerical study of vortex couples in two-dimensional flows. *J. Fluid Mech.* **173**, 225-251 (1986).

4. Flierl, G.R., M.E. Stern & J.A. Whitehead - The physical significance of modons: laboratory experiments and general integral constraints. *Dyn. Atmos. Oceans* **7**, 233-263 (1983).

5. Hopfinger, E.J., & G.J.F. van Heijst - Vortices in rotating fluids. *Ann. Rev. Fluid Mech.* **25**, 241-289 (1989).

6. Heijst, G.J.F. van, & J.B. Flór - Dipole formation and collisions in a stratified fluid. *Nature* **340**, 212-215 (1989).

7. Voropayev, S.I., Y.D. Afanasyev & I.A. Filippov - Horizontal jets and vortex dipoles in a stratified fluid. *J. Fluid Mech.* **227**, 543-566 (1991).

8. Flór, J.B., & G.J.F. van Heijst - An experimental study of dipolar vortex structures in a stratified fluid. *J. Fluid Mech.* **279**, 101-133 (1994).

9. Heijst, G.J.F. van, P.A. Davies & R.G. Davis - Spin-up in a container. *Phys. Fluids* **A2**, 150-159 (1990).

10. Clercx, H.J.H. - A spectral solver for the Navier-Stokes equations in the velocity-vorticity formulation for flows with two nonperiodic directions. *J. Comput. Phys.* **137**, 186-211 (1997).

11. Flór, J.B. - *Coherent Vortex Structures in Stratified Fluids.* PhD thesis, Eindhoven University of Technology (1994).

12. Dalziel, S.B. - *DigImage: Image Processing for Fluid Dynamics.* Cambridge Environmental Research Consultants Ltd. (1992).

13. Konijnenberg, J.A. van de, H.I. Andersson, J.T. Billdal & G.J.F. van Heijst - Spin-up in a rectangular tank with low angular velocity. *Phys. Fluids* **6**, 1168-1176 (1994).

14. Boubnov, B.M., S.B. Dalziel & P.F. Linden - Source-sink turbulence in a stratified fluid. *J. Fluid Mech.* **261**, 273-303 (1994).

15. Geffen, J.H.G.M. van, V.V. Meleshko & G.J.F. van Heijst - Motion of a two-dimensional monopolar vortex in a bounded rectangular domain. *Phys. Fluids* **8**, 2393-2399 (1996).

16. Konijnenberg, J.A. van de - *Spin-up in Non-axisymmetric Containers.* PhD thesis, Eindhoven University of Technology (1995).

17. Gollub, J.P., & T.H. Solomon - Complex particle trajectories and transport in stationary and periodic convective flows. *Physica Scripta* **40**, 430-435 (1989).

Vortex Solitons in Stably Stratified Differentially Rotating Fluid

A. Stegner and V. Zeitlin

LMD, B.P. 99, Université P. et M. Curie, Paris, France

Abstract

Large-scale long-living slowly drifting quasi-axisymmetric vortices are found as solutions of the hydrodynamics equations for a differentially rotating stably stratified fluid layer in the Boussinesq approximation. They are obtained by applying the multiple-scale asymptotic analysis based on perturbative expansions in the amplitude of the pressure perturbation, Rossby number and the gradient of the overall rotation angular velocity. Physically, the appearance of these vortices is due to the mutual compensation of weak nonlinearity and weak dispersion of the long baroclinic Rossby waves. The vortices found have a vertical structure of the linear baroclinic modes. Being intense, they trap fluid particles in the regions (lenses) close to the core and localized in the vertical direction. These lenses may be cyclonic or anticyclonic depending on the given mode.

1 Introduction

In the present paper we address a question of large-scale long-living vortices in a differentially rotating continuously stratified fluid layer. Our basic motivation is provided by numerous astro- and geophysical observations indicating the presence of such structures in planetary atmospheres and in the ocean [1], [2]. We adopt an idea, first formulated in [3] (see also [4] in the oceanic context), that the structures of this kind may be, in principle, explained as solitons arising due to the mutual compensation of (weak) dispersion and (weak) nonlinearity of the long Rossby waves. Although a number of papers developing this approach exist in the literature (for a full list of references cf. [6]), most of the work was limited up to now by the barotropic shallow water. Recent generalizations to a two-layer model [5], [6] indicate that stratification introduces some important new features in the analysis and below we confirm this in the physically and technically much more interesting case of continuous stratification.

The present work is based on our previous studies of the same problem in the barotropic shallow water dynamics [7] and in the two-layer models [6] and uses the nonlinear quasi-geostrophic (NLQG) regime described in these papers. We also use the ideas from the paper [8] where, however, a different asymptotic regime (the intermediate geostrophic one) has been exploited. Let us remind ourselves that the idea underlying the search for solitary vortices in rotating shallow water systems consists in the observation [3] that the Korteveg-de Vries equation in 2 dimensions

$$f_t - aff_x - \Delta f_x = 0 \tag{1.1}$$

has steady-drifting localized solutions (x and y are the coordinates on the plane, Δ is the Laplacian, the subscripts denote the corresponding partial derivatives). They are anticyclones (positive f) or cyclones (negative f) depending on the sign of the constant a. Equations of the type (1.1) arise in perturbation theory in vortex amplitude for slow geostrophically balanced motions in a rapidly rotating fluid layer. To have a consistent perturbation theory and to define slowness properly one has to non-dimensionalize the original set of hydrodynamical equations and to introduce characteristic dimensionless parameters governing the dynamics. They are the Rossby number ϵ, the characteristic pressure perturbation amplitude λ (the nonlinearity parameter), the characteristic stratification parameter n and the characteristic gradient of the overall rotation angular velocity in the y- direction, the β- parameter. Throughout this paper we shall use the β - plane model where the angular velocity is

$$\Omega(y) = \Omega(1 + \beta y). \tag{1.2}$$

All four parameters are supposed to be small, so before developing the physical quantities in perturbation series one should fix their relative values. The relation between ϵ, β and λ together with the first-order balance between the pressure terms and the Coriolis terms in Euler equations (the geostrophic balance) defines a characteristic scale of the structures. The NLQG regime which will be used in this paper consists in choosing

$$\epsilon \sim \beta \sim \lambda^2 \tag{1.3}$$

and describes structures of a characteristic scale larger than the corresponding Rossby radius. It has been shown in [7], [6] that this scaling is the only one consistently leading to (1) because the so-called twisting terms, i.e. the terms with explicit y- dependence inevitably arising in perturbation series, cf. (1.2), appear only in the fourth order in this regime. Below, as usual in multi-scale asymptotic analysis, we shall develop velocity, density and pressure fluctuations in perturbation series in λ, introducing multiple time-scales. The slow-time evolution equations of the type (1.1) for low-order perturbations appear as integrability conditions for dynamical equations governing the dynamics of higher-order corrections.

2   NLQG regime for a weakly stratified fluid layer

We use the Boussinesq approximation for a stratified inviscid fluid layer of depth H_0:

$$Du - 2\Omega v(1 + \beta y) = -\rho^{-1} p_x$$
$$Dv + 2\Omega u(1 + \beta y) = -\rho^{-1} p_y \tag{2.1}$$

$$u_x + v_y + w_z = 0, \quad D\rho = 0, \tag{2.2}$$

together with the hydrostatic equation in vertical direction allowing to relate density and pressure

$$p_z + g\rho = 0. \tag{2.3}$$

Here u, v, w are the three velocity components, $D = \partial_t + u\partial_x + v\partial_y + w\partial_z$. For simplicity we choose rigid lid ($z = H_0$) and rigid bottom ($z = 0$) boundary conditions. The case we consider here is a weakly stratified one. By this we mean that the basic state is characterized by the following density and pressure profiles:

$$\rho = \rho_0(1 + n\rho_s(z))$$
$$p = \rho_0 g H_0((1 - z) + n p_s(z)) \tag{2.4}$$

where ρ_0 is a constant, n is a stratification parameter supposed to be much smaller than ϵ^2 which, roughly , corresponds to the oceanic situation. This will allow us to replace ρ by ρ_0 in the r.h.s. of (2.1) and, thus, simplify considerably the calculations. The p_s and the ρ_s, being still related by the non-dimensionalized version of the hydrostatics Equation (2.3), define some stably stratified profile. We suppose, and this is our second hypothesis as regards to stratification, that the mean stratification length $< l >$, where $l(z) = N^2(z)/g$, N being the Brunt-Vaisala frequency associated with ρ_s, is of the same order as H_0. The full density and pressure profiles, thus, are

$$\rho = \rho_0(1 + n\rho_s + n\lambda\sigma)$$
$$p = \rho_0 g H_0((1 - z) + n p_s + n\lambda\pi) \tag{2.5}$$

where σ, π are dynamical variables characterizing the density and the pressure perturbations. We choose H_0 as a characteristic vertical scale in our problem and get the following hydrostatic equation:

$$\pi_z + \sigma = 0. \tag{2.6}$$

We proceed with non-dimensionalization of our equation by introducing the appropriate scales. The characteristic horizontal scales and velocities of the vortex structures are r_0, v_0, the vertical scale is $< l > \sim H_0 << r_0$. The consistent choice for the vertical velocity scale is $\lambda H_0 r_0^{-1} v_0$. We shall use r_0/v_0 as a time scale, remembering that by definition the Rossby number is $\epsilon = v_0/2\Omega r_0$. It will turn out later that the actual time scale is, in fact, slower by the factor λ, see below. The density and pressure perturbations σ and π are already non-dimensional. The density advection equation in these terms takes the following form

$$D\sigma + \rho'_s w = 0, \tag{2.7}$$

where the prime denotes the z- derivative and one should remember about a factor λ in front of w in the non-dimensionalized Lagrangian derivative D. While non-dimensionalizing the Euler Equations (2.1) we have to take into account our requirement that pressure terms have to equilibrate the Coriolis terms in the zeroth order which gives:

$$\frac{n\lambda g H_0}{2\Omega r_0 v_0} = \frac{\lambda}{\epsilon}\left(\frac{R_0}{r_0}\right)^2 = \mathcal{O}(1), \tag{2.8}$$

where

$$R_0^2 = \frac{n g H_0}{(2\Omega)^2} \tag{2.9}$$

is the baroclinic Rossby radius. We see that by chosing the NLQG regime (1.3) we choose the Burger number to be much smaller than unity. This fixes the characteristic scale of vortices under consideration to be larger than the Rossby radius, cf. (2.8). Note, however, that an Burger number is one order of magnitude smaller than unity when the characteristic scale is only three times larger than Rossby radius.

We, thus, arrive to a following system of dimensionless equations:

$$\begin{aligned}
\epsilon Du - v(1 + \beta y) &= -\pi_x \\
\epsilon Dv + u(1 + \beta y) &= -\pi_y
\end{aligned} \tag{2.10}$$

together with the dimensionless continuity equation

$$u_x + v_y + \lambda w_z = 0 \tag{2.11}$$

(all possible numerical factors of order unity are absorbed into dynamical variables in these equations).

3   Perturbative   analysis   and   axisymmetric   solitonic   solutions

Now our strategy is to develop all the dynamical variables in perturbation series in λ and to get evolution equations for subsequent perturbations. In doing this we use the fact that Equation (2.6) allows to express σ in terms of π and that (2.7) allows to eliminate w. The horizontal velocities are expressed perturbatively in terms of pressure fluctuations via Equations (2.10) and all the results are then substituted in the continuity Equation (2.11), thus, giving a closed equation for dynamical evolution of the pressure.

The perturbative expansions for velocity and pressure fluctuations are:

$$
\begin{aligned}
u &= u^{(0)} + \lambda u^{(1)} + \lambda^2 u^{(2)} + \dots \\
v &= v^{(0)} + \lambda v^{(1)} + \lambda^2 v^{(2)} + \dots \\
\pi &= \pi^{(0)} + \lambda \pi^{(1)} + \lambda^2 \pi^{(2)} + \dots
\end{aligned}
\tag{3.1}
$$

and we have

$$
\sigma^{(i)} = -\pi_z^{(i)}, \quad i = 1, 2, \dots
\tag{3.2}
$$

in the zeroth order in λ the vertical component of velocity is absent and we get the standard geostrophic equations

$$
v^{(0)} = \pi_x^{(0)}; \quad u^{(0)} = -\pi_y^{(0)}.
\tag{3.3}
$$

The continuity equation is trivially satisfied in this order. In the first order in λ we have

$$
v^{(1)} = \pi_x^{(1)}; \quad u^{(1)} = -\pi_y^{(1)}
\tag{3.4}
$$

which gives a zero contribution into the continuity equation, too. However, a first-order w-term appears, cf. (2.7)

$$
w^{(0)} = \rho_s'^{-1} D_h^{(0)} \pi z_z^{(0)}
\tag{3.5}
$$

where D_h^i means a horizontal part of the i-th order Lagrangian derivative

$$
D_h^{(i)} = \partial_{t_i} + u^{(i)} \partial_x + v^{(i)} \partial_y
\tag{3.6}
$$

and we take into account the slow time-scales appearing in higher orders of perturbation theory, $t_i = \lambda^{-i} t$. Note that we are working with stable stratifications only and, therefore, $\rho_s' < 0$. Hence, the continuity equation gives

$$
\left(\frac{D_h^{(0)} \pi_z^{(0)}}{\rho_s'} \right)_z = 0 .
\tag{3.7}
$$

Our first basic hypothesis is that we are looking for a soliton i.e. a packet of long baroclinic Rossby waves which maintains its form due to the mutual compensation of weak nonlinearity and dispersion. As the spectrum of the vertical baroclinic modes is discrete, they propagate with essentially different phase-speeds. Therefore, in order to have a steady drifting structure we have to associate it with a specific vertical mode. Hence, in what follows we are looking for solutions admitting a separation of vertical and horizontal variables with the vertical part corresponding to that of the baroclinic Rossby wave:

$$\pi^{(0)}(x, y, z; t) = f^{(0)}(z)p^{(0)}(x, y; t). \tag{3.8}$$

This implies, after being substituted in (3.7), that

$$\partial_{t_0} p^{(0)}(x, y; t) = 0 \tag{3.9}$$

meaning that the actual time-scale of evolution is not faster than $t_1 = \lambda^{-1} t_0$, i.e. we are dealing with very slow motions. An equation for the baroclinic mode $f^{(0)}(z)$ will appear in the next order, see below.

In the next, second order in λ we get

$$
\begin{aligned}
v^{(2)} &= \pi_x^{(2)} - v^{(0)} y + D_h^{(0)} u^{(0)} \\
u^{(2)} &= -\pi_y^{(2)} - u^{(0)} y - D_h^{(0)} v^{(0)}
\end{aligned}
\tag{3.10}
$$

giving a horizontal divergence

$$u_x^{(2)} + v_y^{(2)} = -\pi_x^{(0)} - J(\pi^{(0)}, \Delta\pi^{(0)}) \; ; \tag{3.11}$$

where $J(..., ...)$ denotes the Jacobian.

The second-order correction to w is

$$w^{(1)} = \rho_s'^{-1}(D_h^{(1)} \pi_z^{(0)} + D_h^{(0)} \pi_z^{(1)}) + \rho_s'^{-2} \pi_{zz}^{(0)} D_h^{(0)} \pi_z^{(0)}. \tag{3.12}$$

Introducing (3.11), (3.12) into the continuity equation and taking into account (3.8) and (3.9) we have

$$\left(\frac{\pi_{zt_1}^{(0)} + J(\pi^{(1)}, \pi_z^{(0)}) + J(\pi^{(0)}, \pi_z^{(1)})}{\rho_s'} \right)_z - \pi_x^{(0)} - J(\pi^{(0)}, \Delta\pi^{(0)}) = 0 . \tag{3.13}$$

This is an inhomogenious partial differential equation for $\pi^{(1)}$ once $\pi^{(0)}$ is known. Here we make our second basic hypothesis assuming that $p^{(0)}(x, y; t)$ is (i) axisymmetric, (ii) steady drifting with some velocity c which means

$$p_{t_1}^{(0)} - cp_x^{(0)} = 0 . \tag{3.14}$$

In this case by introducing polar coordinates r, θ in the moving frame of reference we get from (3.13)

$$\left(\frac{\pi_{\theta z}^{(1)} f^{(0)} - \pi_\theta^{(1)} f_z^{(0)}}{\rho_s'} \right)_z + c\, r \cos\theta \left(\left(\frac{f_z^{(0)}}{\rho_s'} \right)_z - c^{-1} f^{(0)} \right) = 0. \tag{3.15}$$

Now, the equation

$$\left(\frac{f_z^{(0)}}{\rho_s'} \right)_z - c^{-1} f^{(0)} = 0 \tag{3.16}$$

corresponds exactly to that defining vertical mode structure of the baroclinic Rossby waves (cf. for example [9]) which may be easily seen by linearizing the basic Equations (2.6), (2.7), (2.10) and (2.11). Hence, taking into account our preceding remarks we require (3.16) to be valid. Then we get from (3.15), (3.16) exactly the same Equation (3.16) for $\pi_\theta^{(1)}$ which means that either $\pi^{(1)}$ is an axisymmetric function with arbitrary (at this order) dependence on z, or it is of the form $p^{(1)}(x, y; t)f^{(0)}(z)$ where $p^{(1)}(x, y; t)$ is an arbitrary function. Thus, the general solution of (3.13) will be a combination of these two particular solutions with arbitrary coefficients and we can consistently set the total pressure field to be axisymmetric at this order. Possible asymmetric corrections will be discussed later. We, therefore, see that at this order the basic axisymmetric pressure profile has a characteristic vertical structure $f^{(0)}(z)$ of a baroclinic Rossby wave and drifts with the phase speed c of this latter. Both $f^{(0)}$ and c are defined from the eigenvalue problem (3.16) with appropriate boundary conditions for $f^{(0)}(z)$, i.e. they are totally determined by the background stratification profile $\rho_s(z)$. The boundary conditions for $f^{(0)}$ follow from the kinematic boundary condition $w(z = 0) = w(z = 1) = 0$ which gives zero-derivative conditions for $f^{(0)}$. For the linear stratification the eigenfunctions are simple harmonic functions and the eigenvalues are positive for stable stratification meaning a westward drift. In fact, we recover here a dispersionless propagation of long Rossby waves. In the third order in λ we obtain

$$
\begin{aligned}
v^{(3)} &= \pi_x^{(3)} - v^{(1)}y + D_h^{(0)}u^{(1)} + D_h^{(1)}u^{(0)} \\
u^{(3)} &= -\pi_y^{(3)} - u^{(1)}y - D_h^{(0)}v^{(1)} - D_h^{(1)}v^{(0)} .
\end{aligned}
\tag{3.17}
$$

While calculating the horizontal divergence in this order we take into account the axisymmetry and separability of $\pi^{(0,1)}$ and get

$$
u_x^{(3)} + v_y^{(3)} = -\pi_x^{(1)} - \Delta\pi_{t_1}^{(0)} .
\tag{3.18}
$$

The third correction to w is

$$
w^{(2)} = \rho_s'^{-1}(D_h^{(2)}\pi_z^{(0)} + D_h^{(0)}\pi_z^{(2)} + D_h^{(1)}\pi_z^{(1)}) + \rho_s'^{-2}\pi_{zz}^{(0)}(D_h^{(0)}\pi_z^{(1)} + D_h^{(1)}\pi_z^{(0)})
$$

$$
+\rho_s'^{-3}\pi_{zz}^{(0)^2}D_h^{(0)}\pi_z^{(0)}
\tag{3.19}
$$

Using our preceding hypotheses we get from the continuity equation together with (3.18), (3.19)

$$
\left(\frac{\pi_{zt_2}^{(0)} + \pi_{zt_1}^{(1)} + J(\pi^{(2)}, \pi_z^{(0)}) + J(\pi^{(0)}, \pi_z^{(2)})}{\rho_s'} + \frac{\pi_{zz}^{(0)}\pi_{zt_1}^{(0)}}{\rho_s'^2}\right)_z - \pi_x^{(1)} - \Delta\pi_{t_1}^{(0)} = 0 .
\tag{3.20}
$$

As at the previous order, this is an equation for $\pi^{(2)}$ provided $\pi^{(0,1)}$ are known. This is a linear equation and we may, again, consistently set a non-axisymmetric part of $\pi^{(2)}$ to be equal to zero. Then we get a following equation for the yet unknown axisymmetric part of $\pi^{(1)}$

$$\left(\frac{\pi_{zt_1}^{(1)}}{\rho_s'}\right)_z - \pi_x^{(1)} + \left(\frac{f_z^{(0)}}{\rho_s'}\right)_z p_{t_2}^{(0)} + c\left(\frac{f_z^{(0)} f_{zz}^{(0)}}{\rho_s'^2}\right)_z p^{(0)} p_x^{(0)} - cf^{(0)} \Delta p_x^{(0)} = 0 . \quad (3.21)$$

For consistency, $\pi^{(1)}$ has to have the same drift velocity c as $\pi^{(0)}$ and, hence, the same as in (3.16) operator

$$\hat{L}f = \left(\frac{f_z}{\rho_s'}\right)_z - c^{-1}f \quad (3.22)$$

appears in the left-hand side. With the above-mentioned boundary conditions this operator is self-adjoint, so we get an integrability condition for (3.21) by multiplying by $f^{(0)}$ and integrating by z, as $f^{(0)}$ is an eigenfunction of (3.22) with eigenvalue 0. Thus, we arrive to the following evolution equation for $p^{(0)}$.

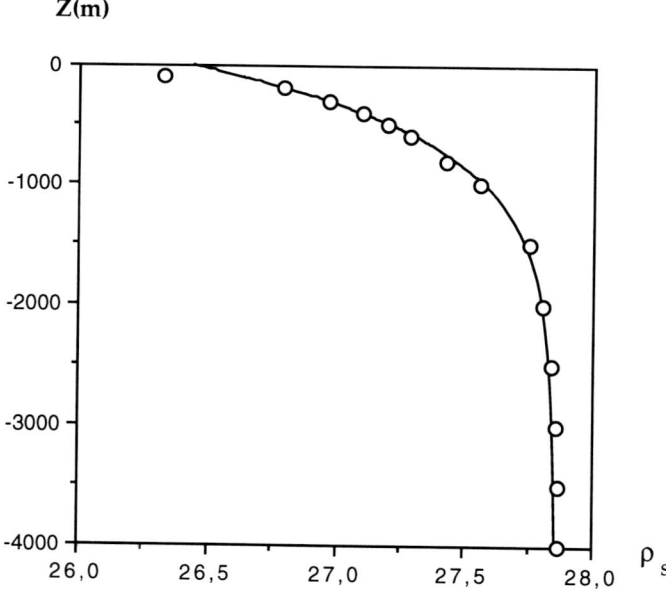

Figure 1. Mean stratification of Atlantic ocean. The solid line is a continuous stratification profile interpolated from observational data (open circles). The corresponding stratification parameter is $n = 10^{-3}$

$$c_1 p_{t_2}^{(0)} - c_2 \Delta p_x^{(0)} - c_3 p^{(0)} p_x^{(0)} = 0 , \tag{3.23}$$

i.e. exactly the 2-d KdV Equation (1.1). The constants c_1, c_2, c_3 are defined as follows:

$$c_1 = \int_0^1 dz\, f^{(0)} \left(\frac{f_z^{(0)}}{\rho_s'} \right)_z , \tag{3.24}$$

$$c_2 = c \int_0^1 dz\, f^{(0)2} = c^2 c_1 , \tag{3.25}$$

$$c_3 = c \int_0^1 dz\, f^{(0)} \left(\frac{f_z^{(0)} f_{zz}^{(0)}}{\rho_s'^2} \right)_z = -c^{-1} \int_0^1 dz\, f^{(0)3} . \tag{3.26}$$

As it was already mentioned, Equation (3.23) admits localized solutions steadily propagating along the x - axis. Just like in the case of the one-dimensional KdV equation they appear as a result of mutual compensation of nonlinearity and dispersion. This means that large-scale steady-drifting solitary axisymmetric vortices whith a sign determined by stratification exist as solutions of our model Equations (2.1–2.3). They are moving slightly faster than dispersionless linear waves (3.14) and rotate rapidly with respect to their drift velocity. Hence, like

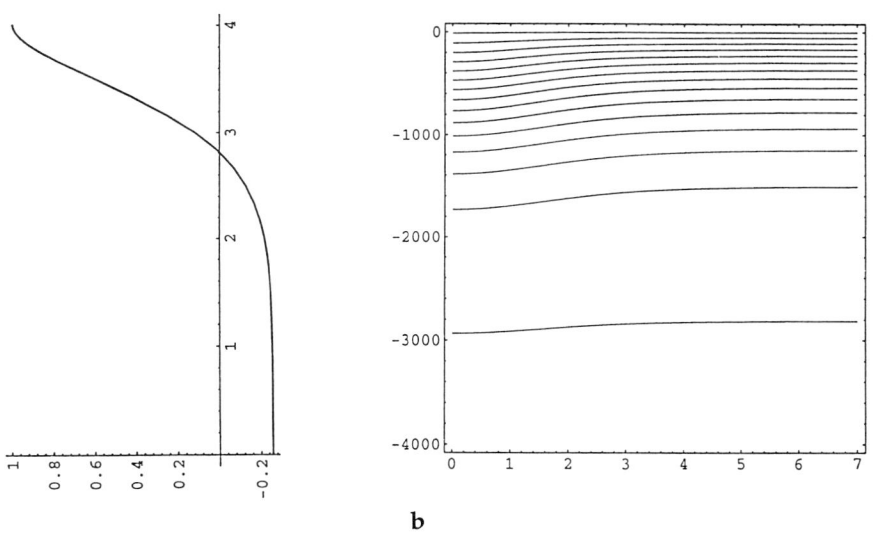

a   b

Figure 2. First vertical eigenmode for pressure pertubation, (a) and the isopycnal lines, and (b) of the corresponding soliton solution of (37) when $\lambda = 0.2$, $\beta = \varepsilon = 0.04$

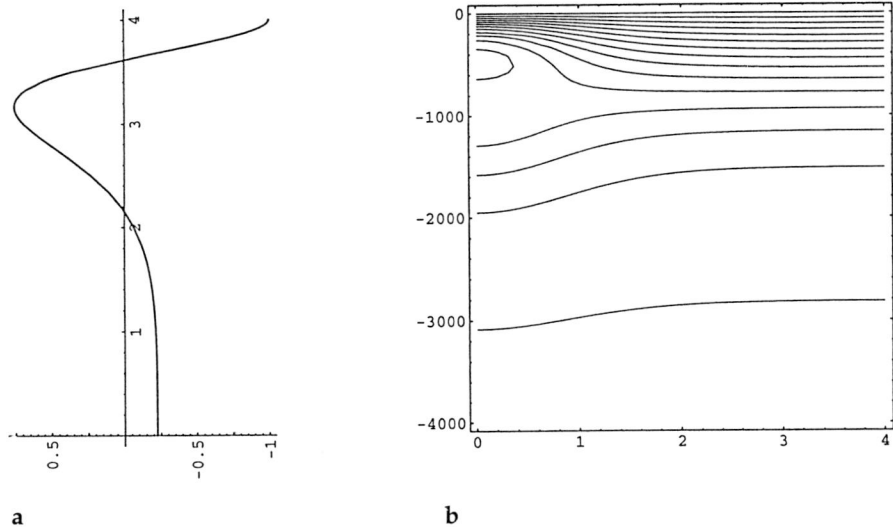

Figure 3. Second vertical eigenmode for pressure perturbation, (a) and the isopycnal lines, and (b) of the corresponding soliton solution (37) when $\lambda = 0.15$, $\beta = \varepsilon = 0.0225$

for the barotropic solutions of this kind, we expect to have regions where fluid particles may be trapped. However, the non-trivial vertical profile of the solutions restricts the vertical extension of these regions thus limiting the transport to lense-like structures.

By integrating by parts it is easy to see that for a stable stratification, $\rho'_s < 0$, the constants c_1 and c_2 are always positive while the constant c_3 may be positive, negative or zero depending on the background stratification. This latter case is degenerate and, indeed, for the linear stratification $\rho'_s = const$ the eigenmodes are harmonic functions, $c_3 = 0$ and there are no solitons. In the general case of non-zero c_3, like for example for the typical oceanic stratification [9] represented in Figure 1, the soliton solutions exist. We present in Figures 2 and 3 the first and the second vertical eigenmodes and the corresponding solitonic solutions for this case.

4   Symmetric and asymmetric corrections to the basic profile

Once the integrability condition for (3.21) is satisfied, one may find the first axisymmetric correction $\pi^{(1)}$ to the basic profile of the vortex given by (3.23). As usual, this correction is not uniquely defined as any function of r multiplied by $f^{(0)}(z)$ may be added to the particular solution of (3.21). This means that *any* axisymmetric correction of order λ respecting the initial vertical structure is admissible which means robustness of the soliton with respect to axisymmetric perturbations. As to the non-axisymmetric ones, their vertical structure was uniquely defined above and if we allow them in (3.20), as well as the corresponding non-axisymmetric part of $\pi^{(2)}$, we get

$$\left(\frac{J(\pi^{(2)}, \pi_z^{(0)}) + J(\pi^{(0)}, \pi_z^{(2)})}{\rho_s'}\right)_z + (p_{t_1}^{(1)} c^{-1} - p_x^{(1)}) f^{(0)} -$$

$$\left((J(p^{(1)}, \Delta p^{(0)}) + J(p^{(0)}, \Delta p^{(1)}))\right) f^{(0)^2} = 0, \tag{4.1}$$

where we used (3.21) and (3.22). As (4.1) is an equation for $\pi^{(2)}$, again, an integrability condition should be satisfied. Taking into account the boundary conditions we see that the terms containing $\pi^{(2)}$ vanish, as well as the second bracket, once the equation is integrated in z over the layer. We, thus, get an equation for $p^{(1)}$

$$J(p^{(1)}, \Delta p^{(0)}) + J(p^{(0)}, \Delta p^{(1)}) = 0. \tag{4.2}$$

Developing $p^{(1)}$ in the Fourier-series in polar angle $p^{(1)} = \sum s_n(r) e^{in\theta}$ and calculating Jacobians in polar coordinates we get the following ordinary differential equation for s_n:

$$s_n'' + \frac{1}{r} s_n' - \left(\frac{n^2}{r^2} + \frac{\Delta p^{(0\prime)}}{p^{(0\prime)}}\right) s_n = 0, \tag{4.3}$$

where prime denotes differentiation with respect to r. A direct numerical study of this equation shows that a localised dipolar ($n = 1$) solution exists while higher n's give solutions growing at infinity. Thus, a dipolar correction of a specific form is also admissible.

5 Discussion

Thus, we have shown that localized steady-drifting solitary vortices exist as asymptotic solutions of the equations of motion of a differentially rotating stably stratified thin fluid layer. These solutions result from the mutual compensation of weak nonlinearity and weak dispersion of the long baroclinic Rossby waves. Their vertical structure is prescribed by solutions of a self-adjoint eigenvalue problem, the eigenvalues corresponding to the phase speeds of the long waves in the direction normal to that of the gradient of the Coriolis force. As the spectrum of phase speeds is discrete, the dispersion in question is in the horizontal while different vertical eigenmodes are travelling with essentially different phase speeds. Hence, our solutions represent weakly nonlinear axisymmetric wave-packets, modulated in the horizontal plane and having, in the first approximation, a fixed vertical structure. In this way they are very similar to solutions found in the barotropic shallow water [7]. As the barotropic solutions they are found in the third order of asymptotic expansion in $\lambda \sim \epsilon^{1/2}$ and, thus, are formally valid for times up to $t_3 = \mathcal{O}(\epsilon^{-3/2})$. We cannot guarantee the coherence of such vortices at longer time-scales, moreover, the explicitly y- dependent twisting terms appear in higher orders of the perturbation theory. The same terms render our asymptotic expansions spatially non-uniform and, thus, formally, we can trust our results only for the vortex core region, while some non-trivial (due to the $x - y$ anisotropy) matching is necessary to get a far-field asymptotics. However, for a rapidly rotating liquid layer the time-scale t_3 may be still long enough and, presumably, the effects of the far field will not influence the solution for shorter time-scales. Hence, it is tempting to interpret the large-scale coherent vortices observed in nature as vortex solitons. The analysis of this kind will be done elsewhere, here we only note that the fact that the structures in question should be larger than the *baroclinic* Rossby radius (2.9) containing a small factor n means that this constraint is much less restrictive here than in the barotropic shallow-water model [7], [6] and may be met, in principle, in the oceanic context. An important property of the solutions found is that they can, as their barotropic counterparts, transport particles trapped in the core region [6].

A crucial question is whether the solitary vortices of this nature are stable. As they are not exact solutions of the equations of motion but, rather, are the asymptotic ones having, by construction, a long but limited life-time, a proper answer to this question can not be given within the framework of the asymptotic expansions and requires a direct numerical analysis of the full system of equations, which represents a separate problem. However, on the basis of the asymptotic analysis we have shown that the vortex is stable with respect to the axisymmetric perturbations of the same vertical structure as the basic profile and that a certain dipolar corrections are also admissible. This indicates that the vortices in question are rather robust. Note also, that our vortex behaves, at the leading order, as a simple linear baroclinic Rossby wave and, hence, no trace of the baroclinic instability is observed in low orders of our asymptotic expansions.

Bibliography

1. Smith B.A. *et al*, 1979, "The Jupiter System Through the Eyes of Voyager 1", Science, **204**, 951 - 972; 1982, "A new look at the Saturn System: the Voyager 2 Images", Science, **215**, 504 - 537; 1989, "Voyager 2 at Neptune: Imaging Science Results", Science, **246**, 1422 - 1449.

2. Ebbesmeyer C. *et al*, 1986, "Detection, Structure and Origin of Extreme Anomalies in a Western Atlantic Oceanographic Section", J. Phys. Oceanogr. **16**, 591 - 612.

3. Petviashvili V.I., 1980, "The Jovian Red Spot and Drift Soliton in Plasma", JETP Letters **32**, 632 - 635.

4. Mikhailova E.I. and Shapiro N.B., 1980, "Two-dimensional model of synoptic disturbances evolution in the ocean", Atm. Ocean Physics - Izvestija, **16**, 823 - 833.

5. Sutyrin G.G. and Dewar W.K., 1992, "Almost symmetric eddies in a two-layer ocean", J. Fluid Mech. **238**, 633 - 656.

6. Stegner A. and Zeitlin V., 1996, "Asymptotic Expansions and Monopolar Solitary Rossby Vortices in Barotropic and Two-Layer Models", Geophys. Astrophys. Fluid Dyn. **83**, 159 - 195.

7. Stegner A. and Zeitlin V., 1995, "What Can Asymptotic Expansions Tell Us About Large-Scale Quasi-Geostophic Anticyclonic Vortices?", Nonlin. Proc. Geophys., **2**, 186 - 193.

8. Romanova N.N. and Tseitlin V., 1985, "Solitary Rossby waves in a weakly stratified medium", Atm. Ocen Physics - Izvestija **21**, 627 - 630.

9. V.M. Kamenkovich *et al* eds "Synoptic eddies in the ocean", Springer, 1986.

An Experimental Investigation of the Entrainment into a Buoyant Jet in a Weak Crossflow

S.J. Gaskin[*] and I.R. Wood[**]

[*]*Department of Civil Engineering and Applied Mechanics, McGill University, Montreal, Canada, and [**]Department of Civil Engineering, University of Canterbury, Christchurch, New Zealand*

Abstract

The entrainment into buoyant jets in a weak crossflow is investigated using a particle image velocimetry (PIV) technique to obtain a detailed picture of the entrainment velocities. The experiments show that the entrainment velocities can be superimposed in the irrotational region. This implies that, given the entrainment velocity for a buoyant jet in a still medium, the entrainment velocity in a moving medium can be calculated by adding the crossflow velocity to the entrainment velocity in a still medium. The implications of this lead to an additional term in the integral momentum equations and show how the normal entrainment assumption changes naturally into the forced entrainment formulation at higher crossflows.

1   Introduction

Buoyant jets and plumes are important in environmental engineering, examples are the release of sewage at deep ocean outfalls and the rise of hot gases from chimney stacks. In these applications a crossflow is often present and an understanding of the effect of the crossflow on the entrainment into the buoyant jet is important in predicting the dilution and trajectory of the buoyant fluid.

The buoyant jet flow is turbulent and in a stationary ambient, after a small region of flow establishment close to the release point, there exists a jet-like region driven by the momentum flux followed by a plume-like region driven by the buoyancy flux. Using the boundary layer assumption, Morton et al. [1] showed that the velocity profiles are self-similar, and that the entrainment into the jet-like flow is proportional to the mean centreline velocity. Later experiments have shown that both the velocity and buoyancy profiles of the jet- and plume-like regions are Gaussian.

In a crossflow, following the zone of flow establishment, are the jet- and plume-like regions where the mean velocity distribution relative to the ambient crossflow is approximately Gaussian. Then, as the crossflow is entrained the mean velocity distribution changes from a Gaussian distribution to that of a distributed vortex pair, in which the horizontal flow velocity approaches the crossflow velocity.

Experiments were performed to investigate the effect of the crossflow on the entrainment into the buoyant jet regions having Gaussian distributions of velocity and buoyancy using a particle image velocimetry (PIV) technique [2]. A detailed picture of the entrainment velocities in the irrotational region outside the jet flow is obtained and indicates that superposition of the entrainment and crossflow velocities occurs resulting in non-axisymmetric entrainment into the buoyant jet. This leads to modifications in the integral equations describing the flow and shows how the normal entrainment assumption changes naturally into the forced entrainment formulation at higher crossflows.

2   Experiments

The experiments were carried out with a negatively buoyant effluent (salt water) released into fresh water contained in a glass sided tank of dimensions 6 m long, 1.5 m wide and 1.1 m deep. The pipe exit releasing the effluent was positioned 0.1 m below the water surface. The crossflow was modelled by towing the pipe through the fresh water. The entrainment velocities were obtained

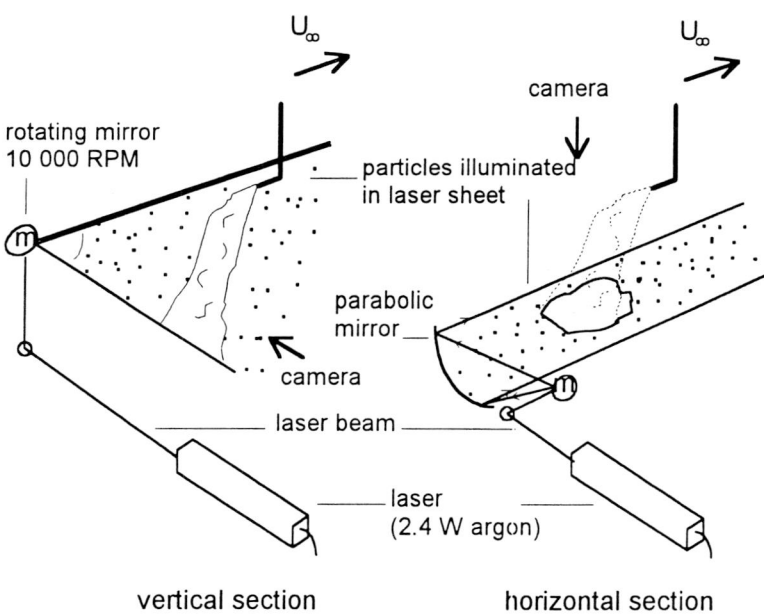

Figure 1. Laser illuminated sections of the buoyant jet flow

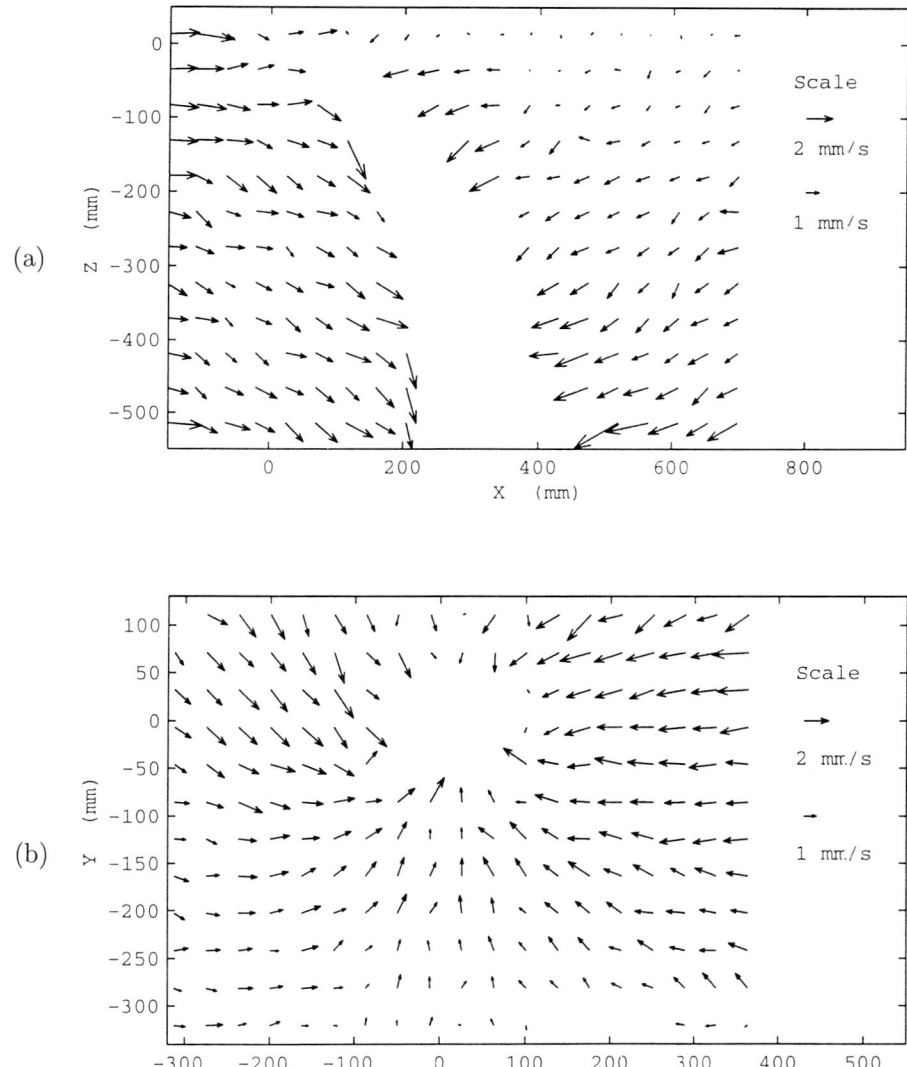

Figure 2. Horizontal buoyant jet $(Fr = 10)$ in a stationary ambient, (a) Vertical centreline section, and (b) Cross-section at $z = 400$ mm

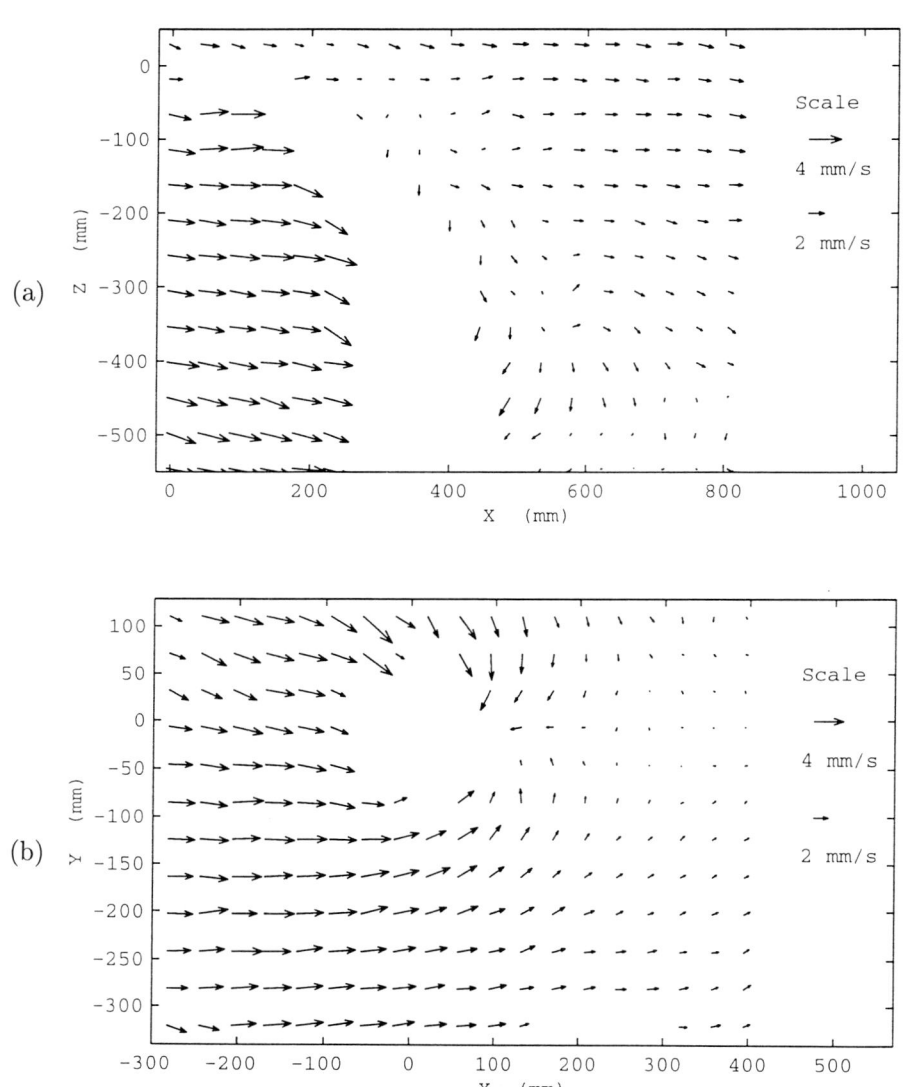

Figure 3. Horizontal buoyant jet ($Fr = 10$) in a crossflow of 2.1 mm/s, (a) Vertical centreline section, and (b) Cross-section at $z = 400$ mm

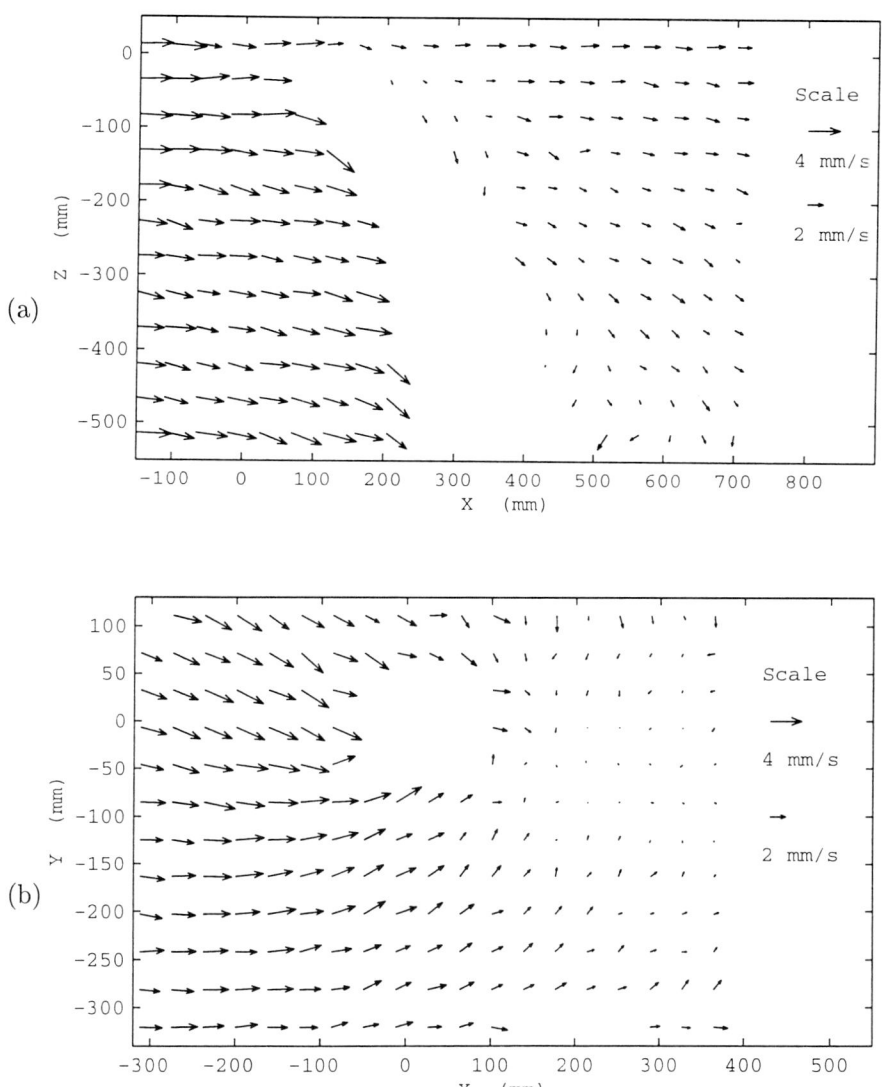

Figure 4. Horizontal buoyant jet $(Fr = 10)$ with a simulated crossflow of 2.1 mm/s added to the stationary ambient entrainment, (a) Vertical centreline section, and (b) Cross-section at $z = 400$ mm

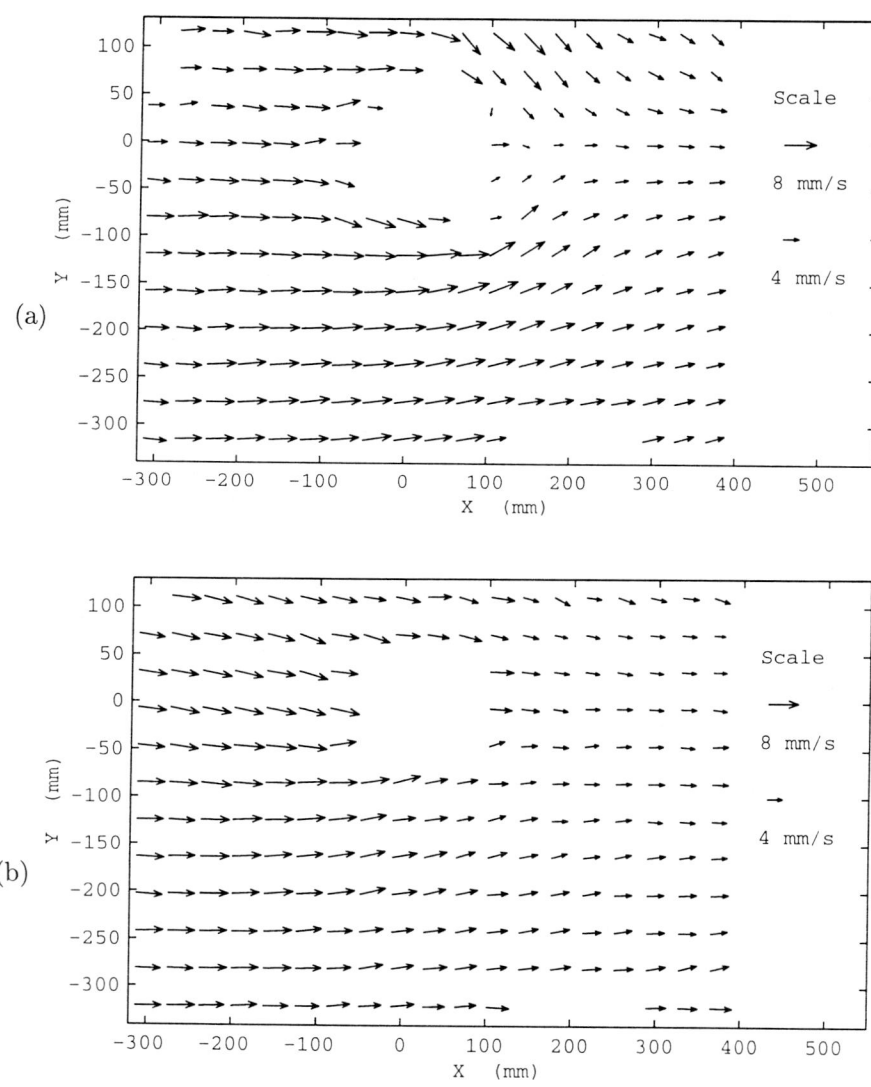

Figure 5. Horizontal buoyant jet ($Fr = 10$, cross-section at $z = 400$ mm), (a) Crossflow of 6.1 mm/s, and (b) Simulated crossflow of 6.1 mm/s added to the stationary ambient entrainment

using a particle image velocimetry (PIV) technique, in which the ambient is seeded with particles and a sheet of laser light illuminates either a vertical or horizontal section through the flow as seen in Figure 1. Pairs of images, separated by a small time interval, were analyzed using a PIV analysis [2]. This uses a pattern matching technique to find the displacement of sub-images during the time interval between the images. The results after analysis and averaging over time are presented as a mean velocity vector map of the entrainment flow.

The first set of experiments looked at a vertical centreline section of a vertical buoyant jet with a densimetric Froude number of 10. As cross-sections could not be obtained with the current equipment, a horizontal buoyant jet was investigated with the horizontal cross-section taken in the near-vertical region of the trajectory. The entrainment flow field of the buoyant jet in a stationary ambient is obtained for the vertical centreline section and a horizontal cross-section as shown in Figure 2. Similar entrainment velocity magnitudes and approximately horizontal streamlines are seen on both sides of the jet. The buoyant jet was then released into a crossflow of 2.1 mm/s (approximately the maximum entrainment flow in a stationary ambient) as seen in Figure 3. The entrainment flow comes predominantly from the upstream side of the jet and in the cross-section a cusp and stagnation point is seen on the centreline of the downstream side. These results can be compared to the simulated velocity field created by adding a velocity vector of 2.1 mm/s to the entrainment flow field in the stationary ambient shown in Figure 4. Neglecting the slight change in the trajectory the entrainment flow fields are very similar in both magnitude and direction. In Figure 5 the cross-section of the jet in a crossflow of 6.1 mm/s (greater than the maximum entrainment velocity in a stationary ambient) is compared to the simulated entrainment field. The comparison is reasonable and it is seen that the entrainment flow comes almost completely from the upstream side.

While there are some differences, particularly close to the port where the buoyant flow feels the wake from the port, the assumption, that the entrainment velocity in a flow is the normal entrainment velocity with the crossflow velocity superimposed, is within experimental errors and it is worth exploring this advected buoyant jet assumption.

3 The entrainment into the buoyant jet

In a stationary ambient earlier experiments show that the time averaged velocity distribution and the time averaged dimensionless buoyancy of the buoyant jet in infinite surroundings can be written respectively as,

$$u_{eg} = U_{eg}e^{-\left(\frac{r}{b}\right)^2} \quad \text{and} \quad \Delta_l = \Delta e^{-\left(\frac{r}{\lambda b}\right)^2} \tag{3.1}$$

where u_{eg} and U_{eg} are the time averaged buoyant jet velocities at r and the centreline respectively, b is the distance where u_{eg}/U_{eg} is $1/e$, Δ_l and Δ are the local and centreline time averaged dimensionless buoyancy $(\Delta\rho g/\rho)$ respectively and λ is the difference in the spread of velocity and buoyancy.

The continuity equation for the flow beyond the zone of flow establishment, [1], is written as,

$$\frac{d}{ds} \int_0^\infty u_{eg} 2\pi r dr = 2\pi \alpha U_{eg} b. \tag{3.2}$$

This states that at any radius the entrainment velocity into the buoyant jet in a stationary ambient is proportional to the maximum buoyant jet velocity, and it implies that the entrainment velocity varies inversely with the radius. It is usual to use the entrainment velocity at a radius of b (where $u_{eg}/U_{eg} = 1/e$), and this velocity is $\alpha U_{eg} (= U_r)$, where α is the traditional entrainment coefficient. The flow in the external field outside the buoyant jet can be modelled as a row of sinks along the axis [3]. For a pure jet the sink strengths are constant and the stream lines are horizontal, and for a pure plume although the sink strengths are not constant (they vary with $s^{2/3}$) and the streamlines not strictly horizontal, the experimental results of [4] show that the departure from the horizontal is not great. Considering a stationary control volume perpendicular to the excess velocity, the assumed Gaussian velocity distribution and the self-similarity of the flow imply an entrainment velocity distribution, graphed in Figure 6, of

$$\frac{u_r}{\alpha U_{eg}} = \frac{1 - e^{-\left(\frac{r}{b}\right)^2}}{\frac{r}{b}}. \tag{3.3}$$

The entrainment velocity tends to zero as the radius tends to infinity and it has a maximum of 0.638 at a dimensionless radius of 1.1.

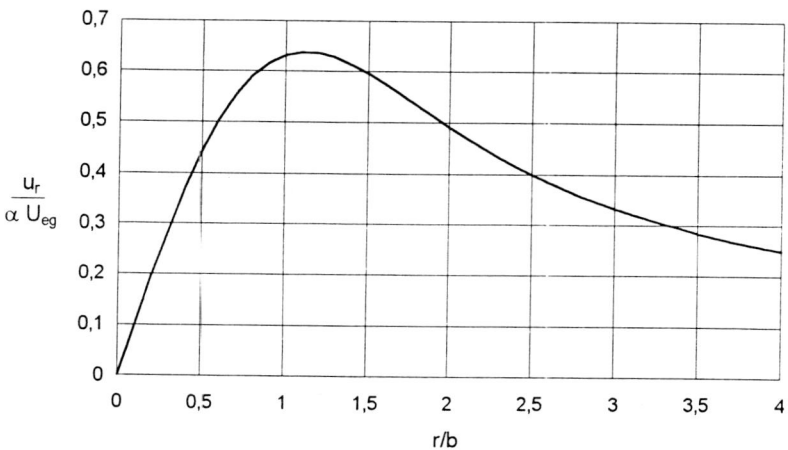

Figure 6. The entrainment velocity implied by the Gaussian excess velocity and the self-similarity of the flow

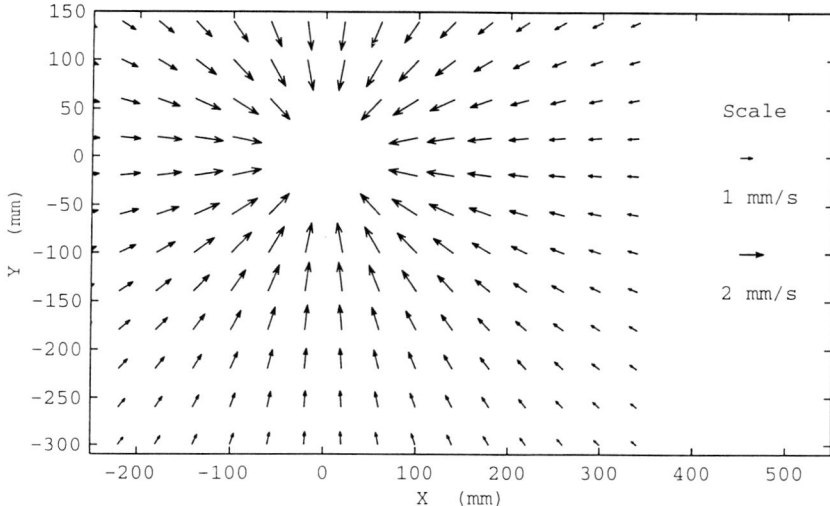

Figure 7. The theoretical entrainment for a buoyant jet $(Fr = 10)$ in a stationary ambient

The stream function approximating this entrainment flow is

$$\Psi = -\alpha U_{eg} b \int_0^\phi \left(1 - e^{-\left(\frac{r}{b}\right)^2}\right) d\phi. \tag{3.4}$$

The velocities predicted by this streamline function, using the sink strength of 2.5×10^{-3} m^3/s estimated from the experimental data of Figure 2(b), are shown in Figure 7.

3.1 The entrainment in a very small crossflow

If it is assumed that, when the buoyant jet is advected, its advected excess velocity distribution (u_{eg}), the turbulence and the pressure distribution are unchanged by the advection, then the entrainment velocity relative to the crossflow velocity (U_∞) will be proportional to the maximum buoyant jet velocity (U_{eg}) measured relative to the crossflow velocity.

Considering the stationary control volume shown in Figure 8 and using the assumption of superposition of the velocities in the irrotational region, the continuity equation becomes

$$d \int_0^R (u_{eg} + U_\infty \cos \alpha_r) 2\pi r dr -$$

$$\left(\int_0^{2\pi} \alpha U_{eg} b \frac{1 - e^{-\left(\frac{R}{b}\right)}}{\left(\frac{R}{b}\right)} \frac{R}{b} d\phi + U_\infty \sin \alpha_r \int_0^{2\pi} \cos \phi b R d\phi \right) ds = 0. \qquad (3.5)$$

Provided R/b is large such that all of the entrainment is included, the above equation reduces to

$$d \int_0^R (u_{eg} + U_\infty \cos \alpha_r) 2\pi r dr - 2\pi \alpha U_{eg} ds = 0. \qquad (3.6)$$

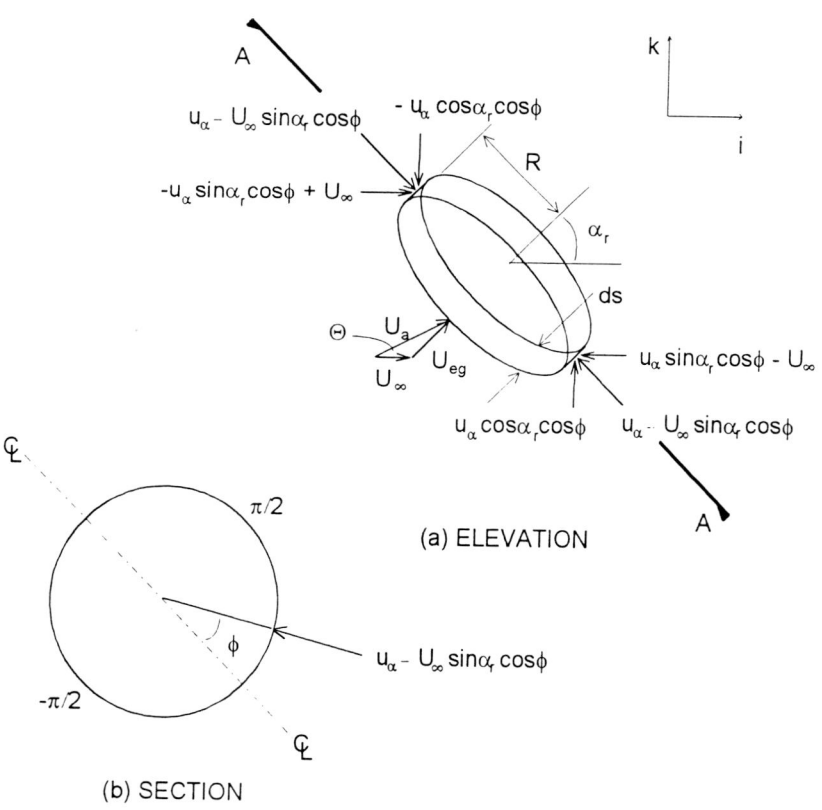

Figure 8. The control volume

It is notable that the term involving the entrainment does not involve U_∞ (when the crossflow velocity is not very small this is not the case). The maximum crossflow for which this equation applies can be determined by calculating the maximum radial velocity in a stationary crossflow (from Equation (3.3)) and setting it equal to the maximum resolved component of the crossflow velocity ($\cos\phi = 1$). The maximum crossflow velocity is then given by $U_\infty \sin\alpha_r / \alpha U_{eg}$ equal to 0.638. For a crossflow less than or equal to this value, the entrainment is described by the normal two dimensional streamfunction, Ψ, for the flow of a sink in a crossflow given by

$$\Psi = -\alpha U_{eg} b \int_0^\phi \left(1 - e^{-\left(\frac{r}{b}\right)^2}\right) d\phi + U_\infty \sin\alpha_r r \sin\phi. \qquad (3.7)$$

For the calculated experimental sink strength of $2.5 \times 10^{-3} m^3/s$ and a crossflow of 2.1 mm/s this gives the velocity vector map in Figure 9, which can be compared to the experimental data of Figure 3. The comparison is good as the velocity magnitudes are close and the same general flow pattern is seen with most of the entrainment occurring on the upstream side. On the downstream side there is little entrainment and again a cusp and a stagnation point are visible.

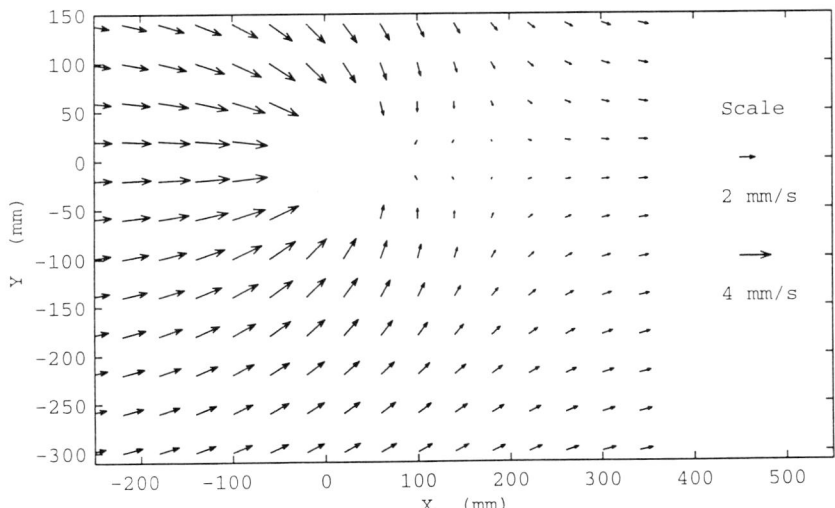

Figure 9. The theoretical entrainment flow for a buoyant jet ($Fr = 10$) in a crossflow of 2.1 mm/s equal to the maximum entrainment flow in a stationary ambient

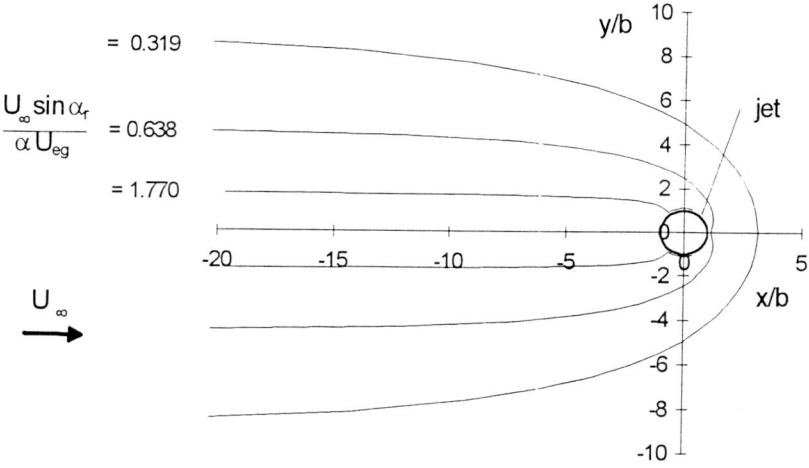

Figure 10. The entrainment bounds for a buoyant jet in a crossflow

The map of the area from which the entrainment originates is obtained by solving for the streamfunction Ψ set to zero. The bounds of the entrainment for the limiting crossflow velocity are plotted as the middle curve in Figure 10. The entrainment function implies a cusp on the centreline similar to that seen in the experimental data of Figure 3. The cusp may be the start of the obvious change from a Gaussian distribution to a vortex pair distribution of excess velocity and buoyancy, which is always observed in a buoyant jet in a crossflow. The outermost curve in Figure 10 is the case for a crossflow velocity of half the limiting velocity ($U_\infty \sin \alpha_r / \alpha U_{eg}$ equal to 0.319), which is the same as for a sink in a uniform flow and the cusp is no longer there.

3.2 The entrainment when the crossflow is greater than the maximum entrainment velocity

When the crossflow is greater than the limiting value, there is a section of the jet on the downstream side over which no entrainment occurs. The point of zero radial velocity can be determined by equating the entrainment velocity to the resolved part of the crossflow velocity along the radius yielding

$$U_\infty \sin \alpha_r \cos \phi = \alpha U_{eg} \frac{1 - e^{-\left(\frac{r}{b}\right)^2}}{\left(\frac{r}{b}\right)}. \tag{3.8}$$

The maximum value of the dimensionless entrainment velocity is 0.638 and occurs at r/b equal to 1.12. Thus the minimum value of ϕ at which the entrainment occurs is ϕ_1 given by

$$\cos \phi_1 = 0.638 \frac{\alpha U_{eg}}{U_\infty \sin \alpha_r}. \tag{3.9}$$

The entrainment is then bounded by the two dimensional streamfunction set to zero, whose trace starts from the point where $\phi = \phi_1$. This yields

$$\Psi = 0 = -\alpha U_{eg} b \int_{\phi_1}^{\phi} \left(1 - e^{-\left(\frac{r}{b}\right)^2}\right) d\phi + U_\infty \sin \alpha_r \int_{y_1}^{y} dy, \quad where \quad y = \frac{r}{b} \sin \phi \tag{3.10}$$

which is plotted for a dimensionless crossflow velocity of 1.77 as the inner curve in Figure 10. Integrating across the limits, this can be expressed for the case of $r/b \to \infty$ as

$$U_\infty \sin \alpha_r y_\infty = \alpha U_{eg} b (\pi - \phi_1) - U_\infty \sin \alpha_r y_1 \tag{3.11}$$

which gives the volume flux at x equals infinity within one arm of the entrainment bounds. The entrainment flow into the buoyant jet can then be written as

$$\frac{d}{ds} \int_0^{r=\infty} \int_0^{2\pi} (U_\infty \sin \alpha_r + u_{eg}) r \, dr \, d\phi = 2(\alpha U_{eg} b (\pi - \phi_1) - U_\infty \sin \alpha_r y_1). \tag{3.12}$$

Then defining

$$q = \frac{d}{ds} \int_0^{r=\infty} \int_0^{2\pi} u_{eg} r \, dr \, d\phi$$

and noting that superposition of the velocities implies

$$\frac{d}{ds} \int_0^{r=\infty} \int_0^{2\pi} (U_\infty \sin \alpha_r) r \, dr \, d\phi = 0$$

and that the maximum natural entrainment velocity occurs at r/b of approximately 1.1, the entrainment flow can be expressed dimensionlessly as either a ratio of the normal entrainment assumption as plotted in Figure 11 and given by

$$\frac{q}{2\pi \alpha U_{eg} b} = \frac{\pi - \phi_1}{\pi} + \frac{U_\infty \sin \alpha_r}{\pi \alpha U_{eg}} 1.1 \sin \phi_1 \tag{3.13}$$

or as plotted in Figure 12 and given by

$$\frac{q}{2 U_\infty b} = \frac{\alpha U_{eg}}{U_\infty} (\pi - \phi_1) + 1.1 \sin \phi_1. \tag{3.14}$$

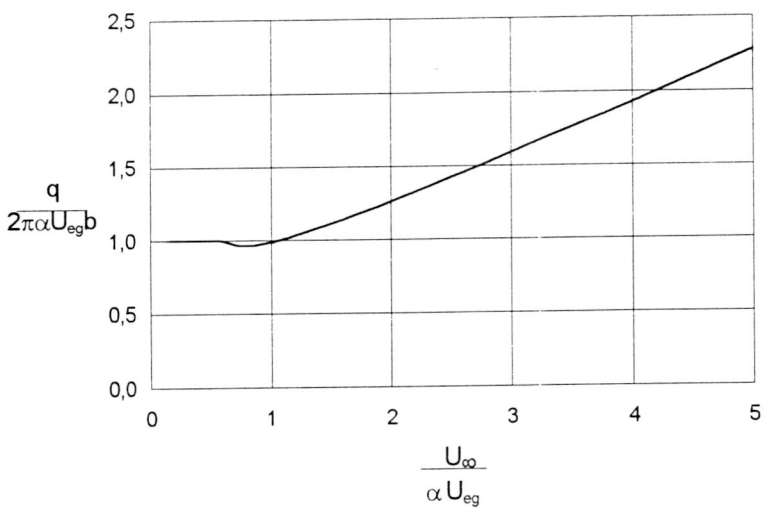

Figure 11. The ratio of the entrainment to the normal entrainment assumption

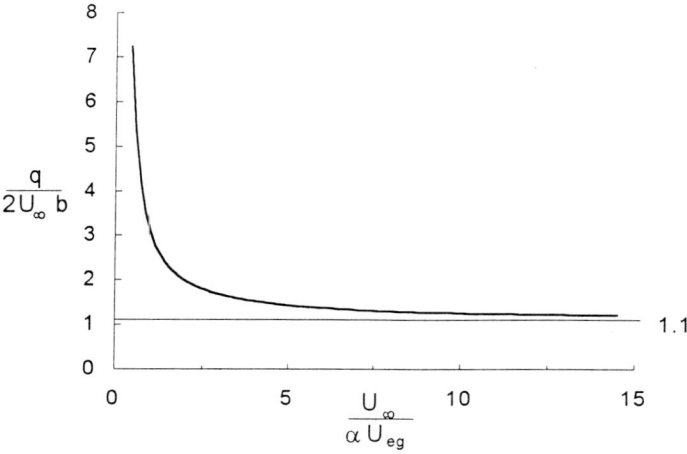

Figure 12. The ratio of the entrainment to the forced entrainment formulation

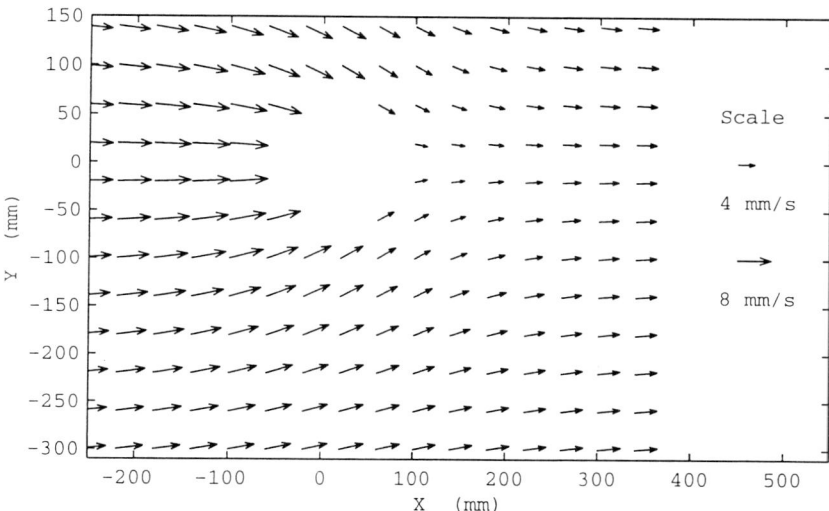

Figure 13. The theoretical entrainment for a buoyant jet ($Fr = 10$) in a crossflow of 6.1 mm/s

These graphs indicate that with increasing crossflow the entrainment flow increases linearly from the normal entrainment assumption and approaches the forced entrainment formulation at higher crossflows. The streamfunction Ψ for a crossflow of any magnitude is then given by

$$\Psi = \frac{\alpha U_{eg} b}{\pi} \left((\pi - \phi_1) + \frac{U_\infty}{\alpha U_{eg}} 1.1 \sin \phi_1 \right) \int_0^\phi \left(1 - e^{-\left(\frac{r}{b}\right)^2} \right) d\phi + U_\infty \sin \alpha_r \sin \phi.$$

$$(3.15)$$

For a crossflow of 6.1 mm/s), this gives the velocity vector map in Figure 13, which can be compared to the experimental data of Figure 5. The general flow pattern and magnitude of the velocities compare well, and both indicate that most of the entrainment comes from the upstream side and that there is a region on the downstream side over which no entrainment occurs.

4   The effect of the entrainment assumption on the momentum equations

The existing integral analysis of the buoyant jet [5] is modified to include the effect of the entrainment assumption on the continuity and momentum equations. The equations are developed considering the control volume shown in Figure 8 and superimposing the resolved part of the crossflow velocity onto the radial velocity. If R is sufficiently large that the integral of the Gaussian distribution of excess velocity and buoyancy become finite, then the equations can be written using Gaussian shape functions (I_q, I_m, $I_{q\Delta}$ and I_Δ for the volume flux, the momentum flux, the turbulent buoyancy and the buoyancy respectively) as follows:

Continuity

$$\frac{d}{ds}(I_q U_{eg} b^2) = 2(\alpha U_{eg} b(\pi - \phi_1) + 1.1 U_\infty b \sin \alpha_r \sin \phi_1). \qquad (4.1)$$

Conservation of horizontal momentum

$$\frac{d}{ds}(I_m U_{eg}^2 b^2 \cos \alpha_r + U_\infty I_q U_{eg} b^2 \cos^2 \alpha_r) = \pi b U_\infty \alpha U_{eg} \sin^2 \alpha_r. \qquad (4.2)$$

Conservation of vertical momentum

$$\frac{d}{ds}(I_m U_{eg}^2 b^2 \sin \alpha_r + U_\infty I_q U_{eg} b^2 \cos \alpha_r \sin \alpha_r) = \pi b U_\infty \alpha U_{eg} \sin \alpha_r \cos \alpha_r + I_\Delta \Delta b^2. \qquad (4.3)$$

Conservation of buoyancy

$$I_{q\Delta} U_{eg} \Delta b^2 + U_\infty \cos \alpha_r I_\Delta b^2 = q_{\Delta 0}. \qquad (4.4)$$

Angle defining the entrainment for crossflows greater than the limiting value

$$\cos \phi_1 = 0.638 \frac{\alpha U_{eg}}{U_\infty \sin \alpha_r}. \qquad (4.5)$$

Geometric equations

$$\frac{dz}{ds} = \frac{U_{eg} \sin \alpha_r}{U_\infty \cos \alpha_r + U_{eg}} \qquad (4.6)$$

$$\frac{dx}{ds} = \frac{U_\infty + U_{eg} \cos \alpha_r}{U_\infty \cos \alpha_r + U_{eg}}. \qquad (4.7)$$

The assumption that the buoyant flow is advected implies that, moving with the crossflow velocity, the radial velocity is the same as for the stationary case. The entrainment constant $\propto$ depends on the type of flow. With a vertical buoyant jet in a stationary ambient, the value of α [5,6] can be written as

$$\alpha = \frac{I_q}{2\pi}\left(0.11 + \frac{b}{2I_m U_{eg}^2 b^2}\frac{d(I_m U_{eg}^2 b^2)}{ds}\right) \qquad (4.8)$$

For a pure jet, it can be shown that α equals $(I_q/2\pi)db/ds$ and for a pure plume α equals $(5I_q/6\pi)db/ds$, where experimental measurements give db/ds as 0.11 for both the jet and plume.

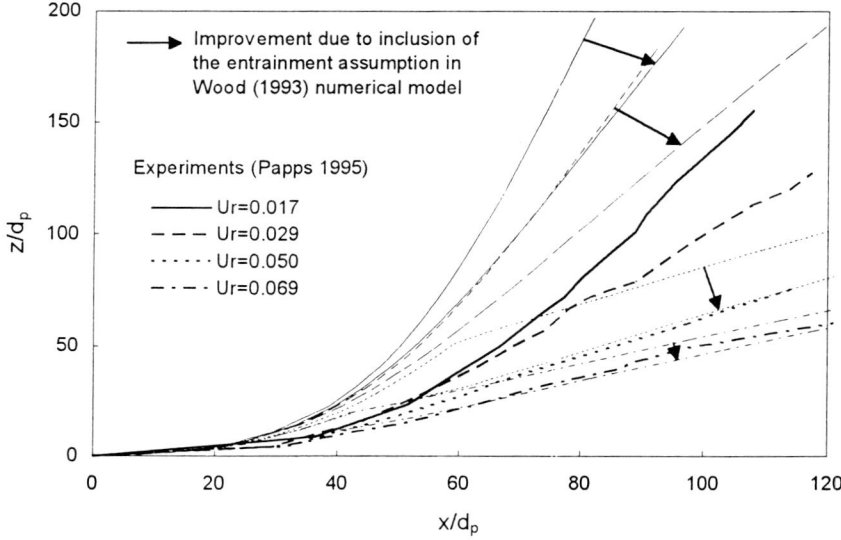

Figure 14. The improvement in the predicted dimensionless trajectory of the modified analysis compared to experimental data

There are sufficient equations to solve for b, U_{eg}, α_r, x and z as a function of s. The equations have been checked analytically for the limiting cases with a crossflow tending to zero of a pure plume and a pure jet and for a more general case a comparison to experimental results follows.

The integral analysis of Wood [5] modified by the inclusion of the entrainment assumption, as described above, has been put into a numerical procedure based on that of Papps [7]. The improvement in the model is shown by comparison of the original model and the present model to experimental data [7] for some horizontal buoyant jets in crossflows of various magnitudes. In Figure 14 the trajectory data is shown and an improvement in the model predictions is seen. The superposition of the crossflow velocity on the radial velocity adds horizontal momentum to the buoyant jet flow which is clearly seen in the trajectory data. The modified model also predicited slightly higher dilutions which improved the fit to the experimental data.

Bibliography

1. Morton, B.R., Taylor, G.I. and Turner, J.S. (1956). Turbulent gravitational convection from maintained and instantaneous sources. *Proceedings of the Royal Society*, **A234**, pp. 1–23.

2. Stevens, C.L. and Coates, M.J. (1994). Applications of maximised cross-correlation technique for resolving velocity fields in laboratory experiments. *International Journal of Hydraulic Research*, **32(2)**, pp. 195–212.

3. Taylor, G.I. (1958). Flow induced by jets. *Journal Aero and Space Sciences*, **25**, pp. 464–465.

4. Rouse, H., Yih, C.S. and Humphries, H.W. (1952). Gravitational convection from a boundary source. *Tellus*, **4**, pp. 201–210.

5. Wood, I.R., Bell, R.G. and Wilkinson, D.W. (1993). Ocean Disposal of Waste Water. World Scientific Publishing Company Limited Singapore.

6. Jirka, G.H. and Harleman, D.R.F. (1979). Stability and mixing of vertical plane jets in a confined depth. *Journal of Fluid Mechanics*, **94**, pp. 275–304.

7. Papps, D.A. (1995). Merging buoyant jets in stationary and flowing ambient fluids. *Ph.D. Thesis*, University of Canterbury, Christchurch, New Zealand.

Three-Dimensional Plume of the Rhône River

Nathalie Durand, Sandrine Arnoux-Chiavassa, Sylvain Ouillon, Vincent Rey and Philippe Fraunié

Laboratoire de Sondages Electromagnétiques de l'Environnement Terrestre, Université de Toulon et du Var, La Garde, France

Abstract

This paper relates to the three-dimensional study of the Rhône river plume hydrodynamics. This work is both experimental and numerical: during a field campaign in the delta area experimental data have been collected and three-dimensional numerical simulations of the dynamical fields and transport of suspended matter have been carried out.

Computational results are presented and their agreement with field measurements is examined.

1 Introduction

The Rhône river mouth is located in the French Mediterranean coast, near Marseille. The discharge of the Rhône fresh water into the ambient sea water creates a very stably stratified flow. Plume dynamics are the result of the wind effects, the shear between riverine water and sea water, the Coriolis effect and the interaction with the general circulation. The studied domain, presented on Figure 1, extends from the river mouth to several kilometres offshore.

A field campaign carried out in November 1994 in the delta area provides us with experimental data that are presented in a first part. In a second part our models are described: a three-dimensional hydrodynamical model used to simulate the dynamics of the plume, and a transport model for the suspended sediment brought by the river. Then, simulation results corresponding to particular conditions encountered during the campaign are presented. The simulated fields are compared from a qualitative point of view to field data.

2 Field campaign

In the aim of studying the Rhône river plume, a field campaign was performed during one month, in November 1994. The data collected are very diverse and provide us with a lot of complementary information [1].

In order to study the plume response to climatic events, we dispose of wind direction and speed measurements, and of the Rhône flow rates during the whole month. The dominant winds during the period are a strong south-easterly wind

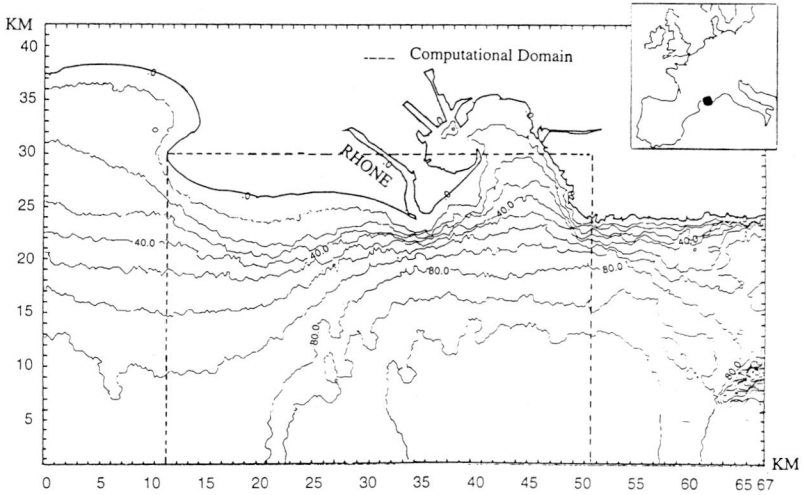

Figure 1. Studied area

blowing with a velocity between 10 and 15 m/s and moderate winds (5 to 10 m/s) from North-West to North-North-East directions. The Rhône flow rate reaches exceptional levels during the campaign. A severe flood occurred on November 7th with a flow rate of almost 9000 m^3/s, when the mean value is 1700 m^3/s.

The vertical hydrologic structure of the plume is investigated using a CTD probe. Temperature and salinity profiles are important in order to evaluate the plume thickness and the rate of vertical mixing.

During the campaign, two VHF radars were situated near the mouth of the river providing surface current maps every 30 minutes.

Three satellite images (SPOT XS and LANDSAT TM) have been processed and maps of suspended matter concentration below surface are obtained [2].

Because of the unusual geophysical and climatic conditions and of the diversity of the measurements, the experimental data collected during the field campaign off the river mouth are of special interest.

3   Three-dimensional modelling

The numerical study of the plume is composed of two parts. The hydrodynamics of the area are simulated for particular conditions (wind, flow rate) encountered during the campaign. Then, the computed velocity fields are used as forcing parameters for the transport of the suspended matter brought by the river into the sea.

3.1 Hydrodynamical model

The three-dimensional dynamics of the Rhône river plume are simulated with the OCKE3D code [3, 4, 5, 6, 7]. It resolves the three-dimensional Reynolds averaged Navier-Stokes equations and the transport equation for the salinity discretized in a Cartesian mesh by a finite volume method. The spatial discretization is done with a second-order centered scheme, except for advection terms which require an upwind scheme; a first order Euler scheme is employed for time-marching procedure. A free-surface elevation is allowed and wind effects are taken into account.

The discharge of the Rhône water into the sea creates a buoyant plume with very strong discontinuities at its limits. The stratification between fresh and saline waters is important and stable, therefore turbulent fluxes through the interface tend to be restrained. The turbulence closure model must take this particular feature into account. The classical k-ε model simulates an isotropic turbulence unadapted to this flow. An anisotropic constant eddy viscosity model gives different values to vertical and horizontal turbulence coefficients, but the turbulence is considered as uniform along each direction. The turbulence closure model developed by Munk and Anderson can easily take into account the effects of a strong stratification; the vertical turbulent coefficients depend, via the local Richardson number, of the density gradients [8]. The vertical coefficients adapt themselves to the strength of turbulence, diminishing in the zones where a strong density gradient inhibits turbulence. These three kinds of closure models have been tested, it clearly appears that the model of Munk-Anderson gives the more realistic results, thus it is chosen for our simulations.

The circulation in the delta area depends on different external forcing parameters. The discharge of fresh water into sea water is the principal one, it is introduced in the model by means of a boundary condition at the river mouth, where an incoming fresh current of three metres depth is imposed. A forcing wind is also taken into account using a classical quadratic law as boundary condition at the sea surface.

3.2 Model of transport of suspended-matter

In order to investigate the carrying of the suspended sediment arriving at the Rhône river mouth, a transport equation for the concentration is resolved [9]. It is a general advection-diffusion equation with a settling term, representing the falling of the particles by gravity. This model is forced by the computed velocity fields. As boundary condition at the mouth is imposed the sediment concentration measured during the field campaign.

The settling velocity of the particles varies with their diameter and density. In our model, this velocity is expressed by the Stokes law, which is adapted to the small particles at low concentrations. Indeed, when the concentration is greater than a certain value, the particles interact with each other during their falling, modifying their settling velocity. At the mouth of the river, the mean diameter of the particles is 10 μm, and their density 2.6.

The diffusivity coefficient for the particles is calculated by a Munk-Anderson law. The turbulent Schmidt number, ratio between turbulent viscosity and sediment diffusivity coefficients, is 0.6 for neutral stratification [10].

4 Presentation of the results

In a first part, a plume test-case is presented in order to examine some vertical profiles.

Then, two simulations are presented, they correspond to conditions encountered during the campaign and for which satellite images are available. We can by this way, compare hydrodynamic and transport modelling results with experimental data.

4.1 Plume test-case

In order to compare several turbulence closure models, some vertical salinity profiles are presented on Figure 2. They have been computed for a plume test-case, meaning a fresh water discharge in a domain with flat bottom and rectilinear coast. Two profiles are simultaneously presented: one of them is computed with a constant eddy viscosity model, and the other one is simulated with a Munk-Anderson model. These profiles have to be compared with CTD profiles collected in November 1994, during the field campaign (Figure 3).

Figure 3 shows a conductivity profile that can be directly linked to salinity profile. It is characterised by a very strong gradient that reveals a very marked interface between two water masses, fresh and saline, that do not mix so much.

The salinity profiles obtained from the numerical simulations are different depending on the chosen turbulence closure model. The Munk-Anderson model restitutes a marked interface separating two water masses of very distinct salinities, when the constant eddy viscosity model creates a wide mixing layer separating very progressively fresh and saline waters.

The CTD profile is clearly better represented by the Munk-Anderson model than by the constant eddy viscosity one. Therefore the Munk-Anderson model is chosen for next simulations.

4.2 26 November 1994: Mean river discharge and northern wind

On November 26th, the Rhône flow rate is 1800 m^3/s and a northerly wind is blowing at 7 to 8 m/s. When the measurements are carried out and the satellite image acquired, climatic conditions are stationary for more than one day.

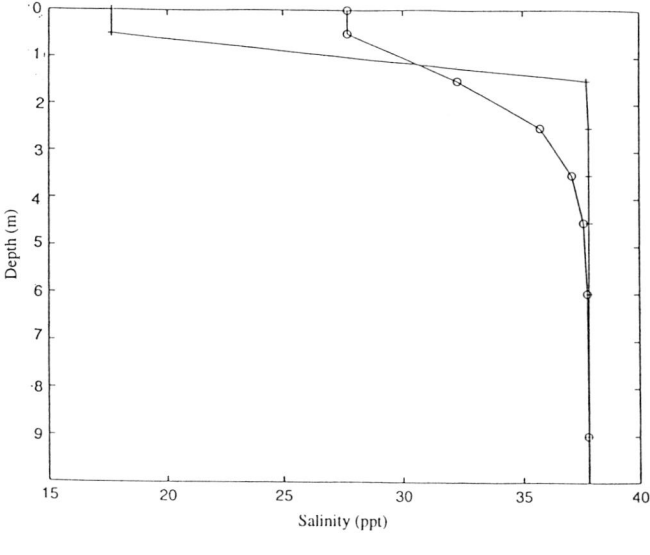

o———o Turbulence Closure Model with Constant Eddy Viscosities

⊢————⊣ Turbulence Closure Model of Munk-Anderson

Figure 2. Simulated vertical profiles of salinity

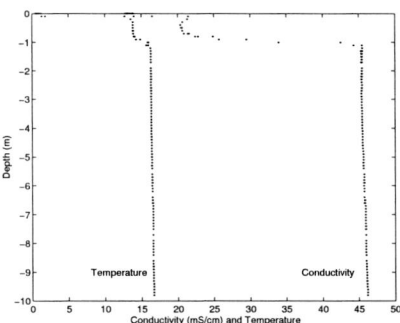

Figure 3. CTD profile

The radar measurements collected the same day allow us to draw a surface current map presented on Figure 4. The velocity discontinuities characterize the lateral limits of the plume, that is clearly south-westward. The horizontal views of velocity and salinity fields at the surface are shown on Figure 5. If we look at the shape, direction and expanse of the plume, we observe a good agreement between experiments and simulation. A deflection of the plume on its right, due to Coriolis effect, is obvious. Indeed, in this case, the Rossby number is about one, thus Coriolis effect is not negligible compared with inertia. On the radar currents map, velocities at 1 km off the mouth are less than 1 m/s. Similar values are provided by simulation. Thus the order of magnitude of the computed fields is correctly represented by the hydrodynamical model.

The results of the transport model are now also presented: Figure 6 presents a surface view of isoconcentrations. On Figure 7 is shown the rough SPOT

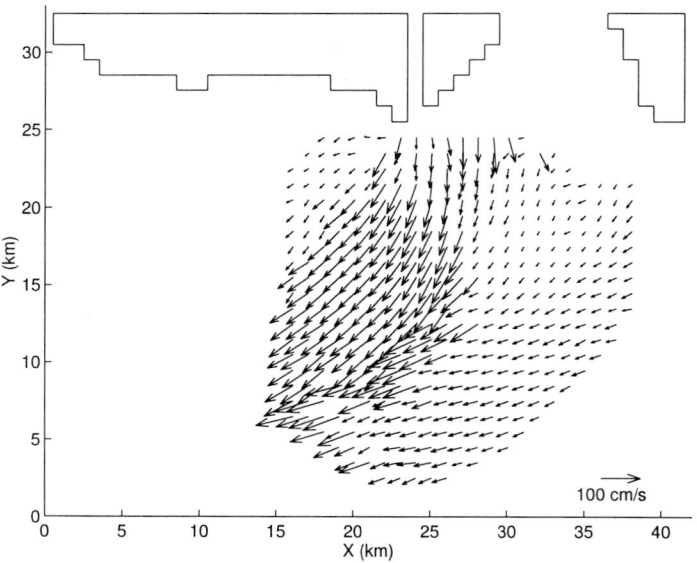

Figure 4. Surface currents mapped by radar on November 26th

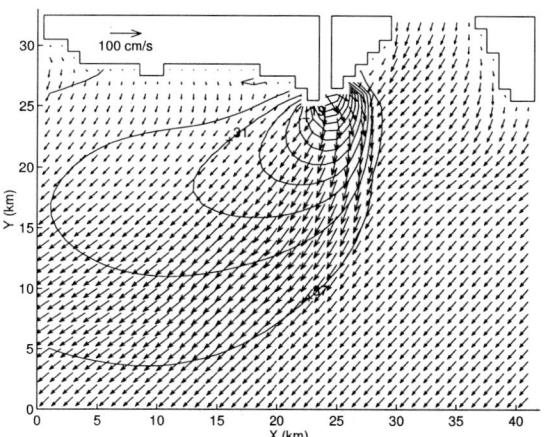

Figure 5. Plane view of the computed flow fields and salinity contours at the sea surface for the November 26th situation

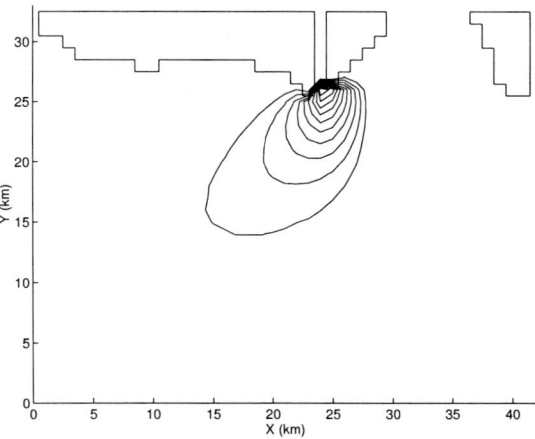

Figure 6. Plane view of the computed sea surface concentrations contours for the November 26th situation

Figure 7. November 26th SPOT satellite rough image

satellite image acquired on November 26th. On this image a turbid plume with strong fronts can be seen. The computed field, as well as the image, shows that the concentration gradients are stronger at the east boundary of the plume. On the satellite image can be seen a zone, near the mouth, where the sediment concentrations are particularly high, then they progressively decrease in the offshore direction. This phenomenon can be explained by the falling of the particles by gravity and the increasing mixing of fresh sediment-charged water and saline water.

4.3   10 November 1994: High river discharge and north-western wind

The climatic and hydrologic conditions encountered on November 10th are a north-westerly wind blowing at 7 to 8 m/s and a Rhône flow rate of 3400 m^3/s. The flow rate has decreased until the severe flood of November 7th but it is still high. Thus the amount of sediment discharging into the sea at the river mouth is still important as it can be seen on processed SPOT satellite image (Figure 8). In this very turbid plume the maximum concentration is about 30 mg/l when it is about 10 mg/l on November 26th.

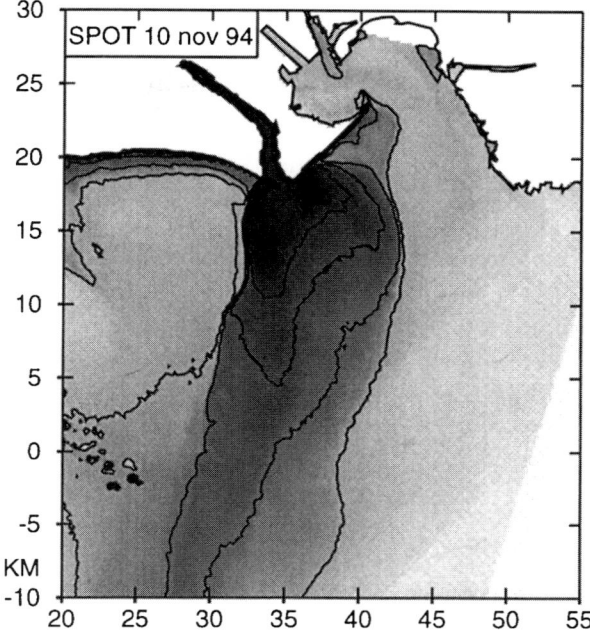

Figure 8. November 10th processed Spot satellite image

If we observe the radar currents map (Figure 9) and the surface horizontal velocity and salinity fields (Figure 10), one can verify that there is a good agreement between experimental and numerical data. As in the previous case, the plume characteristics are quite well represented by the model. The measured surface velocities are higher than these observed on November 26th, from 1.2 to 1.4 m/s at 1 km off the river mouth. The computation correctly reproduces these stronger velocities.

The plume we observe on Figures 8 and 9 is quite different from the November 26th plume. It is strongly southward. The Coriolis effect is not as strong as it is on November 26th. Here is a Rossby number greater than one; the influence of advection on the plume dynamics is more important, in comparison with the Coriolis effect, than in the previous case. This is mainly due to the high flood of the river. The north-westerly wind blowing on November 10th increases this southward plume direction since it opposes to the deflection of the plume on its right.

The experimental data, satellite image and radar surface currents maps, clearly show that very strong fronts exist at the boundaries of the Rhône river plume: the gradients of suspended-matter concentrations are high and the discontinuities in the velocities are important. When we look at the computed fields, we notice that at the lateral boundaries, the isolines of salinity and concentration are relatively distant of each other indicating that for the simulated plume, the gradients at the boundaries are smoother than in the reality. The horizontal mixing is thus probably too strong in the model. The turbulence closure model used takes vertical stratification into account, via the local Richardson number, but in the horizontal plane the turbulent coefficients are uniform. Some improvements have to be done concerning horizontal turbulent coefficients.

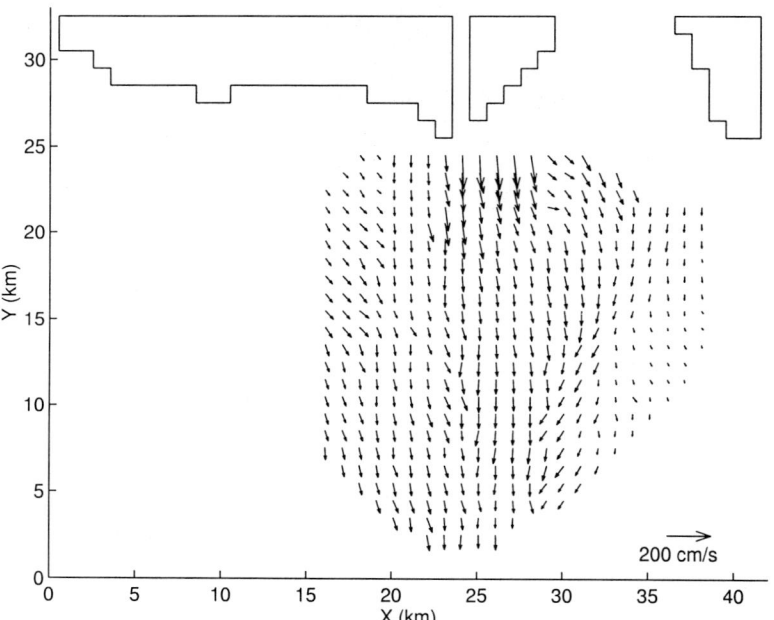

Figure 9. Surface currents mapped by radar on November 10th

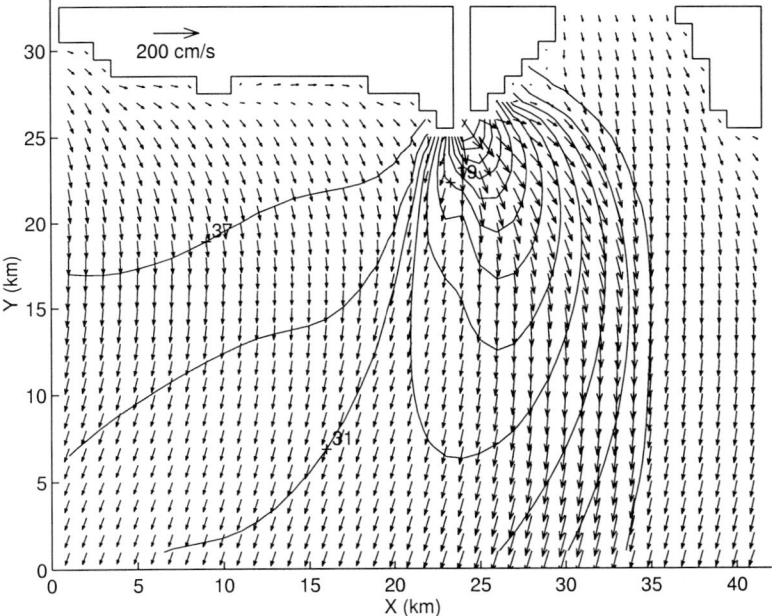

Figure 10. Plane view of the computed flow fields and salinity contours at the sea surface for the November 10th situation

5 Conclusion

The main interest of this three-dimensional study of the Rhône river plume is to associate two kinds of results: experimental data collected during a field campaign and modelling results. A qualitative comparison between these results is encouraging since it shows that our models are adapted to the simulation of stably stratified flows. Indeed the general features of the Rhône plume are quite well represented, although this flow is complex because of the very sharp density gradients in this area and the strong influence of the wind.

Nevertheless, some improvements must be effected. From a numerical point of view, higher order numerical schemes would increase the code accuracy, especially for unsteady situations. Another parametrization of the turbulent coefficients taking into account the strong horizontal density gradients, instead of considering only the vertical ones, may be important. It would allow us to reduce horizontal mixing, which is by now too high, when we compare the computation results and the experimental data. We may introduce an inflow forcing condition more realistic by considering a salt wedge. That would change the inflow depth which is an important parameter of the model. A quantitative refined comparison will be carried out in the aim of calibrating the model. Finally, another external forcing parameter due to the general circulation would be considered: the Liguro-Provençal current that crosses the domain during some periods of the year.

Acknowledgement

This study was supported by the french scientific program PNOC (Programme National d'Océanographie Côtière) and by the European Community Environment Research Program Climatology and Natural Hazard (MEDDELT, contrat EV5V-CT94-0465).

Bibliography

1. Arnoux, S., Baeckeroot, A., Baghdadi, N., Broche, P., Devenon, J.L., Forget, P., Gaggelli, J., De Maistre, J.C., Ouillon, S., Rey, V. and Rougier, G. (1995). Field campaign off the Rhône river mouth. *Proceedings of the 2nd International conference on the mediterranean coastal environment, "MEDCOAST 95"*, Tarragona, Spain.

2. Forget, P. and Ouillon, S. (1996). Surface suspended matter off the Rhône river mouth from visible satellite imagery. *Oceanologica acta*, submitted.

3. Sini, J.F. and Dekeyser, I. (1987). Numerical prediction of turbulent plane jets and forced plumes by use of the k-ϵ model of turbulence. *Int. J. Heat Mass Transfer, 30* (9), 1787–1801.

4. Marcer, R., Fraunié, P., Dekeyser, I. and Andersen, V. (1990). Modélisation numérique d'un couplage physico-biologique en milieu côtier. *Oceanologica acta*, Vol.Sp., *11*, 71–79.

5. Verdier-Bonnet, C., Angot, P. and Fraunié, P. (1995). Paramétrisation de la turbulence pour le phénomène d'upwelling côtier en milieu stratifié. *C. R. Acad. Sci. Paris, Série II*, under press.

6. Arnoux, S., Rey, V., Ouillon, S. and Fraunié, P. (1995). Three-dimensional modelling of the Rhône deltaic coastal zone. *Proceedings of the 2nd International conference of Computer modelling of seas and coastal regions, "COASTAL 95"*, Cancun, Mexico.

7. Verdier-Bonnet, C., Angot, P., Fraunié, P. and Coantic, M. (1997). 3D coastal circulation modelling with a k-ϵ model corrected to account for non-isotropic effects. *29th International Liège Colloquium on Ocean hydrodynamics*, Liège, Belgium.

8. Munk, W.H. and Anderson, E.R. (1948). Notes on a theory of the thermocline. *J. Mar. Res.*, *7*, 276–295.

9. Arnoux-Chiavassa, S., Durand, N., Rey, V. and Fraunié, P. (1995). Three dimensional modelling of the Rhône deltaic fringe. *Proceedings of the 2nd International conference on the mediterranean coastal environment, "MED-COAST 95"*, Tarragona, Spain.

10. Van Rijn, L.C. (1984). Sediment transport, part II: suspended load transport. *J. Hydr.Eng., ASCE*, *110* (11), 1613–1641.

A Heated Vertical Buoyant Jet in a Crossflow

P.N. Papanicolaou[1], J.N.E. Papaspyros, E.G. Kastrinakis and S.G. Nychas

Department of Chemical Engineering, Aristotle University of Thessaloniki, Greece

Abstract

The trajectory and temperature distribution in round vertical heated air jets that discharge into a uniform moving ambient cold air stream, have been studied in the laboratory using hot wire anemometry and a flow visualization technique. The model proposed [1] has been modified by use of two length scales, one to normalize the horizontal and one the vertical distance from the source. The dimensionless trajectory and centerline concentration of round buoyant jets issuing normally into a uniform current are expressed by analytical, non-parametric equations, that fit earlier and the present experimental data very well. The equation of temperature decay at the plume axis and the normalized self-similar profiles of the mean temperature and turbulence properties of plumes in crossflows presented here, can be used in modeling for the calibration and testing of numerical algorithms.

1 Introduction

Round buoyant jets issuing normally into a uniform current of infinite extent are frequently observed above chimney stacks and point fires, and at the disposal sites of cooling water or treated wastewater into rivers. Also, they are observed during accidental buoyant gas leaks, such as ammonia from its storage container in a cross wind, once it has reached the ambient temperature. The trajectory and average dilution of such flows have been investigated by numerous researchers. Several models have been proposed, based on analyses made by the experimenters, who tried to fit the normalized data to some analytical relationship. The trajectory and growth characteristics of non-buoyant round jets in crossflows have been investigated [2], [3] and an attempt has been made [4] to predict the mean flow characteristics of bent-over plumes theoretically. The trajectory of plumes was discussed [5] by field data analysis from bent-over chimney plumes. In relevant studies regarding buoyant jets (forced plumes) [1], [6-8], the investigators have been concerned with the trajectory and dilution of round buoyant jets issuing normally into a uniform cross current. Two

---

[1] Present address: 59 Kefallinias Street, 112 51 Athens, Greece.

length scales based upon the inertial and buoyant characteristics of the jet, which were defined [1], [7] and adopted by relevant bibliography [9] are presented in Table 1. The authors have related the jet to ambient fluid velocity ratio R and the two length scales l_m (or z_m) and l_b (or z_b) to the trajectory and average dilution characteristics, in terms of multi-parametric curves. The main parameter that is the length scale ratio, indicates the relative importance of the initial momentum flux M if compared to the buoyancy flux B of the jet, as a function of some dimensionless distance from the source. In both investigations, the authors have examined the trajectory and average dilution as a function of the horizontal distance from the source x, normalized by either one of the length scales. Since either the inertial or the buoyancy properties of the flow are not present in the normalized distance from the source, their combined effect which can be critical, has been neglected in their analysis.

An alternative length scale could be used, to account for all the initial jet parameters such as the initial momentum M and buoyancy B fluxes of the jet, along with the uniform stream velocity U. In an attempt [8] to have trajectory data in [6] reanalyzed, a length scale l_s and a time scale t_s were used and the dimensionless trajectory was expressed in the form

$$\frac{z}{l_s} = f\left(\frac{x}{Ut_s}\right), \tag{1.1}$$

that fits the data in [6] well. In Equation (1.1) z is the elevation of the jet axis at a horizontal distance x from the source. Also, two simplified relations that predict a jet width and an average dilution as functions of the ratio z/l_s were determined. The length scales l_s and Ut_s used in [8], are the two length scales z_t and x_t respectively (multiplied by some constant), defined in [1] to be the elevation and horizontal distance from the source where the effects of momentum

Table 1. Length scales defined [1], [7]. M and B are the specific momentum and buoyancy fluxes of the jet respectively, and U is the crossflow velocity

Author	Momentum Length Scale	Buoyancy Length Scale
Chu and Goldberg, [1]	$l_m = \left(\dfrac{4M^{1/2}}{\pi U}\right)^{1/2}$	$l_b = \dfrac{4B}{\pi U^3}$
Wright [7]	$z_m = \dfrac{M^{1/2}}{U}$	$z_b = \dfrac{B}{U^3}$

and buoyancy on the flow are in balance. In this article we present a modified Chu and Goldberg [1] (CG hereafter) type of analysis, so that all jet and ambient flow initial parameters are taken into account simultaneously, to make trajectory and dilution data collapse all on a single curve. More specifically, we will be using two length scales similar to x_t and z_t in [1] and used by [8] to determine the buoyant jet trajectory analytically. In this manner, the trajectory and average dilution in buoyant jets, will be normalized and expressed by simple analytical relationships that can be useful in engineering design and model prediction control. The simplified model presented in this study, will be verified with data from earlier experiments [1] and [6,7]. Also, some new data collected for the present work will be analyzed accordingly.

2 Analysis of the modified Chu and Goldberg [1] model

We assume that a vertical buoyant jet shown schematically in Figure 1, with diameter D, initial velocity W, temperature T_o and density ρ_o, issues vertically into a uniform ambient of density ρ_a, temperature T_a and velocity U. The initial jet characteristics are defined [9] as

$$Q = \frac{\pi D^2}{4} W$$

$$M = QW \qquad (2.1)$$

$$B = \frac{\rho_a - \rho_o}{\rho_o} gQ$$

where Q, M and B are the initial specific (per unit mass) volume, vertical momentum and buoyancy fluxes respectively. Two more parameters that characterize the flow, are the initial jet Richardson number Ri defined as

$$Ri = \frac{QB^{1/2}}{M^{5/4}}, \qquad (2.2)$$

and the initial jet to crossflow velocity ratio R defined as

$$R = \frac{W}{U}. \qquad (2.3)$$

A characteristic length scale l_M which describes the relative importance of the inertial and buoyancy characteristics of the jet [9] is

$$l_M = \frac{M^{3/4}}{B^{1/2}}. \qquad (2.4)$$

Following the analysis reported [1], Equation (6) of CG which describes the growth of the specific (per unit mass) vertical momentum flux with x, using Equations (2.2-2.4) above can be written as

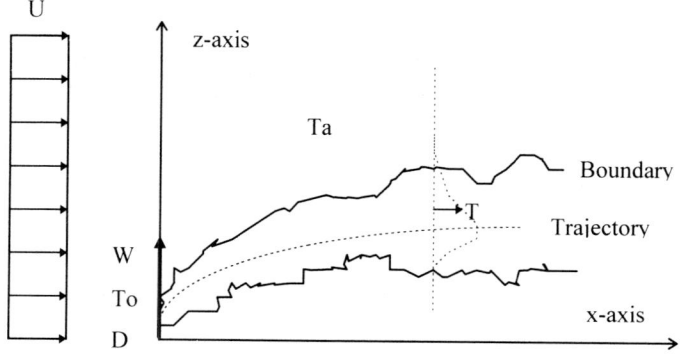

Figure 1. Definition sketch

$$m(x) = \int_{A(x)} uw\,dA = \frac{B}{U}x + M = M\left(1 + RiR\frac{x}{l_M}\right). \qquad (2.5)$$

Following the CG analysis, Equation (11) in [1] with $r = az$, $a = $ constant, can be rewritten as

$$\pi\alpha^2 U^2 z^2 \frac{dz}{dx} = \frac{B}{U}x + M = M\left(1 + RiR\frac{x}{l_M}\right) \qquad (2.6)$$

which upon integration determines the axis elevation of a buoyant jet with distance

$$\frac{z^3}{3} = \frac{M}{\pi\alpha^2 U^2}\left(x + \frac{1}{2}RiR\frac{x^2}{l_M}\right). \qquad (2.7)$$

Multiplying both sides of the equation by RiR/l_M and using Equations (2.2-2.4) to eliminate U and M and replace them with R, Ri and l_M, Equation (2.7) becomes

$$\frac{z}{l_M}(RiR)^{-1/3} = \left(\frac{3}{\pi\alpha^2}\right)^{1/3}\left[RiR\frac{x}{l_M} + \frac{1}{2}\left(RiR\frac{x}{l_M}\right)^2\right]^{1/3}. \qquad (2.8)$$

Equation (2.8) describes the far field elevation of the jet axis normalized by $l_Z = l_M(RiR)^{1/3}$, as a function of the horizontal distance from the nozzle normalized by $l_X = l_M/(RiR)$. Contrary to what earlier investigators have proposed, the dimensionless trajectory becomes a one-parameter curve if the vertical and horizontal distances are normalized by the length scales l_X and l_Z respectively. This relationship is a reformatted Equation (3) of [8], where all the parameters have been determined. In this investigation [8] an added mass coefficient k was determined, and the trajectory data from [6] collapsed on a line for $R > 4$. The author [8], has accepted the value $k = 1$, and stated that "it was not determined with great certainty". If we compare Equation (2.8) with Equation (3) from [8], we see that we do not have to determine such a constant or set restrictions on the initial jet to ambient velocity ratio R.

The average over a cross-section dilution of a round buoyant jet in a crossflow, is defined [9] to be the ratio of the local jet volume flux μ to the initial jet volume flux Q

$$S = \frac{\mu}{Q} = \frac{U\pi r^2}{Q} = \frac{U\pi a^2}{Q}z^2 \tag{2.9}$$

where according to [1], $r = az$. Substituting z from Equation (2.8) into Equation (2.9)

$$SRi(RiR)^{1/3} = (9\pi a^2)^{1/3}\left[RiR\frac{x}{l_M} + \frac{1}{2}\left(RiR\frac{x}{l_M}\right)^2\right]^{2/3} \tag{2.10}$$

that is the dimensionless average dilution over the jet cross section, as a function of the dimensionless distance from the source.

To further connect the present study to earlier investigations, we define the appropriate length scales in Table 2, for the reader's convenience. Equations (2.8) and (2.10) form unique representations of the dimensionless trajectory and average dilution of buoyant jets in a uniform crossflow. The analysis presented here differs from those made earlier in the length scales used. The two length scales of [1], [7] $l_m = M^{1/2}/U$ and $l_b = B/U^3$ are based upon either the inertial or the buoyancy jet characteristics respectively and the crossflow velocity U. Both, the vertical and the horizontal coordinates of the buoyant jet axis had been normalized using either one of those. In this article, following [6] we define two length scales, l_X to normalize the horizontal and l_Z to normalize the vertical distance from the origin. Hence, the parametric relationships that relate the normalized elevation or the average dilution to a normalized horizontal distance from the source with parameter l_m/l_b, are replaced by a single one. The horizontal length scale l_X using Equations (2.2-2.4) may be written as $l_X = MU/B$ and it takes into account the inertial and buoyant behavior of the jet in terms of the initial Ri and l_M, as well as the strength of the jet relative to the crossflow velocity U in terms of R. It is one half of the horizontal distance x_t [1], beyond which the jet buoyancy becomes important if compared to the initial jet momentum. The vertical length scale $l_Z = (M^2/BU)^{1/3}$, is the elevation z_t [1] divided by $(12/\pi a^2)^{1/3}$.

At this point let us look at the asymptotic behavior of the trajectory and average dilution in jets and plumes, and compare the results to earlier investiga-

Table 2. Length scales used in the present (modified CG) analysis

Author	Horizontal Length Scale	Vertical Length Scale
Chu and Goldberg [1]	$x_t = \dfrac{2l_m^2}{l_b} = 2\dfrac{MU}{B}$	$z_t = \left(\dfrac{3l_m^4}{\alpha^2 l_b}\right)^{1/3} = \left(\dfrac{3}{\alpha^2}\dfrac{4}{\pi}\dfrac{M^2}{BU}\right)^{1/3}$
Chu [8]	$Ut_s = \dfrac{MU}{B}$	$l_s = \left(\dfrac{l_m^4}{l_b}\right) = \left(\dfrac{4}{\pi}\dfrac{M^2}{BU}\right)^{1/3}$
Present study	$l_X = \dfrac{l_M}{RiR} = \dfrac{MU}{B}$	$l_Z = l_M(RiR)^{1/3} = \left(\dfrac{M^2}{BU}\right)^{1/3}$

tions. In a pure (momentum driven) jet $Ri = 0$ (small), R is finite, and $l_M = \infty$ (large), hence $RiR(x/l_M) << 1$, and the higher order term on the right hand side of Equations (2.8) and (2.10) can be neglected. Substituting $Ri = l_Q/l_M$, (where l_Q is defined [9] to be the characteristic pure momentum jet length scale $l_Q = Q/M^{1/2}$) into Equation (2.8) it becomes

$$\frac{z}{l_Q R} = \left(\frac{3}{\pi\alpha^2}\right)^{1/3}\left(\frac{x}{l_Q R}\right)^{1/3} \tag{2.11}$$

that is a reformatted Equation (21) of [1] and it is congruent to the trajectory proposed in [2] regarding jets. In a similar manner the corresponding average dilution of a pure jet becomes

$$\frac{S}{R} = (9\pi\alpha^2)^{1/3}\left(\frac{x}{l_Q R}\right)^{2/3}. \tag{2.12}$$

In a pure (buoyancy driven) plume, Ri and R are finite and l_M is small, therefore $RiR(x/l_M) >> 1$ at the far field; keeping the higher order terms in Equations (2.8) and (2.10) one can easily show that

$$\frac{z}{l_M}(RiR)^{-1/3} = \left(\frac{3}{2\pi\alpha^2}\right)^{1/3}\left(RiR\frac{x}{l_M}\right)^{2/3} \tag{2.13}$$

and

$$SRi(RiR)^{1/3} = \left(\frac{9\pi\alpha^2}{4}\right)^{1/3}\left(RiR\frac{x}{l_M}\right)^{4/3}. \qquad (2.14)$$

Equations (2.11-2.14) are the asymptotic jet and plume solutions, and they agree with the far field asymptotes proposed [9] for jets and plumes.

In Figure 2, we have plotted the normalized trajectory Equation (2.8) with $\alpha = 0.5$ and the trajectory data from [1] and [6,7] normalized accordingly. The analysis we have presented is confirmed, since the predicted buoyant jet trajectory fits the experimental data very well. All the data are from experiments that have been performed in water, and are plotted without restrictions regarding

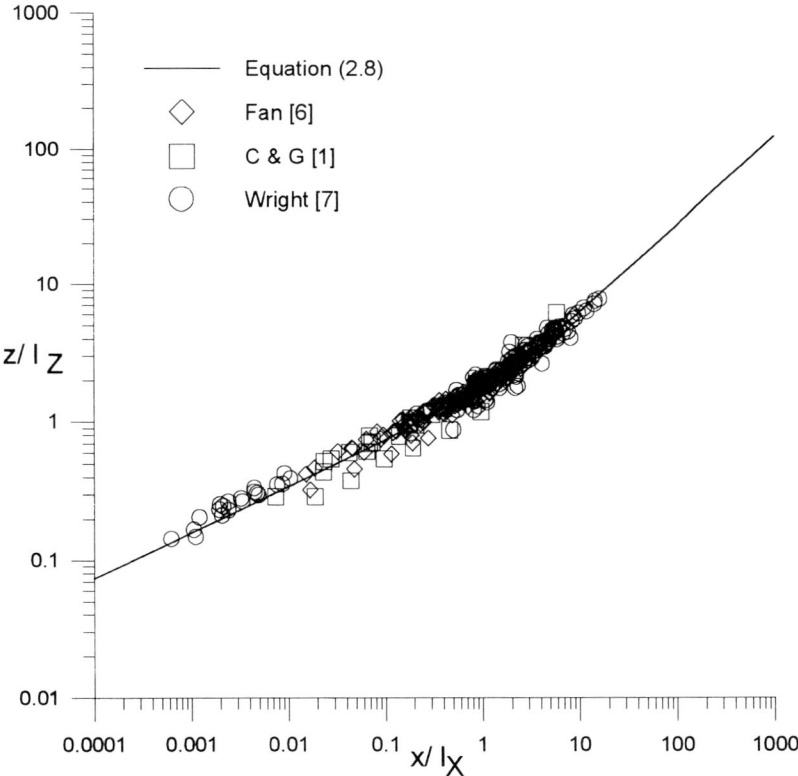

Figure 2. Normalized trajectory of buoyant jets in a crossflow. The data plotted are from earlier investigations [1] and [6,7]

the initial jet and ambient flow parameters R and Ri. In Figure 3 the normalized average dilution Equation (2.10) is plotted as a function of the normalized horizontal distance from the source (dotted line), along with the experimental data [6,7] normalized properly. In the same figure we plotted Equation (2.10) without the coefficient $(9\pi a^2)^{1/3} = 1.92$ (solid line) that fits the normalized centerline dilution data very well. The difference between the two lines plotted in Figure 3 is a factor of about 2, which means that for a jet "radius" $r = az$, $a = 0.5$, the average (over the cross section) dilution is twice the mean centerline dilution $S_c = T_o/T_c$. In summary, from Figures 2 and 3 it is evident that the dimensional arguments made in this paragraph, which emerged from the modified CG model, are congruent with experimental data from water experiments.

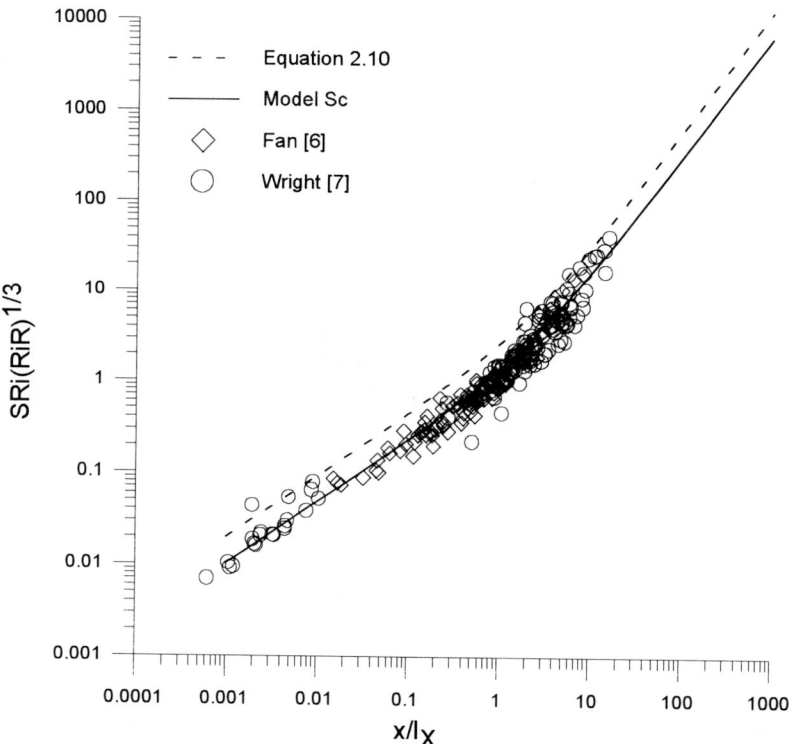

Figure 3. Normalized centerline dilution of buoyant jets in a crossflow. The data plotted are from earlier investigations [6,7]

From the analysis presented above one can determine the behavior of a buoy-ant jet in a crossflow, according to the initial jet characteristics and the crossflow velocity. From the normalized trajectory and dilution Equations (2.8) and (2.10) and the asymptotic jet solution (2.11) and (2.12) a buoyant jet is dominated by the momentum up to a distance x for which $x/l_X < 0.10$ and by the buoyancy ac-cording to Equations (2.13) and (2.14) beyond distances x for which $x/l_X > 10$. In the regime $0.10 < x/l_X < 10$ the flow is in transition from jets to plumes because the effects of the jet momentum and buoyancy are in balance. Further-more, Equations (2.8) and (2.10) dictate that an initially momentum jet with the slightest possible density difference from the uniform ambient, will eventu-ally become a plume. The asymptotic behavior of jets and plumes is confirmed by experimental data in Figures 2 and 3.

3 Experimental

Experiments have been carried out in a suction type wind tunnel with a 2m long octagonal cross-section 30cm × 30cm, model C2-10 manufactured by Armfield Technical Education Company Limited. An arrangement similar to that reported in [3] has been used. The heated jet was placed right upstream from the sharp edge of a 2cm thick flat plate, that was positioned 5cm above the bottom surface of the tunnel. It is made of PVC pipe with i.d. 1.27cm and it was heated via a resistor, supplied by a variable output power supply at a given DC Voltage, in order to obtain the desired initial jet temperature, according to the cold air volumetric flux input. Nozzles with different diameters could be fitted at the end of the pipe with the heating element.

A special smoke generator device was used to mark the jet fluid (see Diagram 1) making possible its visual perception. The mean trajectory of the heated jet was measured on the images (like this in the diagram) of the jet, recorded via a high speed high resolution video tape recorder. Sampling rates varied from 200 or 400 frames per second, and the shutter speed (exposure time) was 0.0001 seconds.

The temperature field of the jet has been measured with a X-wire probe 55P62 manufactured by DISA, whereas the wire in the front was used as a cold-wire. Samples were taken for 20 to 40 seconds at rates 1 or 2 kHz at each location. Measurements were made along vertical lines on the plain of symmetry of the jet and on vertical cross-sectional planes at various distances from the nozzle. The jet diameters used were 2.65mm, 5.4mm, 10.2mm and 11mm. The initial jet temperature T_o was around 80°C, the jet velocity varied from 2.50 to 22.50 m/s and the cross-flow velocity used varied from 1 to 3.3 m/s. The velocity ratio R varied from 2.5 to 9.0 and the jet initial Ri from 5×10^{-4} to 5×10^{-2}. The point where the maximum time-averaged temperature occurred, was found from the vertical mean temperature profile on the plane of symmetry of the jet, thus defining the jet axis elevation. The jet exit velocity was estimated from the cold air volumetric flux measured with a flowmeter, the known jet diameter D and

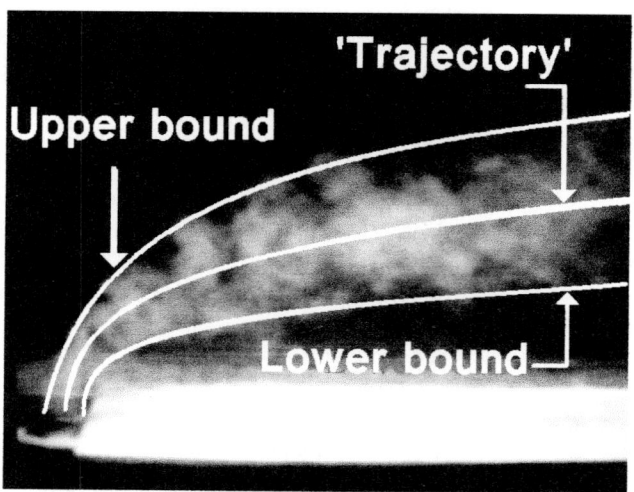

Diagram 1. High speed video camera image from a buoyant jet marked with smoke

the jet exit temperature T_o, using the ideal gas law to account for volumetric changes. The initial jet temperature was measured at the nozzle with a cold wire.

4   Results

In the sections that follow we present the results from our measurements, normalized according to the analysis made earlier. We have assumed that the initial jet velocity and temperature profiles are top-hat for the calculation of the initial jet momentum and buoyancy fluxes. The dimensionless elevation z/l_Z of the jet centerline is plotted versus the normalized horizontal distance x/l_X from the nozzle in Figure 4. Equation (2.8) that is the line with $\alpha = 0.5$ fits the present data well, except for the regime $0.05 < x/l_X < 1$, where the boundary effect of our wind tunnel on the jet is obvious. In fact, we have tried to make measurements in the buoyancy driven, plume regime. Thus we have used a relatively large jet nozzle, at a low velocity W in a cross stream that is moving slowly, to keep the length scale l_X low. We have performed measurements as far downstream from the nozzle as possible, to approach the plume regime ($x/l_X \gg 1$). As the plume grew and rose in the z-direction it was displaced by the wind tunnel top boundary layer, thus keeping its axis at a lower elevation. The asymptotic relation (Equation (2.11)) is verified by the present experimental results. From the trajectory data it is made obvious that a buoyant jet that initially behaves as a momentum jet ($z \propto x^{1/3}$), will eventually become a plume with $z \propto x^{2/3}$.

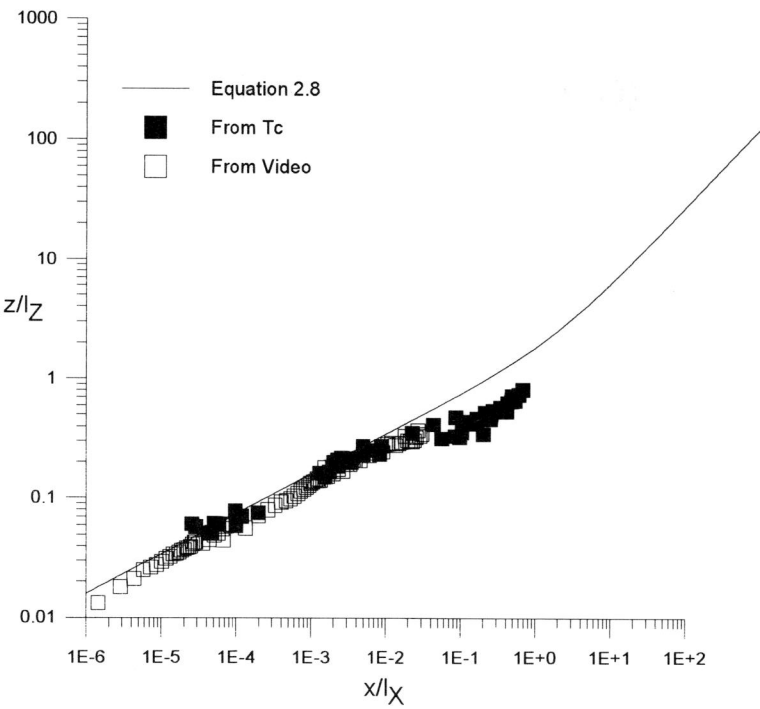

Figure 4. Normalized trajectory of buoyant jets in a crossflow. The data plotted are compared to the jet trajectory Equation (2.8)

In order to determine the temperature distribution in heated jets, we have subtracted the ambient temperature T_a from all temperature data. Therefore, when for example we refer to the mean temperature, we mean the excess (above ambient T_a) temperature. The normalized mean and rms temperatures at the centerline, are plotted versus the dimensionless distance from the nozzle in Figure 5. The average dilution Equation (2.10) with $\alpha = 0.5$ is plotted as a dotted line and the centerline dilution, that is Equation (2.10) without the constant $(9\pi\alpha^2)^{1/3}$ as a solid line. The present dilution data extend from the asymptotic jet regime to the transition from jets to plumes regime, since the normalized distance from the source x/l_X up to which measurements have been performed did not exceed 1. Also, since the open and solid triangles correspond to centerline values of the temperature, the normalized mean centerline dilution $(T_o/T_c)Ri(RiR)^{1/3}$ will be lower from the jet average dilution by a factor of 2 as it was shown in Figure 3. In Figure 5 we have also plotted the dilution with respect to the rms temperature at the axis $(T_o/T_c')Ri(RiR)^{1/3}$. This is greater from the mean centerline dilution as expected since $T' < T_c$ around the axis.

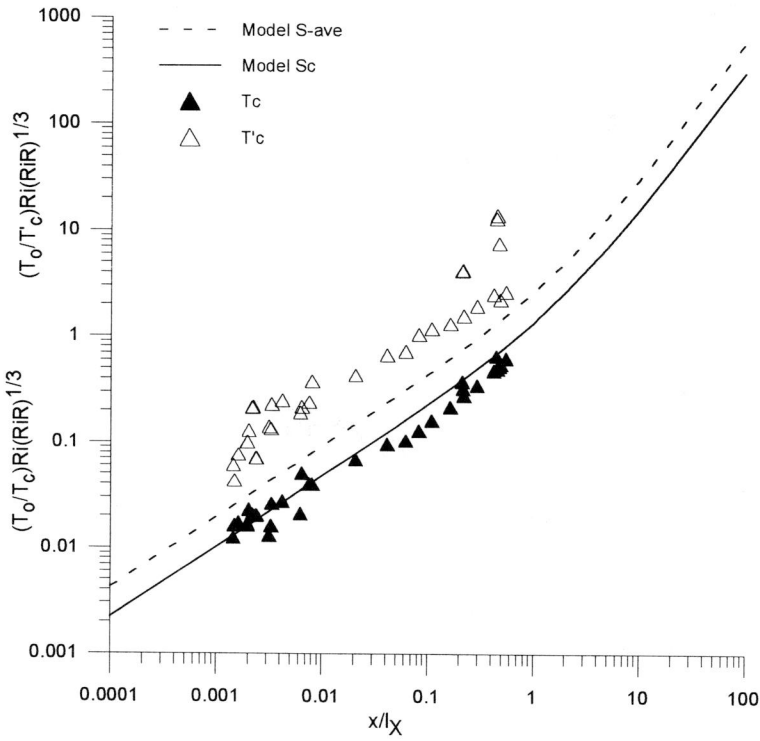

Figure 5. Normalized mean and rms temperature at the centerline of buoyant jets in a crossflow. The data plotted are compared to the average dilution Equation (2.10) (dotted line), and to the mean centerline dilution Equation (solid line)

Also, the rms normalized temperature excess, follows the same asymptotic behavior with the mean temperature, something that has also been observed in round buoyant jets that discharge into calm uniform ambient fluid [10,11].

At this point it is worth looking at the normalized temperature distribution properties in the flow field of the jet besides the centerline values. It is difficult to present some 3-D normalized temperature surfaces. Thus we had to slice the jet with a vertical plane, that is perpendicular to the direction of the crossflow. Then we made measurements along a vertical and a horizontal line on this plane, drawn from the point of intersection with the jet axis. Jet axis is considered to be the point on the plane of symmetry of the jet, where the maximum mean temperature was measured. The coordinates of this point then are defined as (x, z_c). In order to normalize the "radial" distance from the jet axis, we have to determine a jet "radius" r_c, which at a given distance x from the nozzle, can be defined as $r_c = \alpha z_c$, where z_c is calculated from Equation (2.8), and $\alpha = 0.50$ as proposed by CG.

A series of experiments were performed in a jet with diameter $D = 10.2$mm. The initial jet parameters and the distance of the measurements from the source are presented in Table 3. Measurements were taken on the whole plane normal to the crossflow, then the coordinate z_c was determined and the temperature profiles on a horizontal and a vertical line were isolated. The mean normalized temperature excess and turbulent intensity in the horizontal direction are plotted versus y/r_c in Figures 6a and 6b respectively, where y is the horizontal distance from the plane of symmetry of the jet. It is clear that the mean temperature excess horizontal profile is symmetric, and it vanishes beyond one characteristic radius r_c from the jet axis. The intensity of turbulence profile shows two symmetric peaks around $y = \pm 0.5 r_c$, and it is $0.30 T_c$.

The mean normalized temperature and turbulent intensity in the vertical direction are plotted in Figures 7a and 7b respectively versus the normalized distance $(z - z_c)/r_c$. It is clear that the mean temperature excess becomes zero beyond one characteristic radius r_c above the jet axis, while it does not vanish below it.

Table 3. Experimental conditions for cross-sectional temperature measurements

Run	D(mm)	$T_o(^\circ C)$	$T_a(^\circ C)$	W(m/s)	U(m/s)	x(cm)	x/D
D_A	10.2	80.00	24.18	2.83	1.05	10.0	10
D_B	10.2	80.00	24.26	2.83	1.05	17.5	17
D_C	10.2	80.00	24.43	2.83	1.00	25.5	25
D_D	10.2	80.00	24.92	2.83	1.16	40.0	39
D_E	10.2	80.00	25.00	2.83	1.15	53.0	52
D_F	10.2	80.00	23.62	2.83	1.11	71.5	70
D_G	10.2	80.00	24.07	2.83	1.16	86.5	85
D_H	10.2	80.00	24.05	2.83	1.20	100.0	98

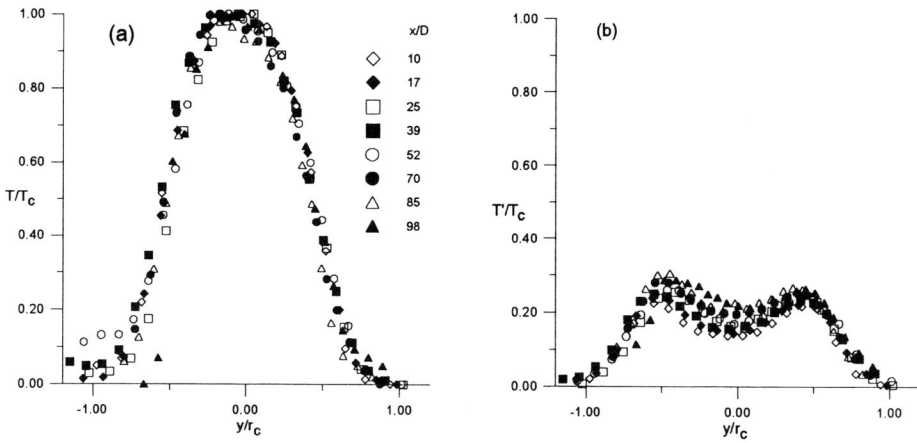

Figure 6. Normalized mean (a) and rms (b) temperature profiles along a horizontal, through the jet axis

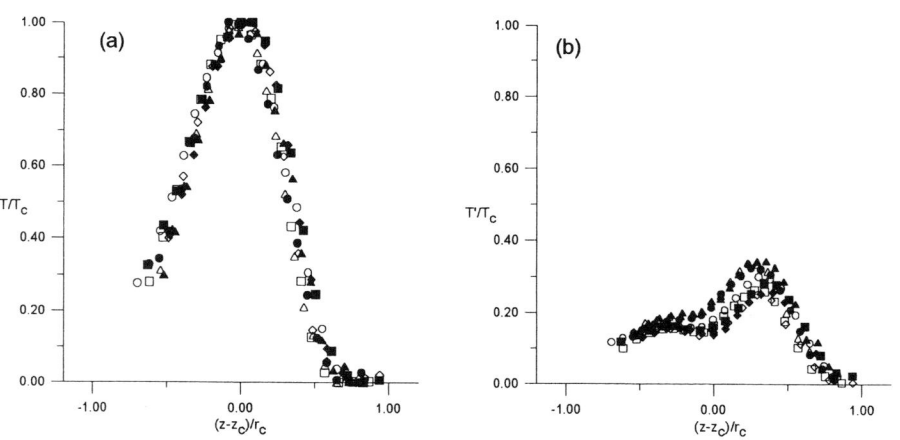

Figure 7. Normalized mean (a) and rms (b) temperature profiles along a vertical, through the jet axis (for legend see Figure 6)

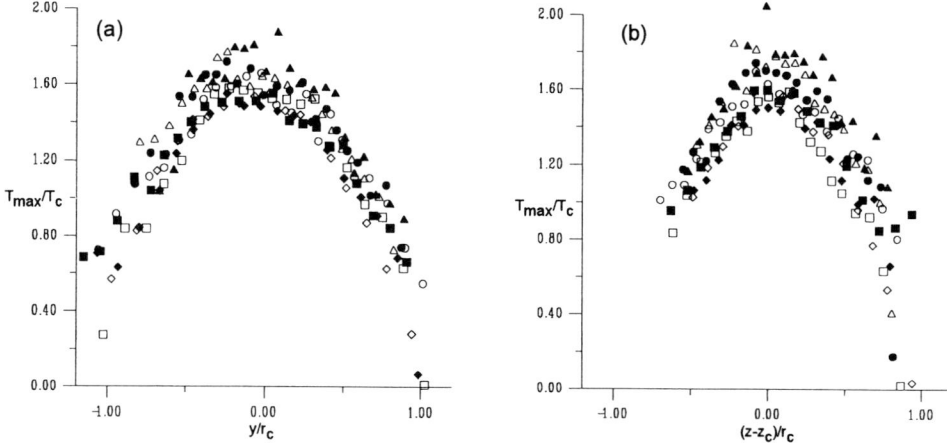

Figure 8. Normalized maximum temperature profiles along a horizontal (a) and a vertical (b) through the jet axis (for legend see Figure 6)

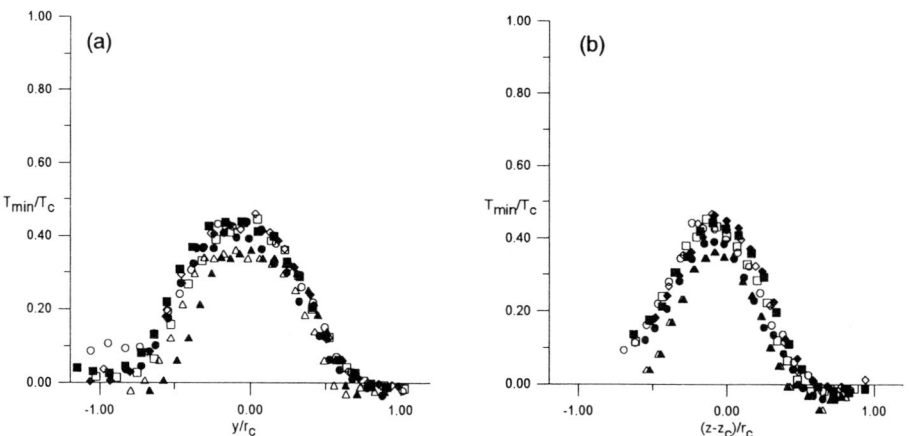

Figure 9. Normalized minimum temperature profiles along a horizontal (a) and a vertical (b) through the jet axis (for legend see Figure 6)

The intensity of turbulence shows a peak of 0.35 around $z - z_c = 0.3r_c$, that is different from the location where T_c peaks.

The mean and rms temperature profiles are not symmetric in the vertical direction as expected, since the wake of the jet must influence the temperature distribution below the jet axis. In Figures 8a and 8b we have plotted the normalized maximum horizontal and vertical temperature profiles, that take a value of 1.6 to 1.8 around the axis. Also, in Figures 9a and 9b we have plotted the normalized minimum horizontal and vertical temperature profiles that take the value 0.45 around the jet axis. From Figures 6-9 it becomes evident that the profiles in the horizontal direction are wider from those in the vertical direction. This means that the jet width defined from the mean temperature profile is wider in the y than in the z - direction. This is in accordance to earlier measurements [6]. Also, the normalized mean temperature profiles using the "jet radius" r_c are self-similar. The mean and rms temperature distributions in Figures 6 and 7 can be used in the prediction of contaminant transport by buoyant jets as well as in numerical model evaluations.

5   Summary and Conclusion

The behavior of round vertical turbulent buoyant jets that discharge normally in a crossflow, can be determined if two length scales, l_X and l_Z are used. The first one has been used to normalize the horizontal distance from the source and the second to normalize the elevation above the source. They take into account both the inertial and buoyant characteristics of the jet, along with the strength of the crossflow. Then, the normalized trajectory and average dilution can be expressed by simple algebraic equations, thus modifying the early CG model. From the normalized average dilution algebraic equation, we can accurately predict the mean centerline temperature decay of buoyant jets. The dimensional analysis reported in earlier investigations regarding the asymptotic jet and plume trajectories and average dilution has been confirmed. Equations (2.8) and (2.10) predict the trajectory and dilution in the transition regime from jets to plumes regardless of the initial jet and ambient fluid characteristics, i.e. R and Ri. From Equations (2.8) and (2.10), the normalized experimental data, and the asymptotic analysis we can state that a buoyant jet is driven by inertial forces (pure jet behavior) up to a horizontal distance equal to $0.1l_X$ from the source, while it is driven by buoyant forces (pure plume behavior) beyond $10l_X$. This is analogous to the jet and plume definitions [10,11] with respect to the characteristic length scale l_M.

The normalized mean centerline dilution and turbulence intensity of the temperature, follow the same decay law with the average dilution, something that has also been observed in both, the velocity and concentration field of a round buoyant jet in calm ambient fluid [10,11]. This means that the 3-D normalized temperature profiles must be self similar. In fact this has been confirmed by measurements made along two characteristic lines that cross the jet axis. The transverse vertical and horizontal distances from the jet axis have been normalized by the "jet radius" proposed by [1], that is one half of the jet axis elevation distance from the nozzle. The mean and rms normalized temperature profiles are self-similar in both directions. The horizontal mean temperature profile is symmetric and wider from the mean vertical one, that is asymmetric. The horizontal turbulent intensity profiles are symmetric across the plane of symmetry of the jet. The asymmetric vertical turbulent intensity profiles peak at $0.25 \, r_c$ above the jet axis. Both normalized mean and rms temperature profiles do not extend beyond one characteristic radius r_c from the axis.

The maximum and minimum temperatures that we have measured, occurred around the jet axis and are $1.7 \, T_c$ and $0.45 \, T_c$ respectively, in both directions. The maximum value measured is low if compared to recent measurements [12] where the minimum measured concentration was zero while the maximum as high as one order of magnitude of T_c. The zero minimum value observed, means that unmixed ambient fluid reaches the jet axis. The reason we have not observed such values here is the spatial resolution ($2 \times 1 \times 1 mm^3$) of the measuring device (cold wire), that is large if compared to size of the probe volume measured by optical techniques [12].

In conclusion, the jet trajectory in a crossflow is given by Equation (2.8). The centerline normalized concentration can be calculated from Equation (2.10) if the constant coefficient is given the value one. This means that the average, over the cross section dilution is twice the dilution at the jet axis. Also, the normalized mean and rms temperature profiles are self - similar.

Acknowledgment

The authors gratefully acknowledge the financial support of EEC STEP Program under Contract No. STEP-PL-900597. The first author acknowledges the Institute of Mathematics and its Applications (IMA) and the Greek Chamber of Engineers (TEE) for their financial support regarding the conference.

Bibliography

1. Chu, V.H. and Goldberg, M.B. (1974). Buoyant forced plumes in cross-flow. *ASCE, J. Hyd. Div.*, **100**, 1203-1214.

2. Keffer, J.F. and Baines, W.D. (1963). The round turbulent jet in a cross-wind. *J. Fluid Mech.*, **15**, 481-496.

3. Pratte, B.D. and Baines, W.D. (1967). Profiles of the round turbulent jet in a cross flow. *ASCE, J. Hyd. Div.*, **93**, 53-64.

4. Priestley, C.H.B. (1956). A working theory of the bent-over plume of hot gas. *Quart. J. Roy. Met. Soc.*, **82**, 165-176.

5. Slawson, P.R. and Csanady, G.T. (1967). On the mean path of buoyant, bent-over chimney plumes. *J. Fluid Mech.*, **28**, 311-322.

6. Fan, L.N. (1967). Turbulent buoyant jets into stratified or flowing ambient fluids. *Report No.* **KH-R-15**, W.M. Keck Laboratory of Hydraulics and Water Resources, California Institute of Technology, Pasadena, California.

7. Wright, S.J. (1977). Mean behavior of buoyant jets in a crossflow. *ASCE, J. Hyd. Div.*, **103**, 499-513.

8. Chu, V.H. (1979). L.N. Fan's data on buoyant jets in crossflow. *ASCE, J. Hyd. Div.*, **105**, 612-617.

9. Fischer, H.B., List, E.J., Koh, R.C.Y., Imberger, J. and Brooks, N.H. (1979). Mixing in inland and coastal waters. Academic Press, New York.

10. Papanicolaou, P.N. and List, E.J. (1987). Statistical and spectral properties of tracer concentration in round buoyant jets. *Int. J. Heat Mass Transfer*, **30**, 2057-71.

11. Papanicolaou, P.N. and List, E.J. (1988). Investigations of round vertical turbulent buoyant jets. *J. Fluid Mech.*, **195**, 341-391.

12. List, E.J. and Dugan, R. (1994). Transition from jet plume dilution to ambient turbulent mixing. NATO ASI, *Recent Research Advances in the Fluid Mechanics of Turbulent Jets and Plumes*, P.A. Davies and M.J. Valente Neves (eds.), Series E, **255**, 1-11, Kluwer Academic Publishers.

Tracer Dispersion by Internal Solitary Waves in Stable Stratifications

William B. Zimmerman[1] and Geert W. Haarlemmer

Department of Chemical Engineering, University of Manchester Institute of Science and Technology

Abstract

Computational fluid dynamics simulations of solitary internal waves propagating in an atmospheric inversion layer are presented. The simulations are initiated by prescribed gusts of cold air at the domain inlet. Properties of the resultant solitary wave evolution are examined. The history of the fluid motion is recorded by either the diffusion of tracer components or the transport of Lagrangian fluid elements. In the case studies presented here, solitary waves of all amplitudes prove to be good at short range mixing and surprisingly poorer than expected at long range transport. When contaminants are introduced after a nocturnal industrial accident they may become trapped in solitary waves triggered by the same event. The upshot is that pollutants are convected away from the source faster then can be expected on the basis of the background wind. When only a few waves are created, mixing is slowed down due to the trapping of fluids by waves which prevents dispersion. When disturbances break up into trains of solitary waves, however, mixing is accelerated. These simulations show that the presence of shear has only little effect on the dynamics of the waves. Both recirculation and pollutant convection still occur.

1 Introduction

As stable stratifications recur in geophysical venues with high frequency, for example nocturnal inversions and ocean thermoclines and pycnoclines, the study of fluid transport mechanisms has wide applicability. Although most studies concentrate on turbulent transport, this is a weak and short range mechanism. Internal gravity waves are also much studied, but if the amplitude is small relative to the waveguide, there is little advection of tracer fluids. Only when the wave is of appreciable amplitude can it trap fluids which it then advects over longer ranges. Indeed, Davis and Acrivos [1] demonstrated that only when the amplitude of localised waves exceeds that of the waveguide does internal recirculation truly trap and advect tracer fluids. These solitary waves maintained

---

[1]Present address: Department of Chemical and Process Engineering, University of Sheffield, Newcastle Street, Sheffield, S1 3JD.

similar shape for long duration in wave tank experiments, but eventually decayed due to dissipative effects. It has been remarked in nearly all wave tank experiments that large amplitude solitary waves are surprisingly easy to generate from practically any big disturbance.

Even though solitary waves have been extensively studied under wave tank experiments [2] [3] to some extent in the field (see for example [4]), and extensively analysed theoretically following on the works of Benney [5], Benjamin [6] [7] and Davis and Acrivos, many questions about their dynamics remain unanswered or only partially addressed. For instance:

- *wavelength*
  Benney's [5] seminal analysis of internal solitary waves assumes the wavelength λ of the disturbance is known in advance so that shallow layer theory can formally be adopted. The *sech*^{2} profile predicted has an explicit wavelength, so that the theory is self consistent. Benjamin's [6] singular perturbation theory makes this assumption. As all subsequent theoretical treatments for more complicated fluid physics and wave guides launch from these pioneering efforts, it is safe to say that there is no theoretical prediction of the wavelength of solitary internal waves.

- *wave structure*
  Essential to shallow layer or singular perturbation theory is that the wave separates into a factor that propagates in the horizontal direction along the waveguide and a factor representing the vertical structure, to leading order in all disturbance field variables – disturbance velocity, disturbance pressure, and disturbance temperature (concentration). This is a strong assumption which one expects must be severely limited in amplitude range. If indeed it is a robust assumption that the wave motion is pseudo-1D, it is likely only in the time asymptotic limit. Otherwise, one would expect the wave structure to be 2-D, not self-similar on each isotherm.

- *permanent form*
  Associated with the wave structure is the wave form. Benjamin [6] sought large amplitude waves of permanent form in an inviscid, dissipation-free fluid, which he found in the steady state limit, after all initial transients have dispersed away. Although the Reynold's numbers of geophysical flows are quite large, dissipative mechanisms are expected to cause wave attenuation only – slow decay. However, there are no estimates of how long the dispersive phase lasts before waves of permanent form result from nonlinearity balancing wave dispersion.

- *advection of tracers*

  As alluded to earlier, given the reported long-life of internal solitary waves, they would be expected to be good long range carriers of tracers, and poor short range mixers. Of course, this expectation depends on the amplitude of the wave. Small amplitude linear waves are not expected to be good mixers, even over short ranges, as the fluid elements themselves are weakly advected. An open question is whether weakly nonlinear waves are capable of either long range transport or short range mixing.

- *wave amplitude*

  Also implicit in the treatments of Benney and Benjamin is the assumption of a well known characteristic amplitude of the wave. As solitary internal waves result from the break up of the initial disturbance, which as a momentum and mass source has quantifiable characteristic scales presumably, it is not the wave amplitude that can be controlled - it is a function of the initial conditions. Finding an *a priori* unambiguous quantification of amplitude nonlinearity $\varepsilon = A_0/H$, where A_0 is the maximum disturbance distance of the isotherm and H is the height of the waveguide, is problematic.

Given the richness of open questions raised above, a flexible approach should be adopted in addressing them. Field observations, though providing invaluable information about the actual application of interest, are usually too sparse to be able to address the above issues in detail. Wave tanks and wind tunnels are excellent visualization venues, but cannot provide the scope of measurements needed without considerable investment. This leaves numerical simulation as the best hope for addressing these questions. It is the approach adopted here.

We have been developing atmospheric simulations of a 2-D domain under idealized conditions as a first approach to the open questions listed above. The first case studies have been reported in the fluid mixing literature in absence of wind shear [8] with an initiation scheme that tweaks the velocity vector of an isolated neighbourhood of grid points. This artificial procedure is especially useful for generating "clean" solitary waves, whose wave dynamics and advecting ability can be closely monitored [9] in the abstract.

In this paper, we are concerned with relieving the ideal initiation by creating an initial disturbance which is both a mass (temperature) and momentum source at the inlet of the domain. These dynamic inlet conditions, performed with a variety of waveguides, permit the study of the effects of generation mechanisms, different wave sizes, stratifications, and the background wind profiles on the wave evolution.

Simulations are performed with either a uniform background velocity or a sheared background. Temperature profiles are only stratified near the bottom surface and left unstratified above 300m. Several types of wave events are simulated. The stratification and the initial disturbances are varied throughout this paper.

Waves are created by blowing a transient cold gust into the waveguide. The scenario is similar to that described in Doviak et al. [10]. They explain how a thunderstorm downdraft can create a train of solitary waves. Other situations where this mechanism may trigger a solitary wave is an explosive release of a cold gas (for example rupture of a tank containing liquid gas) or a sea breeze flowing into a nocturnal inversion layer (gravity current). The origination of waves is influenced by the wind which interacts with the gust. Once the wave develops, the effect of the uniform wind should be negligible.

The method of the simulations is outlined in Section 2. In Sections 3 and 4 the results of the simulations are presented on the propagation of solitary waves in shearless fluids with a density stratification. Two stratifications of different strengths will be compared. The phase speeds of the waves in Sections 3 and 4 can be computed and compared to predictions made by the inviscid weakly nonlinear theories. This is done in Section 5. In Section 6 the effects of shear on the presence of solitary waves is examined more closely.

2   Simulations

The computational fluid dynamics simulation scheme is described in detail in [8] and in the Appendix of [9]. Due to the space constraints neither the finite volume implicit differencing technique nor the model system of partial differential equations appropriate to the simulation of a laminar atmosphere is reproduced here.

The simulations describe the creation of solitary disturbances and their evolution in time. A packet of cold air is released creating a perturbation in the temperature profile and streamlines. No attempt was made to resolve the fine scales of the jet flowing into the stratification. After the initial burst settled down the accuracy increased. The size of the mesh is 40km (200 grid points) by 1000m (50 grid points) and the size of the time steps is 20s. The background wind is set to be either uniformly at 5m/s or sheared.

All of the kinetic energy added is initially localized and then spreads out and evolves into a solitary wave. Potential temperature profile A is linearly stratified from 10 to 11°C in the lower 300m, the upper unstratified layer is set to 11°C (Brunt-Väisälä frequency $1.0 \cdot 10^{-2}$ s^{-1}). Potential temperature profile B is linearly stratified from 10 to 14°C (Brunt-Väisälä frequency $2.2 \cdot 10^{-2}$ s^{-1}) and left unstratified above. The various simulations presented here are listed in Table 1.

The simulations in Sections 3 and 4 show the development of different initial disturbances in two different stratifications. The size of the perturbation is varied in both sections. The influence of the strength of stratification is examined by comparing these two sections. The phase velocities in the simulations are compared to the weakly nonlinear theories in Section 5. Finally in Section 6 the influence of shear in the waveguide is studied.

Table 1. Numerical simulations performed in this paper, with the amplitude amp [m] of the initial disturbance and the particularities of the simulations. () is the time at which the amplitude is measured. v [m/s], time [s], and T[°C] refer to the velocity, duration, and temperature of the initial cold air gust

No	type	v	time	T	amp	ε	aim
1	A	20	40	10	98.8 (1000)	0.33	Lagrangian dispersion
2	A	17	40	10	94.7 (800)	0.31	Lagrangian dispersion
2a	A	15	40	10	58.7 (80)	0.20	Lagrangian dispersion
3	A	5	40	10	30.8 (80)	0.10	Lagrangian dispersion
4	B	20	40	10	129 (400)	0.43	Passive tracer
5	B	10	40	10	85.9 (400)	0.29	2D structure examined
6	B	5	40	10	42.4 (400)	0.14	2D structure examined
7	B	20	40	10	154.9 (200)	0.52	$Ri = 1.9 \cdot 10^2$
8	A	15	20	10	68.2 (200)	0.22	$Ri = 1.2$

3 Light stratification

The simulations in this section focus on the evolution of solitary disturbances in weak stratifications. Tracers and Lagrangian fluid elements show the history of the fluid motion and the mixing properties of solitary waves. How the size of the waves influences fluid trapping is investigated. The stratification is very light. First, simulations of relatively large amplitude waves are presented (simulations 1, 2 and 2a). Then an example of a small wave is given in simulation 3.

3.1 Large waves

3.1.1 Simulation 1: $\varepsilon = 0.33$

In this first simulation a large volume of air is released from the bottom boundary, in the centre of the mesh. The air was released in 40s with a velocity of 20m/s over the length of 200m (concentration tracer is unity). This disturbance alters the previously steady temperature profile and streamlines. A set of 10,000 Lagrangian particles is followed in this simulation. The temperature profile directly after the introduction of the disturbance is shown in Figure 1. The frame of reference moves with the background wind, isolating the convection due to wave propagation.

After the initial gust, the vertical velocities directed downward are, after 80s, already larger than upward wind velocities. Since there is still, after 80s, much vertical momentum, the wave amplitude continues to grow. As the perturbation develops, an evolving wave propagates to the right in the frame of reference, i.e. faster than the wind.

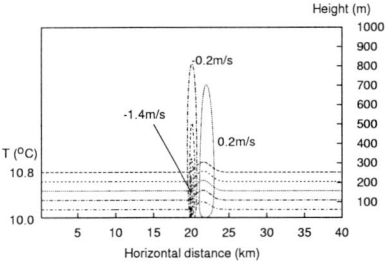

Figure 1. Potential temperature and the vertical velocity profiles in simulation 1 after 80s

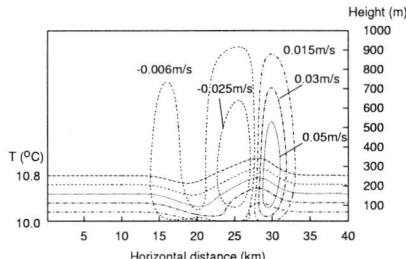

Figure 2. Potential temperature and vertical velocity profiles in simulation 1 after 2000s

Figure 2 presents the solution after 2000s. The wave has gained in amplitude at the expense of kinetic energy. The initial disturbance developed into an elevation followed by a depression. Weakly nonlinear theory predicts the development of a train of solitons. It appears however that dispersive and dissipative effects prevent this from happening. Wave dispersion is strong since the stratification is weak and the wave amplitude is relatively large.

As time progresses (Figure 3) the wave slowly disperses and becomes more shallow. Eventually it will run off the scale. The evolution of the 10.9°C isotherm is plotted in Figure 5. The wave first grows until it reaches a maximum amplitude and than it slowly decays. Over the first 4000s the overall velocity of the wave is 3.1m/s. Compared to the overall velocity the background flow of 5m/s this phase speed is significant.

A non-buoyant chemical tracer was introduced with the initial cold gust. The concentration in the introduced air was unity. Figure 4 displays the concentration immediately after introduction and after 4000s. The tracer was convected along with the wave and pushed forward. Part of the tracer slips out of the back of the wave which gives strong horizontal mixing effects. Vertical mixing appears to be blocked by the stratification.

The history of fluid transport can be visualised by introducing Lagrangian fluid elements and tracing their position in time (Figure 6). The fluid elements chosen are situated in a column of 100m×90m. As the wave evolves the fluid elements are advected. The fluid elements remain near the lower boundary. Buoyancy effects prevent rapid vertical mixing.

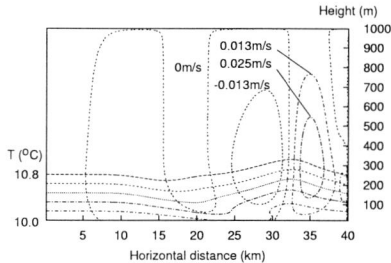

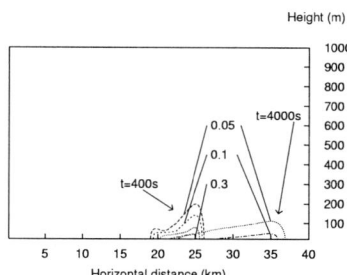

Figure 3. The potential temperature profile and vertical velocity in simulation 1 after 4000s

Figure 4. Concentration of the tracer after 400s and 4000s in simulation 1

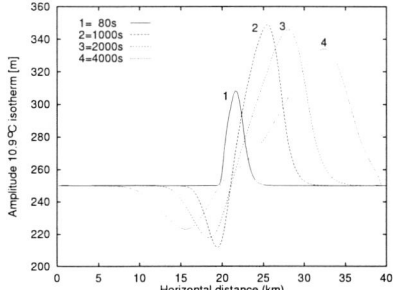

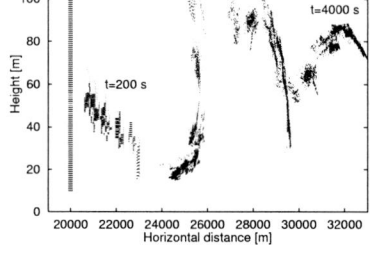

Figure 5. The position of the 10.9°C isotherm as a function of time in simulation 1

Figure 6. The positions of 10,000 particles as a function of time in simulation 1

3.1.2   Simulation 2 and 2a: $\varepsilon = 0.31$ and $\varepsilon = 0.20$

The waves created in these simulations are smaller than the wave in the simulation 1. The gust was let in with 17m/s for 40s. Again the background wind is 5m/s and the grid moves at the same velocity. Figure 7 shows the temperature and vertical velocity profile after 800s of simulation time. Clearly the evolution is very similar to simulation 1. The perturbation created a wave with an amplitude of 94.7m after 800s. Simulation 2a was started with a gust of 15m/s for 40s. The results of these simulations are qualitatively very similar, therefore in this section only the figures for simulation 2 are given.

Although the wave is smaller, the effects found in simulation 2 are similar to the simulation 1. The fluid with the tracer is trapped and convected away from the source for a considerable distance. It is clear that both mixing and convection play an important role in these wave evolution simulations. As can be seen in Figure 8, the concentration of the contaminants decreases in time due to diffusion and mixing. Convection is caused by the wave propagation. A similar result can be seen from the particles traced in (Figure 9). Although the effectiveness of the wave propagation is less than in simulation 1 there is still a considerable effect. One particle is labelled with a ◇ to show how the cloud rotates.

3.2   Small waves

3.2.1   Simulation 3: $\varepsilon = 0.10$

The gust that caused the disturbance in this simulation is now much smaller than in simulations 1, 2 and 2a (5m/s). The disturbance develops in a similar way although the maximum amplitude is reached much faster. In Figure 10 the location of the 10.9°C isotherm is depicted. The most striking feature of the simulation is that the wave moves to the left, upwind, rather than to the right.

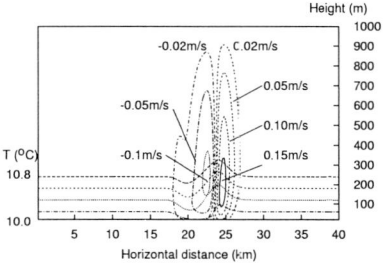
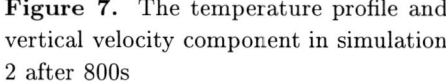
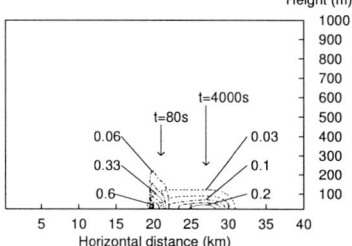

Figure 7. The temperature profile and vertical velocity component in simulation 2 after 800s

Figure 8. Concentration of the tracer component after 80s (left) and 4000s (right) in simulation 2

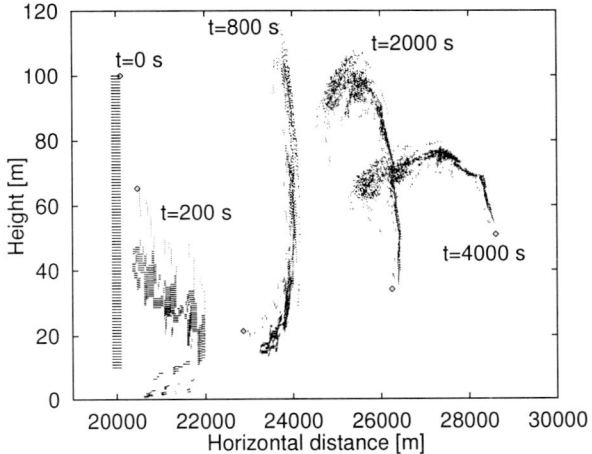

Figure 9. The positions of 5,000 particles as a function of time in simulation 2. One particle is marked with a ◇

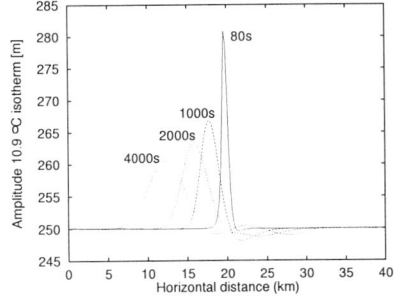

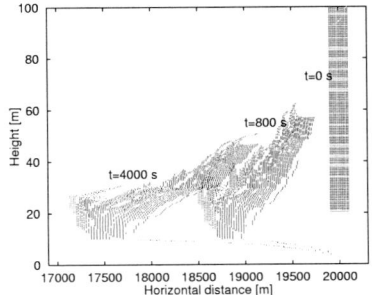

Figure 10. The position of the 10.9°C isotherm at different times in simulation 3

Figure 11. The positions of 10,000 particles as a function of time in simulation 3

Apparently there is a mechanism determining the direction the wave travels based on amplitude. Small waves move left; large waves move right.

The classical weakly nonlinear theories (KdV and BDO) allow waves only to travel in one direction. The phase velocity is dependent on the amplitude. In the weakly nonlinear theories the wave velocity is linear in the amplitude. The small waves found in this simulation move significantly slower than the large waves in the previous sections.

For simulation 3 the history of the particles is followed and depicted in (Figure 11). Obviously, in this simulation advection is less pronounced than in the previous simulations with larger waves. The convective effects are weaker and disappear once the wave becomes too small.

3.3   Discussion of the results

The simulations in this section show that large waves play an important role in the convection of fluids. Smaller waves are less effective in this respect. Waves travel in both directions in these simulations, convecting the trapped fluids either upwind or downwind. All waves in this section reveal a complex 2-D structure with a wavy tail. The waves do not take a simple $sech^2$ shape.

4   Strong stratification

In a stronger stratification it is expected that nonlinearity will play a bigger role, and the phase speed of the waves will increase. Some simulations in this section exhibit these expectations. The size of the disturbance is varied.

4.1   Large waves

4.1.1   Simulation 4: $\varepsilon = 0.43$

Compared to simulations 1–3 a stronger stratification is used. The lower 300m has a linear temperature profile increasing from 10°C to 14°C. The top 700m was left unstratified at 14°C. The grid was fixed. There was no need artificially to extend the domain using moving grids. Due to greater wave velocities in stronger stratifications, the events and the evolution take place faster. A disturbance was released with a velocity of 20m/s for 40s (temperature 10°C, concentration tracer unity). This creates a wave with an amplitude of 129m (after 400s).

The disturbance propagates away from the source and develops into a train of two solitary waves, Figures 12 and 13. Eventually (after 3600s) the wave runs off the grid. The gap between the two waves is clearly seen from the location of the 13.8°C isotherm (Figure 14). The velocity of the wave is significantly higher than the background wind. The disturbance initially travels at 9.5m/s (later decreased to 8.7m/s). This is considerably faster than the background wind (5m/s).

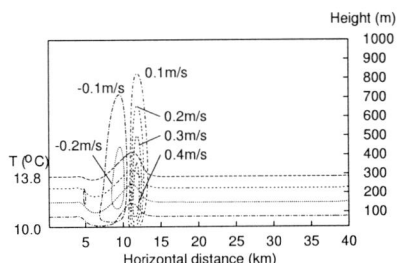

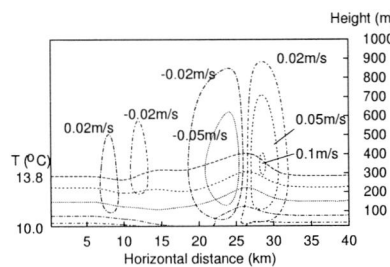

Figure 12. The potential temperature profile and vertical velocity component in simulation 4 after 400s

Figure 13. The temperature profile and vertical velocity in simulation 4 after 2000s

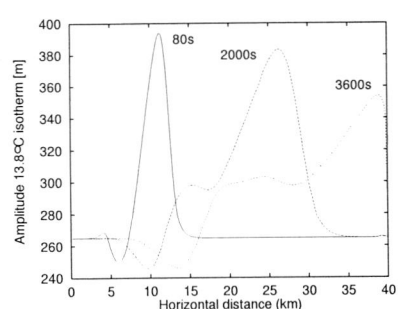

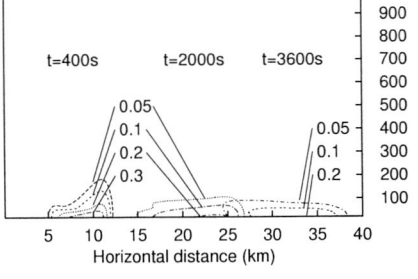

Figure 14. Location of the 13.8°C isotherm for three times during simulation 4

Figure 15. Concentration profiles after 400s, 2000s and 3600s in simulation 4

The tracer is convected faster as well. Although there is strong horizontal mixing, there is not much vertical mixing. (See Figure 15). The tracer, and therefore the fluids in the lower layers of the inversion layer, are pushed forward by the wave. Some tracer escapes during this convection, hence the long tail. The wave provides an effective horizontal mixing mechanism.

4.2   Smaller waves

4.2.1   Simulation 5 and 6: $\varepsilon = 0.29$ and $\varepsilon = 0.14$

These simulations use the same background conditions as simulation 4, but the waves are considerably smaller. Simulation 5 consists of a disturbance introduced at a velocity of 10m/s (amplitude initial wave 85.9m) and simulation 6 at a velocity of 5m/s (amplitude initial wave 42.4m). For clarity, the modulation of a single isotherm is shown.

Clearly, small waves are not very effective in convection of tracers. This is very similar to the results found in simulations 3. The location of the 13.8°C isotherms are depicted in Figures 16 and 17. The vertical scale applies to the lowest lines. Each additional line has 25m added to its vertical location to allow a clearer picture.

The disturbance travels much slower than in simulation 4. The waves propagate upwind in contrast to the larger wave in simulation 4 that moved downwind. It is clear that the waves separate into a larger and steeper solitary wave and a more shallow, smaller solitary wave. This agrees with weakly nonlinear theory. These waves are strictly speaking not solitons because they are not of permanent form.

The weakly nonlinear theories assume that the waveform has the self-similar form at every streamline. This means that the disturbance can be separated into a disturbance field $A(x,t)$ and a vertical structure of the waveguide $\Psi(z)$. For the simulations in this paper, coordinate separation does not occur. The locations of the 13°C and the 11°C isotherms in simulation 5 are shown in Figures 18 and 19. The disturbance takes a different form at every streamline and is therefore not separable. It may take a very long time for the energy to spread evenly on the streamlines. A more sophisticated 2-D model is needed to describe these waves.

4.3   Discussion of the results

An increase in the strength of the stratification clearly shows that the wave breaks into a train of solitary waves. This is what would be expected on the basis of the 1-D weakly nonlinear theories. The 2-D wave structure of strongly nonlinear wave propagation, however, is revealed.

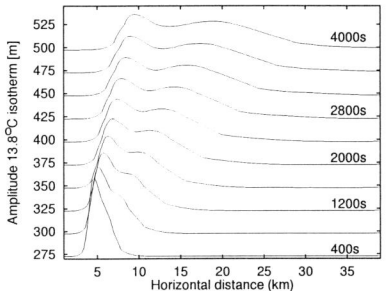

Figure 16. The position of the 13.8°C isotherm at different times in simulation 5. Each additional line is shifted 25m vertically

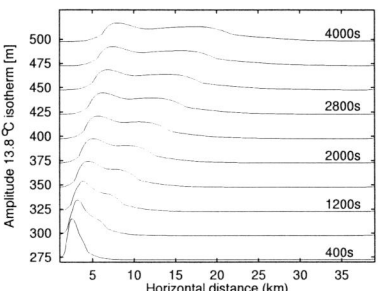

Figure 17. The position of the 13.8°C isotherm at different times in simulation 6. Each additional line is shifted 25m vertically

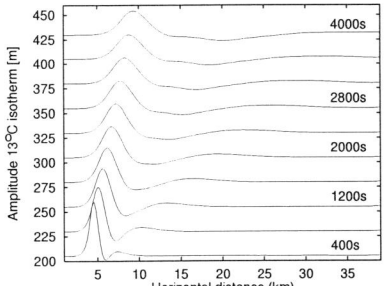

Figure 18. The position of the 13°C isotherm at different times in simulation 5. Each additional line is shifted 25m vertically

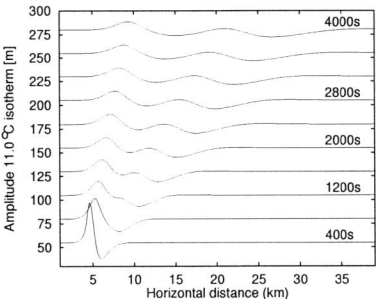

Figure 19. The position of the 11°C isotherm at different times in simulation 5. Each additional line is shifted 25m vertically

5   Phase velocities

An increase in strength of stratification causes in increase in phase speed (c) of the wave. It is possible to predict the phase speed in the simulations by solving the Taylor-Goldstein eigenvalue problem [11]. The eigenvalue problem arises from the leading order in the perturbation analysis. The eigenvalue found is the phase speed (c_0) of an infinitely long wave (wavenumber $k = 0$).

For the waveguides used in this paper various assumptions can be made. When it is assumed that both top and bottom boundaries are rigid lids, the shallow water theory of Korteweg and de Vries (KdV) applies [5]. When the stratified waveguide is embedded in deep fluid, the Benjamin-Davis-Ono [1][7] (BDO) theory applies. The long wave phase speeds for both theories are shown as a function of the strength of the stratification in Figure 20. It can be seen that the differences between the two theories are small. The deep fluid equations predict a slightly higher phase speed than the shallow fluid equations. The constraining rigid lid has a retarding effect.

The phase velocity is predicted below with corrections due to small, but finite amplitude [11]. Here α is the coefficient of the nonlinear term in the evolution equation.

$$\text{KdV}: \quad c \;=\; c_0 + \frac{1}{3}\varepsilon\alpha H$$

$$\text{BDO}: \quad c \;=\; c_0 + \frac{1}{4}\varepsilon\alpha H. \tag{5.1}$$

The coefficients α and the long wave phase speeds c_0 above are computed and displayed in Table 2.

The phase velocities as a function of the amplitude are graphed for the simulations 1 to 6 in Figures 21 and 22. A comparison is made to the weakly nonlinear theories in each of the graphs. In some cases the initial velocity of the wave is very large although the velocities invariably decrease quickly to the level predicted by the weakly nonlinear theories. From a physical point of view the

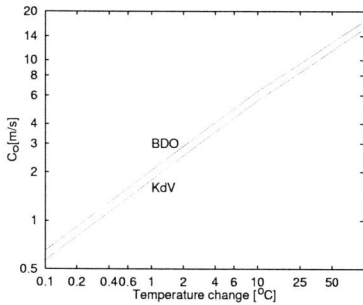

Figure 20. Long wave phase speed c_0 for a linearly stratified waveguide of 300m height, predicted by the KdV and BDO theory. Both axes are logarithmic

Table 2. Long wave phase speed and coefficients of the nonlinear term for the KdV and BDO equations in the simulations of this paper

Stratification	c_0 (KdV) [m/s]	α (KdV) [s^{-1}]	c_0 (BDO) [m/s]	α (BDO) [s^{-1}]
A	1.78	0.00978	2.05	0.0137
B	3.55	0.0195	4.07	0.0271

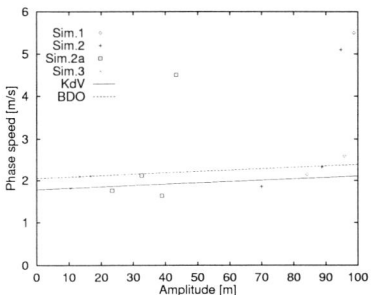

Figure 21. Wave velocities as a function of amplitude found in simulations 1–3 compared to weakly nonlinear theories

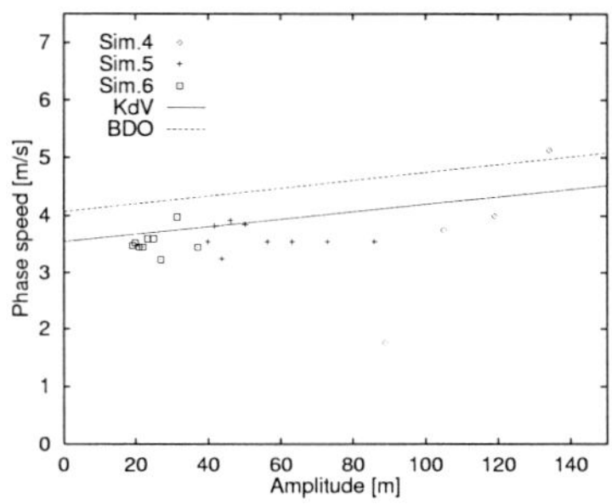

Figure 22. Wave velocities as a function of amplitude found in simulations 4–6 compared to weakly nonlinear theories

shallow water theory (KdV) should be used because the the wavelength is considerably larger than the height of the waveguide. There is good agreement with the KdV theory for strong stratifications in Figure 22. For the weak stratification in Figure 21 there is no reason to assume that the KdV theory is more accurate than the BDO theory. The trend of the phase speed increasing with the wave amplitude as suggested by (5.1) is clear from the graphs for both stratifications.

6   Sheared waveguides

Two simulations here examine how shear influences wave propagation and convection. The inclusion of shear in numerical and mathematical models for wave guides can be separated into two areas. Shear profiles can be characterised by the Richardson number as defined by: $Ri = N^2/\bar{u}_z^2$, where N the Brunt-Väisälä frequency and $\bar{u}$ the background velocity profile. When the Richardson number anywhere in the flow is larger than 0.25 it is expected that the dynamics of the wave motion is very similar to that of un-sheared waveguides. If there are areas of high shear gradients in the flow the Richardson number may locally be smaller then 0.25. Thus the flow is susceptible to instabilities and critical layers. In a critical layer the phase velocity of a wave is equal to the background wind velocity. This acts as a capping mechanism on a waveguide in much the same way as a rigid lid.

The scientific interest in sheared waveguides and critical layers is very high. Various authors (for example [12]) treat trapping mechanisms that can be attributed to shear layers. The simulations presented here concentrate on how disturbances evolve into solitary waves. Section 6.1 presents a simulation with weak shear, while a simulation with a strong shear layer is presented in Section 6.2. Both simulations have Richardson number higher than 0.25 to avoid shear instabilities.

6.1 High Richardson number

Flows with a high Richardson numbers have relatively strong buoyancy forces compared to shear forces, so buoyancy opposes vertical motion, and turbulence is suppressed. The simulation in this section has a background $Ri = 1.9 \times 10^2$. The background velocity linearly increases from 5 to 5.5m/s over 1000m. The potential temperature stratification increases linearly from 10 to 14°C over the first 300m and is left unstratified above. The wave was created by blowing in air with a tracer at a velocity of 20m/s for 20s. The disturbance was then left to evolve.

The temperature profile 200s and 1000s after the start of the simulation are graphed in Figure 23. It is clear that a large amplitude wave arises from the disturbance. When the wave evolves further it retains its coherent structure despite the shear. The wave develops the well known wavy tail seen in previous simulations.

The tracer appears to be indifferent to the presence of the wave in Figure 24. It is not trapped but slowly disperses. The shear profile is clearly visible from the shape of the cloud of tracer. The vertical velocity profile (Figure 25) shows that the wave remains one coherent structure and does not align itself to the flow.

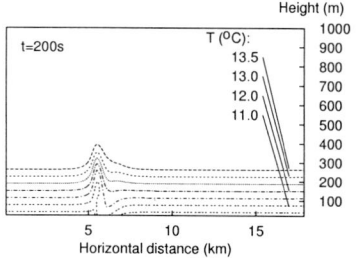 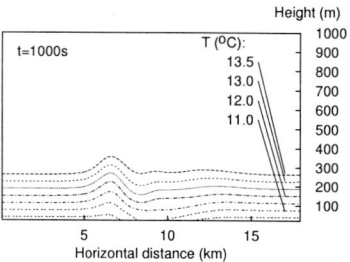

Figure 23. The temperature profiles at t=200s (left) and t=1000s (right) in simulation 7

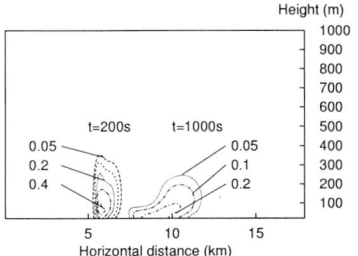

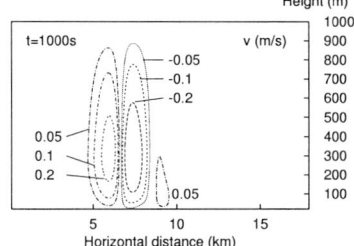

Figure 24. The concentration profiles at t=200s and t=1000s in simulation 7

Figure 25. The vertical velocity profile at t=1000s in simulation 7

6.2 Low Richardson number

When the Richardson number decreases to $O(1)$ shear forces become increasingly important. The simulation in this section has a potential temperature stratification that increases linearly from 10 to 11°C in the first 300m and is left unstratified at 11°C above 300m. The background velocity increases linearly from 2m/s at the lower boundary to 12m/s at the top ($Ri = 1.2$). The wave was created by blowing in air (10°C) with a tracer at a velocity of 15m/s for 20s.

The temperature profiles after 200s and 1600s in the simulation are given in Figure 26. The most striking feature of these graphs is that the wave form appears to be restricted towards the top of the stratified layer. The amplitude of the streamlines decreases with increasing height. The phase velocity of the leading right-moving wave in the simulation (after 1600s at approximately 15km) is $c = 5.6$m/s. This is the background velocity at 350m altitude, indicating a critical layer in the waveguide. The capping mechanism present is the critical layer.

The centre of the tracer in simulation 8 is convected at a velocity of circa 2.5m/s, Figure 27. This cannot be attributed to the presence of the wave. The vertical velocity profile of the wave (Figure 28) shows that the wave does not align itself to the shear profile but remains a coherent structure as in the previous simulation.

6.3 Discussion of the results

The simulations show that the waves treated here are not very successful in trapping fluids. While the wave retains a coherent structure, the tracer aligns itself to the flow. There are some interesting dynamics observed in simulation 8. There is a critical layer in the waveguide. In general these critical layers are assumed to exist only when the local $Ri < 1/4$. The velocity profile does not experience a strong capping mechanism, indicating that a critical layer is a pseudo-slip boundary. No further detailed investigation of critical layer properties is permitted by the simulations here.

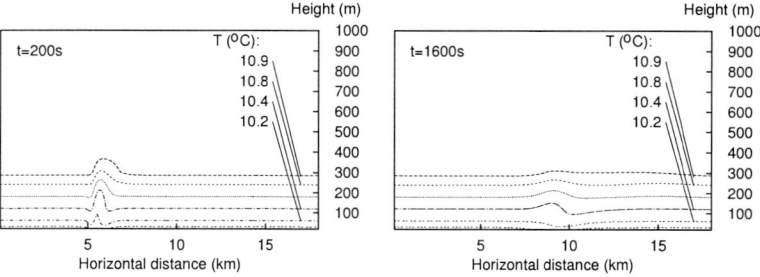

Figure 26. The temperature profiles in °C, Left: t=200s, Right: t=1600s. Simulation 8

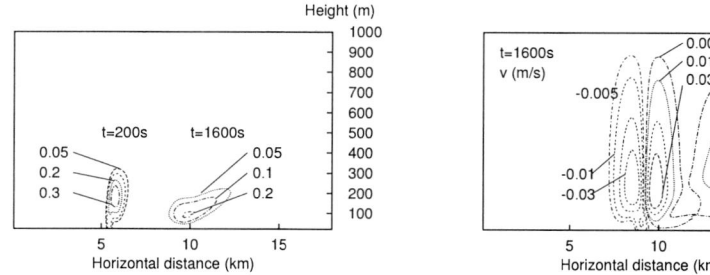

Figure 27. The concentration profiles at t=200s and t=1600s. Simulation 8

Figure 28. The vertical velocity profile at t=1600s. Simulation 8

7   Discussion and conclusions

The simulations in this paper reveal a number of properties of solitary waves. It was shown that the weakly nonlinear theories are partly successful in describing and predicting the dynamics. Due to the complex nature of the wave motion some properties cannot be explained by the classical one dimensional theories.

- *Fluid convection*
  The waves under study are very efficient in transporting fluids. The waves are transient in nature and evolve due to dissipation and dispersion. When the wave amplitude decreases the trapped fluids are lost. This means that this type of solitary wave is an efficient mechanism for horizontal transport and mixing. There are no clear recirculations observable in the temperature profiles. In the way the Lagrangian fluid elements are convected some rotation is visible in the waves. There are clear effects of the presence of the waves on the introduced fluids. Even though nonlinearity is not very strong the waves convect some of the fluids over large distances. When the waves are small, or become small, this effect decreases.

- *Break up of the disturbance into a train of solitary waves*
  A disturbance in weak stratifications (simulations 1–3) does not produce a train of solitary waves in these simulations. When the stratification becomes stronger (simulations 4–6) the nonlinear effects become more important and the disturbance is more likely to break up than in simulations 1–3. Viscous dissipation and dispersion can prevent waves from breaking up. Christie (1989) showed this effect for the dissipative BDO-Burgers equation.

- *Waveforms*
  All the simulations initially reveal solitary waves with a true two dimensional structure. The wave structure does not separate. The top streamline in the inversion layer is perturbed in a $sech^2$-like form. Lower streamlines however reveal different wave forms. When a wave passes, the fluid is lifted, then moves down and overshoots its old position. This behaviour can be accounted for by the KdV equation with sufficient dispersion. The solutions of the KdV-KS Equation [13] however have wavy tails on one side of the wave and therefore may provide a more suitable description.

- *Phase velocities*
  It is possible to predict the phase velocities of the observed waves reliably by either KdV or BDO theory. Thus the speed at which the pollutants are convected is predictable as well, provided there is sufficient information about the waveguide.

- *Inclusion of shear*
  The simulation in sheared waveguides show that the fluids trapped in a wave quickly escape, even in weak shear. It was also shown that even at very large Richardson numbers (weak shear) the evolution of the waves is not hampered by the presence of shear; the wave retains a coherent structure. Simulation 8 showed that high shear ($Ri = 1.2$) may produce capping mechanisms for waveguides as would be expected for Richardson numbers smaller than $1/4$.

Acknowledgements

The first author thanks NATO for support under crg 940242 for discussions with Professor M.G. Velarde of Complutense University of Madrid. The second author is grateful for a UMIST postgraduate fellowship. Use of the PHOENICS CFD engine is acknowledged.

Bibliography

1. Davis, R.E. and Acrivos, A. (1967). "Solitary waves in deep water." *J. Fluid Mech.*, *29*:593–607.

2. Maxworthy, T. (1976). "Experiments on collisions between solitary waves." *J. Fluid Mech.*, *76*:177–185.

3. Amen, R. and Maxworthy, T. (1980). "The gravitational collapse of a mixed region into a linearly stratified fluid." *J. Fluid Mech.*, *96*:65–80.

4. Rees, J.M. and Rottman, J.W. (1994). "Analysis of solitary disturbances over an antarctic ice shelf." *Boundary-Layer Met.*, *69*:285–310.

5. Benney, D.J. (1966). "Long nonlinear waves in fluid flows." *J. Math. Phys.*, *45*:52–63.

6. Benjamin, T.B. (1966). "Internal waves of finite amplitude and permanent form." *J. Fluid Mech.*, *25*:241–270.

7. Benjamin, T.B. (1967). "Internal waves of permanent form in fluids of great depth." *J. Fluid Mech.*, *29*:559–592.

8. Haarlemmer, G.W. and Zimmerman, W.B. (1996). "Enhancement of mixing by internal solitary waves in stratified flows." Mixing V, *I Chem E Symposiums Series*, H. Benkreira ed. *140*:225-235.

9. Haarlemmer, G.W. and Zimmerman, W.B. (1997). "Convection of pollutants by internal solitary waves." *Nonlinear Processes in Geophysics*, in press.

10. Doviak, R.J., Chen, S.S., and Christie, D.R. (1991). "A thunderstorm generated solitary wave observation compared with theory for nonlinear waves in a sheared atmosphere." *J. Atmos. Sci.*, *48*(1):87–111.

11. Rottman, J.W. and Einaudi, F. (1993). "Solitary waves in the atmosphere." *J. Atmos. Sci.*, *50*(14):2116–2136.

12. Skyllingstad, E.D. (1991). "Critical layer effects on atmospheric solitary and cnoidal waves." *J. Atmos. Sci.*, *48*(14):1613–1624.

13. Zimmerman, W. B. and Velarde, M.G. (1996). "A centre manifold approach to solitary internal waves in a sheared, stably stratified fluid layer." *Nonlinear Processes in Geophysics*, *3*:110–114.

Vertical Dispersion of Passive Scalars Within and Across Stably Stratified Inversion Layers

V. Schilling and D. Etling

Institute of Meteorology and Climatology, University Hannover, Germany

Abstract

Transport of substances into and across inversion layers has been simulated numerically by means of a two-dimensional finite difference model. Two simulation experiments have been selected in order to demonstrate dispersion of passive scalars caused by Kelvin-Helmholtz instabilities and breaking gravity waves, respectively.

From simulated results it can be gained that during a period of $60min$ enhanced turbulence within the temperature inversion yields mass fluxes of $\overline{C'w'} \leq 7.6cms^{-1}$ (C: normalized concentration). An effective mass flux, averaged horizontally and in time, amounts to $2.1cms^{-1}$ in the case of Kelvin-Helmholtz instabilities and $1.6cms^{-1}$ resulting from breaking gravity waves. At the final stage a normalized concentration of about 10% the initial concentration below the inversion is found at a level just above the inversion's top.

1 Introduction

Elevated shallow layers with strong stable stratification (inversion layers), often topping a near neutral or even unstably stratified layer below, can be found quite often in the atmosphere and oceans. These layers, wherein the temperature increases with height, are usually considered as effective barriers to the vertical transport of passive scalars. But on occasion, transport of momentum, heat and mass can be observed within and across these inversion layers as shown, for example, by Miller [1], Goodman and Miller [2] or Readings et al. [3]. It is of some importance to know some details about the occurrence and strength of those unusual transport events in cases of strong smog situations or possible accidents occurring in industrial plants.

The mixing events may be due to overshooting thermal plumes from an underlying convective boundary layer, due to breaks caused by fronts or other external phenomena or due to unstable internal gravity waves and dynamical instabilities in the presence of additional windshear across the inversion layer.

There are only a few measurements indicating a vertical transport both into and across inversion layers topping a boundary layer. The most intensive measurements have been done at the west coast of the United States [1], [2]. They indicate that, although the coastal inversion layer was both statically and dynamically stable, there was a vertical flux of substances (Na, Cl, ozone, condensation nuclei) across the inversion. This transport, approximately $15 \, \text{cm} \, \text{s}^{-1}$, is ascribed to the action of stable internal gravity waves moving through a layer of strong vertical shear of the horizontal wind. Horizontal mixing near their crest would result in a net vertical pumping of mass.

Another phenomenon could be observed by Readings et al. [3] using a high power Doppler radar. At the top of a convective boundary layer they found convective penetration into the topping inversion layer yielding small scale dynamical (Kelvin-Helmholtz) instabilities just above penetrating convective plumes. These instabilities may cause a possible flux of heat, mass and momentum, but representative values could not be quantified because of the absence of knowledge of the inversion's overall structure.

Other studies concerning this problem have dealt with stable layers at very high levels in the atmosphere (tropopause) and more larger scale instabilities than we will discuss in this paper (for example [4], [5]).

In order to get a more detailed impression and knowledge of the phenomena and mechanisms yielding a vertical flux of substances within and across inversion layers we carried out a number of numerical simulation experiments by using a two-dimensional LES-(Large Eddy Simulation-) model and a Lagrangian dispersion model. As a first step of our investigations we have focussed our interest to phenomena caused by the internal structure of the inversion layer itself. This means, that we discuss vertical dispersion of passive scalars due to Kelvin-Helmholtz instabilities and breaking gravity waves caused by the static stability and the vertical windshear within the inversion layer in contrast to phenomena, as breaks or generally destruction of the inversion, caused by fronts, advection of strong turbulence, enhanced convective boundary layers or other external processes.

Theoretically, the instabilities discussed in this paper are based on the Taylor-Goldstein equation

$$[\bar{u}(z) - c]^2 \left[\frac{\partial^2 w'}{\partial z^2} - k^2 w' \right] - [\bar{u}(z) - c] \frac{\partial^2 \bar{u}}{\partial z^2} w' + N^2(z) w' = 0, \qquad (1.1)$$

where $\bar{u}$, $c = c_r + i c_i$, w' and k denote the mean horizontal wind velocity, the complex phase speed, the vertical perturbation velocity and the horizontal wavenumber, respectively. N is the Brunt-Vaisala frequency

$$N^2 = \frac{g}{\widetilde{\Theta}} \frac{\partial \Theta}{\partial z}, \qquad (1.2)$$

with Θ and $\widetilde{\Theta}$ as the local and the horizontal mean of the potential temperature, respectively.

The Kelvin-Helmholtz instability as well as the breaking of gravity waves are solutions of the Taylor-Goldstein equation, i.e. combinations of $\bar{u}(z)$, k and c yielding an imaginary part of the phase velocity ($c_i \neq 0$). It can be shown that

Kelvin-Helmholtz instabilities occur if there exists an inflexion point within the velocity profile and the local Richardson number Ri is smaller than the critical Richardson number Ri_{cr}:

$$\frac{\partial^2 \bar{u}}{\partial z^2} = 0 \quad \text{and} \quad Ri = N^2 \left(\frac{\partial \bar{u}}{\partial z}\right)^{-2} < Ri_{cr} = 0.25. \tag{1.3}$$

Gravity waves will break (producing enhanced turbulence via convective instabilities) if, for example, their phase velocity c equals the horizontal wind velocity $\bar{u}$ at some height (gravity wave-critical level interaction) or if their amplitudes A will become maximum ($A \sim U/N$).

In this paper we will describe the concept of the Eulerian dynamic model and the Lagrangian dispersion model in Section 2. We have selected two exemplary numerical experiments, which will be presented in Section 3 (initial conditions) and Section 4 (results and discussion) giving an impression of the general development and magnitude of the simulated vertical dispersion of passive scalars within and across inversion layers, vicariously.

2 Model concept

Numerical simulation of dispersion processes parallel to dynamical phenomena require the use of two models in combination. In our studies an Eulerian flow model is used to simulate the dynamical system of the instability processes while a Lagrangian (Monte Carlo) model describes the dispersion of substances simulated as passive scalars. Both model parts are operated simultaneously and are coupled by using the dynamically simulated wind velocity components $\bar{u}$ and subgrid turbulent kinetic energy E as input values required for the dispersion model. During the first phase of our investigations, which is described partly in this paper, we use the two-dimensional version of the model to study a number of different situations and to get an impression of the structure and magnitude of those dispersion processes. A complete description of this model version is given by Schilling [6] (see also [7], [8], [9]).

2.1 Flow model

Our experiments are based on a model which originally was developed to study the diffusion of aircraft emissions due to Kelvin-Helmholtz instabilities [9] and breaking gravity waves [7]. It is based on the primitive equations of motion using the Boussinesq approximation, Reynolds-averaging and gradient relationship for Reynolds flux terms. The system of equations consists of the equation of motion

$$\frac{\partial u_i}{\partial t} = -u_k \frac{\partial u_i}{\partial x_k} + \delta_{i3} g \frac{\Theta^*}{\widetilde{\Theta}} - \frac{1}{\widetilde{\rho}} \frac{\partial p^*}{\partial x_i} + \frac{\partial}{\partial x_k} \left[K_m \left(\frac{\partial u_i}{\partial x_k} + \frac{\partial u_k}{\partial x_i} \right) \right] \tag{2.1}$$

with $i, k \in \{1, 3\}$, the continuity equation for anelastic media

$$0 = \frac{\partial}{\partial x_i} \left(\widetilde{\rho} u_i \right), \tag{2.2}$$

and a heat transfer equation

$$\frac{\partial \Theta}{\partial t} = -u_k \frac{\partial \Theta}{\partial x_k} + \frac{\partial}{\partial x_k} \left[K_h \frac{\partial \Theta}{\partial x_k} \right]. \tag{2.3}$$

Here u_i denotes the Reynolds-averaged velocity vector; $\widetilde{\Theta}$ and $\widetilde{\rho}$ are the mesoscale averaged potential temperature and air density, respectively, and Θ^* and p^* are the deviations from the mesoscale mean values of potential temperature and pressure, respectively. These equations are solved numerically at a finite difference grid. The subgrid eddy diffusion coefficients (for example, K_m for momentum) are determined by a one-equation turbulence model using the Prandtl-Kolmogorov relation

$$K_m = c_m l \sqrt{E}, \tag{2.4}$$

where $c_m = 0.2$ is an empirical constant, $E = 0.5 \overline{u_i'^2}$ is the turbulent kinetic energy, and l is a mixing length related to the grid size by

$$l = 0.2 \sqrt{\Delta x \cdot \Delta z}. \tag{2.5}$$

The turbulent kinetic energy is obtained from the equation

$$\frac{\partial E}{\partial t} = -u_k \frac{\partial E}{\partial x_k} + K_m \frac{\partial u_j}{\partial x_k} \left(\frac{\partial u_k}{\partial x_j} + \frac{\partial u_j}{\partial x_k} \right) - K_h N^2 + \tag{2.6}$$

$$\frac{\partial}{\partial x_k} \left(2 K_E \frac{\partial E}{\partial x_k} \right) - \frac{c_\epsilon}{l} E^{3/2},$$

with $c_\epsilon = 0.064 = \alpha_0^3$ (α_0 is Kolmogorov constant). The diffusion coefficients for heat K_h and turbulent kinetic energy K_E are set equal to that for momentum

$$K_h = K_E = K_m;$$

that is, the turbulent Prandtl number $Pr_t = K_m / K_h = 1$.

The complete set of equations has been solved numerically on a staggered grid of resolution ($\Delta x = 20$ m, $\Delta z \geq 10$ m) by using a simple Euler forward scheme for integration in time. To calculate the nonlinear advection terms special techniques are required owing to the sharp velocity gradients in the vicinity of the inversion centre. In this model we used an "upstream spline" technique to minimize numerical diffusion effects while yielding results of great accuracy [10], [11]. In addition, it should be mentioned that for the pressure perturbations a Poisson equation is solved by using a fast "one-step" solver [12].

2.2   Dispersion model

Transport and dispersion of passive scalars due to mean flow and turbulence is simulated by means of a Lagrangian particle model (for example, [13]). Several thousands of weightless particles are set free below the inversion layer, and their trajectories are followed by

$$x_i(t + \Delta t) = x_i(t) + [\overline{u_i}(t) + u_i''(t)] \Delta t. \tag{2.7}$$

While the mean velocity components $\bar{u}_i$ are available from the dynamic model, the turbulent velocity fluctuations u'_i are obtained from

$$u''_i(t) = R_L u''_i(t - \Delta t) + u^*_i(t), \qquad (2.8)$$

where R_L is the Lagrangian autocorrelation given by

$$R_L = \exp(-\frac{\Delta t}{\tau_L}). \qquad (2.9)$$

u^*_i is a random velocity, which can be obtained from a stochastic process via the Monte Carlo method:

$$u^*_i = \sqrt{1 - R_L^2}\, \sigma_{u_i} \Omega + (1 - R_L)\tau_L \frac{\partial \sigma^2_{u_k}}{\partial x_k} \delta_{ik}. \qquad (2.10)$$

Ω is a random number with normal distribution and σ_{u_i} are the velocity standard deviations, which are obtained from the turbulent kinetic energy by

$$\sigma_{u_i} = \sqrt{E}. \qquad (2.11)$$

The Lagrangian timescale τ_L is given by

$$\tau_L = \frac{K_m}{E}. \qquad (2.12)$$

By solving the equations numerically, it is possible to obtain particle positions for every time step Δt of the flow development. Concentrations can then be calculated by counting the number of particles contained in a box of an adapted numerical grid.

3 Geometry and initial conditions

Two simulation experiments shall represent the dispersion of passive scalars within and across inversion layers due to Kelvin-Helmholtz instabilities and breaking gravity waves. For the first experiment we have chosen a more simple atmospheric situation. A near neutral stratified boundary layer is topped by an inversion layer (index i) of thickness $\Delta z_i = 100m$ and temperature increase $\Delta T_i = 0.2K$ ($\Delta \Theta_i = 1.2K$, see Figure 1). Above the inversion the atmosphere is slightly stable stratified ($\partial \Theta / \partial z = 0.0035 K m^{-1}$). Throughout the inversion layer the horizontal wind velocity increases from about $\bar{u}_b = 2ms^{-1}$ at the inversion base ($z_b = 550m$) up to $\bar{u}_t = 10ms^{-1}$ at the inversion top ($z_t = 650m$), so that the minimum Richardson number $Ri_i \approx 0.05$. From that, we aspect the formation of Kelvin-Helmholtz instabilities within this dynamically unstable inversion layer.

The second case is characterized initially by temperature and velocity profiles similar to those observed by Goodman and Miller [2] in the area of the U.S. west coast inversion (Figure 2). The inversion layer (centered at $z_i = 400m$) is found dynamically stable stratified ($Ri \approx 20$). In order to study dispersion processes due

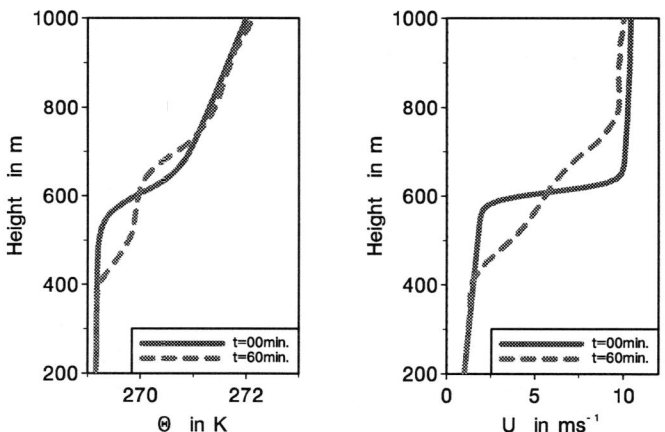

Figure 1. Initial ($t=0min$) and final ($t=60min$) profiles of horizontally averaged potential temperature (left) and horizontal wind velocity (right). Simulation: Dispersion due to Kelvin-Helmholtz instabilities

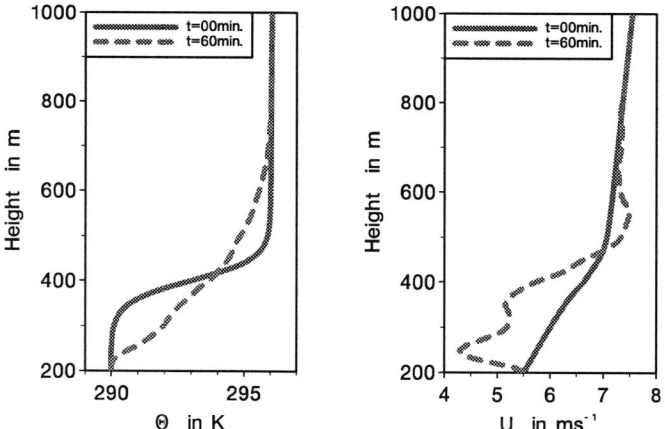

Figure 2. Same as in Figure 1, but showing the dynamically stable inversion layer of simulation experiment "breaking gravity waves"

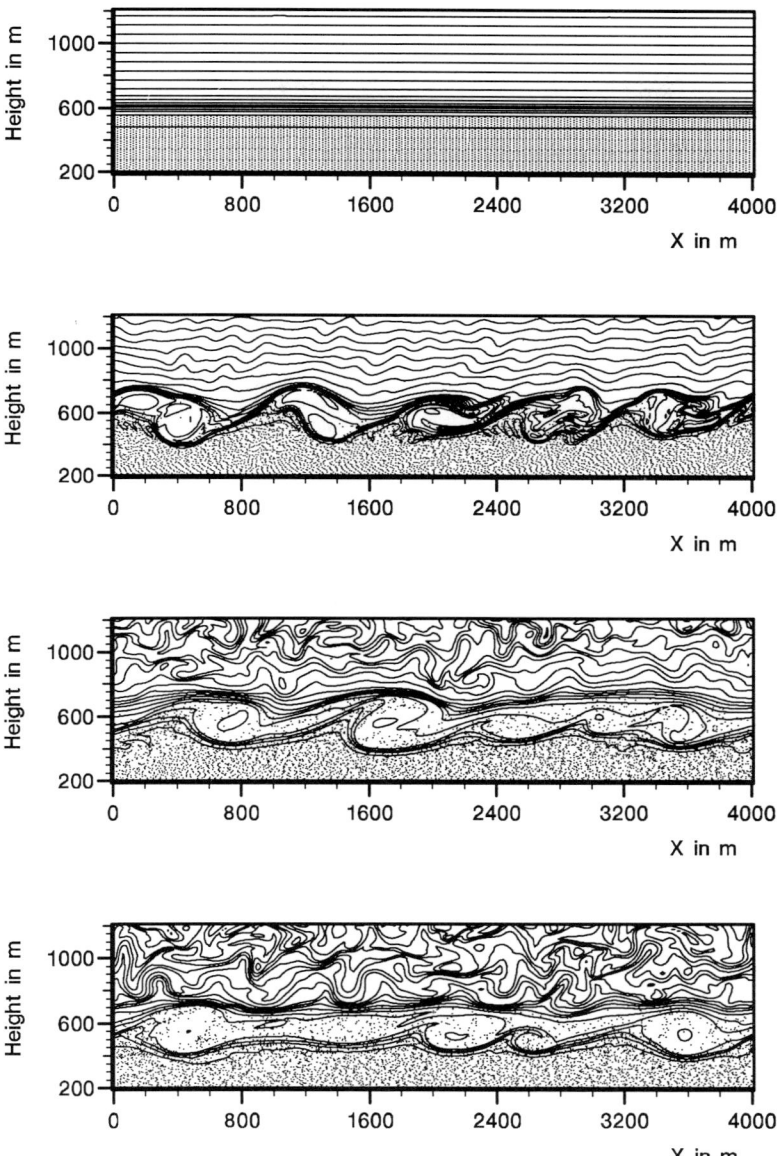

Figure 3. Isotherms of potential temperature Θ (contour interval $\Delta\Theta = 0.5K$) and particle distribution in vertical cross sections at simulated times $t = 0$, $10min$, $30min$ and $45min$ (from top to bottom) showing dispersion due to Kelvin-Helmholtz instabilities

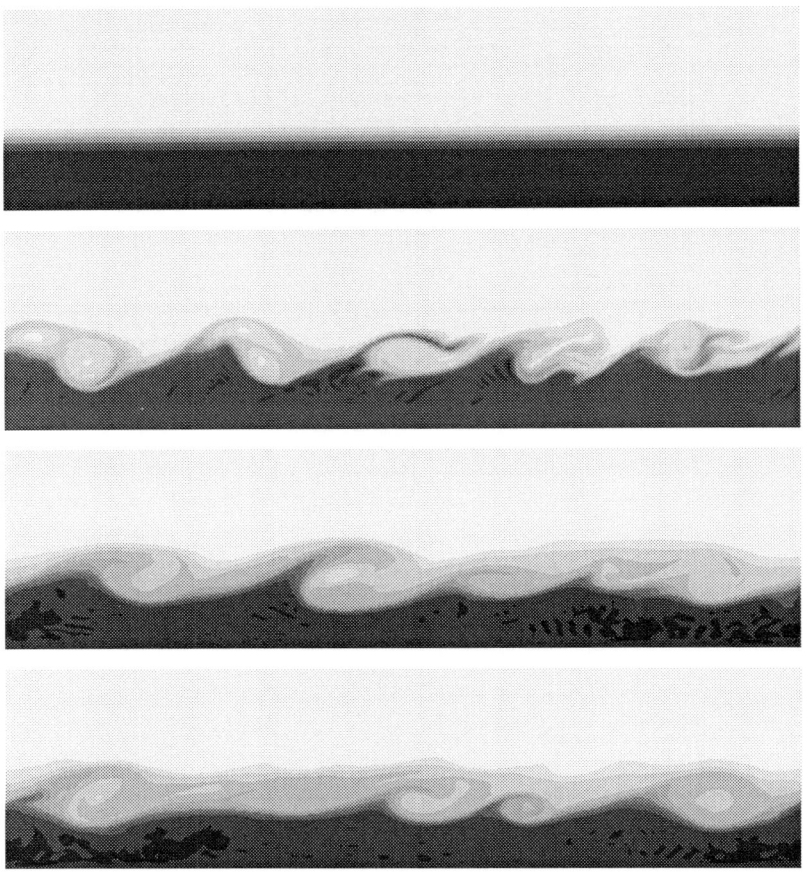

Figure 4. Same cross sections and times as in Figure 3, but representing normalized concentration of passive scalars (each grey shade represents a range of concentration $\Delta C = 0.1$)

to breaking gravity waves we initialized (following Fritts [14]) a set of four gravity waves of phase velocity equal to the horizontal wind velocity at the inversion's top ($c = u_t$) and of different wavelength ($100m$, $200m$, $400m$ and $1000m$) just above the inversion base.

Both numerical experiments are simulated within a model area, which extends $4km$ in horizontal and from $z = 200m$ to $z = 1500m$ in vertical direction. For the lateral boundaries we have chosen cyclic conditions, while at the lower and upper boundaries the background quantities are assumed to remain constant and all deviations from the mean state vanish.

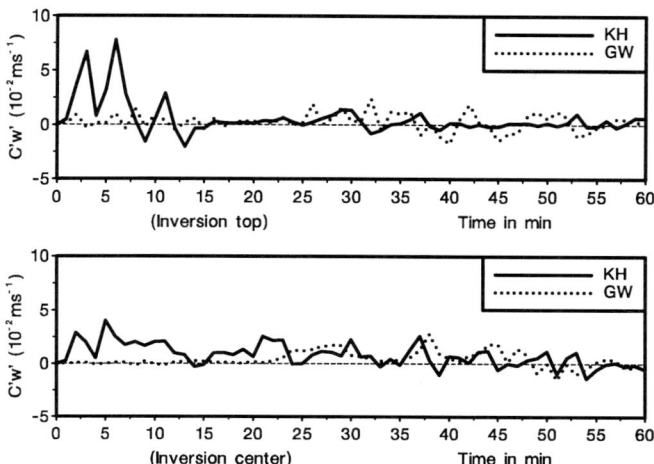

Figure 5. Horizontally averaged turbulent fluxes of passive scalars (C: normalized concentration) at the middle (bottom) and at the top of the inversion layer from simulations of "Kelvin-Helmholtz instability" (KH) and "breaking gravity waves" (GW)

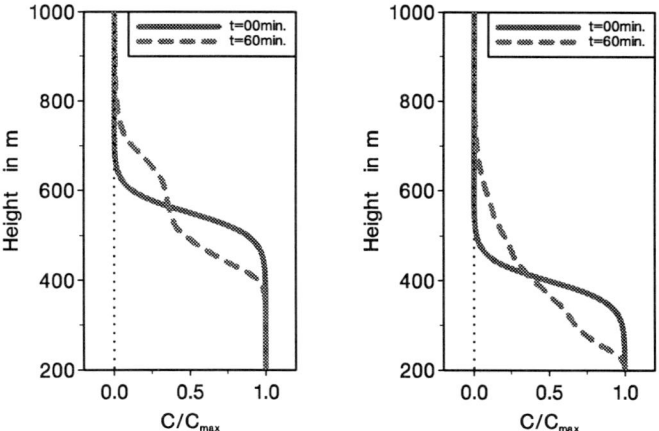

Figure 6. Initial ($t = 0min$) and final ($t = 60min$) profiles of horizontally averaged normalized concentration from simulation of occuring Kelvin-Helmholtz instabilities (left) and breaking gravity waves (right)

At the beginning of the simulations we set free a number of 10 000 weight-less particles below the respective inversion layer giving a maximum normalized concentration $C = 1$ (see, for example, Figures 3 and 6).

In addition to the mean state, a resolved background perturbation velocity is added to the flow field in order to take into account the effect of ambient turbulence [15]. This perturbation velocity ($|\mathbf{v}'| < 0.2ms^{-1}$), however, corresponds to a resolved turbulent energy spectrum limited by wavelenght $\lambda_E = 200m$ at the lower and $\lambda_E = 40m$ at the higher frequency end.

4   Dispersion within and across inversion layers

Significant transport of substances into and across inversion layers is "per se" possible due to enhanced turbulence, which can be produced by dynamical insta-bilities or by breaking gravity waves via convective instabilities. Both phenomena have been simulated numerically, and the results of two of these simulations may show the development of the dynamical processes as well as typical magnitudes of fluxes within and across inversion layers.

Partly, it is possible that inversion layers can become dynamically unstable, sometimes. This requires the existence of an inflexion point within the vertical profile of the horizontal wind velocity ($\partial^2 u/\partial z^2 = 0$, Figure 1) and the local Richardson number has to drop below a critical value of about 1/4. Conse-quently, Kelvin-Helmholtz instabilities occur in order to stabilize the flow by turbulent mixing. In Figure 3 combinations of isentropes and single particles in x-z cross sections show different stages of both the dynamical as well as the dispersion process.

At the beginning ($t = 0$) a more slightly stable stratified inversion layer ($\partial\Theta/\partial z = 0.012 Km^{-1}, N \approx 2 \cdot 10^{-2}s^{-1}$) may become dynamically unstable due to the increase of vertical windshear. Below the inversion a number of particles represent an unknown substance of normalized concentration $C = 1$. During a few minutes an instability grows building the wave-like structure of Kelvin-Helmholtz billows. As it can be expected, following previous studies (for example [6], [9], [16], [17]), the "wave"-length λ_{KH} is found about 7.5 times the inver-sion depth Δz_i and the maximum aspect ration (amplitude/λ_{KH}) becomes 1/5. The subsequent phase of pairing and breaking of the Kelvin-Helmholtz billows is characterized by an intensive mixing of momentum, heat and mass (see also Figure 4). As a result, the intensive windshear has been reduced so that the flow becomes dynamically stable. On the other hand, the same mixing mechanism reduces the stable stratification so that the temperature gradient is no longer positive throughout the "inversion" ($\partial\Theta/\partial z < 0.01 Km^{-1}$) and the layer itself has been broadened (Figure 1). It has to be noticed that, actually, this and the further development are three-dimensional processes, which can be simulated only in approximation by using a two-dimensional model.

Mixing and dispersion of the passive scalars is caused by the dynamical pro-cesses and the initiated background turbulence (Figure 4). Breaking Kelvin-

Helmholtz vortices mix the particles into and within the inversion layer (Figure 5), while fluxes across the inversion into the air above are due to the background turbulence, additionally. The effect of the latter process was analysed, but it can not be identified from Figures 3 or 4 because the simulated time of one hour is found too short for particle sampling above the "inversion" as analysed by, for example, Goodman and Miller [2]. The simulated mass fluxes were averaged horizontally at heights of inversion center and inversion top (Figure 5). The latter was identified as the height at which very stable stratification below changes to standard stratification above. During the first fifteen minutes, approximately, the Kelvin-Helmholtz instability causes a maximum flux of $\overline{C'w'} = 7.8cms^{-1}$ upwards through the inversion layer, while later on only weak fluxes both upwards and downwards through the respective levels can be observed (Figure 5). From the resulting vertical distribution of the horizontally averaged concentration (Figure 6) a representative vertical flux of $\widetilde{C'w'} = 2.1cms^{-1}$ can be calculated, which corresponds to the magnitude of mass flux velocity $w' \approx 0.15ms^{-1}$ found by Goodman and Miller ([2]) although the responsible processes differ. A further good agreement concerns to the heat fluxes. The simulated maximum of turbulent temperature fluxes $\overline{T'w'} = 0.23Kms^{-1}$ is found in accordance to the values of $\overline{T'w'} \leq 0.25Kms^{-1}$ measured by Readings et al. [3].

The second simulation experiment demonstrates the effect of breaking gravity waves on the dispersion of passive scalars. The situation is characterized by a combination of four gravity waves occuring within a very stable stratified inversion layer similar to the one observed at the U.S. west coast [2]. The phase velocity of the gravity waves is assumed to be equal the horizontal wind velocity at the inversion top $(c = u_t = 7ms^{-1})$ so that, at this height, a "critical level" causes breaking and reflexion of upward propagating gravity waves [14]. The different wave-lengths represent both "trapped" and "untrapped" gravity waves, which means that waves of frequencies larger than the Brunt-Vaisala frequency of the layers outside will be trapped within the inversion layer by reflexions at the respective boundaries (see, for example, [18]).

As a consequence of propagating wave-critical level interaction the simulated gravity waves become convectively unstable causing the production of enhanced turbulence (Figure 7). It has to be mentioned again that the latter process is of more three-dimensional nature so that our simulations have to regard as approximations with respect to the final processes of turbulence production. The increased turbulent activity within the inversion layer is quite strong enough to realize a positive upward flux of momentum, heat and mass into and, together with the background turbulence, across the inversion (Figure 5). Due to the strong stratification these fluxes are of lower intensity than those simulated in the case of Kelvin-Helmholtz instabilities $(\overline{T'w'} \leq 0.18Kms^{-1}, \overline{C'w'} \leq 2.6cms^{-1})$. Vertical profiles of the horizontally averaged quantities (Figures 2 and 6) indicate that the strong stratification is weakened and the thickness of the inversion layer increases up to about three times its initial value. In contrast, the windshear is

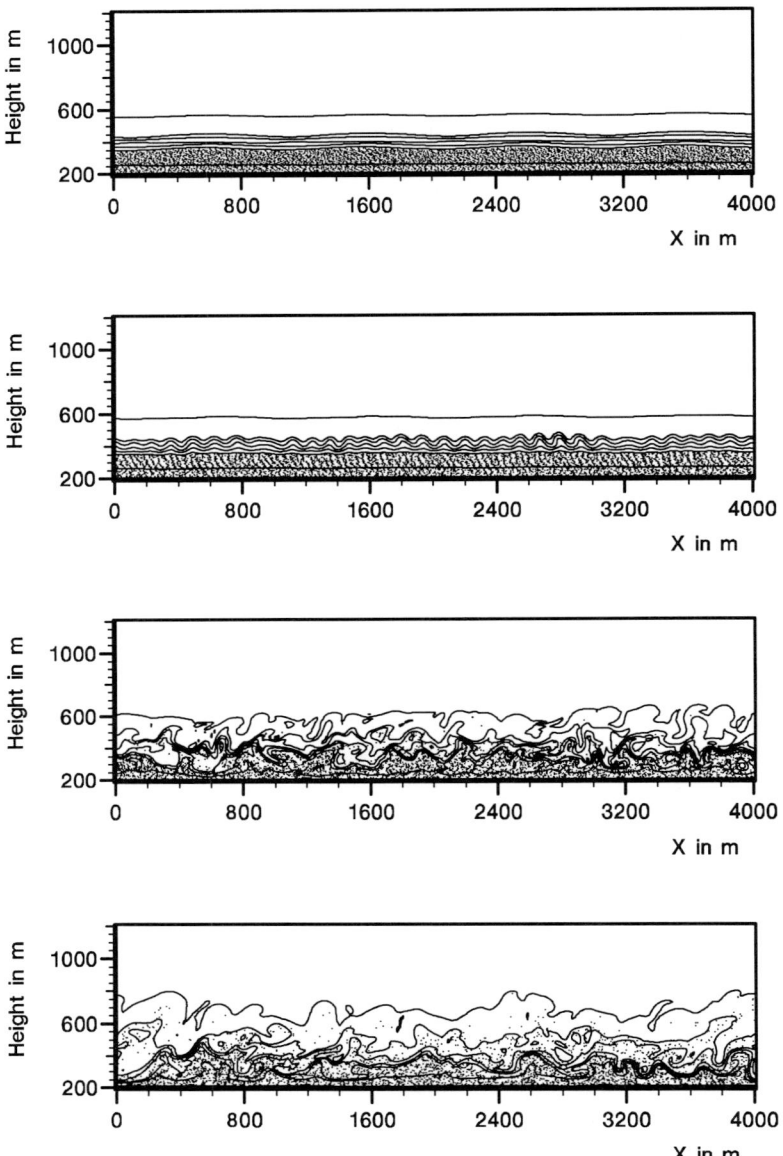

Figure 7. Isotherms of potential temperature (contour interval $\Delta\Theta = 0.5K$) and particle distribution in vertical cross sections at simulated times $t = 10min$, $20min$, $30min$, $45min$ (from top to bottom) showing dispersion due to breaking gravity waves

found strengthened within a layer below the critical level. This can be explained as a result of a momentum flux upwards caused by interaction of upward propagating gravity waves and their respective reflexions at the critical level yielding an acceleration of the flow [6], [7], [14].

The final ($t = 60min$) vertical distribution of passive scalars is found similar to the changed temperature distribution (Figures 2 and 6). During the simulated period an averaged effective flux of $\widetilde{C'w'} = 1.7cms^{-1}$ at the inversion's top causes a final concentration of $C \approx 0.1$ just above the inversion layer.

5 Conclusion

Results of two-dimensional numerical simulations have shown that dispersion of passive scalars within and across inversion layers is possible due to dynamical Kelvin-Helmholtz instabilities and breaking gravity waves. Fluxes of momentum, heat and mass into and across the inversion are caused by enhanced turbulence produced by the dynamical instabilities or by breaking gravity waves via convective instabilities.

Horizontally averaged mass fluxes of about $7.8cms^{-1}$ in maximum could be found resulting from Kelvin-Helmholtz instabilities while breaking gravity waves yield maximum fluxes of $2.6cms^{-1}$. During the simulated period ($t = 60min$) an effective flux of passive scalars of about $2cms^{-1}$ cause a final concentration of 10% of the initial concentration at a level just above the inversion layer in both cases, approximately.

Finally it has to be mentioned that, although simulation experiments indicate possible mass fluxes across inversion layers due to single phenomena, transport across real inversions during a longer period can be quantified only if it will be possible to quantify the temporal and spatial occurence of those phenomena within inversion layers.

Bibliography

1. Miller, A. (1968). Wind profiles in west coast temperature inversions. *Rep.* 4, 57 pp., Meteorol. Dep., San Jose State Coll., San Jose, Calif.

2. Goodman, J. K. and Miller A. (1977). Mass transport across a temperature inversion. *J. Geophys. Res.*, *82*, 3463-3471.

3. Readings, C. J., Golton, E. and Browning, K. A. (1973). Fine-scale structure and mixing within an inversion. *Boundary-Layer Meteorol.*, *4*, 275-287.

4. Danielsen, E. F. (1968). Stratospheric-tropospheric exchange based on radioactivity, ozone, and potential vorticity. *J. Atmos. Sci.*, *25*, 502-518.

5. Reiter, E. R. (1975). Stratospheric-tropospheric exchange processes. *Rev. Geophys. Space Phys.*, *13*, 459-474.

6. Schilling, V. (1993). Effective Diffusion von Luftbeimengungen in der Stratosphäre verursacht durch Kelvin-Helmholtz Instabilitäten und brechende Schwerewellen. *Rep. 44, Inst. f. Meteorol. u. Klimatologie Univ. Hannover*, D-30455 Hannover, Germany (in German).

7. Schilling, V. and Etling, D. (1996). Vertical mixing of passive scalars owing to breaking gravity waves. *Dyn. Atmos. Oceans*, *23*, 371-378.

8. Schilling, V., Siano, S. and Etling, D. (1996). Dispersion of aircraft emissions due to wake vortices in stratified shear flows: a two-dimensional numerical study. *J. Geophys. Res.*, *101* (D15), 20965-20974.

9. Schilling, V. and Jannsen, U. (1992). Particle dispersion due to dynamical instabilities in the lower stratosphere. *Contrib. Atmos. Phys.*, *65*, 259-273.

10. Purnell, D. K. (1976). Solution of the advective equation by up-stream interpolation with a cubic spline. *Mon. Weather Rev.*, *104*, 42-48.

11. Mahrer, Y. and Pielke, R. A. (1978). A test of the up-stream spline interpolation technique for the advective terms in a numerical model. *Mon. Weather Rev.*, *106*, 818-830.

12. Sweet, R. (1977). A cyclic reduction algorithm for solving block tridiagonal systems of arbitrary dimensions. *SIAM J. Numer. Anal.*, *14*, 706-720.

13. Legg, B. J. and Raupach, M. R. (1982). Markov-chain simulation of particle dispersion in inhomogeneous flows: The main drift velocity induced by a gradient in Eulerian velocity variance. *Boundary Layer Meteorol.*, *24*, 3-13.

14. Fritts, D. C. (1985). A numerical study of gravity wave saturation: Nonlinear and multiple wave effects. *J. Atmos. Sci.*, *42*, 2043-2058.

15. Fung, J. C. H., Hunt, J. C. R., Malik, N. A. and Perkins, R. J. (1992). Kinematik simulation of homogeneous turbulence by unsteady random Fourier modes. *J. Fluid Mech.*, *236*, 281-318.

16. Patnaik, P. C., Sherman, F. S. and Corcos, G. M. (1976). A numerical simulation of Kelvin-Helmholtz waves of finite amplitude. *J. Fluid Mech.*, *73*, 215-240.

17. Davis, P. A. and Peltier, W. R. (1979). Some characteristics of the Kelvin-Helmholtz and resonant overreflection modes of shear flow instability and of their interaction through vortex pairing. *J. Atmos. Sci.*, *36*, 2394-2412.

18. Metcalf, J. I. (1975). Gravity waves in a low-level inversion. *J. Atmos. Sci.*, *32*, 351-361.

Surface and Planetary Boundary-Layer Modelling of Stably Stratified Atmospheric Boundary Layers

Dapeng Xu[1], Wensong Weng, Lucilla Chan and Peter Taylor

Department of Earth and Atmospheric Science, York University, Ontario, Canada

Abstract

For surface layer (0-100m height) studies of stable stratified flow over topography, the conventional constant flux layer flows with log-linear profiles for velocity and potential temperature or density are unsuitable as they stand, since velocity linearly increases with height. If they are applied throughout too deep a layer, internal gravity waves excited by the flow are trapped by the Froude number increasing above 1 at some height. Artificial sources/sinks of heat and momentum can be introduced as a device to replace slowly evolving background boundary-layer profiles by steady ones for studies of short time scale perturbations.

With this artifact we can obtain realistic background profiles of velocity and density or buoyancy frequency for linear, mixed spectral finite-difference (MSFD) numerical model studies of stably stratified boundary-layer flows over topography over a range of Froude numbers with both trapped and vertically propagating waves. Some results will be presented and compared to inviscid and other solutions.

As one step in the development of models of stably stratified boundary-layer flow over hills and other topography a robust, high-order closure model of stratified boundary-layer flow over flat terrain is developed. Issues associated with the production of dissipation and model constants are addressed.

1 Introduction

One advantage that a surface-layer model has over a planetary boundary-layer (PBL) model is that it has fewer controlling parameters. This is of especial significance in parameterization of drag and other boundary-layer physical quantities over small scale topography. In a neutrally stratified boundary layer, no problems arise when a surface-layer model, applying a conventional constant stress (a logarithmic wind profile) concept, is used to calculate drag over subgrid scale topography [1]. In a stably stratified boundary layer, however, the conventional

---

[1]Present address: Certicom Corporation, 200 Matheson Boulevard West, Mississauga, Ontario, Canada, L5R 3L7.

constant flux layer flow with log-linear profiles for velocity and potential temperature or density, causes trouble because the linearly increasing velocity in the outer layer would lead to a Froude number ($Fr = Uk/N$, where U is horizontal wind component, k is a horizontal wavenumber and $N^2 = (g/\Theta_s)(\partial\Theta/\partial z)$ is the buoyancy frequency) greater than 1 and would thus trap internal gravity waves excited by topography with wavenumber k. A way to work around this is to introduce artificial sources or sinks of momentum and heat to modify the slowly evolving background boundary-layer profiles, in the belief that they have no significant impact on the short time scale perturbations that we are studying.

A PBL model is however more applicable than a surface-layer model for wider horizontal domains. This is of importance in the context of parameterizations of boundary-layer physics over subgrid scale topography in a global or large scale model, since the subgrid scale is, under this situation, well beyond 10^4m. A mile stone towards our goal of a robust, second order turbulence closure, PBL model for stably stratified flow over complex terrain is to develop a one-dimensional PBL model for flow over flat terrain. It turned out to be more challenging than we expected to develop such a model. Firstly, we came across a problem that is quite similar to what [2] has been described with the $E - \epsilon$ turbulence closure model. The cause of the problem was traced to the production term in the dissipation rate, ϵ, equation. We will discuss a way to re-parameterize the dissipation production term in Section 4.1. Secondly, we found, in the process of developing this one-dimensional PBL model, that constants now in use, in the parameterizations of pressure-strain and pressure-correlation terms in the shear stress and heat flux equations, are inconsistent with atmospheric observations and fail to satisfy some constraints provided by relatively simple flow situations. We will re-evaluate these constants via atmospheric observational data in Section 4.2.

2   Governing equations and turbulence closure schemes

In the Cartesian co-ordinate system, equations for the conservation of momentum, heat and mass, under the Bussinesq assumption, are

$$\frac{DU_i}{Dt} = f(U_k - U_{gk})\epsilon_{3ik} + \frac{g}{\Theta_s}(\Theta - \Theta_s)\delta_{3i} - \frac{1}{\rho}\frac{\partial P}{\partial x_i} - \frac{\partial \overline{u_i u_k}}{\partial x_k}, \qquad (2.1)$$

$$\frac{D\Theta}{Dt} = -\frac{\partial \overline{u_k\theta}}{\partial x_k} \qquad (2.2)$$

and

$$\frac{\partial U_k}{\partial x_k} = 0, \qquad (2.3)$$

where upper-case and lower-case letters represent mean and fluctuation quantities respectively; summation is implied whenever an index repeats in the same term; f is the Coriolis parameter, Θ_s is a reference temperature; $\overline{u_i u_k}$ and $\overline{u_k\theta}$ are shear stresses and heat fluxes that we will discuss in detail in the next two subsections. In surface-layer models, the first term on the right hand side of Equation (2.1) is neglected while in the PBL model that term is retained.

2.1 LRR turbulence closure

Here LRR refers to Launder, Reece and Rodi's [5] second order turbulence closure. Note that in their original paper, Launder et al. constructed the closure only for neutral stratification. The second order turbulence closure we list below combines the work of [6] and [7]. The equations for Reynolds stresses, heat fluxes and the dissipation rate of turbulent kinetic energy read

$$\frac{D\overline{u_i u_j}}{Dt} = P_{ij} - \frac{2}{3}\delta_{ij}\epsilon + \Phi_{ij} + G_{ij} + \frac{\partial}{\partial x_k}d_{ijk}, \tag{2.4}$$

$$\frac{D\overline{u_i \theta}}{Dt} = -\overline{u_i u_k}\frac{\partial\Theta}{\partial x_k} + P_{i\theta} + \Phi_{i\theta} + G_{i\theta} + \frac{\partial}{\partial x_k}d_{ik\theta}, \tag{2.5}$$

$$\overline{\theta^2} = -C_{\theta\theta}\frac{E}{\epsilon}\overline{u_k\theta}\frac{\partial\Theta}{\partial x_k} \tag{2.6}$$

and

$$\frac{D\epsilon}{Dt} = \frac{C_{\epsilon 1}}{2}(P_{kk} + G_{kk})\frac{\epsilon}{E} - C_{\epsilon 2}\frac{\epsilon^2}{E} + C_\epsilon\frac{\partial}{\partial x_k}\left(\frac{E}{\epsilon}\overline{u_k u_\ell}\frac{\partial\epsilon}{\partial x_\ell}\right), \tag{2.7}$$

where

$$\begin{aligned}\Phi_{ij} = & -C_1\frac{\epsilon}{E}(\overline{u_i u_j} - \frac{2}{3}\delta_{ij}E) - D_1(P_{ij} - \frac{1}{3}P_{kk}\delta_{ij}) \\ & -D_2 E(\frac{\partial U_i}{\partial x_j} + \frac{\partial U_j}{\partial x_i}) - D_3(D_{ij} - \frac{1}{3}P_{kk}\delta_{ij}) \\ & -C_3(G_{ij} - \frac{1}{3}G_{kk}\delta_{ij}),\end{aligned} \tag{2.8}$$

$$G_{ij} = -\frac{1}{\Theta_s}(g_j\overline{u_i\theta} + g_i\overline{u_j\theta}), \tag{2.9}$$

$$d_{ijk} = C_s\frac{E}{\epsilon}\left(\overline{u_i u_\ell}\frac{\partial\overline{u_j u_k}}{\partial x_\ell} + \overline{u_j u_\ell}\frac{\partial\overline{u_k u_i}}{\partial x_\ell} + \overline{u_k u_\ell}\frac{\partial\overline{u_i u_j}}{\partial x_\ell}\right), \tag{2.10}$$

$$P_{ij} = -\left(\overline{u_i u_k}\frac{\partial U_j}{\partial x_k} + \overline{u_j u_k}\frac{\partial U_i}{\partial x_k}\right), \tag{2.11}$$

$$D_{ij} = -\left(\overline{u_i u_k}\frac{\partial U_k}{\partial x_j} + \overline{u_j u_k}\frac{\partial U_k}{\partial x_i}\right), \tag{2.12}$$

$$\begin{aligned}\Phi_{i\theta} = & -C_{\theta 1}\frac{\epsilon}{E}\overline{u_i\theta} - C_{\theta 2}P_{i\theta} + C'_{\theta 2}\overline{u_k\theta}\frac{\partial U_k}{\partial x_i} \\ & +C''_{\theta 2}(\overline{u_i u_j} - \frac{2}{3}\delta_{ij}E)\frac{\partial\Theta}{\partial x_j} - C_{\theta 3}G_{i\theta},\end{aligned} \tag{2.13}$$

$$d_{ik\theta} = C_\theta \overline{u_k u_\ell} \frac{E}{\epsilon} \frac{\partial \overline{u_i \theta}}{\partial x_\ell}, \tag{2.14}$$

$$P_{i\theta} = -\overline{u_k \theta} \frac{\partial U_i}{\partial x_k} \tag{2.15}$$

and

$$G_{i\theta} = -\frac{g_i}{\Theta_s} \overline{\theta^2}. \tag{2.16}$$

In above formulae, $C_{\theta\theta}$, $C_{\epsilon 1}$, $C_{\epsilon 2}$, C_ϵ, C_1, C_2, $D_1 = (C_2 + 8)/11$, $D_2 = (30C_2 - 2)/55$, $D_3 = (8C_2 - 2)/11$, C_3, C_s, $C_{\theta 1}$, $C_{\theta 2}$, $C'_{\theta 2}$, $C''_{\theta 2}$, $C_{\theta 3}$ and C_θ are constants and will be discussed in Section 4.2. $g_i = (0, 0, -g)$ is the gravity vector.

2.2   E-ℓ turbulence closure

E-ℓ turbulence closure is a $1\frac{1}{2}$ order turbulence closure scheme, widely used in atmospheric modelling. The turbulent kinetic energy (TKE), E, is controlled by

$$\frac{DE}{Dt} = \frac{1}{2}(P_{kk} + G_{kk}) - \epsilon + \frac{\partial}{\partial x_k}\left(K \frac{\partial E}{\partial x_k}\right), \tag{2.17}$$

where the eddy viscosity K is defined as

$$K = (\alpha E)^{1/2} \ell_m, \tag{2.18}$$

ϵ is the dissipation rate of TKE and is represented by

$$\epsilon = \frac{(\alpha E)^{3/2}}{\ell_d}. \tag{2.19}$$

The α is the ratio of surface shear stress and surface TKE. In the surface-layer models we use $\alpha = 0.25$, largely for historical reasons, while in the PBL model we use $\alpha = 0.17$, which matches atmospheric observations. The mixing length, ℓ_m, and the dissipation length ℓ_d are prescribed as

$$\frac{1}{\ell_m} = \frac{1}{\kappa}\left(\frac{1}{z + z_0} + \frac{\beta}{L}\right) \tag{2.20}$$

and

$$\frac{1}{\ell_d} = \frac{1}{\kappa}\left(\frac{1}{z + z_0} + \frac{\beta - 1}{L}\right), \tag{2.21}$$

where κ (taken to be 0.4) is the Karman constant, z_0 is the surface roughness length, β (taken to be 5.0) is a constant and L is the Obukhov length.

The shear stresses and heat fluxes are calculated with

$$\overline{u_i u_j} = \frac{2}{3}E\delta_{ij} - K\left(\frac{\partial U_i}{\partial x_j} + \frac{\partial U_j}{\partial x_i}\right) \tag{2.22}$$

and

$$\overline{u_i \theta} = -\frac{K}{Pr} \frac{\partial \Theta}{\partial x_i}, \tag{2.23}$$

where Pr is Prandtl number. In the surface layer models we use $Pr = 1$ for simplicity and in the PBL model we use $Pr = 0.74$ following [4].

3 Surface layer models

3.1 Background profiles

$Fr = 1$ is a critical value determining whether internal gravity waves can propagate ($Fr < 1$) or not ($Fr > 1$). A constant flux layer with log-linear profiles for velocity and potential temperature, i.e.

$$U(z) = \frac{u_*}{\kappa} \left(\ln \frac{z + z_0}{z_0} + \frac{\beta z}{L} \right) \tag{3.1}$$

and

$$\Theta(z) - \Theta_s = \frac{\theta_*}{\kappa} \left(\ln \frac{z + z_0}{z_0} + \frac{\beta z}{L} \right), \tag{3.2}$$

where u_* is the friction velocity and θ_* is a temperature scale, defined by $u_* \theta_* = -\overline{w \theta_0}$, generates a constant temperature gradient, thus a constant buoyancy frequency and a increasing wind, U, with height for large z. This implies that somewhere along the height the Froude number would be greater than 1. Therefore, a constant flux layer actually serves as a rigid lid at some height of the boundary layer and prevents gravity waves from propagating upwards.

In reality there is perhaps no such thing as a steady state stably stratified boundary layer. What we usually see in nature is a slowly evolving (on an order of 10 hours) nocturnal boundary layer. We may ignore the varying of the boundary layer within a short time period, say one hour, and view the boundary layer as in a quasi-steady state. This allows us to modify the slowly evolving background boundary-layer profiles without worrying about their impacts on the short time scale perturbations that we are interested in.

After some trials, we have eventually adopted a velocity profile that reduces to a log-linear form close to the surface and matches a constant free stream velocity in the outer layer, it reads,

$$U(z) = \frac{1}{2} \left(g_1(z) + g_2(z) + U_\infty - \sqrt{(g_1(z) + g_2(z) - U_\infty)^2 + c_0^2} \right), \tag{3.3}$$

where $g_1(z)$ is the usual log-linear profile (3.1) and $g_2(z)$ is defined as

$$g_2(z) = u_* \left(c_1 + c_2 \ln \frac{z + z_0}{z_0} + c_3 \left(\frac{z + z_0}{z_0} \right)^\gamma \right). \tag{3.4}$$

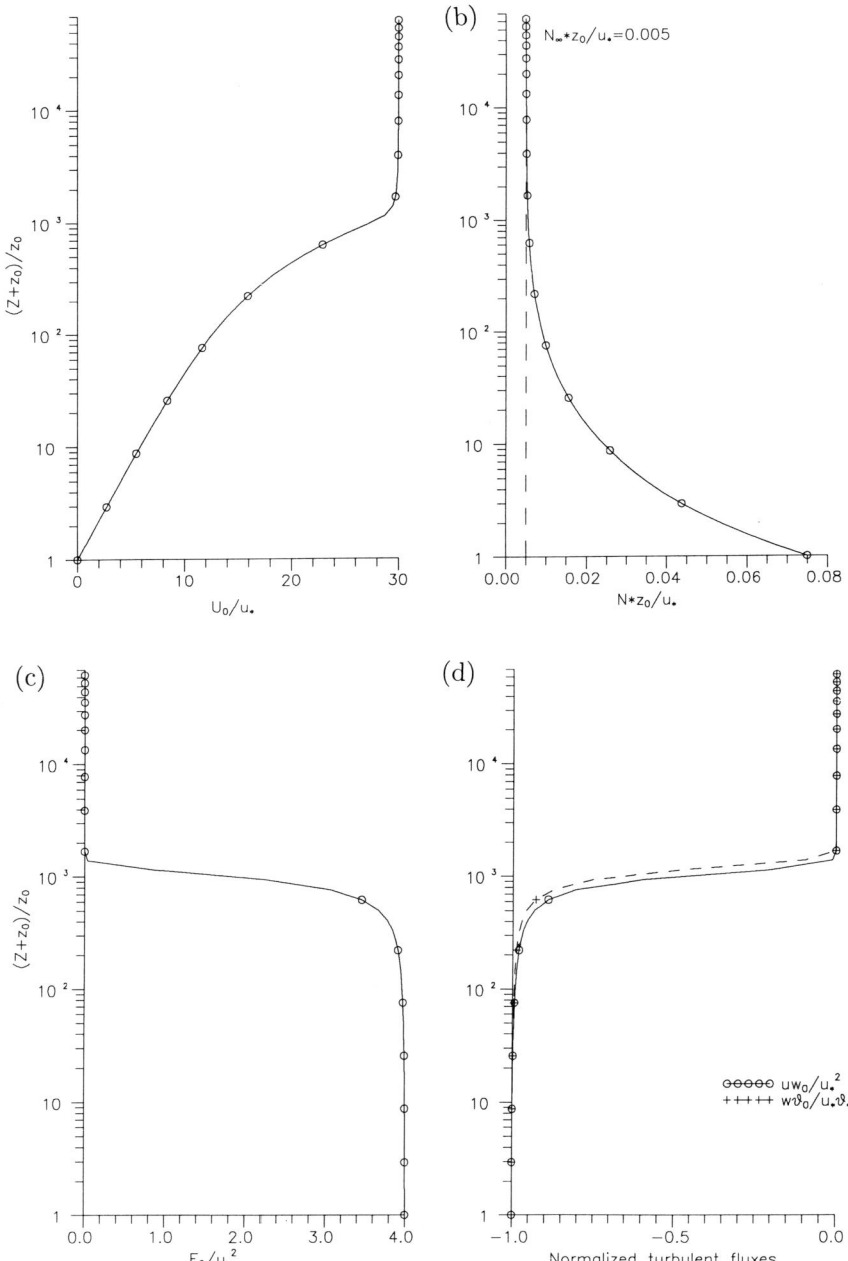

Figure 1. Background profiles for $Fr = 0.6$ with $z_0 = 0.1m$, $u_* = 0.2ms^{-1}$, $U_\infty = 6ms^{-1}$, $\theta_* = 0.0265°K$, $\Theta_s = 290°K$, $N_\infty = 0.01s^{-1}$ and $L = 112m$. (a) dimensionless velocity, (b) dimensionless buoyancy frequency, (c) dimensionless TKE, and (d) dimensionless shear stress and heat flux

The constants c_1, c_2 and c_3 are given as follows

$$c_3 = \frac{c_0^2}{4U_\infty^2 \kappa \gamma^2} \left(\frac{\beta z_0}{L} + \frac{2u_*}{\kappa} \left(1 + \frac{\beta z_0}{L} \right)^2 \right),$$ (3.5)

$$c_2 = \frac{c_0^2}{4U_\infty^2 \kappa} \left(1 + \frac{\beta z_0}{L} \right) - \gamma c_3$$ (3.6)

and

$$c_1 = \frac{c_0^2}{4U_\infty u_*} - c_3.$$ (3.7)

The shape of the profile can be slightly changed by adjusting the two empirical constants, c_0 and γ. In the following calculations we use $c_0 = 0.15U_\infty$ and $\gamma = 0.1$.

Once the profiles for velocity and temperature along with the controlling parameters, z_0, u_*, L, U_∞, Θ_s and θ_* are given, other profiles, such as TKE, can be calculated from the governing equations and turbulence closure scheme. In Figure 1 we show the background profiles, obtained via the method described herein, that we are going to use in the surface layer models in the following two subsections.

3.2 The MSFD-STAB model

The MSFD-STAB model is a stable stratification extension of the mixed spectral finite difference (MSFD) model for neutrally stratified turbulent flow over complex terrain [3]. The MSFD model works in a terrain-following co-ordinate and decomposes each variable into a known background part and an unknown perturbation. The model applies spectral methods in the horizontal directions and employs a finite difference scheme in the vertical. The discretized equations for the unknown perturbations are solved by using a block LU facterization algorithm.

To handle the possible upward internal gravity wave propagation, Klemp and Durran's [8] radiation condition is used at the upper boundary. In spectral space, the radiation condition gives,

$$\tilde{p} - \frac{N_\infty}{|k|} \sqrt{1 - \frac{U_\infty^2 k^2}{N_\infty^2}} \, \tilde{w} = 0,$$ (3.8)

where U_∞ and N_∞ represent free stream velocity and buoyancy frequency respectively, and ˜ implies a Fourier transformed variable.

In Figure 2 we show some comparisons of drag coefficient versus Froude number from our MSFD-STAB model and those from an inviscid flow model. Figure 2(a) demonstrates results for $Fr < 1$ while Figure 2(b) covers the results for $Fr > 1$. We note that turbulence has a tendency to reduce drag, by an order of 20% in the small Froude number regime. This implies that the parameterization of drag over subscale topography that in use in the global and regional scale

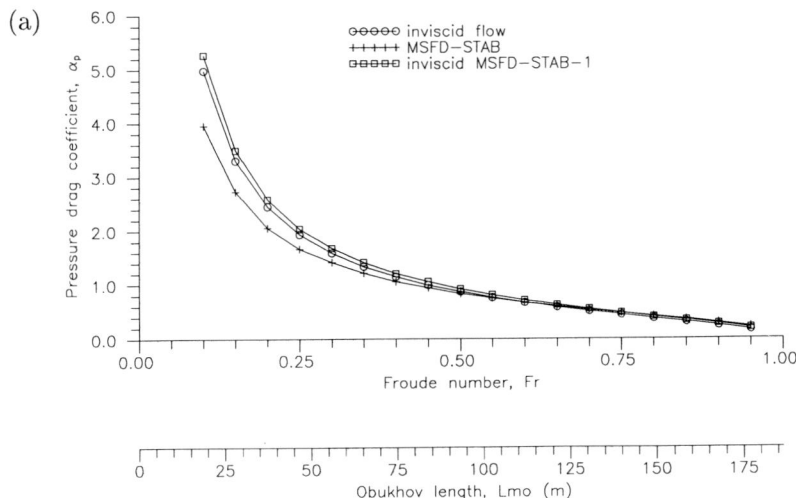

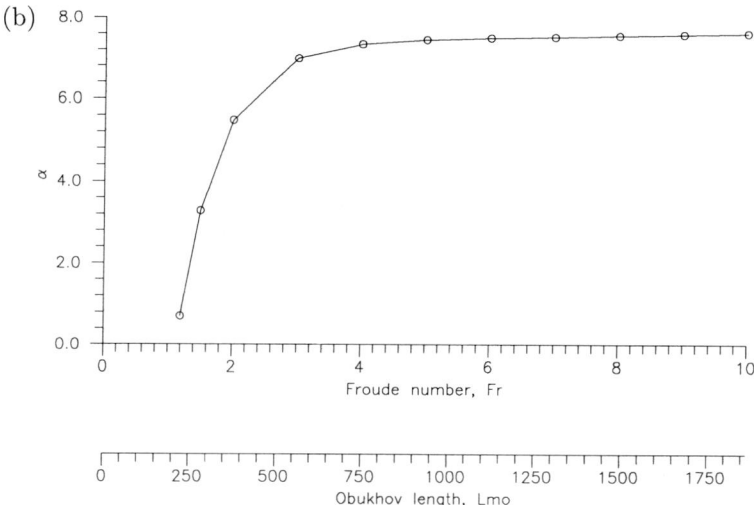

Figure 2. Pressure drag coefficient variation with Froude number for stable stratified flow over sinusoidal terrain with $2\pi a/\lambda = 0.15$. (a) $Fr < 1$, and (b) $Fr > 1$

models are reasonable. A more detailed discussion of these results is presented in [9].

3.3 A 2D, non-linear finite difference model

This model is a stable stratification extension of the finite difference model for neutrally stratified flow over hilly terrain, originally developed by Taylor [10]. The model uses a artificial compressibility algorithm and a staggered grid finite difference scheme. The model adopts the same background flow treatment and the same radiation upper boundary condition as we described earlier. Results have been obtained with log-linear profiles and $Fr > 1$ but implementation of the upper radiation boundary condition in conjunction with the method of artificial compressibility is proving difficult.

4 PBL model

4.1 Re-parameterization of the dissipation production

When applying the LRR model listed in Section 2.1 to a 1D, neutrally stratified atmospheric boundary layer, we find that the model generates a much too deep boundary layer. This is very similar to the result Detering and Etling [11] obtained from the standard $E - \epsilon$ model. The cause of this can be easily traced, by replacing the dissipation equation with a specified turbulence length scale, to the dissipation equation. We consider that the dissipation production term in the equation, $C_{\epsilon 1}/2(P_{kk} + G_{kk})\epsilon/E$, should be re-parameterized as (see [12])

$$\frac{C_{\epsilon 1}\alpha^3 E^2}{\ell_\epsilon^2}, \tag{4.1}$$

where

$$\frac{1}{\ell_\epsilon} = \frac{1}{\kappa(z + z_0)} + \frac{\beta}{\kappa L} + \frac{f}{0.00027|U_g|}, \tag{4.2}$$

based on the concept that, in a spectral cascade model, dissipation production is dependent on input to the cascade from the energy containing part of the spectrum, with a characteristic length scale ℓ_ϵ.

4.2 Constants in the LRR model

To limit the length of this paper, we omit the details how the constants are determined but instead briefly describe the procedures involved.

Applying the LRR model to a simple flow situation such as a constant flux layer, the model would reduce to a very simple form. This provides self consistency constraints. Once some of the constants are determined by experimental or observational data, the remainder should derived from these constraints.

Turbulence models are generally developed in the mechanical engineering community and the constants within the model are evaluated with laboratory

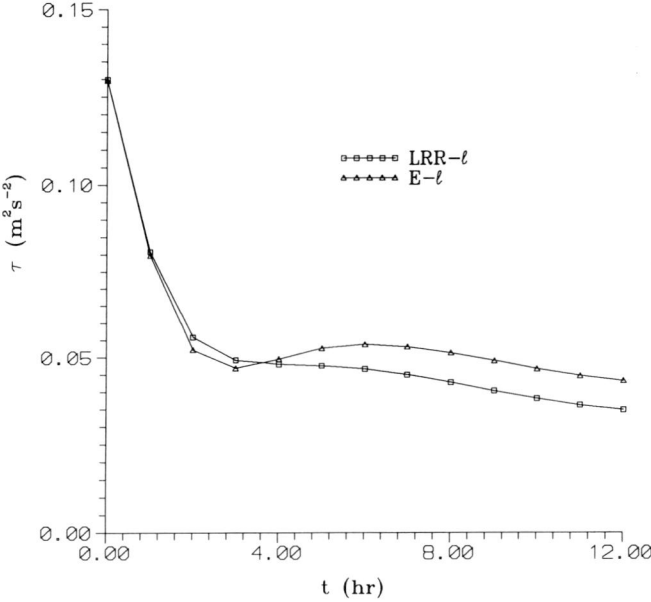

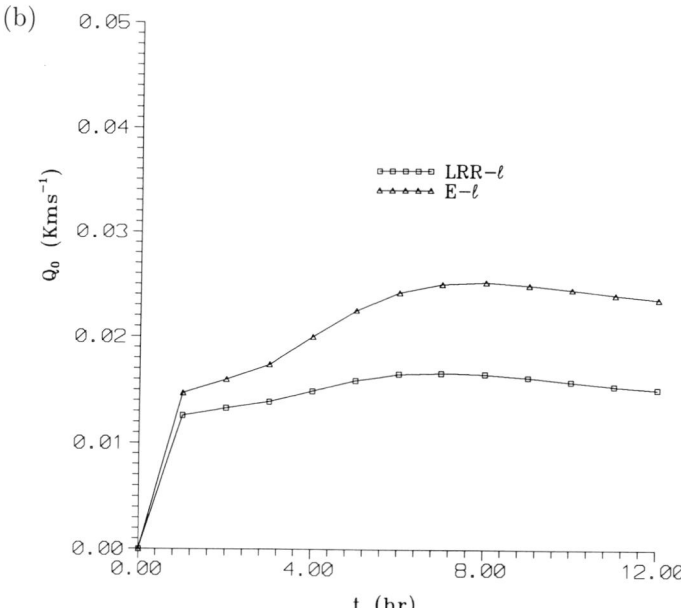

Figure 3. Evolution of surface shear stress and heat flux with a cooling rate of $1°C$/hour for $Ro = 10^6$. (a) surface shear stress, and (b) surface heat flux

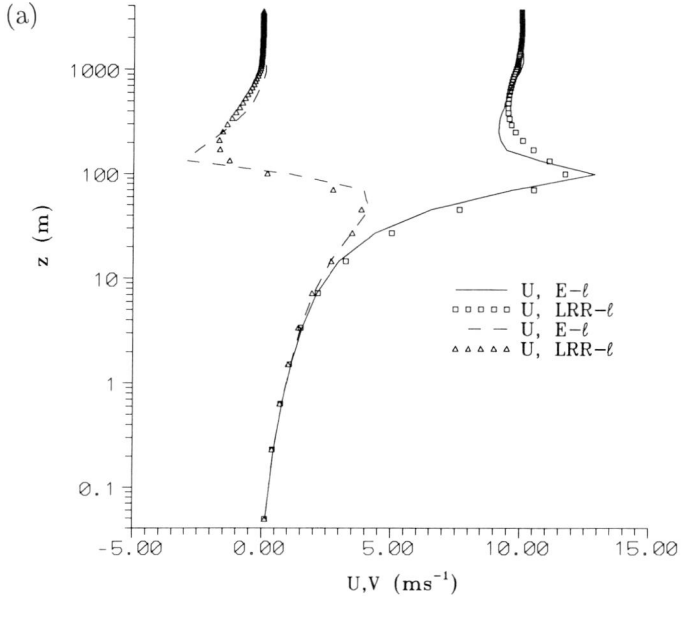

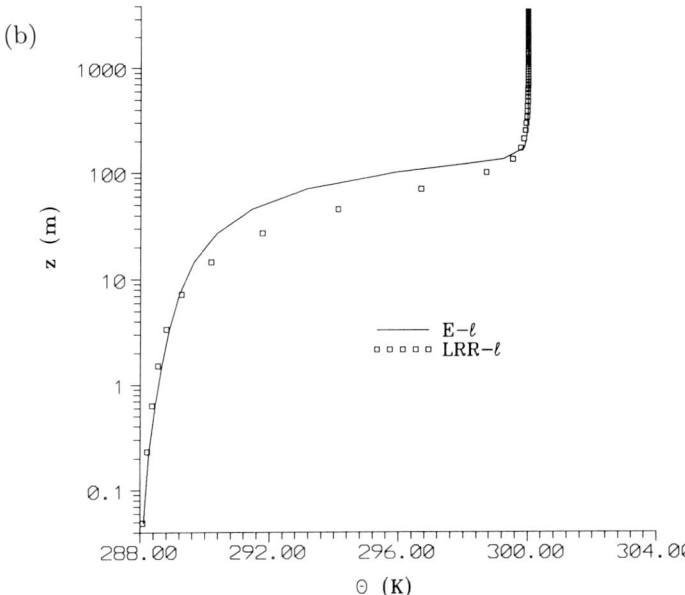

Figure 4. Wind and potential temperature profiles in a stable boundary layer with $Ro = 10^6$, after 12 hours cooling at the surface at a rate of $1°C/$hour. (a) Wind components, and (b) potential temperature

(a)

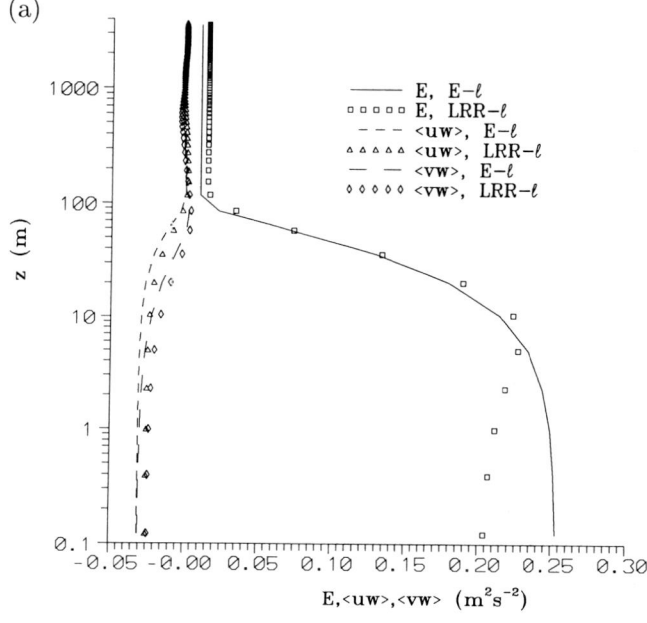

(b)

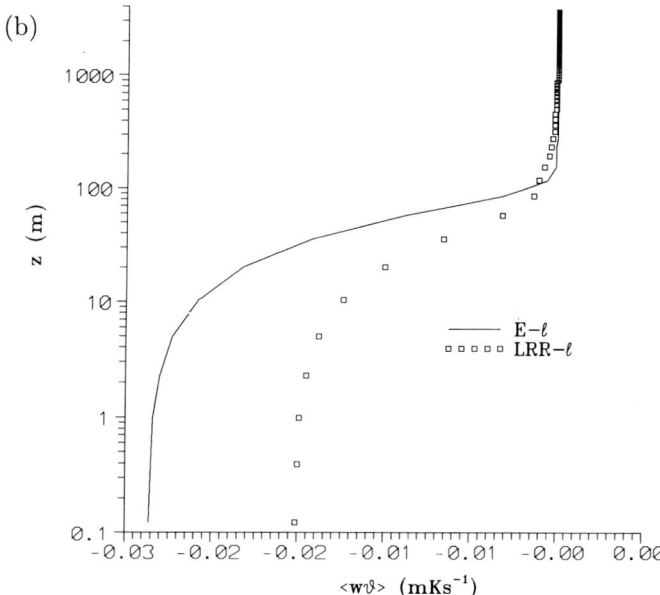

Figure 5. Turbulence quantities in a stable boundary layer with $Ro = 10^6$, after 12 hours cooling at the surface at a rate of $1°C$/hour. (a) TKE and shear stresses, and (b) heat flux

Table 1. Constants used in atmospheric and engineering modelling

	C_s	C_1	C_2	C_3	C_ϵ	$C_{\epsilon 1}$	$C_{\epsilon 2}$
Atmospheric Modelling	0.7	2.0	0.9	0.9	0.05	1.44	1.92
Engineering Modelling	0.22	1.8	0.6	0.6	0.15	1.44	1.92
	C_θ	$C_{\theta 1}$	$C_{\theta 2}$	$C_{\theta 3}$	$C'_{\theta 2}$	$C''_{\theta 2}$	$C_{\theta\theta}$
Atmospheric Modelling	0.13	2.9	0.4	0.4	0.15	-0.19	0.96
Engineering Modelling	0.15	2.9	0.4	0.4	0	0	1.6

experiment. Unfortunately, there are significant differences in turbulence statistics between laboratory experiment and atmospheric observation [13]. It seems we have to re-evaluate the constants with atmospheric data before we can actually apply these turbulence models to atmospheric applications. In Table 1 we list the constants obtained from atmospheric data along with those currently in use in the mechanical engineering community.

4.3 Sample results

We simulate a 1D stable boundary layer with a Rossby number of 10^6. We start with a neutrally stratified boundary layer, then cool the surface at the cooling rate of $1°C$/hour and integrate the model for 12 hours. To make comparisons we also make calculations with the $E - \ell$ model for the same flow situation. The prescribed turbulence length used in the $E - \ell$ model is the same as that in Equation 4.2, i.e., $\ell = \ell_\epsilon$.

In Figure 3, we plot surface shear stress and surface heat flux evolution during the 12 hour period. We see that after one hour or two of rapid adjustment, the surface stress and flux then evolve slowly. The flow seems to be in a quasi-steady state if we look at it on a time scale of order of an hour. Velocity and potential temperature profiles after 12 hour of integration are plotted in Figure 4. Overall agreement between different closure models is fairly good though there are differences in detail. Turbulence quantities are presented in Figure 5. Other than the differences between different closure models in surface shear stress and surface heat flux, which are also reflected in Figure 3, the results from the two models agree reasonably well.

5   Summary

For surface-layer models, the traditional constant flux layer with log-linear wind and temperature profiles behaves, at the height where $Fr > 1$, like a rigid lid that keep internal gravity waves from escaping. Artificial sources or sinks can be added to the slowly evolving background profiles to work around the trapping gravity wave problem without invoking disturbances to the perturbations we are interested in.

Second order turbulence closure schemes do not seem to be at a "ready-to-use" stage for atmospheric applications. We believe that one should re-parameterize the production term in the dissipation equation and customize the constants in the model for atmospheric applications.

Bibliography

1. Xu, D. and Taylor, P. A. (1995) Boundary-layer parameterization of drag over small-scale topography. *Quart. J. R. Meteorol. Soc.*, **121**, 433–443.

2. Detering, H. W. and Etling, D. (1987) Application of the $E - \epsilon$ turbulence model to the atmospheric boundary layer. *Boundary-Layer Meteorol.*, **33**, 113–133.

3. Beljaars, A. C. M., Walmsley, J. L. and Taylor, P. A. (1987) A mixed spectral finite difference model for neutrally stratified boundary-layer flow over roughness changes and topography. *Boundary-Layer Meteorol.*, **38**, 273–303.

4. Wyngaard, J. C. (1975) Modelling the planetary boundary layer—extension to the stable case. *Boundary-Layer Meteorol.*, **9**, 441–460.

5. Launder, B. E., Reece, G. J. and Rodi, W. (1975) Progress in the development of a Reynolds stress turbulence closure. *J. Fluid. Mech.*, **68**, 537–566.

6. Gibson, M. M. and Launder, B. E. (1978) Ground effects on pressure fluctuations in the atmospheric boundary layer. *J. Fluid. Mech.*, **86**, 491–511.

7. Jones, S. and Musonge, P. (1983) Modelling of scalar transport in homogenous turbulence flows. *Proc. 4th Symp. on Turbulent Shear Flows*, 17.18–17.24, Karlsruhe.

8. Klemp, J. B. and Durran, D. R. (1983) An upper boundary condition permitting internal gravity wave radiation in numerical mesoscale models. *Monthly Weather Review*, **111**, 430–444.

9. Weng, W., Chan, L., Taylor, P. and Xu, D. (1996) Modelling stably stratified boundary-layer flow over low hills. *Submitted to Quart. J. R. Meteorol. Soc.*

10. Taylor, P. A. (1977) Some numerical studies of surface boundary-layer flow above gentle topography. *Boundary-Layer Meteorol.*, **11**, 439–465.

11. Detering, H. W. and Etling, D. (1987) Application of the $E - \epsilon$ turbulence model to the atmospheric boundary layer. *Boundary-Layer Meteorol.*, **33**, 113–133.

12. Xu, D. and Taylor, P. A. (1997) An $E - \epsilon - \ell$ turbulence closure scheme for PBL models: the neutrally stratified case. *Boundary-Layer Meteorol.*, **84**, 247–266.

13. Xu, D. and Taylor, P. A. (1997) On turbulence closure constants for atmospheric boundary-layer modelling: neutral stratification. *Submitted to Boundary-Layer Meteorol.*, **84**, 267–287.

Beamsteering Analysis of Internal Gravity Waves in the Stable Atmospheric Boundary Layer

J.C.W. Denholm-Price*, J.M. Rees*[1], P.S. Anderson** and J.C. King**

*School of Mathematics and Statistics, University of Sheffield, and
**British Antarctic Survey, Cambridge

Abstract

In this paper we present some preliminary findings from the preparation of a gravity wave climatology at Halley IV station, Antarctica, using data gathered during the second phase of the British Antarctic Survey's Stable Antarctic Boundary Layer Experiment in 1991 (STABLE II). The method of "beamsteering" is used to evaluate wave parameters from data gathered using an array of six microbarographs. Results are presented from artificial data designed to test the beamsteering code and also to compare it with the "correlation" method used in a previous study. Two case studies which illustrate the kind of wave events observed during the STABLE II programme are described.

1 Introduction

Studies of the stable atmospheric boundary layer (SABL) have shown that the transport of momentum and generation of turbulence due to gravity waves can be important [2, 4]. The SABL is less well understood than the convective boundary layer due to its relative complexity, shallowness and its unsteady nature (typified by the ephemeral nature of internal gravity waves). By developing a climatology of gravity waves at Halley IV station, Antarctica (75.6°S, 26.7°W), we aim to increase the understanding of the importance of these waves and aid the development of models and parametrizations of the SABL. Previous studies (for example [5]) have used data gathered from mid-latitude sites where the SABL is destroyed and recreated in the diurnal cycle. The data used in this study were gathered by the British Antarctic Survey during the second intensive phase of the Stable Antarctic Boundary Layer Experiment (STABLE II) at Halley, which is

---

[1] Author for correspondence.

situated on the Brunt Ice Shelf (a map of which can be found in [13], page 62, or [11]). An array of six microbarographs was installed to detect waves alongside a $32m$ instrumented mast which is used to study the SABL.

A small part of the STABLE II data has been analysed previously using an old version of the beamsteering code [13]. The STABLE II dataset is further analyzed herein using a revised beamsteering program. The beamsteering method uses an array of instruments distributed spatially, represented in Figure 1 (North is towards the top), to analyze the direction and speed of wave-like features traversing the array. Essentially the method convolves the data from the instruments with spatial information from the array (the "array transfer function") to yield an estimate for the discrete frequency-wavenumber spectrum [3]. The shape of the array is chosen so that the array transfer function approximates a delta function [7], with a maximum at the origin in wavenumber space. When a wave crosses the array the maximum in the wavenumber plane moves. The magnitude and direction of this shift indicates the direction and speed of the wave. The beamsteering algorithm searches for the displaced maximum and uses its position to derive estimates for phase speed and propagation direction for a range of frequency bins (see [1, 13] for further details of the beamsteering method. It was also used by [5]).

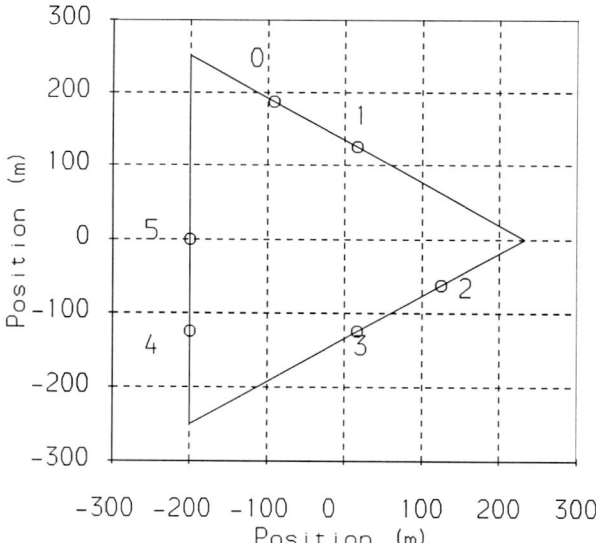

Figure 1. Schematic of the layout of the microbarograph array

Before analyzing the STABLE II data the beamsteering program was tested in detail using synthetic data. Results from this analysis are presented in Section 2 where the performance of the beamsteering program is compared with a different method based on correlations (as used by [11]). The method of beamsteering is preferred over the simpler correlation method as it is more robust. Detailed testing of large programs like the beamsteering package is essential before they can be used reliably. A number of problems were identified and corrected during the testing phase of the work before analysing the STABLE II dataset. Results from the analysis of a small section of the dataset are presented in Section 3.

In Section 3.1 a large amplitude wave event from the STABLE II dataset is analyzed. This event, observed in October 1991, provides further verification of the beamsteering program because it is possible to see "by eye" its propagation direction. It also provides the impetus for future work analyzing the behaviour of groups of atmospheric waves.

The analysis of 5 days of data from the 1991 STABLE II observational programme is presented in Section 3.2. Using this limited section of data we are able to demonstrate some of the variety of gravity waves that can be observed at Halley. This also aids the development of tools for the final climatology using the entire 227 day STABLE II dataset. In keeping with the climatological theme, only broad conclusions are drawn from these data. Discussion of the results are presented in Section 4.

2   Testing the beamsteering program

In order to test the beamsteering program and gauge its ability to extract information from "noisy" data containing waves, synthetic data were generated and processed using the beamsteering program. The same synthetic data were also processed using the correlation method, as used by [11], for comparison. This method uses the correlation between pairs of instruments in the microbarograph array to calculate the time-lag between instruments as the wave passes over the array. By using the two independent lags between three instruments it is possible to estimate the wave speed and direction. Therefore in the current array of six microbarographs we used the instruments in two sub arrays (grouping microbarographs 1,3,5 and 0,2,4) to evaluate the wave direction and speed twice. This redundancy could be used as a measure of self checking in the analysis, however in the discussion that follows the estimates from the two sub arrays are not differentiated.

2.1   The beamsteering program and array sensitivity

The sensitivity of the array can be estimated by considering various physical characteristics of the microbarograph array and beamsteering programs. A wave must not travel so fast that it traverses the array in less than twice the sampling rate of the microbarographs. Since the array size is $\sim 500m$ and the microbarographs were sampled at a fixed rate of $10s$ then the wave transit time must be less than $\sim 20s$. Thus phase speeds of up to $50ms^{-1}$ may be observable. This exceeds the maximum phase speed observed by [11] and is comparable with that found by [6].

The array size also affects the wavelength of waves that may be observed. For example, if the wavelength is less than the spacing between "neighbouring" instruments then the maxima may be misinterpreted and the wave aliased to a longer wavelength. Since the beamsteering algorithm considers the array as a whole, the minimum wavelength must be in the range $125 - 500m$. The minimum wavelength found by [6] was $\sim 5km$ which suggests that the current array is of an appropriate size.

A low pass filter is used to detrend the microbarograph data, with a cutoff frequency of $f_{cut} = 0.0006Hz$, which corresponds to a period of $\sim 30min$. This limits the range of frequencies to which the beamsteering array will be sensitive. The maximum frequency considered by the beamsteering is $f_{max} = 0.02Hz$ (periods of $\sim 50s$). The range $f_{cut}-f_{max}$ is sufficient to cover the expected range of frequencies of gravity waves [5]. The beamsteering program then produces estimates of wave parameters for a variable number of frequency bins in the range $f = f_{cut}-f_{max}$. In the current analysis 32 linearly spaced bins were used, derived from 512-point blocks of data. This corresponds to 42 minutes of data from the microbarographs which were sampled at $0.1Hz$.

The beamsteering algorithm itself imposes a restriction on the range of phase speeds that can be estimated. In order to find efficiently the wave parameter estimates for a particular frequency bin the beamsteering program searches for peaks over a limited range of wavenumbers, from 0 to a maximum wavenumber $k_{max} = 0.04m^{-1}$. Thus at a given frequency, f, the *minimum* phase speed c that the beamsteering algorithm is capable of detecting is $c=c_{min} = \frac{2\pi f}{k_{max}}$. This corresponds to phase speeds of $\sim 0.09 - 3ms^{-1}$ and this is again sufficient to encompass all of the waves observed by [6, 11].

2.2 Synthetic data

Synthetic data were generated by first picking a random phase direction (from 0–360°), frequency ($f = 0.0006$–$0.02Hz$) and phase speed (in the range $c = 0$–$60ms^{-1}$ subject to $c > c_{min}$). To a pure sinusoidal wave generated with these characteristics, lagged appropriately for the six microbarographs, was added a random amount of noise. The synthetic wave was (arbitrarily) assigned an amplitude of $20\mu bars$ and the random noise varied between 0 and 200% (0–$40\mu bars$). Typically waves observed in the STABLE II dataset have root mean square amplitudes between 4 and $40\mu bars$. Due to the design of the microbarographs turbulent fluctuations

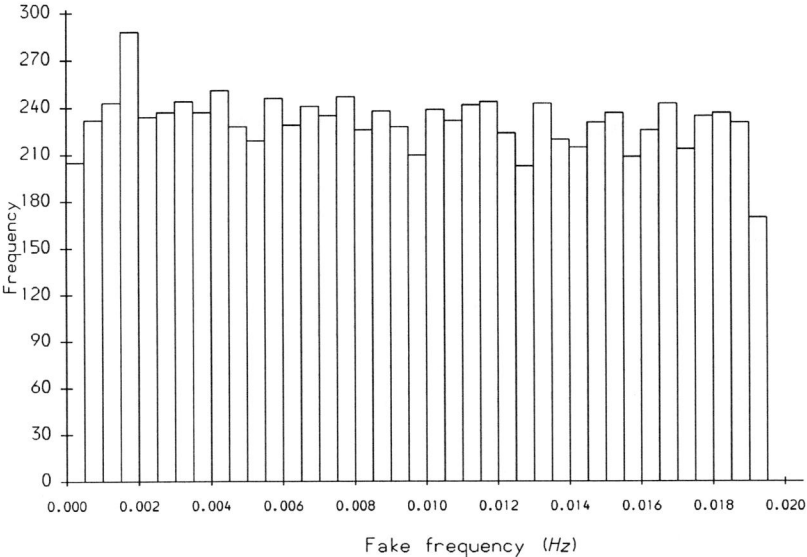

Figure 2. Histogram illustrating the distribution of waves

in the boundary layer are filtered out of the microbarograph data [1] so that the "random" signal is in all cases much less than 100% of the wave amplitude.

A total of 9100 random waves were generated in this way to obtain good coverage of the desired parameter space. This is illustrated in Figure 2 where the number of synthetic waves analysed per beamsteering frequency bin is plotted. Histograms of frequency versus phase speed and random amplitude are broadly similar. If the beamsteering program processes successfully such a wide range of synthetic data we have confidence that it will be effective when dealing with "real" data such as from the STABLE II dataset.

2.3   Results from testing with synthetic data

The approximate percentage frequency of "correct" estimates from the beamsteering and correlation methods in Figure 3 are plotted against the synthetic wave frequency. The curves show the percentage of estimates with wave direction predicted to within 30° of the value imposed on the synthetic data. The solid lines represent synthetic waves with root mean square noise amplitude in the range 0–100% of the basic wave amplitude ($20\mu bars$) and the dashed lines 100–200%. Ideally the frequency of oc-

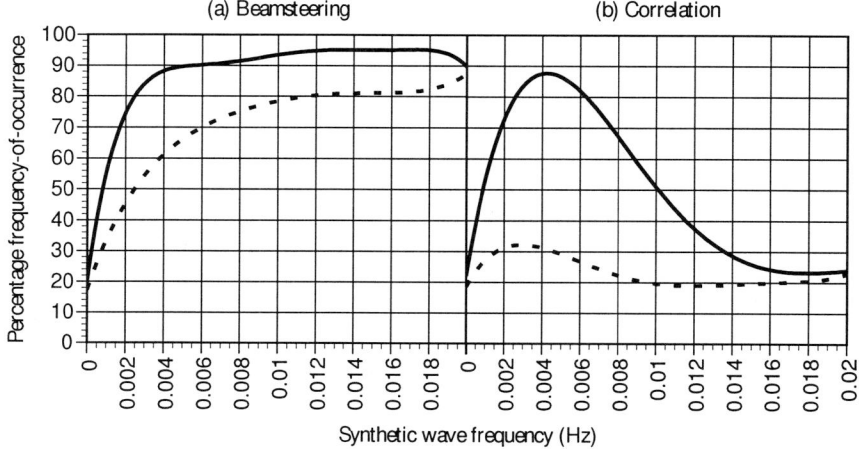

Figure 3. Beamsteering and correlation directional performance

currence should be around 100% for all wave frequencies[2]. We can see immediately that the beamsteering performs well over a wide range of frequencies and generally better than the correlation method. The drop in sensitivity as $f \to 0Hz$ can be explained in both cases by the low pass trend removal that is applied to the synthetic data and the increased wavelength for such waves.

We can see that the beamsteering method produces estimates of wave direction within 30° for almost 100% of the synthetic data with random noise between 0 and $20\mu bars$, over most frequencies. This contrasts strongly with the correlation method which peaks near 100% for $f \sim 0.004Hz$ only for low noise amplitudes, with reduced sensitivity elsewhere.

The beamsteering program shows a similar frequency response for synthetic phase speed and wavelength and so we conclude that it performs well, better than the correlation method which it is intended to replace. Therefore in the next section we present results from beamsteering analysis of a small part of the STABLE II data.

3   Analysis of STABLE II data

In the following two subsections we present a brief analysis of some of the STABLE II dataset using the beamsteering program described above. The testing performed in Section 2 using a single monochromatic wave simplifies

---

[2]Small variations around 100% are to be expected since the number of synthetic waves generated with a given noise amplitude within a given frequency band is not constant.

the beamsteering analysis because there is only ever *one* correct answer. When considering real data we do not know *a priori* how many waves there may be, which obviously complicates matters. The beamsteering is incapable of distinguishing between two monochromatic waves at the same frequency – the beamsteering algorithm "sees" the two waves as one with a different direction entirely, depending on their relative amplitude. However it is considered extremely unlikely that such a situation could arise.

Difficulties may occur when there is more than one wave at different frequencies. The beamsteering program was designed to identify the most likely correct estimates. Due to the leakage of energy from side lobes in the Fourier transform on which the beamsteering is based, a single peak in spectral power due to a monochromatic wave from a given direction will produce estimates with similar directions in neighbouring frequency bins. When there are two waves at frequencies separated by a number of frequency bins, multiple estimates for different frequencies will be produced with directions determined by the two waves. The beamsteering program therefore accepts wave estimates that have neighbouring estimates with similar direction. Further tests based on the spectral power, via a Fisher test, and coherence are also performed for increased confidence. This procedure was shown to work well with single, monochromatic waves in Section 2 and was also tested successfully using synthetic data with more than one wave source at different frequencies.

3.1 A large amplitude event

On 12 October 1991 a large amplitude wave event was recorded in the STABLE II dataset. This wave event is unusual in being of a large amplitude ($\sim 160\mu bars$) and exhibits high coherence across the microbarograph array. It serves as a further test of the beamsteering program since it is easy to identify visually the direction of travel from the microbarograph signal. In Figure 4a five hours of data from one microbarograph is displayed and the large amplitude event is clearly visible between hours 15.5 and 16.5. In Figure 4b time series data from microbarographs 1, 3 and 5 are expanded from hours 15.54 to 15.65. We can see clearly that the wave reaches microbarograph 3 first and microbarographs 1 and 5 almost simultaneously. From Figure 1 this implies that the wave is travelling from $\sim 90°$.

Spectral analysis of this wave packet reveals that there are four peaks in the power spectrum, as shown in Figure 4c. The beamsteering program produces four wave parameter estimates corresponding to these peaks, which are shown in Table 1. In the first column of the table, wave estimates 1–4 correspond to the beamsteering estimates in Figure 4c. The result derived using the correlation method is shown in the last line. We can see that the beamsteering analysis suggests that the large amplitude wave is made up of three components propagating from an easterly direction, $\sim 106°$, which

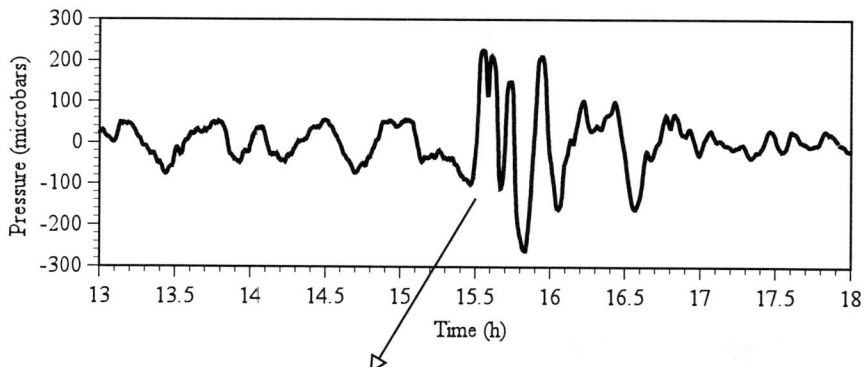

(a) Time-series of microbarograph 1 output.

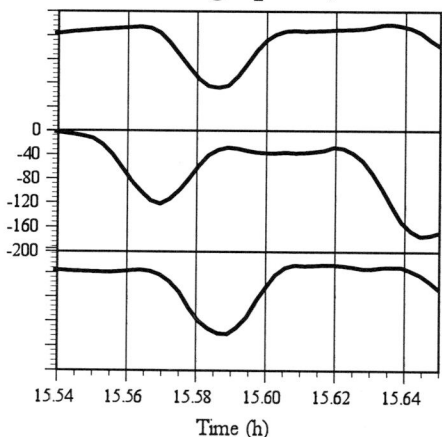

(b) Expansion of time-series
from microbarographs 1, 3 and 5.

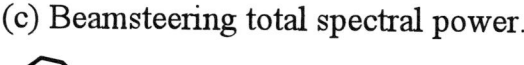

(c) Beamsteering total spectral power.

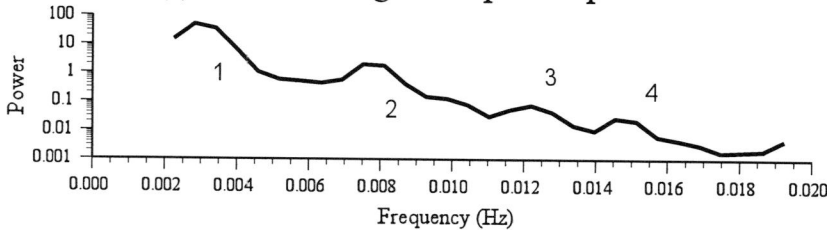

Figure 4. Beamsteering analysis of large amplitude wave event

Table 1. Wave estimates from analysis of data in Figure 4

Wave estimate	Wave parameters			
	Frequency	Direction	Speed	Length
1	$0.0028Hz$	$106.49°$	$4.14ms^{-1}$	$1463.4km$
2	$0.0075Hz$	$103.75°$	$4.69ms^{-1}$	$623.2km$
3	$0.0122Hz$	$112.04°$	$6.02ms^{-1}$	$493.1km$
4	$0.0146Hz$	$201.04°$	$8.65ms^{-1}$	$594.5km$
correlation method	$\sim 0.01Hz$	$73.02°$	$5.34ms^{-1}$	$534km$

agrees with the simple observation that the wave first reaches microbarograph 3 in Figure 4b and agrees reasonably with the single result from the correlation analysis. The beamsteering analysis suggests that the fourth peak in Figure 4b represents a higher frequency wave propagating from a southerly direction, $\sim 200°$.

Further information can be derived concerning the waves by considering cross-spectra. [12] has shown that cross-spectra can be used to help identify different types of atmospheric motion. The cross-spectral phase is used to differentiate between neutral waves that propogate horizontally, vertically propogating waves and Kelvin-Helmhotz "billows" [12, 13]. The cross-spectrum between two time series (in this case the time series correspond to output from two instruments) can be defined as $C_{xy}(f) = E[X^*(f)Y(f)]$ where $X(f)$ and $Y(f)$ are the Fourier transforms of time series $x(t)$ and $y(t)$, respectively. "*" denotes complex conjugate and E expectation which is taken to be an average over a section of data. Then the cross-spectral phase is defined as

$$ph(xy) = tan^{-1}\left(\frac{\mathcal{I}m[C_{xy}(f)]}{\mathcal{R}e[C_{xy}(f)]}\right)$$

where $\mathcal{I}m$ and $\mathcal{R}e$ denote the imaginary and real parts, respectively.

By analysing the cross-spectral phases from the mast data and microbarograph 0 it is possible to gain some insight into the origin of the waves. The cross-spectral phase between pressure, p, from microbarograph 0 and the vertical velocity at $4m$ (w_{4m}) is $ph(pw_{4m}) \sim \pm\frac{\pi}{2}$. Similarly for pressure and temperature at $2m$ (T_{2m}), $ph(pT_{2m}) \sim (0, \pi)$. These suggest the presence of trapped, neutral waves.

The phase from vertical velocity and temperature at $4m$ is also consistent with this interpretation, being $ph(wT)_{4m} \sim \frac{\pi}{2}$. [12] suggests that the sign of $ph(wT)$ should depend upon the relative magnitudes of the wave phase speed, c, and mean wind speed, $\overline{u}$. The wave phase speeds in Table 1 are between 4 and $9ms^{-1}$ and the near surface mean wind speed at that time was $\overline{u} \sim 4ms^{-1}$. $Ph(wT)$ is positive for the wave estimates in Table 1 which agrees with the observation that $c > \overline{u}$.

Thus it seems at first that there are *two* separate waves which coincide at the location of the microbarograph array, propagating from widely separated directions. An alternative explanation could be proposed, of which only a brief description will be given here as it is the subject of further research. It is anticipated from previous studies, such as [11], that the direction of gravity wave propagation should generally be veered slightly from the mean wind direction when the waves are generated topographically. During most of the day under consideration the mean wind was from $\sim 230°$ which is close to the propagation direction of the fourth wave in Figure 4c and Table 1. It may in fact be the case that only a single group of waves was generated, propagating from an easterly direction as the first three estimates in Table 1 show. The high frequency component, that is peak 4 in Figure 4c, has veered parallel to the mean wind. It is suggested that the wave corresponding to the fourth estimate was generated concurrently with estimates 1–3 and that its behaviour is analogous to surface water ripples which propagate in the wind direction. This interaction appears to disrupt the high frequency component of the wave as its array-averaged coherence is less than 0.25, compared to > 0.5 for estimates 1–3. Evidently further study is required to investigate the circumstances under which this mechanism might occur and its significance in the SABL.

It is interesting to note that the majority of the components of this wave event propagate from the direction of the Brunt Ice Shelf. In their analysis of the STABLE I data [11] found that waves propagating from this direction are commonplace but were usually associated with mean winds from a similar direction and were therefore interpreted as being generated by the topography of the ice shelf. The event considered here has a particularly large amplitude and propagates in a direction opposed to the surface wind direction. This suggests that the wave may not have a strictly topographic source. An alternative explanation could be that the wave event was generated at the Hinge Zone where there is a sharp change in the topography from the Antarctic interior to the ice shelf. Katabatic winds are observed to be prevalent in the Antarctic interior [14] but are *not* observed to propagate far onto the ice shelf [10] where the slope of the surface is shallow $(\sim \frac{1}{2000})$ compared with the slope from the continental plateau towards the ice shelf where it is $\sim \frac{1}{100}$. Therefore it is possible that this large amplitude event was generated in the region where the katabatic winds decay and its large amplitude enabled it to be observed at Halley station which is $\sim 50km$ from the Hinge Zone. Further investigation of this type of event is warranted as there are several large amplitude events present in the STABLE II dataset.

3.2   Results from several days data

Our second case study involves the beamsteering analysis of 5 day's data, taken from 4 – 8 May 1991. During this period the near surface mean wind direction is approximately constant, being from an easterly direction of $\sim 80°$, which implies a long upwind fetch across the Brunt Ice Shelf. Vertical profiles of wind direction from daily radiosonde ascents show that the wind direction is remarkably constant with height, varying by less than $60°$ from the surface direction over $2500m$ above the ground. Climatological studies (for example [9]) show that winds from this direction are the most common at Halley and analysis of wind speed and direction from the STABLE II data confirms this.

In Figure 5 histograms of the beamsteering estimates for wave direction and phase speed are shown for the 5 days' data. We can see that the majority of waves propagate from directions close to the mean wind, with the majority being veered from the mean wind by less than $90°$. Such directional consistency and alignment with the mean wind direction suggests that the waves are likely to be generated by topography, similar to the

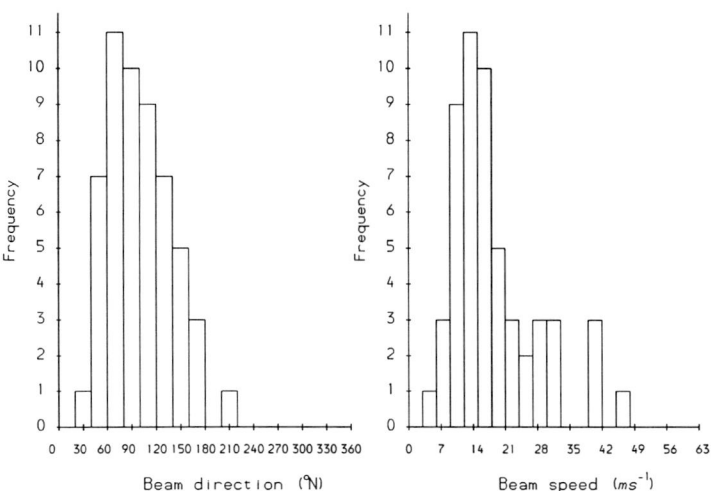

Figure 5. Beamsteering estimates for wave direction and phase speed

results found by [11]. In contrast to [11], however, we can see that there are a significant number of waves whose direction is *backed* from the mean wind direction.

The wave phase speeds in Figure 5 are also comparable with the surface mean wind speed, which varies between 4 and $15ms^{-1}$ at $4m$. Using the beamsteering estimates for phase speed and the frequency bin in which they are derived we can estimate the dispersion relation for the waves. It is assumed that the dispersion relation takes the form $\lambda \propto f^{-a}$, where λ is the wavelength. Using the method of least squares to fit this power law to the beamsteering estimates we find that a lies between 0 and 4.5, with the majority in the range $0 < a \leq 2$. A histogram of a is shown in Figure 6 where we can see that the majority of waves have dispersion relations with $a > 1$ but that there are a significant number with $a < 1$.

Evidently given this form for the dispersion relation the phase speed c is then directly related to frequency f by $c \propto f^{1-a}$. If $a = 1$, c is constant. Such a borderline case is unlikely to be found as a is unlikely to be equal to exactly 1. If $a > 1$, c increases with frequency. This is the most common, stable form for a group of waves.

When $a < 1$, c decreases with frequency and this situation is less common. Waves with such a dispersion relation were not observed by [11] who found $a = 1.15 \pm 0.1$. If c is inversely proportional to f then waves with

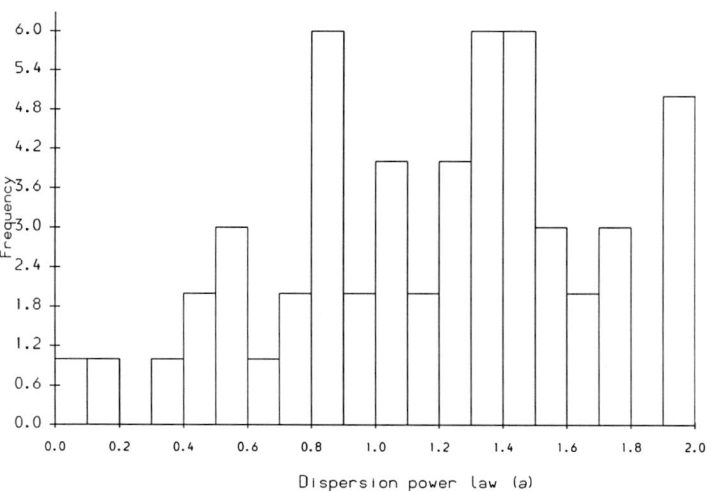

Figure 6. Distribution of the dispersion parameter, a

shorter wavelengths are overtaken by longer waves and it is difficult to envisage the formation of a "wave packet". Instead the generation of these waves must continue for longer so that such related groups of waves could be observed simultaneously at Halley, which is probably some distance from the source.

It seems that such a fundamental change in the wave dispersion relation must be associated with changes in the waves' generation. In Figure 7a the mean wind speed, $\bar{c}$, at $32m$ is plotted against time for the 5 days under consideration (day 0 corresponds to 1 January 1991). We can see that $\bar{c}$ increases up to the fourth day and decreases during the fifth. In Figure 7b a similar time series of a is plotted. To highlight the different regimes, values of $a \geq 1$ are plotted as "+" symbols and $a < 1$ as "x". We can see that, apart from four isolated events, $a > 1$ for days 123-126. During day 126 a group of waves is detected with $a < 1$ and afterwards a is once again greater than 1. It is evident from Figure 7 that the onset of these "unstable" waves ($a < 1$) from day 126 coincides with a maximum in $\bar{c}$. The wind speed also changes most rapidly with time during day 126, although the wind direction and temperature do not vary in such a marked

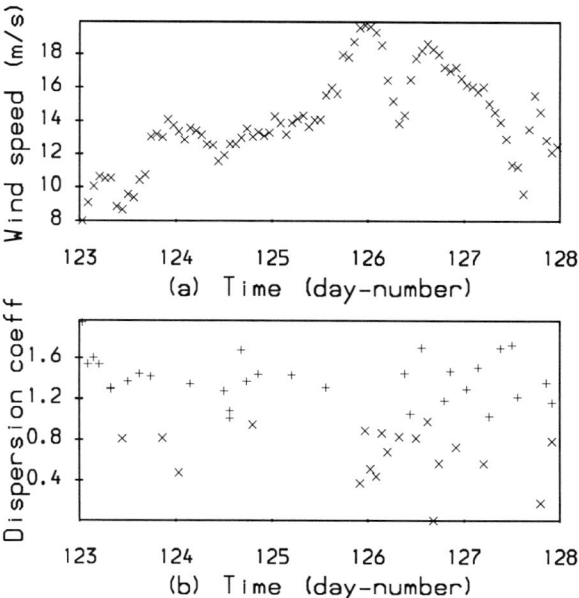

Figure 7. Time series of (a) $\bar{c}$, and (b) a

way. Such rapid shearing results in a minimum in the gradient Richardson number below $32m$.

We could use the cross-spectral phase to differentiate between horizontally propagating neutral waves and vertically propagating waves. However the cross-spectral phase data were incomplete at the time of publication so a definitive answer is difficult. The "unstable" waves detected on day 126 have directions that are closest to the wind direction as a whole and are all quite similar in this regard. In contrast, the "stable" waves ($a > 1$) from days 123–125 and 127 show much greater directional scatter and are furthest from the wind direction. It is possible, therefore, that the majority of the waves detected in this 5 day sample are generated by shear aloft and that the group of waves detected on day 126 with $a < 1$ are topographically generated. Further analysis of this data is required.

4   Discussion and conclusions

In this paper we have demonstrated the ability of the beamsteering program to identify wave parameters effectively using the array of six microbarographs depicted in Figure 1. We have identified and analysed a large amplitude wave event and discussed possible mechanisms for its generation. We have also indicated the direction future work on the climatology will take by analysing a subset of the STABLE II dataset. The two case studies have shown that a variety of waves are observed at Halley and that their behaviour and source mechanisms are varied.

Acknowledgement

This work was completed under grant GR3/9321 from the U.K. Natural Environment Research Council. The authors would also like to acknowledge the previous work of Ian McConnell on the beamsteering program (as described in [13]).

Bibliography

1. Anderson, P.S., Mobbs, S.D., King, J.C., McConnell, I. and Rees, J.M. (1992). A microbarograph for internal gravity wave studies in Antarctica. *J. Antarc. Sci.*, **4**(2), 241-248.

2. Belcher, S.E. and Wood, N. (1996). Form and wave drag due to stably-stratified turbulent-flow over low ridges. *Quart. J. R. Met. Soc.*, **122**, 863-902.

3. Capon, J. (1969). High-resolution frequency-wavenumber spectrum analysis. *Proc. of the IEEE*, **57**(8), 1408-1418.

4. Derbyshire, S.H. and Wood, N. (1992). The sensitivity of stable boundary layers to small slopes and other influences. *IMA Conf. Proc. "Stably Stratified Flows"*. Ed's. Castro, I.P. & Rockliff, N.J., Oxford, 105-118.

5. Einaudi, F., Bedard Jr., A.J. and Finnigan, J.J. (1989). A climatology of gravity waves and other coherent disturbances at Boulder Atmospheric Observatory during March-April 1984. *J. Atmos. Sci.*, **46**(3), 303-329.

6. Gedzelman, S.D. and Rilling, R.A. (1978). Short-period atmospheric gravity waves: A study of their dynamic and synoptic features. *Mon. Wea. Rev.*, **106**, 196-210.

7. Haubrich, R. (1968). Array design. *Bull. of the Seismological Soc. of America*, **58**(3), 977-991.

8. King, J.C., Mobbs, S.D., Rees, J.M., Anderson, P.S. and Culf, A.D. (1989). The stable Antarctic boundary-layer experiment at Halley station. *Weather*, **44**, 398-405.

9. King J.C. (1989). Low-level wind profiles at an Antarctic coastal station. *J. Antarc. Sci.*, **112**, 169-178.

10. King, J.C. (1993). Control of near-surface winds over an Antarctic ice shelf. *J. Geophys. Res. Atmos.*, **98**, 12949-12953.

11. Rees, J.M. and Mobbs, S.D. (1988). Studies of internal gravity waves at Halley Base, Antarctica, using wind observations. *Quart. J. Roy. Meteor. Soc.*, **114**, 939-966.

12. Rees, J.M. (1991). Studies of internal gravity waves at Halley base, Antarctica, using wind observations. *Boundary-Lay. Met.*, **55**(4), 325-343.

13. Rees, J.M., McConnell, I., King, J.C. and Anderson P.S. (1992). Observations of internal gravity waves over an Antarctic ice shelf using a microbarograph array. *IMA Conf. Proc. "Stably Stratified Flows"*. Ed's. Castro, I.P. & Rockliff, N.J., Oxford, 61-79.

14. Parish, T.R., Bromwich, D.H. and Tzeng, R.Y. (1994). On the role of the Antarctic continent in forcing large-scale circulations in the high southern latitudes. *J. Atmos. Sci.*, **51**(24), 3566-3579.

A Case Study of Plume Dispersion During the Evolution of a Stable Nocturnal Boundary Layer

M. Bennett*, H.E. Jørgensen**, E. Lyck†, P. Løfstrøm†,
T. Mikkelsen** and S. Ott**

*Environmental Technology Centre, Department of Chemical Engineering,
University of Manchester Institute of Science and Technology, **Risø National
Laboratory, Roskilde, Denmark, and †National Environmental Research
Institute, Roskilde, Denmark

Abstract

In July 1995, a series of measurements (BOREX '95) were made of the dispersion of a tracer plume from a 21 m instrumented tower in Borris Heath in Jutland. Facilities available included two scanning Lidars, SF_6 tracer samplers and full micrometeorological instrumentation. Between 2250 and 0030 DST on the night of 11-12/7/95, the tracer plume was monitored over 3 series each of 30 min duration. Repeated Lidar scans were taken at travel distances of 600, 2000 and 2400 m; SF_6 tracer was monitored with a single bag sampler at 600 m and an arc of samplers and a fast-response instrument at ~2600 m; and wind, temperature and turbulence profiles were measured up to a height of 21 m at the source. There was a moderate E airflow over the period, with very dry air allowing an intense stable stratification to develop over the dry heathland. On the first two runs, the plume direction at 20 m remained steady to within $1°$, though the Lidar cross-sections showed a veering of the wind with height, amounting to $\sim 6°$ over 35 m. On the plume centreline, non-zero concentrations were observed on over 80% of scans at 2.4 km travel. In the final series, however, the plume backed by $\sim 8°$ in 30 min, with centreline intermittency being around 50% at 2.4 km. Despite the steadiness of the driving conditions, turbulence intensities oscillated widely over the period of the measurements. This seems to be due to the intensification and subsequent break-down of wind speed and temperature gradients around 10 m. Attempts to model dispersion in the nocturnal boundary layer should take account of the unsteadiness of flow and turbulence in stable conditions.

1 Introduction

Between 7th–13th July 1995, a series of dispersion measurements were made in Borris Heath in Jutland (cf. Figures 1 and 2). This measurement campaign (BOREX '95) followed previous campaigns (BOREX '92, BOREX '94) organized by the Danish National Laboratory at Risø and National Environmental

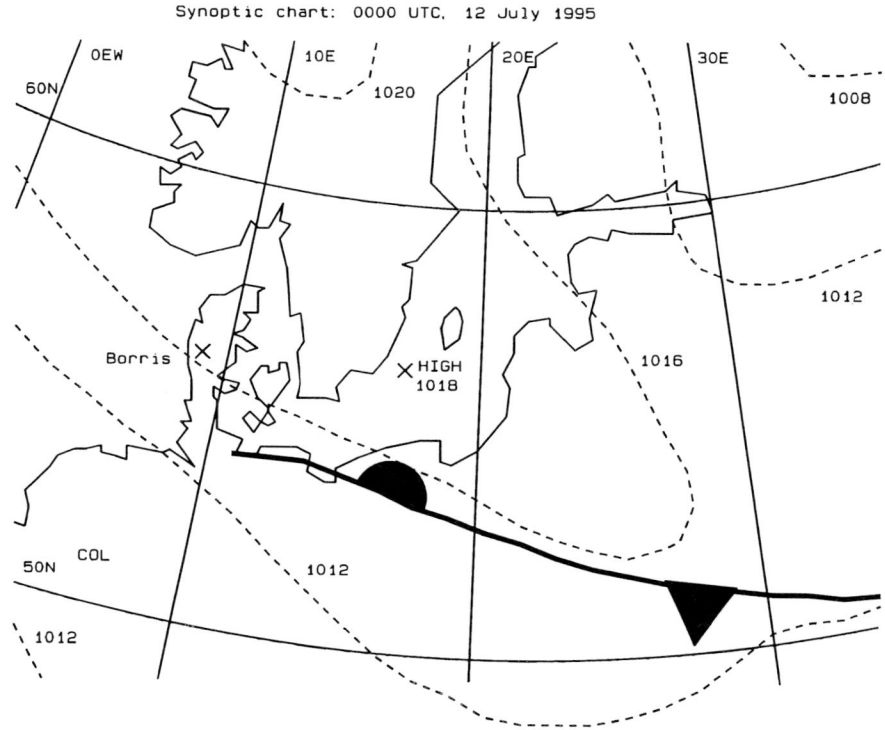

Figure 1. Synoptic situation at 0000 UTC on 12th July 1995

Research Institute at Roskilde [1–3]. These campaigns were intended to provide data for the validation of regulatory dispersion models.

The study area is a Danish military training ground. It is a lowland heath, extending roughly 6 km EW by 4 km NS. The overall topography is gentle, with heights above sea level ranging between 11 m and 25 m, though locally moraines give the impression of a rugged landscape. Vegetation cover is mostly rough grass and heather over a very light sandy soil. The heath is largely surrounded by pine plantations, beyond which there is farmland.

In the Borris '95 campaign, smoke and SF_6 tracer were released from a 21 m mast in the middle of the heath (Figure 2). The smoke was generated from $SiCl_4$ and aqueous NH_3: these liquids were pumped to the top of the mast and released from adjacent nozzles. The procedure gives a dense smoke from a modest emission. Propylene tracer was also used on some trials, in collaboration with Dr. C.D. Jones of CBDE Porton.

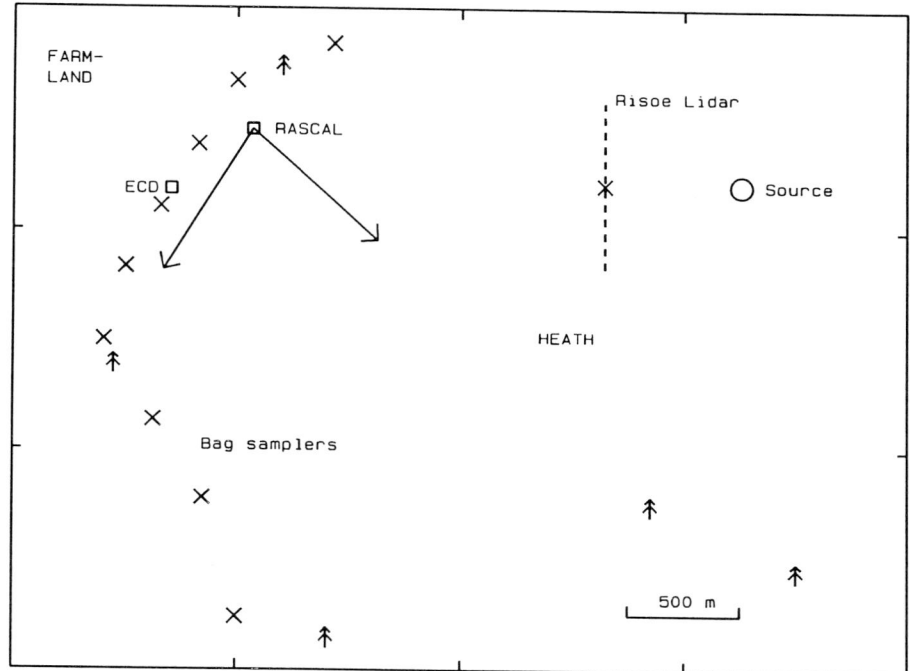

Figure 2. Plan of the experimental area for the night of 11th–12th July 1995. Trees denote the positions of pine plantations. Crosses denote bag samplers. "ECD" denotes the fast-response tracer detector

Two Lidars were available. Risø supplied an infra-red Lidar with a range-resolution of 1 m. UMIST supplied the Rapid-Scanning Lidar (RASCAL) originally developed by the Central Electricity Generating Board [4,5]. This had recently been upgraded through a grant from the Natural Environment Research Council. Its range resolution in normal operation remains 5 m but it is now controlled by a PC, enabling much preliminary analysis to be carried out during the course of a field trial.

NERI provided SF_6 sampling. This consisted both of bag samplers, which were analyzed subsequently on site using a gas chromatograph with electron capture detector, and a real-time ECD [6].

Meteorological measurements were made on the 20 m mast. Cup and vane anemometers were situated at 2, 10 and 21 m; thermocouples at 0.23, 0.88, 1.93, 3.37, 5.20, 7.41, 10.00, 12.96, 16.31 and 20.00 m; and ultrasonic anemometers at 2, 7, 10 and 20 m. The Solent anemometer at 7 m was considered to be the most reliable.

In normal daytime operation, arcs of bag samplers were disposed around the source at ranges of 200, 400 and 600 m and at increments of 5°. The NERI vehicle with its fast-response ECD would site itself close to the plume centreline along one of these arcs, while the Risø Lidar would scan tangentially to the arc. RASCAL would be sited ~800 m away and would make alternate tangential vertical scans at two downwind ranges. The scanning schedule was predicated by the programme of bag sampling: typically, this involved three series each of 30 min duration separated by a 10 min pause.

Nighttime measurements, in stable conditions, were made on three evenings during the campaign. On the evening of 10/7/95, the above procedure was continued until 2330, i.e. long after the material had ceased to disperse to the ground; on the evening of 8/7/95, horizontal scans were made over a surface release; and the measurements for the night of 11–12/7/95 are discussed below.

2   Measurements: 11th–12th July 1995

The experimental lay-out for the night of 11th–12th July 1995 is shown in Figure 2, while the synoptic situation is shown in Figure 1. High pressure over southern Scandinavia drew warm dry air from central Russia over the Baltic and Jutland. Analysis of the synoptic chart for 0200 DST suggests a surface geostrophic wind of $G \approx 13.8$ m s^{-1}. (It should be noted that at 850 mb a thermal wind of about 4 m s^{-1} opposed the geostrophic wind so the net wind above the boundary layer fell steadily with height). With clear skies, a stable boundary layer developed and a well-defined, steady plume was advected to the western boundary of the heath.

Bag samplers were distributed along the lane on the West side of the heath (i.e. at ranges of 2400–3000 from the source), with a single sampler at a range of 600 m. The Risø Lidar was also deployed at this intermediate range. RASCAL was sited to the North of the far-field arc and made alternate scans on azimuths of 132° and 212°, thereby obtaining plume cross-sections at downwind distances of 2000 m and 2400 m.

SF$_6$ tracer was released at a rate of 0.141 g s^{-1} from 2220. Three 30 min bag samples were taken, starting at 2250, 2325 and 0000; scans were made simultaneously with RASCAL. Data were unfortunately lost from the first RASCAL series, initially because the laser had to be stopped to allow NERI personnel access to their samplers and later because too distant a range window had been set. A typical plume cross-section is shown in Figure 3.

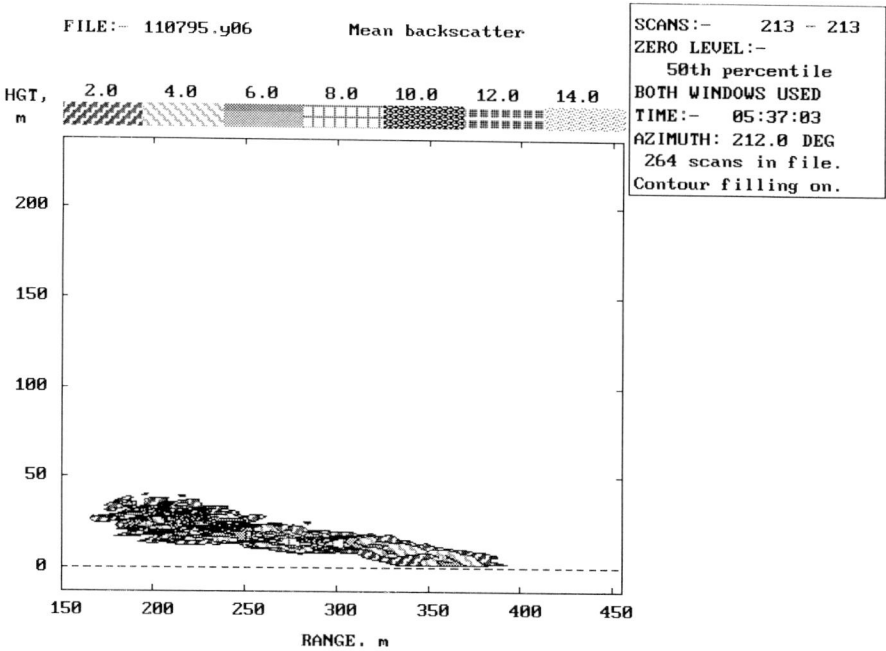

Figure 3. Instantaneous plume cross-section at 23:49:19 at 2400 m travel. The scanning plane crosses the plume centreline at an angle of about 120°. (A software error has led to the time being in error by 65536 s.) This is a single scan out of a series of 264

3 Evolution of stable boundary layer

Figure 4 illustrates the evolution of the stable boundary layer between 2100 and 0300 over the night of 11–12/7/95. In particular, we show the bulk Richardson number (Ri) between 10 m and 21 m, the sensible heat flux (Q_S) at 7 m and the Monin-Obukhov length (L_{MO}) at 7 m. It is the period before 0030 which is of most interest to us in interpreting our dispersion measurements and we see here that the behaviour is anomalous. Despite the steady driving conditions (clear sky, defined gradient wind) the stability oscillates with a time period of about 1 h. We may characterize an "on" phase as having $Ri < 0.2$, $L_{MO} > 10$ m and $Q_S < -15$ W m^{-2}, while an "off" phase is associated with larger values of Ri and small values of sensible heat flux and Monin-Obukhov length.

Turbulent intermittency is, of course, a well recognized phenomenon. For large values of Ri, turbulent transfers die away, allowing the wind aloft to

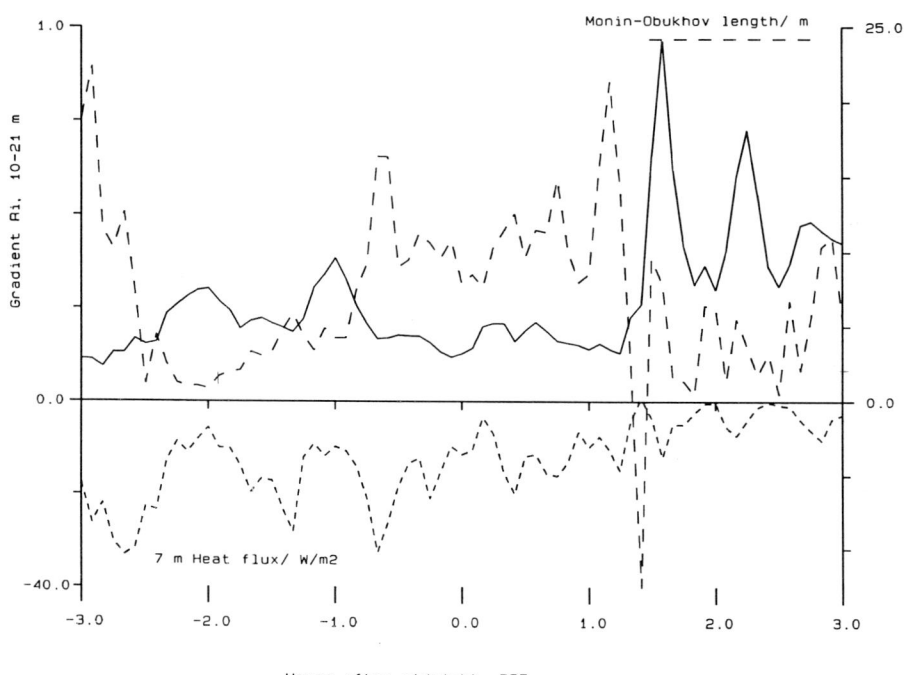

Figure 4. Evolution of surface stability between 2100, 11/7/95 and 0300, 12/7/95. The positive left-hand scale shows the gradient Richardson number between 0 and 1; the negative left-hand scale shows the sensible heat flux down to -40 W m^{-2}; the right-hand scale shows the Monin-Obukhov length: this changes sign briefly when cold surface air is advected at 0130

accelerate under Coriolis forces. Eventually the wind shear becomes sufficient to break down the stable layer, giving a burst of turbulence. In this case, however, the gradient wind seems to be too large for such turbulence breaking to take place. Derbyshire and Wood [7] propose that turbulence becomes intermittent if the magnitude of the surface sensible heat flux exceeds

$$Q_{\max} = \frac{T \rho c_\rho}{g}(R_{fc}/\sqrt{3})G^2 f, \tag{3.1}$$

where T is the absolute air temperature and the critical Richardson number is about 0.25. This formula suggests that the moderately large geostrophic wind on the night of 11–12/7/95 should have been adequate to maintain a sensible heat flux down to -100 W m^{-2}. In fact, it oscillated around a mean value of about -15 W m^{-2}.

Figure 5 gives perhaps a hint as to what is going on. This figure shows the wind and temperature profiles for successive off (High Ri) and on (Low Ri) phases. It appears that the off phase is associated with a weak velocity gradient and relatively high temperature at heights between 7 m and 15 m. With the following on phase, the surface wind speed has dropped, while the air at intermediate heights has cooled more rapidly than that at the surface or at 20 m. We propose that this be interpreted in terms of a "two-switch valve" model. The weak velocity gradient aloft at the end of the on phase leads to a high value of Ri between 10 m and 20 m and a decay of turbulence in that layer. The "valve" at 15 m then closes and allows no further momentum flux to the surface layer. This layer, however, is still turbulent and in contact with the surface so it decelerates. Assuming a logarithmic wind-speed profile of depth Δz, the layer would halve its initial velocity in a time

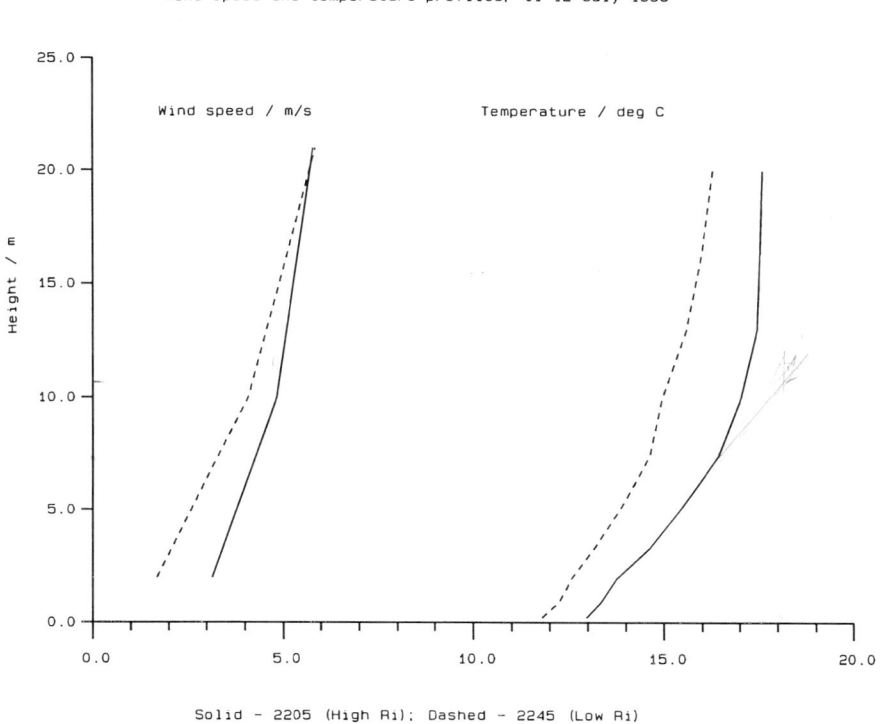

Figure 5. Wind and temperature profiles for "on" (low Ri, high turbulence) and "off" (high Ri, low turbulence) phases

$$t_{1/2} = \frac{1}{k^2} \frac{\Delta z}{u(\Delta z)} \log\left(\frac{\Delta z}{z_0}\right) \left(\log\left(\frac{\Delta z}{z_0}\right) - 1\right), \qquad (3.2)$$

which in this case comes to about 10 min. Of course, since in practice the valve does not completely close, the surface layer would not decelerate as quickly as this. As the surface layer decelerates, the velocity gradient increases and the turbulence will eventually be restimulated. The valve will therefore turn itself on again some 15–20 min after it has been switched off.

Meanwhile, the surface has continued to radiate heat. During the off phase this can only come from the ground and from the layer of air close to the surface. This deficit accumulates while turbulence is weak. When the on phase returns, sensible heat is transported downwards, initially to the surface air and subsequently to the vegetation and sub-surface layer. Conduction of excess heat flux to the surface will persist for a period similar to the off phase. The ultimate effect is to cool the air around 10 m, increase Ri and switch the valve off again.

We suggest that under these conditions the layer of air around 10 m has acted as a valve. When closed (by high Ri) it automatically reopens when the surface air has decelerated. When opened (by a large velocity gradient) it automatically recloses when it has been cooled by the surface heat deficit.

If the driving forces, G and Q_s remained fixed, it is possible that these oscillations would eventually be damped out. On the night in question, however, there was a qualitative change about 0130. A pool of cold surface air was advected, associated with a brief positive sensible heat flux. The wind dropped and backed. This change may be associated with the front marked in Figure 1.

4 Nighttime dispersion: results

Software has been written [5] to analyze cross-sections of the kind shown in Figure 3. Standard plume parameters: (height, plume direction, σ_y, σ_z, total backscatter, total area etc.) may be calculated for each instantaneous cross-section. In the course of a 30 min series, several hundred such scans may be performed; means and variances of the above parameters may thus equally be calculated. We show in Figure 6, the time-series of σ_z measured by RASCAL at 2400 m travel. For comparison, we have also plotted the predictions of the diffusion equation,

$$\sigma_z(x) = 2Kx/u, \qquad (4.1)$$

where we have assumed that the eddy diffusivity, K, may be approximated by the eddy viscosity, K_M, predicted by the Businger-Dyer flux-profile relationships,

$$K_M(z) = \frac{ku_* z}{1 + 5\frac{z}{L_{MO}}}. \qquad (4.2)$$

For a source height of $z = 20$ m and the range of L_{MO} experienced, this reduces approximately to

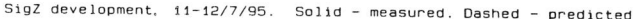

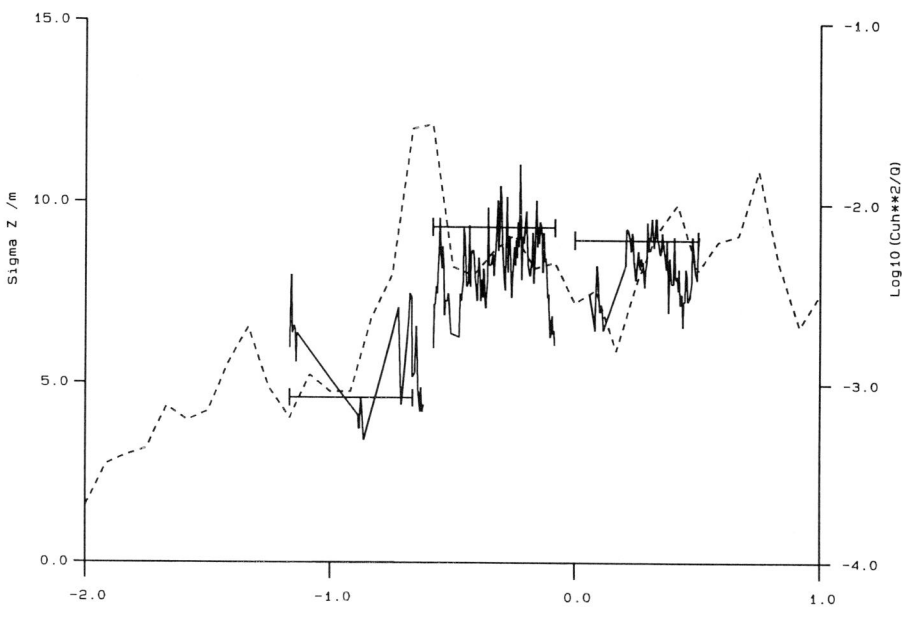

Figure 6. Predicted and observed instantaneous vertical plume spread (σ_z) at 2400 m travel. The bars show the normalized surface peak tracer concentrations at 2600 m travel

$$K_M = 0.08 u_* L_{MO}. \qquad (4.3)$$

The absolute agreement between measurement and simple theory in Figure 6 is surprisingly good. To a limited extent, the time-dependences of measurement and theory are also similar: the slower dispersion rates around 2300 and 0000 have been picked up, but the active turbulence around 2325 has apparently not affected the dispersion.

It is arguable that these instantaneous plume statistics are easier to predict than time-mean statistics. An instantaneous passive plume spreads as a results of statistically well-defined processes; in this case, diffusive eddies of scale $L_{MO}/5$. Analogously, instantaneous buoyant plumes were found to behave in a self-similar pattern with their spread being proportional to their buoyant rise relative to the surrounding air [8]. It is, on the other hand, much more difficult to predict time-mean plume spreads since these depend on the statistics of large-scale motions in the boundary layer.

We have also, in Figure 6, shown the normalized peak SF_6 concentrations from the bag samplers. These show the same general behaviour as the Lidar estimates of dispersion. In a simple Gaussian plume model, the normalized centreline ground-level concentration is given by

$$Ln\left(\frac{Cuh^2}{Q}\right) = Ln\left(\frac{h^2}{\pi\sigma_y\sigma_z}\right) - \frac{h^2}{2\sigma_z^2} \qquad (4.4)$$

and thus $\log_{10}(Cuh^2/Q) = -4.53$ for $\sigma_y \approx \sigma_z = 0.2h = 4$ m, and $\log_{10}(Cuh^2/Q) = -1.86$ for $\sigma_y \approx \sigma_z = 6$ m. The SF_6 measurements thus suggest a plume spread similar to that measured with the Lidar. Since the plume was much narrower than the spacing between the samplers, there is a wide degree of uncertainty in the estimated plume spread, with the measured concentration representing a lower limit to the centreline concentration and hence to σ_z. We should, of course, expect the 30 min mean values of σ_z to be greater than the instantaneous values measured by Lidar. A more precise estimate may be made by noting that at 0013 the surface plume swept over the fast-response SF_6 detector and a peak concentration of 2400 ng m^{-3} was then measured. This corresponds to a normalized concentration of $\log_{10}(Cuh^2/Q) = -1.39$, or $\sigma_z \approx 7$ m. Again, this is in satisfactory agreement with the plume spread measured by the Lidar.

5   Variation in wind direction

The RASCAL software allows us to calculate the mean direction of advection of the plume for each instantaneous cross-section. The time series of plume direction is shown in Figure 7, together with the direction measured at 21 m on the mast. It may be seen that there is very good correlation between the two, though with an offset of a few degrees. Comparison with other measurements of wind direction suggested that this small offset arose from difficulties in aligning the vane at this height. The bag samplers at 2600 m travel indicated that the surface plume lay between 85° and 91° throughout the period of measurements, tending to the lower value by 0030. The slight time offset apparent in Figure 7 is consistent with the plume advection time.

Two features of the direction development may be noted. Firstly, the direction backed very sharply at 0130 as the cooler air mass arrived. The direction then continued to vary widely.

Secondly, the wind direction prior to this event, and especially before midnight, was extremely steady. This may be seen from Figure 8, which displays the intermittency (defined as the fraction of time that tracer is present) of the plume observed between 2325 and 2355. We have here taken the 132 scans in the series at this azimuth and, for each pixel, counted the number of scans for which the measured backscatter is above background. The non-zero proportion has then been plotted as a percentage. (The streak above the plume arises from a single low-energy shot. Such glitches have been identified and removed in the analysis presented in Figure 6.)

BL development, 11-12 July 1995

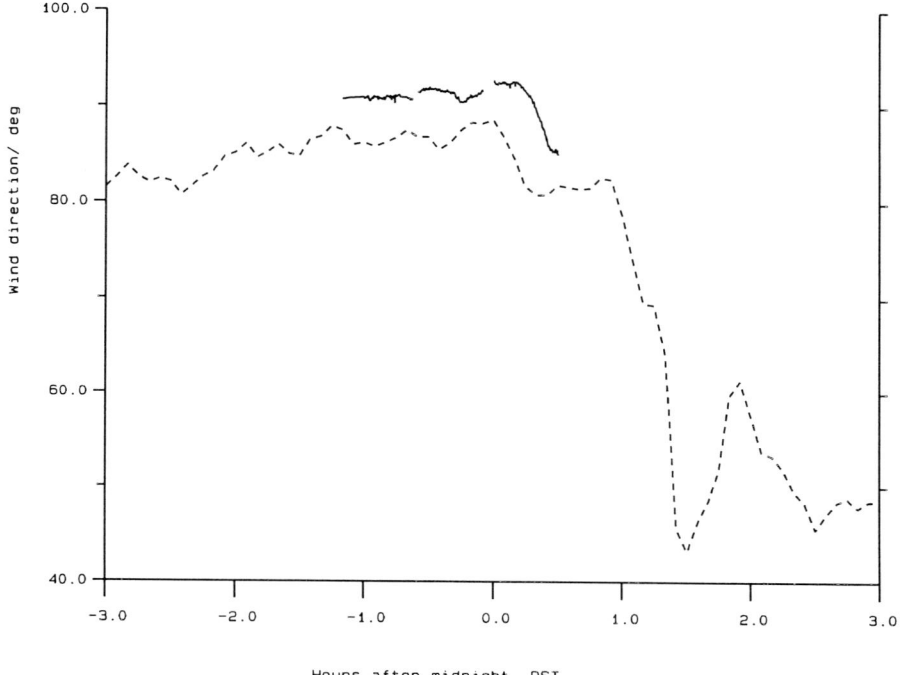

Figure 7. Evolution of wind direction. The dashed line is the direction measured with the wind vane at 21 m; the solid line is calculated from the plume cross-section at 2400 m travel

The peak intermittency in the plume of Figure 8 may be seen to be 90%, while at the surface, the peak intermittency exceeded 80%. This is most unusual. Peak intermittencies observed in the daytime boundary layer at only 200 or 400 m from the source struggled to exceed 10%: here we are looking at the plume 2400 m from the source.

After midnight, the plume started to back and this inevitably lead to a reduction in the intermittency. The peak value in the final Lidar series was less than 50%.

Figure 8 allows us to make a sensitive estimate of the mean veer in wind direction with height. Geometrically the mean difference in plume direction between the surface and 35 m may be seen to be about 3.1°. Since we have no reason for believing that a given smoke particle at a given measurement height has passed other than equal times at all heights between the source and this height, this implies a wind direction difference of 6.2° between the surface and

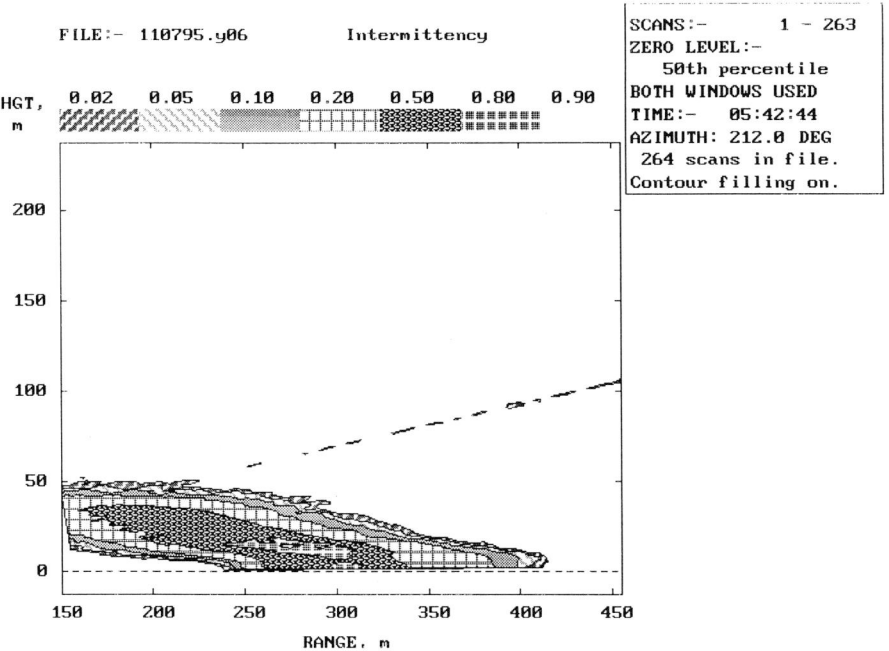

Figure 8. Intermittency cross-section of plume between 2325 and 2355

35 m. The expected direction difference may be calculated from the Ekman formula, which predicts

$$\frac{\partial \psi}{\partial z} = \sqrt{\frac{f}{8K_M}} \tag{5.1}$$

at the surface. Using Equation (4.3) for the eddy viscosity and the measured values of u_* and L_{MO} at 2325 for the on phase and 2305 for the off phase we estimate that the wind direction difference between the surface and 20 m should vary between 16.1^o and 48.3^o. (The latter is, of course, well beyond the linear régime of Equation (5.1)). It is thus clear that, although the veered plume in Figures 3 and 8 may be spectacular, its slope is far less than would have developed had the stable boundary layer remained steady for long enough for the surface wind to come to equilibrium with the Coriolis forces.

Finally, we may note that it is also possible to use near-horizontal Lidar scans to observe the effects of such vertical shears in wind speed and direction. Figure 9 shows such a scan, obtained at 2245 on 8/7/95. Eastings and Northings

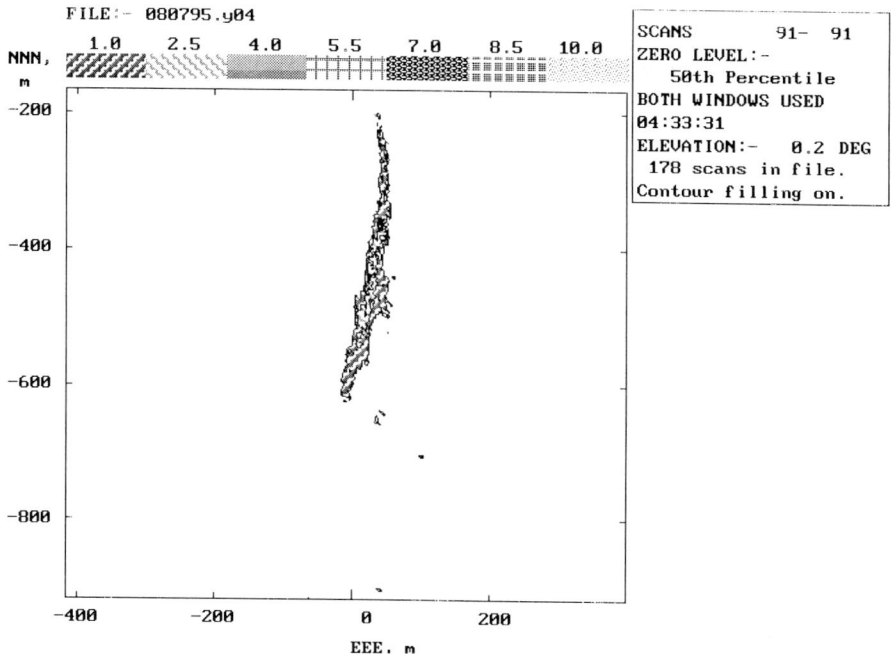

Figure 9. Near-horizontal scan through puff at 2246 on 8/7/95. Note the curvature which arises as a result of wind direction shear

are in m relative both to the source and the Lidar, which were co-located. A surface puff had been released some 220 s earlier, and had drifted southwards in a light Northerly wind. For reasons of laser safety, the Lidar could not be scanned absolutely horizontally; allowing for the terrain, the section rises from a height of about 4 m at 200 m to about 6 m at 500 m. It may be seen that the mean drift direction increases from 171° at a range of 250 m to 181° at 500 m. There is no suggestion here that the mean wind direction has increased by 10° over a height difference of 2 m. Rather, the particulate detected at 250 m has spent most of the last 220 s at heights below the scanning plane and been thus both delayed and backed; the particulate at 500 m has travelled above the scanning plane and been accelerated and veered. The net effect is to provide the hodograph shown in Figure 9.

6 Conclusions

The analysis we have presented has not broken any new ground theoretically. Rather, we have shown how our measurements of plume dispersion in a stable boundary layer are quantitatively in broad agreement with some very basic theoretical concepts.

The external conditions in this experiment were as steady as could reasonably be expected in stable conditions: a moderate, steady wind, a cloudless sky and flat terrain. Nevertheless, the observed turbulence parameters were far from steady, with (for example) the Monin-Obukhov length varying more than 10-fold over an hourly cycle. Plume advection and dispersion reflected this unsteadiness. Realistic models of plume dispersion in the stable boundary layer should take such unsteadiness into account.

Acknowledgements

This work was carried out under the Danish Strategy Environmental Research Programme. We are grateful to Mr. Des Doocey, Mr. Søren Lund and Mr. Jan Nielsen for technical assistance.

Bibliography

1. Mikkelsen, T., Jørgensen, H.E., Løfstrøm, P. and Lyck, E. (1996). *Borex '92: Atmospheric Dispersion Experiment on Concentration Fluctuations - Data Report Risø-R-925(EN)*.

2. Mikkelsen, T., Jørgensen, H.E., Løfstrøm, P. and Lyck, E. (1996). *Borex '94: Atmospheric Dispersion Experiment on Concentration Fluctuations - Data Report Risø-R-926(EN)*.

3. Mikkelsen, T., Jørgensen, H.E., Løfstrøm, P., Lyck, E. and Bennett, M. (1998). *Borex '95: Atmospheric Dispersion Experiment on Concentration Fluctuations - Data Report Risø-R-927(EN)*.

4. Bennett, M., Sutton, S. and Gardiner, D.R.C. (1992). "Measurements of wind speed and plume rise with a rapid-scanning Lidar". *Atmos. Environ.*, **26A**, pp. 1675–1688.

5. Carruthers, D.J., Edmunds, H., Bennett, M., Woods, P.T., Milton, M.J.T, Robinson, R., Underwood, B.Y. and Franklin, C.J. (1996). "Validation of the UK-ADMS dispersion model and assessment of its performance relative to R-91 and ISC using archived Lidar data". *Department of the Environment, Research Report Number: DOE/HMIP/RR/95/022*.

6. Lyck, E. (1990). "A tracer technique for full-scale atmospheric dispersion experiments". *Ninth Symposium on Turbulence and Diffusion, American Meteorological Society, Boston*, pp. 127–129.

7. Derbyshire, S.H. and Wood, N. (1994). "The sensitivity of stable boundary layers to small slopes and other influences". *Stably Stratified Flows: Flow and Dispersion over Topography*, Clarendon Press, Oxford, pp. 105–118.

8. Bennett, M., Sutton, S. and Gardiner, D.R.C. (1992). "An analysis of Lidar measurements of buoyant plume rise and dispersion at five power stations". *Atmos. Environ.*, **26A**, pp. 3249–3263.

Forecasting of Plume Dispersion around a Power Plant

M.J. Souto*, V. Pérez-Muñuzuri*, J.A. Souto**, M. de Castro** and J.J. Casares**

*Group of Nonlinear Physics, Faculty of Physics, University of Santiago de Compostela, Spain, and **Department of Chemical Engineering, University of Santiago de Compostela, Spain

Abstract

Transport and dispersion of pollutants in the lower atmosphere are predicted using a mesoscale Lagrangian particle model and a mesoscale Lagrangian plume model, coupled with a prediction of buoyant plume rise, which allow us to compare their results under stable and convective conditions. A time dependent hydrostatic meteorological model has been developed to predict wind and temperature fields and to infer needed subgrid scale turbulence quantities.

The models are applied to the problem of predicting the ground level concentration (glc) of the SO_2 emitted by As Pontes Power Plant, in northwestern Spain. The numerical results obtained by the meteorological and dispersion model are compared with real data from seventeen fully automated remote stations (for ground level concentration of SO_2, NO_x and particles), nine meteorological towers and one Remtech PA-3 sodar located within an area of 3600 km^2 around the Power Plant.

1   Introduction

The legal limit of emissions guarantee, for the majority of locations and under most meteorological conditions, the observance of the law for ground level concentrations. However, the existence of single sources of significant magnitude, or the danger of fugitive emissions to the atmosphere, and also specific meteorological conditions may cause an impact of the plume on the ground that may go beyond the legal limits for the pollutant. To avoid the occurrence of these rare episodes, it is necessary to predetermine the maximum emission allowed for any meteorological condition [1].

Two combined models, meteorological and diffusion, for the simulation of the plume transport at mesoscale distance are developed [2]. The system is based on a three dimensional time dependent meso-β-hydrostatic meteorological model and two types of Lagrangian diffusion model: an adaptive plume model and a particle model. In the first case, the conservation of aerosol material χ in the atmosphere, expressed mathematically as

$$\frac{\partial \overline{\chi}}{\partial t} = \overline{u}^j \frac{\partial \overline{\chi}}{\partial x^j} + \frac{\partial}{\partial x^j} \overline{u'_j \chi'} + S_x \quad j = 1, 2, 3 \tag{1.1}$$

is solved using one form of the Gaussian plume model, where advection by the grid-volume averaged velocity and spatial variations of density are ignored. In the second case, the advection and diffusion are estimated by a stochastic surrogate to (1.1) where the mesoscale model predictions are used to determinate the statistical properties of the transport and mixing of the pollutants [3].

2   Meteorological model

The meteorological dynamical model is a three dimensional time-dependent mesoscale model based on finite difference solution of the hydro-thermodynamical equations. A terrain-following coordinate system is used to introduce the topography in the model. The vertical turbulent fluxes, or diffusion terms, account for the vertical mixing at the atmosphere and their definition depends on the stability of the layer being simulated. They are obtained as a function of surface turbulent parameters, that have been calculated using an iterative procedure based on wind speed and temperature values at two different levels near the ground [4]. In this case, the lowest level was chosen at 3 meters, the first vertical layer of the grid.

The depth of the planetary boundary layer (PBL) is usually associated with an inversion and it is calculated by a prognostic equation mainly depending on the surface heating [5]. The Bussinger-Dyer relationships between turbulent parameters are used to solve the prognostic equation for the PBL. During the transition from convective to stable conditions, the PBL tends to adjust exponentially toward an equilibrium depth. In this case, the height produced during transition times can be considered as a fictitious height during which the stable layer near the surface develops and becomes well established [6]. The parameterizations for boundary layer structure allow us to calculate the standard deviations of the subgrid scale velocity components needed for the Draxler method application in the plume modeling and for the Lagrangian autocorrelation calculation in the particle model.

The first-order closure of the meteorological model does not allow to obtain high-order turbulent parameters as σ_u, σ_v, σ_w the standard deviations of the subgrid scale velocity components. Then, variables in the meteorological model and planetary boundary layer parameterizations were used to obtain them by using the following expressions [3]:

$$\sigma_w = \begin{cases} K_m / A\lambda_{M_w} & z/L \leq 0 \\ 1.2l \left(\frac{Ri_c - Ri}{Ri_c} \right)^{0.58} \left[\left(\frac{\partial u}{\partial z} \right)^2 + \left(\frac{\partial v}{\partial z} \right)^2 \right]^{1/2} & z/L > 0 \end{cases} \tag{2.1}$$

with l the mixing length including a buoyancy length scale that indicates the degree of suppression of vertical motions by the static stability [7].

$$\sigma_u = \sigma_v = \begin{cases} u_*(12 + 0.5z_i/\mid L \mid) & z/L \leq 0 \\ \\ 2.3u_* & z/L > 0. \end{cases} \qquad (2.2)$$

The meteorological model equations are solved by a finite difference method. A forward-in-time, upstream in space scheme is used for the advection terms. For the diffusion terms a semi-implicit scheme with weight of 75% on a future time step is used and a constant time step of 30 seconds. The rest of spatial derivatives are solved by a forward-in-time, centered in space scheme. For Coriolis terms, as well as for radiative terms, an explicit scheme is used.

Forty vertical levels beginning at 3 meters above the ground spaced logarithmically until reaching the top at 7000 meters and 31 × 31 grid points on each vertical level with a grid mesh of 2000 meters have been used. The area under study is localized in northwestern Spain between 7° 36' and 8° 12' (W) and between 43° 9' and 43° 40' (N). To avoid the first spurious effects due to initialization the model is run for 3 h with the requirement that time-dependent forcing terms are not permitted to occur.

3   Dispersion models

3.1   Lagrangian particle model

In this model the number of emitted particles is a constant value of 10 particles per time step and reaches a maximum of 20000. They are emitted with an initial random horizontal dispersion that situates them in the area corresponding to chimney top. The location of individual particles in the terrain following coordinate system is given by:

$$x_i(t + \Delta t) = x_i(t) + [\overline{u^i}(t) + u'^i(t)]\Delta t \quad i = 1, 2, 3. \qquad (3.1)$$

The variables $\overline{u^i}$ represent the mean values of wind velocity components obtained directly from the mesoscale model calculation and u'^i represent the turbulent fluctuations whose values are evaluated statistically from the boundary layer formulation used in the mesoscale model. According to the statistical approach we assume that u'^i is a semirandom variable composed of two components, one correlated component and the other a purely random component obtained by manipulating computer-generated random numbers [8]. These subgrid scale components can be written as [9,10]:

$$u'^i(t) = u'^i(t - \Delta t)R_{u_i}(\Delta t) + u''^i(t) \qquad (3.2)$$

where R_{u_i} are the Lagrangian autocorrelations for each velocity component as a function of the time step in the numerical model, $\Delta t = 30s$. The second term on the right can be obtained randomly from a normal probability distribution with a zero mean and standard deviations given by:

$$\sigma'_{u'} = \sigma_{u^i}[1 - R^2_{u^i}(\Delta t)]^{1/2} \tag{3.3}$$

where σ_{u^i} are the standard deviations of the subgrid scale velocities.

The correlations needed in the last expression are determined from:

$$R_{u^i}(\Delta t) = exp(-\Delta t/T_{L_{u^i}}) \tag{3.4}$$

where the Lagrangian time scale T_L is determined for each component from the scale of turbulence as evaluated from the turbulence spectra [11] $T_{L_{u^i}} = 0.2\beta_{u^i}\lambda_{mu^i}/\overline{V}$ where λ_{mu^i} is the peak wavelength in the spectra for each component and its value depends on atmospheric stability [12,13], and $\overline{V} = [(u^i)^2]^{1/2}$. The ratio of Lagrangian to Eulerian time scale β for each component is given by $\beta_{u^i} = 0.6\overline{V}/\sigma_{u^i}$.

3.2   Puff model

As a solution for the simulation of the plume transport, a Lagrangian adaptative model [14] is used with a discrete representation of the plume by puffs described by five points at different heights. The equation for the distribution of the concentration C for the whole puff, from the total amount of pollutants, Q_c, assigned is given by:

$$C = \frac{Q_c}{^3\sqrt{(2\pi)^2\sigma_y^2\sigma_z}} exp\left[-\frac{1}{2}\left(\frac{y}{\sigma_y}\right)^2\right] exp\left[-\frac{\overline{Z}^2}{2}\right] \tag{3.5}$$

and $\overline{Z}$ is a non-dimensional vertical coordinate defined as, $\overline{Z}(z) = \int_{z_0}^z \frac{d\xi}{\overline{V}_z(\xi)}$.

The vertical movement for each of the points is obtained as the sum of the vertical movement of the air, the plume rise and the vertical diffusion. This contribution of the turbulent diffusion to the vertical movement of the plume rise is a function of σ_z using the virtual distance X_v [15]. Draxler [16] equations, completed with Irwin [17] results, allow to calculate σ_y, and σ_z as:

$$\sigma_z = \sigma_w\frac{X_v}{\overline{V}} \qquad \left[\frac{d\sigma_z}{dX}\right]_{x=x_v} = 0 \tag{3.6}$$

$$\sigma_z = \frac{\sigma_w}{\overline{V}}\frac{X_v}{1 + 0.9(0.02X_v/\overline{V})^{0.5}} \qquad \left[\frac{d\sigma_z}{dX}\right]_{x=x_v} = \frac{\sigma_w}{\overline{V}}\frac{1 + 0.45(0.02X_v/\overline{V})}{(1 + 0.9(0.02X_v/\overline{V})^{0.5})^2} \tag{3.7}$$

$$\sigma_y = \sigma_\theta\frac{X_v}{1 + 0.9(0.001X_v/\overline{V})^{0.5}} \qquad \left[\frac{d\sigma_y}{dX}\right]_{x=x_v} = \sigma_\theta\frac{1 + 0.45(0.001X_v/\overline{V})^{0.5}}{(1 + 0.9(0.001X_v/\overline{V})^{0.5})^2}. \tag{3.8}$$

σ_w and σ_θ are obtained from the meteorological model. σ_z can be obtained directly for convective conditions from (3.5). The calculation of σ_z for stable and neutral conditions (3.6) and the calculation of σ_θ (3.7) require a Newton-Raphson iterative procedure and in this case it has been applied with a 0.1 m of accuracy.

3.3   Plume rise

A plume rise calculation routine is included, since buoyancy represents one of the primary mechanisms governing the motion of plumes generated from sources as the tall industry stack studied here [18]. The buoyancy flux is calculated using temperature differences between stack and ambient air. The emitted gases temperature is considered constant and equal to 450 K.

Vertical potential temperature gradient indicates the atmospheric stability at each level. In both cases (stable and unstable) a derivation of the conventional two-thirds power law [22] is applied to obtain the plume rise. Since we are considering non-homogeneous atmospheric conditions, a double calculation of plume rise is needed for each time step. One at the initial time and another at the end of the time step, i.e., thirty seconds later. So, the plume or particle rise, $\triangle h$, (it depends on what kind of model is being used) will be the difference between the rise for the final position (X_{end}) and the corresponding to the initial position (X_{ini}). The expressions used are

$$\triangle h = 2.04 \left[\left(\frac{0.86F(1 - cos(Nx_{end}/\overline{V}))}{N^2\overline{V}} + R_0^3 \right)^{1/3} \right.$$
$$\left. - \left(\frac{0.86F(1 - cos(Nx_{ini}/\overline{V}))}{N^2\overline{V}} + R_0^3 \right)^{1/3} \right] \tag{3.9}$$

under convective conditions while, for neutral and stable conditions:

$$\triangle h = 1.35F^{1/3} \left[\frac{x_{end}^{0.58} - x_{ini}^{0.58})}{V} \right]. \tag{3.10}$$

The constants that appear in these formulae have been used from laboratory experiments and direct measurements of wind speed, plume rise and dispersion [20]. F is the buoyancy parameter, $\overline{V}$ the velocity at the stack, R_0 is the stack radius, N the Brunt-Väisäilä buoyant frequency and $\overline{V}$ is, at least, equal to the critical velocity $\overline{V}_c = 5.3 \cdot 10^{-4} F^{1/3} Z_s^{2/3}$, where Z_s is the stack height equal to 350 meters.

4   Results and discussion

As Pontes Power Plant is located in the north west of Spain. This area is characterized by steep hills and sea inlets bathed by the Atlantic Ocean. Both

M.J. Souto et al.

(a)

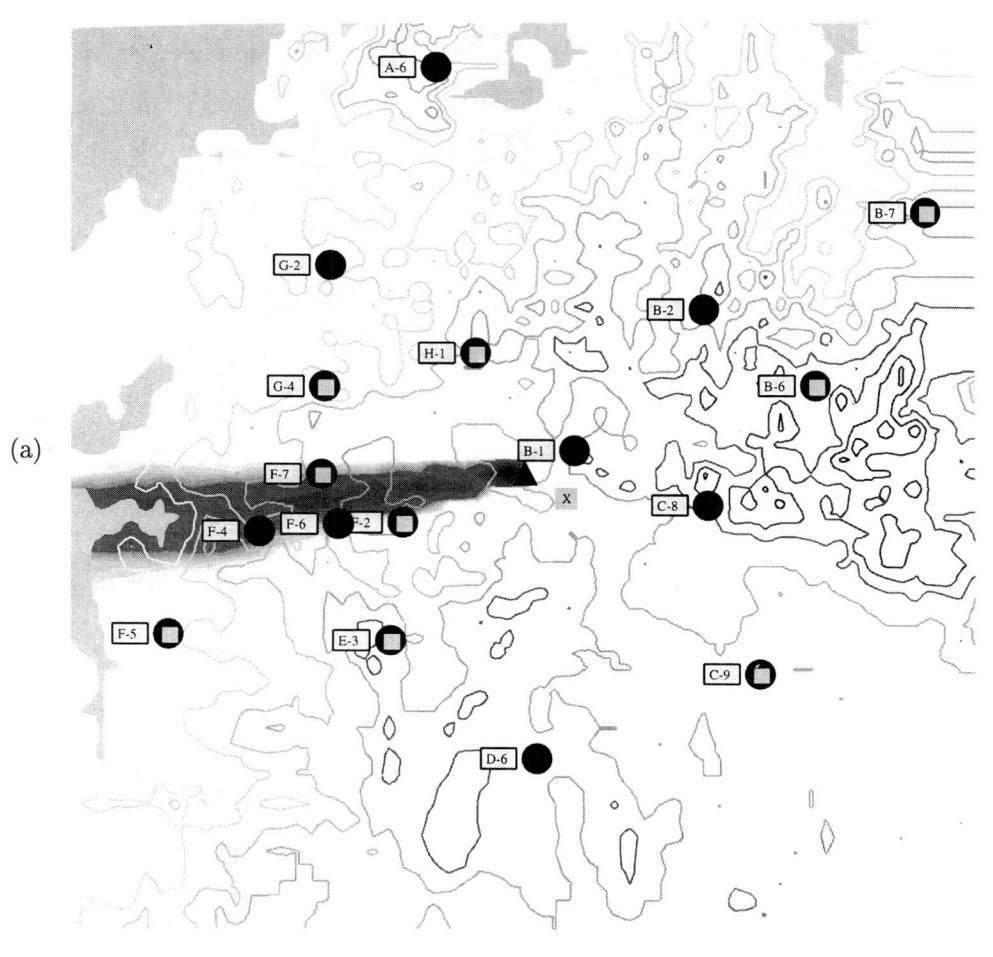

Figure 1.

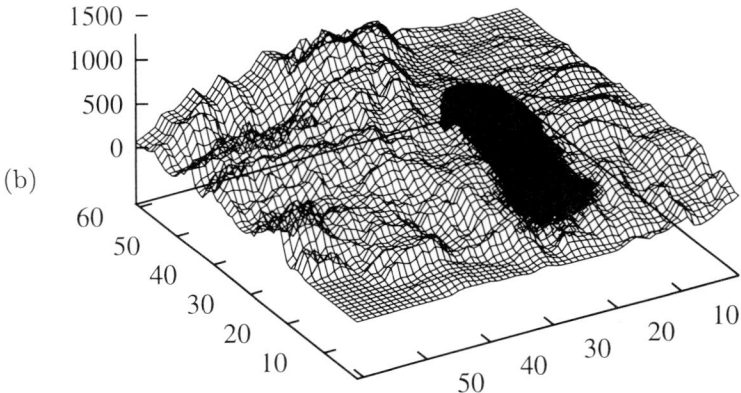

1500 –
1000 –
500 –
0 –

(b)

60
50
40
30
20
10

50
40
30
20
10

Figure 1. Horizontal plume dispersion calculated, (a) with the puff model, and (b) with the particle model, under convective conditions (17:00 LST)

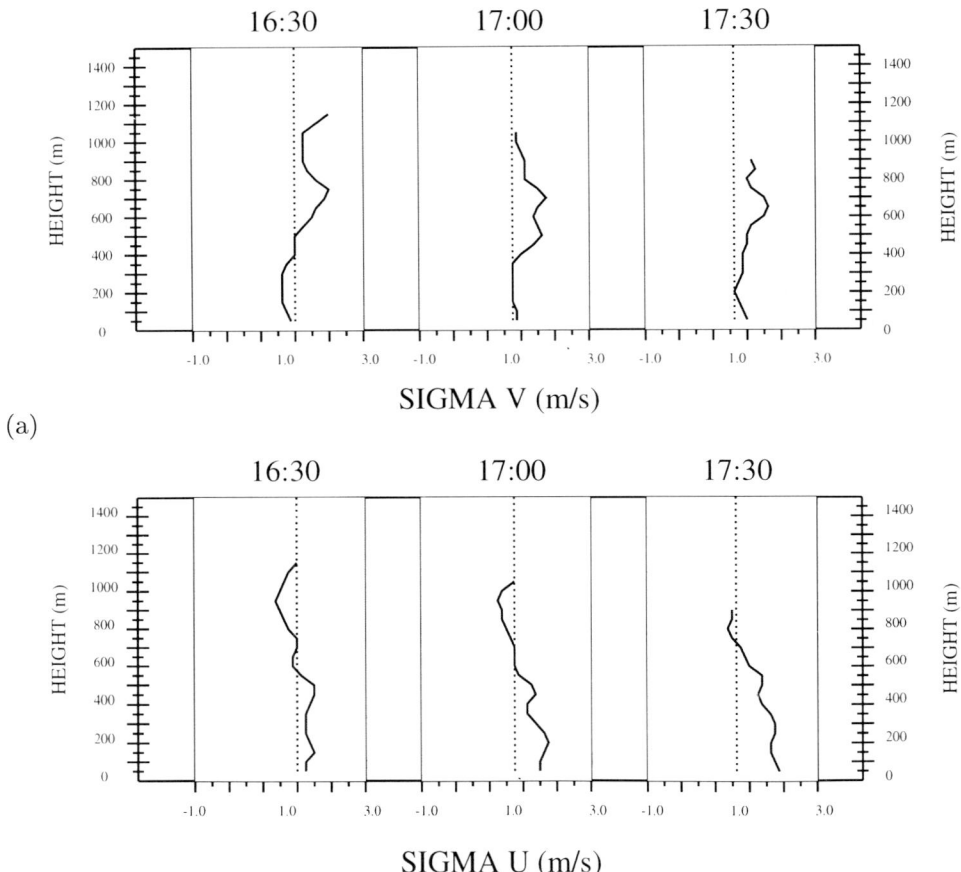

(a)

Figure 2.

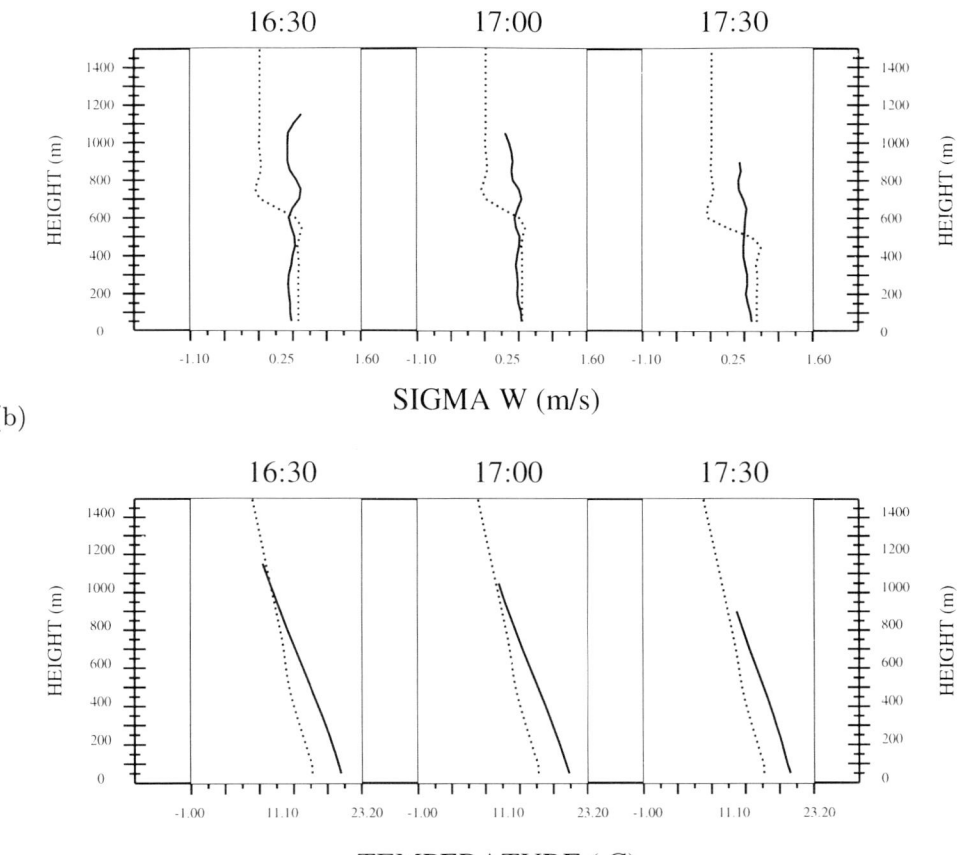

(b)

Figure 2. Comparison between model results (dots) and sodar measurements (continuous line) of standard deviations and temperature under convective conditions. Vertical-axis represents height in meters

M.J. Souto et al.

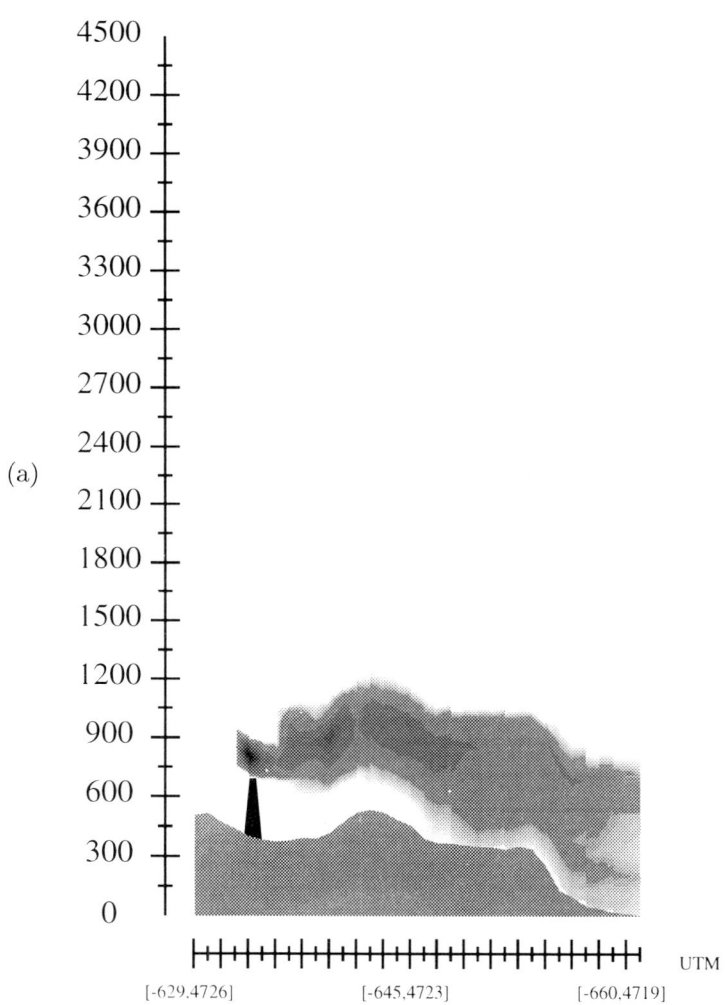

(a)

Figure 3.

(b)

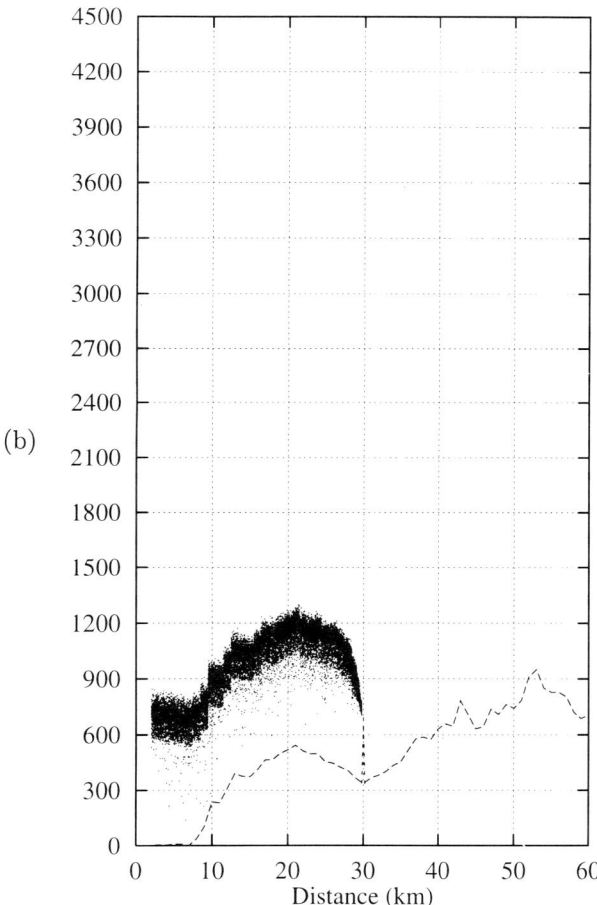

Figure 3. Vertical plume profile obtained at 17:00, (a) with the puff model, and (b) with the particle model. Note that while the x-axis in (a) is shown in UTM coordinates, in (b) the x-axis is shown in our local grid coordinates. Height is in meters for both representations

M.J. Souto et al.

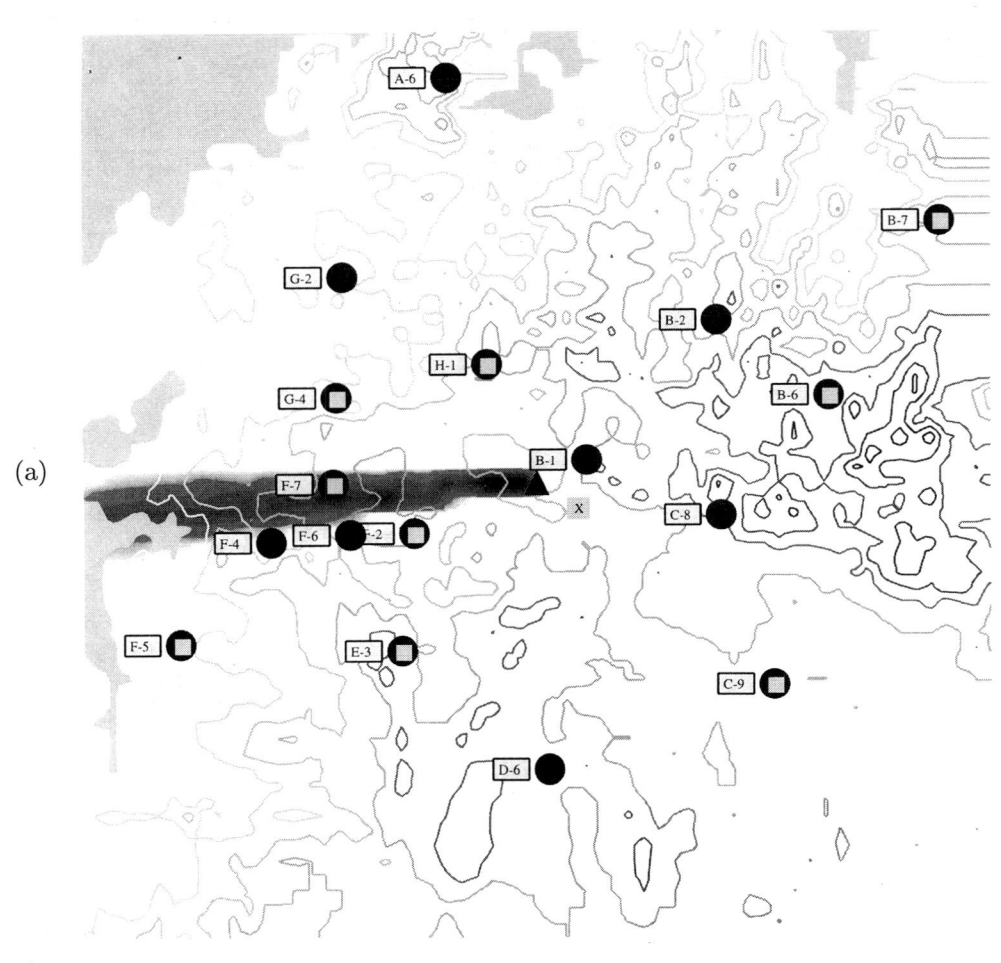

(a)

Figure 4.

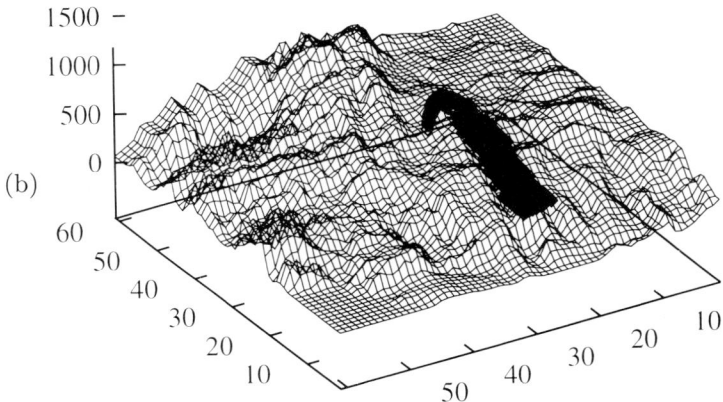

(b)

Figure 4. Horizontal plume dispersion calculated, (a) with the puff model, and (b) with the particle model, under stable conditions (23:00 h)

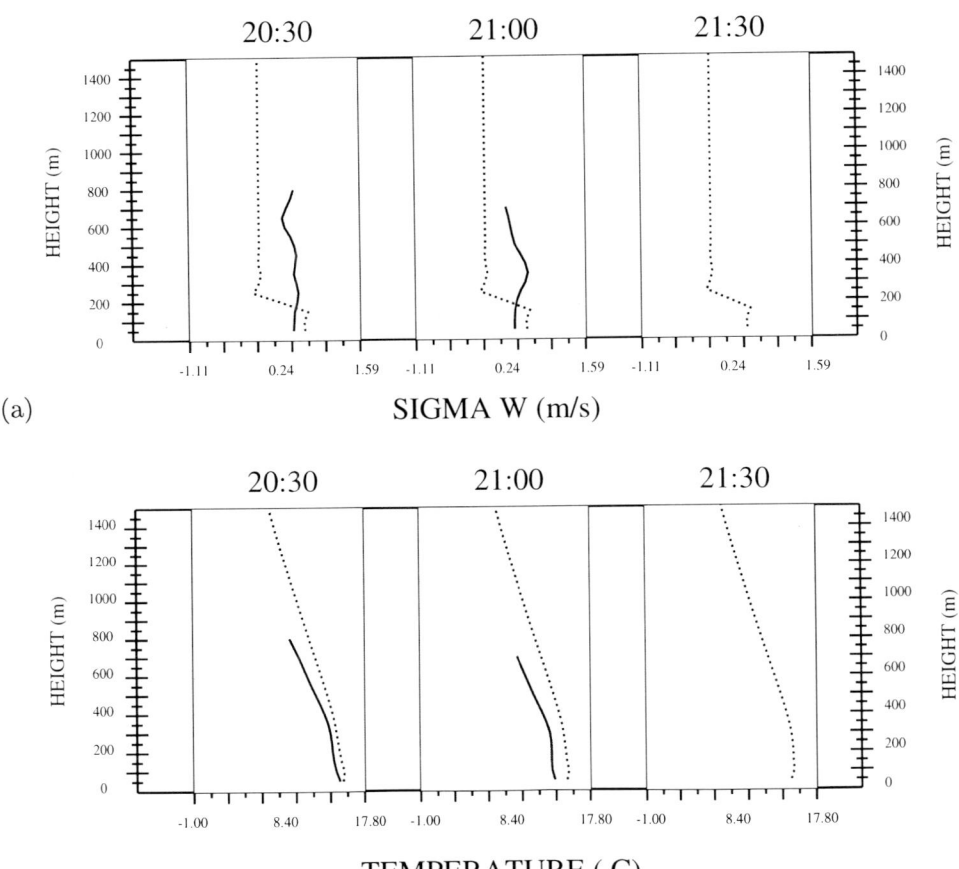

(a)

Figure 5.

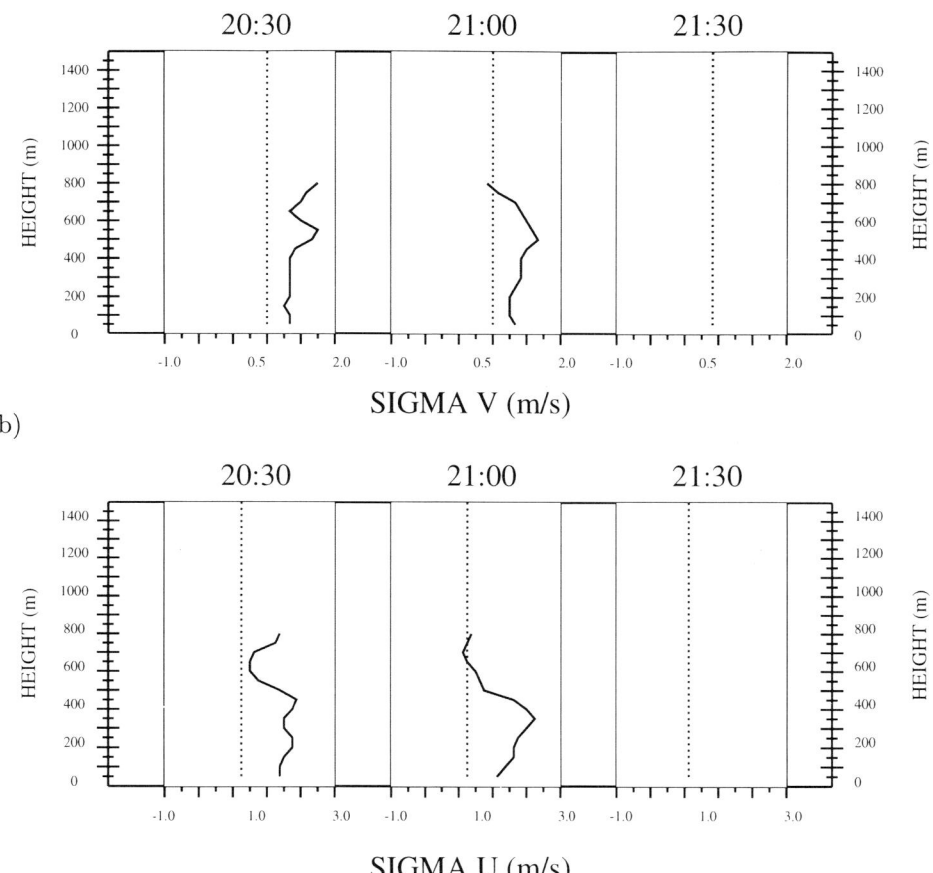

Figure 5. Comparison between model results (dots) and sodar measurements (continuous) of standard deviations and temperature. Vertical-axis represents height in meters

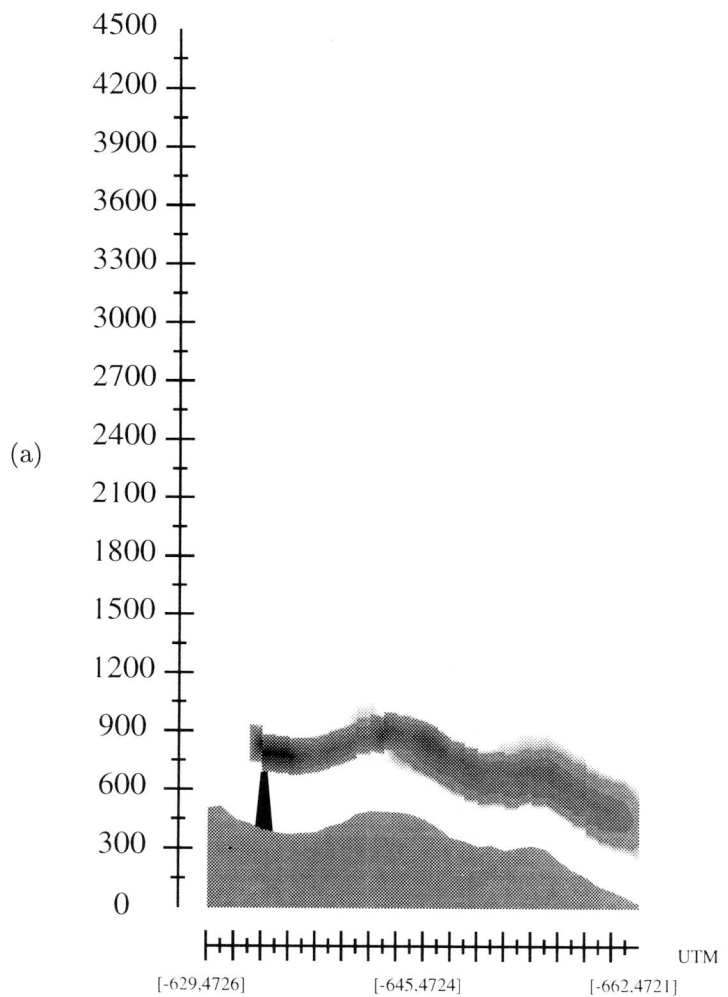

Figure 6.

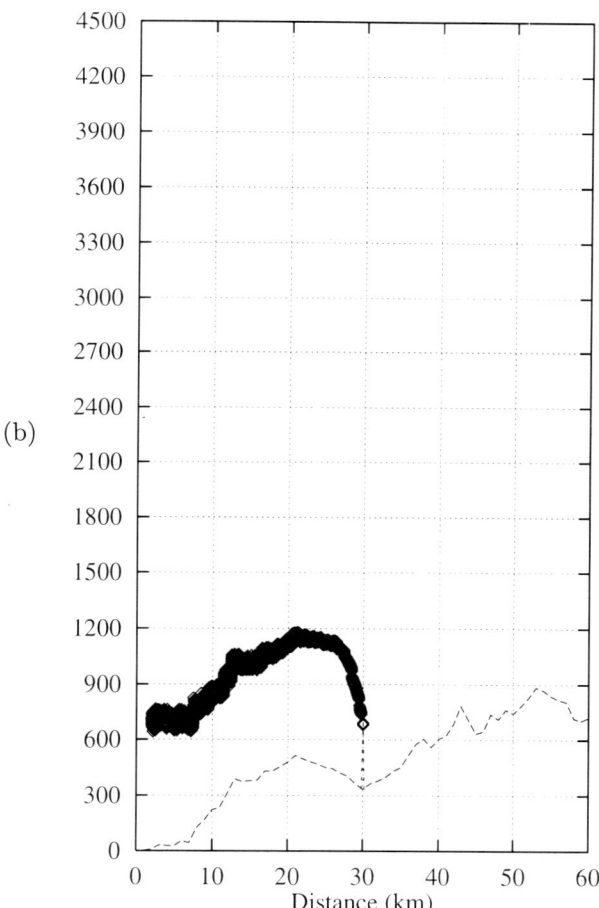

Figure 6. Vertical plume profile obtained at 23:00, (a) with the puff model, and (b) with the particle model

the meteorological and Lagrangian dispersion models are run for forecasting SO_2 glc around this power plant, within a radius of 30 km. A grid of 9 meteorological towers, one Remtech PA-3 sodar and 17 SO_2 glc remote stations provide 5-minutes average data for validation of models results. Here in this study, we have selected September 3rd, 1996, for research purposes. During this day, middle clouds, temperatures ranging from 11 to 20 degrees and easterly winds were observed. Results will be presented both for the puff and particle model under unstable and stable conditions for comparison.

Figure 1 shows the horizontal plume dispersion obtained with the puff model (a) and with the particle model (b). The horizontal area represented is 60×60 km. Both models agree quite well, with the maximum of the horizontal dispersion of about 12 km in both cases.

The parameterizations used for horizontal wind speed standard deviations as well as the vertical profile of temperature are compared with measurements obtained with the Sodar in Figure 2. Here, a nearly unstable layer grows from the surface up to 500 meters. A stable layer follows from 500 m to 1000 m where the plume is embedded. A nearly adiabatic layer appears at the upper levels. The lack of a parameterization for vertical wind speed standard deviation at levels higher than the mixing height is the cause of the important differences observed between measurements and model predictions. Note the high accuracy of the forecasted value of σ_w below the PBL height (about 600 meters at 17:00 LST). Above this level, the predicted values are forced to tend to zero, while measured values do not show this behavior.

The predicted values of σ_w are used both by the puff and particle models to predict the ground impact observed around 17:00 LST. Figure 3 shows both vertical plume profiles simulated under convective conditions. Puff model and particle model can channel the plume in the stable layer and predict its impact on the ground, specially important about 20 km far from the stack. These results can be compared with the measurements obtained at F-stations, and allow us to confirm the correct trajectory and impact of the plume simulated.

Under stable conditions (21:00 LST), Figure 4 shows the horizontal plume dispersion obtained with both models. They are very similar again, but in this case the plume is less developed than for convective conditions, with the maximum value about 8 km. The plume has not rotated since the afternoon and the horizontal projection remains over F-stations.

Figure 5 shows predicted and measured values of σ_u and σ_v for stable conditions as well as for temperature and σ_w. Here the horizontal dispersion could not be reproduced to match the measured values so well as under unstable conditions and the model obtain higher values, specially near the ground. The model reproduces the surface inversion generated due to radiation that grows as the night goes by. At 21:00 LST it is situated at about 200 meters. From 200 m to the upper levels a stable temperature profile is predicted. Due to the small value of the PBL height under this stable conditions, the σ_w values are different from sodar measurements almost at all heights due again to the lack of a parameterization for vertical wind speed standard deviation at levels higher than the PBL.

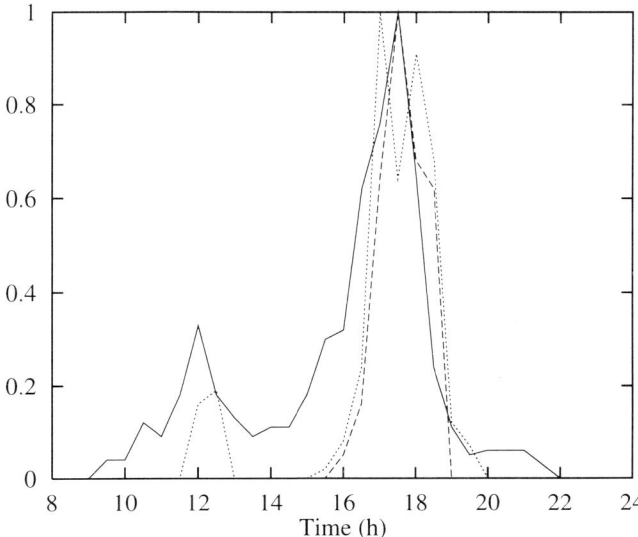

Figure 7. Time evolution of SO_2 glc measured (continuous line) and predicted with puff model (dashed line) and particle model (dots)

The plume shows in this case less vertical dispersion than in convective situation and no ground impact occurred, as shown in Figure 6, neither with the puff model (a) nor with the particle model (b), in accordance to the measurements obtained in the glc stations.

To summarize, in Figure 7 we show the percentage of glc measured at all the F-stations during the whole day in comparison with that predicted by both dispersion models: the puff and particle models. Note, that while both models reproduce the afternoon impact (around 17:00 LST), only the particle model is able to reproduce the small impact observed at noon. Under stable conditions, no important impacts were recorded and the models also reproduce this situation.

5   Conclusions

An air pollution package, based in the use of models that allow the prediction of the impact of the plume on the ground 24 hours in advance, is presented. The results show a good agreement between glc predictions and measurements. But, more effort is required to improve and calibrate the models.

This package is being applied, on a trial basis, at As Pontes Power Plant to optimize the consumption of imported sub-bituminous coal to be mixed with the local lignite, for enviromental reasons.

Acknowledgements

This work is financially supported by Endesa.

Bibliography

1. Pérez-Muñuzuri, V., Miguez, G., Souto, M.J., Souto, J.A. and Casares J.J. (1996) "A micro-mesoscale meteorological model for prediction of ground level concentration of pollutants near a power plant", Anales de Fisica, **92**, 38-48.

2. Souto, J.A., Pérez-Muñuzuri, V., Ludwig, F.L. and Casares, J.J. (1994) "Meteorological and atmospheric diffusion model for air pollution forecasting" In *Computer Techniques in Environmental Studies V*, Pollution Modeling, I, Zanetti, Computational Mechanics Publications, UK, pp. 281-288.

3. Pielke, R. A. (1981). "Mesoscale meteorological modeling", Academic Press, NY.

4. San José, R. (1991). "A simple approach to evaluate mixed layer depths", *Env. Software*, **6**, 161-167.

5. Pérez-Muñuzuri, V., Souto, M.J., Casares J.J. and Pérez-Villar, V. (1996) "Terrain-induced focusing of wind fields in the mesoscale", *Chaos, Solitons and Fractals*, **9**, 1479-1494.

6. Smeda, M.S. (1979) "Incorporation of planetary boundary-layer processes into numerical forecasting models". *Bound. Layer Meteorol.*, **16**, 115-129.

7. Stull, R.B. (1991) "An introduction of boundary layer meteorology", Kluwer, Dordrecht.

8. Fernández, J.F., Cremades, L. and Baldasano, J.M. (1994). "Dispersion modelling of a tall stack plume in the Spanish mediterranean coast by a particle model", *Atmos. Env.*, **11**, 1331-1341.

9. Hanna, S.R. (1979). "Some statistics of Lagrangian and eulerian wind fluctuation", *J. Applied Met.*, **18**, 518-531.

10. Smith, F.B. (1968). "Conditioned particle motion in a homogeneous turbulent fields", *Atmos. Env.*, **2**, 491-508.

11. Pasquill, F. (1974). "Atmospheric Diffusion", Halsted Press, Wiley, NY.

12. Kaimal, J.C., Wyngaard, J.C., Haugen, D.A., Cote, O.R., Izumi, Y., Caughey, S. J., and Readings, C.J. (1976). "Turbulence structure in the convective boundary layer", *J. Atmos. Sci.*, **33**, 2152-2168.

13. Caughey, S.J., Wyngaard, J.C. and Kaimal, J.C. (1979). "Turbulence in the evolving stable boundary layer", *J. Atmos. Sci.*, **36**, 1041-1052.

14. Ludwig, F.L., Salvador, R. and Bornstein, R. (1989). "An adaptative volume plume model", *Atmos. Env.*, **23**, 127-138.

15. Ludwig, F.L. (1982). "Effect of a change of atmospheric stability on the growth rate of puffs used in plume simulation models", *J. Applied Met.*, **21**, 1371-1374.

16. Draxler, R.R. (1976). "Determination of atmospheric diffusion parameters", *Atmos. Env.*, **10**, 99-105.

17. Irwin, J.S. (1979). "Estimating plume dispersion -A comparison of several sigma schemes", *J. Applied Met.*, **2**, 92-114.

18. Zhang, X. and Ghoniem, A. (1994) "A computational model for the rise and dispersion of wind-blown, bouyancy-driven plumes. II.-Linearly stratified atmosphere", *Atmos. Env.*, **18**, 3005-3018.

19. Briggs, G.A. (1975). "Plume rise predictions", In *Lectures on air pollution and environmental impact analysis*, Haugen D. A., Am. Meteorol. Soc., Boston, pp. 59-111.

20. Bennet, M., Sutton, S. and Gardiner, D.R.C. (1992) "An analysis of lidar measurements of buoyant plume rise and dispersion at five power station", *Atmos. Env.*, **18**, 3249-3263.

Separated Flow Around Bluff Obstacles at Low Froude Number; Vortex Shedding and Estimates of Drag

J.C.R. Hunt* and H.J.S Fernando**

*Department of Applied Mathematics and Theoretical Physics, University of Cambridge, and ** Arizona State University, USA

Abstract

For stably stratified flows around bluff obstacles we review different concepts for how sloping vortices are shed from the obstacle surfaces and develop in the regions of separated flow.

Approximate analyses are developed for the vortices and the disturbances to the density field that they generate. New experiments were undertaken to study how these shed vortices oscillate, distort and break up; results are broadly consistent with the analysis. Thence a tentative synthesis is proposed for the separated wake structure downstream of bluff obstacles and of mountains in stable flows. This explains the observations that the separated regions have a minimum thickness at one particular Froude number which, it is suggested, also implies a minimum in the drag.

1 Introduction

Three hypotheses have been suggested about the form of the flow downstream of bluff bodies (hills of height H) placed in uniform flow at low Froude number, ($F = U/NH$). See Figure 1.

- (Hyp. 1) Consider inviscid flow in the middle layer [M] below the level $z = H_{ds}$ of the dividing streamline height above which is the "top layer" [T] (recently reviewed and analysed by Hunt et al 1997). Note that $H_{ds} = H - \alpha U/N$ where α is a coefficient equal to about 1.0 ± 0.3. By assuming no coupling between the flow in these two layers [1] deduced that [M] is a region in which the flow moves in horizontal planes with zero vertical vorticity, so that the flow does not separate.

- (Hyp. 2) (a) For most viscous flow, experiments on bluff bodies (where heights H are of the same order as the diameter L and the value of $F \gtrsim 0.02$) the interpretation has been that, below [T], separation is **similar** to that around cylindrical bodies with a cross section corresponding to that of the hill at some heights within [M] because the flow moves in approximately horizontal planes.

(b) The y co-ordinate of the separation location $y_s(z)$ is determined quantitatively by the analogous cylindrical body **at every level**, i.e. $y_s(z) = y_s(R(z))_{\text{cyl}}$. It is observed that there are some variations of separation location with Froude number and Reynolds number that are not explained by this analogy. (See [2] and references therein.)

- (Hyp. 3) (a) For viscous flow, an alternative concept has been proposed by [3], based on their experiments of flow around spheres, that, when $F \ll 1$, the sides of the separated flow regions are approximately vertical, and its upper surface (presumably also below $[T]$) is approximately horizontal. This implies that the y co-ordinate of the separation locations $y_s(z)$ on a bluff body are approximately in the same vertical plane parallel to the flow ($y_s(z) = $ constant).

(b) Hyp. (2b) and (3) are mutually contradictory, but Hyps. (2a) and (3) could be consistent if there is one level (z_s) at which the separation position for all z (below z_s) corresponds to that of two dimensional flow around a cylinder having the equivalent shape as the cross section of the hill at height z_s, (i.e. $y_s(z) = y_s(R(z))_{\text{cyl}}$).

(c) If hypothesis (2a) is valid, hypothesis (3b) implies (though [3] did not state as much) that the separation position is defined by the equivalent cylinder at the level just below the top layer $[T]$, i.e. $z_s = H_{ds}$.

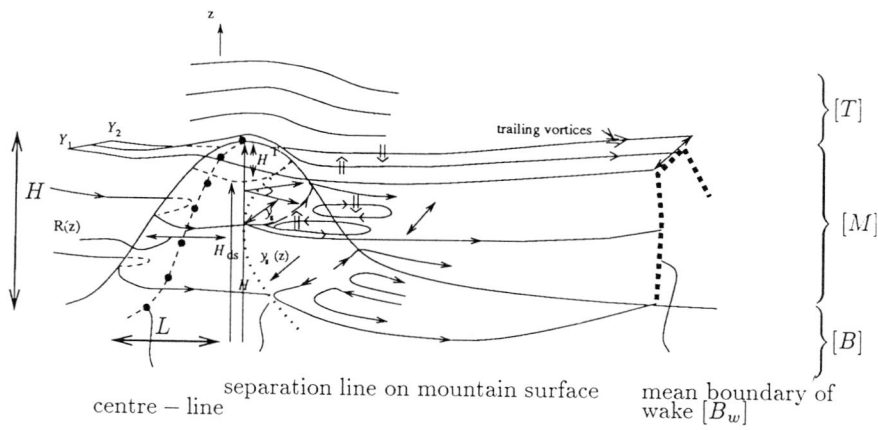

separation line on mountain surface   mean boundary of
centre − line   wake $[B_w]$

Figure 1. Schematic of side view / rear view of wake for low Froude number flow over a mountain

No theoretical reasons have been advanced for the third hypothesis, nor have any suggestions been made about its limitations. Detailed measurements of the shape of the separated flow region and the variation with height of the separation position have not been made by those advocating hypothesis (2). If the separation position is only measured at one level (as in [2]), the data is not detailed enough to distinguish between hypotheses (2b) and (3).

The purpose of this note is to develop some physical concepts and a brief flow visualisation experiment on shed vortices, and then propose a new hypothesis (N) for the separated flows at low F, based on hypotheses (2) and (3).

2 Wake vortices in stratified flow

2.1 Concepts on sloping vortices

If the hypothesis (2) is valid, the surface defined by the streamlines of the separated flow is sloping in the sense that there is a finite angle $\bar{\theta}$ between the tangent to the surface and the horizontal (see Figure 2). This implies that axes of the vortices that form in the separated shear layer are also sloping at an angle $\bar{\theta}$ to the horizontal or $\theta(= \pi/2 - \bar{\theta})$ to the vertical. We make an approximate theoretical analysis of their dynamics using two different approaches.

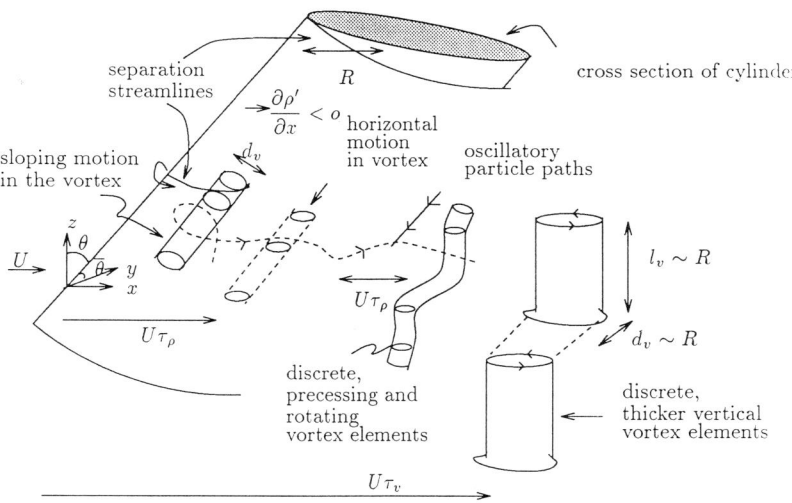

Figure 2. Sketch how sloping vortices falling up in the separated shear layer of a sloping bluff obstacle becomes a line of discrete fat vertical vortices as the wake develops downstream

2.1.1 Vortex dynamics

First consider how the vortex is formed in the separated flow and thence the perturbation density distribution that it causes. If the Froude number of the vortex is $F_v = \frac{\Gamma}{d_v^2 N} \lesssim 1$, where Γ is the circulation, the vertical velocities in the vortex are of the order $\frac{\Gamma}{d_v} \sin\theta$. [If $F_v \ll 1$, low Froude number analysis is required.] This induces density fluctuations denoted by ρ_v'. Since the vortex is generated by the "rolling up" (i.e. large scale instability) of the shear layer shed from the sloping separation line, there is a net rise in density in the separated layer so that the perturbation in density around the vortex is not axisymmetrical. As the vortex moves downstream it entrains fluid from outside the shear layer, so that ρ' decreases. Thus there is a weak overall "shear layer" gradient $(\partial\rho'/\partial x)_{sl} < 0$ over the cross section of the vortex. Since the vortices as they are shed vary in strength and orientation along their length, ρ_v' and $(\partial\rho'/\partial x)_{sl}$ vary along the lengths of the vortices.

The vertical velocity within the vortex induces a density gradient $\partial\rho'/\partial y \sim -\left(\frac{\partial w}{\partial y}\right)t\partial\rho/\partial z \sim -\sin\theta.t.(\Gamma/d_v^2)\partial\overline{\rho}/\partial z$ and a small corresponding buoyancy force, which generates a horizontal component of *vorticity*, so that $d\omega_x/dt \sim -(N^2 t\Gamma/d_v^2)\sin\theta$. This tends to reduce the magnitude of the vertical *velocity* in the vortex (i.e. $\partial w/\partial y$) and the density fluctuations on a time scale τ_ρ of the order N^{-1} but does not immediately affect the overall alignment of the vortex (i.e. the angle θ).

If $F_v \ll 1$, then $\sin\theta \sim F_v$ (based on the calculation of mean flow around a bluff body at low Froude number ([4]). In that case the density fluctuations are even weaker and any vertical velocity fluctuations are damped out even sooner. Consequently $(\partial\rho'/\partial x)_{sl}$ is also reduced by a factor of the order of F_v.

Once the density fluctuations have decayed on the time scale of N^{-1} or less, non-linear interactions between the residual vertical vorticity at different horizontal levels in a sloping vortex begin to be significant. These tend to cause vortices at different levels to induce circulating motions by vortices at other levels and thereby to increase the overall width of the vortices and to reduce their slope. This process occurs on a larger time scale of the order d^2/Γ. The sloping regions of vorticity are unstable to small disturbances, tending to create discrete organised vertical vortices. See Figure 2.

The weak larger scale density gradients in the rolling up shear layer also affect the overall alignment of the vortex. The effect of the streamwise density gradient $(\partial\rho'/\partial x)_{sl} < 0$ leads to $d\omega_y/dt < 0$ and a precession of the vortex as a whole on a time scale N^{-1}.

Thus although

$$\tau_\rho \sim N^{-1}, \tag{2.1}$$

the time τ over which the vortex oscillates, thickens and tends to become vertical is given by

$$\tau\Gamma/d_v^2 \sim 1. \tag{2.2}$$

Typically the fully developed vortices have a scale of the order of the radius $R(z)$, of the cross section of the obstacle at that height (Hyp. 2a), and a circulation of the order of $\lambda U_0 R$, where the factor of λ is about 0.3 [5]. It follows that

$$\frac{\tau U}{R} \sim 1. \tag{2.3}$$

Therefore on this argument at low Froude number the vortices should become vertical within a distance of their formation that is comparable with R, although within the vortex the motions become horizontal on the shorter time scale of the order $F_v R/U$.

2.1.2 Vorticity analysis using RDT

Our second approach is to use linear "rapid distortion" theory (for example [6]) to examine in more detail how the perturbations of density caused by the sloping vortices evolve with time. The limitation of this approach is that it is based on the assumption that these perturbations are generated instantaneously, whereas in fact they grow with the development of the vortices. On the other hand the analysis does lead to a more detailed study of the evolution of the internal structure of the vortices and shows how this depends on the initial form and orientation of the vortices.

The detailed analysis given in the appendix focusses on the development of a density perturbation, $\rho(\mathbf{x}, t)$ whose initial form $\rho_o(\mathbf{x})$ at $t = 0$ is oscillatory in the x direction. But it is confined within a strip of thickness d in the yz plane, lying at an angle $\bar{\theta} = (\pi/2 - \theta)$ to the y-axis where $\alpha = \tan\bar{\theta}$. Thus

$$\rho_o = \rho_{o_{mx}} e^{-n^2/d^2} e^{ik_1 x} \quad \text{where} \quad n^2 = (z - \alpha y)^2/(1 + \alpha^2). \tag{2.4}$$

The buoyancy frequency of the background stratification is N.

It is found that when $Nt \gg 1$ (using the method of stationary phase)

$$\rho \sim \rho_{o_{mx}} \frac{d}{l(t)} (\sqrt{1 + \alpha^2}) e^{i(Nt + k_1 x + n^2(1 + \alpha^2)/d^2)} e^{-n^2(1 + \alpha^2)d^2/l^4} \tag{2.5}$$

where the scale length $l(t) = \sqrt{Nt}/k_1$, increases with time. Because the potential energy of the fluctuations propagates outwards, the density perturbations decay (with the inverse square root of time) on a time scale $(1/N)$ or $(\tau U/R) \sim FH/R$. This is the same as the time scale τ_ρ indicated by (2.1).

Note that this time scale is smaller than that at which density perturbations in homogeneous 3-dimensional turbulence decay because in such a turbulent flow at any point an equal amount of energy is propagating away as is received! This explains why the lateral variation with the normal distance n from the vortex axis of density oscillates with a wave length proportional to $\sqrt{Nt}$. However the perturbation **decreases** and decays over a lateral distance of order $l^2/((\sqrt{1 + \alpha^2})d)$, which increases with time and decreases as α tends to $\pi/2$. Thus the more vertical the initial orientation of the vortices the **narrower** they remain as they develop; correspondingly the more horizontal vortices widen faster. The phase variation is significant physically because it implies that some of the vorticity changes at the outer edge of the vortices, especially those at shallower angles, which would include some rapid mixing in these cases.

Our physical analysis of how the vortex distorts the density contour in Section 2.1.1, implies that the initial density perturbation is proportional to $\cos\bar{\theta}$, i.e. $\rho_{o_{mx}} \propto \frac{1}{\sqrt{1+\alpha^2}}$. This suggests that in (2.3), the maximum amplitude at time t is independent of the slope. However, since when vortices are shed from a separation line there is some variation in their orientation, α is not a fixed quantity. Now (2.4) shows that there is a selection mechanism to preserve the more vertically orientated disturbances further downstream and to ensure that they spread outwards less rapidly.

In this analysis the effects of any initial horizontal velocity fluctuations are neglected because the model is linear and these fluctuations are not affected by the density gradient. Although the shed vortices have a component of velocity that is horizontal, in the linear analysis this motion does not change downstream. Therefore it is only the vertical motions and density perturbations of the vortices that need to be analysed when considering the linear aspects of the selection mechanism that controls the development of vortices downstream.

2.2   Flow visualisation study on wake vortices

A simple flow visualisation experiment was performed in the tow tank of the Environmental Fluid Mechanics laboratories at Arizona State University to examine this concept. A circular cylinder with a diameter of 4 cm and length of about 0.3 m were towed at 0.5 cm sec in a stratification of about 2 rad s^{-1} so that $F_R \simeq 0.06$. The axis was 30° to the vertical. Fluorescein dye was introduced upstream of the obstacles at two levels separated by about 3 cm. It was found that the flow separated at the edges of the cylinders and that sloping vortices were shed parallel to the edges. The orientation of the vortices became approximately vertical within about 1.5 diameter downstream. In most cases they precessed as their orientation changed. This is consistent with the order of magnitude estimate (2.4).

In a second set of experiments (with the axis of a circular cylinder at angle (θ) of 15° to the vertical) we also studied how the vortices broke up into discrete vertical lengths so that the vortex lines connecting these discrete vortices (and the dye lines that marked them) became horizontal at the levels of the "tops" and "bottoms" of each vertical vortex. There was also some evidence of the development of a complex radial structure of the vortices indicated by the analysis, such as reversal of vorticity and widening of the vortices. (See Figure 3.)

In a third set of experiments the axis of the cylinder was 45° to the vertical. There was a striking difference in that the vortices widened markedly to about 4 times the width of the cylinder; within the vortices the motion was horizontal

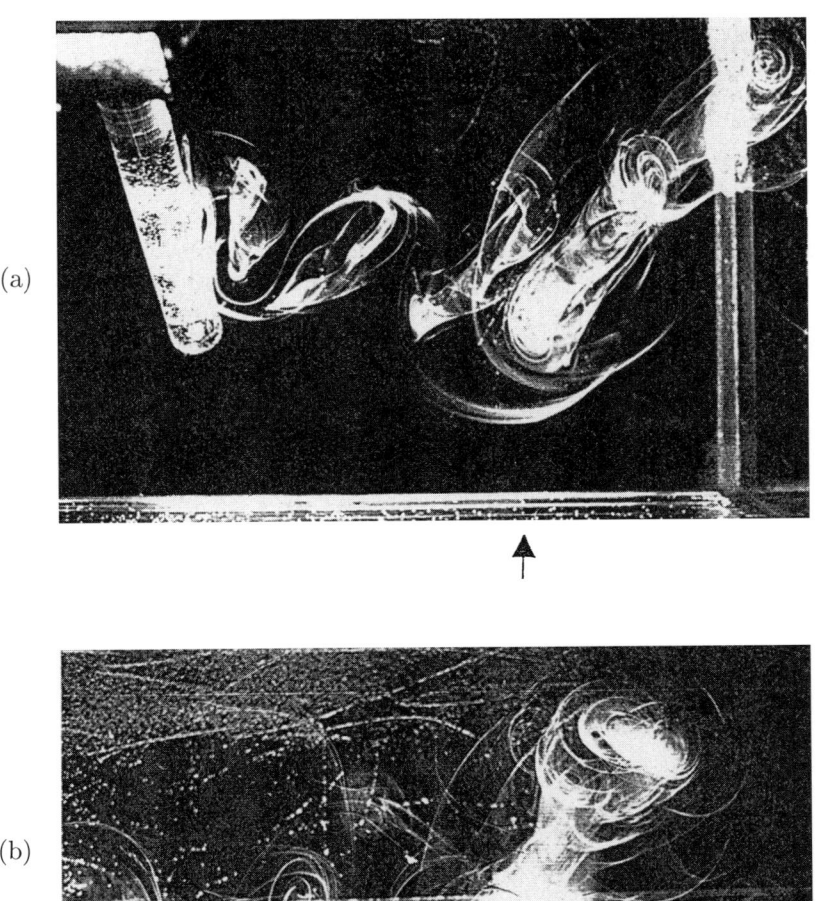

(a)

(b)

Figure 3. Photographs of experiments of vortex shedding behind a circular cylinder placed at 15° to the vertical in a plane perpendicular to the direction of motion in a stratified flow with $F \simeq 0.3$. (a) Close to the cylinder. (The arrow indicates how two vortices form), and (b) about 5 diameter downstream. (Note the complex structure and fattening of the vortices)

but their axes were highly tilted rather than having a cellular structure as is evident in the case where $\theta = 15°$.

2.3   Vertical scale of shed vortices

As the boundary layer separates at each level, the instabilities in the boundary layer trigger growing disturbances in the separated free shear layer (see for example the flow visualisation photographs in Hunt & Snyder 1980). These distances are on the scale of the boundary layer (and therefore much less than $R(z)$), and initially are only correlated vertically over this scale. As the sloping vortices grow in the separated layer, they induce vertical motion and displace denser fluid upwards above lighter fluid; in other words they induce within the vortices some hydrostatically unstable motion. This triggers some disturbances along the vortices on the scale of the diameter d_v. This may explain why long vortices do **not** form as they do in unstratified flow where in cylinder wakes they are well correlated over about 5 diameters.

Hence in our sloping cylinder experiments the shed vortices were correlated over a distance of the order R. (This is consistent with the results of [2] (for example their Figure 10d)). Because they are only correlated over a distance of the order R the vortices are shed downstream from each cross section of the sloping cylinder. This explained why the wakes of the sloping cylinders were effectively parallel to the cylinder but at the same time the vortices were vertical.

The experiments also showed, as did the earlier experiments of [7] and [8], that near the body the shed vortices were correlated in the vertical direction over a distance of about one diameter (Figure 3). (Far downstream when vortices are weaker and are advected by the mean flow, vertical velocity perturbations are weak, which leads to the vertical correlation distance of their horizontal velocity being much less than their diameter.)

3   Separated wakes in stably stratified viscous flows

The Hypotheses 2 and 3 set out in Section 1 and the concepts and experimental results reviewed in Section 2 suggest the following new hypothesis (N) for wakes when $F \ll 1$.

3.1   New hypothesis (N)

(a) For obstacles, hills etc. whose height H are of the order of or less than their radii L (defined, say at the half height $z \simeq H/2$ for a hill), then the shed vortices become approximately vertical within a radius R at each height. See Figure 4. For steep sided obstacles such as hemispheres, steep cones also, the effective radius for the application of this theory is larger than L.

The external and separated flow is approximately horizontal below the dividing streamline height $z = H_{ds}$ (the vertical velocity components are at most

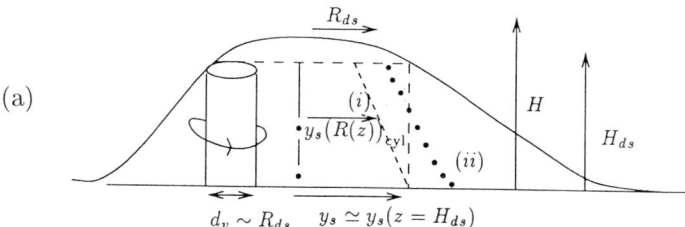

(a)

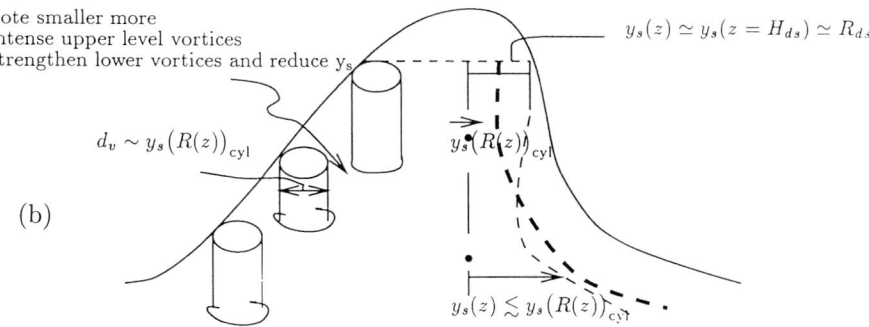

(b)

Figure 4. Sketch of the developing vortices in the wakes of hills in stably stratified flows. (a) Low aspect ratio hills: $H/L \lesssim F^{1/P}$; approximately vertical sides to separated region, (i) is typical separation line for very high Reynolds number curve; and (ii) for moderate Reynolds number, and (b) high aspect ratio hills: $H/L \gtrsim F^{1/P}$ sloping sides to separated region

of order FU_0, the numerical factor being a small fraction). The effective radius of the hill at the height $z = H_{ds}$ is R_{ds}, and the location of separation is defined as $y = \pm y_{ds}$ (if the shape is axisymmetric). Since at this height vortices are approximately vertical (within a distance of R_{ds} downstream) and have a diameter of the order R_{ds}, they are also vertically correlated over a distance of the order R_{ds}. It is implied that the vortices shed at height H_{ds} control the separation over a height R_{ds} below this level. This is only possible if $y_s(R(z))_{cyl}$ is less than or of the order of R_{ds}. (This control operates through the stronger upper vorticity intensifying the vorticity in lower layers.)

Therefore, if

$$H_{ds} \lesssim R_{ds}, \tag{3.1}$$

the lateral boundary of the separated wake region is approximately vertical. Using the mathematical description for the shape of rounded (axisymmetric) hills, namely $z = H/(1 + (R/L)^P)$ the criterion (3.1) can be re-expressed as $(H/L) < F^{1/P}$ where $P \simeq 2$. Thus to satisfy (3.1) as $F \to 0$, H/L must decrease i.e vertical wakes boundaries only occur for hills with low slope when F becomes very small.

Note that this hypothesis is consistent with the approximate calculation of [4] for matching the top layer with the flow below $z = H_{ds}$ in the middle layer. In particular the low pressure associated with downflow on the lee side of the hill in [T] matches the low pressure in the horizontal separated flow below H_{ds}.
(b) If the criterion (3.1) is not satisfied, i.e. for high narrow hills or very low Froude number

$$H_{ds} > R_{ds} \quad \text{or} \quad H/L > F^{1/P}, \tag{3.2}$$

then the vertical correlation scale of vortices in the separated shear layer (of R_{ds}) is less than the height H_{ds} of the wake. Consequently decorrelated vortices are shed at lower levels and, as with the sloping cylinder, the position $y_s(z)$ of the lateral boundaries of the separated wake corresponds approximately to that of the wake of the equivalent cylinder at each level $y_s(R(z))_{\text{cyl}}$. Near the level $z = H_{ds}$, the separation is driven by the pressure in [T] and this means that $y_s \simeq R_{ds}$. At lower levels $y_s(R(z))_{\text{cyl}} > R_{ds}$.

The implications of (3.2) on the separation position leads to two hypotheses:

Hyp N(a) for $H/L < F^{1/P}, y_s(z) \simeq y_s(z = H_{ds})$,   if $y_s(R(z))_{\text{cyl}} < 2R_{ds}$

$$\tag{3.3}$$

and

Hyp N(b) for $H/L > F^{1/P}, y_s(z) \simeq mx\{y_s(z = H_{ds}), y_s((R_s(z))_{\text{cyl}}\}$.   (3.4)

Note that as H/L changes from less than to greater than $F^{1/P}$, these hypotheses imply a sharp rise in y_s, provided $H/L \ll 1$. In reality there is a smooth transition because the vortex dynamics change smoothly from one region to another.

3.2   Comparison with experiments

The hypothesis N can be tested against the experiments of [2] who measured y_s near $z = 0$ (on the centreplane of a sphere of radius L for which $H/L = 1$, as a function of F and Reynolds number. According to (3.3) and (3.4) (which for a sphere become $F^{1/2} < 1/12$ or $F^{1/2} > 1/12$), the value of $y_s(z = 0)$ should be approximately $y_s(R(0))_{\text{cyl}} \simeq L \sin 110°$ for $F \lesssim 0.5$ and equal to $R(z = H_{ds}) \sin 110°$ for $1 > F \gtrsim 0.5$. Thus near $F \simeq 0.5$, there should be marked reduction in the width of the separated region from $0.9L$ to $0.6L$, (as separation angle θ_s changes from $110°$ to $140°$). The experiments at $Re = 10^2$ showed a rapid increase in θ_s from $110°$ to $150°$ as F increased from 0.2 to 0.5.

This is a satisfactory level of agreement, eventhough the critical value of F is not very small. Note that in the experiments of [8], in the cases where $F \simeq 0.2$ and 0.4, $H/L \sim F^{1/2}$, there is evidence from flow visualisation that θ_s is maximum at $z = H_{ds}$ and decreases as z decreases but because $R(z)$ rapidly increases, $y_s(z)$ increases slightly as z decreases.

Vosper et al [9] have measured the Strouhal number of vortices in the wakes of hemispheres and cones and thence have inferred the variation of the wake widths. They conclude that the wake widths decrease as F decreases and that, for the hemisphere, the hypothesis N (a) (Equation 3.3) is approximately satisfied by the data for $F \gtrsim 0.1$, even though $H/L > 1$. (This is because the theory for mountains with low slope does not exactly apply to hills with such high slopes.) However, for steep sided cones the estimate of y in (3.3) appears to be an underestimate.

There are many photographs of snow being blown around in the wakes of steep mountains and yet being confined within the separation lines; this corresponds to the situation defined by (3.4).

3.3 Review of the hypothesis

We can now comment on how our hypotheses N (a) and (b) relate to the Hypotheses 1, 2a, 2b and 3 described in Section 1. It is clear that N (a) corresponds to Hyp. 3 and to Hyp. (2a) for the range of H/L and F defined by (3.1), and that N (b) corresponds to Hyp. (2a) and (2b) for the range defined by (3.2).

It also follows that we can consider Hyp. (1) as a limiting case for a range of values of H/L and F, since, as $F \to 0$, the depth of the layer [T] and the radius at H_{ds} both tend to zero, i.e. $R_{ds} \to 0$. Therefore from (3.3) $y_s/L \to 0$, if $(H/L)F^{-1/P} \to 0$ and $F \to 0$. This is likely to be the only practical situation in which a viscous flow could approximately satisfy the Drazin limit. For example if $F = 0.05$, and $(H/L) = 0.1$, then $y_s/L \sim 0.2$. This corresponds approximately to the **inviscid flows** simulated numerically by [10], when the width of the separated region was very small.

4 Implications for the drag

This note is a tentative synthesis of previous research about separated and non-separated flows above hills and bluff bodies in stably stratified flow at low Froude number. We have used experimental results, physical order of magnitude arguments and idealised model calculations.

The results show why previous concepts for wakes may be valid, but for restricted ranges of hill slope and Froude number. This synthesis is an important element in the conceptual model and approximate calculations of strongly stratified flow over hills by [4].

A particularly important application of this synthesis is the estimation of the drag of the mountain in strongly stratified flow which consists of the wave drag from the $[T]$ region

$$D^{(W)} \sim \rho U^2 HLF^{3/2} \text{ (for a rounded hill)} \tag{4.1}$$

and the "form" drag $D^{(F)}$ caused by the flow below the dividing streamline height H_{ds}.

By considering the wake width y_s,

$$D^{(F)} \sim \rho U^2 H_{ds} y_s. \tag{4.2}$$

Thus from (3.3), for $H < LF^{1/P}$,

$$y_s \sim L_{ds}, \quad D^{(F)} \sim \rho U^2 H_{ds} L (1 - F) F^{1/2} \tag{4.3}$$

while from (3.4), for $H \gtrsim LF^{1/P}$,

$$y_s \sim L, \quad \text{and} \quad D^{(F)} \sim \rho U^2 HL; \quad \text{if} \quad (H/L) \ll 1. \tag{4.4}$$

Note that there is a discontinuity between the expressions (4.2) when $H = LF^{1/p}$. This would be a smooth transition in reality as the vortices are adjusted from the vertically correlated to the decorrelated form.

Thus for a rounded hill the analysis reveals that the total drag is of the order $\rho U^2 HL$ when $F \ll (H/L)^2$ and when F rises to about 1.0, but D has a minimum value of the order $\rho U^2 H^2$ when $F \sim (H/L)^2$. The expression in (4.3) has a **maximum** when $F \simeq 1/3$ (when the effect of the decreasing depth of region (M) is balanced by the increasing wake width.) Figure 5 shows a graph of the drag components and total drag derived from (4.1), (4.3) and (4.4).

The results developed here indicate that when the condition (3.1) is satisfied, the form drag $D^{(F)}$ may be significantly reduced. This sensitivity of form drag to variation in F is a surprising result which has not yet been examined experimentally. Its only experimental support comes from the laboratory flow visualisation results of [2] and [9]. Although their hills had finite slopes, the vortex dynamics should be *similar* to that over hills with low slope at low Froude number. It is not consistent with the model of [11] which was used to estimate the mountain drag in the ECMWF forecasting system; but in that case Coriolis effects need to be considered.

If the same arguments are applied to a steep hill where $H/L \gtrsim 1$, such as a cone, the expression (4.1) for the wave drag (from region $[T]$) is replaced by $D^{(W)} \sim \rho U^2 HLF^2$, and the form drag affecting the region $[M]$ is now given by $D^{(F)} \sim \rho U^2 H_{ds} L \sim \rho U^2 HL(1 - \alpha F)$ because the height of the shed vortices and the wake is still limited by the height H_{ds}. In this case the total drag decreases as F increases until the wave drag becomes significant. This is consistent with the experiments on the drag of cones and hemispheres of [9].

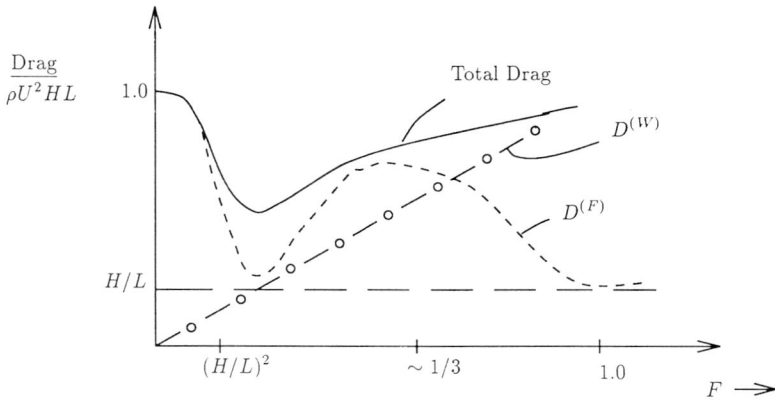

Figure 5. Sketch of the wave-drag and form-drag components and the total drag of a rounded hill with *low slope* (i.e. $H/L \ll 1$) in stable flow

Bibliography

1. Drazin, P.G. (1961). On the steady flow of a fluid of variable density past an obstacle. *Tellus*, **13**, 239-251.

2. Lin, Q., Lindberg, W.R., Boyer, D.L. & Fernando, H.J.S. (1992). Stratified flow past a sphere. *Journal of Fluid Mechanics*, **240**, 315-354.

3. Sysoeva, E.Y. & Chashechkin, Y.D. (1988). Spatial structure of a wake behind a sphere in a stratified liquid. *Journal of Applied Mechanics and Technical Physics*, **5**, 655-660.

4. Hunt, J.C.R., Feng, Y., Linden, P.F., Greenslade, M.D. & Mobbs, S.D. (1997). Low Froude Number Stable Flows Past Mountains. To appear in Il Nuovo Cimento.

5. Roshko, A. (1954). On the drag and shedding frequency of two-dimensional bluff bodies. NACA Technical Note 3169 (also *Journal of Aeronautical Science*).

6. Hanazaki, H & Hunt, J C R (1996). Linear processes in unsteady stably stratified turbulence. *Journal of Fluid Mechanics*, **318**, 303-337.

7. Brighton, P.W.M. (1978). Strongly stratified flow past three-dimensional obstacles. *Quarterly Journal of the Royal Meteorological Society*, **104**, 289-307.

8. Hunt, J.C.R. & Snyder, W.H. (1980). Experiments on stably and neutrally stratified flow over a model three-dimensional hill. *Journal of Fluid Mechanics*, **96**, 671-704.

9. Vosper, S.B., Castro, I.P. & Mobbs, S. (1997). Experimental studies of stratified flow over orography. (Presented at the 5th IMA Conference on Stably Stratified Flows; to be published.)

10. Smolarkiewicz, P.K. & Rotunno, R. (1989). Low Froude number flow past three-dimensional obstacles. Part I: Baroclinically generated lee vortices. *Journal of Atmospheric Sciences*, **46**, 1154-1164.

11. Lott, F. & Miller, M. (1997). A new sub-grid scale orographic drag parametrization: Its formulation and testing. *Quarterly Journal of the Royal Meteorological Society*, **537**, 101-127.

Appendix

Linear (Rapid Distortion Theory) calculation of the perturbation to a stably stratified inviscid fluid produced by an initially sloping perturbation to the density field, based on the analysis of [6].

Linearised governing equations

$$\frac{du_i}{dt} = \frac{1}{\rho}\frac{\partial}{\partial x_i}p - g\delta_{3i}\rho' \tag{4.5}$$

$$d\rho'/dt = -u_3\partial\bar\rho/\partial x_3 \tag{4.6}$$

$$\partial u_i/\partial x_i = 0. \tag{4.7}$$

Let $x_i \equiv (x, y, z)$.

Initial condition

$$u_i(\mathbf{x}, t=0) = 0; \rho'(\mathbf{x}, t=0) = \rho_{o_{mx}}e^{-n^2/d^2}e^{ik_1 x}$$
$$\text{where} \quad n^2 = (z - \alpha y)^2/(1+\alpha^2). \tag{4.8}$$

The perturbation slopes at an angle $\tan^{-1}\alpha$ to the y axis.

Solution: Taking the Fourier Transform, denoted by $\tilde\rho$, in y, z,

$$\tilde\rho(k_1; k_2, k_3; t=0) = \tilde\rho_o 2\pi\sqrt\pi \quad \rho_{o_{mx}}.\overline{d_v}.\delta(k_2 + k, \alpha) \times \exp(-k_3^2\overline{d}^2/4), \tag{4.9}$$

where $\overline{d} = d\sqrt{(1+\alpha^2)}$, and $\delta(\ )$ is a one-dimensional delta function.

From (4.5)-(4.7), it follows that $\tilde\rho$ satisfies

$$\frac{d^2\tilde\rho}{dt^2} + N^2\sin^2\theta\tilde\rho = 0, \tag{4.10}$$

where $\sin\theta = \sqrt{k_1^2 + k_2^2}/k^2$.

From (4.8) it follows that at $t = 0$, $d\tilde\rho/dt = 0$. Thence from (4.9)

$$\tilde\rho(k_1, k_2, k_3; t) = \tilde\rho_o \cos(Nt\sin\theta). \tag{4.11}$$

Taking the inverse F.T. in k_2, k_3 and using the delta function in (4.9) to simplify the integral in k_2, yields

$$\rho(\dot{\mathbf{x}}, t) = e^{ik_1 x}\rho_{o_{mx}}.\sqrt\pi.\overline{d}.I(y, z)$$

$$\text{where} \quad I = \int_{-\infty}^{\infty}\sqrt\pi e^{-k_3^2 d^2/4 + i(k_3\alpha y - k_3^2)}\cos(Nt\sin\theta).dk_3 \tag{4.12}$$

and now $\sin\theta = [1 - k_3^2/(2k_3^2 + k_1^2)]^{\frac12}$.

Taking the asymptotic value of $Nt \to \infty$, and using the method of stationary phase (which implies that the main contribution to the integral occurs where $\theta \to \pi/2$ or $k_3/k_1 \to 0$), leads to

$$I \propto \sqrt{2\pi} \frac{e^{iNt} k_1}{\sqrt{Nt}} e^{i(\alpha y - z)^2 k_1^2/(Nt)} e^{-(\alpha y - z)^2 k^4 \overline{d}^2/(Nt)^2}. \tag{4.13}$$

The implications of the results are discussed in Section 2.1.2.

Investigation of the Effect of Atmospheric Stability on Characteristic Decay Times of Tracer Concentrations in the Wake of an Isolated Obstacle

I. Mavroidis and R.F. Griffiths

Environmental Technology Centre, UMIST, Manchester

Abstract

The effect of atmospheric stability on characteristic decay times of a gas tracer in the near-wake of an isolated obstacle has been investigated. Field experiments have been conducted in flat terrain using a model 2 m cube. Two cube orientations were considered, with the mean wind direction either normal or at 45 degrees to the cube. The experiments involved the release of a tracer gas upwind of the obstacle, with fast-response detectors measuring concentrations at various points in the obstacle wake, at a frequency of 100 Hz. Tracer gas was released continuously for a limited period in order to fill the wake. Thereafter, the source was switched off using a solenoid valve, and the concentration in the wake was found to decay in an exponential manner. The residence time, which is defined as the time it takes for the concentration to decay to $1/e$ of its original value, was measured. This procedure was repeated several times in order to provide confidence in statistics. The atmospheric stability conditions ranged from very stable to very unstable. The period required for the detrainment of gas is much longer in stable than in unstable conditions, mainly due to the lower wind speeds and higher concentrations observed in stable conditions. The residence time is found to depend mainly on the wind speed and not on the atmospheric stability.

1 Introduction

The investigation of building influenced dispersion is very important for estimating concentrations or doses of contaminants in the case of accidental releases or in the cases of stacks located near obstacles. A substantial body of literature exists on building influenced flows, and two very good reviews of such flows are given by Meroney [1] and Hosker [2]. The phenomenon of gas residence in the wakes of obstacles is very important in pollution dispersion problems, since it results in the creation of secondary sources in the near-wake of obstacles. The detrainment behaviour in the wake of an obstacle can be determined from the decay times of concentrations measured in the wake region. Characteristic decay times in the wakes of obstacles have been investigated mainly in the wind tunnel [3-6], while research in the field is very limited (Drivas and Shair [7] and Higson et al [8]). In order to describe the phenomenon of gas residence in the wake of an obstacle a number of characteristic decay times can be considered. Firstly, one may consider the actual time that it takes for the gas concentration to fall below a threshold level from the instant that the gas source is switched off (i.e. the time required for the recirculation region to become effectively free of gas), which is referred to here as the decay duration (t). The most commonly used characteristic decay time in the literature is the residence time, which is a measure of the rate of concentration decay. The residence time (T_d) is defined as the time it takes for the concentration to decay to $1/e$ of its original value. This definition arises naturally, since the concentration has been observed to fall exponentially once the source has been switched off [3-6]. A non-dimensional residence time (τ) is derived from the residence time using the equation:

$$\tau = \frac{UT_d}{H} \tag{1.1}$$

where U is the wind speed (m/s) at obstacle height and H is the height of the obstacle (m).

Although the significance of the effects of atmospheric stability on the dispersion of pollutants is well recognized, the literature on the experimental examination of flow and dispersion around buildings in different stability conditions is quite limited. Investigations of the effect of atmospheric stability on pollution dispersion near obstacles is mainly concentrated on stable flows over hills. Yang and Meroney [9] and Kothari, Peterka and Meroney [10] have examined gaseous dispersion into stratified building wakes in the wind tunnel and Higson et al [8] have examined the effect of atmospheric stability on concentration fluctuations and wake retention times in the field. Castro [11] examined the flow around a two-dimensional obstacle in a towing tank and compared the results with numerical computations. Zhang et al [12] have compared results of experiments conducted in a towing-tank with results from numerical modelling of stable atmospheric flow around a cubical building. Robins [13] presented a review of flow and dispersion around buildings in light wind conditions and Snyder [14] has investigated the influence of stratification on building wakes in a salt-water stratified tank.

It should be emphasized that despite its importance for the dispersion of pollutants in built-up areas, research on the effect of atmospheric stability on characteristic decay times in the wake of obstacles is limited to a few observations in the field by Higson et al [8]. The main findings of Higson et al were that in stable conditions (a) long periods of high concentrations are experienced which are interspersed by long periods of zero concentrations and (b) the duration of gas detrainment is much longer than in unstable conditions and that even the non-dimensional residence time is approximately 30% longer than in unstable conditions. It must be noted, however, that the experiments described by Higson et al were not designed specifically to measure characteristic decay times, since the gas was not switched off at the source. The residence time was measured in cases where the entrainment of gas in the near-wake ceased due to a temporary shift in the wind direction. The experiments described in this paper were specifically designed to investigate the effect of atmospheric stability on characteristic decay times in the near-wake of a cube.

2 Experimental setup

The field experiments described here were conducted between July and August 1995 at Dugway Proving Ground, in Utah, USA. The model building used for the experiments consisted of a 2 m wooden cube. The tracer gas source was located 2.5 H upwind of the cube and at the cube height (H). Two orientations of the cube were investigated, such that the mean wind direction was approximately normal to or at 45 degrees to the leading face of the cube. The surrounding terrain was flat and the roughness length of the site was approximately 16 mm.

Since the main purpose of the experiments was to investigate the characteristic decay times in the wake of the model building under different atmospheric stability conditions, experiments were performed at different times of the day to cover a variety of stability conditions, ranging from unstable to stable. Unstable conditions were observed during the day with the sun high in the sky, whilst stable conditions were observed during the night, with neutral conditions existing in the period around sunset or sunrise, as well as on the limited occasions when clouds covered the sky and the wind speed was high. However, in the latter case experiments were not conducted since the danger of a storm was high. In order to conduct experiments under stable conditions trials were performed in the early morning hours, starting before 4:00 a.m. and continuing until sunrise (at around 7:15 a.m.), when the conditions changed to near neutral and then to unstable. Experiments were also conducted in the daytime, when the conditions were unstable.

Tracer gas was released continuously for a minimum period of 30s in order to fill the wake region with gas. A bivane anemometer was used to measure the wind direction. The wind direction signal then passed through a low pass filter which effectively averaged the data over a period of approximately 30s. If the average horizontal wind direction lay within ±10 degrees of that desired, an indicator light was triggered and the tracer was then switched off using a solenoid valve. Meteorological measurements were carried out simultaneously with the gas concentration measurements using an Ultrasonic anemometer. This provided three orthogonal components of the wind vector and the speed of sound in air, which is used for calculating the temperature, all of which were sampled at a frequency of 21 Hz.

The tracer gas used was propylene and the detectors deployed were UVIC®
detectors (Ultra Violet Ion Collectors). The UVIC® detector is a fast response
gas monitoring instrument which provides a useful calibratable range from about
0.01 to 1000 ppm by volume, with a response time of about 0.02 seconds. An
account of the development of the UVIC® detector is given by Griffiths [15].
The principle of operation of the UVIC® detector is that of photo-ionization of
gases. A stream of air is drawn through a tube by a small fan and passes across
the face of a lamp emitting UV (Ultra-Violet) radiation. The photon energy of
the UV radiation emitted is approximately 10.6 eV. This can ionize gases with
ionization potentials of this value or less. The normal components of air have
higher ionization potentials, and are therefore not detected. The ions produced
when ionizable species are present are collected by a suitably biased co-axial
electrode system downstream of the UV lamp, and the current is converted to
a voltage signal by an electrometer. The detectors were calibrated at the time
of use in the field using the purpose-built gas calibration system. Data were
acquired at a frequency of 100 Hz.

Seven UVIC® detectors were used in total during the experiments. Five detectors were located across the width of the wake, with a separation of 0.375H
(0.75 m) between them, at a height of 0.5H and at a distance of 0.75H downwind of the rear face of the cube. These detector locations, as well as the
position of the source, allow comparisons with previous wind tunnel work [5-6].

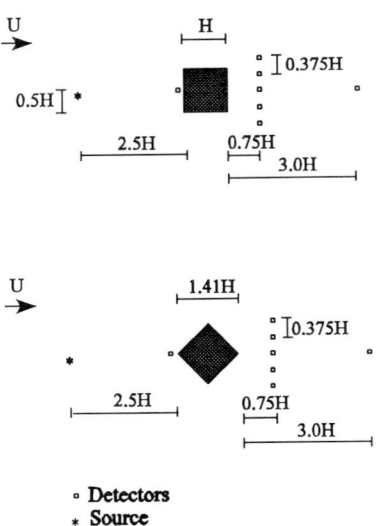

Figure 1. Plan view of the experimental configurations for the cube normal and at
45 degrees to the flow

Two additional detectors were deployed, one located at the centre of the upwind face of the cube and the other 3.0H downwind of the rear face of the cube, on the same vertical plane with the source and the centre of the cube, and at height of 0.5H. The two experimental configurations are presented in Figure 1.

3 Analysis and results

3.1 Data analysis

In order to derive values of the characteristic decay times, concentrations are averaged over the lateral array of five detectors located 0.75H downwind of the cube. The advantage of this arrangement is that it allows direct comparison with results from wind tunnel investigations [3-6]. Furthermore, it reduces the effect of concentration variability in the near-wake since concentrations are averaged over the lateral array of detectors. Concentration results are presented either in ppm or in the form of the non-dimensional concentration K_c:

$$K_c = \frac{CUH^2}{Q} \tag{3.1}$$

where C is the measured mean concentration in ppm (volume) multiplied by 10^{-6} (for pure undiluted gas $C = 1$), U is the mean wind speed in m/sec, H is the height of the cube in m and Q is the volumetric flow rate of the gas source in m^3/sec. The wind speed is averaged over the period of the decay duration (t) for each individual trial.

Concentrations are initially averaged over a period of 1 second and then the residence time (T_d) is calculated using a least squares fit. An example of how concentrations in the wake of a cube normal to the flow are affected when the gas is switched off at the source is presented in Figure 2 (similar behaviour is observed when the cube is oblique to the flow). The trial described here was performed under stable atmospheric conditions (Pasquill stability class F) and the gas was switched off at the source at 1063sec. Figure 2 presents concentrations versus time in a semi-logarithmic plot. Concentrations are averaged over the lateral array of five detectors located 0.75H downwind of the rear face of the cube (at a height of 0.5H). The concentration is falling exponentially, with the decaying concentration appearing as a steady decline with superimposed fluctuations. The value of the residence time (the time it takes for the concentration to decay to 1/e of its original value) is approximately 10.2sec. The duration of the concentration decay is approximately 50sec.

The trials investigating characteristic decay times were repeated several times under different atmospheric stability conditions in order to derive statistically stable averages and to explore the effect of stability on the results. In total 60 trials were undertaken with the cube normal to the flow and 58 trials with the cube at 45 degrees to the flow. Individual residence time measurements were taken using a least squares fit to the concentration data from each trial. The residence time was non-dimensionalized with the obstacle height and the wind speed averaged over the period t for each individual trial. Then all the individual values were averaged to give the final ensemble average value. The variability of the individual values of the characteristic decay times mainly depended on the variability of the wind speed and the other meteorological conditions. In the case of the non-dimensional residence time, where the effect of the wind direction is excluded, the variability was very small. Figure 3 shows the number of occurrences for different values of the non-dimensional residence time in the case of a cube normal to the flow. There is a very clear peak around the mean value.

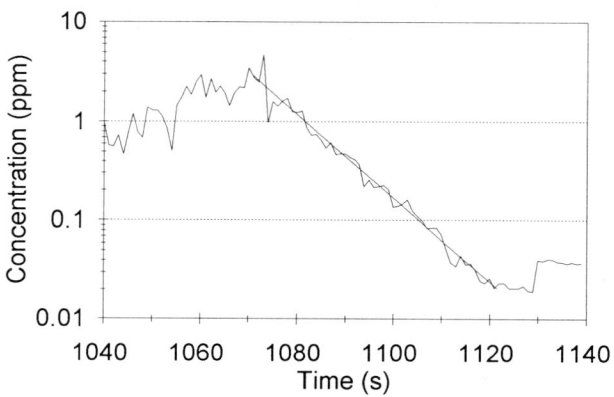

Figure 2. Concentration time series averaged over five detectors located in a lateral array 0.75H downwind of the cube, presented in a semi-logarithmic graph. Gas switched off at 1063sec. The straight line shows the exponential decay of concentration

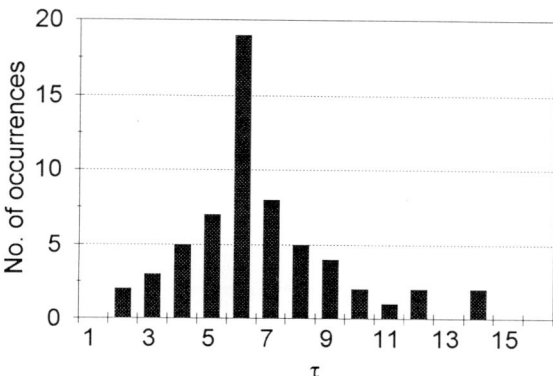

Figure 3. Histogram of values of the non-dimensional residence time for the cube normal to the flow

3.2   Effect of stability on characteristic decay times

Table 1 and Table 2 show the average values of t, T_d and τ in the case of the cube normal and at 45 degrees to the wind respectively. Ensemble averages are presented for all the trials, for those conducted under stable conditions and those conducted under unstable conditions.

The results presented in Tables 1 and 2 confirm the finding of Higson et al [8] that the duration of concentration decay is much longer in stable than in unstable conditions. This is mainly due to the fact that the wind speeds observed during stable conditions are lower and the concentrations observed are higher, and accordingly the period required for the concentration to fall below a threshold level is longer. It should be noted here that higher initial concentrations are observed in this case in stable conditions only because the gas was switched off at the source when the wind direction was within a window of ±10 degrees from that required. If concentrations are averaged over a long period in stable conditions, the resulting values might be lower than in unstable conditions, due to the continuous plume meandering observed under stable conditions and low wind speeds.

The residence time and the non-dimensional residence time, however, do not depend on the initial concentration (i.e. the concentration observed at the nominal zero time), since they are a measure of the rate of concentration decay.

Table 1. Cube normal to the flow. Mean values of t, T_d, τ, K_c and U for all the trials, those performed under stable conditions and those performed under unstable conditions

Conditions	$t(s)$	$T_d(s)$	τ	K_c	$U(m/s)$
All conditions	37.65	11.31	6.18	0.69	1.31
Stable conditions	47.68	10.88	5.97	0.93	1.37
Unstable conditions	30.49	11.62	6.33	0.5	1.27

Table 2. Cube at 45 degrees to the flow. Mean values of t, T_d, τ, K_c and U for all the trials, those performed under stable conditions and those performed under unstable conditions

Conditions	$t(s)$	$T_d(s)$	τ	K_c	$U(m/s)$
All conditions	41.43	9.88	9.53	1.41	3.10
Stable conditions	53.77	13.00	9.52	2.18	1.61
Unstable conditions	31.41	6.97	9.54	0.77	2.43

The residence time depends mainly on the wind speed observed during the decay episode. The dependence of the residence time on wind speed is clearly shown in Figure 4, where the residence time is plotted against the wind speed averaged over the period t for each individual trial, in the case of the cube normal to the flow. For higher wind speeds there is a slight decrease of the residence time as the wind speed decreases, but as it decreases to values lower than 1 m/s the residence time increases rapidly, reaching values up to 5-7 times higher than those observed for higher wind speeds. Since the residence time depends mainly on the wind speed, similar residence times are observed under stable and unstable conditions in the case of the cube normal to the flow, where the average wind speeds between stable and unstable conditions do not differ much. On the other hand, in the case of the cube at 45 degrees to the flow, where the wind speed is higher in the case of unstable conditions, the residence time is clearly longer in the case of stable conditions. This is the first indication that, outside of the effect of wind speed, atmospheric stability does not affect the residence time significantly.

If this is the case, and since the residence time is non-dimensionalized using the mean wind speed for the period of concentration decay, the non-dimensional residence time should be the same for stable and unstable conditions. This is confirmed by the values of the non-dimensional residence time presented in Tables 1 and 2, in the case of the cube normal and at 45 degrees to the flow respectively. Thus it is apparent that atmospheric stability has a negligible overall effect on the value of the non-dimensional residence time. This appears

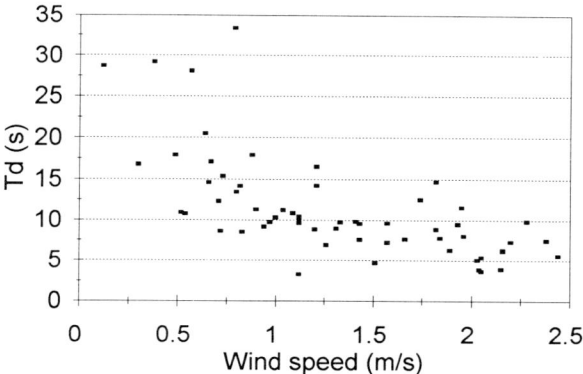

Figure 4. Residence time as a function of the wind speed in the case of a cube normal to the flow

to be in contrast with the results of Higson et al [8], who found that the non-dimensional residence time was approximately 30% longer in stable conditions. However, it should be noted that during the experiments of Higson et al the gas was not actually switched off at the source, but the residence time was calculated from continuous source experiments, during which the wind direction changed suddenly, transporting the gas away from the obstacle. It was possible though that - in some of their trials - a proportion of the plume was still entrained in the wake after the nominal zero time. Furthermore, during the experiments described by Higson et al only a small number of residence time measurements were made, and the differences observed between stable and unstable conditions are within the margin of uncertainty expected.

3.3 Comparison of the non-dimensional residence time with the empirical formula of Fackrell

The non-dimensional residence time has a more universal significance than either the decay duration or the residence time, since it is a measure of the rate of concentration decay after removing the effect of the wind speed and thus it can be applied universally for modelling purposes. This feature of the non-dimensional residence time is further enhanced by the fact that it is independent of atmospheric stability, as was revealed in the previous section. Fackrell [5] performed a number of experiments in the wind tunnel, under neutral stability conditions, and combined the results in order to derive an empirical relation for the non-dimensional residence time. This relation is given by (3.2):

$$\tau = \frac{11(W/H)^{1.5}}{1 + 0.6(W/H)^{1.5}} \tag{3.2}$$

where W is the width of the obstacle. For buildings oblique to the flow, although less information is available, Fackrell [5] suggested that the above relation is valid, provided that the effective building dimensions are used. For a cube normal to the flow this relation gives a value of $\tau = 6.9$, while for a cube at 45 degrees to the flow $\tau = 9.2$. Wind tunnel data of Vincent [3,4] give a value of τ between 5 and 8 for a cube normal to the flow in near uniform flow.

The results from the field trials presented here show that the non-dimensional residence time averaged over all the trials in the case of the cube normal to the flow has a value of approximately 6.2 (Table 1), while in the case of a cube at 45 degrees to the flow it has a value of approximately 9.5 (Table 2). The results of the present field measurements agree with the values obtained using the empirical formula of Fackrell (3.2). In the case of a cube normal to the flow the difference between the two values is approximately 10%, while in the case of the cube at 45 degrees to the flow the difference between the two values is approximately 3%. This is a surprisingly good agreement between the field results presented here and the wind tunnel results described by the empirical formula of Fackrell. This agreement indicates that, at least for the limited cases examined here, the formula proposed by Fackrell gives a very good estimate of the residence time in the wakes of obstacles in the atmosphere, irrespective of the atmospheric stability conditions.

4 Conclusions

The characteristic decay times in the wake of a cube normal and at 45 degrees to the flow were investigated in the field. Experiments were performed under a variety of stability conditions ranging from very unstable to very stable. The main findings are summarised below:

- The duration of the concentration decay is longer in the case of stable conditions than in unstable conditions. This is attributed mainly to the lower wind speeds occurring under stable conditions and to the fact that concentrations in the wake of a cube are in general higher under stable conditions.

- The residence time is mainly affected by the wind speed. As a result longer residence times are observed in stable conditions, when the wind speed is in general lower. Apart from the effect of the wind speed the residence time does not appear to be affected by stability. As a result the values of the non-dimensional residence time do not differ between stable and unstable conditions.

- The values of the non-dimensional residence time observed in the field for a cube normal ($\tau = 6.2$) or at 45 degrees to the flow ($\tau = 9.5$) agree very well with the values calculated using the empirical formula derived by Fackrell [5] from experiments performed in a wind tunnel under neutral stability conditions ($\tau = 6.9$ and $\tau = 9.2$ respectively). This indicates that at least for simple structures Fackrell's formula adequately predicts the value of the non-dimensional residence time.

5   Acknowledgements

This work has been carried out with the support of PLSD, UK Ministry of Defence, under agreement 2044/004/CBDE. The authors would like particularly to acknowledge the help of Dr. C.D. Jones and Dr. S.C. Cheah of PLSD. The contribution of Dr. H.L. Higson in planning the experiments is gratefully acknowledged. Many thanks are also due to Mr. W. Evans, and to the staff of the Dugway Proving Ground, Utah, USA, and especially to Mr. C. Biltoft.

Bibliography

1. Meroney, R.N. (1982). Turbulent diffusion near buildings. *Engineering Meteorology*, (Editor: E. Plate), Elsevier Scientific Publishing Company, Amsterdam, The Netherlands, 481–525.

2. Hosker, R.P., Jr. (1984). Flow and diffusion near obstacles. *Atmospheric Science And Power Production*, (Editor: D. Randerson), U.S. Department of Energy, 241–326.

3. Vincent, J.H. (1977). Model experiments on the nature of air pollution transport near buildings. *Atmospheric Environment*, **11**, Elsevier Science Ltd, 765–774.

4. Vincent, J.H. (1978). Scalar transport in the near aerodynamic wakes of surface mounted cubes. *Atmospheric Environment*, **12**, Elsevier Science Ltd, 1319–1322.

5. Fackrell, J.E. (1984). Parameters characterizing dispersion in the near wakes of buildings. *J. Wind Eng. Ind. Aero.*, **16**, Elsevier Science Publishers B.V., Amsterdam, The Netherlands, 97–118.

6. Hunt, A. and Castro, I.P. (1984). Scalar dispersion in model building wakes. *J. Wind Eng. Ind. Aero.*, **17**, Elsevier Science Publishers B.V., Amsterdam, The Netherlands, 89–115.

7. Drivas, P.J. and Shair, F.H. (1974). Probing the airflow within the wake downwind of a building by means of a tracer technique. *Atmospheric Environment*, **8**, Elsevier Science Ltd, 1165–1175.

8. Higson, H.L., Griffiths, R.F., Jones, C.D. and Biltoft, C. (1995). Effect of atmospheric stability on concentration fluctuations and wake retention times for dispersion in the vicinity of an isolated building. *Environmetrics*, **6**, John Wiley & Sons Ltd, 571–581.

9. Yang, B.T. and Meroney, R.N. (1970). Gaseous dispersion into stratified building wakes. Technical Report for the U.S Atomic Energy Commission, AEC Report No. C00-2053-3, USA.

10. Kothari, K.M., Peterka, J.A. and Meroney, R.N. (1980). Stably stratified building wakes. Report NUREG/CR-1247, U.S. Nuclear Regulatory Commission, USA.

11. Castro, I.P. (1993). Effects of stratification on separated wakes: Part 1, weak static stability. *IMA Conf. Proc. on Waves and Turbulence in Stably Stratified Flows*, **40**, Clarendon Press, Oxford, 323–346.

12. Zhang, Y.Q., Arya, S.P. and Snyder, W.H. (1996). A comparison of numerical and physical modeling of stable atmospheric flow and dispersion around a cubical building. *Atmospheric Environment*, **30(8)**, Elsevier Science Ltd, 1327–1345.

13. Robins, A. (1994). Flow and dispersion around buildings in light wind conditions. *IMA Conf. Proc. on Stably Stratified Flows: Flow and Dispersion over Topography*, **52**, Clarendon Press, Oxford, 325–358.

14. Snyder, W.H. (1994). Some observations of the influence of stratification on diffusion in building wakes. *IMA Conf. Proc. on Stably Stratified Flows: Flow and Dispersion over Topography*, **52**, Clarendon Press, Oxford, 301–324.

15. Griffiths, R.F. (1993). Emissions and environmental monitoring using energetic UV radiation: a new development in portable ambient monitoring. *Proceedings of Monitor '93 Conference*, Spring Innovations, Manchester, 57–62.

Experiments on Two-Dimensional Lee-Wave Breaking in Stratified Flow

P. Bonneton*, O. Auban** and M. Perrier**

*D.G.O., Université Bordeaux 1, Talence, France, and **Meteo-France CNRM, Toulouse, France

Abstract

This paper describes the results of an experimental study on lee wave breaking over two-dimensional obstacles in linearly stratified flow. Two main dimensionless parameters govern the flow dynamics : F, the internal Froude number based on the obstacle height, h, and $F_L = U/NL$, where L is the horizontal scale of the obstacle. Particular emphasis is given on the influence of F_L (which characterises the non-hydrostatic effects) with respect to the lee-wave dynamics and the wave-overturning condition.

1 Introduction

In the past several years, a large research effort has been devoted to the understanding of internal waves generated by stratified flows over topography. Extensive reviews may be found in Smith [1] and Baines [2]. This research is motivated by the importance of the lee-wave dynamics on a wide range of spatial scales in the atmosphere and also in the ocean. On the mesoscale, Clark and Peltier [3] and Peltier and Clark [4] have demonstrated, from numerical simulations, that lee-wave breaking plays an important role in the development of severe atmospheric downslope winds. Subsequently, lee-wave breaking was studied theoretically by Smith [5] and Laprise and Peltier [6] and experimentally by Rottman and Smith [7] and Castro and Snyder [8]. The main conclusion of these studies is that when the Froude number F ($F = U/Nh$, where U is the upstream velocity, h is the obstacle height and $N = (-g/\rho_0 \, d\rho/dz)^{1/2}$ is the Brunt-Väisälä frequency) is smaller than the critical value F_c for which streamlines locally overturn, wave breaking occurs and leads to a very different flow configuration. This configuration is characterised by an acceleration of the flow near the surface, on the lee side of the obstacle, with an associated increase in the surface drag. In this regime, a turbulent region is located above the lee side of the mountain and acts as a wave reflector.

The present paper describes new experimental results on lee-wave breaking. In Section 2, the experimental facilities and procedures are described. In Sections 3 and 4, the influence of the non-hydrostatic effects on wave breaking over two-dimensional obstacles is analysed.

2 Experimental procedure

Experiments were conducted in two different water-towing tanks of respective sizes ($\mathcal{H}$ x $\mathcal{W}$ x $\mathcal{L}$, see Figure 1) 0.5 x 0.5 x 4 m^3 and 0.7 x 0.8 x 7 m^3. These two tanks are of interest to test confinement effects on the wave field. Many experiments dealing with lee waves in stratified fluid adopted a towing carriage technique where the hill was mounted on a flat baseplate suspended from the carriage and towed upside-down across the stationary fluid (for example Hunt and Snyder [9]). To avoid perturbations which could be generated by the baseplate, we have prefered to choose the upright towing configuration presented in Chomaz et al. [10] and Kadri et al. [11]. The obstacle was suspended by eight nylon threads of 0.25 mm in diameter. Impulsively started, the obstacle, was towed tangentially along the bottom of the tank. Careful image analysis showed that, in the velocity range studied ($U \in [0.25, 5 \text{ cm/s}]$), no vibrations or oscillations were detectable. Reynolds numbers associated with the supporting threads range from 0.5 to 10 at which the downstream disturbances are negligible. Physical constraints do not allow the mountain to be towed with a gap of less than 2 mm between the baseplate and the tank floor. However, experiments with increasing gap spacings have not revealed significant flow perturbations up to 4 mm. We therefore chose to work with a gap of 3 mm.

The experiments were performed with a two-dimensional Gaussian obstacle. Its cross-sectional profile is given by the expression: $h(x) = h\exp(-x^2/2L^2)$, where $h = 1.6$ cm and $L = 2.83$ cm ($r = h/L = 0.57$). This Gaussian shape was chosen because of its particularly simple Fourier transform, useful for theoretical analysis (discussed in Section 3), and also because, contrary to the classical Witch of Agnesi profile, the same profile can be used for axisymmetric three-dimensional obstacles. The obstacle was uniform across the spanwise section of the tank, with a spacing of 2 mm at each side.

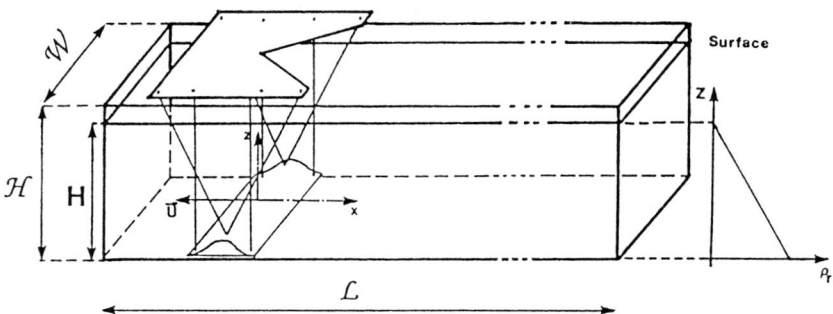

Figure 1. Experimental setup

The stratification (NaCl solution) was obtained by a computer-monitored filling process. In all the experiments described here, the density profiles were linear with a Brunt-Väisälä frequency N in the range of $N \in [0.8, 1.3 \text{ rad/s}]$. Due to the cumulative erosion of the density profile at the bottom of the tank, a given fluid stratification was used at most four times. This ensured that the maximum deviation of the linear density profile was less than 1 kg m^{-3}.

Visualisation and measurement techniques used in the experiments have been described in Bonneton et al. [12] and Chomaz et al. [10]. In all the flow visualisation results presented, the flow is from left to right.

The ratio h/H, where H is the water depth, ranges from $h/H = 0.025$ to 0.035, such that the vertical confinement effects may be considered negligible. Tests have shown that even at ratios as high as $h/H = 0.1$, the same

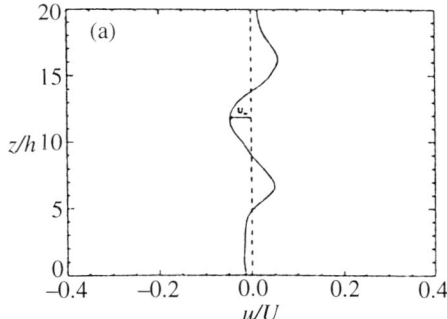

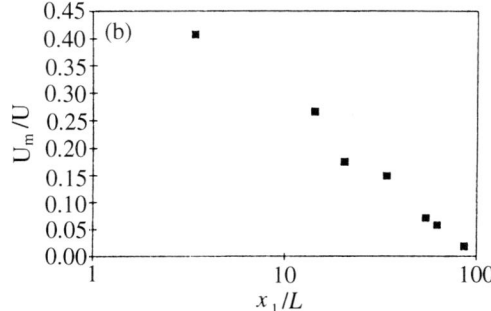

Figure 2. Upstream velocity perturbations in the laboratory frame, measured at the point ($x_1 = 0$) located 80 cm from the end of the tank, for $F = 0.6$. (a) horizontal velocity profile at $x_1/L = 62$; and (b) maximum horizontal velocity perturbation as a function of the distance x_1, between the obstacle and the measurement location

wave-breaking regime is obtained - independently of the vertical confinement. However, the reflection of upstream perturbations at the end of the tank can perturb the upstream flow. To quantify this phenomenon, we have measured, from particle displacements, the horizontal upstream velocity profile (in the laboratory frame) 80 cm from the end of the tank. This velocity profile, when the distance between this fixed location and the mountain is $x_1/L = 62$, is shown in Figure 2(a). In Figure 2(b), we plot the maximum velocity U_m for $x_1/L \in [2, 85]$. It can be seen that, at a distance $x_1 \sim 1m$ ($x_1/L \sim 40$), upstream the obstacle, the maximum velocity perturbation is less than 10% of the towing velocity U. We therefore consider that the reflection of upstream perturbations is negligible until the obstacle is less than 1 m from the end of the tank.

The flow is characterised by three non-dimensional parameters: the Froude number F ($F \in [0.5, 3]$), $F_L = U/NL$ ($F_L \in [0.28, 0.62]$) and the Reynolds number $Re = Uh/\nu$ ($Re \in [120, 690]$). For these values of Re, the viscous effects on the lee-wave dynamics can be considered to be negligible, as suggested by the following results. Figures 3(a) and 3(b) show a comparison between the observed

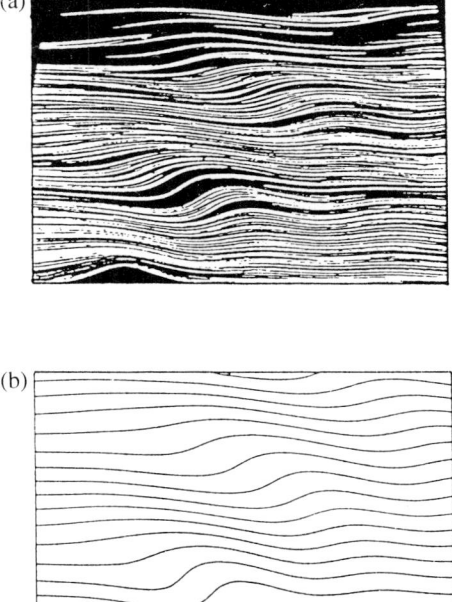

Figure 3. Comparison between the observed flow, (a) ($Nt = 150$) and Long's model solution; and (b) for $F = 1.2$

flow pattern and streamlines calculated from Long's model [13] (presented in Section 3), for $F = 1.2$. Good agreement can be observed between the predicted and measured phase structure and only a small discrepancy of the maximum amplitude of the lee wave A ($A/h = 1.5$ for the observed flow and $A/h = 1.75$ for Long's solution) is detected. This implies that the flow may be adequately described by Long's inviscid and unbounded model and suggests that viscous and confinement effects are only of minor importance in these experiments. The motions within the breaking zone, however, are expected to be controlled by viscous effects.

3 Preliminary considerations on the lee-wave dynamics over two-dimensional obstacles

Lee-wave fields over two-dimensional obstacles are mainly controlled by the two non-dimensional parameters F and F_L, which characterise, respectively, the degree of linearity and the degree of non-hydrostaticity of the flow. Our purpose, in this section, is to underline the importance of the non-hydrostatic effects on the dynamics of the lee waves observed experimentally.

3.1 Horizontal length scale definition

To study the lee-wave dynamics it is paramount to use a universally applicable definition of the horizontal length scale L which is used in the non-dimensional parameter F_L. This scale is generally defined as the streamwise length of the obstacle at the half-height elevation. But the physical meaning of this definition is not evident and it does not represent a scale which allows all shapes to be scaled correctly. We therefore propose a new definition for L (see also Bonneton et al. [14]), which enables us to compare results obtained from different topography shapes.

Lee waves are controlled by the lower boundary condition of the vertical velocity component w. For smooth topographies, this condition can be simplified and replaced by the linear free-slip boundary condition:

$$w(x, z = h(x)) \simeq w(x, 0) = U \frac{\partial h}{\partial x}$$

which can be written in Fourier space as

$$\| \hat{w}(k, z = 0) \| = U \| k\hat{h} \|$$

where $\hat{w}$ and $\hat{h}$ are the Fourier transforms of w and h, and k is the horizontal wavenumber. The dominant mode of the wave field is associated with the wavenumber k_m for which $\| k\hat{h} \|$ reaches a maximum. This allow us to define L as $L \equiv 1/k_m$. For $F_L < 1$ the dominant mode k_m belongs to the propagative part of the wave spectrum ($\omega/N(= Uk_m/N = F_L) < 1$, where ω is the wave frequency) and for $F_L > 1$ the dominant mode k_m corresponds to the evanescent part of the wave spectrum ($\omega/N > 1$). For three-dimensional topographies, Bonneton et al. [14] and Kadri et al. [11] have shown that, $F_L = 1$, marks the transition between a flow dominated by a lee wave of maximum amplitude (which suppresses the separation of the boundary layer) and a three-dimensional turbulent wake regime.

3.2 Influence of F_L on the lee-wave structure

The steepening of the lee wave is controlled by the Froude number F. Long's model [13] shows that when F decreases the steepening increases and for a critical value, F_c, close to 1, the streamlines are vertical (i.e., the horizontal velocity is zero). For $F < F_c$, there exists a region with a statically unstable density gradient, which marks the limit of applicability of Long's model. It was generally assumed that when the streamlines reach this critical (vertical) condition, lee-wave breaking is just about to occur.

The influence of F_L, on the other hand, has not been examined in detail yet. Here, we analyse its influence on lee-wave overturning using Long's non-hydrostatic and unbounded model. To obtain the exact solution to Long's model, it needs to be solved numerically. We have used the algorithm of Laprise and Peltier [15], where the governing Helmholtz' equation is solved using Fourier transform techniques and where the correct nonlinear lower boundary condition is implemented iteratively. Figures 4(a) and 4(b) show the wave fields calculated for two different values of F_L ($F = F_c$). We note that the location of maximum steepening of the streamlines is located at about $3/4\lambda$ ($\lambda = 2\pi U/N$) from the centre of the obstacle for all F_L. It can also be observed that when F_L increases, the zone of wave overturning moves downstream and approaches the ground. This phenomenon is related to the non-hydrostatic propagation of wave energy. The location of maximum steepening is mainly determined by the direction of the group velocity (in the obstacle frame) which is associated with the dominant mode of the wave field. From the linear theory for dispersive waves (Lighthill [16]) we can easily show that this group velocity increasingly tilts downstream as F_L increases. The angle θ, between the vertical and the direction of maximum local wave energy, increases with F_L and reaches an asymptotic value of $45°$. It should be noted that this asymptotic value differs from the value of $90°$ obtained for lee waves generated by three-dimensional axisymmetric obstacles (Bonneton et al. [14] and Kadri et al. [11]). We will see in the next section that these observations on the location of maximum steepening as a function of F_L (for Long's steady solutions), can be applied to determine the wave-overturning location for $F < F_c$.

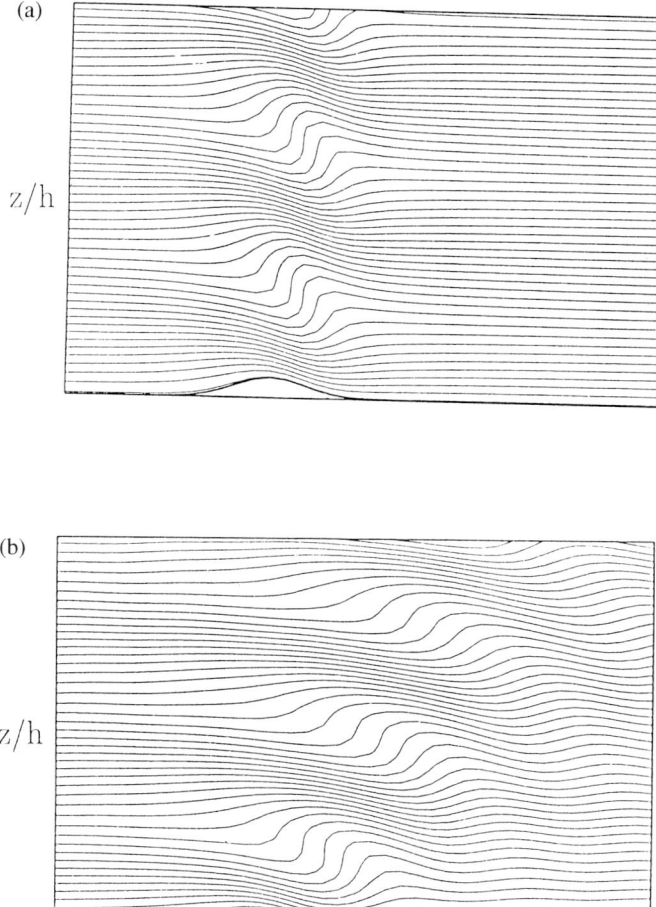

Figure 4. Streamlines for steady lee waves over a Gaussian obstacle calculated from Long's model, for $F = F_c$. (a) $F_L = 0.1$ ($F_c = 1.21$); and (b) $F_L = 0.54$ ($F_c = 0.96$)

4 Experimental results

In the previous studies by Castro [17], Rottman and Smith [7] and Castro and Snyder [8], which examined lee-wave breaking, the flow was visualised by emitting dye at various positions near the obstacle. This technique allows mixing events to be detected, but does not give information about the dynamical processes which lead to mixing. In this study, we have used particle-streak techniques, which reveal the dynamical processes as well as the transient evolution of the flows.

4.1 Qualitative description of the flow

Figures 5(a)–(d) present particle-streak visualisations for different Froude numbers around the critical value predicted by Long's model, $F_c = 0.96$. For $F = 1.2$ (Figure 5(a)) a steady lee-wave field well predicted by Long's theory (as in Figures 3(a) and 3(b)) can be observed. For $F = 1$ (for example, Figure 5(b) at $Nt = 150$), the flow fluctuates about Long's (steady) solution over a long time scale ($Nt \simeq 50$), with no transient breaking. This observation is in contrast to the results of earlier investigations (for example, Clark and Peltier [3], Laprise and Peltier [6]), where it was found that the criteria for static and dynamic instabilities are simultaneously satisfied. It is, however, in agreement with the recently developed asymptotic theory (in the hydrostatic limit) by Prasad et al. [18]. For $F \in]0.8, 1[$, we observed intermittent wave-breaking. Each "period" consisted of a phase of lee-wave steepening followed by a local overturning phase (which implies a decrease in lee-wave amplitude). Continuous wave breaking is observed for $F \leq 0.8$ (Figures 5(c) and 5(d)).

The transient evolution of the lee-wave breaking (leading up to continuous breaking) is presented in Figures 6(a)–(d), for $F = 0.6$ ($F_L = 0.34$). Just after the impulsive start of the obstacle, the lee-wave energy freely propagates with an angle in agreement with the value calculated from Long's model for $F_L = 0.34$. Near $Nt \sim 35$, we observed a transition between a freely propagating internal wave flow and a flow characterised by the existence of a mixed region which traps the internal wave field close to the ground (for example Figures 6(c) and 6(d)).

4.2 Overturning condition

From visualisations we have determined the wave overturning location, allowing us to compare it with Smith's theory. Smith [5] developed a nonlinear hydrostatic theory to predict the height of the well-mixed region and the associated increase of the wind speed and drag. Unfortunately, this theory does not produce severe wind state solutions for $F < 1.015$, which is the main range of wave breaking experimentally observed. Smith attempted to overcome this problem by taking into account the effect of upstream blocking. He thus obtained a

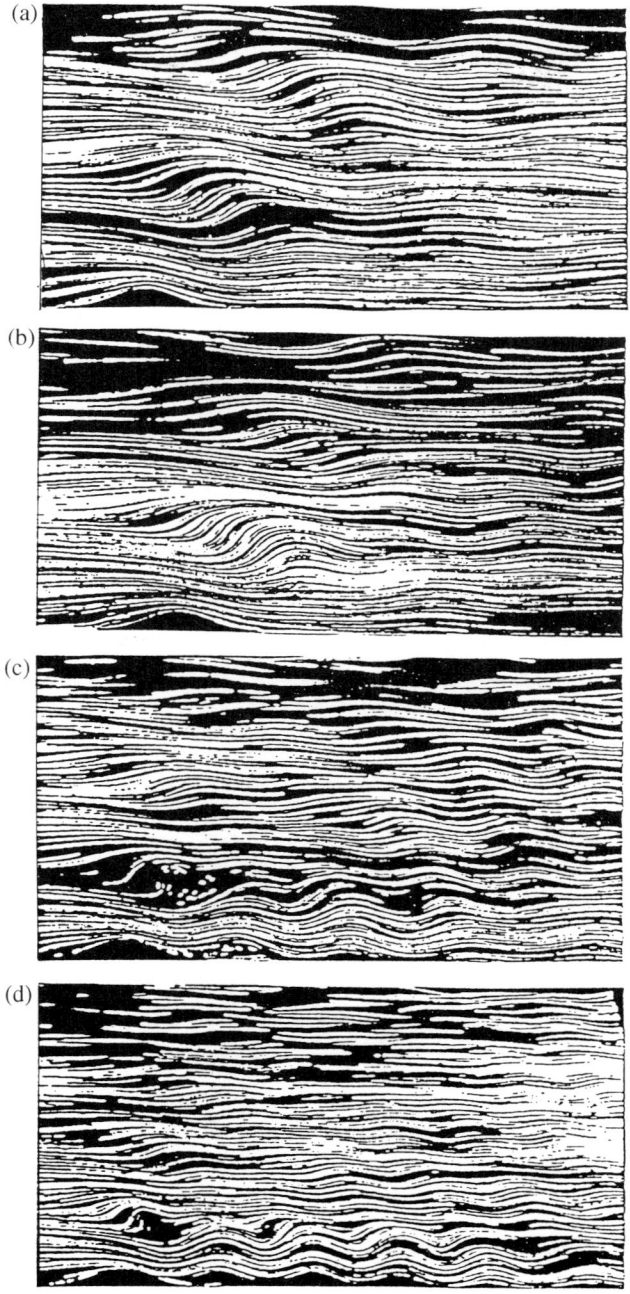

Figure 5. Particle streak trajectories at $Nt = 150$. (a) $F = 1.2$; (b) $F = 1$; (c) $F = 0.8$; and (d) $F = 0.6$

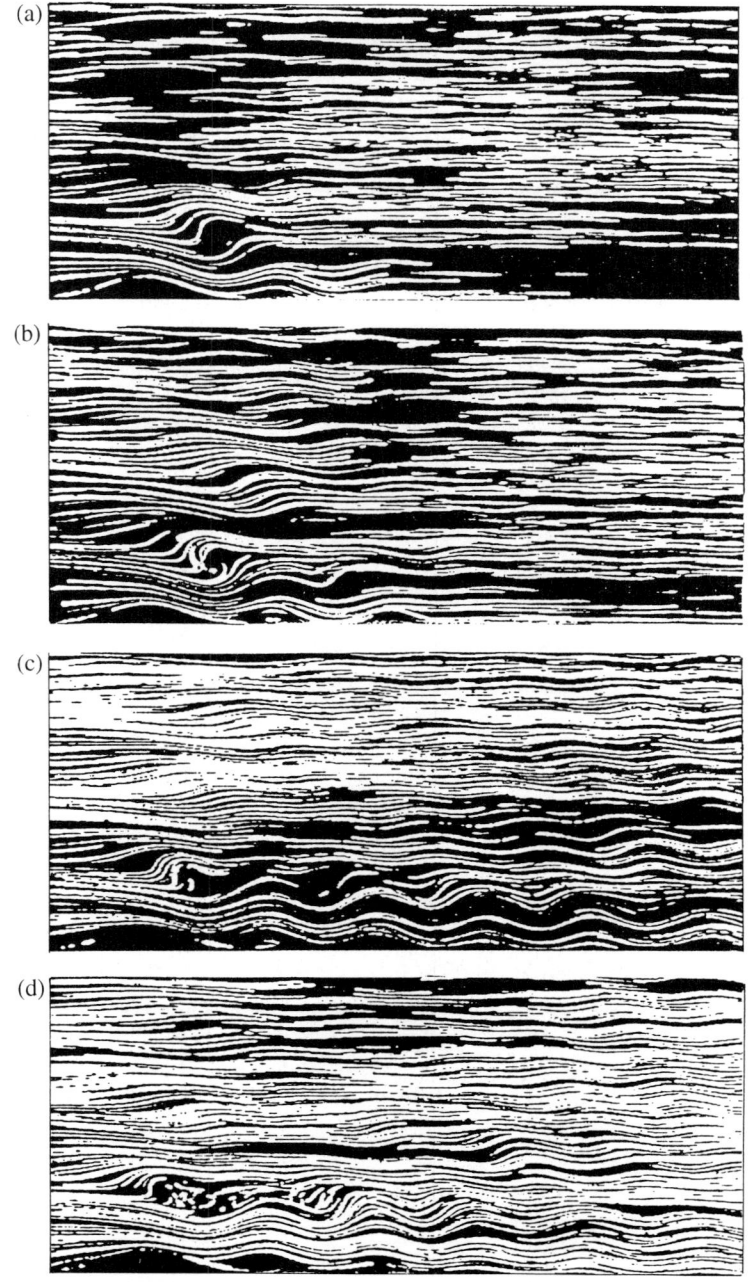

Figure 6. Particle streak trajectories for $F = 0.6$. (a) $Nt = 21$; (b) $Nt = 39$; (c) $Nt = 150$; and (d) $Nt = 250$

semi-empirical law relating the non-dimensional critical streamline height $H_d N/U$ to the Froude number F,

$$\frac{H_d N}{U} = \frac{3\pi}{2} + \frac{bN}{U},$$ (4.1)

where b is the depth of the blocked layer upstream the obstacle. The general form of this law can be established from simple physical arguments. As discussed in Section 3, the height of the critical steepening of the streamlines, for hydrostatic lee waves, is equal to $3/4\lambda$ (see also Queney [19]). When the blocking phenomenon is present, it is necessary to add the depth of the blocked layer to obtain the effective height, i.e. $H_d = 3/4\lambda + b$, and which normalised by U/N leads the non-dimensional law (Equation (4.1)). Following the experimental work of Snyder et al. [20] (and, implicitly, Sheppard's [21] energy argument), Smith suggested that $b/h = 1 - \alpha F$, with $\alpha = 0.985$. It should be noted that α is an empirical value. It follows that

$$\frac{H_d N}{U} = \frac{3\pi}{2} - 0.985 + \frac{1}{F}.$$ (4.2)

Rottman and Smith [7] have measured $H_d N/U$ experimentally for three different cosine-shaped topographies, and have found good agreement with Equation (4.2). However, contrary to what Rottman and Smith assert, these experimental results do not confirm the nonlinear hydrostatic theory of Smith, but only the physical arguments developed by these authors (as recalled above).

In Figure 7, the non-dimensional critical streamline height $H_d N/U$ measured for a Gaussian obstacle ($r = h/L = 0.57$) and a cylinder (complementary experiments have been done with a cylinder of radius $R=1$ cm, towed at mid-hight of the tank, $F \in [0.18,0.8]$), is plotted as a function of F, with the values obtained by Rottman and Smith. To be consistent with the definition used in Smith's theory, we define H_d as the upstream height of the streamline that marks the top of the wave breaking region. In this manner we observe, like Rottman and Smith, that $H_d N/U$ essentially evolves like $1/F$ as a result of the blocking phenomenon. The high values of $H_d N/U$ obtained for the cylinder indicate, as already discussed by Snyder et al. [20], that the depth of the blocked layer depends on the topographical shape, i.e., the α coefficient is not a constant. For Gaussian and cosine-shaped obstacles it can be observed that $H_d N/U$ tends to decrease as the steepness of the topography increases. This is related to non-hydrostatic effects which are not taking into account by Equation (4.2). To emphasise this effect by eliminating the blocking contribution, we have plotted $H_d N/U - 1/F$ as a function of F_L (for a given obstacle, F and F_L vary together: $F_L = rF$, where $r = h/L$), and shown in Figure 8. It can be seen that $H_d N/U - 1/F$ decreases when F_L increases, in agreement with the observations of Section 3, concerning the location of maximum steepening of the streamlines. It can also be observed in Figure 8 that the hydrostatic theory of Smith underestimates H_d for small F_L. The implication is that the very good agreement between Equation (4.2) and Rottman and Smith's measurements for non-hydrostatic flows (see Figure 7), appears to be related to the choice of α ($\alpha = 0.985$).

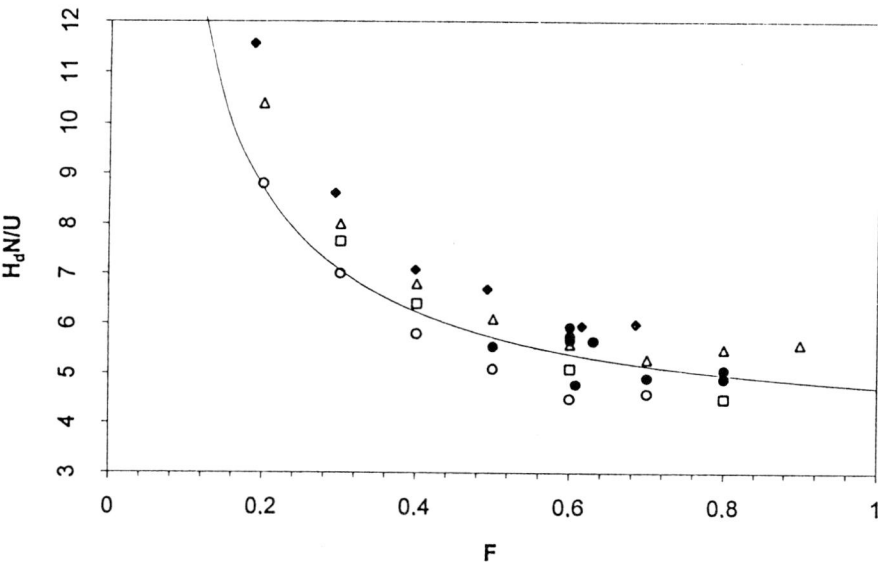

Figure 7. Non-dimensional critical streamline height $H_d N/U$ as a function of F. Experimental measurements: •, Gaussian obstacle ($r = 0.57$), ◇, cylinder; cosine-shaped ridges (Rottman and Smith [7]): ○, steep ridge ($r = 1.47$); □, intermediate ridge ($r = 0.89$); △, gentle ridge ($r = 0.42$); ——, Smith's theory

Figure 9 presents a sketch of non-hydrostatic lee-wave overturning, which includes the blocking phenomenon, as suggested by Smith, and the effect of the shift of the breaking zone due to non-hydrostatic effects, as described in Section 3. In Figure 10, the horizontal distance between the point of maximum steepening (just before the lee wave overturns) and the obstacle's crest, L_d, is plotted as a function of F_L. We find a good agreement between the measured values and those calculated with Long's model. We observe in Figures 6 that H_d does not vary with time, but that the overturning region moves upstream and reaches a fixed location just above the obstacle at $Nt \sim 150$.

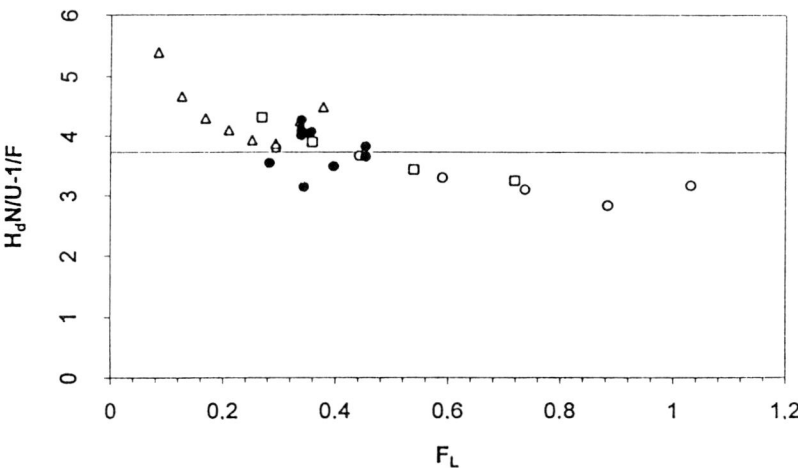

Figure 8. No n-dimensional critical streamline height $H_d N/U - 1/F$ as a function of F_L. Experimental measurements: •, Gaussian obstacle ($r = 0.57$); cosine-shaped ridges (Rottman and Smith [7]): ○, steep ridge ($r = 1.47$); □, intermediate ridge ($r = 0.89$); △, gentle ridge ($r = 0.42$); ——, Smith's theory

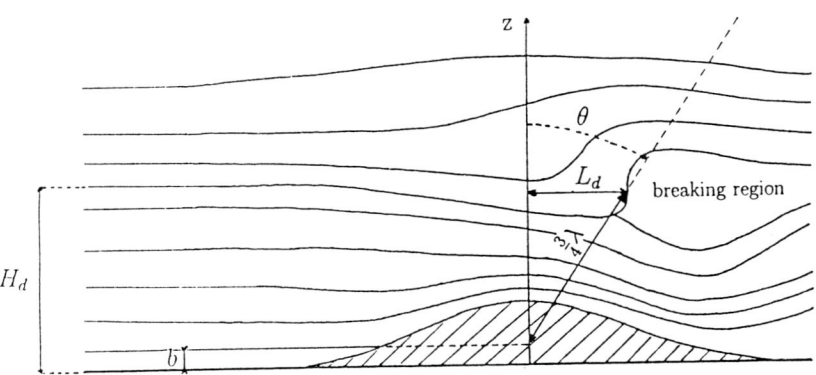

Figure 9. Sketch of lee-wave overturning

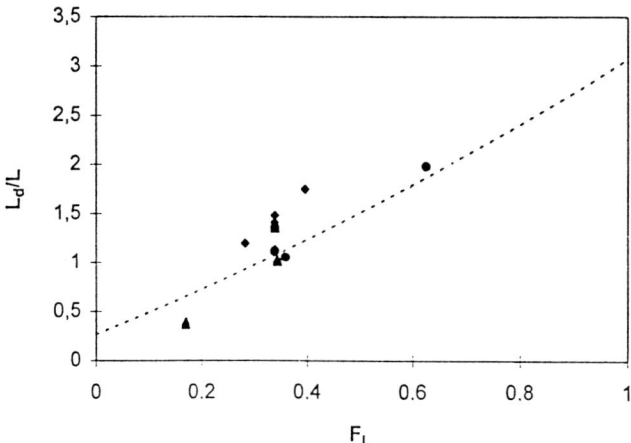

Figure 10. Horizontal distance between the point of maximum steepening and the obstacle's crest, as a function of F_L. $\triangle$, experimental measurements; ———, Long's model solution

5 Conclusion

On the basis of flow visualisations we have demonstrated the influence of non-hydrostatic effects on the location of lee-wave breaking. The measured values of the upstream height of the streamline that marks the top of the well-mixed region $H_d N/U$, were shown to increase as F_L decreases, in agreement with theoretical arguments based on non-hydrostatic steady Long's model. This corresponds to an extension of the work of Smith [5] and Rottman and Smith [7].

The experiments described here are part of a program to validate the MESONH and SUBMESO non-hydrostatic numerical models, developed in France. In this context, the extension of this study to large Reynolds numbers ($Re \sim 10000$) has recently been undertaken in the facilities of Météo-France to characterise the turbulence induced by lee-wave overturning.

Acknowledgement

This work was financially supported by the CNRS program PATOM. We are grateful to each member of the "Simulation Physique des Ecoulements Atmospheriques" team from the Centre National de Recherches Mtorologiques for their help.

Bibliography

1. Smith, R.B. (1989) Hydrostatic airflow over mountains. *Advances in Geophysics.* **31**, 1-41.

2. Baines, P.G. (1995) *Topographic effects in stratified flows.* Cambridge University Press.

3. Clark, T.L. & Peltier, W.R. (1977) On the evolution and stability of finite-amplitude mountain waves. *J. Atmos. Sci.* **34**, 1715-1730.

4. Peltier, W.R. & Clark, T.L. (1979) The evolution and stability of finite-amplitude mountain waves. Part 2: Surface wave drag and severe downslope windstorms. *J. Atmos. Sci.,* **36**, 1498-1529.

5. Smith, R.B. (1985) On severe downslope winds. *J. Atmos. Sci..* **42**, 2597-2603.

6. Laprise, R. & Peltier, W.R. (1989) The linear stability of nonlinear mountain waves: implications for the understanding of severe downslope windstorms. *J. Atmos. Sci.,* **46**, 545-564.

7. Rottman, J.W. & Smith, R.B. (1989) A laboratory model of severe downslope winds. *Tellus,* **41A**, 401-415.

8. Castro, I.P. & Snyder, W.H. (1993) Experiments on wave breaking in stratified flow over obstacles. *J. Fluid Mech.* **225**, 195-211.

9. Hunt, J.C.R. & Snyder, W.H. (1980) Experiments on stably and neutrally stratified flow over a model three-dimensional hill. *J. Fluid Mech.* **96**, 671-704.

10. Chomaz, J.M., Bonneton, P. & Hopfinger, E.J. (1993) The structure of the near wake of a sphere moving in a stratified fluid. *J. Fluid Mech.,* **254**, 1-21.

11. Kadri, Y., Bonneton, P., Chomaz, J.M. & Perrier, M. (1996) Stratified flow over three-dimensional topography. *Dyn. Atmos. Oceans,* **23** , 321-334.

12. Bonneton, P., Chomaz, J.M. & Hopfinger, E.J. (1993) Internal waves produced by the turbulent wake of a sphere moving horizontally in a stratified fluid. *J. Fluid Mech.,* **254**, 23-40.

13. Long, R.R. (1953) Some aspects of the flow of stratified fluids. A theoretical investigation. *Tellus,* **5**, 42-58.

14. Bonneton, P., Roux, N. & Perrier, M. (1995) Etude exprimentale de la dynamique des ondes de relief. *Note du CNRM, Mto-France,* **12**, 1-19.

15. Laprise, R. & Peltier, W.R. (1989) On the structural characteristics of steady finite-amplitude mountain waves over bell-shaped topography. *J. Atmos. Sci.,* **46**, 586-595.

16. Lighthill, M.J. (1978) *Waves in fluids.* Cambridge University Press.

17. Castro, I.P. (1987) A note on lee wave structures in stratified flow over three-dimensional obstacles. *Tellus* **39A**, 72-81.

18. Prasad, D., Ramirez, J. & Akylas, T.R. (1996) Stability of stratified flow of large depth over finite-amplitude topography. *J. Fluid Mech.* **320**, 369-394.

19. Queney, P. (1948) The problem of air flow over mountains : a summary of theoretical studies. *Bull. Amer. Meteor. Soc.* **29**, 16-26.

20. Snyder, W.H, Thompson, R.S., Eskridge, R.E., Lawson, R.E., Castro, I.P., Lee, J.T., Hunt, J.C.R. & Ogawa, Y. (1985) The structure of strongly stratified flow over hills : dividing-streamline concept. *J. Fluid Mech.* **152**, 249.

21. Sheppard, P.A. (1956) Airflow over mountains, *Q. J. R. Met. Soc.* **82**, 528.

The Effects of Stable Stratification on the Flow Around a Horizontal Circular Cylinder in a Channel of Finite Depth

Shigehira Ozono*, Noboru Aota** and Yuji Ohya*

*Research Institute for Applied Mechanics, Kyushu University, Kasuga, Japan, and **Sumitomo Heavy Industries Limited, Tokyo, Japan

Abstract

This paper describes a numerical study of the two dimensional flow of linearly stably-stratified Boussinesq fluid around a circular cylinder in a channel of finite depth. The Reynolds number used is high enough for the vortex shedding to occur behind the cylinder in neutral flows. Attention was focused upon the interaction between the vortex shedding and internal gravity waves. The columnar disturbances and lee waves appear when parameter K is close to integers in conformity with linear theory. Here, K is defined by $NH/\pi U$, where N is the Brunt-Väisälä frequency, H the channel depth, and U the basic flow speed. As K increases, the symmetric configuration with the cylinder placed at the mid-depth first stimulates the symmetric vertical dominant mode of the columnar disturbance at $K \sim 2$, but the non-symmetric cases with even a slight displacement of the cylinder stimulate the non-symmetric mode at $K \sim 1$. A flow alteration accompanied by an abrupt change in the Strouhal number occurs shortly after appearance of the lee wave. This alteration is thought to be caused when the lee wave emerges and its characteristics like amplitude and wavelength satisfy some conditions under which the vortex shedding is significantly affected. There is a crucial difference in the way the lee wave affects the vortex shedding between the symmetric and non-symmetric cases. In the symmetric case, the wake cavity is suppressed symmetrically by the first lee wave troughs. On the other hand, in the non-symmetric cases, the wake cavity is blocked by the first trough from one side and biased toward a nearby wall.

1 Introduction

The flow of a uniformly stratified fluid past a two-dimensional obstacle in a channel of finite depth has been the subject of extensive study. Most of the works were concerned with the flow over surface-mounted obstacles, in which no vortex shedding due to the interaction between the separated shear layers could be introduced, but there are only a few studies as to the stratified flow in the presence of the vortex shedding.

Droughton and Chen [1] reported that the flow was observed to be very sensitive to the relative position of a circular cylinder and a slender body in a towing tank for the blockage ratio from 0.05 to 0.10. More recently, Boyer et al. [2] observed the flows past a circular cylinder in stably-stratified fluid under different conditions and depicted a flow regime diagram in a Froude- vs. Reynolds-number plane. But the effects of the boundaries were not taken into account, although the blockage ratio ranges from 0.03 to 0.20.

Hanazaki [3] carried out a numerical simulation of the flow past a plate in a channel of finite depth. Although the Reynolds number used is 20, so that no vortex shedding is supposed to be created, the results may be even profitable for the explanation as to the high-Reynolds number flow with vortex shedding in that it gives a clear view of the long internal waves trapped in a channel. According to his work, there are time-dependent oscillations in each vertical mode of the upstream advancing columnar disturbances, which are related to unsteadiness in the drag coefficient. To separate it from blocking, the columnar disturbance is defined as the long internal waves causing an almost horizontal motion relative to the basic flow. Based on a linear dispersion relation by assuming $k_x \to 0$, where k_x is the horizontal wavenumber, linear theory (for example, Wei et al.[4]) claims that the columnar disturbance of mode n propagates upstream at the speed of

$$(K/n - 1)U \qquad (1.1)$$

with reference to the obstacle. In fact, Hanazaki [3] showed that the upstream advancing wave speeds in a linearly stratified flow were in conformity with (1.1).

By using a density-stratified wind tunnel, we have done an experimental work on the flow around a circular cylinder with a blockage ratio of 0.17 [5]. The Reynolds number used was of the order of 10^3 to 10^4. As expected from the neutral flow, the vortex shedding was formed in a usual way for weak stratification. But as stratification was raised, critically from $K \sim 1.0$ the roll-up of separated shear layers was seen to be suppressed and the position of their interaction shifted farther downstream in the wake. This flow alteration resulted in an abrupt change in the vortex shedding frequency. This phenomenon suggests that the internal wave of first mode (i.e., non-symmetric mode) appears first. The phenomenon has been an intriguing problem to be solved.

Now let us specify the goal of the present work. The main purpose of the paper is to examine the effects of asymmetry on the flow. To this end we numerically simulate the flows around a circular cylinder with its relative placement to the channel varied. Through numerical analysis, it is possible to obtain detailed information about the flow structure, in particular, the behavior of the upstream influences caused by internal waves. Along with the present numerical study, we conducted a water tank test [6] for a few cases using a circular cylinder, where much care was paid to realize symmetry for the symmetric configuration. Thus, some computed results are compared with the experimental ones available. Major attention is focused on the relation between the flow alteration tied to the critical change in the Strouhal number and the internal gravity waves.

2 Simulation details

2.1 Flow configurations

A typical flow configuration is presented in Figure 1a. The channel depth H is held 5d, where d is the diameter of the circular cylinder, and hence the blockage ratio d/H is 0.2. To examine the effects of asymmetry on the flow of interest, we vary the placement of the cylinder. The cases are specified by the displacement ΔH from the mid-depth of the channel. As a basic case, the cylinder is settled at the mid-depth of the channel. This is referred to as the "symmetric case" hereafter ($\Delta H = 0$). Apart from the symmetric case, two "non-symmetric

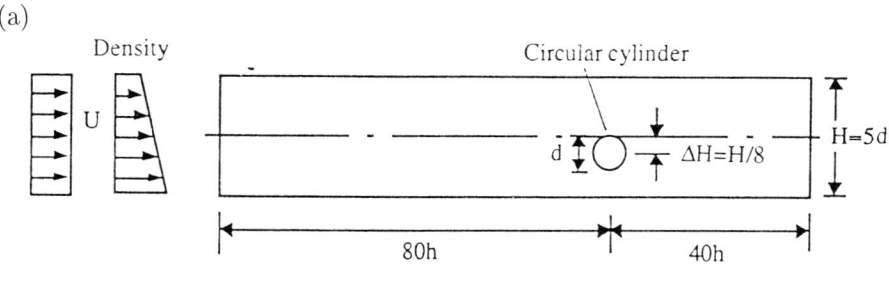

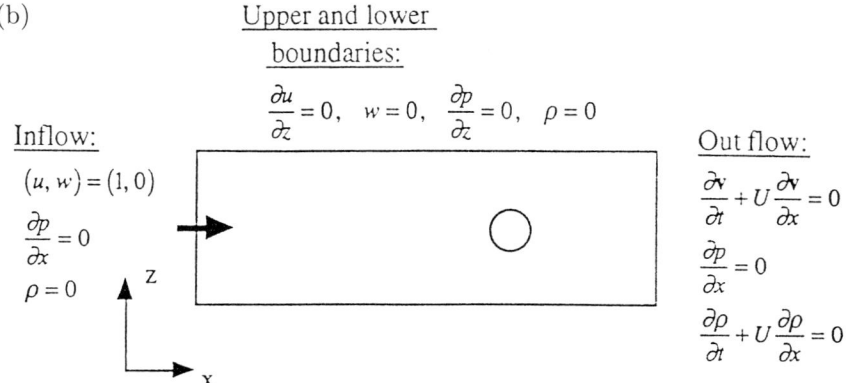

Figure 1. (a) Configuration of the non-symmetric case with $\Delta H = H/8$; and (b) outer boundary conditions

cases" with $\Delta H = H/8$ and $H/32$ are used. The slightly displaced case with $\Delta H = H/32$ is introduced to confirm the sensitivity of the flow to asymmetry. The parameter K ranges from 0 to 3.0 and the Reynolds number based on d is held 10^3.

2.2 Discretization and solution

Under the Boussinesq approximation, the two-dimensional governing equations for the incompressible viscous flow can be written in a dimensionless form as:

$$\frac{D\mathbf{v}}{Dt} = -gradp + \frac{1}{Re}\Delta\mathbf{v} - \frac{\rho}{Fr^2}\hat{i} \tag{2.1}$$

$$div\mathbf{v} = 0 \tag{2.2}$$

$$\frac{D\rho}{Dt} = w \tag{2.3}$$

where $\mathbf{v} = (u, w)$ is the velocity, p and ρ are the perturbed pressure and density respectively, and $\hat{i} = (0, 1)$ is the unit vector. The Reynolds number Re and the Froude number Fr are defined by $Re = Ud/\nu$, and $Fr = U/Nd$ respectively, where $N^2 = -(g/\rho_0)(d\rho'/dz)$. Here, g is the gravitational acceleration, ρ' is the dimensional basic density, and ρ_0 is the mean density at the mid-depth.

The outer boundary conditions are presented in Figure 1b. The top and bottom boundary conditions require some attention. To isolate pure behaviors of the columnar disturbances, the free-slip condition is assumed. The boundary conditions on the body are as follows:

- velocity: $(u, w) = 0$,

- pressure: calculated from imposing $(u, w) = 0$ within the body region,

- density: linearly extrapolated.

To solve the equations, we adopt a finite difference method. In our preliminary computations, we used a conventional finite difference scheme based on a single grid system. But as stratification was raised, the calculations mostly broke down. Hence, we applied a difference scheme based on a combination of two grids proposed by Suito et al. [7] to the spatial term and hence attained highly stratified flows. According to that paper, this scheme allows the spatial terms to conserve their leading error terms irrespective of the rotation of coordinate.

A skew grid is superposed on an original one. In the computational plane, the skew grid is rotated to the original one by 45 degrees as presented in Figure 2a. Two coordinate transformations are introduced as:

$$x = x(\xi, \zeta), z = z(\xi, \zeta) \tag{2.4}$$

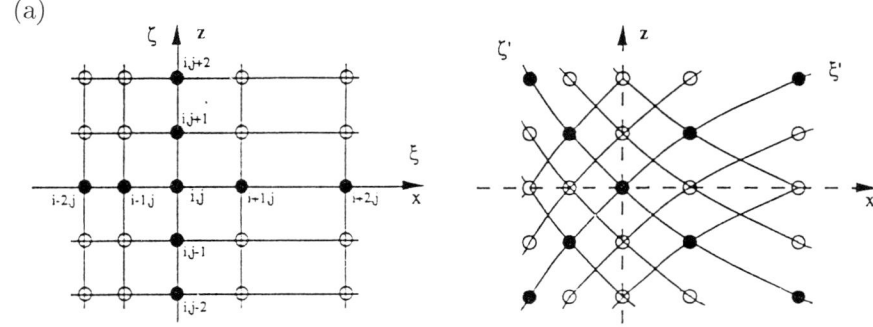

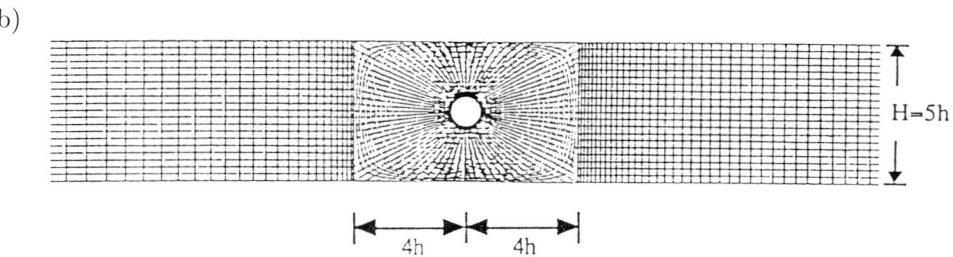

Figure 2. (a) Original and skew grids; and (b) part of the grid system used

$$x = x(\xi', \zeta'), z = z(\xi', \zeta') \tag{2.5}$$

where (ξ, ζ) and (ξ', ζ') denote the computational plane variables of the original and skew grid respectively. All the spatial terms of the original and skew grids are discretized in accordance with each transformation rule on each grid. Then the original and skew grids are averaged by a weight of 2:1 respectively.

The procedure to solve the equations is based on the marker-and-cell (MAC) method [8]. The space derivatives except the advective terms are approximated with the second-order central differences. The advective terms are approximated with a third-order upwind scheme. A Poisson equation is solved with the successive over-relaxation (SOR) method. The time evolution equations are solved with the second-order Runge-Kutta method.

The grid system used is presented in Figure 2b. The region around the cylinder is covered by an O-type grid with 81 (tangential direction) ×60 (radial) grid points. Circumferential grid lines are clustered toward the body surface to capture finer scale fluid motions. The up- and downstream regions outside the O-type grid consist of rectangular grids with 81 (horizontal) ×21 (vertical) grid points in common. The time increment is set to 0.002.

3   Results and discussion

3.1   Streamlines

Figure 3 shows flow patterns for the symmetric case. For the neutral flow ($K = 0$), the classical Kármán vortex street is formed in the wake. The vortex shedding survives up to a value of $K = 1.5$, although the wake cavity seems to be suppressed gradually. At $K = 1.6$, the vortex shedding disappears. When $K = 2.0$, the lee wave troughs are evidently observed downstream at $x \sim 10d$. As K increases further beyond $K = 2.0$, the wavelength of the lee waves is gradually shortened. As shown unambiguously for $K = 2.5$, the wake cavity is impeded symmetrically by the lee waves and shrinks extremely.

Flow patterns for the non-symmetric case with $\Delta H = H/8$ are presented in Figure 4. For the neutral flow ($K = 0$), the vortex street is formed normally. From $K = 1.0$, the lee wave appears with the first trough downstream at $x \sim 25d$, but its amplitude is very small. When $K = 1.1$, the lee wave can be seen clearly. As K increases further, the wavelength of the lee wave becomes shorter, so that the first trough approaches the cylinder and seems to deflect the wake cavity downward forcibly. The vortex shedding seems to vanish at $K = 2.0$. Note that when K is larger than 2.0, the lee waves of symmetric mode dominate, as clearly seen from the pattern for $K = 2.5$.

Flow patterns for the non-symmetric case with $\Delta H = H/32$ are shown in Figure 5. Overall, the variation of patterns with K is closely similar to the case with $\Delta H = H/8$ rather than the symmetric case. It is interesting to note that this feature results from introduction of slight asymmetry. Comparing the respective patterns with a certain K, we can see that the lee wave amplitude of the case with $\Delta H = H/32$ is smaller than that of the case with $\Delta H = H/8$. This difference in amplitude may be attributed to the responsibility of each mode to the relative placement of the cylinder. More specifically we can say that the larger ΔH is, the larger the lee wave amplitude will be. For $K < 1.6$, the amplitude of the lee waves is too small for its appearance to be identified clearly, although very slight undulation can be observed far downstream for $K = 1.3$. However, at least when K reaches 1.6, the lee wave is created significantly. As K increases further, the wavelength of the lee wave becomes shorter in a similar manner to the case with $\Delta H = H/8$. The vortex shedding seems to be totally inhibited for $K \geq 1.8$.

3.2   Perturbed streamlines

Figures 6–8 show the perturbed streamlines at a dimensionless time $t = Ut'/d = 100$, where t' is the dimensional time. They can be obtained by subtracting the basic flow from a flow field.

In the symmetric case (Figure 6), as K becomes greater than 2.0, disturbances with elongated closed streamlines like vortices advance upstream. The disturbance amplitude varies in space and the eddy-like disturbances are shed

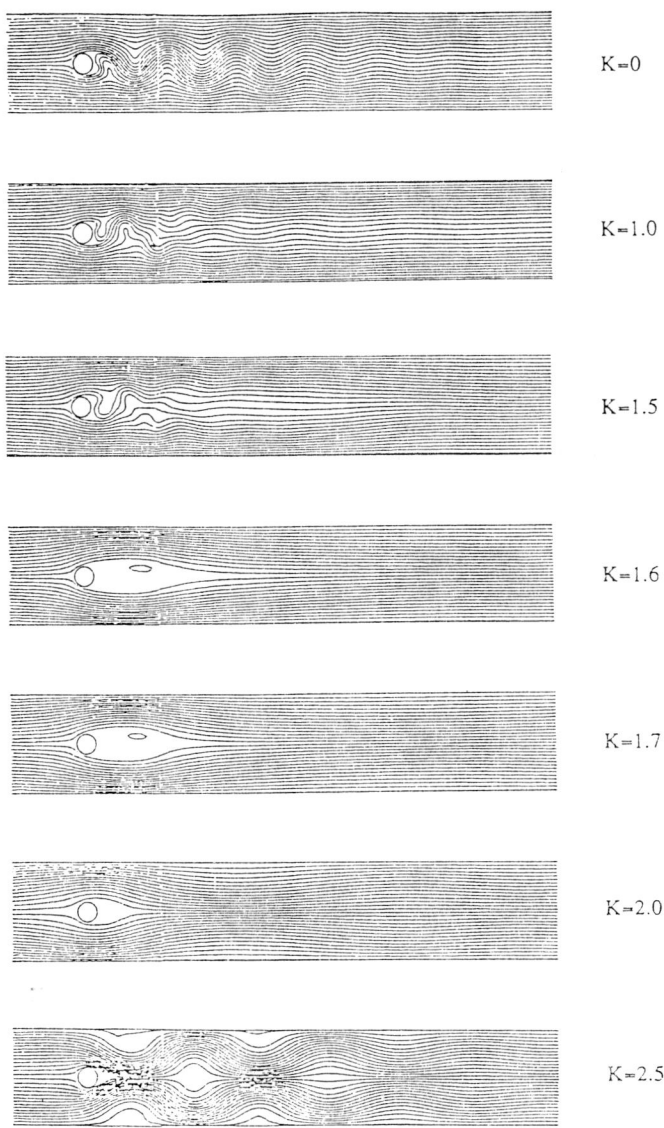

Figure 3. Streamlines for the symmetric case: $Ut'/h = 100$

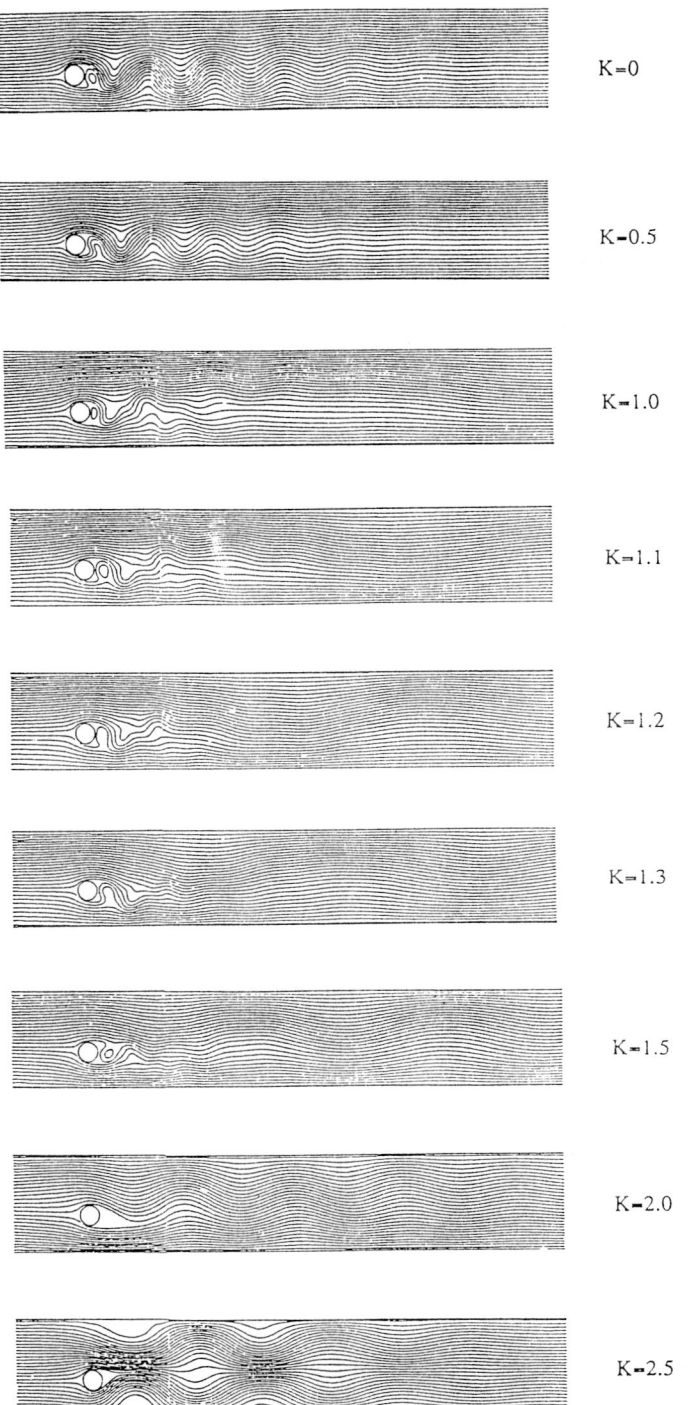

K=0

K=0.5

K=1.0

K=1.1

K=1.2

K=1.3

K=1.5

K=2.0

K=2.5

Figure 4. Streamlines for the non-symmetric case with $\Delta H = H/8 : Ut'/h = 100$

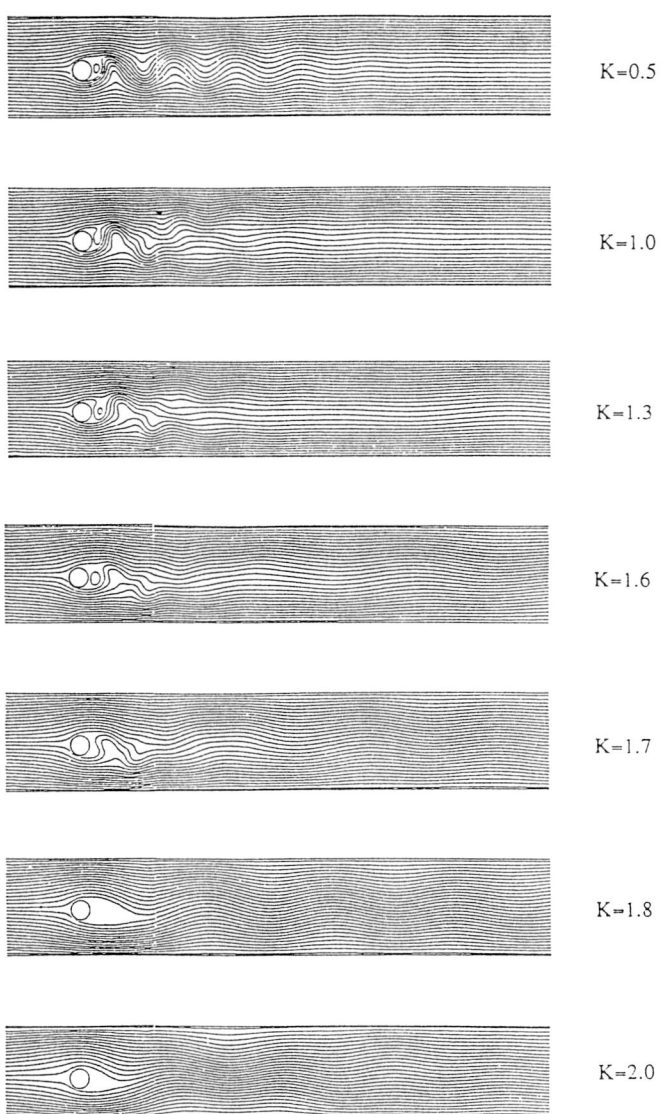

K=0.5

K=1.0

K=1.3

K=1.6

K=1.7

K=1.8

K=2.0

Figure 5. Streamlines for the non-symmetric case with $\Delta H = H/32 : Ut'/h = 100$

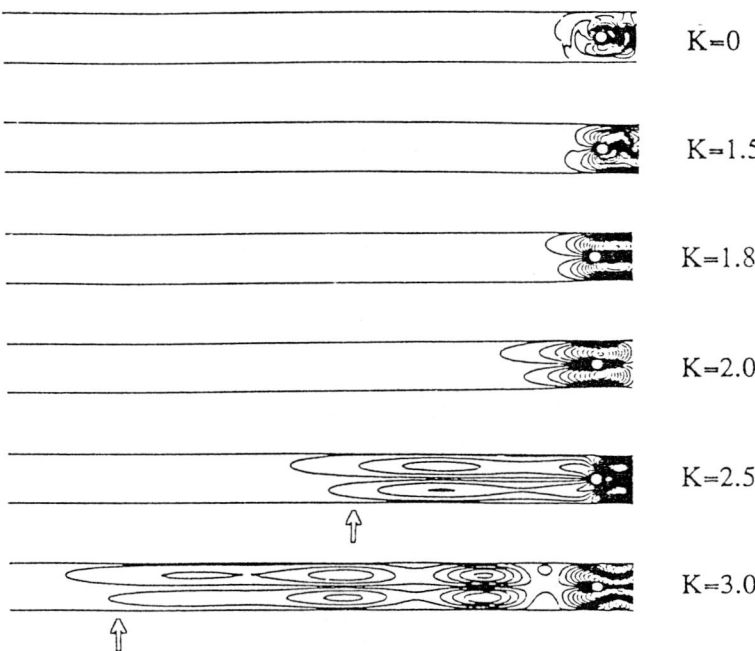

Figure 6. Perturbation streamlines for the symmetric case: $Ut'/h = 100$. The arrows indicate the locations of the fronts of columnar disturbances of second mode predicted by linear theory

upstream from the body repeatedly as found by Hanazaki [3]. From the vertically symmetric shape of the disturbances, this is thought to be the columnar disturbances of second mode.

Unlike the symmetric case, it is for K greater than 1.0 that the upstream advancing columnar disturbances are created in the non-symmetric case (Figures 7 and 8). Since the disturbances are shed upstream from the upper side of the cylinder, this is thought to be the columnar disturbance of first mode. We can see that the disturbance of the case with $\Delta H = H/32$ is significantly weaker than that of the case with $\Delta H = H/8$ in magnitude. It is of particular interest to note that although, in the non-symmetric case with $\Delta H = H/32$, only slight asymmetry is introduced to the flow configuration (i.e., 3.1% of the channel depth), the value of K for which the columnar disturbance appears jumps from ~ 2.0 to ~ 1.0.

The upstream fronts of columnar disturbances are in good agreement with those predicted by the formula (1.1) corresponding mode $n = 2$ for the symmetric case and mode $n = 1$ for the non-symmetric cases in Figures 6-8, although

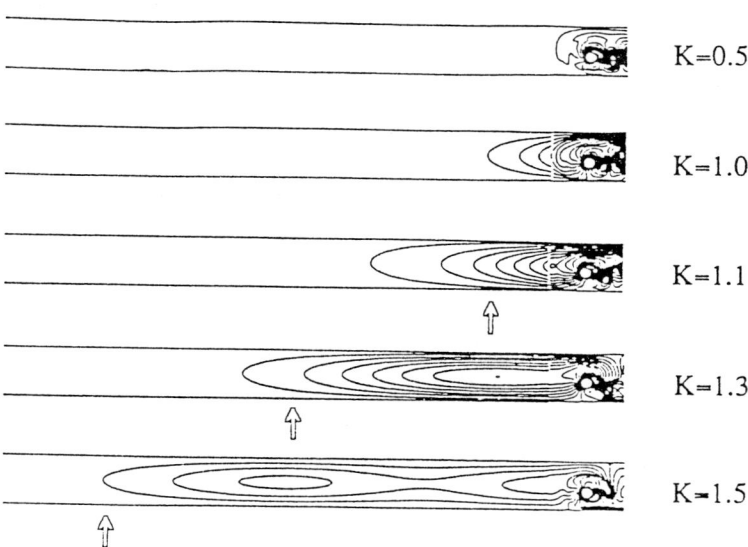

Figure 7. Perturbation streamlines for the non-symmetric case with $\Delta H = H/8 : Ut'/h = 100$. The arrows indicate the locations of the fronts of columnar disturbances of first mode predicted by linear theory

the wave speeds at $K \sim 1$ for the non-symmetric cases are somewhat inconsistent. Comparing Figures 7 and 8, we can find close similarity between them. This fact shows that only if a pair of non-symmetric cases have an equal value of K, they will have an equal wave speed. Moreover, this provides us the following understanding. Although the dominant mode of the disturbance is determined alternatively by the placement of the obstacle itself (i.e., whether the configuration is symmetric or not), once the disturbances of a certain mode are created, the wave speeds are not characterized by obstacle conditions, but by linear theory.

Here we should mention an exceptional manner. According to the linear theory, the columnar disturbances of mode $n = 1$ cannot propagate upstream for $K \leq 1$ (see (1.1)). But in practice, even at $K = 1.0$ (so-called "resonance") certain disturbances do stretch upstream as shown in Figures 7 and 8. Likewise, for the symmetric mode $n = 2$, they cannot propagate upstream for $K \leq 2$ in theory, but as shown in Figure 6, some disturbances of symmetric mode stretch significantly at $K = 2.0$.

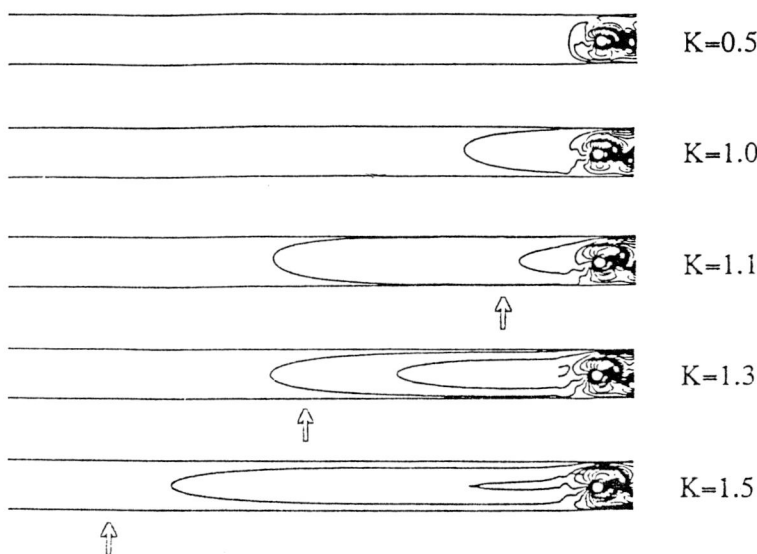

Figure 8. Perturbation streamlines for the non-symmetric case with $\Delta H = H/32$: see Figure 7 for legend

3.3   Strouhal number

Figure 9 shows the variation of the Strouhal number St (St is defined by fd/U, where f is the dominant frequency of vortex shedding) *vs.* K together with the experimental results [6]. From both the numerical and experimental results for the symmetric case, St is decreased with K increasing up to 1.5. In the numerical analysis, the vortex shedding is totally inhibited when K exceeds 1.5, so it is unable to define the Strouhal number. But according to the experiment, the vortex shedding can be observed even for $K > 1.5$. We observed that when K exceeded 1.7, the Strouhal number increased abruptly and at the same time the roll-up of separated shear layers was suppressed and the position of their interaction was shifted farther downstream in the wake. It should be noted that despite this flow alteration, the vortex shedding itself was retained throughout the experiment.

   For the non-symmetric case with $\Delta H = H/8$, St is decreased with K increasing up to 1.1, at which the lee wave appears clearly. But with further increase in K, St increases again until the vortex shedding is inhibited totally at $K = 2.0$. In the experiment also, when K is larger than 1.3 at which the lee wave emerges, the Strouhal number shows an abrupt increase. In accordance with this abrupt change in St, we observed a flow alteration similar to the symmetric case. From

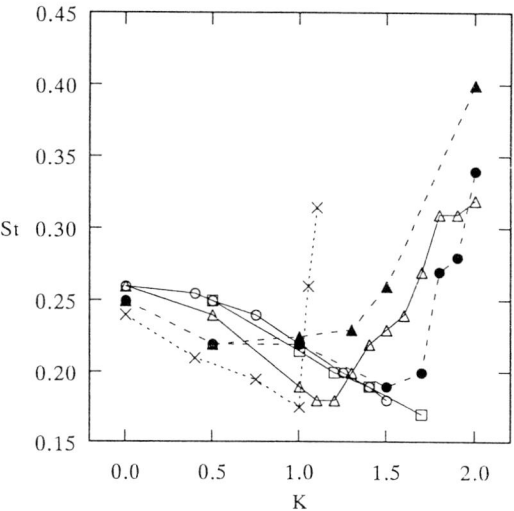

Figure 9. Variation of Strouhal number with K: ○, symmetric case (numerical); △, non-symmetric case with $\Delta H = H/8$ (numerical); □, non-symmetric case with $\Delta H = H/32$ (numerical), ●, symmetric case (experimental); ▲, non-symmetric case with $\Delta H = H/8$ (experimental). ● and ▲ are taken from reference [6]. × denotes the data from reference [5] using a wind tunnel

Figure 4 the abrupt flow alteration at around $K = 1.1$ and 1.2 is not so clear, but the critical behavior of St taking a minimum at around 1.1 suggests a qualitative change in the dynamics of vortex formation.

For the non-symmetric case with $\Delta H = H/32$, St decreases up to $K = 1.7$ in a similar manner to the symmetric case. This is probably because the amplitude of the lee wave is so small that St is influenced very little by the lee wave.

3.4 Flow patterns in the $K - \Delta H$ plane

Figure 10 summarizes the variation of the flow patterns with K. Overall, the columnar disturbances and lee waves appear at integers of parameter K (i.e., $K \sim 1$ for the non-symmetric cases and $K \sim 2$ for the symmetric case). In the non-symmetric case with $\Delta H = H/32$, exceptionally, the value of K for which the lee wave appears falls around 1.3. This is because the lee wave is too small in amplitude for its value to be identified precisely. But at least as far as the columnar disturbance is concerned, the discrete modal structure as predicted by linear theory is properly realized in the present flow situation. Note that even a slight displacement stimulates the internal waves of non-symmetric mode.

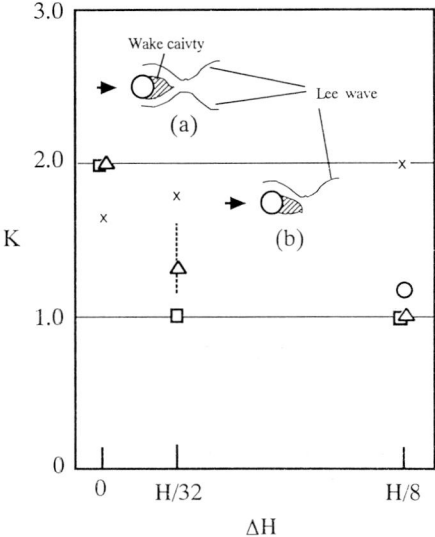

Figure 10. Schematic representation of the variation of flow patterns: ○, the flow alteration accompanied by an abrupt change in the Strouhal number occurs; △, the lee wave appears; □, columnar disturbance appears. The dotted line at $\Delta H = H/32$ denotes that the lee wave should exist, but not clearly identified; ×, the vortex shedding is totally inhibited in the simulation

On the basis of these linear aspects, we can understand the flow alteration accompanied by an abrupt change in the Strouhal number as seen in Figure 9 for the non-symmetric case with $\Delta H = H/8$, although the flow alteration is not so apparent from the flow patterns in Figure 4. The implicit flow alteration occurs shortly after appearance of the lee wave at $K = 1.0$. This alteration is thought to be caused when the lee wave emerges and its characteristics like amplitude and wavelength satisfy some conditions under which the vortex shedding is significantly affected. For the symmetric case and non-symmetric case with $\Delta H = H/32$ in the numerical analysis, the vortex shedding vanishes before the Strouhal number reaches a minimum, so it is impossible to specify the critical Strouhal number.

On the contrary, the dynamics of the so-called "vortex suppression" is not straightforward. As a matter of fact, the vortex shedding was observed even at relatively large K in the laboratory experiment. To discuss the vortex suppression thoroughly, other factors like the Reynolds number should be taken into consideration. We put such discussion outside the scope of this paper. Nonetheless, we can point out the following characteristics. There is a crucial difference in the way the lee wave affects the vortex shedding between the symmetric and non-symmetric cases. The flow patterns at higher K when the wavelength is shortened enough disclose this difference clearly. In the symmetric case, the wake cavity is suppressed symmetrically by the first lee wave troughs (see $K = 2.0$ in Figure 3, or a descriptive sketch (a) in Figure 10). On the other hand, for the non-symmetric cases, the wake cavity is blocked by the lee wave from one side and biased toward a nearby wall (see $K = 2.0$ in Figure 4 and $K = 1.8$ in Figure 5, or a descriptive sketch (b) in Figure 10).

As mentioned before in Figure 3 for $K = 1.6$, there seems to be no lee wave for the symmetric case. One possible explanation is that although the lee wave emerges for that value, its amplitude is too weak for its appearance to be identified only by inspection. According to our numerical analysis [9] under similar conditions to the present study except for using a rectangular cylinder, for the symmetric configuration, when $K = 1.8$, the vortex shedding vanishes and at the same time the lee waves are observed. Although the body shape is different, it seems reasonable to suggest that this feature should also hold true for the circular cylinder case. However, further studies may be required to confirm this inference.

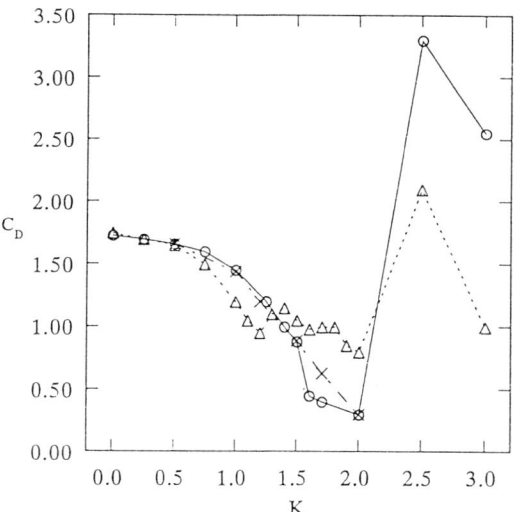

Figure 11. Variation of the drag coefficient with K: ◯, symmetric case; △, non-symmetric case with $\Delta H = H/8$; ×, non-symmetric case with $\Delta H = H/32$

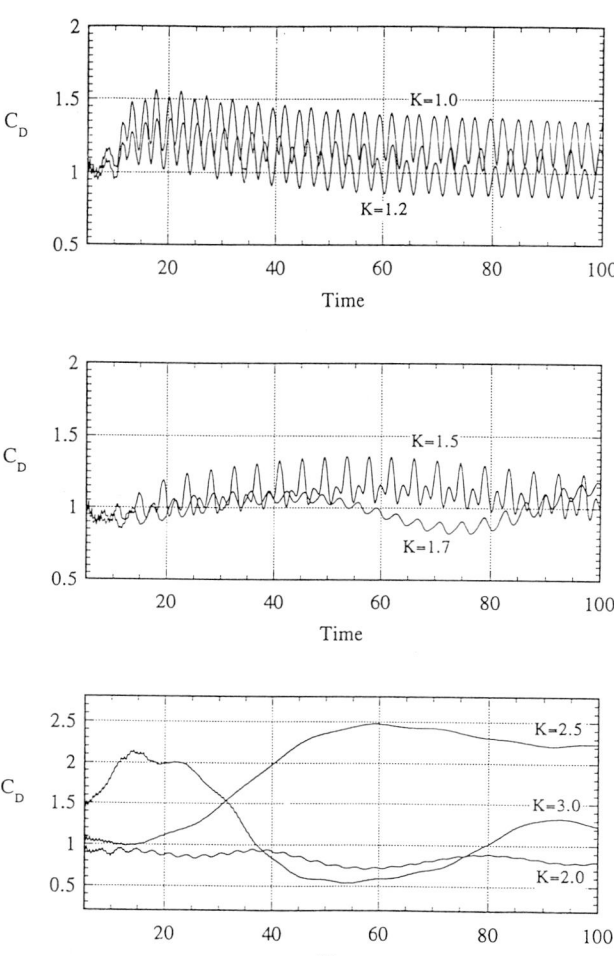

Figure 12. Time development of the drag coefficient for the non-symmetric case with $\Delta H = H/8$

3.5 Drag coefficient

Figure 11 shows the variation of the drag coefficient C_D with K. Here, C_D is defined by $C_D = F_x/\{(1/2)\rho_0 U^2 d\}$, where F_x is the drag force. For the symmetric case, the drag coefficient decreases with K increasing up to 2.0. However, with further increase in K, C_D increases rapidly. For the non-symmetric case with $\Delta H = H/8$, roughly speaking, C_D is reduced at integers of K as predicted by linear theory for the surface-mounted configurations [10]. One of the minima occurring at $K = 1.1$ corresponds to that of the Strouhal number and this feature supports the qualitative change in the dynamics of vortex shedding again. For the non-symmetric case with $\Delta H = H/32$, C_D decreases up to 2.0 in a similar manner to the symmetric case. This is probably because the amplitude is so weak that the slight asymmetry has little effect on the drag coefficient in common with the Strouhal number.

The time development of the drag coefficient for the non-symmetric case with $\Delta H = H/8$ is shown in Figure 12. The relatively high-frequency oscillations with time for $K \leq 1.7$ are probably due to vortex shedding. This is proved by the fact that these oscillations disappear when K becomes 2.0, when the vortex shedding vanishes. A closer inspection reveals that the time history is not exactly monoharmonic, but large- and small-amplitude waves appear by turns. This feature may reflect the non-symmetric structure in the non-symmetric case, i.e., the wake cavity being blocked by the lee wave from one side. It is interesting that for $K = 1.5$ and 1.7 some long-period unsteadiness exists. Such unsteadiness has been a major topic in the study as to the flow over surface-mounted obstacles these days and several explanations were proposed [11].

4 Remarks on a previous experimental result

In our previous experiment using a wind tunnel [5], the circular cylinder was supposed to be settled in a symmetric configuration. As K increases from zero, vortex shedding appeared to be intensified up to $K \sim 1.0$. From a certain value of $K \sim 1.0$, however, the roll-up of the separated shear layers was critically .suppressed and the position of their interaction shifted farther downstream in the wake. This flow alteration resulted in an abrupt change in the vortex shedding frequency (Figure 9). Although this abrupt flow change may be tied to the generation of the internal waves, the odd integer of $K \sim 1.0$ is not consistent to the symmetric configuration, because $K = 1$ is a critical value beyond which the first vertical mode appears according to linear theory. For the symmetric case, it is natural that the internal waves of the vertical symmetric (second) mode should occur from $K \sim 2$.

A plausible explanation for this apparent inconsistency is as follows. The experimental arrangement probably had some imperfections such as the placement of the circular cylinder and the velocity and/or density in the approaching flow. As a matter of fact, the cylinder was not positioned precisely at the mid-depth for a technical reason. Then some of them may have broken the symmetry in combination and driven the non-symmetric first mode at $K \sim 1.0$. In support of this scenario, Figure 8 evidently shows that introduction of slight asymmetry to the flow configuration causes the columnar disturbance of first mode at $K \sim 1.0$ unlike the "symmetric case". Furthermore, as shown in Figure 9, close similarity between the wind tunnel test and the present computation for the non-symmetric case with $\Delta H = H/8$ can be seen, although the computed St does not show so drastic change.

5 Conclusions

A finite difference simulation was conducted to investigate the two dimensional flow of linearly stably-stratified Boussinesq fluid around a circular cylinder in a channel of finite depth. The Reynolds number was taken to be 10^3, so that the vortex shedding behind the cylinder was formed normally for neutral flows. Parameter K ranged from 0 to 3.0. The major results obtained are as follows:

1. The columnar disturbances and lee waves appear when parameter K is close to integers in conformity with linear theory. The perturbed streamlines reveal that as K increases, the symmetric configuration first stimulates the symmetric vertical dominant mode (second mode) of the columnar disturbance at $K \sim 2$, but the non-symmetric configurations with even a slight displacement of the cylinder stimulates the non-symmetric mode (first mode) at $K \sim 1$.

2. The flow alteration accompanied by an abrupt change in the Strouhal number occurs shortly after appearance of the lee wave. This alteration is thought to be caused when the lee wave emerges and its characteristics like amplitude and wavelength satisfy some conditions under which the vortex shedding is significantly affected. There is a crucial difference in the way the lee wave affects the vortex shedding between the symmetric and non-symmetric cases. In the symmetric case, the wake cavity is suppressed symmetrically by the first lee wave troughs. On the other hand, in the non-symmetric cases, the wake cavity is blocked by the first trough from one side and biased toward a nearby wall.

3. The flow is sensitive to asymmetry. Introduction of even slight asymmetry could cause internal waves of asymmetric mode. In conducting a laboratory experiment much care should be paid for the effects of the wave reflection on the walls.

Bibliography

1. J.V. Droughton and C.F. Chen. The channel flow of a density-stratified fluid about immersed bodies. *Transaction of the ASME*, 122–130, 1972.

2. D.L. Boyer, P.A. Davies, H.J.S. Fernando and X-H. Zhang. Linearly stratified flow past a horizontal circular cylinder. *Phil. Trans. R. Soc. Lond.*, **A328**, 501–528, 1989.

3. H. Hanazaki. Upstream advancing columnar disturbances in two-dimensional stratified flow of finite depth. *Phys. Fluids A*, 1–12, 1976-1987, 1989.

4. S.N. Wei, T.W. Kao and H.P. Pao. Experimental study of upstream influence in the two-dimensional flow of a stratified fluid over an obstacle. *Geophys. Fluid Dyn.*, **6**, 315, 1975.

5. Y. Ohya and Y. Nakamura. Near wakes of a circular cylinder in stratified flows. *Phys. Fluids A Letters*, **2-4**, 481–483, 1990.

6. Y. Ohya and N. Aota. (Unpublished).

7. H. Suito, K. Ishii and K. Kuwahara. Simulation of dynamic stall by multi-directional finite-difference method. 26th AIAA Fluid Dynamic Conference, 19-22/San Diego, CA, 1995.

8. F.H. Harlow and J.E. Welch. Numerical calculation of time-dependent viscous incompressible flow of fluid with free surface. *Phys. FLuids*, **8**, 2182–2189, 1965.

9. S. Ozono, N. Aota and Y. Ohya. Stably stratified flow around a horizontal rectangular cylinder in a channel of finite depth. *Proc. of 2nd Int. Symposium on Comp. Wind Engng.*, (CWE96), Fort Collins, 1996.

10. K. Trustrum. An Oseen model of the two-dimensional flow of a stratified fluid over an obstacle. *J. Fluid Mech.*, **50**, 177–188, 1971.

11. I.P. Castro, W.H. Snyder and P.G. Baines. Obstacle drag in stratified flow. *Proc. R. Soc. Lond.*, **A429**, 119–140, 1990.

Mixing in Stratified Flow Over Three-Dimensional Obstacles

M.F. Paisley* and I.P. Castro**

*School of Computing, Staffordshire University, and **Department of Mechanical Engineering, University of Surrey

Abstract

Numerical computations of viscous stratified flows over three-dimensional obstacles are presented. Attention is concentrated on the localised regions where density gradients are reversed and vertical mixing occurs, namely the upstream flow reversal zone, downstream separated flow under a lee wave crest and the region surrounding a breaking lee wave. A particular case is examined in detail, and data relating to the evolution in time of mixed flow in each of these regions is considered. Modelling such phenomena requires the use of a turbulence model, and the question of the adequacy of either a mixing length or a one-equation turbulence model is addressed. Flow gradients are high in mixing regions and the issue of sufficient grid resolution is discussed.

1 Introduction

The interaction of the flow of a stratified fluid with topography can be highly complex, even for relatively simple geometries, and a wide range of phenomena can occur. Implicit in some of these phenomena is flow recirculation and/or streamline overturning, where density gradients are locally reversed. Heavier fluid overlies lighter and the resulting static instability causes vertical mixing. Such regions are an important feature of the dynamics of atmospheric flow over terrain and can be influential in the context of local and mesoscale meteorology and the transport of atmospheric contaminants.

In the presence of a stable density gradient the flow of a stratified fluid over topography is characterised by the Froude number, $F_h = U/Nh$, where U, N and h are the upstream velocity, buoyancy frequency and obstacle height respectively. At high Froude numbers when lee waves are absent, recirculation and mixing would occur in the region of separated flow behind a sufficiently bluff obstacle. At lower Froude numbers, however, the dynamics of the flow may lead to the occurrence of mixing in at least three other regions. When lee waves are present, for example, separation on the lee slope may be suppressed, and may occur instead under lee wave crests further downstream. Separation and mixing may also occur at low levels in the upstream flow, due to flow stagnation. In addition, downstream lee waves may break, in which case streamline overturning leads to static instability and mixing aloft.

Numerous experimental and numerical studies of stratified flow over obstacles have been carried out in recent years. Experimental work with three-dimensional obstacles includes [1-3], while the numerical work in three dimensions includes [4-6]. Realistic modelling of mixing effects requires the use of viscous equations and some form of turbulence parametrisation. This paper describes recent high Reynolds number viscous computations in three dimensions modelled by the unsteady incompressible Navier Stokes equations, with either a mixing length or a one-equation turbulence model. Computations have been performed for three obstacles of cosine-shaped cross-section used in the experimental study [3]. Some of the results were described in [7], where it was shown that conditions for the onset of wave-breaking in the downstream flow for the wider obstacles, and vortex-shedding for the narrowest, agreed well with the experimental data.

This paper is concerned with the time-development of the flow field and particularly the regions of the flow where mixing occurs. Two aspects are in mind: firstly the adequacy of the turbulence model and, secondly (because flow gradients are neccessarily high in such regions) the effect of grid refinement.

2   Numerical model

Much of the detail of the numerical method has been omitted here, and can be found in previous work, for example [7,8].

2.1   Geometry

A schematic diagram of the computational domain and surface plots of the obstacles used are shown in Figures 1 and 2. The boundary conditions on the computational domain correspond to those of a towing tank, with uniform flow and linear density gradient specified at inflow and zero-gradient conditions at outflow. At the bottom of the domain no-slip conditions are applied for $|x| < 8$ and symmetry conditions elsewhere. The conditions at the top of the domain and the far side wall correspond either to a moving wall or a symmetry plane. Only half the domain is used, with symmetry imposed along the centre plane, except in the low Froude number cases for the axisymmetric obstacle. Three grids have been used, of sizes $64 \times 24 \times 32$, $96 \times 32 \times 48$ and $128 \times 48 \times 64$. The grid spacings near the hill for the medium grid are approximately $\Delta x/h = 0.2$, $\Delta y/h = 0.2$, $\Delta z/h = 0.05$ (with the first grid point above the surface at $z/h = 0.025$), with the respective dimensions for the coarser and finer grids approximately double and half these values.

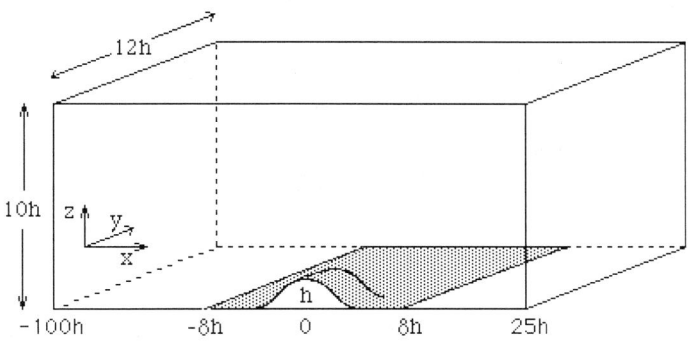

Figure 1. Typical computational domain

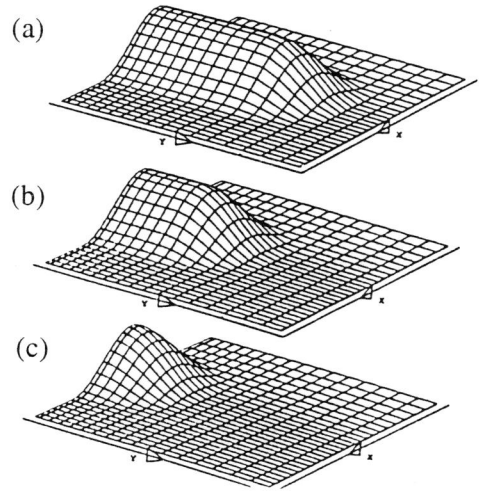

Figure 2. Topography used. (a) COS3, (b) COS2, and (c) COS1

2.2 Discretisation and solution procedure

The equations for turbulent stratified flow to be solved are

$$\frac{Du_i}{Dt} = -\frac{\partial p}{\partial x_i} - \frac{1}{F_h^2}\theta\hat{z} + \frac{\partial}{\partial x_j}\left\{\frac{1}{Re}\frac{\partial u_i}{\partial x_j} + \tau_{ij}\right\}$$

$$\frac{D\theta}{Dt} = \frac{\partial}{\partial x_j}\left\{\frac{1}{Re.Pr}\frac{\partial \theta}{\partial x_j} + H_j\right\}$$

$$\frac{\partial u_i}{\partial x_i} = 0.$$

Following a boundary conforming transformation, the equations are discretised on staggered grids using a finite volume method. Standard interpolation is used for the viscous terms, while the treatment of the advective terms guarantees monotonicity, [9,10]. The scheme has been demonstrated to be effective in the prevention of density overshoots in mixing regions in flows over two-dimensional obstacles [11]. Coupling between the momentum equations and the continuity equation is achieved using a pressure correction method. Second order backward timestepping and "impulsive start" initial conditions were used in all cases with a nondimensional timestep of $U\Delta t/h = 0.5$ or 1.0.

2.3 Turbulence parametrisation

The components of the Reynolds stress tensor and the turbulent contribution to the diffusive term in the scaler transport equation are given by

$$\tau_{ij} = \nu_t\left\{\frac{\partial u_i}{\partial x_j} + \frac{\partial u_j}{\partial x_i}\right\}, \quad H_j = (\nu_t/Pr)\frac{\partial \theta}{\partial x_j},$$

where ν_t is an eddy viscosity and Pr is the turbulent Prandlt number. In the mixing length model the eddy viscosity is determined as $\nu_t = l^2 S$, where the strain rate S is given by

$$S^2 = \frac{1}{2}\left\{\frac{\partial u_i}{\partial x_j} + \frac{\partial u_j}{\partial x_i}\right\}^2,$$

and the mixing length l is given by $1/l = 1/l_o + 1/\kappa z$, where l_o is a constant.

In the one-equation model, a transport equation of the form

$$\frac{Dk}{Dt} = \frac{\partial}{\partial x_j}\left\{\left(\nu + \frac{\nu_t}{\sigma}\right)\frac{\partial k}{\partial x_j}\right\} + P_k - \epsilon + G_k$$

is solved for the turbulent kinetic energy, k, where P_k, G_k and ϵ are terms representing shear production, buoyancy production and dissipation rate. The eddy viscosity is now $\nu_t = C_\mu k^{1/2}l$ where C_μ is a standard constant.

In stratified flow the eddy viscosity can be modified according to the local conditions of stability. Defining the value of the local gradient Richardson number as

$$Ri = \frac{1}{F_h^2} \frac{\partial \theta / \partial z}{S^2}$$

and setting a value for the critical value Ri_c (say 0.25), the eddy viscosity can be modelled by:

Stable $(Ri > 0)$:
$$\begin{cases} \nu_t (1 - Ri/Ri_c)^2, & 0 \leq Ri \leq Ri_c \\ 0, & Ri_c \leq Ri \end{cases}$$

Unstable $(Ri < 0)$: $\quad \nu_t (1 - Ri)^{1/2}$

Solid wall boundary conditions for both models are dealt with using the usual wall function treatment, implemented as a modification to the eddy viscosity immediately adjacent to the boundary. There are inconsistencies in this whole approach, for the near-wall conditions in these flows are far from the conditions under which the log law applies, particularly in regions of recirculation. Notwithstanding this, an estimate of the local friction velocity, u_τ, is obtained and used to calculate the nondimensional height of the first grid point above the surface, z^+, which, for strict validity, should be in the turbulent region with $z^+ > 30$. The two models differ in the method used to deduce the friction velocity: the one-equation model uses the square-root of the turbulent kinetic energy, while the mixing length model uses the logarithmic profile. The disavantage in the latter case is that u_τ vanishes at stagnation, while in the former it need not.

The implication for grid refinement is that care needs to be taken to ensure that the near-surface grid points remain sufficiently far from the wall that they continue to lie in the turbulent layer. On the other hand, good resolution of flow features close to the surface demands that grid spacings (determined in the vertical direction largely by where the near-surface grid point is) be as small as possible. The heights of the near-surface points in the grids used here are 0.05, 0.025 and 0.01 (ie 5%, 2.5% and 1% of the hill height), and tests showed that for these to lie consistently in the turbulent layer (except at stagnation), the Reynolds number needed to be of the order of 10^5. $Re = 10^4$ was found to be too low, and computations with this Reynolds number with both models show an undue degree of grid dependence, where, typically, a breaking wave present in the flow computed on a coarse grid might be absent on a finer grid. The computations described here were thus performed with $Re = 10^5$, and although this is somewhat higher than a typical value in a towing tank experiment, at least some measure of grid independence could be found.

The one-equation model also requires an inflow boundary condition for k; this was set so that free stream levels of turbulence were approximately 0.025%. The constant l_o controls the overall level of eddy viscosity, and tests with a vertical barrier showed that the results of computations with the value of $l_o = 0.1h$ reproduced behaviour such as unsteadiness and wave-breaking seen in experiments. Tests also showed that for stratified flow over a flat plate in two dimensions, computations with both models gave similar vertical profiles of flow quantities. Further computations of breaking waves over a two-dimensional obstacle of the same cross-sectional shape as the three-dimensional obstacles used here indicated only minor differences.

3   Results

We discuss initially one of the general difficulties of computing stratified flows over obstacles, namely that of dealing with upstream propagating modes. Computing times (and towing-times in experimental tanks) are limited by the arrival time of the reflection of the fastest mode from the upstream boundary. Linear theory indicates that the amplitude of such modes is proportional to the lee-wave amplitudes, and hence the strength of upstream modes is expected to decay as the obstacle width decreases. This is certainly the case for the obstacles here as shown in Figure 3, where the horizontal velocity perturbation is shown at $x/h = -10$ at a nondimensional time of $Ut/h = 25$ for each obstacle, and

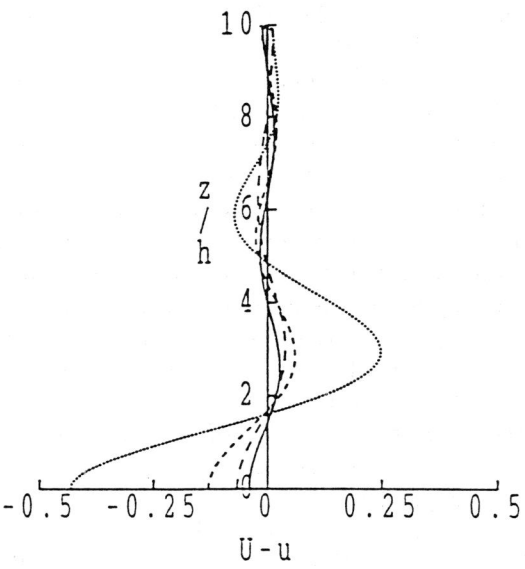

Figure 3. Vertical profile of upstream modes at $x/h = -10$ and $Ut/h = 25$. COS1, COS2, COS3, 2D

$F_h = 0.6$. Since $K = D/(\pi h F_h) = 5.31$, five modes are expected, but by this nondimensional time only the fastest three have arrived. The two-dimensional case is shown for comparison and indicates the dramatic reduction in amplitude for the three-dimensional problem. In these computations we have been careful to extend the upstream boundary sufficiently far to delay the arrival of reflections from upstream. For Froude numbers in the range $0.6 \leq F_h \leq 0.8$ an upstream boundary set at $x/h = -100$ suffices to allow computation up to at least $Ut/h = 35$.

Having established that the downstream flow (and the flow immediately upstream) can evolve free from the influence of upstream reflections, we turn now to our objective of investigating the typical mixing processes present in these flows. A particular case is chosen, that of the widest cosine hill (COS3) at $F_h = 0.7$, indicated on the regime diagram in Figure 4. Wave-breaking is known to occur under these conditions experimentally at a Reynolds number of around 10^4, without the additional complication of merging between the breaking region and the region of separated flow. The time evolution of the flow field for computations using the two turbulence models on the $96 \times 32 \times 48$ grid is shown in Figure 5. The flow fields are qualitatively similar while streamlines are overturning and wave-breaking occurs. The longer time behaviour is rather different, however, for the breaking region has diminished almost completely by $Ut/h = 30$ in the flow obtained with the mixing length model, whereas it is still present in

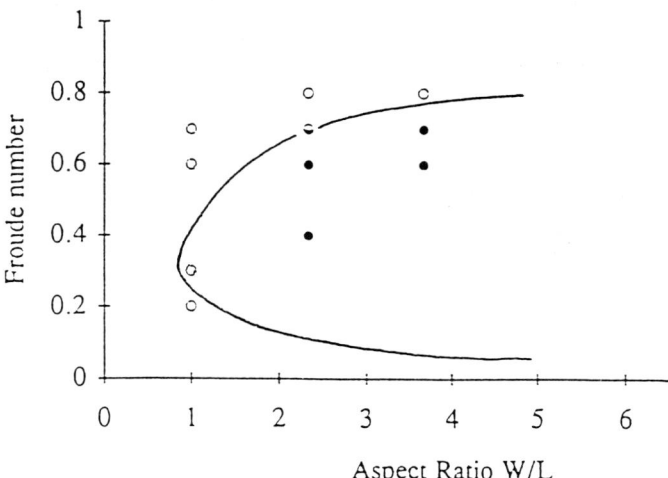

Figure 4. Aspect ratio versus Froude number regime diagram. Solid symbols indicate wave-breaking. Solid line is critical Froude number found experimentally (Castro and Snyder, 1993)

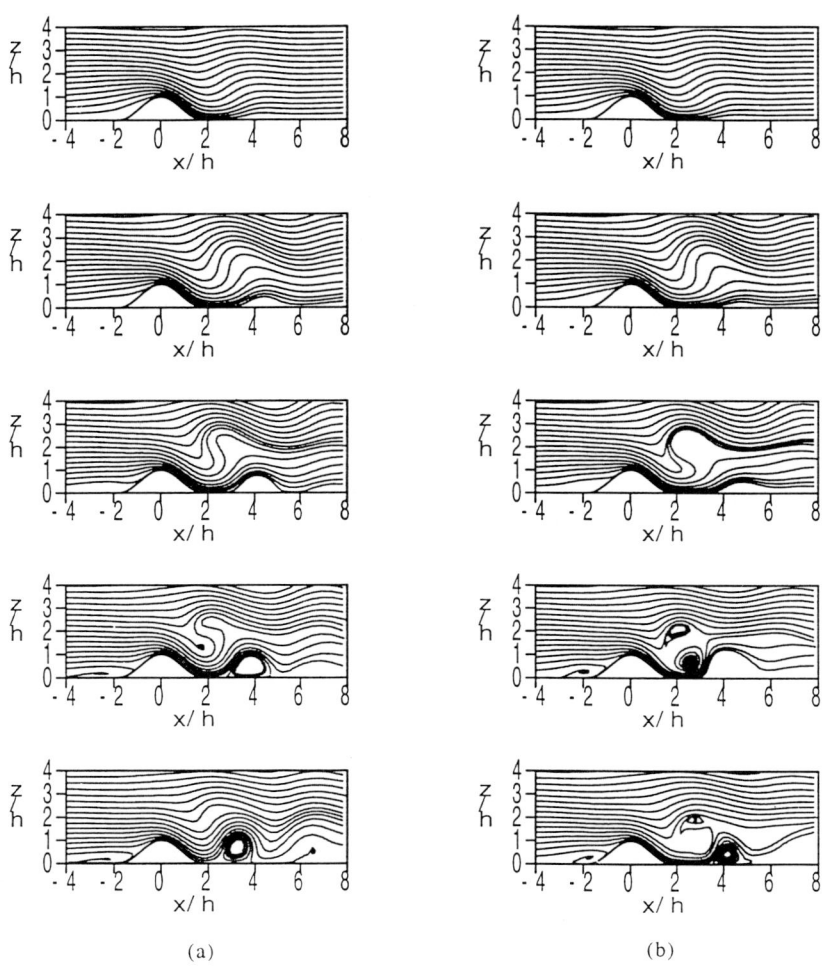

Figure 5. Time development of flow over the COS3 obstacle with $F_h = 0.7$ at nondimensional times (from top) $Ut/h = 5$, 10, 15, 20 and 30. (a) Mixing length model, and (b) one-equation model

the flow obtained with the one-equation model. At present it is not known whether this is simply due to the value of l_o being too high, or, more fundamentally, whether it relates to the deficiency of the wall function treatment for the mixing length model near stagnation points. In what follows, however, we shall concentrate on the development of the flow field and examine the regions where mixing occurs in more detail, in turn, starting with the zone of reversed flow upstream, followed by the separated region downstream and finally the breaking region.

It is well-known that a region of reversed flow appears on the upstream surface of an obstacle if the Froude number is sufficiently low. The reversed flow is accompanied by two stagnation points - the furthest upstream of which is a saddle and the other a node. The location of the former point moves upstream as obstacle width increases and Froude number decreases. These trends are shown in Figure 6 and agree qualitatively with inviscid computations [6] and experiments [1]. The latter point is the attachment point of the "dividing streamline" [12]. Streamlines originating above the dividing streamline pass over the obstacle and those below pass around the obstacle. The energy argument in [12] gives the upstream height H_s of the dividing streamline as $H_s/h = 1 - F_h$ independent of obstacle shape. More recent analysis [13,14], however, suggests that flow stagnation upstream is rather due to changes in the density field, which is dependent on obstacle shape. This leads to a dependence on the obstacle shape of the dividing streamline height, a view which the (admittedly limited) results, Figure 7, tend to support.

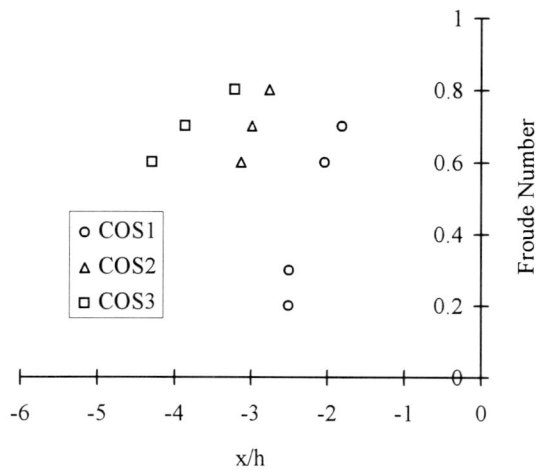

Figure 6. Location of upstream stagnation point

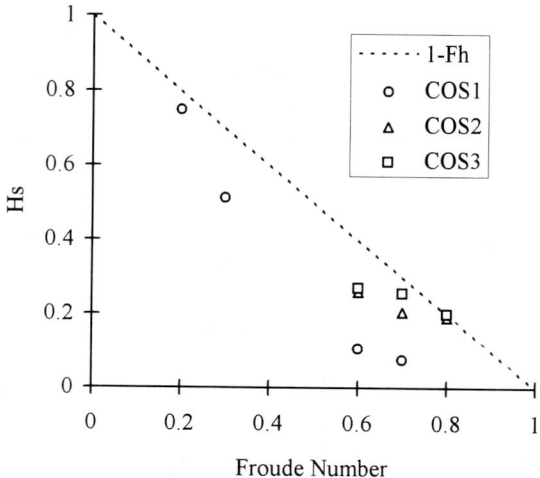

Figure 7. Dividing streamline heights

Figure 8 shows vertical profiles of flow quantities through the upstream flow reversal zone for the computations with the two turbulence models. The horizontal location of the vertical profiles is $x/h = -2.0$, which approximately corresponds to the point of maximum vertical deviation of the streamlines near the base of the obstacle. Figure 8(a) shows the time development of the horizontal velocity profiles. Reversed flow is established rapidly with the mixing length model, with the flow reaching essentially a steady state by around $Ut/h = 15$. The development of the flow obtained with the one-equation model is not so rapid, however, and a steady state does not appear to be reached by $Ut/h = 30$. The profiles given by the two turbulence models are similar in terms of the vertical height at which reversed flow starts, although rather different in terms of the peak negative velocity achieved. Figure 8(b) compares the horizontal velocity profiles at $Ut/h = 20$ calculated on the medium and the fine grids, where very little difference is seen in the case of the mixing length model. The rather larger difference for the one-equation model is attributed to the wall function treatment, for inspection showed that on the fine grid $z^+ < 30$ in this region (in contrast to the mixing length model), and grid independence in the near-wall region might not be expected. These two profiles taken together, however, suggest that the resolution of the medium grid in this region is quite adequate. The time-development of the corresponding density profiles on the medium grid is given in Figure 8(c). The approach to almost steady state occurs rapidly with the mixing length model, with the one-equation model not achieving steady state, in keeping with the velocity profiles. The profiles given by the two models are qualitatively

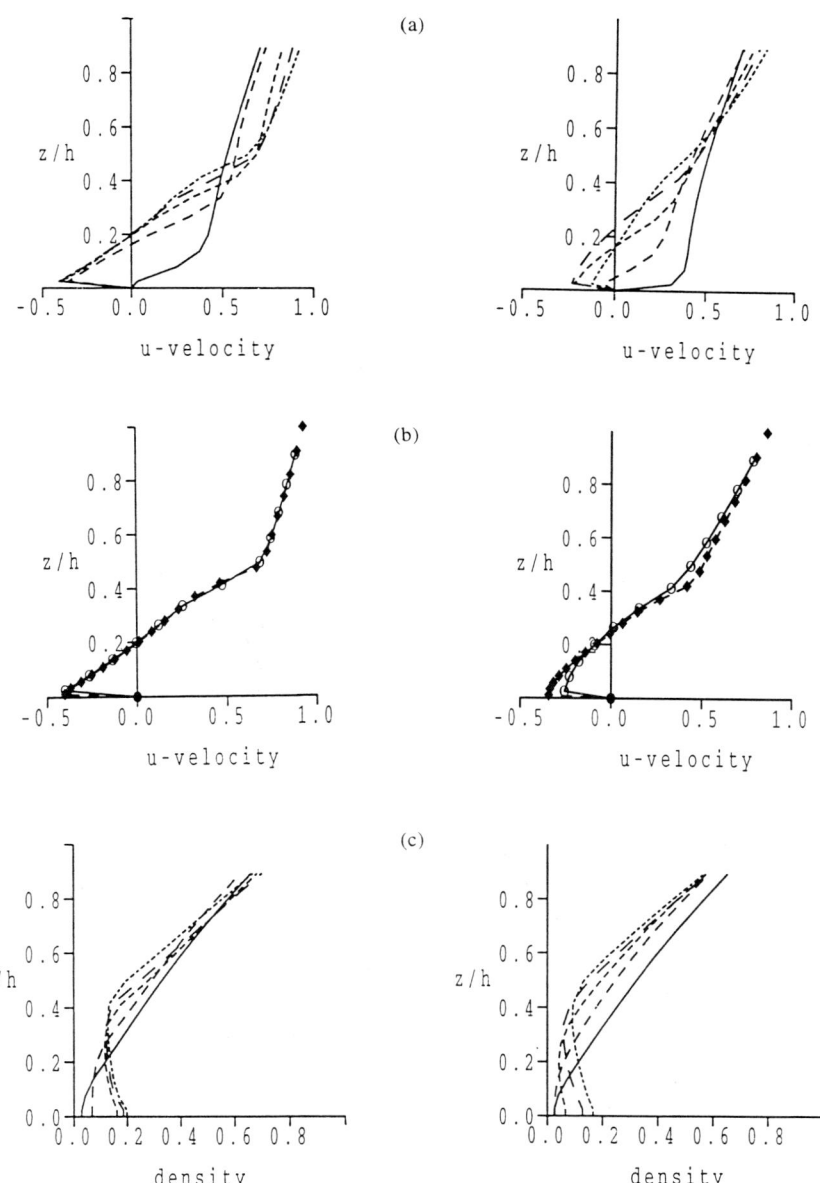

Figure 8. Vertical profiles through upstream flow reversal zone ($x/h = -2.0$) obtained with mixing length (left) and one-equation (right) models. (a) Time development of horizontal velocity ——— $Ut/h = 5$, $- - -$ 10, - - - 15, $——$ 20, $- \cdot - \cdot -$ 30, (b) effect of grid refinement on horizontal velocity at $Ut/h = 20$: ◯ $96 \times 32 \times 48$, ◆$128 \times 48 \times 64$, and (c) time development of density - same key as (a)

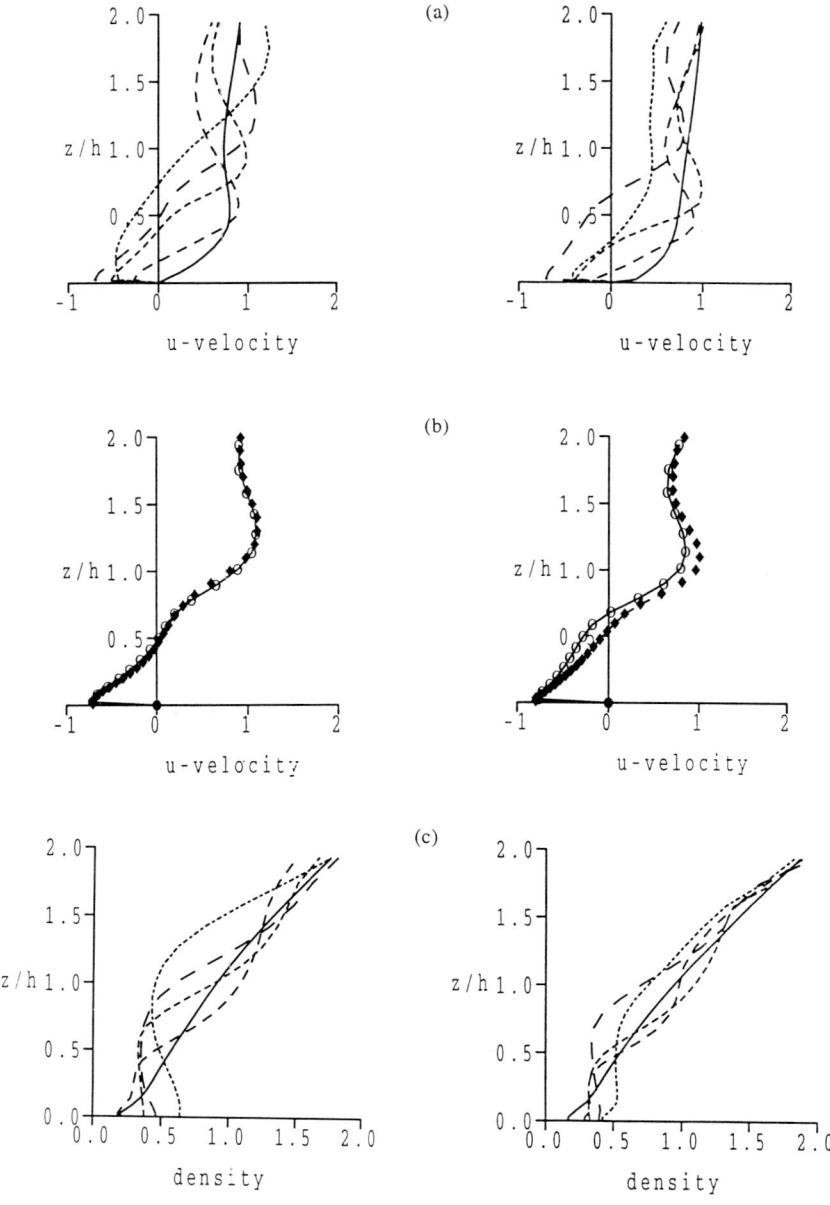

Figure 9. Vertical profiles through downstream recirculation region ($x/h = 4.0$) obtained with mixing length (left) and one-equation (right) models. (a) Time development of horizontal velocity ——— $Ut/h = 5$, $- - -$ 10, $- \cdot - \cdot$ 15, $— — —$ 20, $- \cdot - \cdot -$ 30, (b) effect of grid refinement on horizontal velocity at $Ut/h = 20$: ○ $96 \times 32 \times 48$, ◆$128 \times 48 \times 64$, and (c) time development of density - same key as (a)

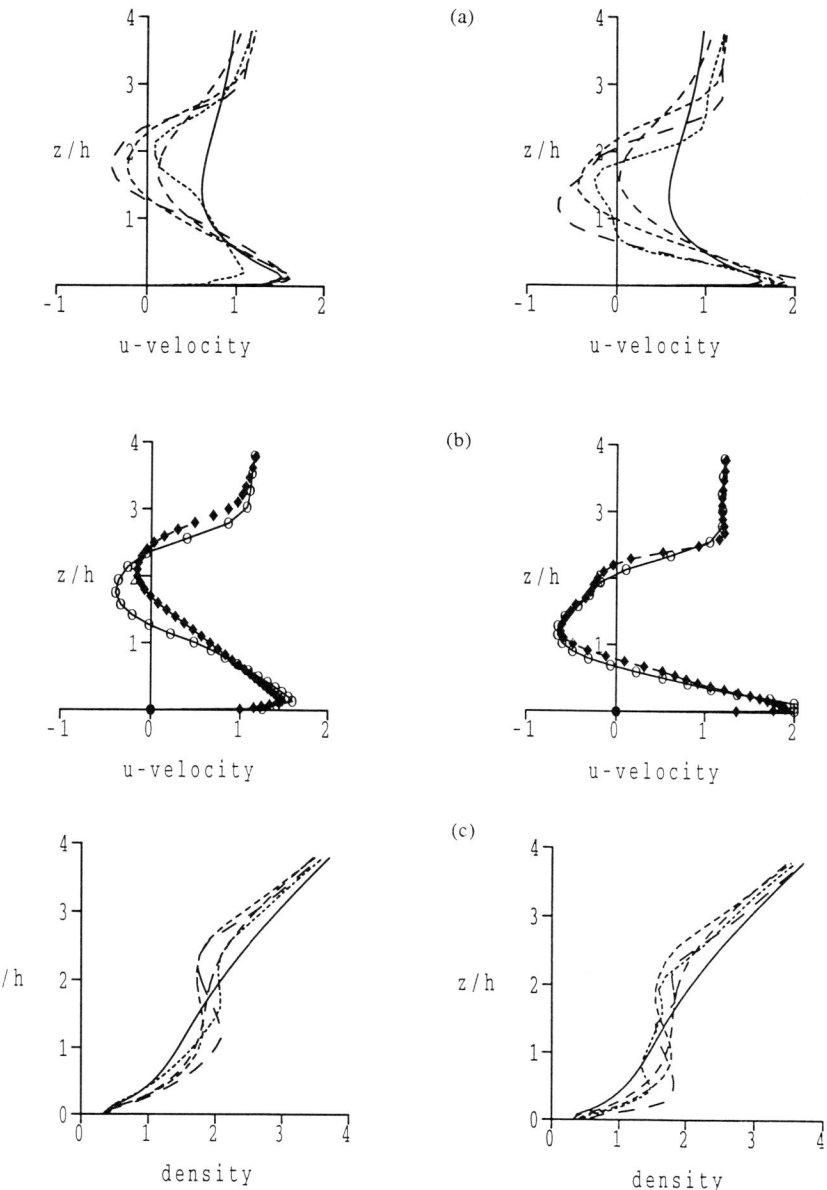

Figure 10. Vertical profiles through downstream breaking region ($x/h = 2.5$) obtained with mixing length (left) and one-equation (right) models. (a) Time development of horizontal velocity ———— $Ut/h = 5$, – – – 10, - - - 15, —— —— 20, – · — · — 30, (b) effect of grid refinement on horizontal velocity at $Ut/h = 20$: ○ 96 × 32 × 48, ♦ 128 × 48 × 64, and (c) time development of density - same key as (a)

similar, and both indicate a region of mixed flow to a depth of around 40% of the hill height.

Figure 9 shows vertical profiles through the downstream recirculation region under the first lee wave for the two computations. The location of the region is approximately constant throughout the computation with the one-equation model, while for the mixing length model it shifts upstream slightly once the breaking region disappears. The time development of the velocity profile, Figure 9(a), differs somewhat according to turbulence model, but is qualitatively similar up to and including $Ut/h = 20$ with divergence thereafter. Grid refinement has little effect with either turbulence model, Figure 9(b), indicating that the grid is quite adequate for this region too. Density profiles up to and including $Ut/h = 20$, Figure 9(c), show only minor differences in the time development of the mixed region.

Perhaps the most interesting profiles are those of Figure 10, which relate to the wave-breaking region. All profiles are taken at $x/h = 2.5$, except that at $Ut/h = 20$ for the mixing length model which is at $x/h = 2.0$. The development of the velocity profile, Figure 10(a), shows that the breaking region is of rather greater extent with the one-equation model, with reversed flow occuring much nearer the surface than it does with the mixing length model. By the time $Ut/h = 30$ is reached, the breaking region with the one-equation model has reduced in vertical extent, while it has disappeared altogether with the mixing length model. The effect of grid refinement is now quite dramatic with the mixing length model, Figure 10(b), with significantly less reversed flow on the fine grid, giving much worse agreement than in either of the other two regions, and streamline plots show that the breaking region on the fine grid is of much more limited extent. It would appear that this grid dependence is a direct result of the failure of the wall function treatment. Inspection of the values of z^+ along the surface certainly shows that validity is lost at several grid points in the neighbourhood of stagnation, and further runs with grid refinement everywhere except at the wall region indicated much closer agreement. The agreement for the one-equation model is much better, on the other hand, and a similar inspection reveals only minor losses of validity near downstream stagnation points. The density profiles, Figure 10(c), are consistent with the velocity profiles, and indicate for the one-equation model the general reduction in vertical extent of the breaking region.

4   Conclusions

Computations of high Reynolds number stratified flows over three-dimensional obstacles have been presented. Two turbulence models, a mixing length model and a one-equation model, have been used to allow computation of flows with regions of streamline overturning, recirculation and mixing. Both models have been demonstrated to be capable of representing flows with breaking lee waves, although with the same value of the mixing length parameter a breaking wave only persists in the flow computed with the one-equation model. Vertical profiles of flow quantities through the recirculation zones upstream and downstream reveal only minor differences according to turbulence model, although there are rather larger differences in the breaking regions. The effects of grid refinement have been assessed, and for the flow reversal zones upstream and downstream little dependence on the grid is observed provided the validity of the wall function treatment is maintained. For the mixing length model this validity cannot be maintained near stagnation points, and this is thought to be the explanation for much better grid independence observed in the wave-breaking region in the case of the one-equation model than the mixing length model.

Bibliography

1. J.C.R. Hunt and W.H. Snyder (1980). Experiments on stably and neutrally stratified flow over a model three-dimensional hill. *J. Fluid Mech.*, **96**, 671–704.

2. J.W. Rottman and R.B. Smith (1989). A laboratory model of severe downslope winds. *Tellus*, **41A**, 401–415.

3. I.P. Castro and W.H. Snyder (1993). Experiments on wave-breaking in stratified flow over obstacles. *J.Fluid Mech.*, **255**, 195–211.

4. P.K. Smolarkiewicz and R. Rotunno (1989). Low Froude number flow past a three-dimensional obstacle. Part I: Baroclinically Generated Lee Vortices. *J. Atmos. Sci.*, **46**, 1154–1164.

5. P.K. Smolarkiewicz and R. Rotunno (1989b). Low Froude number flow past a three-dimensional obstacle. Part II: Upwind Flow Reversal Zone. *J. Atmos. Sci.*, **47**, 1498–1511.

6. H. Hanazaki (1988). A numerical study of three-dimensional stratified flow past a sphere. *J. Fluid Mech.*, **192**, 393–419.

7. M.F. Paisley and I.P. Castro (1995). Numerical computations of stratified flow over three-dimensional obstacles. *Numerical Methods for Fluid Dynamics V*, Editors: Morton and Baines, Clarendon Press, Oxford, 523–532.

8. M.F. Paisley and I.P. Castro (1996). A numerical study of wave-breaking in stratified flow over obstacles. *Dyn. Atms. Oceans*, **23**, 309–319.

9. B. Van Leer (1974). Towards the ultimate conservative difference scheme II. *Monotonicity and conservation combined in a second order scheme*, **JCP 14**, 361–370.

10. B.P. Leonard and S. Moktari (1990). Beyond first-order upinding: the ultrasharp alternative for non-oscillatory steady state simulation of convection. *Int. J. Numer. Methods Eng.*, **30**, 729–766.

11. M.F. Paisley, I.P. Castro and N.J. Rockliff (1994). Steady and unsteady computations of strongly stratified flow over a vertical barrier. *Stably Stratified Flows: Flow and Dispersion over Topography*, Editors: Castro and Rockliff, Clarendon Press, Oxford, 39–59.

12. P.A. Sheppard (1956). Airflow over mountains. *Quart. J. R. Met. Soc.*, **82**, 528–529.

13. R.B. Smith (1989). Mountain-induced stagnation points in hydrostatic flow. *Tellus*, **41A**, 270–274.

14. J.C.R. Hunt, Y. Feng, P.F. Linden, M.D. Greendslade and S.D. Mobbs (1997). Low Froude number stable flows past mountains. *Nuovo Cimento Della Societa Italiana Di Fisica C - Geophysics and Space Physics*, **20(3)**, 261–272.

Boundary Effects on Buoyant Cloud Discharges

Peter A. Davies, Ian R. Coulbourn, Helen Newton, Guoliang Yu and Judith Doorschott

Department of Civil Engineering, University of Dundee

Abstract

Experiments are described in which surface and subsurface buoyant clouds are discharged from a circular orifice placed near a vertical wall. The characteristic dimensions of the cloud are measured as a function of (i) the density difference $\Delta\rho/\rho_o$ between the source and receiving fluid, (ii) the source discharge velocity U_0, (iii) the time τ of discharge, and (iv) the distance y_0 of the vertical wall from the discharge source. The results are analysed to show the time development of the along- and cross-stream dimensions of the spreading surface cloud and attempts are made to scale these dimensions appropriately to reveal the effects of the side wall boundary.

1 Background

Though many marine outfall systems operate continuously with either gravity-fed or pumped discharge, several are designed to store effluent at an intermediate stage before discharging into coastal waters (Davies and Neves [1]). Such systems usually operate intermittently, with the intervals between separate discharges being determined by, for example, the supply volume of available wastewater and/or the need to synchronise discharge with tidal cycles or diurnal periods of low energy cost. The relationship between the intervals separating individual discharge events and typical dilution times in the receiving waters control the extent to which individual discharges may be considered to behave independently. A programme of modelling experiments has been conducted to investigate this problem, with attention being paid not only to the behaviour of buoyant clouds of effluent in effectively-unconfined receiving waters but also to the role of side and bottom boundaries in affecting the dilution of an individual cloud. Initial consideration has been given to horizontal discharges (surface and submerged) from a circular source into quiescent receiving waters but experiments are underway with cross flows.

2 Definition

We consider the horizontal discharge Q of fluid of kinematic viscosity ν and density ρ_o into a fluid of density $(\rho_o + \Delta\rho)$. The discharge, having source velocity

$U_0 = 4Q/\pi d^2$, takes place for time τ from a circular source of diameter d and the receiving waters are contained within a large rectangular box of depth H_0, width W and length L. The source is situated at a level $z = z_*$ in the tank, relative to the free surface $z = H_0$ and the bottom $z = 0$ and at a lateral location $y = y_*$ relative to the centreline $y = 0$.

3   Experiments

The experiments were carried out using salt solutions to generate the density differences between the source and receiving fluids. A pumping arrangement was installed to provide a prescribed discharge rate from an external reservoir in which the density of the fluid was maintained at a constant value; an electronically-controlled valve was used to provide a discharge for a prescribed time. Dye was added to the source fluid and the motion and structure of the resulting buoyant jet was monitored using a video recorder. Subsequent analysis of the video record was made using the image processing facilities of the *DigImage* system (Dalziel [2]) and by direct measurements from individual frames. Observations were made from the surface (to determine the surface spreading) and, in the cases of subsurface discharge, the side (in order to determine the trajectory of the buoyant cloud).

3.1   Observables

Figure 1 shows a definition sketch appropriate for the surface and subsurface discharge flows, in which the key surface observables are shown to be the

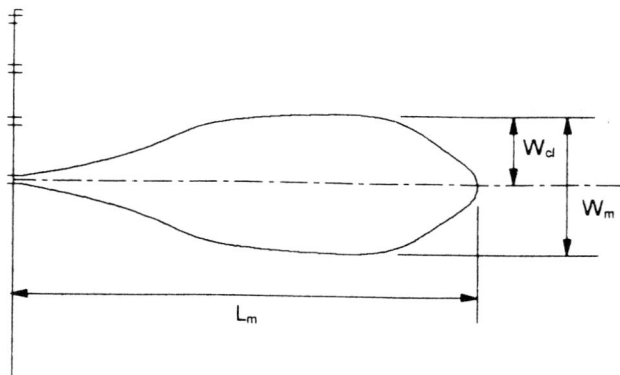

Figure 1. Definition sketch showing dimensions L_m, W_m and W_{cl} of spreading surface cloud

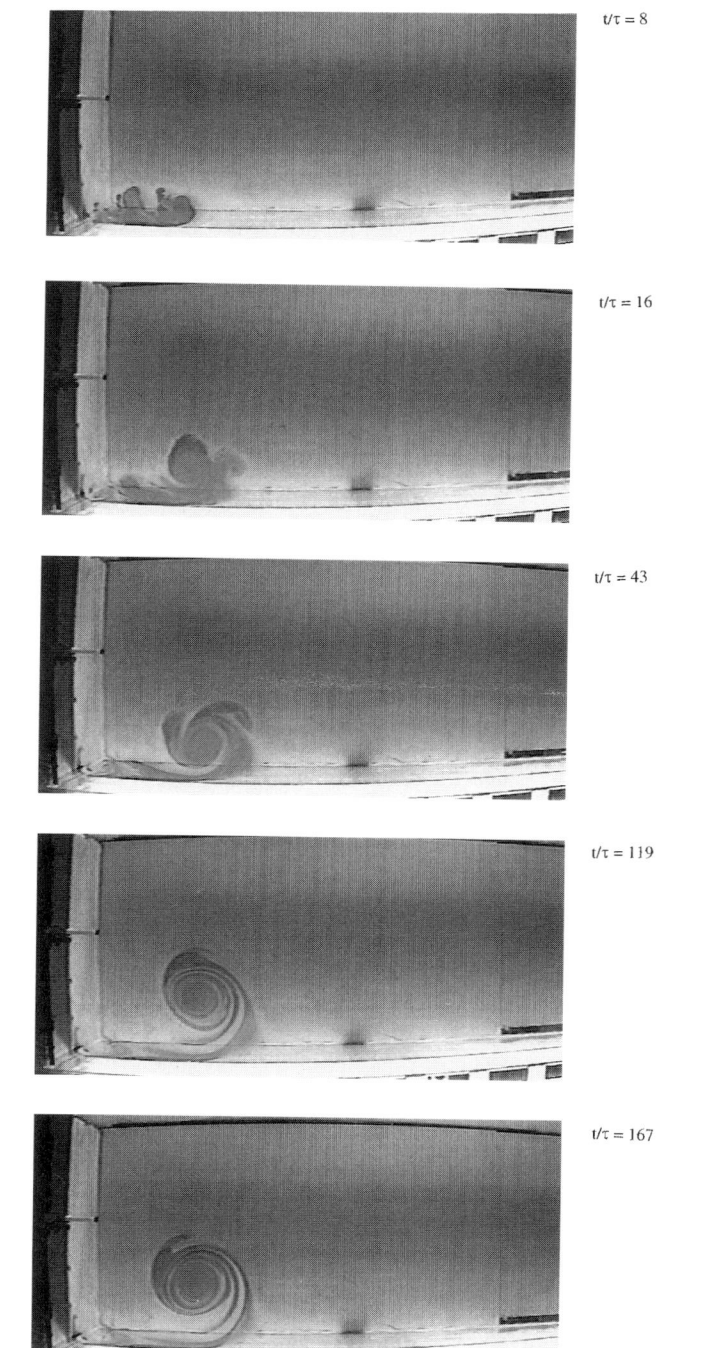

$t/\tau = 8$

$t/\tau = 16$

$t/\tau = 43$

$t/\tau = 119$

$t/\tau = 167$

Davies et al

Fig 2

Figure 2. Sequences of instantaneous cloud patterns for surface $(y/W = 0$ and $0.5)$ and subsurface $(y/W = 0$ and $y/W = 0.5)$ discharges

Davies et al.

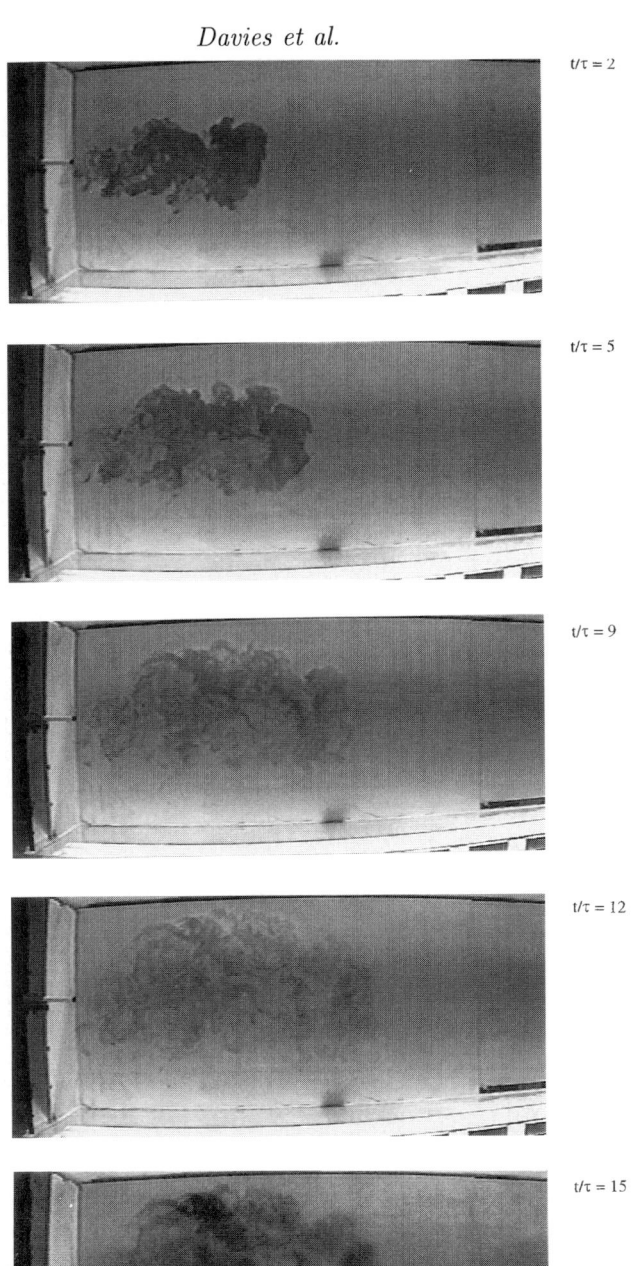

Figure 3. Sequences of instantaneous cloud patterns for surface ($y/W = 0$ and 0.5) and subsurface ($y/W = 0$ and $y/W = 0.5$) discharges

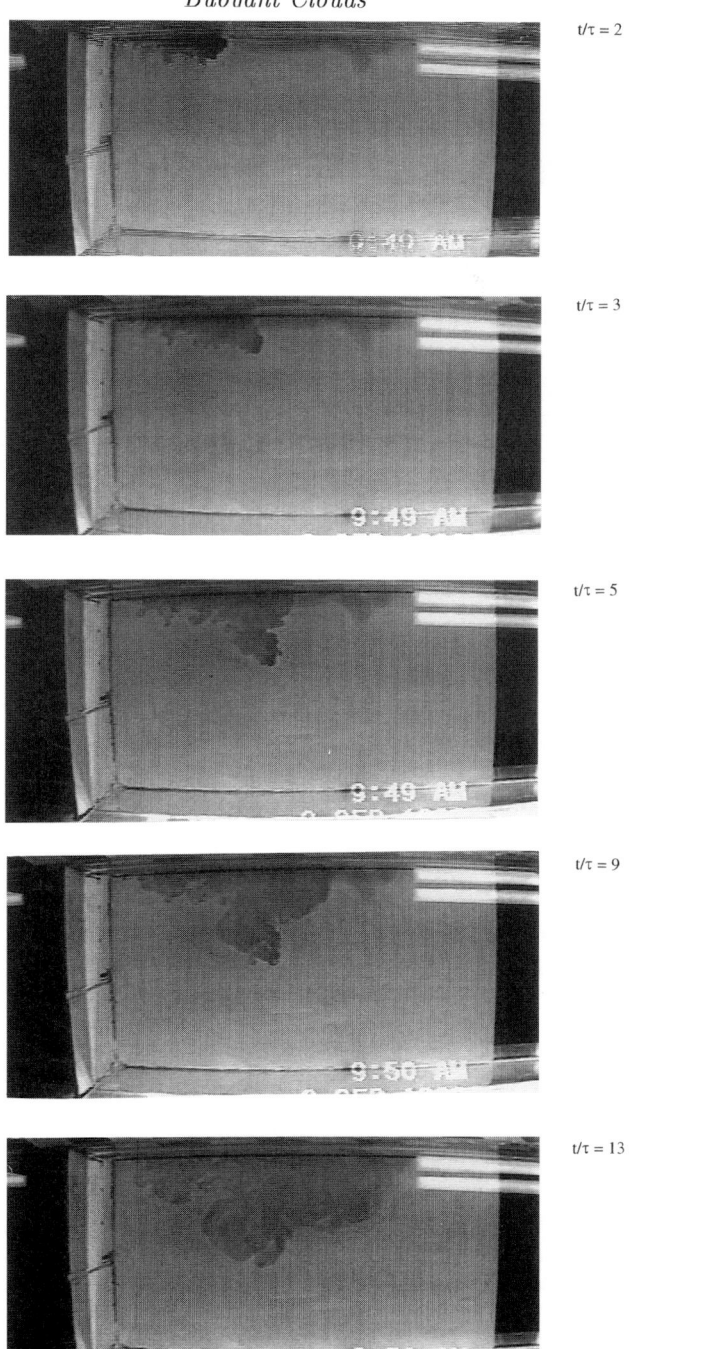

Figure 4. Sequences of instantaneous cloud patterns for surface ($y/W = 0$ and 0.5) and subsurface ($y/W = 0$ and $y/W = 0.5$) discharges

Davies et al.

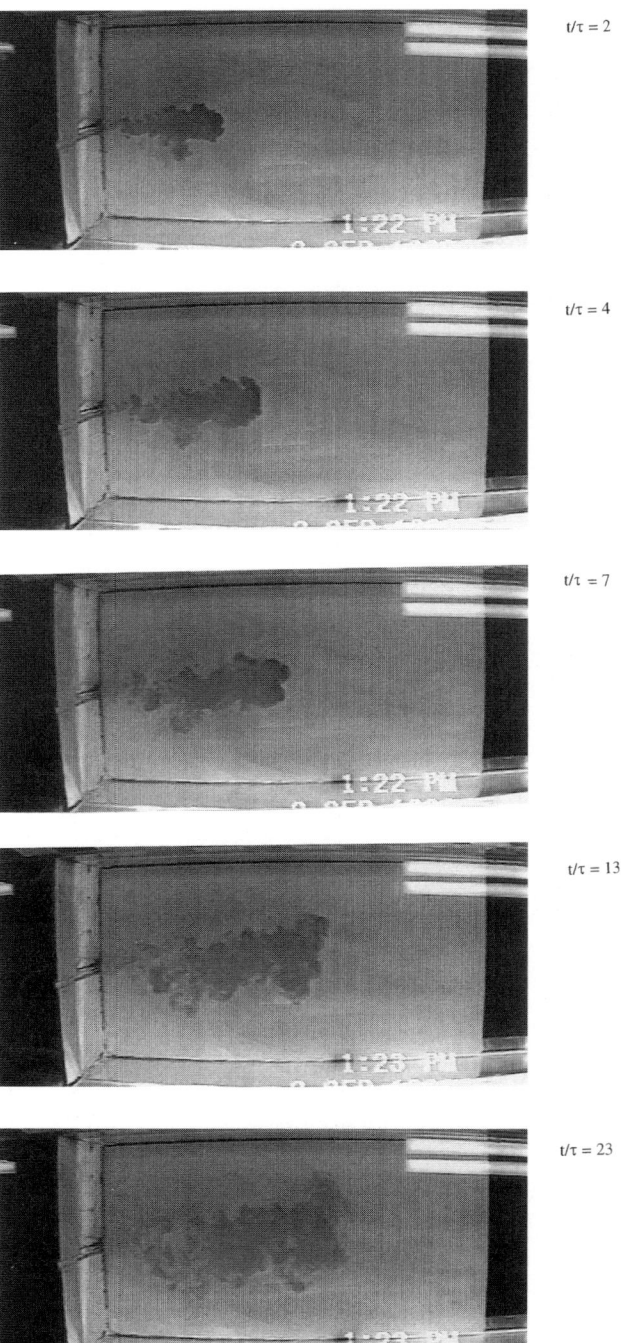

Figure 5. Sequences of instantaneous cloud patterns for surface ($y/W = 0$ and 0.5) and subsurface ($y/W = 0$ and $y/W = 0.5$) discharges

dimensions L_m (the downstream distance from the source to the head of the cloud), W_m (the maximum width of the cloud) and W_{cl} (the maximum half width of the cloud, as measured from the centre-line $y = 0$), all of which are time-dependent.

For subsurface discharges, the additional observables utilised in the analyses of the plume trajectories were the coordinates (x_f, y_f) of the front and the top (x_t, y_t) of the rising cloud, as seen in a vertical plane $(x - z)$ aligned with the source location. As with the observables from the surface discharge cases, all dimensions were time-dependent.

3.2 Results

Figures 2–5 show sequences of instantaneous cloud patterns for different elapsed time after initiation of the discharge and for different positions of the source with respect to the centreline $(y/W = 0)$ and the wall $(y/W = 0.5)$. Both surface and subsurface discharges are displayed and (at least for the centreline configuration) the surface manifestations of both modes of discharge are seen

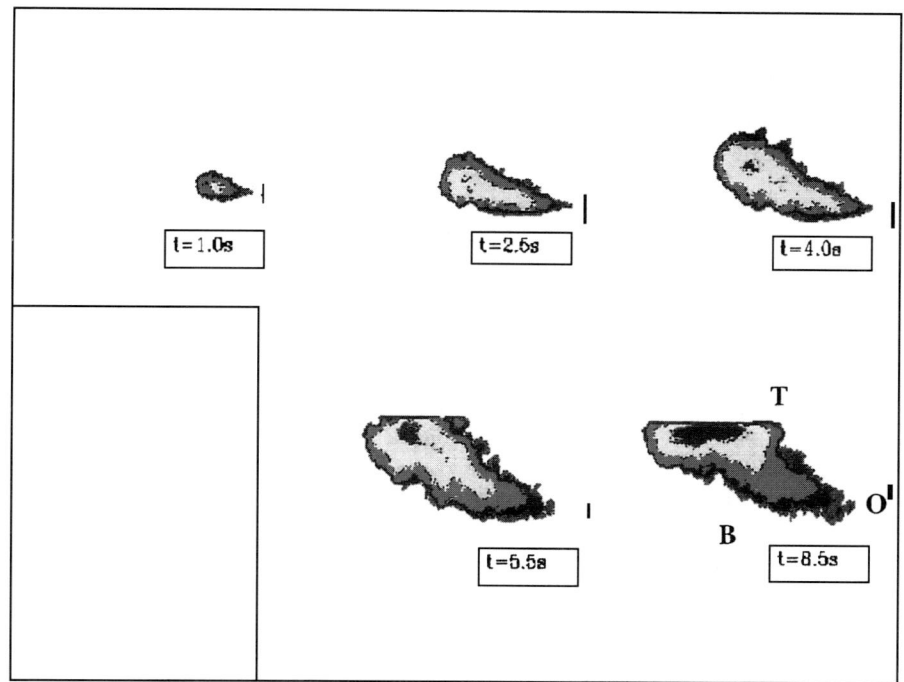

Figure 6. Colour sequences of processed images

to be qualitatively similar. However, for a given depth of discharge, the effect of the wall is seen to be dramatic for discharges taking place adjacent close to this boundary, with asymmetrical entrainment processes causing important structural changes to occur with the spreading cloud.

Figures 6 and 7 show side elevation time sequences of processed images for individual clouds released from subsurface centreline discharges, with the forward part of the cloud showing a behaviour that is essentially that observed with a steady discharge under otherwise-identical initial conditions. The base and rear of the cloud behave quite differently from the steady discharge counterparts, as can be expected when (i) the source momentum and buoyancy driving the flow cease to be maintained during the buoyancy rise of this part of the cloud, and (ii) the density field in the receiving waters is changing with time following the initial release. Figure 8 shows the time dependent growth (in dimensional time) of both x_f and y_f.

In order to investigate further the apparent close relationship between the trajectories of the front of the cloud for the steady and intermittent discharge cases, the trajectory coordinates (x_f, y_f) for the single cloud are shown in

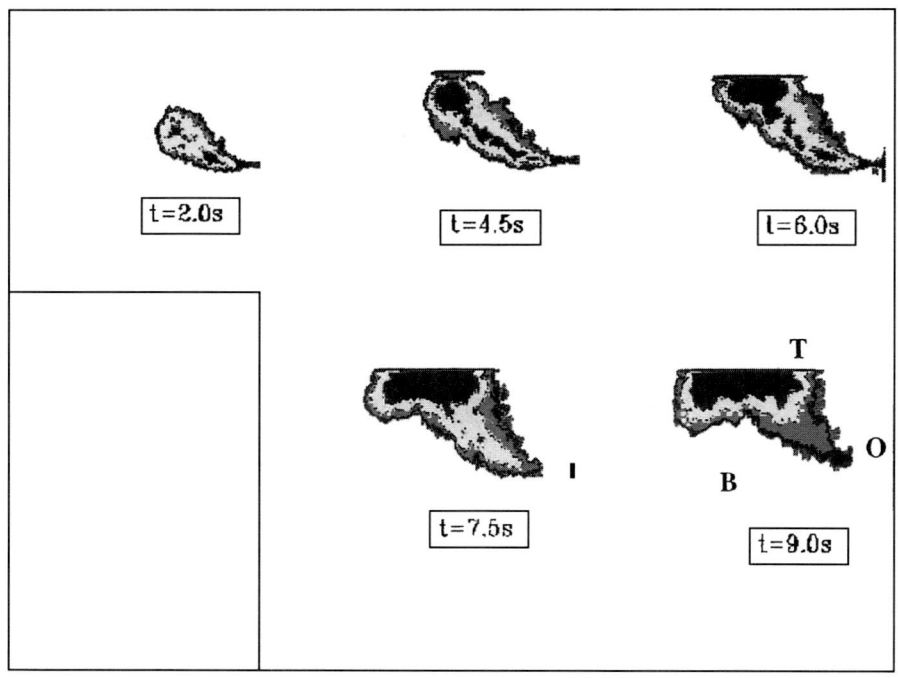

Figure 7. Colour sequences of processed images

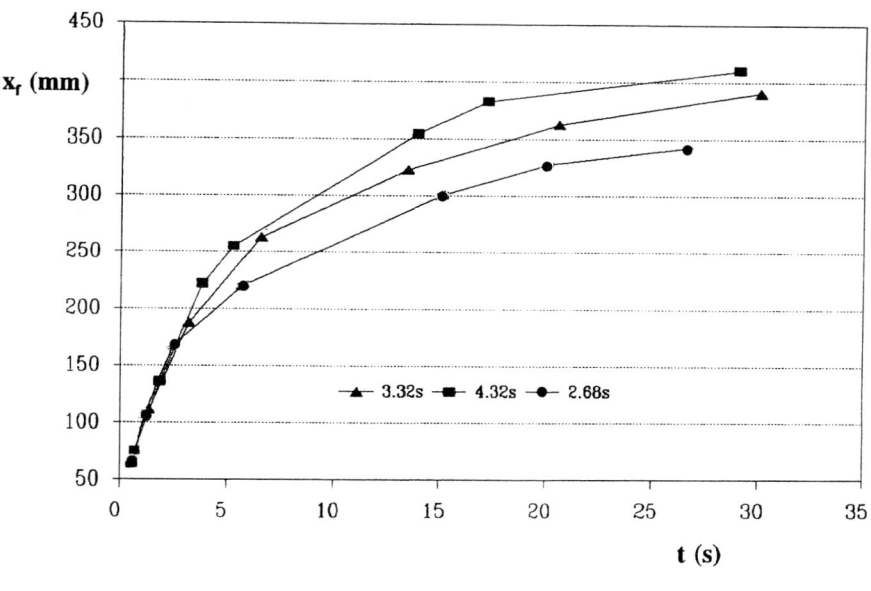

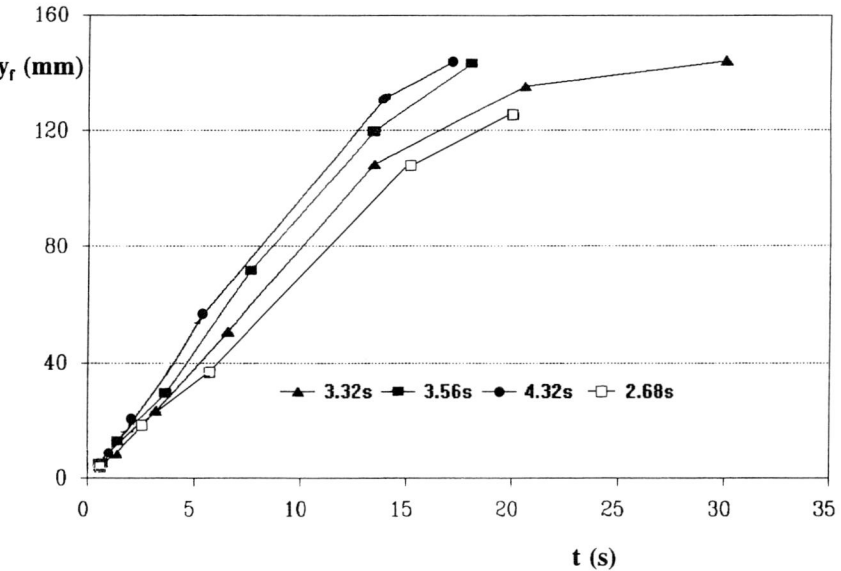

Figure 8. Time dependent growth (in dimensional time) of both x_f and y_f

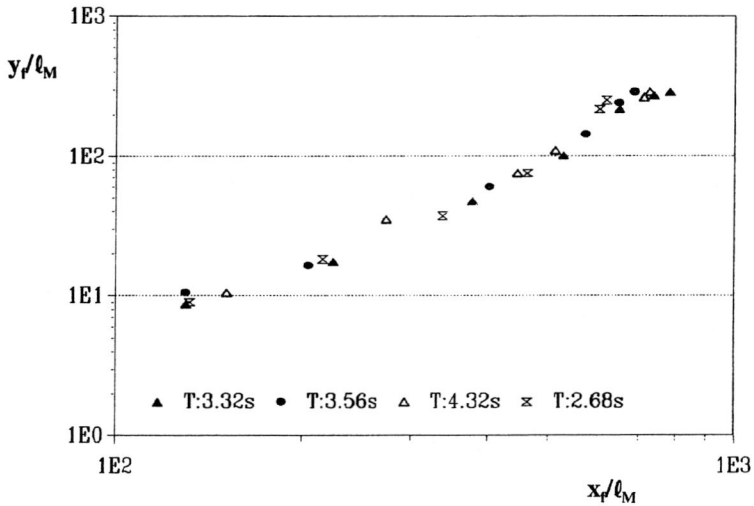

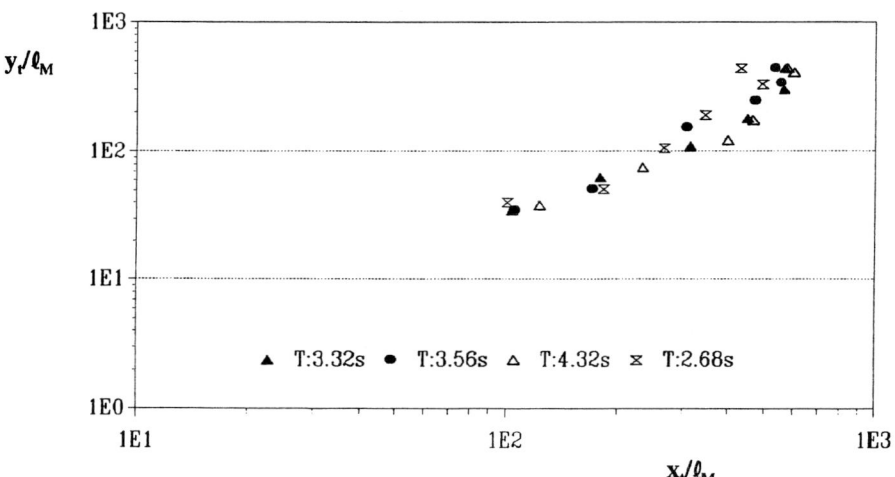

Figure 9. Graphs showing l_m scalings for trajectories

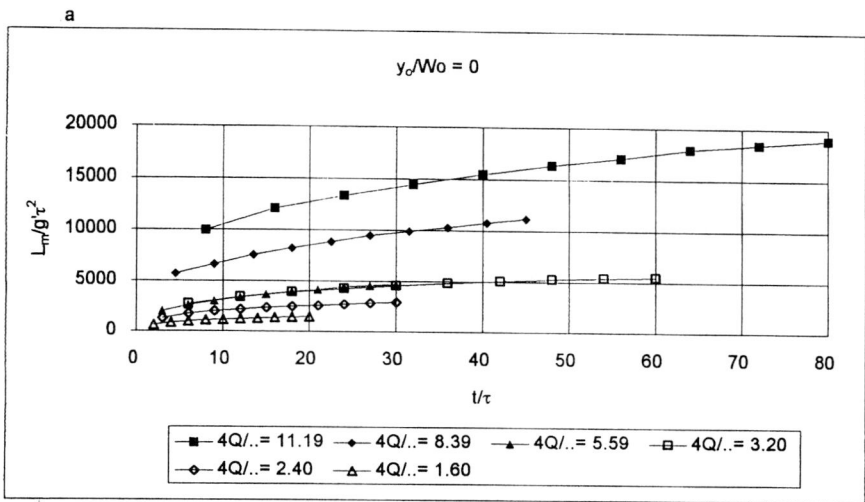

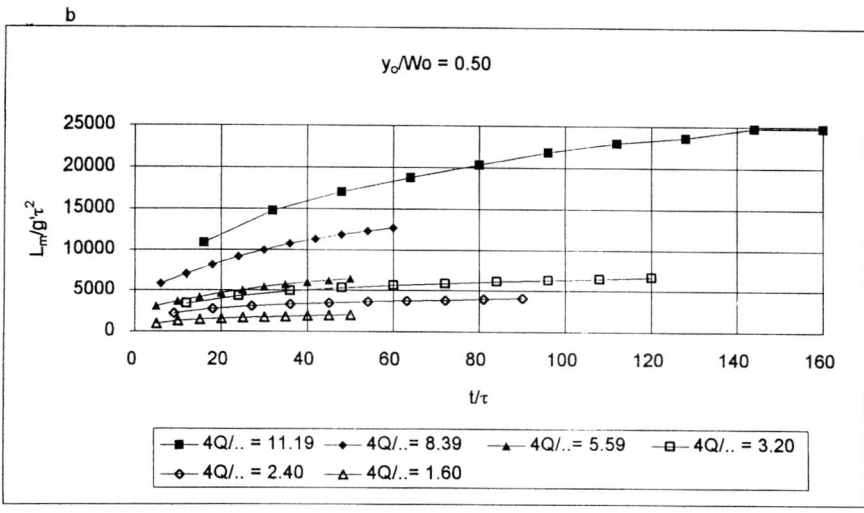

Figure 10. Measurements of the surface axial dimensions L_m of the spreading clouds are shown in dimensionless form in Figures 10 (surface discharge) and 11 (subsurface discharge) for two different values of y_*/W and for different values of $U_0/g(\Delta\rho/\rho_o)\tau(= 4Q/\pi d^2 g(\Delta\rho/\rho_o)\tau)$

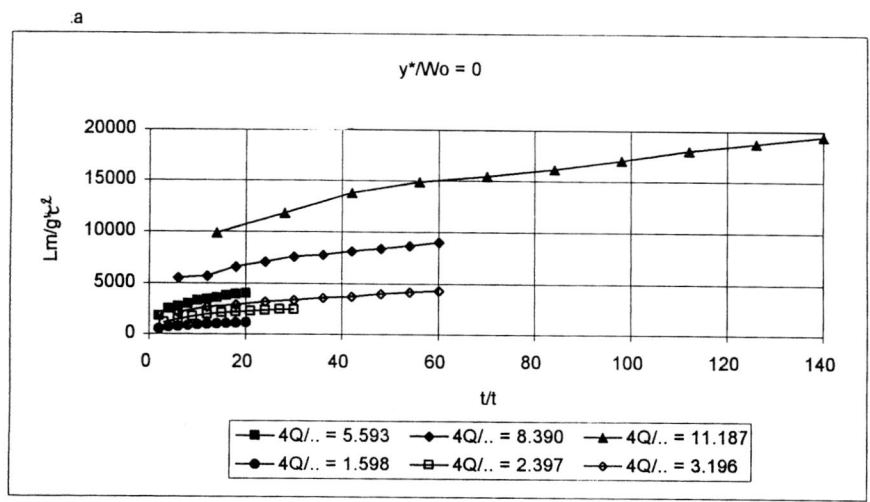

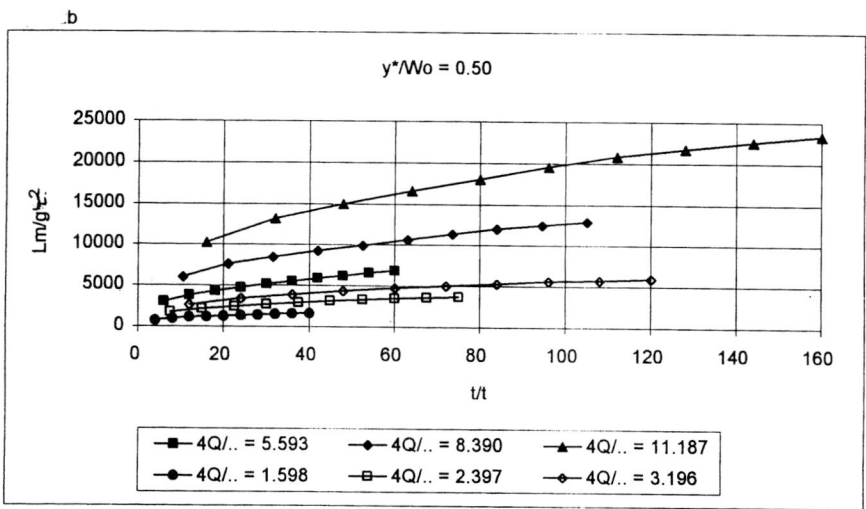

Figure 11. Measurements of the surface axial dimensions L_m of the spreading clouds are shown in dimensionless form in Figures 10 (surface discharge) and 11 (subsurface discharge) for two different values of y_*/W and for different values of $U_0/g(\Delta\rho/\rho_o)\tau(= 4Q/\pi d^2 g(\Delta\rho/\rho_o)\tau)$

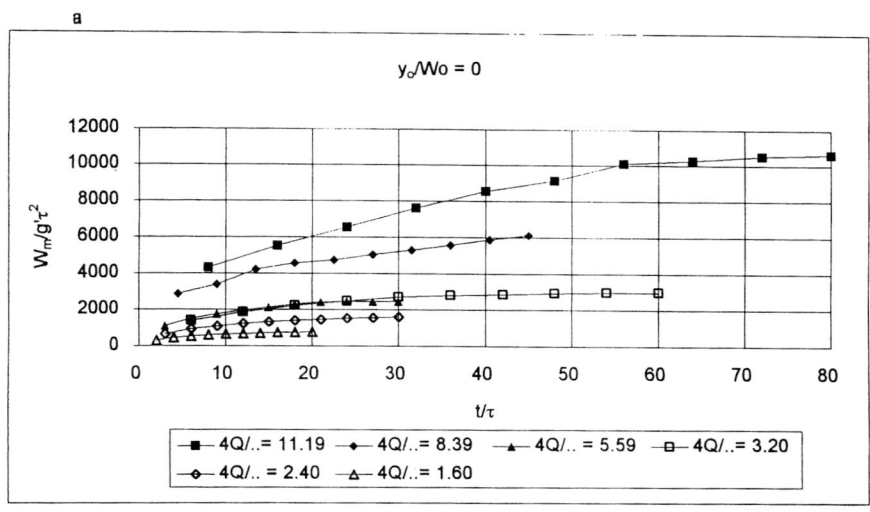

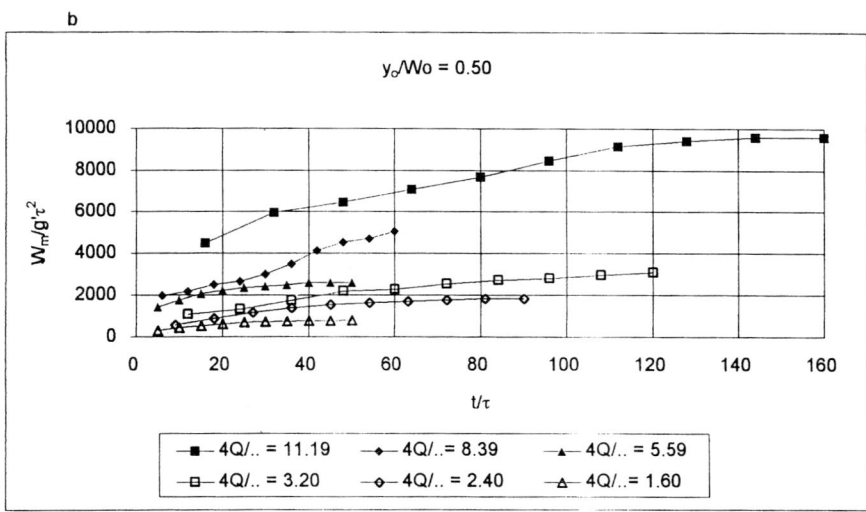

Figure 12. Measurements of the dimensionless surface width $W_m/g(\Delta\rho/\rho_o)\tau^2$ versus time for two different values of y_*/W and for different values of $U_0/g(\Delta\rho/\rho_o)\tau(= \ 4Q/\pi d^2 g(\Delta\rho/\rho_o)\tau)$ for surface (Figure 12) and subsurface (Figure 13)

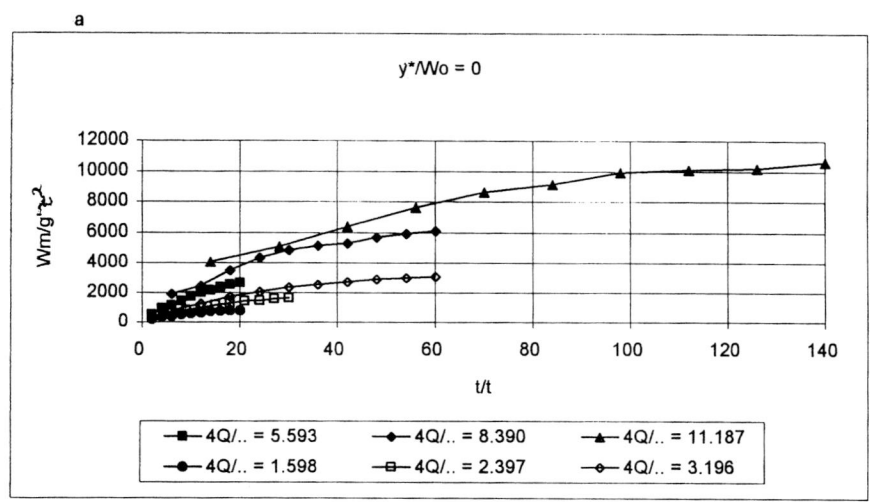

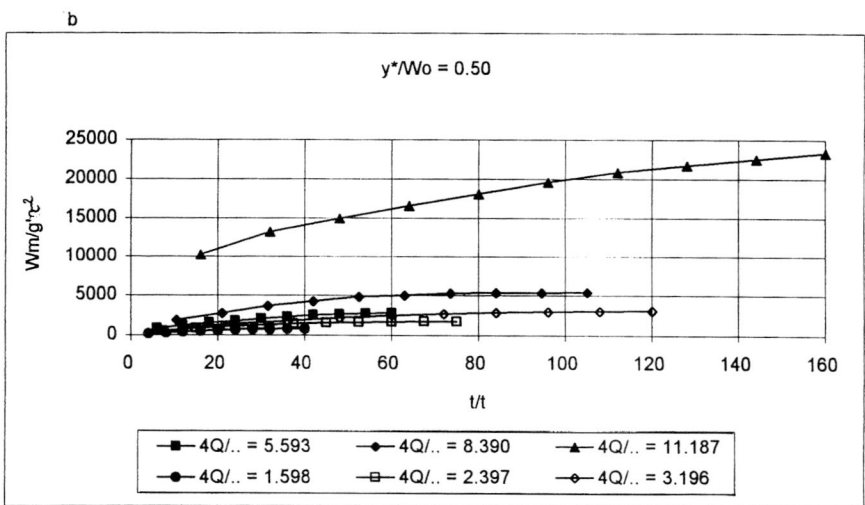

Figure 13. Measurements of the dimensionless surface width $W_m/g(\Delta\rho/\rho_o)\tau^2$ versus time for two different values of y_*/W and for different values of $U_0/g(\Delta\rho/\rho_o)\tau (= 4Q/\pi d^2 g(\Delta\rho/\rho_o)\tau)$ for surface (Figure 12) and subsurface (Figure 13)

dimensionless form in Figure 9, by normalising with the quantity l_m $(= M^{3/4}/B^{1/2})$ – a length scale known to control the trajectories of steady buoyant jets (List [3]). Here, the specific momentum and buoyancy fluxes M and B respectively are defined in the conventional manner as $M = 4Q^2/\pi d^2$ and $B = Qg\Delta\rho/\rho_o$. At least for these centreline discharges, the l_m scaling seems to represent the trajectory date well for the front and top parts of a given rising individual cloud.

Measurements of the surface axial dimensions L_m of the spreading clouds are shown in dimensionless form in Figures 10 (surface discharge) and 11 (sub-surface discharge) for two different values of y_*/W and for different imposed conditions $(U_0, \Delta\rho/\rho_o, \tau)$. Dimensional analysis indicates that a convenient way of representing the data is in terms of length scales $g(\Delta\rho/\rho_0)\tau^2$ and time scales $U_0/g(\Delta\rho/\rho_o)$; accordingly, each graph in Figures 10 and 11 is presented as $L_m/g(\Delta\rho/\rho_o)\tau^2$ versus the dimensionless elapsed time t/τ, for different values of the independent parameters $U_0/g(\Delta\rho/\rho_o)\tau(= 4Q/\pi d^2 g(\Delta\rho/\rho_o)\tau)$ and y_*/W. In these plots, the value of the Reynolds number $Re = U_0 d/\nu$ is assumed to be suffi-ciently high for the effects of viscosity on the flow to be negligible. As expected, the axial dimension of the cloud increases with time to reach an asymptotic value determined by the imposed values of $U_0/g(\Delta\rho/\rho_o)\tau(= 4Q/\pi d^2 g(\Delta\rho/\rho_o)\tau)$ and y_*/W. A similar behaviour is seen with the transverse dimension W_m (see Figures 12 and 13), though the dependence of the dimensionless trans-verse dimension $W_m/g(\Delta\rho/\rho_o)\tau^2$ upon y_*/W is seen to be much stronger than that of the counterpart axial dimension L_m. This can be seen by comparing these values at a reference elapsed time for the same values of $U_0/g(\Delta\rho/\rho_o)\tau$ $(= 4Q/\pi d^2 g(\Delta\rho/\rho_o)\tau)$.

4 Conclusions

These preliminary studies of turbulent buoyant clouds show that the surface dimensions can be suitably described in terms of the length and time scales $g(\Delta\rho/\rho_o)\tau^2$ and $U_0/g(\Delta\rho/\rho_o)$ respectively and that the proximity of a vertical wall to the source of discharge has a significant effect upon these dimensions through its effect upon the symmetry of lateral entrainment into the turbulent jet. This effect is seen most noticeably in the effect upon the transverse dimension of the buoyant surface cloud, with this dimension showing a significant reduction as the value of y_*/W increases, for otherwise identical conditions. The effects of varying y_*/W upon the axial dimensions of the cloud are relatively weak.

For subsurface discharges, the scaling of the buoyant cloud trajectory with the length scale l_m is seen to provide a satisfactory representation of the data for the forward and upper regions of the plume.

Acknowledgements

The work presented above has been funded by the UK Engineering and Physical Sciences Research Council and by the EU Erasmus Programme. The authors acknowledge this support with gratitude.

Bibliography

1. Davies, P.A., and Neves, M.J. (eds), (1994) Recent research advances in the fluid mechanics of turbulent jets and plumes. *NATO ASI series, Series E: Applied Sciences, Kluwer Academic Publishers, Dordrecht, NL* **225**.

2. Dalziel, S.B., (1992) Decay of rotating turbulence; some particle tracking experiments. *Appl. Scient. Res.* **49**, 217-244.

3. List, E.J., (1982) Turbulent jets and plumes *Ann Rev. Fluid Mech.* **14**, 189-212.

The Effect of the Tidal Barrage on the Flows and Mixing in the Tees Estuary

A.M. Riddle and R.E. Lewis

Brixham Environmental Laboratory, Zeneca Limited, Brixham, Devon

Abstract

The Tees estuary has recently been transformed by the construction of a barrage which has reduced the tidal length from 44 km to 18 km. The barrage now restricts the upstream movement of the salt resulting in the formation of a new freshwater region to landward. Surveys undertaken before and after construction of the barrage, in 1990 and 1995 respectively, have demonstrated changes to the hydrographic conditions in the estuarial waters.

The blocking of the tidal flow at the barrage has resulted in a marked increase in stratification due to a reduction in tidal current speed and in the rate of vertical transfer of salt. A feature of the results is that the time at which the most intensive mixing takes place has changed from the ebb to the flood tide. This shift in the timing is apparently due to a different mechanism controlling the mixing in the post-barrage estuary. Bulk Richardson numbers (R_{ib}) computed from 1995 data for Billingham Reach show a similar pattern, but with reduced variability, to the corresponding values for 1990 on neap tides. A distinct change was found on spring tides, with R_{ib} values falling below 1.0, indicating the potential for unstable conditions and increased vertical mixing. Values of R_{ib} below 1.0 occurred for approximately 4 hours on the flood tide in 1995 compared with a similar duration but on the ebb tide in 1990.

1 Introduction

The Tees estuary is situated in the north east of England (Figure 1) receiving waters from the rivers Tees and Leven. An average flow of 20 m^3 s^{-1} of freshwater discharges through the estuary, although the summer flow drops to 2 m^3 s^{-1} and winter flows may reach 565 m^3 s^{-1}. The estuary passes through the towns of Stockton, Billingham and Middlesbrough and has been subject to many man made changes since the industrial revolution. Major sections have been straightened (Figure 1) and much of the lower estuary has been reclaimed confining the estuary to a narrow channel which is now regularly dredged to allow the entrance of large ore carrying ships and many other forms of shipping. Effluent discharge from industry and an increasing population have caused gross pollution of the waterway, which reached its worst in the early 1970s.

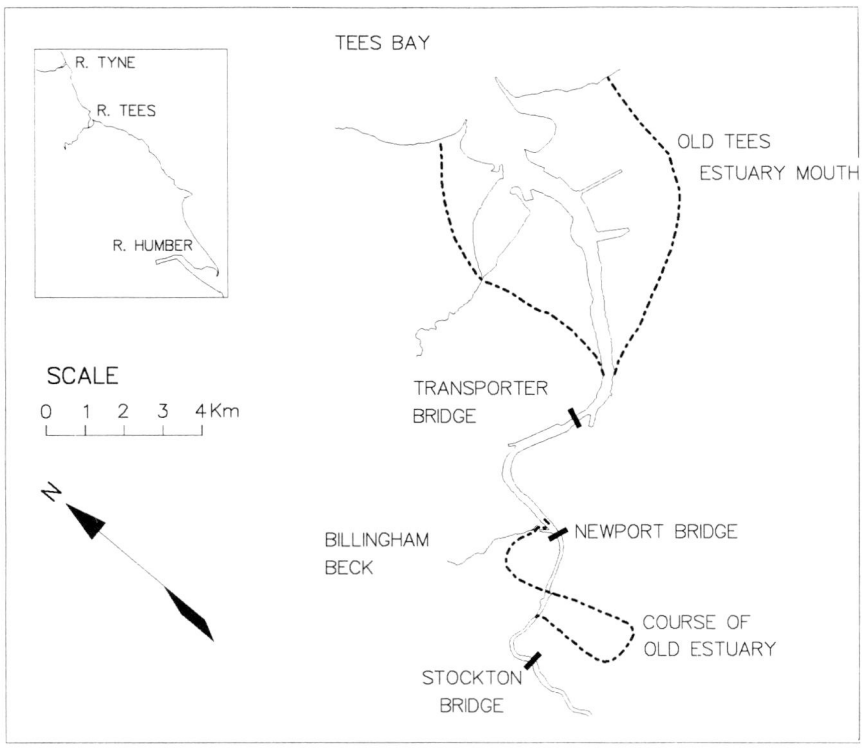

Figure 1. Course of the old and current Tees estuaries

A recent major change to the estuary has been the construction of a barrage across the estuary at a position approximately 2 km downstream of Stockton (Figure 2). The barrage was completed in late 1994 and brought into operation in January 1995. The barrage is built to a height above peak high tides, thus excluding the tide from the upstream section and reducing the tidal estuary from the former 44 km to only 18 km, so that the area above the barrage is now freshwater.

Major surveys of the hydrography and water quality of the estuary have been carried out at 5 year intervals since 1970. These surveys have been used to monitor improvements in water quality with time and to provide data for validation of mathematical models of the estuary. The surveys have been run as joint ventures supported by ICI and the Environment Agency (EA - formerly the National Rivers Authority and before that the Northumbrian Water Authority) and other industries on Teesside. Many other smaller surveys have been carried out in intermediate years by ICI and the EA.

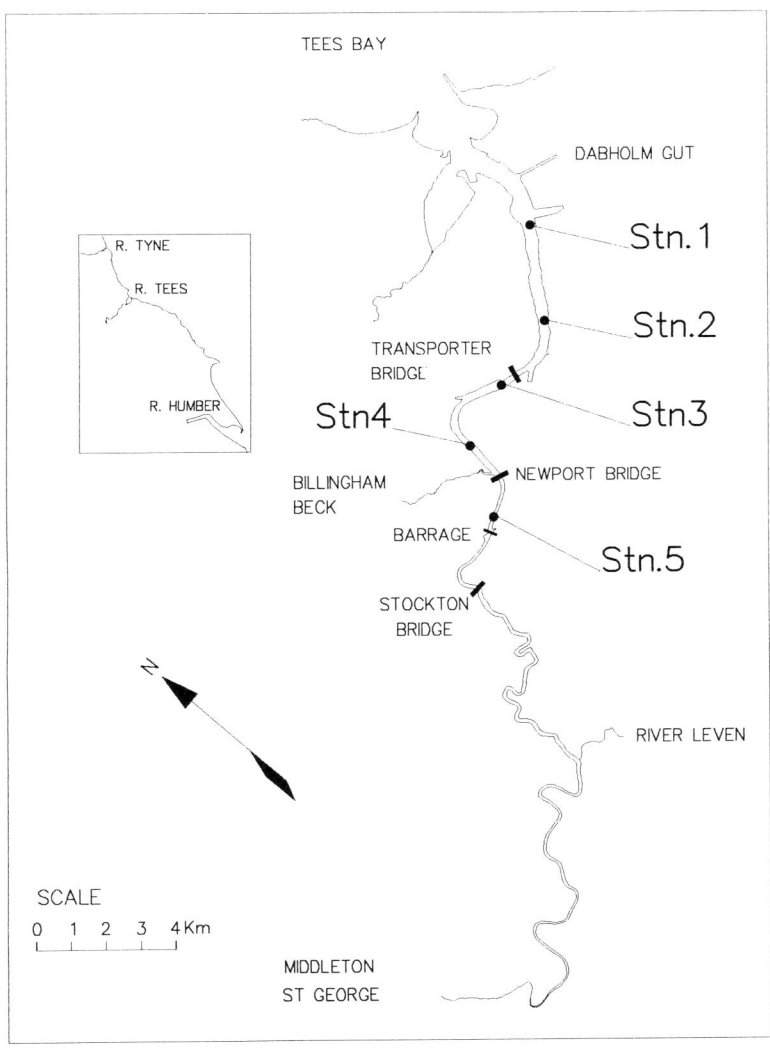

Figure 2. Positions of the survey stations and the barrage

The plans for the barrage were known prior to the 1990 estuary survey and the survey was designed so that the same stations could be re-surveyed in 1995 to provide a direct comparison of the estuary pre- and post-barrage situations. Five stations were used at the positions shown on Figure 2. The stations were surveyed for 3 days on neap tides and 3 days on spring tides on both years (Table 1 gives details of the dates, tidal ranges and river flows during the surveys). On each day the survey vessels were in position in the centre of the channel and sampling from 0600 BST to 1830 BST. Vertical profiles of current speed and

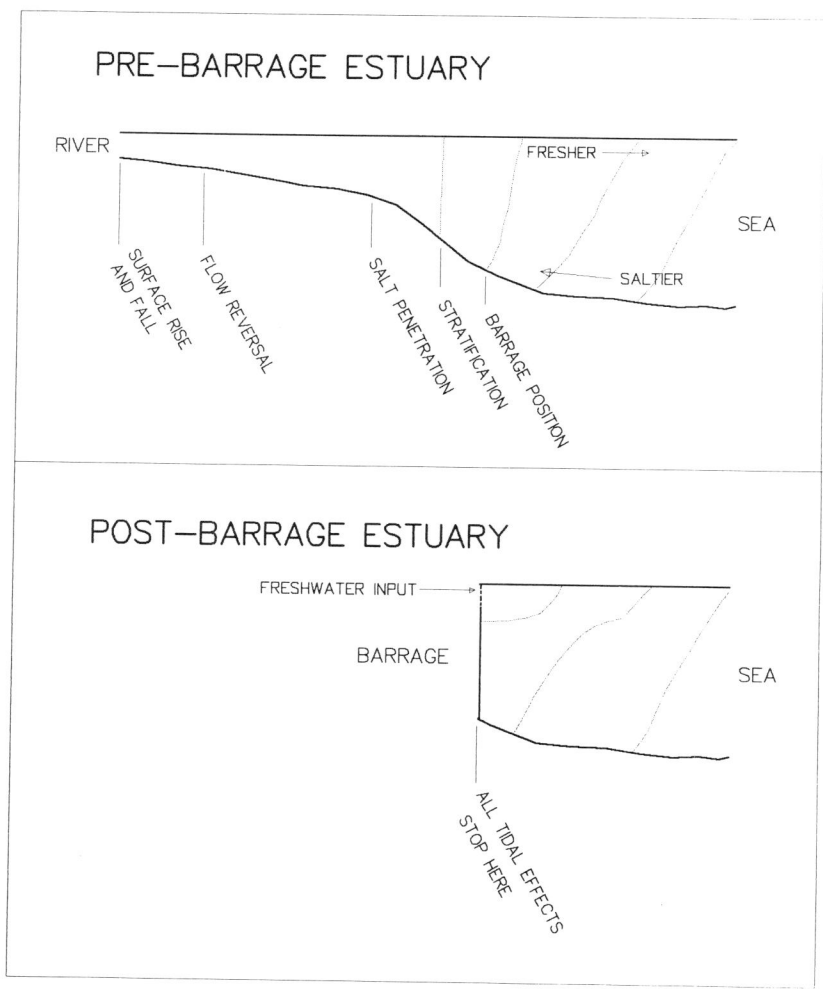

Figure 3. Tidal intrusion pre- and post-barrage

Table 1. Tidal and river flow conditions during the surveys

1990			1995		
Date	Tidal Range (m)	River Flow ($m^3\ s^{-1}$)	Date	Tidal Range (m)	River Flow ($m^3\ s^{-1}$)
2 June	2.5	2.03	6 June	2.9	5.98
3 June	2.4	2.41	7 June	2.7	6.49
4 June	2.3	3.86	8 June	2.8	6.80
9 June	3.8	6.11	13 June	4.7	5.03
10 June	4.3	3.24	14 June	5.0	4.60
11 June	4.1	2.89	15 June	4.9	4.82

direction, salinity and temperature were taken every half hour throughout the tide by sampling at the water surface, 0.5 m, 1 m and 1 m intervals to the bed; many chemical parameters were also measured, however this paper only deals with the physical data.

Figure 3 shows a schematic representation of the estuary before and after the barrage, summarising the influence of the barrage on the water surface, tidal flow and salinity intrusion. Tidal variation of the water surface is now curtailed by the barrage instead of extending to Middleton St. George (Figure 2), reversal of the flow direction formerly occurred up to 38 km from the estuary mouth and salt intrusion occured up to 26 km from the estuary mouth.

This paper documents the changes that have occurred to the hydrodynamics, salt balance and mixing in the estuary due to the barrage; Rowland [1] has also presented some analyses of the survey data in his thesis. The surveys were carried out on 2–4 June (neap) and 9–11 June (spring) in 1990, and on 6–8 June (neap) and 13–15 June (spring) in 1995.

2 Flow

Water flow is the dominant process in an estuary where freshwater from the river meets and mixes with salt water from the sea. Tidal flows dominate the lower estuary, but further upstream the river flow can become dominant at times of heavy or prolonged rainfall. At times of high river flow the estuary is well flushed and water quality improved; the joint surveys have all been carried out during the summer, normally in June or July when river flows have been relatively low.

2.1 River flow

The river water now enters the estuary at the barrage and is controlled to avoid upstream flooding in winter and to provide a freshwater resource for the new canoe slalom in the summer. Due to this, the flow of freshwater to the estuary

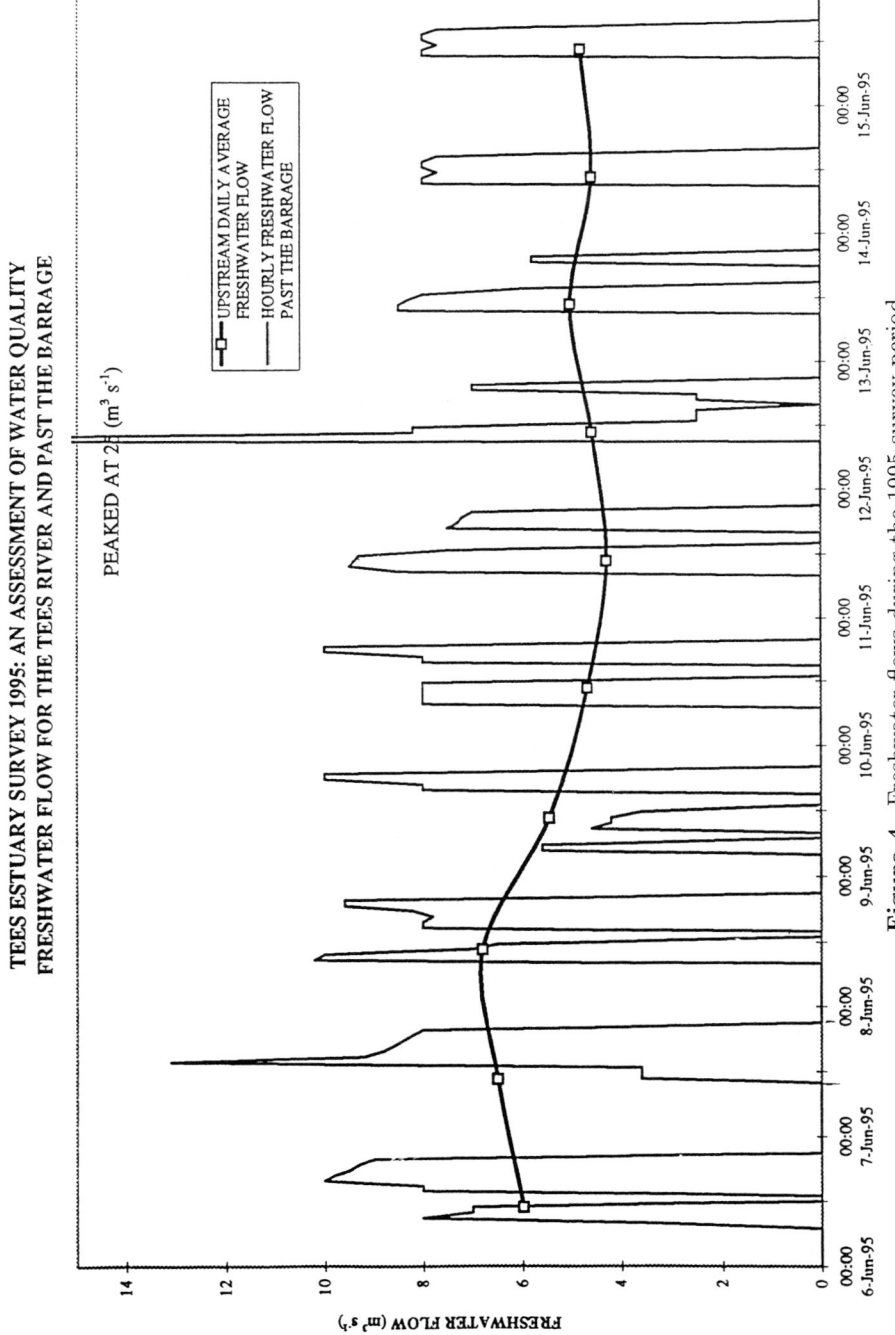

Figure 4. Freshwater flows during the 1995 survey period

is no longer continuous as shown on Figure 4 for the period of the 1995 survey; this plot also shows the daily average river flow.

The annual average flows of the Tees and Leven total 20 m^3 s^{-1}, with a minimum of 2 m^3 s^{-1} and a 10% exceedance flow of 47 m^3 s^{-1} as measured at the Low Moor and Leven Bridge gauging stations [2].

2.2 Tidal flow

The tidal flows in the estuary have been significantly affected by the barrage. Of the survey stations only position 1 was unaffected. Table 2 gives the station positions and peak flood and ebb currents for spring tides in 1995; the percentage reduction in peak flow from the values in 1990 is also tabulated. The variation in current speed over a tidal period on a spring tide is shown for Station 4 (Figure 5(a)) and Station 5 (Figure 5(b)) for the pre-barrage (1990) and post-barrage (1995) situations; near surface and near bed values are plotted. The velocity of water flowing past these stations has significantly reduced and there is little movement of the water in the lower layer at Stations 4 or 5. As the barrage has reduced the strength of the tidal currents so much, surface currents at the upstream station are now largely dependent on the supply of freshwater over the barrage.

Figure 6 shows plots of the tidal average current, which is given by the constant term computed by fitting the following harmonic equation to the survey data at each station and sampling depth:

$$Cur = A + B^* sin(\omega t) + C^* cos(\omega t) + D^* sin(2\omega t) + E^* cos(2\omega t)$$

where A, B, C, D and E are constants, ω is the angular frequency of the semi-diurnal tide and t the time.

It can be seen from the figure that the tidal average flows in the surface layers, in the upper and middle estuary have reduced since the construction of the

Table 2. Tidal flow amplitude at the survey stations

Station	Distance From Barrage (km)	Peak Flood Flow ($m\ s^{-1}$)	% Reduction	Peak ebb Flow ($m\ s^{-1}$)	% Reduction
1	11.6	0.69	5	0.49	7
2	9.5	0.73	11	0.40	25
3	6.0	0.57	12	0.35	42
4	3.3	0.51	17	0.37	52
5	0.8	0.34	75	0.19	78

(a) Station 4

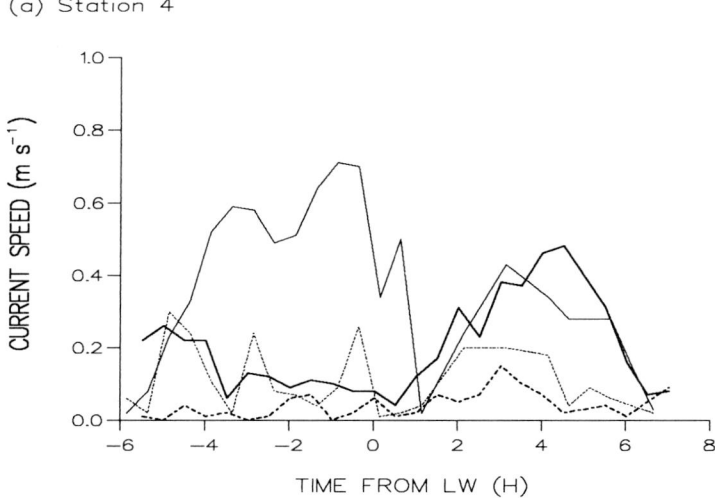

(b) Station 5

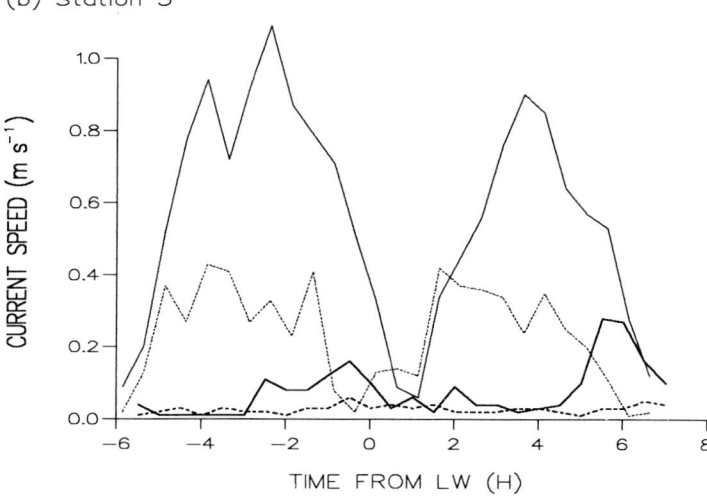

Figure 5. Tidal currents for, (a) Station 4, and (b) Station 5; solid lines - surface current, dashed lines - near bed current. 1990 data - normal lines, 1995 data - bold lines

(a)   (b)

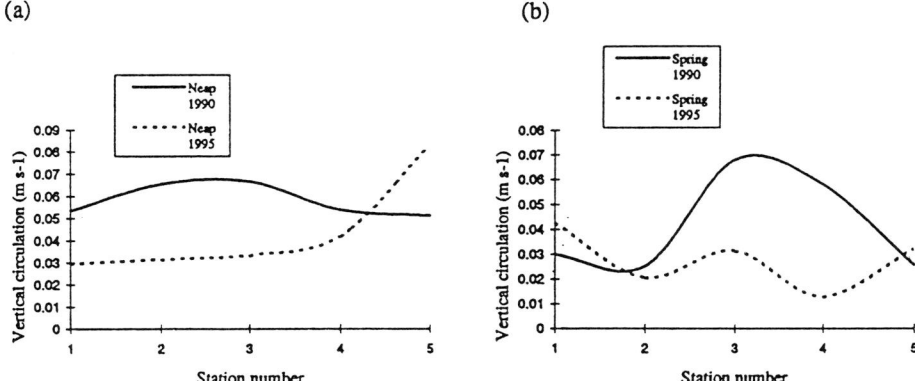

Figure 6. Tidal average surface current for 1990 and 1995; (a) neap tide, and (b) spring tide

barrage with the exception of the area very close to the barrage (Station 5) where the river input dominates the surface flow. The surface currents carry material towards the sea, on tidal average, and thus tidal average flows are important for flushing the estuary. The effect of the barrage has been to increase the flushing time from Billingham Reach from approximately 4 days at low river flows to 6-7 days.

3 Salinity and temperature

The salt distribution in the estuary has been significantly affected by the barrage (see Lewis and Lewis, 1983 for a detailed analysis of the salt balance in the estuary pre-barrage), and now full salt water (34 psu) penetrates the estuary in the lower layer right up to the barrage, whereas pre-barrage the maximum salinity at Station 5 was 30 psu.

The barrage resulted in strong layering of salinity between the waters originating from the river and the sea, forming an interface at a depth of approximately 1.5 m over the 11.5 km long section surveyed. This layering was generally stronger than in the pre-barrage situation, with a salinity difference between the surface and bottom layers of up to 27 psu in the upper estuary as compared with a difference of 8 psu formerly. Figure 7 shows the change in stratification caused by the barrage at mid-flood and mid-ebb tides. There is now little variation in the lower layer salinity throughout the tide at all the survey stations so that variations in stratification are principally due to changes in the surface salinity.

The temperature measurements in 1995 showed a slight increase in surface to bottom temperature differences compared to 1990, but the significance of this

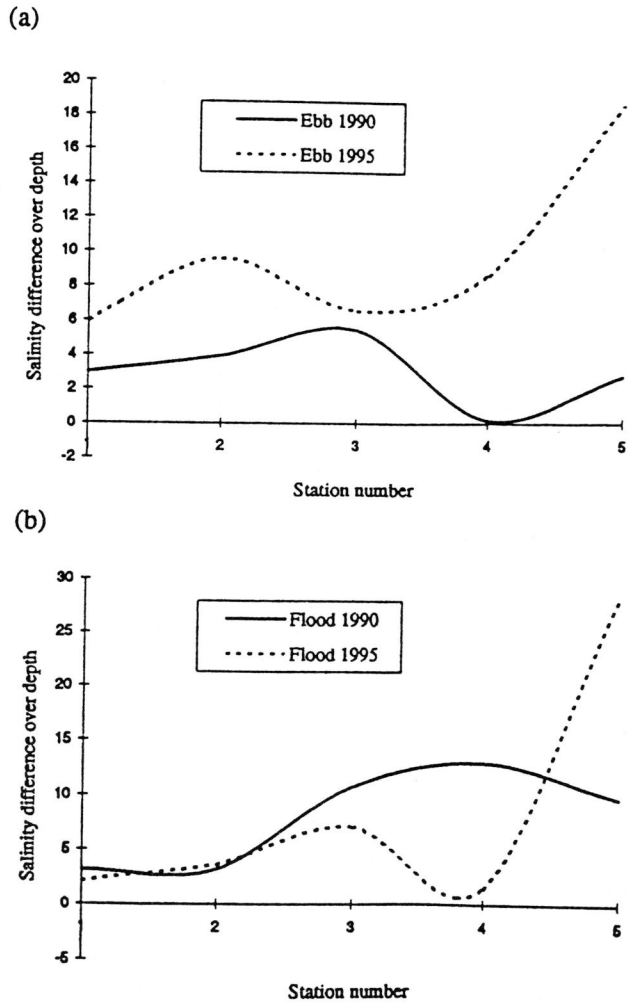

Figure 7. Stratification for 1990 and 1995; (a) mid-ebb tide, and (b) mid-flood tide

is more difficult to assess because it is influenced by the weather conditions prior to the survey periods.

4   Mixing

Simulating the salinity distribution in the estuary using a $2 - D$ laterally integrated model has shown that increasing the vertical mixing for a two hour period, centred on half flood, improves the match with observations [3]. The enhancement in mixing represented a very significant increase over that estimated for the same reach prior to construction of the barrage. The model also indicated that the effect of the barrage has been to appreciably reduce the vertical mixing of salt on the ebb tide. Vertical profiles of the observed salinity at Station 4 in 1990 and 1995 are plotted at hourly intervals over the tidal period in Figure 8; comparison of the profiles illustrates the decreasing difference between the surface salinity and the values in the lower layer during the flood in the post-barrage estuary. The profiles also illustrate the marked increase in stratification on the ebb due to the barrage. Contours of salinity along the estuary and through depth (Figure 9) indicate that the main mixing was observed at Station 4 in Billingham Reach, located at 3.3 km, and to a smaller extent at Station 3, located at 6.0 km.

A bulk Richardson number, R_{ib}, can be defined by:

$$R_{ib} = \frac{gh\triangle\rho}{\rho_m u_m^2}$$

where u_m and ρ_m represent the depth mean current and density over the total depth h, and $\triangle\rho$ is the density difference between the surface and bottom [4]. Values for R_{ib}, computed for Station 4 in Billingham Reach from the survey data for 1990 and 1995, are shown on Figure 10(a) for neap tides and Figure 10(b) for spring tides. The neap tide values show similar patterns for both years, but generally suggest greater stability (i.e. greater R_{ib} values) in the post-barrage estuary, with the higher values occurring near times of high and low water. No periods of instability ($R_{ib} < 1.0$) occurred in the 1995 data compared with short periods (up to 1 hour) at mid flood and ebb tides in 1990. On spring tides (Figure 10(b)), the R_{ib} values were generally lower and showed a distinctly different pattern from that on neap tides; in the pre-barrage situation R_{ib} values less than 1.0 occurred for approximately 4 hours on the ebb tide but remained above 1.0 on the flood. However, after construction of the barrage, the R_{ib} values can be seen to be below 1.0 for approximately 4 hours on the flood tide, but most of the values were above 1.0 during the ebb.

The modelling studies have indicated that the observed changes in the stratification on the flood tide for the post barrage situation are principally associated with vertical mixing, rather than differential advection of the longitudinal distribution of salt [5]. The cause of this mixing remains unclear, but it appears

A.M. Riddle and R.E. Lewis

(a) 3 JUNE 1990

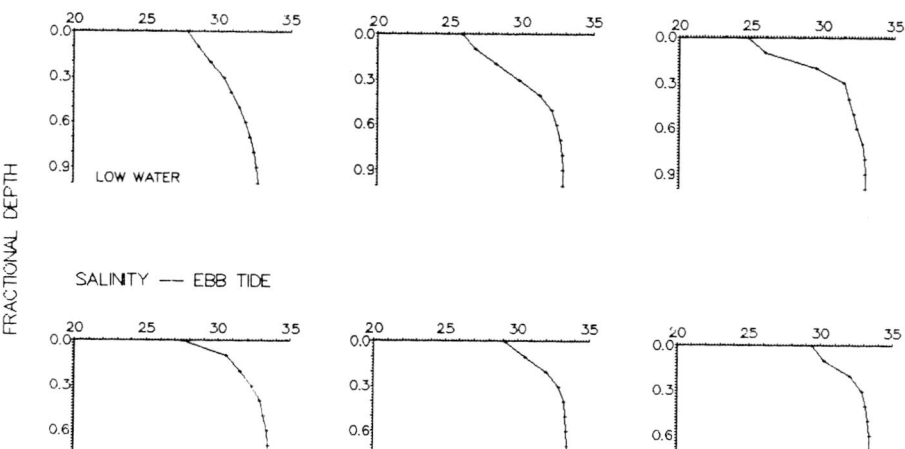

(b) 6 JUNE 1995

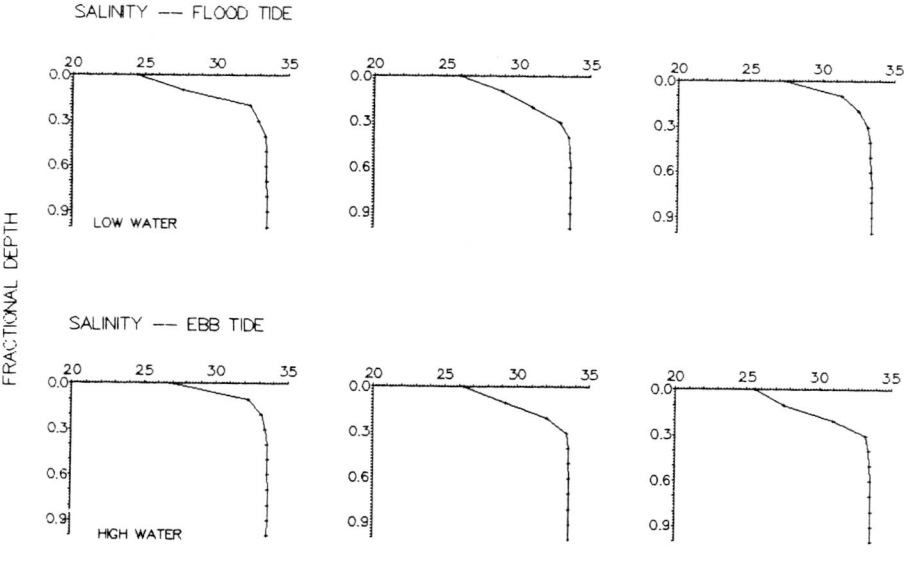

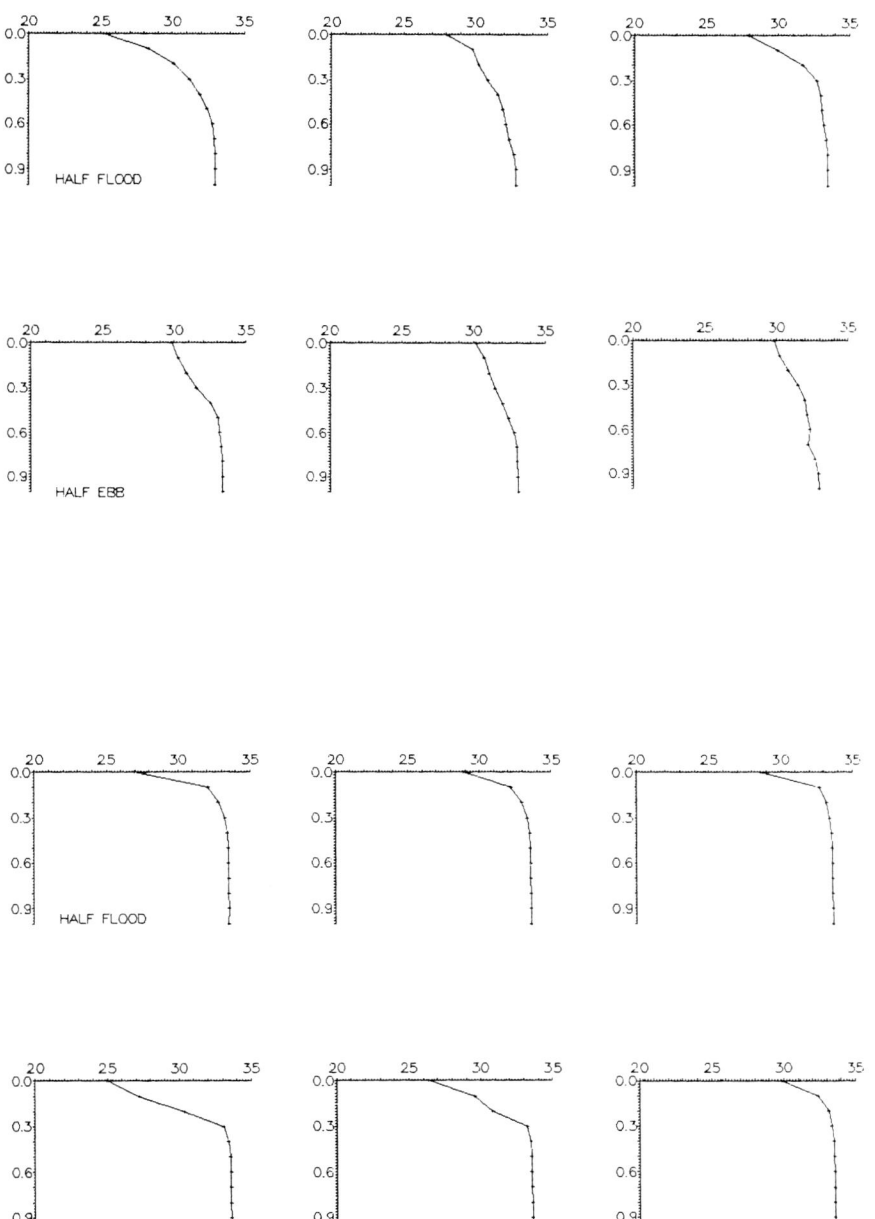

Figure 8. Vertical salinity profiles at hourly intervals through the tide; (a) 3 June 1990, and (b) 6 June 1995

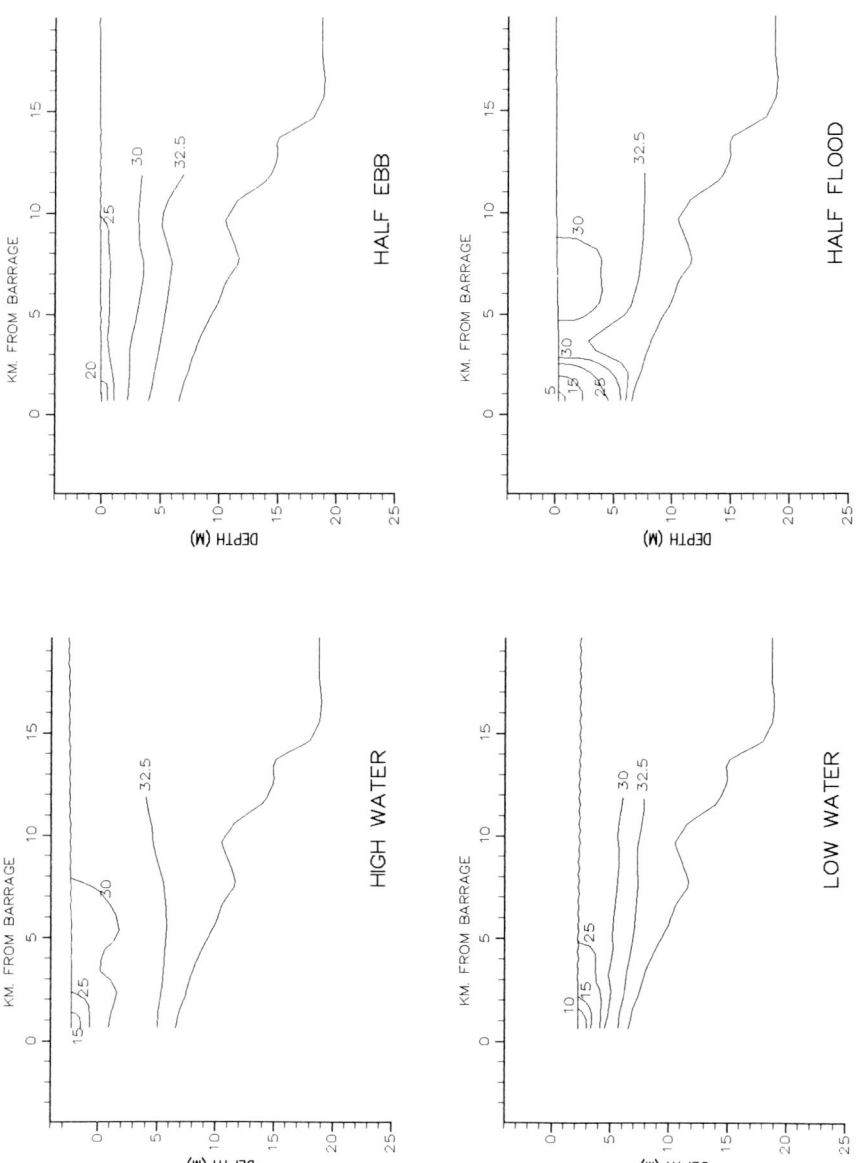

Figure 9. Salinity contours, low water, half flood, high water and half ebb

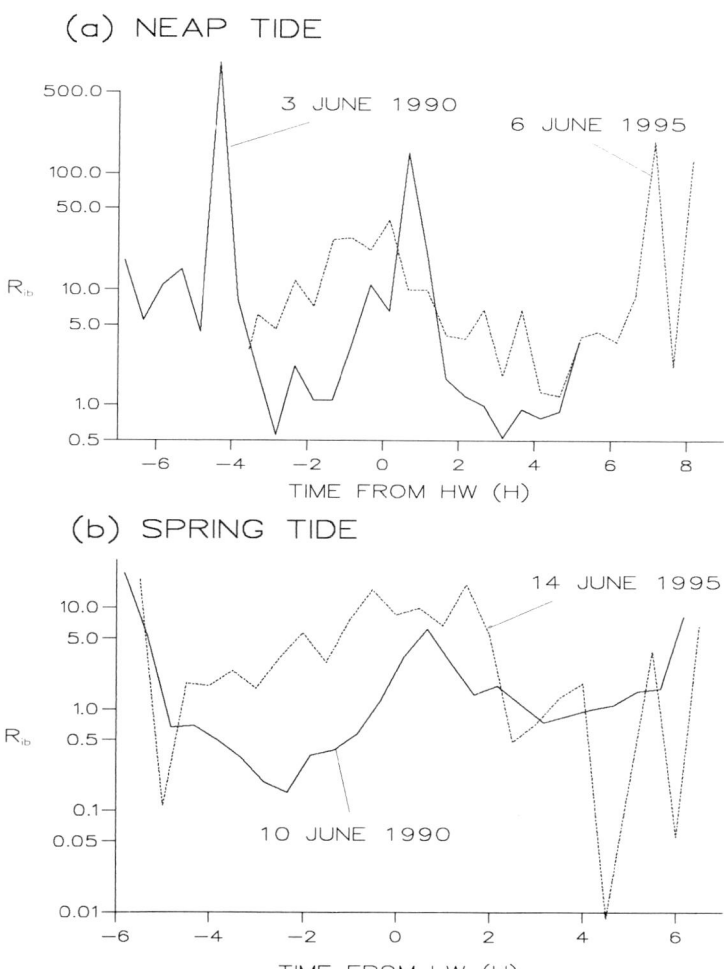

Figure 10. Bulk Richardson numbers at Station 4 1990 and 1995; (a) neap tide, and (b) spring tide

to be associated with interfacial instabilities between the upper and lower layers
arising from shear across this region during intrusion of salt on the flood tide [6].

5   Conclusions

The construction of a barrage across the Tees estuary, reducing the tidal length
from 44 km to 18 km, has had a profound effect on the flows, salinity distribution
and mixing characteristics of the area. Surveys of the estuary, jointly sponsored
by the Environment Agency, ICI and other Teesside industries, were carried out
before construction of the barrage in 1990, and after barrage completion in 1995.

Freshwater from the Tees and Leven rivers now enters the estuary over the
barrage and can be controlled, especially during the lower flow periods of the
summer; the freshwater inflow to the estuary is now not necessarily continuous.
Tidal flows have been significantly reduced for a distance of approximately 9 km
downstream from the barrage with reductions of 80% and 50% at distances of
0.8 km and 3.3 km from the barrage on the ebb tide.

The salinity distribution in the estuary has changed since the completion of
the barrage; stratification has strengthened, raising the peak surface to bottom
salinity difference from about 8 psu in 1990 to 27 psu at station 5 in 1995. Full
strength seawater now reaches up to the barrage in the lower layer of the estuary.

Mixing characteristics in the estuary have changed with the major part of
the vertical mixing now occuring at mid-flood tide instead at the early part of
the ebb tide. Bulk Richardson (R_{ib}) numbers for Station 4 in Billingham Reach
show that the barrage has increased the stability on neap tides. On spring tides,
the ebb tide stability has also been increased by the barrage but on the flood a
period of instability now occurs ($R_{ib} < 1.0$), and this appears to be responsible
for enhanced vertical mixing during the flood. This instability seems to arise at
the density interface between the upper and lower layers and be associated with
shear across this region produced by the flood tide intrusion of salt.

Acknowledgements

We would like to thank ICI and the Environment Agency for permission to
publish these data.

Bibliography

1. P.A. Rowland (1996). Vertical mixing mechanisms over a tidal cycle and
   the effect of a tidal barrage on a partially stratified estuary. *M.Sc. Thesis*,
   University of Wales, Bangor.

2. Environmental Agency (1996). Data sheets on the flows in the river Tees at
   Low Moor and the river Leven at Leven Bridge.

3. R.E. Lewis, A.M. Riddle and J.O. Lewis (1997). Effect of a tidal barrage on currents and density structure in the Tees estuary. *8th Biennial Conference on Physics of Estuaries and Coastal Seas.* In press.

4. R.E. Lewis and J.O. Lewis (1997). The onset and effect of intermittent buoyancy changes in a partially stratified estuary. *7th Biennial Conference on Physics of Estuaries and Coastal Seas - Mixing in Estuaries and Coastal Seas, Buoyancy Effects on Coastal Dynamics*, Editors: D.G. Aubrey and C.T. Friedrichs, *Coastal and Estuarine Studies*, **53**, American Geophysical Union, pp. 331-340.

5. H.M. Nepf and W.R. Geyer (1996). Intratidal variations in stratification and mixing in the Hudson estuary. *J. Geophys. Res.*, **101(C5)**, pp. 12,079-12,086.

6. W.R. Geyer and D.M. Farmer (1989). Tide induced variation of the dynamics of a salt wedge estuary. *J. Phys. Ocean*, **19(8)**, pp. 1060-1072.

Entrainment and Mixing in Long Sea Outfalls

K.H.M. Ali*, R. Burrows*, K. Spence** and P.A. Davies[†]

*Department of Civil Engineering, University of Liverpool, **Department of Civil Engineering, University of Sheffield, and [†]Department of Civil Engineering, University of Dundee

Abstract

The paper presents an experimental investigation into the characteristics of salt wedges in long sea outfalls.

Several high-level instrumentation systems were employed to study the various aspects of the problem. Salt wedge lengths and profiles were obtained for various densimetric Froude numbers. Velocity distributions of the freshwater layer of the primary salt wedges were obtained and used to calculate wall and interfacial shear stresses.

Detailed salinity readings were obtained using a ten-channel conductivity meter system. Velocities in the risers were estimated using a four-unit CDD video camera system and pump-pulsed dye injection. The salinity and velocity results were used to calculate entrainment rates.

Dispersion and mixing in model and prototype outfalls were also studies using the results of salt-water purging experiments. Entrainment was also considered.

1 Introduction

A considerable amount of work has been conducted into the characteristics of salt wedges in open channels [8]. Studies of the intrusion of such salt wedges into enclosed conduits have not received similar attention.

Sewerage outfalls to the marine environment, discharging treated or untreated effluents of approximately "freshwater density", have traditionally been extended into shallow coastal waters. Although recent advances in environmental legislation have led to simple open-ended short outfall pipes being replaced by much longer "long-sea" outfalls with closed ends and multiple vertical discharges through risers and ports, to aid dilution and dispersal of pollutants, the relevance of the phenomenon remains.

The continued presence of salt wedges in outfall pipes can cause problems of blockage and reduced hydraulic performance as reported by the UK Water Research Centre [15] (WRc), in connection with the apparent malfunction of a number of modern "long-sea" outfalls. Over the long-term, existence of the flow stratification and salt wedge in the pipe invert allows the settlement of particles,

591

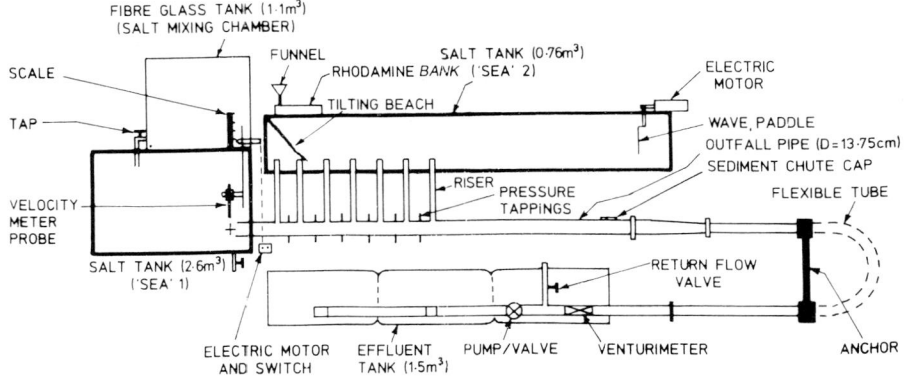

Figure 1. Simplified schematic representation of experimental rig

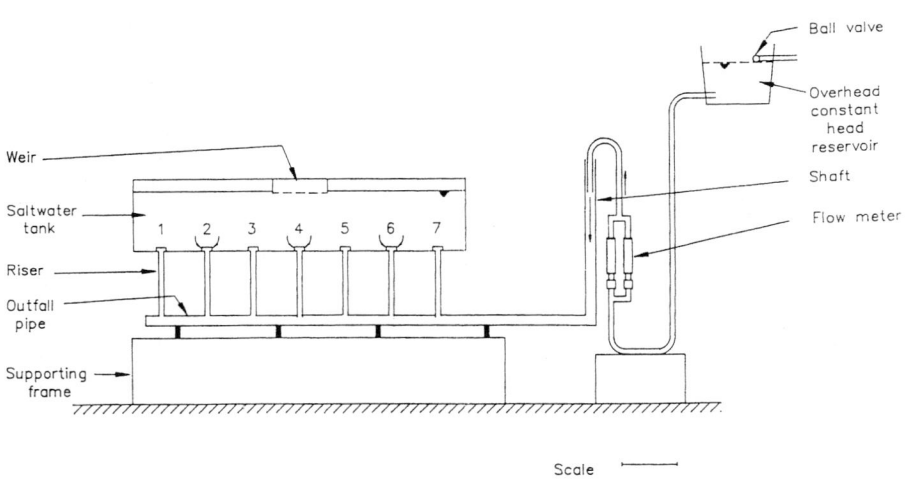

Figure 2. Schematic presentation of the Dundee outfall model

both those transported by the effluent flow which falls through the interfacial boundary, and also sea bed sediment whipped up by storm wave activity and drawn in with the saltwater intrusions. Entrainment of salt water into the effluent flow by turbulence at the interface encourages flocculation and may cause settlement of fine silts and clay which would otherwise be discharged. Once settled, the salt water in the near stationary overlying layer inhibits subsequent re-suspension of sediment and reduces the duration of exposure when freshwater flows are sufficiently large to purge out the salt wedge.

Extensive laboratory studies have been conducted on this topic by the authors at the University of Dundee [5] and Liverpool [3] over the past decade using four different outfall models. The Liverpool group have focused on the characteristics of salt wedges and on the influence of flow variability and wave activity on the performance of diffuser outfalls.

The paper describes detailed velocity and salinity measurements obtained using different numbers of risers. Dispersion, mixing and entrainment phenomena are considered. Boundary and interfacial shear stresses are calculated. Changes in salinity and velocity-distributions resulting from a stepped discharge are also considered.

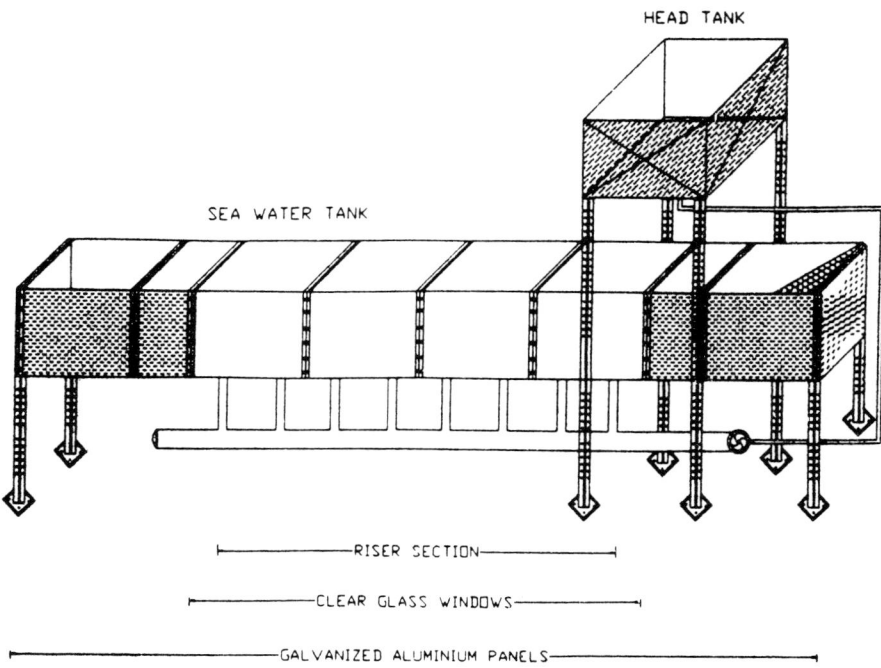

Figure 3. Layout of the lastest Liverpool outfall modl arrangement

1.1 Experimental arrangements

Three different experimental set-ups were used in this investigation.

1. The original Liverpool outfall pipe arrangement

    The apparatus consisted of a clear perspex pipe, 13.75 cm in diameter. The outfall pipe, approximately 6 m in length, incorporated a diffuser section towards its downstream end, consisting of seven risers at spacings of 25 cm. The risers were fabricated from clear perspex so that observations can be made throughout the experiments. Each riser had a length of 0.5 m and in internal diameter of 5.0 cm. The total length of the diffuser section was about 2.0 m.

    The apparatus incorporates three tanks (Figure 1), all of which were used in the storage of the prepared brine solution. The first tank was positioned at the closed end of the outfall pipe and had an approximate capacity of 2.6 m^3. The third tank was positioned directly above the outfall pipe and its capacity was 0.76 cm^3.

    The freshwater mains provided a direct supply of water to the storage tanks throughout the experiment, simply by operating a valve. The valve was connected directly to the mains supply pipe, which branched out to two different storage areas at the valve position; one branch supplying the storage tank positioned below the outfall pipe whilst the other supplied the large fibre-glass tank, positioned directly above the closed end of the outfall pipe. The direction in which the valve was opened (i.e. clockwise or anti-clockwise) determined the direction of flow of the freshwater from the mains. The water supply was conveyed from the storage tanks into the outfall pipe itself by means of the supply pump.

    A constant-speed centrifugal pump was used to control the supply of freshwater to the outfall pipe. A venturi-meter was incorporated into the pipe system controlling the supply of water to the outfall pipe.

    A miniature propeller meter, 16 mm in diameter was used to measure the velocity distributions in some of the experiments. Dye injection was also used for velocity measurement. A hand-held Paar density meter was also used. This hand-held meter was equipped with a clear digital display which provided simultaneous readings of local density and temperature.

2. The Dundee outfall arrangement

    The experimental apparatus is shown in Figure 2.

    The sea tank was 1.23 m wide, 6.11 m long and had a working depth of 0.92 m. The outfall pipe was 8.0 long and had an average diameter of 8.8 cm. The risers were 7 in number and were 1.30 long. Their internal diameter was 3.5 cm and their spacing was 0.85.

A multi-channel conductivity meter was used to monitor continuously the density saltwater in the main outfall pipe and in the risers.

A dye injector, a video recorder and four video cameras were used in the tests.

Two salinity probes were installed (at 10 cm apart), in each riser as a precaution to ensure that at least one of them would provide salinity records should the other become faulty.

The salinity records were analysed and graphical hard copies of the variation of riser salinity with time were produced automatically at the end of each test.

Video film records of dye movements in the riser columns were analysed at a later time. Respective riser velocities were estimated by timing the movements on the fronts of the dye pulses between 5cm marks that were engraved on the outside of each riser column on either side of the point of discharge of the hypodermic needle.

3. Latest Liverpool arrangement

This new experimental set-up is shown in Figure 3 and it depicted a tunnelled sea outfall system. A large 13 m long galvanised aluminium tank represented the marine environment and a perspex pipe system attached underneath the tank modelled the outfall. Figure 3 illustrates the main components of the model set-up including an overhead storage tank as headworks. The set-up was made as versatile as possible to allow for various adjustments such as riser combinations, riser length, outfall pipe slope, height of overflow weir, etc.

A clear perspex pipe constructed with rubber-flanged joints and marked at various sections was used as the main outfall pipe. From the extreme downstream end of the outfall pipe a ball valve was fitted to allow for the discharge of stagnant or unwanted water from the system. This was followed by eight risers connected perpendicular to the outfall pipe to meet the sea tank above. Along the outfall pipe and at various riser sections on the risers acetate sheets grid marked with gauging scales were attached to facilitate velocity measurements. In likewise manner pinholes were also filled to enable the insertion of probes and dye injection needles.

The head tank was made of aluminium and was 1.62 by 1.50 by 0.7 m and stands 3.31 metres above the outfall invert level and is linked to an electronic gate valve by a PVC pipe with a ball valve at the upper end. Freshwater flow into the head tank from the mains was controlled by a ballcock valve. In the various experiments, monitoring the flow rate from the head tank was by means of an electronic gate valve system. By means of an orifice plate with differential pressure tap points, the electronic gate valve system. By means of an orifice plate with differential pressure tap points, the electronic gate valve was regulated using a control unit. From this point, freshwater flowing into the outfall is fed initially down a slope of 1:500 within the first 3 m of the outfall pipe length then into the riser section.

At the sea end a wooden wave paddle system connected to a KEELAVITE WAVE GENERATOR was installed into the sea tank.

The electronic gate valve (Sensycon 23/16-12) is the main source for regulating freshwater inflow into the outfall system. With the outlet valve of the head tank on full bore, the electronic gate valve can be manipulated from a CMI microprocessor control unit to achieve the desired flowrate. The control unit is fed from a pressure transducer (Ashdown Process Control Limited Pressure Transducer: 0-25" WG), which comprises an orifice plate with upstream and downstream tapping points.

Flow velocities in the risers were obtained by the monitoring of the movement of injected Rhodamine B dye in the risers. With the aid of CCD cameras, the dye movement within the risers over the scaled acetate tape section is recorded. Salinity measurements were obtained using a system using 10 conductivity probes. A peristaltic pump was used to suck continuous liquid through the probes.

2 Experimental results

2.1 Flow visualisation tests

Dye injection was used to study mixing and entrainment in the outfall pipes. Dye was injected upstream of the risers and was observed to be expelled through one riser (two risers in operation) [4]. The dye trace entered riser 2 in vertical streak lines and progressively mixed fully across the full width of the riser (see Figure 4(a)). The dye slug in the salt wedge moved upstream at a steady rate. Thus the salt wedge was continuously being replenished from the saltwater tank. Since the saltwater was flowing upstream and the thickness of the wedge remained constant, then it is anticipated that the inflowing salt water was being dispersed through the surface area of the wedge into the flowing freshwater.

The dye injected in the freshwater, upstream of riser 2, moved in a continuous stream through the interface layer into the salt wedge. Mixing took place and the mechanism of mixing was an injection of eddies from the rest of the salt wedge into the moving freshwater current (Keulegan, 1949 [10] (see Figure 4(b)). An interesting point was noted between the flow paths of the dye injected at Ch 110 and at Ch 75 cm. Unlike the path followed by the dye injected at Ch 110 cm, for the same flowrate, the dye injected at Ch 75 traced a different pattern. Dye was injected at chainage 75 cm and the dye trace in the fresh water was expelled through riser 2 (see Figure 4(c)). Dye in the salt layer moved upstream and downstream, as expected (see Figure 4(d)). Traces of dye were

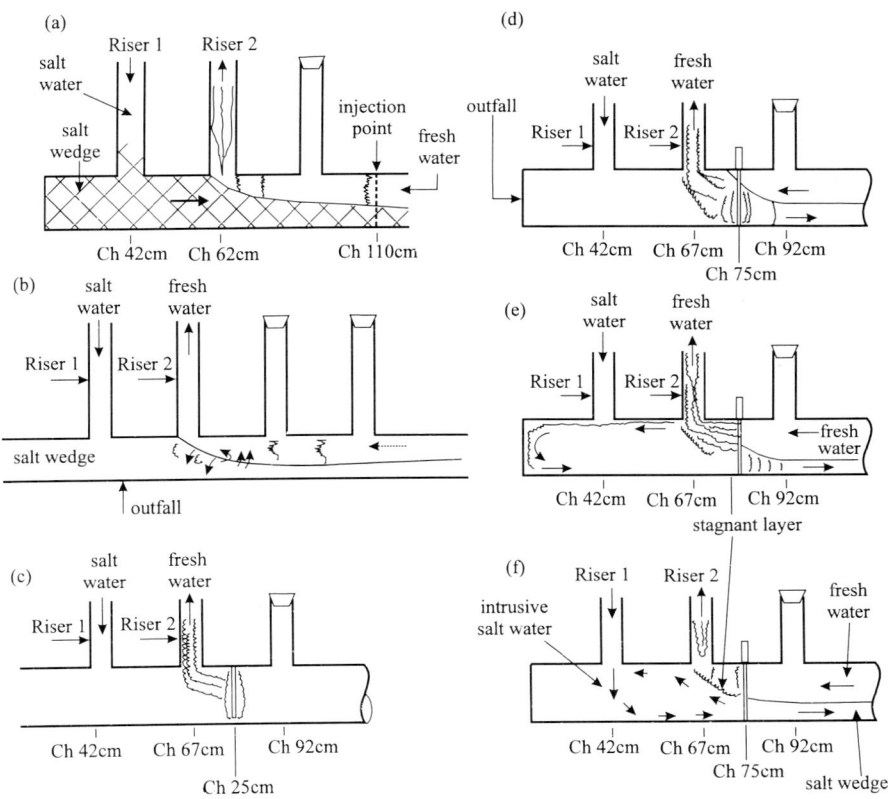

Figure 4. Results of some of the dye injection tests

transported downstream along the top edge of the outfall pipe and it followed a return path at the invert (see Figure 4(e)).

At Ch 75 cm, the dye was injected into quiescent layers, and these dye particles were engulfed by the circulatory motion of the intrusive saline water (see Figure 4(f)).

The intrusive salt water strikes the invert of the pipe and changes direction. This induces a circulatory motion in the most downstream salt wedge layer, as shown in Figure 4(f). The eddies in this circulatory motion engulfed the dyed freshwater and the dye was transported in the salt water upstream to riser 2.

Systematic experiments were conducted in horizontal outfall pipes of various diameters to study the variation of salt-wedge length with densimetric Froude number. Dye injection was used to identify the wedge. Purging experiments

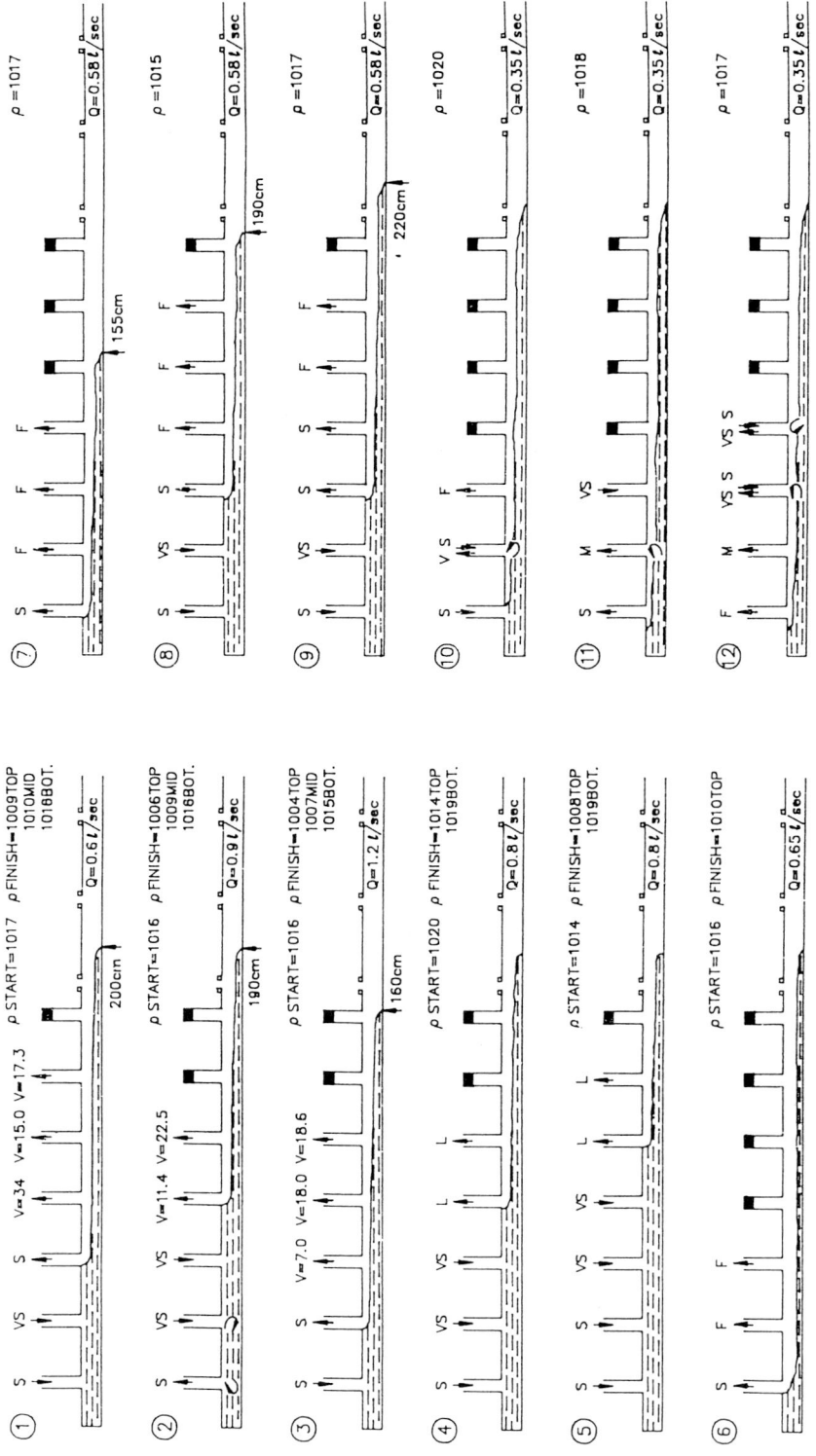

Figure 5. Spencer's outfall visualisation experiments. Velocities are classified between very slow (VS) and fast (F)

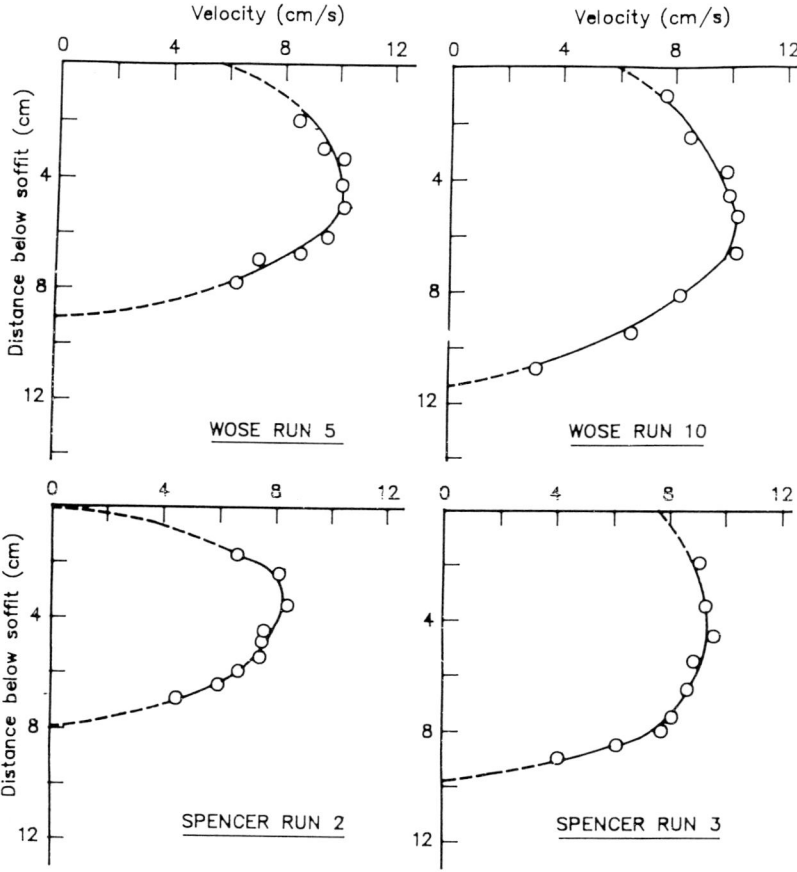

Figure 6. Experimental velocities in a pipe with a salt wedge ($D = 13.75$ cm)

were also carried out using different numbers of risers. Positions and profiles of the wedges were obtained and some of the results are shown in Figure 5.

2.2 Velocity distributions

Figure 6 shows velocity distributions obtained by Wose [19] for the freshwater flow above arrested salt wedges ($D = 13.75$ cm). Figure 7 shows velocity distributions obtained by Mort [14] at various sections along his outfall pipe ($D = 10.5$ cm).

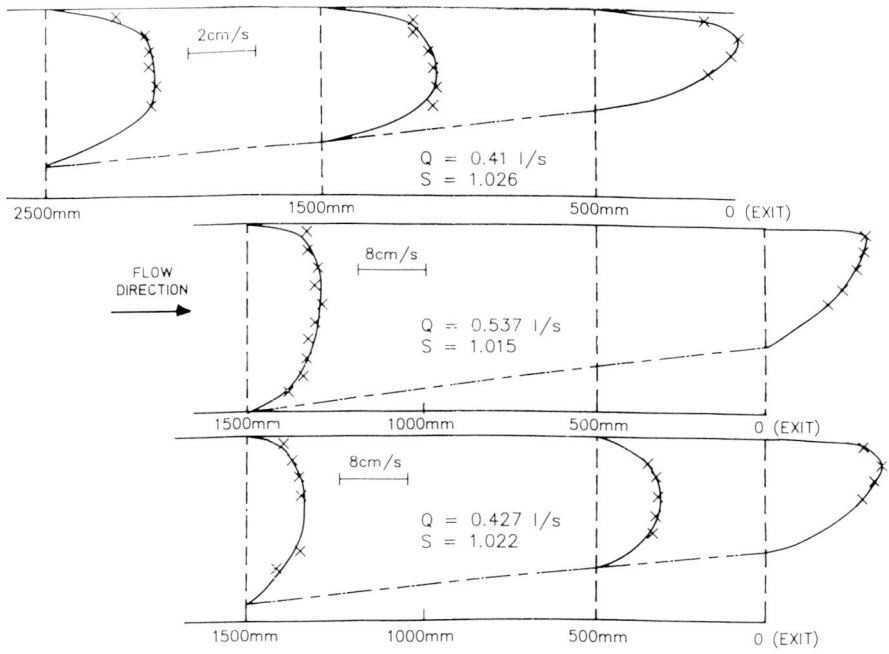

Figure 7. Velocity distributions along an outfall pipe with a salt wedge

2.3 Vertical density profiles in the presence of a salt-wedge

Fine bore brass tubes were fixed at different depths from the pipes soffit in order to sample fluid over the depth of the pipe. This system was used at different stations along the pipe. Water was expelled from the tubes when opened, because of the positive pipe pressure. Fluid samples were collected, at the same time, from all these tubes and the fluid densities were obtained.

Figure 8 shows typical density profiles with risers 1 and 4 in operation. This figure shows the presence of a thick brackish layer indicating that the interface is not sharp and distinct.

3 Theoretical considerations

3.1 Mixing in sea outfalls

Longitudinal variations of concentrations in outfall pipes can be approximately obtained using the following methods:

1. The method of Masters [11]

Following Masters, it is assumed that all changes in concentration are caused by advection. In Figure 9 the environmental system to be modelled has a volume, V, and has equal flows, Q, into and out of the box. Masters [2] assumed that the contents of the box are at all times completely mixed so that the pollutants concentration, C, in the box is the same as the concentration leaving the box. We can write the following relationship:

Accumulation rate = Input rate - Output rate,

or

$$V\frac{dC}{dt} = S - QC \tag{3.1}$$

where:

- V = box volume (m^3); C = concentration in the box and in the existing waste stream (g/m^3); S = total rate at which pollutants enter the box (g/hr) and Q = total flow rate into and out of the box (m^3/hr).

Integrating Equation (3.1) and simplifying, Masters obtained:

$$C = (C_0 - C_\infty)\text{Exp}\left[-Qt/V\right] + C_\infty \tag{3.2}$$

where:

- C_0 = concentration at $t = 0$ and C_∞ = concentration at $t \to \infty$.

2. The method of Fisher et al. [8]

Fischer et al. [8] give the following relationship for the approximate solution for the spreading from a continuous point source in two dimensions:

$$C = \frac{\dot{M}}{U\sqrt{4\pi Et}}\text{Exp}\left[-\frac{y^2}{4Et}\right] \tag{3.3}$$

where $\dot{M}$ = strength of line source; y = normal distance; E = dispersion coefficient; U = mean velocity and t = time. A line source of $\dot{M}$ units into a flow of depth, d, is equivalent to a point source of strength, $\dot{M}/d$ in a two-dimensional flow for which Equation (3.3) becomes:

$$C = \frac{\dot{M}}{Ud\sqrt{4\pi Et}}\mathrm{Exp}\left[-\frac{y^2}{4Et}\right] \tag{3.4}$$

where $x = Ut$ and $\dot{M} = QC$.

Equations (3.2) and (3.4) were applied to some model and field results. Figure 10 shows variations of salinity with time obtained by Wose et al. [20] using their Dundee outfall. These results are given using only one riser for two different locations along it. If, during purging, the changes in salinity were caused mainly by advection, then Equation (3.2) shows that a plot of $ln[(C - C)/(C_0 - C_\infty)]$ against t should result in a straight line. Figure 11 confirms this.

If mixing in the riser was mainly caused by dispersion at the later stages of purging (large values of t) then Equation (3.4) shows that a plot of the maximum concentration ($y = 0$) against $1/\sqrt{t}$ should result in a straight line. This is demonstrated clearly in Figure 12. Values of the dispersion coefficient, E, were calculated and were found to be extremely small.

Figure 13 shows the results of a purging experiment conducted by WRc [15] on the Eastbourne outfall. Using the results for time periods later than about 13:24 BST (constant inflow), Equation (3.2) was applied and the results are given in Figure 14. Again, straight lines result. The slopes of the straight lines for the different risers are different because of the different values of Q and V.

3.2   Entrainment

A general description of salt water entrainment by an upper layer of fresh water flowing at higher velocity can be found in Dyer [7]. When shear becomes sufficiently intense, the interface between layers becomes disturbed by waves, which grow higher with increasing shear [6]. Eventually the waves break and globules of the lower, denser, layer are ejected from the wave crests into the lighter fluid above, in a strictly one-way process.

The entrainment flux is given by [6]:

$$\frac{dm}{dt} = V_e C \quad R_i < R_{ic}$$

$$\frac{dm}{dt} = 0 \quad R_i < R_{ic}$$

where dm/dt = rate of mass entrainment of salt, V_e = entrainment velocity which is given by:

$$V_e = \frac{0.1\Delta U}{(1 + 63R_i^2)^{3/4}} + E\Delta U \tag{3.5}$$

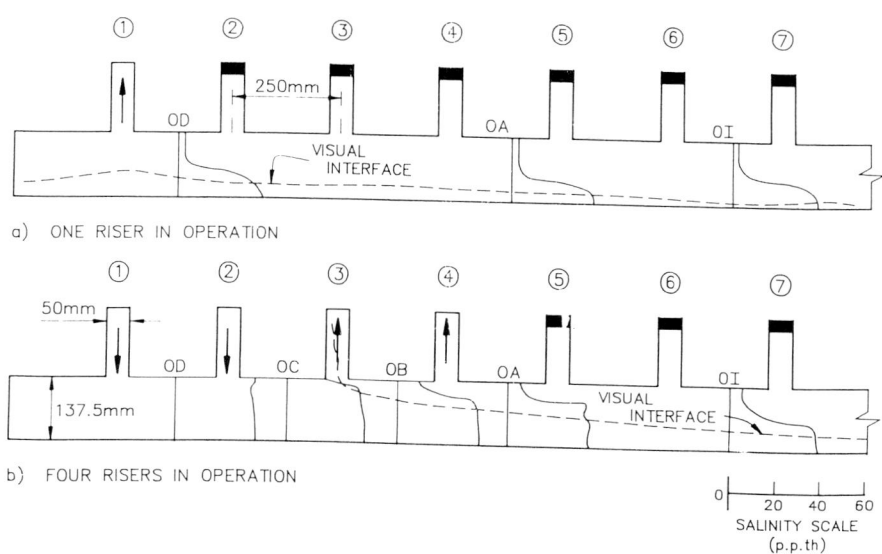

Figure 8. Density distribution along outfall using one and four risers

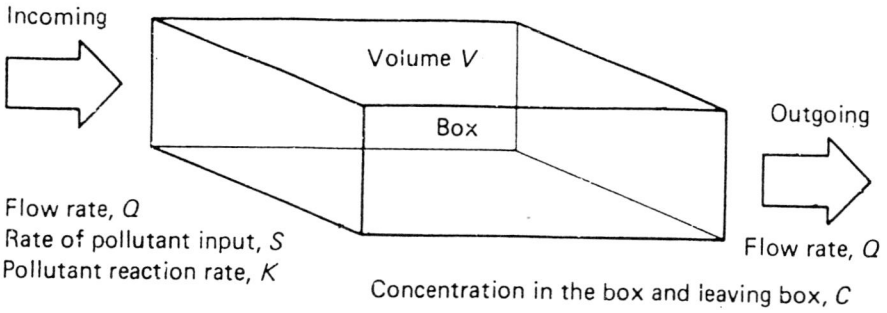

Figure 9. A box model for a transient analysis

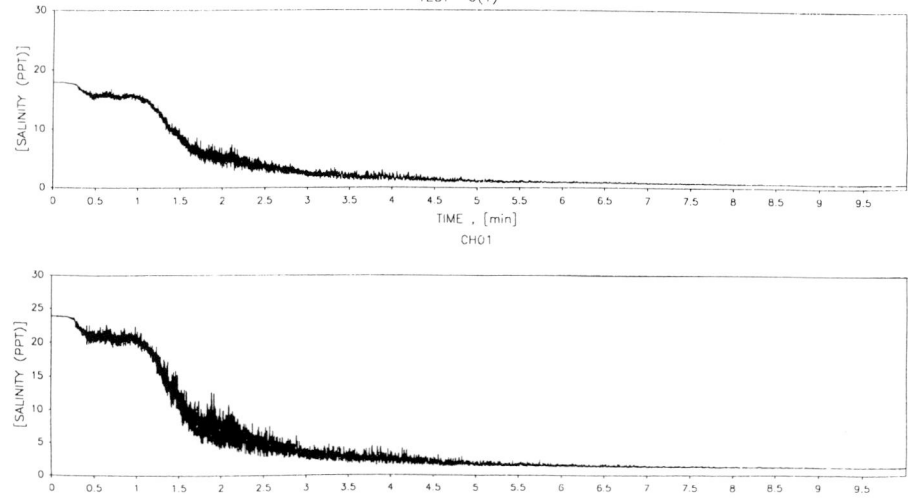

Figure 10. Outfall flow = 0.40 1/s, saltwater density = 1.025 g/cc

where

$$
\begin{aligned}
E \;\; &= \;\; \text{entrainment coefficient} \\
\Delta U \;\; &= \;\; [(u_o - u_s)^2 + (v_o - v_s)^2]^{1/2} \\
&= \;\; u \text{ and } v \text{ are the horizontal depth-averaged velocities} \\
&\quad\;\; (\text{subscripts } o \text{ and } s \text{ denoting saltwater and freshwater, respectively}) \\
C \;\; &= \;\; \text{mass concentration of saline water} \\
Ri \;\; &= \;\; \text{is the bulk Richardson number, defined as:}
\end{aligned}
$$

$$R_i = \frac{\Delta \rho g d}{\rho_o (\Delta U)^2} \tag{3.6}$$

where $\Delta\rho = \rho_s - \rho_o$ is the density difference and d = thickness of the saltwater layer.

According to the above formulation, entrainment can only occur at a sufficiently low value of Ri. The critical value for $Ri(= R_{ic})$ is usually assumed to be 10.

A plot of the entrainment coefficient versus the Richardson number is presented in Figure 15 for salt (see Reference [6]).

3.3   Entrainment in outfall pipes

An attempt was made to calculate the discharge entrained across the interface in the outfall pipe. Using the method of Ali and Jaefar-Zadeh [13]:

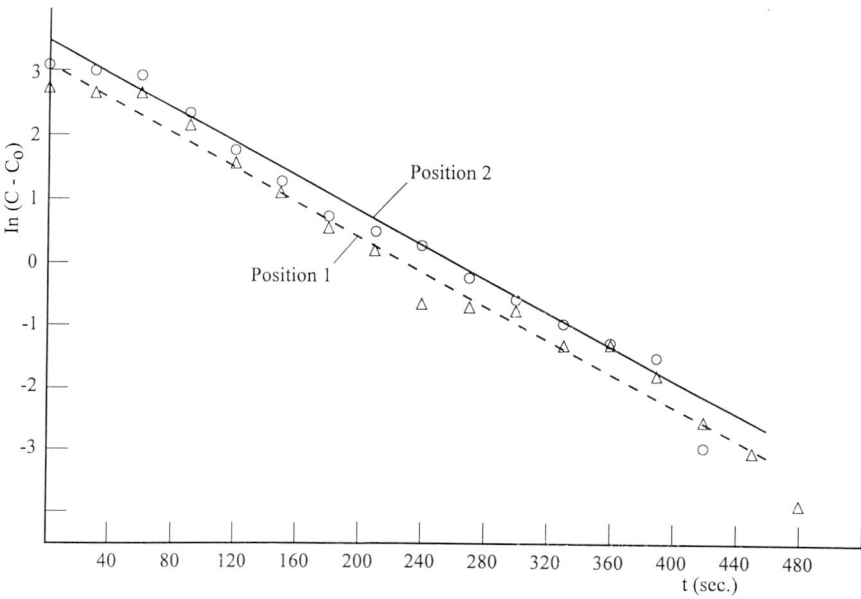

Figure 11. Variation of $In(C - C_0)$ with t for a single riser

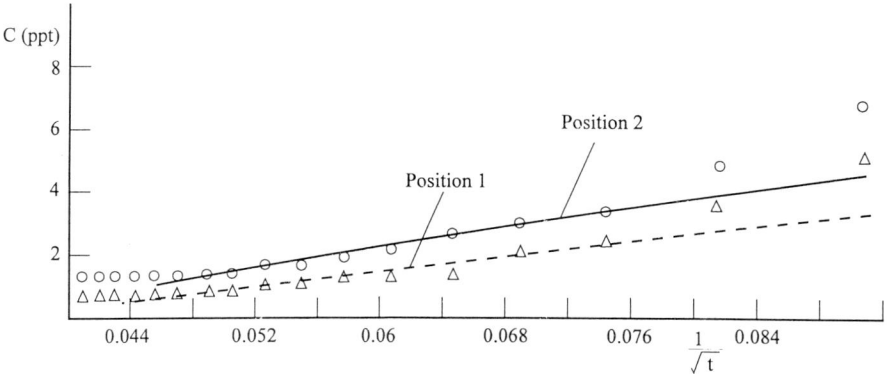

Figure 12. Variation of C with $1/\sqrt{t}$ for a single riser

(a)
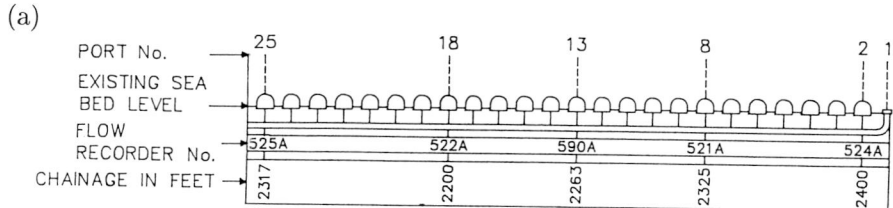

(b)
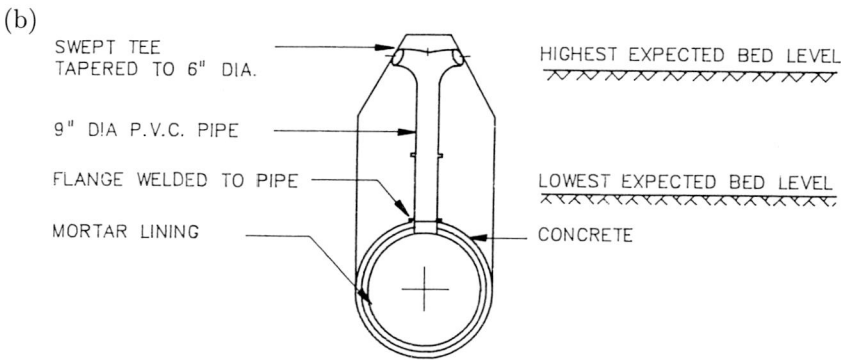

(c)
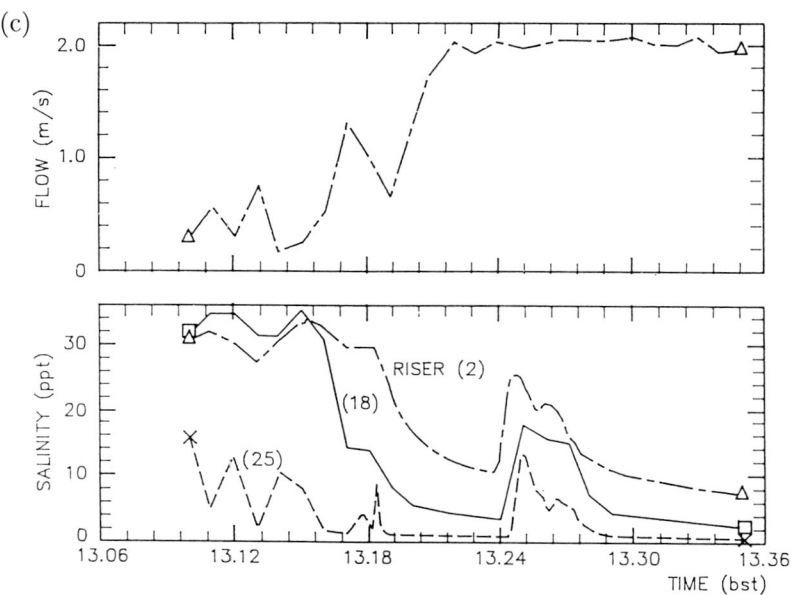

Figure 13. (a) Eastbourne outfall, diffusser section, (b) Eastbourne outfall, section at riser, and (c) Eastbourne outfall, salt water purging

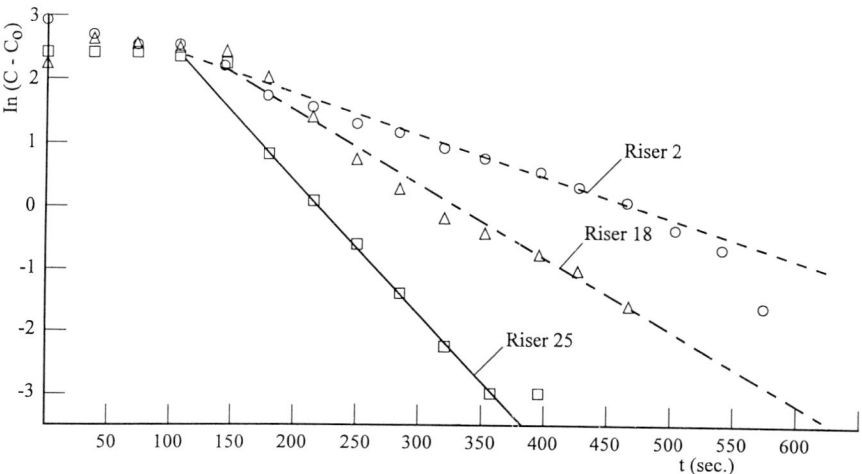

Figure 14. Variation of $In(C - C_0)$ with t for the Eastbourne outfall

$$E = \frac{V_e}{U} - cR_i^{-1}. \tag{3.7}$$

The average entrained discharge, Q_e, is given by:

$$Q_e = V_e \overline{T}\, \overline{L} \tag{3.8}$$

where $\overline{T}$ and $\overline{L}$ are the average width and length of the interface.

Table 1 shows that the entrained discharge can be significant for small values of R_i.

3.4 Calculation of interfacial shear stresses

Extensive velocity distributions were obtained along the outfall pipes in the presence of salt wedges. Examples are given in Figures 6 and 7. The velocities were obtained for different diameter, flow rate and saline density. Dimensionless velocities of the form $v/V_{\max}$ against y/h_1, for elevations above the location of maximum velocity, were obtained. Similar results were obtained for the elevations below the maximum velocity ($v/V_{\max}$ against y_2/h_1). The results are shown in Figure 16. Also given are the theoretical power law velocity distribution curves for various values of n:

$$\frac{\nu}{\nu_*} = C_1 \left(\frac{y\nu_*}{\nu}\right)^{1/n} \tag{3.9}$$

or

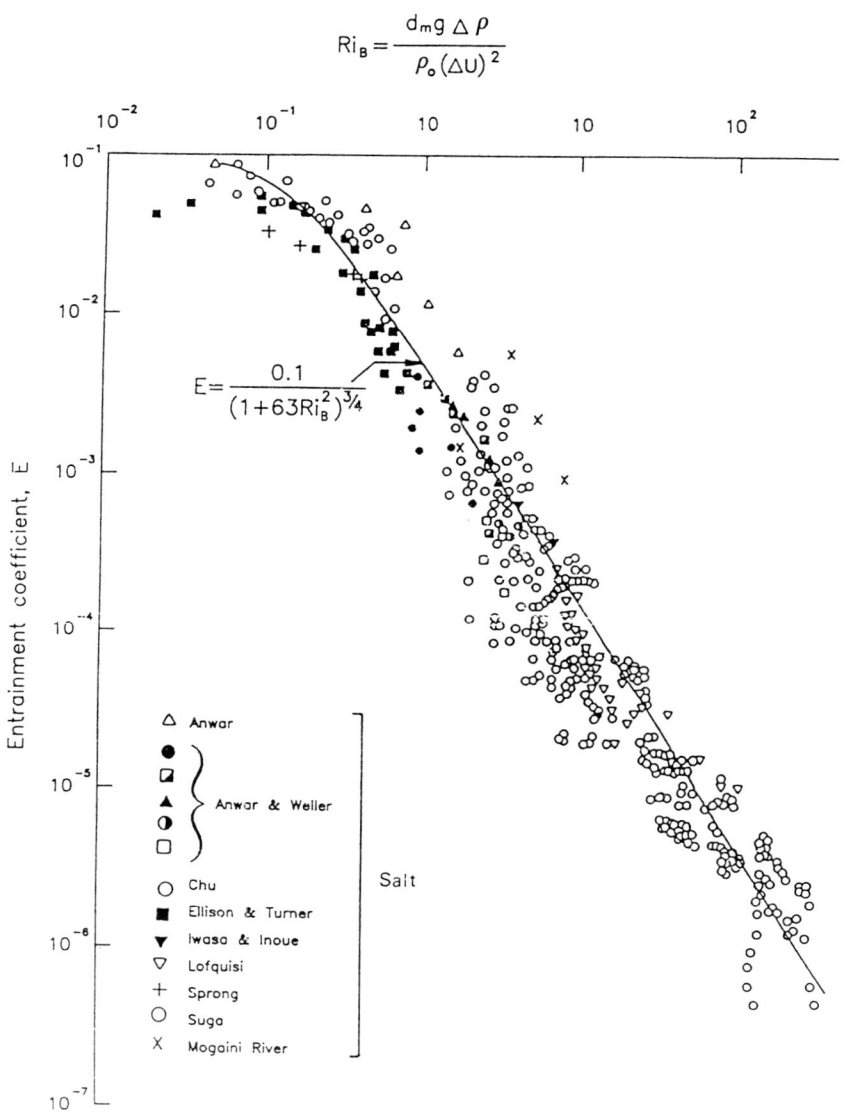

Figure 15. Entrainment function for the case of an arrested salt wedge

Table 1. Entrained discharge for various outfalls

Researcher	Number of Risers	R_i	Q_e (cc/sec)	Q (cc/sec)
Cornick (Liverpool)	One	5.17	2.0	1111
Cornick (Liverpool)	Four	1.50	11.1	702
Wose (Dundee)	One	0.29	163.0	400

$$\frac{\nu}{V_{\max}} = \left(\frac{y}{h_1}\right)^{1/n} \tag{3.10}$$

where:

- h_1 = thickness of freshwater layer; v_* = shear velocity; n and C_1 are parameters strongly dependent on the roughness height but weakly dependent on the Reynolds number [1]. For a given Reynolds number, an increase in roughness height corresponds to a reduction in n and C_1.

Applying Equation (3.9) the zones above and obtaining the values of n and C_1 from Reference [1], centreline interfacial and wall shear velocities were calculated. These results show that, generally, the centreline shear velocities for the interface are higher than those for the solid boundary. It must be stressed, however, that these shear velocities are very approximate because of the considerable scatter in the experimental velocity-distributions as well as the uncertainty regarding the position and thickness of the salt wedge. Also, these results refer only to the centreline distributions.

Figure 17 is reproduced from Rouse [16] and gives the velocity distribution in a rectangular pipe having a smooth boundary on one side and a rough one on the opposite side. This figure shows that the logarithmic velocity distribution law applies very well using different coefficients for the two boundaries. It can be shown that the power law given by Equation (3.9) also applies giving values of $n = 6.06$ and 3.96 for the smooth and rough boundaries respectively. The similarity between Figures 16 and 17 is very evident.

3.5 Purging tests using the new Liverpool outfall (four risers)

A typical purging test is described. This was named test T7 and was divided into four sub-tests T7A to T7D [18] using the last four seaward risers only. The four sub-divisions in the flow lasted for a twenty-minute duration. The fresh water in-flow introduction was done in a stepwise manner up to a maximum flow of about $1.0 l/s$ as shown in Figure 18.

In the tests, a reduction in salinity was observed in all risers at quite different rates within the sub-tests but by similar stepwise increases as the discharge was

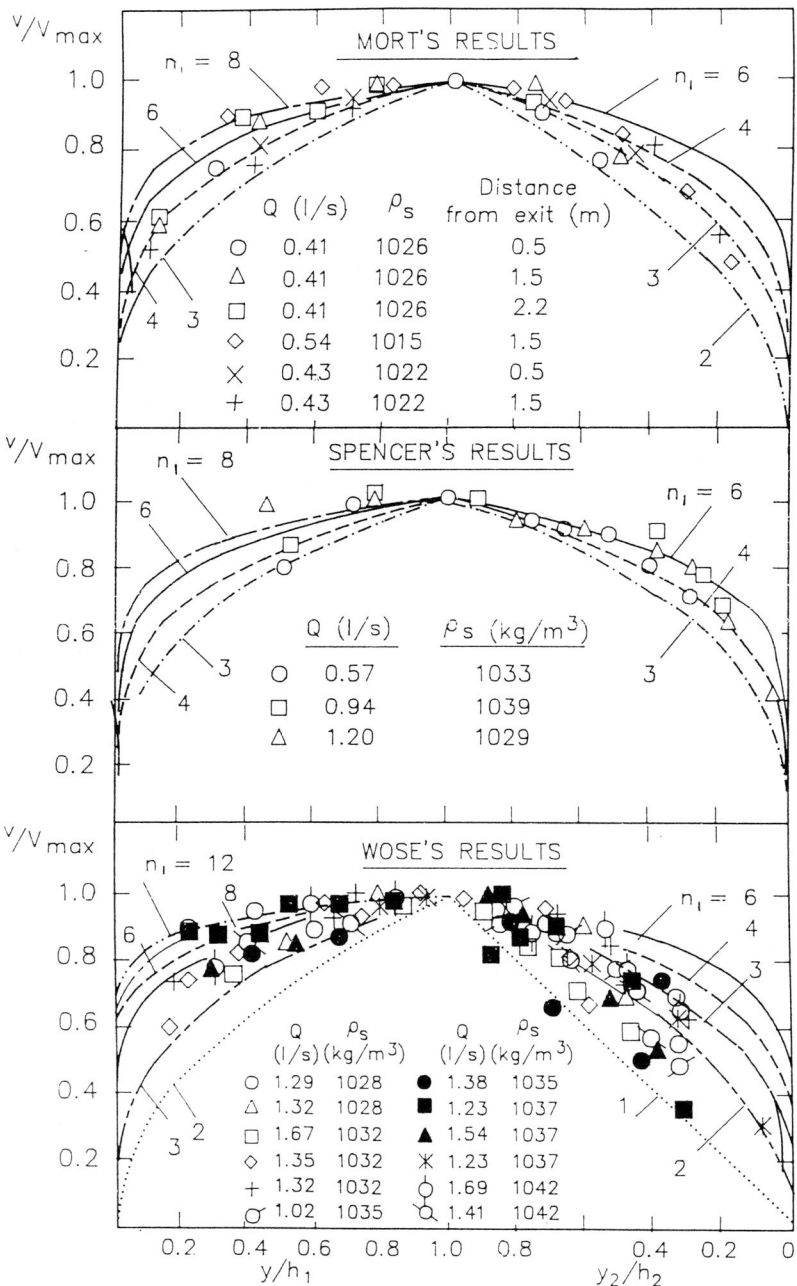

Figure 16. Dimensionless velocity-distributions

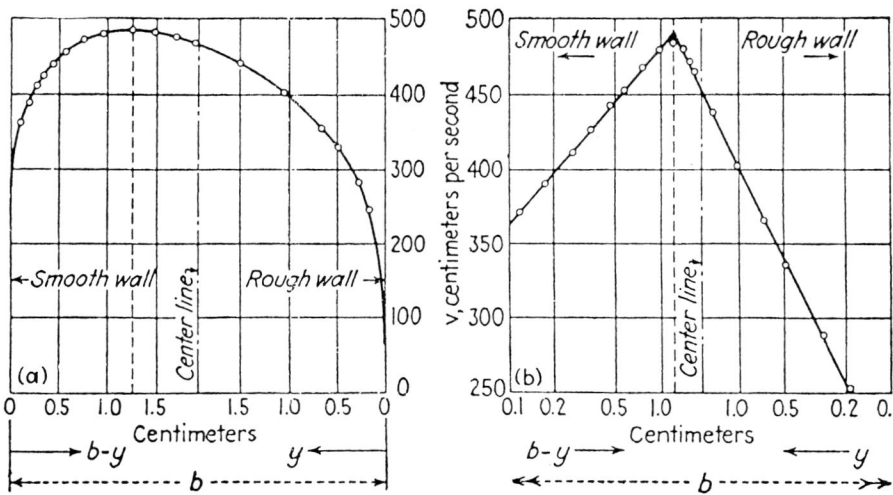

Figure 17. Velocity traverses between smooth and rough boundaries

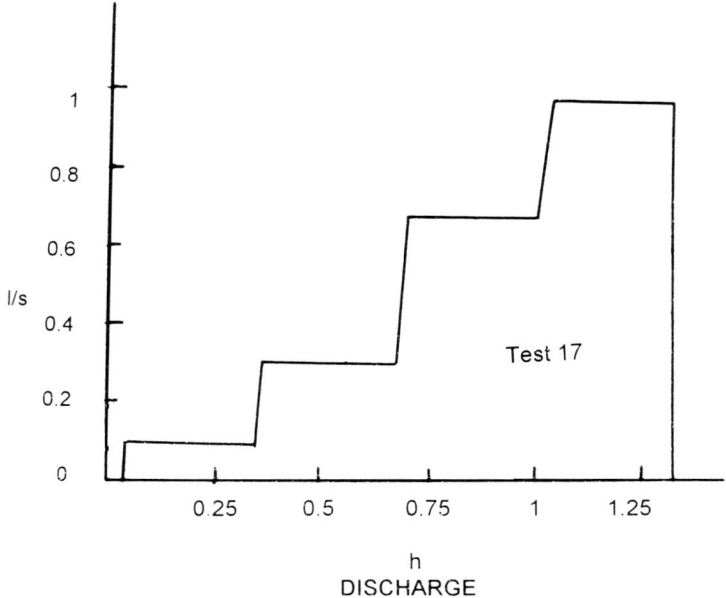

Figure 18. Gate valve discharges

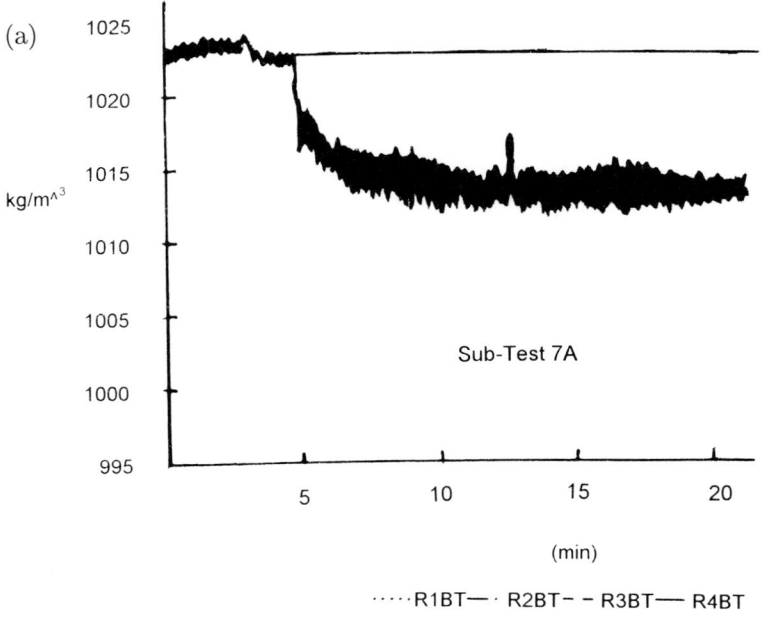

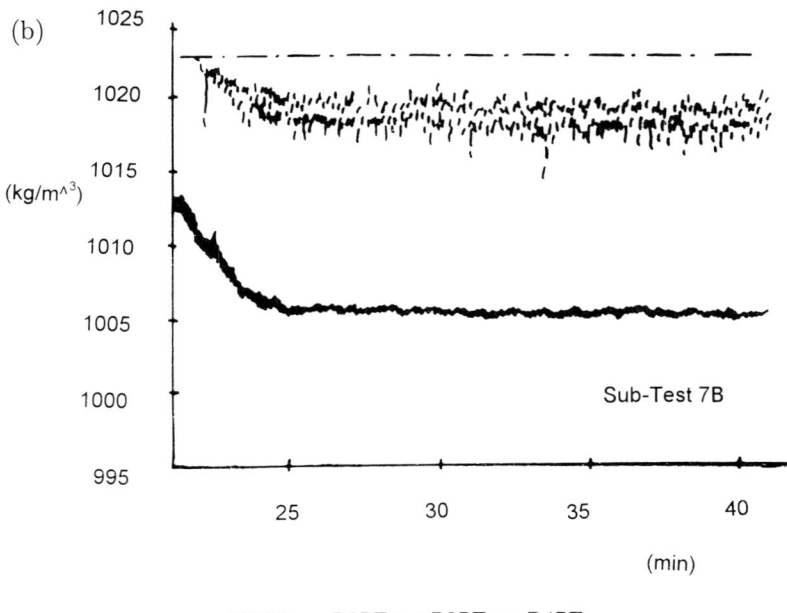

Figure 19. (a) Density changes for channels R1BT-R4BT during sub-test 7A, and (b) Density charges for channels R1BT-R4BT during sub-test 7B

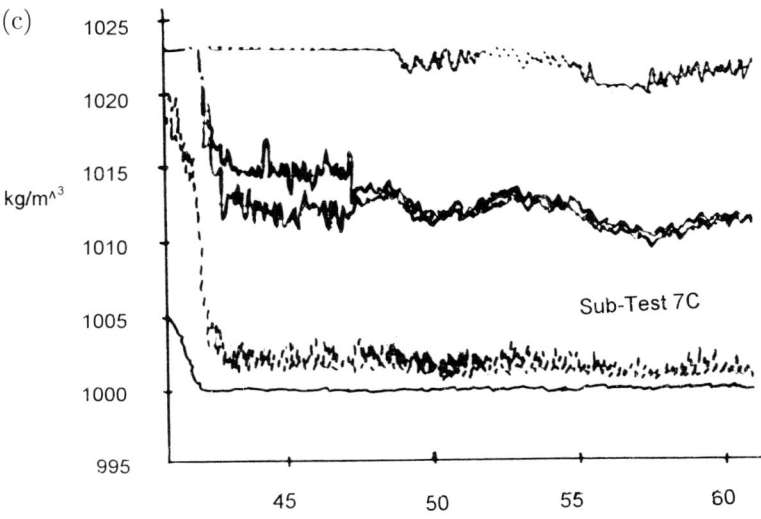

····· R1BT——· R2BT— —R3BT —— R4BT

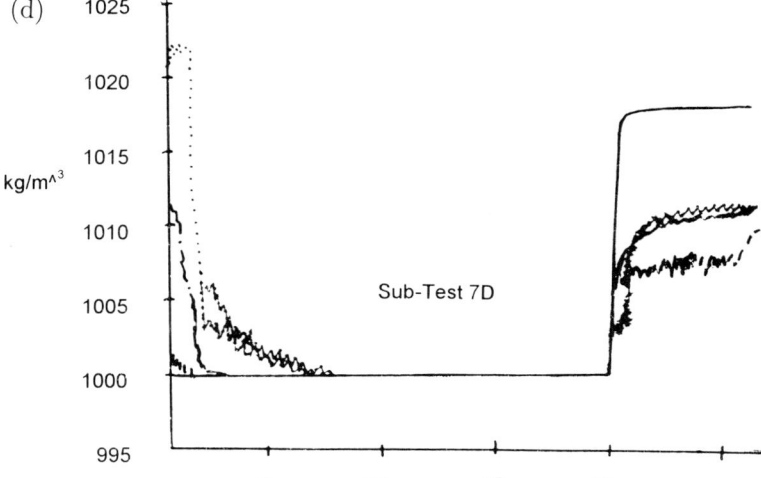

(min)

····· R1BT——· R2BT— — R3BT——R4BT

Figure 19. (c) Density charges for channels R1BT-R4BT during sub-test 7C, and (d) Density charges for channels R1BT-R4BT during sub-test 7D

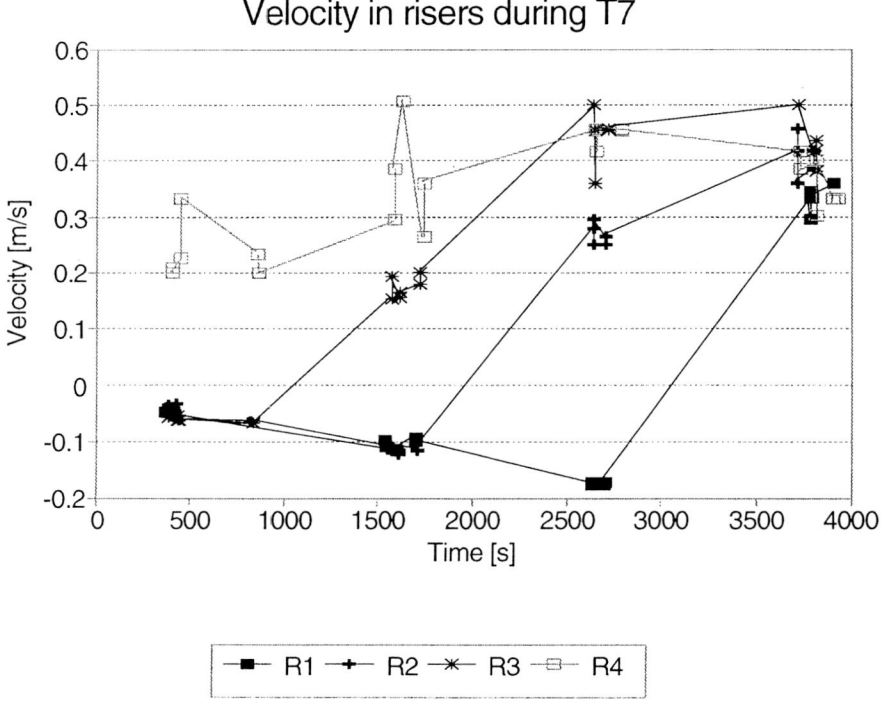

Figure 20. Riser velocities during test number T7

increased with the increase in sub-test progression. For sub-test T7A which was started with a gradual release of freshwater, risers R_1, R_2 and R_3 stayed intrusive with only R_4, the most landward riser, discharging. As shown in Figure 19, it did maintain a mixed flow from around the seventh minute and its velocity was very high.

The freshwater input was increased to $0.318l/s$ and after a flow readjustment period, R_3 slowly began to discharge but with a solution of very high seawater concentration whilst R_4 almost gained full fresh water discharging condiitions. The velocity plot for sub-test 7B shows an increase for R_3 whilst R_1 and R_2 showed a gradual decrease (Figure 20). The arrested wedge reduced in height and volume as expected and it took a steeper profile with more turbulent flow at the interface. Profiles are plotted in Figure 21.

The third increase in fresh water in-flow reversed the intensive flow in R_2 (Figure 20). When the peak flow was applied in sub-test 7D, all intrusion was cleared as well as the stratification.

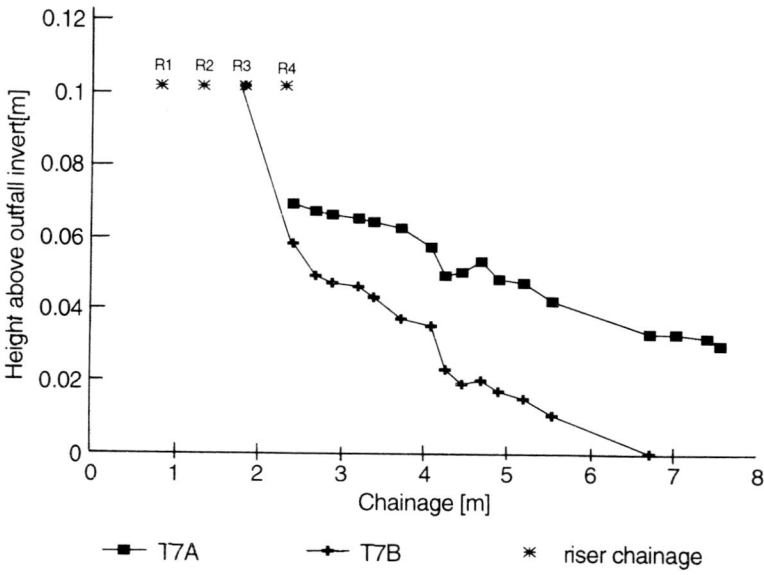

Figure 21. Wedge profiles during test number T7

4 Conclusions

The following conclusions were obtained from the work reported herein.

1. The presence of a stationary salt wedge significantly changes the velocity distribution within the overlaying freshwater flow.

2. In many experiments the interface was not sharp or distinct, in some cases covering a considerable proportion of the flow. This is an illustration of the entrainment mechanism at work in interfacial regions of the two fluids.

3. Mixing within the "freshwater" layers in outfalls seems to be caused mainly by advection.

4. Entrainment at the interface can be considerable for small values of the bulk Richardson number.

5. Approximate calculations show that the centreline interfacial shear stresses are bigger than those at the solid boundary.

6. Under positive flow increments, risers are systematically purged in the seawards direction.

Acknowledgements

The authors would like to acknowledge the financial support of the Science and Engineering Research Council (SERC) in providing Research Grant No. GR/G21841, under which many of the reported experiments were conducted. Thanks are also due to K. Whyte, E.V. Wellen and S. Charavanamuttu for conducting some of the experiments as part of their Masters studies.

Bibliography

1. Ali, K.H.M. (1976). "Prediction of flow development on spillways". *Journal of the Hydraulics Div.*, pp. 1401–1404.

2. Ali, K.H.M. and Jaefar-Zadeh, M.R. (1989). "Circulation and mixing in a stratified reservoir". *Journal Hydraulic Research*, **27**, p. 683.

3. Burrows, R., Ali, K.H.M., Davies, P.A. and Wose, A.E. (1996). "Studies of saltwater purging from a model sea outfall diffuser". *Proc. Instn. Civ. Engrs. Wat. Marit. and Energy*, **118**, pp. 77–87.

4. Charavanamuttu, S. (1994). "Salt intrusion and energy losses in long sea outfalls". *M.Sc. Thesis*, University of Liverpool.

5. Charlton, J.A., Davies, P.A. and Bethune G.H.M. (1987). "Sea water intrusion and purging in multi-port sea outfalls". *Proc. Instn. Civ. Engrs.*, Part 2, pp. 263–274.

6. Costa, M.V. (1995). "Three-dimensional modelling of cohesive sediment transport in estuarine environments". *Ph.D. Thesis*, University of Liverpool.

7. Dyer, K.R. (1973). "Estuaries, a physical introduction". John Wiley and Sons, London.

8. Fischer, H.B., List, E.J., Koh, R.C.Y., Imberger, J. and Brooks, N.H. (1979). "Mixing in inland and coastal waters". Academic Press, New York, pp. 28–54 and 104–139.

9. Harleman, D.R.F. (1961). "Stratified flows". *Handbook of Fluid Dynamics*, Section 26, Editor: V. Streeter, McGraw-Hill.

10. Keulegan, G.H. (1949). "Interfacial instabilities and mixing in stratified flows". *Journal of Research of the Natural Bureau of Standards*, **43**, pp. 487–500.

11. Masters, G.M. (1991). "Introduction to environmental engineering and science". Prentice Hall, Englewood Cliffs, New Jersey, pp. 1–34.

12. Metcalf and Eddy Inc. (1991). "Wastewater engineering". Third Edition, McGraw-Hill Inc., pp. 1213–1240 and 1265–1273.

13. Moore, M.N. and Long R. (1970). "An experimental investigation of turbulent stratified shearing flows". *Journal of Fluid Mechanics*, **49(4)**, pp. 635–655.

14. Mort, R.B. (1989). "Investigation into the effects of wave action on long sea outfalls". *Ph.D. Thesis*, University of Liverpool.

15. Neville-Jones, P.J.D, Darling, C. and McNamara, M. (1987). "Hydraulic performance of long sea outfalls". *Report ER 216E*, Water Research Centre, Swindon.

16. Rouse, H. (1961). "Fluid mechanics for hydraulic engineers". Dover Publications Inc., New York, p. 1.

17. Spencer, A.P. (1990). "An experimental study of saline intrusion and sediment transport in long sea tunnelled outfalls". *B.Eng. Thesis*, University of Liverpool.

18. Whyte, K. (1995). "Hydraulics of flow in marine outfall diffuser manifolds". *M.Sc. Thesis*, University of Liverpool.

19. Wose, A.E. (1992). "Saline intrusion and sedimentation in sea outfalls". *Ph.D. Thesis*, University of Liverpool.

20. Wose, A.E., Burrows, R., Davies, P.A. and Ali, K.H.M. (1934). "Purging studies on sea outfall diffuser system". *Report*, University of Liverpool.

Fluid Mud Transport in the Laboratory and in the Field

K.H.M. Ali*, M. Crapper** and B.A. O'Connor*

**Department of Civil Engineering, University of Liverpool, and* ***Department of Civil and Environmental Engineering, University of Edinburgh*

Abstract

The paper gives the results of an experimental and theoretical investigation into the formation and transport of fluid mud. Various initial mud concentrations, bed slopes and ambient fluid velocities and directions were investigated. Fluid mud velocity, thickness, elevation and density were measured at different longitudinal sections for various times.

The dependence of fluid mud viscosity on concentration was established. Also, entrainment of fluid mud into the ambient fluid was considered.

The effect of wave action on the transport of cohesive sediment was also studied.

1 Introduction

Studies of sediment dynamics in estuaries have commonly been performed in the past through field studies and physical and numerical modelling. Such studies have generally been carried out using an appropriate combination of the above methods, their choice being conditioned by the purpose of this study, the costs involved and, in the case of engineering studies, the desired degree of accuracy of the project.

Estuarine cohesive sediment, commonly called mud, is composed primarily of silt and clay. Mud contains a large proportion of very small particles which have a large specific area such that the effect of the surface physico-chemical forces becomes as important as the effect of gravity forces [9].

The ability to predict the movement of cohesive sediments has a significant economical and ecological importance in the development of new engineering works and the maintenance of existing installations [1]. Furthermore, this ability is crucial in the understanding of the distribution of certain pollutants, in particular heavy metals which are absorbed on to the clay and silt particles.

Fluid mud is a dense suspension containing a concentration of mud flocs which is high enough to change significantly the physical properties of the mud/water mixture compared with that of clear water with the same salinity and temperature [9]. Fluid mud is formed under a variety of conditions. Once formed it may flow under the influence of gravity and hydrostatic forces and then settle and dewater in navigation channels and berths, causing significant rates of siltation over and above that due to settlement directly from suspension.

The paper describes an experimental and theoretical investigation into the formation and transport of fluid mud. The effects of bed slope, initial mud concentration and ambient current speed and direction were investigated. Wave action on cohesive sediment transport was also studied and entrainment of fluid mud into the ambient fluid was also considered.

2   Experimental models

Two experimental models were used to study the formation and transport of fluid mud:

1. An open channel 0.15 m wide, 1.2 m high and 2.4 m long was used. This narrow channel was pivoted at one end, with a mechanical jack attached to the other end. This channel was constructed of perspex to help visual observation. For several initial concentrations, bed slopes of 1/20, 1/10 and 1/5 were used. Fluid mud velocity, depth, concentration and elevation were obtained for each slope and initial concentration. All of these results were obtained for an initially still water conditions. Details of the experimental procedure and the results are given in Reference [1].

2. The Race Track Flume (RTF) was used for most of the detailed experiments (Figures 1 and 2). This is a recirculating flume consisting of two semi-circular bends of internal radius of 0.75 m joined by 4 m long straight sections. The flume's width is 0.305 m and the maximum working depth of water when the flume is horizontal is 0.575 m. The flow is driven by a toothed belt which rests just in the water surface in the rear straight section of the flume. The straight working sections in the race-track flume allow secondary flows which are developed in the bends to decay considerably. The Race Track Flume can also be tilted to a maximum slope of 1/15.61 in the working section. This feature is essential for the study of the flow of fluid mud down a slope under the influence of gravity.

   In some of the experiments, an artificial channel was built inside the RTF to give a bed slope of 1/10. The flume was modified to give a reasonable approach to the upstream section of the new channel.

A considerable number of experiments were carried out using the Race Track Flume. The effects of current speed, initial mud concentration and channel slope on the thickness, velocity and elevation of the fluid mud layer were investigated. The Race Track Flume was prepared by filling it with saline water of the appropriate density. The correct amount of mud was introduced so as to give the desired average concentration when in suspension. The motor of the toothed belt was then set to full speed and the flume was run for about half an hour to ensure

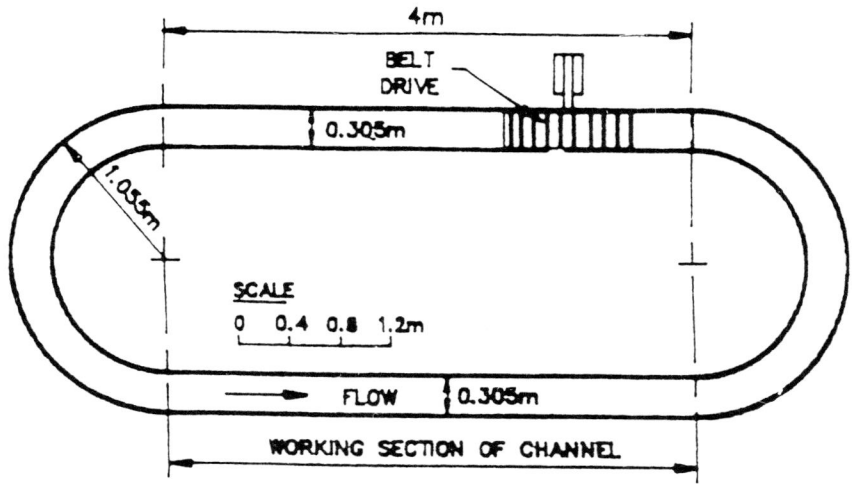

Figure 1. Plan of the Race Track Flume

Figure 2. Side view of the Race Track Flume in tilted position

that all the mud was in suspension and to give it enough time to flocculate. The motor speed would finally be reduced to the desired current value for the fluid mud test.

Measurement of fluid mud characteristics commenced as soon as the mud flocs began to settle. Rhodamine-B dye was injected in vertical lines to enable the fluid mud velocity profiles to be measured and a CCD video camera with a zoom lens and SVHS recorder were used to permanently record the process for later analysis. Further measurements of fluid mud velocity, thickness and elevation were made during the course of each experiment.

Concentrations of fluid mud were determined by sucking samples from the fluid mud layer using a syringe and a thin glass tube connected to a vernier and set at the height of the centre of the moving fluid mud layer. Samples were either stored in bottles and later weighed using a specific gravity bottle or passed through a Paar oscillating-U-tube density meter.

Some experiments were also conducted to study the interaction between surface water waves and a soft mud bed. A wave generating paddle was fitted to the RTF and the toothed belt drive was removed. Reasonably regular waves were obtained. These retained their identity over a considerable period of time.

Wave records were obtained using a series of capacitance wave gauges located along the flume. Continuous signals corresponding to surface water levels were recorded and digitized using a data acquisition system. The FOSLIM [3] probe was used to measure the concentration of the suspended sediment. Concentrations higher than the FOSLIM limit were obtained by passing samples collected using a syringe through a PAAR oscillating-U-tube density meter.

China, clay and Mersey mud were used in these experiments. The mud beds were subjected to different frequencies and wave heights.

3 Theoretical considerations

3.1 Velocity of fluid mud

The approximate analysis described here was obtained by Ali and Georgiadis and is a modification of the work of Harleman [5] and Odd and Rodger [9]. This analysis applies to the unsteady flow of a thin layer of liquid mud which has the physical properties of a Bingham Plastic Fluid.

It is assumed that the depth of the fluid above the liquid-mud layer is large compared with that of the liquid-mud layer. Therefore, the induced velocities in the upper layer may be neglected. The assumed velocity and shear stress distribution are shown in Figure 3. It is also assumed that the region of interest is far enough from the upstream end of the channel to ensure that the velocity distribution is fully developed.

Following the analysis of Viessmall et al. [12] and using Newton's second law, Ali and Georgiadis [1] obtained for $t \geq t_o$ to

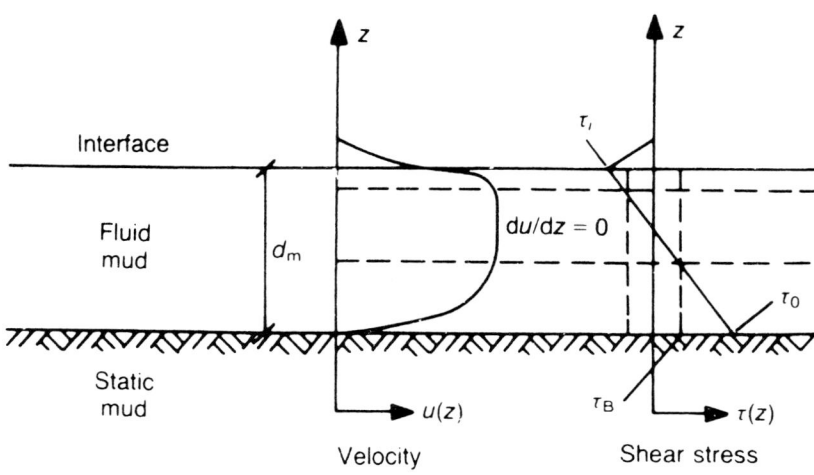

Figure 3. Assumed velocity and shear stress distribution of a layer of fluid mud flowing down a slope in still water

$$\Delta \rho_m g d_m \Delta x B S - (\tau_o + \tau_i) B \Delta x = - \left(\rho_m d_m B \Delta x \frac{d u_m}{d t} + \rho_m u_m i \Delta x \right) \quad (3.1)$$

where d_m = depth of fluid-mud layer; u_m = velocity of fluid-mud layer; ρ_m = density of fluid-mud; g = acceleration due to gravity; S = channel slope; Δx = length of section; $\Delta \rho_m$ = density difference between the liquid-mud layer and the surrounding fluid; $d u_m / d t$ = acceleration of the mud-layer flow; τ_o = boundary shear stress; τ_i = interfacial shear stress; B = channel width and i = the spatial and time variation of the lateral flow of mud into the liquid-mud layer (which equals the vertical settling of suspension into the fluid-mud layer). It was assumed that this lateral inflow enters the moving liquid mud with no velocity component in the direction of flow. In Equation (3.1), it was also assumed that $\delta u_m / \delta t \geq u_m \ \delta u_m / \delta x$.

The boundary shear stress, τ_o, was assumed to be given by

$$\tau_o = \tau_B + \mu_m \frac{d u_m}{d} = \tau_B + \frac{\rho_m f u_m^2}{8} \quad (3.2)$$

where τ_B = Bingham yield strength; f = friction factor and μ_m = dynamic viscosity of fluid mud.

For steady uniform motion of fluid-mud, $d u_m / d t = 0$ and $i = 0$. Equation (3.1) becomes

$$u_m = \left[\frac{8\Delta\rho_m g d_m S - 8\tau_B}{\rho_m (1 + d) f} \right]^{1/2}.$$ (3.3)

Later, Equation (3.1) will be used to calculate the viscosity of fluid mud in the laboratory and in the field.

3.2 Distribution of concentration of fluid mud

A study of the velocity-distributions for fluid mud formed in the laboratory and in the field reveals that, often, the flow is similar to that for a Newtonian fluid between parallel plates [5]. An approach for the distribution of fluid-mud concentration can be obtained using the velocity-profile obtained by Ippen and Horleman [7].

Newtonian Velocity Distribution

Ippen and Harleman [7] developed the following relationships (Figure 4):

$$\frac{u}{\bar{u}} = 1 + 2\frac{z}{h_2} - \frac{1}{2J} \left[\left(\frac{z}{h_2} \right)^2 + \frac{1}{3}\left(\frac{z}{h_2} \right) - \frac{1}{12} \right]$$ (3.4)

$$\frac{u_i}{u_{\max}} = \frac{12J - 1}{12J^2 + 4J + \frac{1}{3}}$$ (3.5)

$$\frac{u_i}{\bar{u}} = 2 - \frac{1}{6J}$$ (3.6)

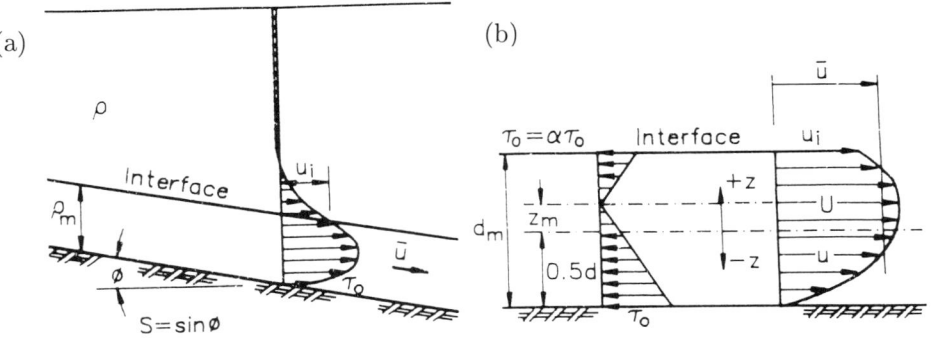

Figure 4. (a) Steady, uniform flow in lower-layer fluid, and (b) Shear velocity distribution in lower-layer fluid

$$\frac{z_m}{h_2} = 2J - \frac{1}{6}. \tag{3.7}$$

(Note: $\bar{u} = u_m$; $h_2 = d_m$)

where

$$J = \frac{F^2}{RS}; \quad F = \frac{\bar{u}}{\sqrt{g'h_2}}$$

and

$$R = \frac{\bar{u}h_2}{\nu}.$$

$$g' = g\Delta\rho/\rho; \quad S = \text{bed slope.} \tag{3.8}$$

For laminar flow, $J = 0.14$, which gives:

$$\frac{u}{\bar{u}} = 1.2976 + 0.8095 \left(\frac{z}{h_2}\right) - 3.5714 \left(\frac{z}{h_2^2}\right). \tag{3.9}$$

$$\frac{u_i}{u_{\max}} = 0.6026 \tag{3.10}$$

$$\frac{z_m}{h_2} = 0.1133 \tag{3.11}$$

$$\frac{u_i}{\bar{u}} = 0.8095. \tag{3.12}$$

Note:

z is measured upwards from mid-section.
We will transform the axis and use y measured upwards from the bed, i.e.:

$$y = z + \frac{h_2}{2}. \tag{3.13}$$

Substituting for z from Equation (3.13) into Equation (3.9) gives:

$$\frac{u}{\bar{u}} = 4.3809 \left(\frac{y}{h_2}\right) - 3.5714 \left(\frac{y}{h_2}\right)^2. \tag{3.14}$$

Obtaining an Expression for the Mixing Coefficient

We will assume that the shear stress for fluid mud is given by:

$$\tau = \epsilon\rho\frac{du}{dy} \tag{3.15}$$

where ϵ = momentum transfer coefficient and du/dy is the velocity gradient.

The shear stress distribution is represented in Figure 4. Similar triangles give:

$$\tau = \frac{(0.6133 \, h_2 - y)}{0.6133 \, h_2} \tau_o. \tag{3.16}$$

From Equation (3.14), the velocity gradient is given by:

$$\frac{du}{dy} = 4.3809 \frac{\bar{u}}{h_2} - 7.1428 \frac{\bar{u} y}{h_2^2}$$

or

$$\frac{du}{dy} = \frac{4.3809 \, \bar{u}}{h_2^2} (h_2 - 1.6304 \, y). \tag{3.17}$$

Following Hydraulics Research Limited [6], the bulk density of a suspension of muddy saline water in terms of its temperature, salinity and mud concentration is given by:

$$\rho = 1000 + (0.797 - 0.001875 \, T)S - 1000 \left[\frac{A(T-4)}{277} \right]^B + 0.62 \, C \tag{3.18}$$

where:

$$
\begin{aligned}
T &= \text{water temperature } (^\circ C) \\
S &= \text{salinity (ppt)} \\
C &= \text{suspended solids concentration (ppt).}
\end{aligned}
$$

For $12^\circ C < T < 34^\circ C$

$$A = 0.562, \, B = 1.85.$$

For a given T and S, Equation (3.18) becomes:

$$\rho = (\alpha_1 + \beta_1 C)\rho_o. \tag{3.19}$$

Substituting for τ, du/dy and ρ in Equation (3.15) gives:

$$\epsilon = 0.2283 \frac{u_*^2 h_2}{\bar{u}(\alpha_1 + \beta_1 C)} \tag{3.20}$$

where:

$$u_* = \sqrt{\frac{\tau_o}{\rho_o}}. \tag{3.21}$$

We will assume that the sediment transfer coefficient, ϵ_s, is equal to the momentum transfer coefficient, ϵ.

The steady state sediment transport equation for a suspension of fluid mud is:

$$CW + \epsilon_g \frac{dC}{dy} = 0. \tag{3.22}$$

The terminal velocity, W, is assumed to be given by:

$$W = W_o(1 - k_2 C)^{\beta_2} \tag{3.23}$$

where $k_2 = 0.008$, $W_o = 2.6$ mm/s, $\beta_2 = 4.65$ and C is measured in g/ℓ.

Substituting for ϵ_s from Equation (3.20) and for W from Equation (3.23) into Equation (3.22) and integrating, gives:

$$\lambda y = -\int \frac{dC}{C(\alpha_1 + \beta_1 c)(1 - k_2 C)^{\beta_2}} + K \tag{3.24}$$

where:

$$\lambda = +4.3802 \frac{W_o \bar{u}}{u_*^2 h_2}. \tag{3.25}$$

We integrate the RHS of Equation (3.24) by using:

$$\phi = 1 - k_2 C; \quad C = \frac{(1 - \phi)}{k_2}; \quad dC = -\frac{1}{k_2} d\phi.$$

Substituting and simplifying, we obtain:

$$\int \frac{dC}{C(\alpha_1 + \beta_1 C)(1 - k_2 C)^{\beta_2}} = \alpha_5 \int \frac{d\phi}{1 + \alpha_7 \phi - \alpha_6 \phi^2)\phi^{\beta_2}} \tag{3.26}$$

where:

$$\alpha_3 = \alpha_1 + \frac{\beta_1}{k_2}$$

$$\alpha_4 = -\frac{\beta_1}{k_2}$$

$$\alpha_5 = -\frac{1}{\alpha_3}; \quad \alpha_6 = \alpha_4/\alpha_3 \quad \text{and} \quad \alpha_7 = \alpha_6 - 1.$$

The integration of Equation (3.26) can be obtained using a standard solution for the following function:

$$\int \frac{dx}{X^{n+1} x^m}$$

where:

$$X = a + bx + cx^2, \quad q = 4ac - b^2.$$

In our case $n = 0$ and $m = 4.65$.

We obtain the solutions form $= 4$ and $m = 5$ and interpolate for $m = 4.65$.

Solution

From Equation (3.24):

$$\lambda_y = -F(\phi) + K \qquad (3.27)$$

where:

$$F(\phi) = \int \frac{d\phi}{(1 + \alpha_7\phi - \alpha_6\phi^2)\phi^{4.65}} \qquad (3.28)$$

when $y = a$, $C = C_a$ and $F(\phi) = F(\phi_a)$, and when $y = H$, $C = C_H$ and $F(\phi) = F(\phi_H)$.

Equation (3.28) finally becomes:

$$\frac{y - a}{H - a} = \frac{F(\phi) - F(\phi_a)}{F(\Phi_H) - F(\phi_a)}. \qquad (3.29)$$

Equation (3.27) can also be written in the following form:

$$\lambda y = -F(\phi) + [\lambda a + F(\phi_a)]. \qquad (3.30)$$

λ can be obtained from the graph of $F(\phi)$ against y.

We used $\rho_o = 1000$ kg/m^3, $\alpha_1 = 1.024$, $\beta_1 = 0.00052$, $k_2 = 0.008$.

4   Results

Figure 5 shows fluid mud velocity distributions for various times, motor speeds and current directions. These results were obtained using the Race Track Flume with a slope of 1/15.61. These distributions are very similar to those for flow between parallel plates (see Ippen and Harleman [7] and Harleman [5]).

Figures 6–9 show the variation of fluid mud mean velocity, thickness, elevation and density with time. These results were obtained for a Race Track Flume slope of 1/10 and are given for four longitudinal sections.

Figures 10 and 11 show some of the fluid mud velocity distribution obtained by H.R. Wallingford [6] from Parrett Estuary. Also shown in Figure 11 is the theoretical curve for laminar flow between parallel plates.

Using the experimental fluid mud results, kinematic fluid mud viscosities were calculated using Equation (3.1) and are given in Figure 12. This figure shows considerable scatter but it does reveal a strong dependence of the kinematic viscosity on fluid mud concentration.

These results can be expressed by the following relationship:

$$\nu = 4.256 \times 10^{-4} C^{2.405} \qquad (4.1)$$

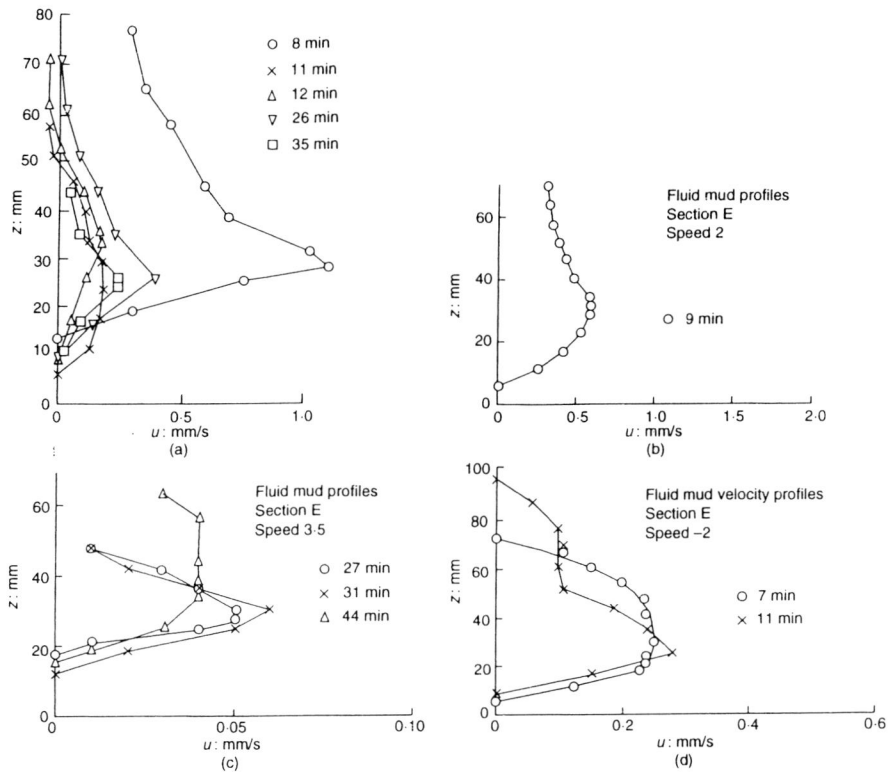

Figure 5. Fluid mud velocities for various paddle speeds

(ν is in cm^2/s and C in g/ℓ).

Entrainment

Costa [2] stated that the physical process leading to mass removal from fluid mud layers by turbulent shear (entrainment) has been recognised to be different from those leading to mass removal from a settled bed showing a structured solids matrix (surface erosion or mass erosion, the latter following bed liquefaction, fluidisation or bed failure). In estuaries, after slack water, rapid flow reversal causes the top layer of fluid mud, not sufficiently dewatered to form a cohesive bed, to be entrained by turbulent flow in a similar manner to the entrainment of a layer of salt water underneath flowing fresh water [8].

Costa [2] stated that a general description of salt water entrainment by an upper layer of fresh water flowing at higher velocity can be found in Dyer [4].

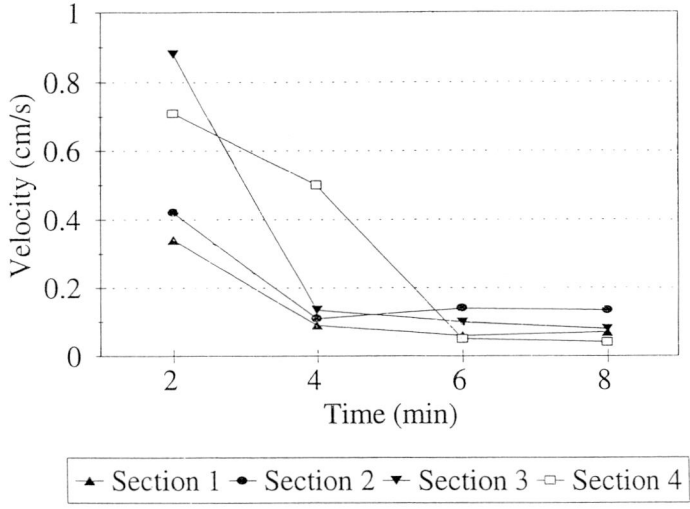

Figure 6. Variation of fluid mud velocity with time at four longitudinal sections (1/10 slope)

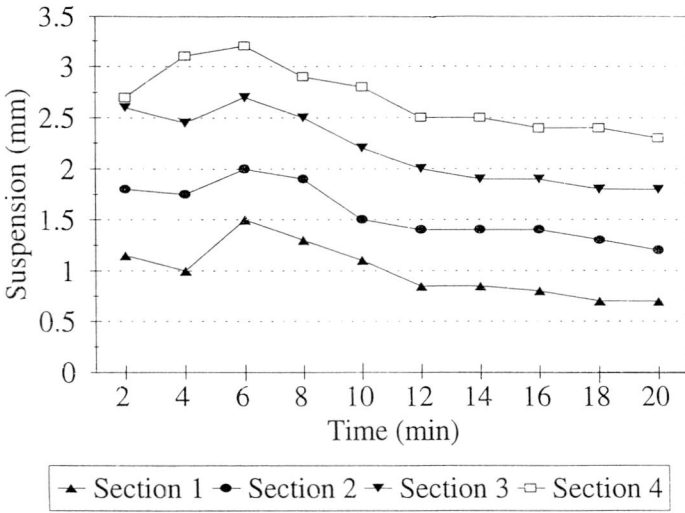

Figure 7. Variation of fluid mud thickness with time (slope = 1/10)

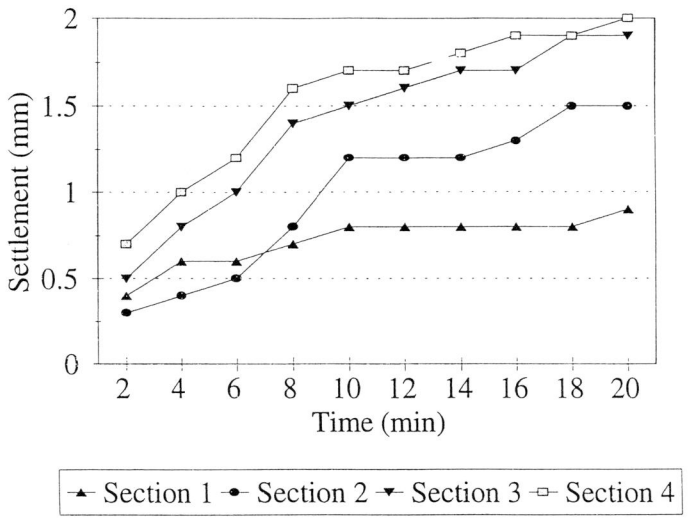

Figure 8. Variation of depth of settled bed thickness with time

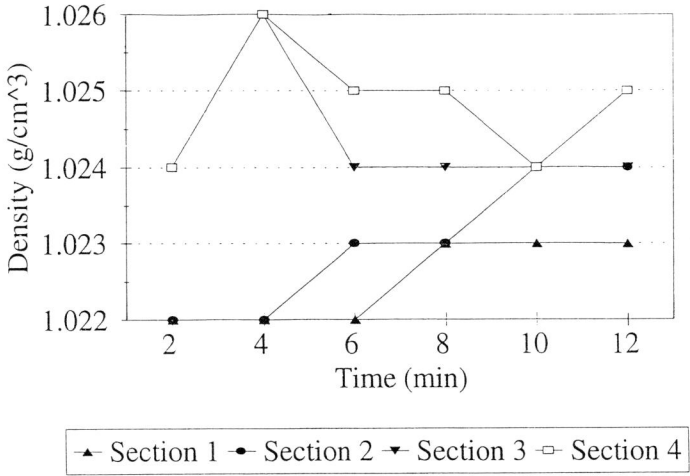

Figure 9. Variation of fluid mud density with time for four longitudinal sections (slope = 1/10)

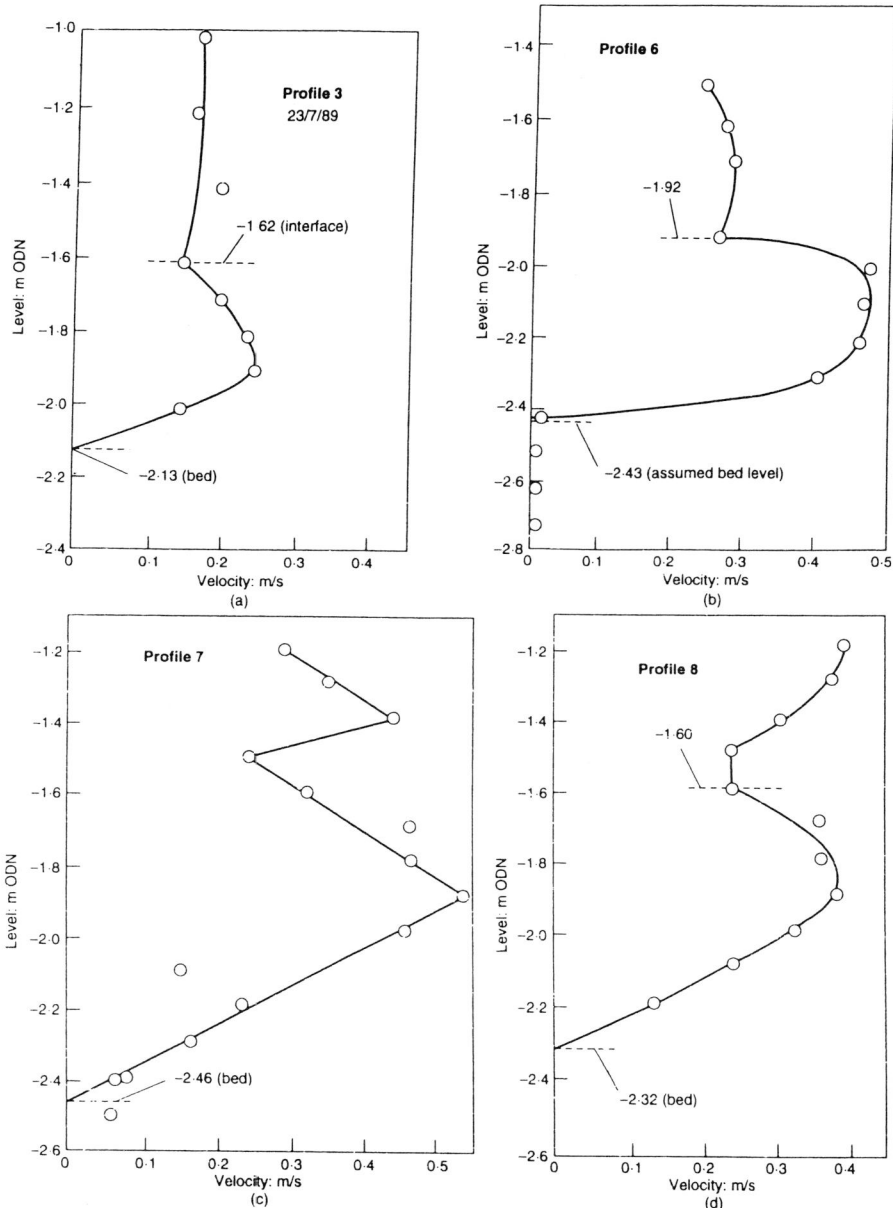

Figure 10. Field velocity distributions

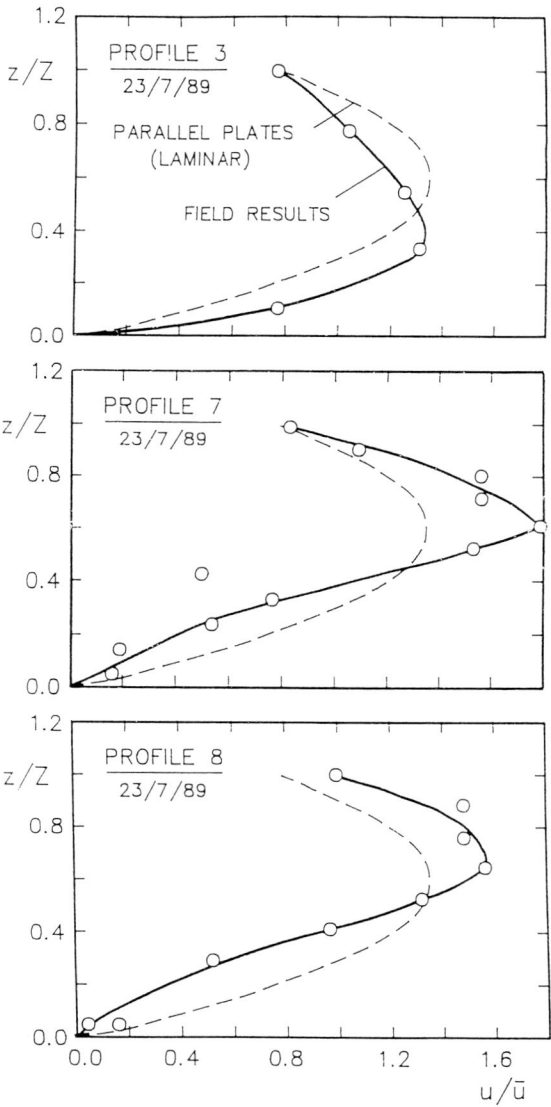

Figure 11. Comparison between field and Harleman's velocity distributions

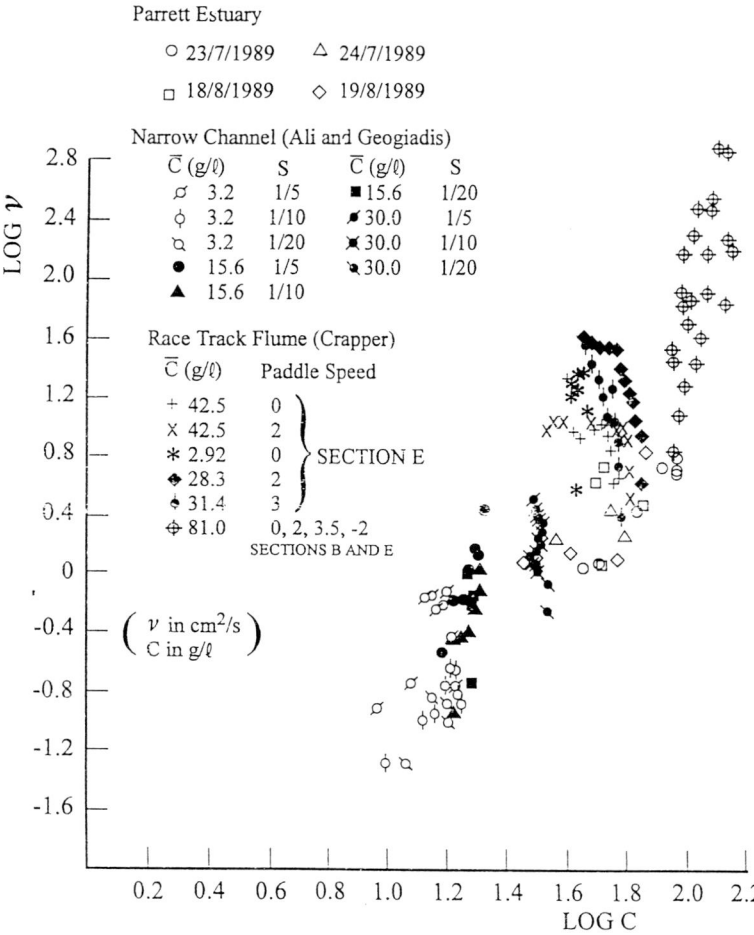

Figure 12. Variation of log ν with log C

When shear becomes sufficiently intense, the interface between layers becomes disturbed by waves, which grow higher with increasing shear. Eventually the waves break and globules of the lower, denser layer are ejected from the wave crests into the lighter fluid above, in a strictly on way process. This characterisation of the phenomenon generally matches the description of the same phenomena observed by Mehta and Srinivas [8] for the case of fluid mud and overlying low concentration layers, in a race-track type flume. However, the latter authors were able to give a more detailed description of the phenomenon for the case of their experimental conditions. At low Richardson numbers ($R_i = (\Delta\rho_m/\rho)gd_m/u_m^2$), $R_i < 5$, the authors found that entrainment appeared to be turbulence domi-

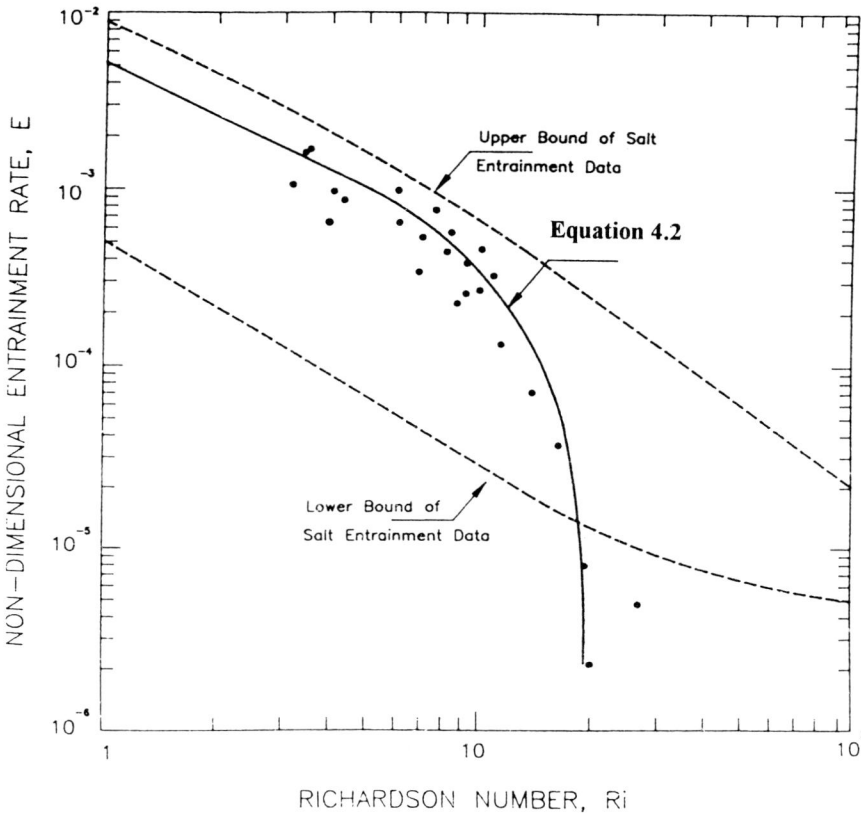

Figure 13. Non-dimensional entrainment versus Richardson number (adapted from Costa)

nated, with a rather diffuse and highly irregular interface. As R_i increased above 5 the interface became better defined and convoluted with large, irregular, undulations, entrainment being dominated by wave breaking, while interspersed, solitary type waves (which seemed to decay without breaking) were also observed. The frequency and amplitude of the disturbances decreased with increasing R_i and, as it increased above 10, the disturbances could be seen to grow slightly in amplitude, sharpen into non-linear crests and disappear suddenly, as the *roller-action* of eddies sheared off the crests. For values of R_i beyond 20, the intensity of entrainment appeared to taper off rapidly with increasing Richardson number. Differences between the entrainment of salt water and cohesive sediment are also noted by Mehta and Srinivas [8]. These are related to the effects of particle settling, cohesion and viscosity of the lower layer, although quantita-

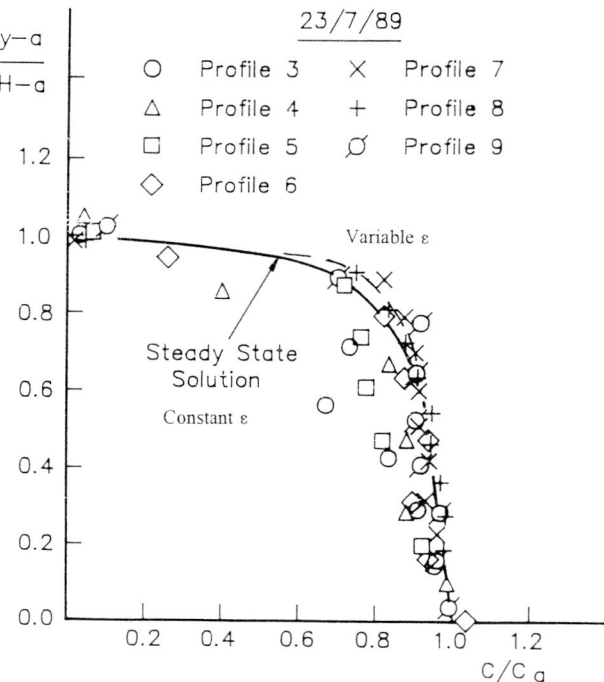

Figure 14. Dimensioness distribution of fluid mud concentration using field data

tive information about the effects of such parameters is scarce. It is concluded by the same authors that mud and salt are entrained at similar rates for low Richardson numbers but that, for increasing R_i (i.e. greater than about 10), mud entrainment rates decrease rapidly relative to those of salt (see Figure 13).

Mehta and Srinivas [8] conducted experiments using three types of sediment; kaolinite, a kaolinite/volclay bentonite mixture and natural mud from a lake. Entrainment was observed to fall rapidly for R_i greater than 10 and the non-dimensional entrainment rate E became almost independent of the Richardson number for R greater than about 15. The best fit resulting from the improved data set was now (see Figure 13)L

$$E = AR_t^{-1} - DR_t \tag{4.2}$$

where $A = 0.0052$ and $D = 0.000016$ (according to the same authors coefficient D reflects mud properties, i.e. settling and cohesion).

Most of the values of R_i for the present experiments were much bigger than 10. The Parrett estuary results, however, resulted in values of about 2–8.

Figure 14 shows typical dimensionless concentration distributions. These values are obtained from the field measurements obtained by Hydraulic Research Limited from the Parrett Estuary [6]. Also given are the theoretical curves for constant and variable mixing coefficient ϵ.

Clearly, there is excellent agreement between theory and experiment for large values of t (high profile numbers). Also, the assumption of a variable ϵ has little effect.

4.1 Experiments with waves

Several experiments were conducted for different wave periods, lengths and water depths. These were carried out using a rigid bed. Velocities near the bed were calculated using linear wave theory.

Using these results an optimum water depth and wave period were selected for maximum sediment movement.

China clay was next added to the saline water solution. It was allowed to settle along the bed of the channel for such a time that ensured there wasn't any in suspension (usually left overnight before commencement of experiment). Before setting the wave paddle in motion density samples of the saline water were obtained at different depths above the bed.

For the selected wave paddle setting, waves were started and observations were taken of the effect of the wave action in stirring the china clay from the bed. Density readings were taken at different time periods and elevations above the bed.

Once the wave paddle was set in motion, there was an initial period of about three minutes where the china clay didn't show any movement from the bed. Gradually, the surface motion caused the clay particles to "rise". This was noticeable throughout the whole length of the working section.

After approximately ten minutes, some layered bands of china clay were discernible within the depth of saline water.

1. Immediately below the water surface, there was little evidence of china clay particles as the water hadn't changed colour since before the wave machine had been started. This band was only 1–2 cm in depth.

2. Below this, and about 3/4 up from the channel bed, a very well defined band of china clay was visible. This layer was 3–4 cm in depth and its intensity developed as the time increased.

3. A final layer developed from the channel bed to the underside of band 2. China clay particles were clearly visible for the whole depth, with the concentration gradually increasing from top to bottom.

There was lateral movement of the china clay particles. Their movement was confined to the oscillatory motion of the surface waves and the particles moved in the direction of wave propagation.

Despite the rise of china clay particles, there always remained a shallow depth of chin clay on the channel bed.

Sleath [11] stated that because of the very low fall velocities associated with cohesive sediments, it is usual to assume that, for steady flows, the concentration is more or less uniform over the whole depth of flow over the bed layer. Owen [10] suggested the following formula for the concentration C of sediment in suspension at any time t:

$$C = C_0 \exp\left[-\frac{AW}{d}\left(1 - \frac{\tau_o}{\tau_d}\right)t\right] \qquad (4.3)$$

where C_0 = concentration at time $t = 0$; A = constant; W = fall velocity; d = mean depth of water; τ_o = shear stress on the bed; τ_d = critical shear stress for deposition. It is usually accepted that the critical shear stress at which particles are deposited on the bed is less than that required for their erosion [10].

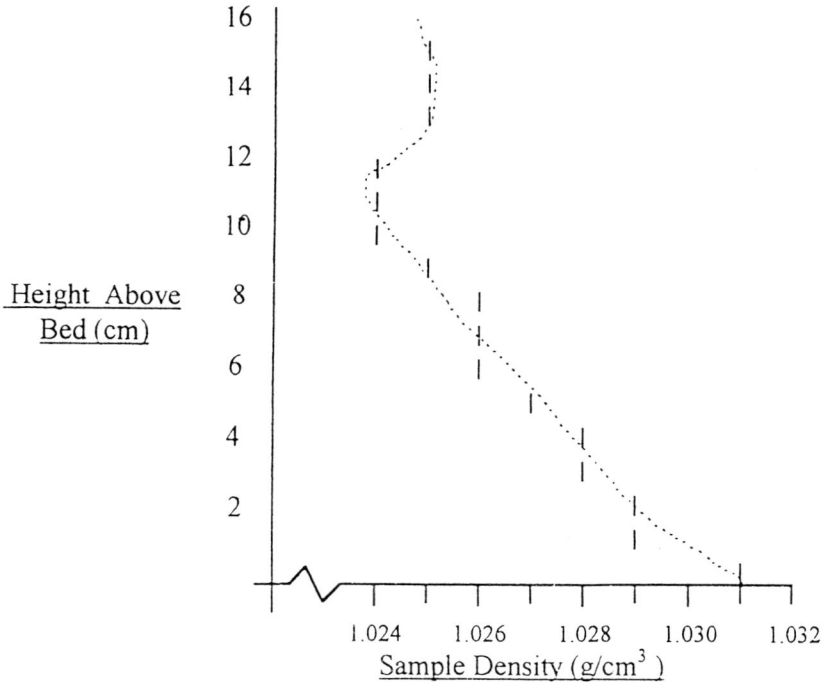

Figure 15. Changes in mud density caused by wave action (paddle setting 4, water depth 0.175 m)

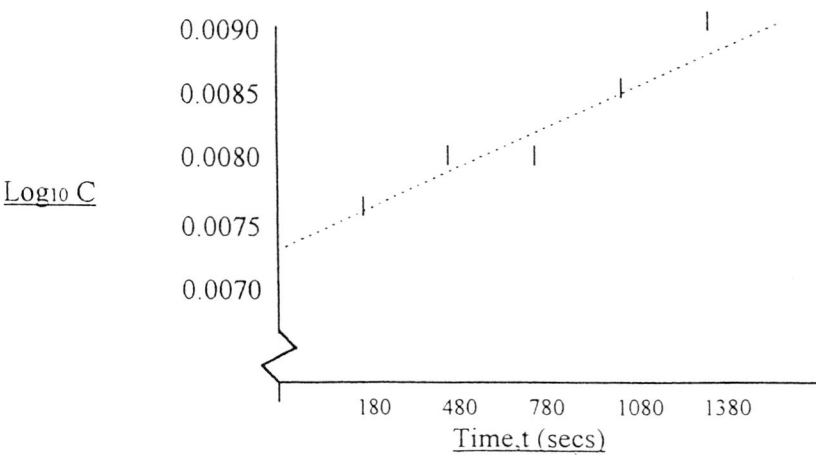

Figure 16. Variation of log (against t for wave condition given in Figure 15 (height above the bed = 4 cm)

Figure 15 shows a typical plot of the variation of density with height above the bed. Figure 16 shows a plot of log C against t for concentrations measured 4 cm above the original bed. This plot results in a reasonable straight line thus verifying Equation (4.3).

5 Conclusions

The main conclusions arising from the consideration of fluid mud flow are as follows:

1. The clear water, hindered settling fluid mud and settled bed layers identified by Ali and Georgiadis and other researchers are also discernible in the Race Track Flume Experiments. It was, however, clear that the water column in the RTF was in fact a single entity with continuously varying properties.

2. The fluid mud flow pattern identified by Ali and Georgiadis of early peaks in velocity and thickness followed by a gradual decline was also present in the Race Track Flume experiments.

3. Fluid mud flow in the Race Track Flume appeared to be entirely laminar in nature.

4. Fluid mud viscosity determined from the narrow channel, Race Track Flume and the field measurements were found to increase with the increase in concentration.

5. Newtonian fluid velocity profiles give a reasonable estimate of fluid mud velocity profiles measured in the Race Track Flume with low concentration. Agreement was also reasonable with the Parrett estuary field results.

6. Calculations show that entrainment was very unlikely for the laboratory results.

7. Wave action on mud beds was well illustrated in the present experiments.

Acknowledgements

The work described in this paper was funded under the UK Science and Engineering Research Council Grant No. GR/H14991 for which the authors express their gratitude.

Thanks are also due to W. Roberts, A.J. Cooper and N.V.M. Odd of Hydraulic Research Limited, Wallingford, United Kingdom.

We would also like to thank E. Mahoney and J. Dilks for conducting the wave experiments and some of the fluid mud tests.

Bibliography

1. Ali, K.H.M. and Georgiadis, K. (1991). "Laminar motion of fluid mud". *Proc. Instn. Civ. Engrs.*, Part 2, **91**, pp. 795–821.

2. Costa, R.G. (1995). "Three-dimensional modelling of cohesive sediment transport in estuarine environments". Ph.D. Thesis, University of Liverpool.

3. Delft Hydraulics Laboratory (1991). *Manual on Optical Silt Measuring Type OSLIM*, Delft Hydraulics Laboratory, The Netherlands.

4. Dyer, K.R. (1973). "Estuaries, a physical introduction". John Wiley and Sons, London.

5. Harleman, D.R.F. (1961). "Stratified flow". *Handbook of Fluid Dynamics*, Editor: V.L. Street, McGraw-Hill, New York, **26**.

6. Hydraulics Research Wallingford Limited (1992). "Fluid mud in estuaries". *ETSUTID 4084*, Energy Technology Support Unit, Department of Energy, United Kingdom.

7. Ippen, A.T. and Harleman, D.R.F. (1952). "Steady-state characteristics of sub-surface flow". Circular No. 521, US National Bureau of Standards, pp. 79–93.

8. Mehta, A.J. and Srinivas, R. (1993). "Observations on the entrainment of fluid mud by shear flow". *Nearshore and Estuarine Cohesive Sediment Transport*, Editor: A.J. Mehte, Coastal and Estuarine Studies Series, **41**, AGU, Washington DC, USA.

9. Odd, N.V.M. and Rodger, J.G. (1986). "An analysis of the behaviour of fluid mud in estuaries". Report No. SR84, Hydraulics Research Wallingford Limited.

10. Owen, M.W. (1977). "Problems in the modeling of transport, erosion, and deposition of cohesive sediments". *The Sea*, Editors: E.D. Goldberg, I.N. McCave, J.J. O'Brien and J.H. Steele, **6**, John Wiley and Sons, New York, pp. 515–537.

11. Sleath, J.F.A. (1984). "Sea bed mechanics". John Wiley and Sons, New York, pp. 279–291.

12. Viessman, W.J. et al. (1972). "Introduction to hydrology". Intext Educational Publishers, pp. 188–2070.

Table 1.

C (g/ℓ)	φ	F(φ) m = 5	F(φ) m = 4	F(φ) m = 4.65	C/C_a	$\frac{y-a}{H-a}$ Constant ε	Variable ε
3	0.976	1.385	0.763	1.164	0.03	1.000	1.000
5	0.960	1.376	0.762	1.150	0.05	0.996	0.981
10	0.920	1.349	0.758	1.142	0.10	0.989	0.980
15	0.880	1.315	0.748	1.100	0.15	0.985	0.979
20	0.840	1.271	0.731	1.080	0.20	0.981	0.979
25	0.800	1.215	0.706	1.000	0.25	0.977	0.978
30	0.760	1.140	0.670	0.974	0.30	0.972	0.978
35	0.720	1.040	0.619	0.900	0.35	0.968	0.977
40	0.680	0.905	0.548	0.780	0.40	0.963	0.976
45	0.640	0.719	0.449	0.650	0.45	0.957	0.974
50	0.600	0.458	0.312	0.405	0.50	0.949	0.972
55	0.560	0.082	0.120	0	0.55	0.941	0.967
60	0.520	-0.470	-0.152	-0.359	0.60	0.930	0.962
65	0.480	-1.306	-0.542	-0.500	0.65	0.916	0.963
70	0.440	-2.615	-1.114	-2.095	0.70	0.896	0.949
75	0.400	-4.744	-1.977	-3.100	0.75	0.868	0.933
80	0.360	-8.376	-3.321	-6.600	0.80	0.825	0.895
85	0.320	-14.939	-5.509	-13.200	0.85	0.759	0.823
90	0.280	-27.683	-9.271	-21.250	0.90	0.646	0.736
95	0.240	-54.846	-16.229	-55.000	0.95	0.436	0.368
100	0.200	-120.476	-30.429	-88.700	1.00	0	0

NOTE
We used $C_a = 100g/\ell$ and $C_H = 3g/\ell$.
The last column in the above table is $(y - a)/(H - a)$.